Space, Time, Place

Third International Conference on Remote Sensing in Archaeology

17th-21st August 2009
Tiruchirappalli, Tamil Nadu, India

Edited by

Maurizio Forte
Stefano Campana
Claudia Liuzza

BAR International Series 2118
2010

Published in 2016 by
BAR Publishing, Oxford

BAR International Series 2118

Space, Time, Place

ISBN 978 1 4073 0659 9

BAR Publishing is the trading name of British Archaeological Reports (Oxford) Ltd. British Archaeological Reports was first incorporated in 1974 to publish the BAR Series, International and British. In 1992 Hadrian Books Ltd became part of the BAR group. This volume was originally published by Archaeopress in conjunction with British Archaeological Reports (Oxford) Ltd / Hadrian Books Ltd, the Series principal publisher, in 2010. This present volume is published by BAR Publishing, 2016.

Printed in England

BAR
PUBLISHING

BAR titles are available from:

BAR Publishing
122 Banbury Rd, Oxford, OX2 7BP, UK
EMAIL info@barpublishing.com
PHONE +44 (0)1865 310431
FAX +44 (0)1865 316916
www.barpublishing.com

Contents

Session 5
INTEGRATED REMOTE SENSING TECHNOLOGIES FOR THE INTERPRETATION OF LANDSCAPE ECOSYSTEMS: CONFRONTING SCALES OF DETAIL

Session 6
DIGITAL CULTURAL ATLASES (PROJECTS AND PROTOTYPES) AND CULTURAL ATLAS COMPONENTS (GAZETTEERS, TIME-PERIODS, BIOGRAPHY AND SOCIAL NETWORKS, THESAURI, TECHNICAL INFRASTRUCTURE, CONTENT SOURCES AND DISPLAY MODELS)

Introduction

S. Campana,[1] M. Forte[2] and C. Liuzza[3]

[1]University of Siena, ITALY
[2]University of Merced, USA
[3] University of Stanford, USA

Some years ago in Rome in occasion of the IInd International Conference on Remote Sensing in Archaeology we focused the introduction of the proceedings on the development of remote sensing **from the perspective of the availability of different tools** – we may consider that for a long time, **RS consisted almost entirely of aerial** archaeology, but from the later part of 1980s and early 1990s the Remote Sensing toolkit became enriched by the advent of satellite imagery, airborne multispectral and hyperspectral techniques, radar data and LiDAR, along with ground-based geophysical methods – **into a form of survey and exploration** in which tools and techniques are critically involved form the very beginning to the end of the archaeological process, with a significant impact on the methodology of the investigation and the character of the archaeological interpretation which can be achieved.

Three years after the second conference we see that this trend continues to grow, with an ever increasing recognition of the positive interaction between archaeologists whose aim is to address archaeological and historical questions and a variety of techniques developed by technicians, engineering and physicians but now applied ever more frequently in the field by archaeologists themselves, still of course in many case in partnership with technical colleagues.

At the IIIrd International Conference on Remote Sensing in Archaeology Conference has been presented about seventy papers, dived in fourteen sessions, with speakers coming from almost every part of the world.

Generally speaking we believe that the schedule and the sessions of the meeting are broadly representative of the key topics currently particularly concern in discussion amongst the scientific community. We are thinking here of such topics as the automatic extraction of archaeological features as well as data integration, archives and the global atlas, 3D reconstruction and last but not least data interpretation. Concerning this last point it is regrettable that out of hundred papers submitted only two were concerned with this fundamental aspect of the archaeological process! We feel very strongly that the question of interpretation is one to which we should return in the next conferences.

To conclude this short introduction I would like to add only one or two more remarks, based on my personal experience in working within the University of Siena. Most of the projects that the University has carried out in Italy and elsewhere over the past thirty years have been fundamentally **aimed at academic research**. At the beginning of 2009 however, we established within the University a spin-off company to bring the know-how of the Laboratory of Landscape Archaeology and Remote Sensing to the national and hopefully international market of archaeological prospection. In the first six months of our operations we have been brought in to contact with construction firms whose interests definitively do not lie in academic research but are focused instead exclusively **on what we have come to know broadly as rescue archaeology**. So, what differences have we noticed and why should these be interesting to the worldwide audience? Some years ago in Italy at an international summer school in archaeology, Prof. Dominic Powlesland, Chris Musson and myself started a discussion on what we termed "emptiness" in landscape archaeology. As is well known, in academic research we try to fill or explain emptiness by an iterative process of formulating archaeological questions, trying to detect coherence or conflict in the available evidence and then redefining the questions which we wish to ask and to answer. Within the academic sphere fieldwork is largely based on remote sensing and surface collection survey. But before we can commit ourselves to the inevitably destructive process of excavation we need a firm indication of the existence of buried archaeology: **as result within academic archaeology we do not excavate where the prospection data is mute**.

Rescue archaeology starts from a completely different perspective. We can of course apply the same kinds of strategies and methods that we use in academic research but there is a fundamental difference in that **the whole of the endangered area will come under excavation of a sort** even if the archaeologists usual survey tools have failed to reveal positive evidence of buried archaeology.

The difference in a few words is that academics excavate only where some specific evidence is available with the result that they almost never systematically test areas where the basic survey methods have failed to produce positive evidence of settlement or other kinds of human activity.

In our particular case the spin-off company stated work recently a major prospection project along the line of a new motorway in northern Italy where the environmental situation relatively favorable for archaeological work, consisting of a flat plain overlying soils of sand and gravel. The financial provision is sufficient for us to develop a highly detailed and multifaceted prospection program, involving the integrated use of archaeological and historical literature, maps, documentary sources, place name evidences, historical and newly commissioned air photography, LiDAR and systematic and high speed geophysical survey, field by field so that every square meter of the endangered area is examined by both magnetic and geoelectrical techniques in particular the Geocarta systems.

Without going into the detail, even in this case - where we are probably applying the highest intensity of survey techniques available within rescue archaeology - we may inevitably find that some important evidence of human activity escape our search, as might happened with some forms of burials or other ephemeral activities.
The challenge and the great opportunity should be recognized with the systematical verification of the prospection dataset. It is possible for instance to go back to the data and to check whether such problem of apparent non-detection are related to instrument sensitivity or resolution, or perhaps derive instead from some misunderstanding or omission in data interpretation.

We have to give serious attention to both of these possibilities, and the opportunity for systematic and large scale ground-truthing which this kind of rescue work gives to the archaeologist at the same time **offers us the chance** to learn and understand more about **issues of archaeological visibility**, **emptiness** and the inevitable but sometimes unacknowledged **limitation of our current remote sensing techniques**.

Preface

D. Dayalan

Archaeological Survey of India, Chennai, India

Remote sensing is now commonly refers to the science of identifying, observing, interpreting and measuring objects or surface, without coming into direct contact with them. By this definition remote sensing encompasses the field of aerial photography. The term, "remote sensing," was first introduced in 1960 by Evelyn L. Pruitt of the U. S. Office of Naval Research. However, the first aerial photograph was taken in 1858. Long before satellites and computers started dominating the field of remote sensing, people were taking pictures of the earth's surface from afar with the help of balloons and kites. In 1858, Gaspard Felix Tournachon (later known as "Nadar") captured the first recorded aerial photograph from a balloon tethered over the Bievre Valley and M. Arthur Batut took the first aerial photographs of Labruguiere, France in the late 1880s by using a kite.

Notwithstanding the application of remote sensing for various purposes in sporadic manner is reported elsewhere, the field of remote sensing took shape only during the 1960s as an outgrowth of aerial photography. William Pecora, Director of the Geological Survey of United States of America, took the lead in pushing for a spacecraft that could cover the entire world with false colour images. The first of them is Landsat 1 by the National Aeronautics and Space Administration (NASA) which reached the orbit in 1972. The French government launched the SPOT program, the Satellite Pour I'Observation de la Terre in 1978 and in 1982 established the firm of SPOT image to market its photos. Lockheed was the first company which obtained the license from the USA authorities for building commercial remote sensing satellite system and launched its IKONOS satellite in 1994. The other system followed this are: Orbview of Orbital Sciences, Quickbird for the firm of DigitalGlobe, etc.

Remote sensing and GIS are complementary technologies. The remote sensing systems are used to observe the earth's surface through the sensors mounted on the satellite and aircraft and to collect information of an object, area or phenomenon. The geographic information systems (GIS) can be used to interpret remote sensing data more accurately by integrating them with data derived from various other sources. The new development in the technology like, microwave remote sensing, GPS, GPR, etc are also being utilized for various studies. As the satellites images are obtained in digital form it can be processed directly in the computer. Moreover the satellite data are more reliable, repetitive, non-invasive, rapid and cost-effective information. All information is exactly localized and gathered in one tool. The direct extraction of topographic and thematic maps for terrain use is also possible with this technology.

Notwithstanding the remote sensing, GIS and related geo-information technologies are used in the fields such as town and country planning, watershed management, agriculture, forestry, mineral resource development and prevention and mitigation of natural disasters, the necessity for the utilization of them for archaeological purposes and the conservation and maintenance of the cultural and natural properties has been felt only in the recent years. In fact the GIS and remote sensing are the excellent tools for the documentation of archaeological sites and monitoring of cultural and natural heritage properties. Some of the well known early attempts of application of remote sensing for archaeological pursuit are the famous shots of Foro Romano taken in 1899 by Giacomo Boni and the photographs of Pompei in 1910. The fortified wall of the ancient historical city of Sisupalgarh, Orissa, India was identified properly by using aerial photography taken in 1948. The other classic example of the use of remote sensing is the discovery of Gallo-Roman villa in 1979. The site was beneath a farm and was identified on the basis of the crop pattern through aerial photographs. Some of the most prominent remote sensing researches are the study of Maya cultural properties in Mesoamerica and the Petén region of northern Guatemala. The Petén is a densely forested region and it is extremely difficult to survey and because of this remote sensing offer a solution to this research problem.

The growing diffusion of ideas and application of spatial approaches using GIS, digital mapping, digitalization in archaeology, opened a new line of intellectual enquiries and insights into the past of the humanity and therefore remote sensing had gained immense application value in understanding and interpreting the remains of the past. In recognition of this trend, UNESCO and the European Space Agency (ESA) launched the Open Initiative on the Use of Space Technologies to monitor natural and Cultural Heritages in 2001. The "Open Initiative" is a framework of co-operation to assist countries to improve the observation, monitoring and management of natural and cultural sites as well as of their surroundings, through space technologies.

First International Conference on Remote Sensing Archaeology was organized by the Chinese Academy of Sciences in October 2004 at Beijing. The research papers and the discussions held there evidently exhibited that the concept of remote sensing was significantly wider than in the past and involved the integration of numerous different technologies and field of application. The Second International Conference on Remote Sensing Archaeology held at Rome in December 2006, in continuation of the initiative at Beijing and fruitful discussion was held on promoting the use of integrated methodologies in remote sensing archaeology so as to help in the creation of new and sustainable policies in the monitoring, interpretation, fruition and communication of the cultural heritages.

Followed this, the III International Conference on Remote Sensing Archaeology- **SPACE TIME AND PLACE 2009** was held at Tiruchirappalli, Tamil Nadu, India in between 17th - 21st August 2009 in association with the Bharathidasan University, Tiruchirappalli, the University of California, Merced and Berkeley, the University of Siena (Italy) and by the REACH Foundation, Chennai. More than 60 research papers were presented in the conference by the delegates from 17 counties. The wide range of perspectives, approaches and issues on the use of remote sensing, GIS, GPS, GPR and digital technologies in documenting, analyzing and interpreting archaeological and anthropological contexts was discussed vividly in the conference and significance and application of remote sensing and digital and numerous other technologies in the field of archaeology was well recognized. Special sessions were also held to discuss about the archaeology in the digital age, digital cultural atlas and heritage management, conservation and sustainable utility of heritage properties. ECAI organized a session dealing with the issues related to heritage sites, especially urban centres. After the conference the delegates were taken to the world heritages monuments at Tanjavur, Darasuram and Mamallapuram to have glimpses of the kaleidoscopic variety of rich and magnificent cultural heritages in India.

The conference was organised with the advisory, technical and monitory support from various institutions and well wishers. We are extremely beholden to Shri, V, Narayanasamy, Hon'ble Minister for State for Planning, Parliamentary Affairs and Culture, Government of India for sparing his valuable time for inaugurating the conference. We also take this opportunity to express our gratitude to Dr. M. Ponnavaikko, Vice-Chancellor, Bharathidasan University, Tiruchirappalli for kindly agreed to organise the conference at Tiruchirappalli. We gratefully acknowledge the valuable guidance, co-operation and encouragement received from Prof. Lewis Lancaster, University of California, Berkeley, Prof. Maurizio Forte, University of California, Merced, Dr. T, Satyamurthy, REACH Foundation, Chennai and Dr. Stefano Campana, University of Siena, Siena, Italy at various stages in organizing the conference. We are equally thankful to Claudia Liuzza, University of California, Merced and Shri. A. Anil Kumar, Archaeological Survey of India, TSP, Chennai for making extensive correspondence with all the delegates and extend all co-operation in arranging the conference and collection of papers from the delegates. We should not fail to thank Prof. K. Nedunchezhean, Bharathidasan University, Tiruchirappalli for taking care of local hospitality of all the delegates at Tiruchirappalli and Mr. Madan, Chennai for creating the web site for the conference. Thanks in ample measure are no less due to the authorities of British Archaeological Reports (BAR), Oxford, England for the excellent publication of the Proceedings of the International Conference. Our sincere and heartfelt thanks are due to the staffs of the Archaeological Survey of India, TSP, for arranging Photo Exhibition on "Monument Heritages of Tamil Nadu" in the conference venue for the delegates, scholars and public.

We hope that the Proceeding of the III International Conference on Remote Sensing Archaeology will be useful to understand the emerging trends in Remote Sensing, promising new research pathways along with multi-disciplinary interaction, conservation issues, Digital Cultural Atlases, Cultural and natural sites resource management and so on and read with avidity by students and scholars alike all over.

Acknowledgements

We would like to take this opportunity to thank all of the people, institutions, association and companies whose help has permitted us to bring this international event to a successful conclusion.

First of all we feel pleasantly obliged to thank prof. Lewis Lancaster of the University of California, Berkeley who gave a valuable contribution to the successful organization of the conference.

We are grateful to Dr. D. Dayalan of the Archaeological Survey of India for the logistical support and the hard work achieved to organize locally the conference.

We are grateful to all the institutions that permitted us to bring the workshop to a successful conclusion and particularly we should remember: University of California, Berkeley and Merced, University of Siena (laboratory of Landscape Archaeology and Remote Sensing, www.lapetlab.it), Archaeological Survey of India, Aerial Archaeology Research Group.

A special debt of gratitude is due to ECAI which made the greatest contribution to securing financially the publication of the proceedings.

This workshop has received generous support from the National Science Foundation Award # IIS-0929851 for travel of delegates. We also wish to acknowledge additional support from the Henry Luce Foundation's Award to ECAI for Cultural Atlas development.

Some of the editorial costs of this volume as well as some grants to attend the workshop has been covered from Leica Geosystems (world leader in the development of high end topographic instruments, www.leica-geosystems.com) and Archaeolandscapes Tech & Survey (company specialized in technological services for archaeology, www.atsenterprise.com)

A special thanks, however, is due to Dr. Francesco Fontanelli (University of Siena) whose efforts helped the set up of the conference proceeding.

Remote sensing and the humanities

Lewis Lancaster

University of California, Berkeley

Abstract

There is a crisis looming on the horizon for the Humanities. In recent years, the rapid pace of digitization of the world's libraries and archives has made an unprecedented amount of information widely available. While this access to increasingly well produced digital collections is a boon to Humanities and Humanistic Social Sciences research (e.g. history, anthropology, cultural geography), it presents Humanities scholars with certain non trivial problems. As the amount of data increases, the need for such visualizations will become more pressing. Geo-registration of data allows us to put it in mapping that includes remote sensing. Surrendering the word to image can allow us to study heritage sites, objective works of art and craftsmanship having diverse parts or elements in a manner that presents us with an intelligible and distinctive whole.

Keywords Remote sensing, humanities, GIS, visualization

The rapid changes in technology, research, and data management, has made us aware that constant adjustments are required to keep abreast of new developments. It is disconcerting to find that our previous information can be unreliable when we incorporate the current data. In the face of this deconstruction of received narratives about history and causality, I take heart in the statement of Tom Stoppard in his play *Arcadia*.

"It makes me so happy. To be at the beginning again, knowing almost nothing.... A door like this has cracked open five or six times since we got up on our hind legs. It's the best possible time of being alive, when almost everything you thought you knew is wrong"

I like the statement not because of the prediction of being "wrong" but because of the wonder of being alive in a time of great change. Being able to identify the "wrong" has come to most of us through the writings of Karl Popper and the search for falsification as a principle of scientific method. But this method does not tell me which of the options is "right", it only identifies the one which is to be rejected. Grobstein says it very well when he states, we are always finding what is "less wrong" rather than what is "right." If we give up trying to find the "best" or the "right" from among a multitude of variables, then our strategy of inquiry changes dramatically. We will be less concerned with the former subjects of inquiry. Instead, our questions will be ones that could not have been asked previously.

Rene Girard, a literature specialist at Stanford University and one of the forty members of the Academie Francaise, has generated a great deal of criticism and praise for his evaluation of the state of Humanities.
He says:

"I think the very notion of humanities is at stake today because of this insistence that they not be touched by religion or science... That's why I think the humanities are withering on the vine."

I am not so pessimistic about the future of the Humanities. The rather dramatic shift that is needed and is actually occurring in the use of technology within Humanistic study concerns the limits on "what can be considered". We now have expanded access to data and the ability to slip from one source and one disciplinary field to another. Consequently, the goals and strategies of traditional Humanities scholarship have begun to falter. I do not say that Humanities fields are disappearing but rather that the traditional methodology is at question. This is naturally a discomfiting situation for scholars who have operated under the norms of "properties and rules." In my field, we have a long history of how to study what is called "India." But what if the "properties and rules" that I have come to follow for understanding Indian Buddhism are not the only characteristics we find in the expanded world of data and spatial analysis. A similar point can be made for the changes in sciences. For example, development of "Complex Systems" in the 1980s has expanded "what can be considered" in many fields. Thus, the emphasis of specialization and a closed focus that characterized science, with the subsequent barrier to looking at other fields of inquiry is giving way and scholars have begun to look for patterns that go across multiple fields. Such patterns have often been missed when the more focused research limited the range of "what can be considered."

We have seen the struggle between the arts and science play out in Europe for centuries. Ernst Gombrich went so far as to say "art is altogether different from science." However, when we look back to the 15th century and the work of Leonardo da Vinci, we see an artist struggling to convince people that painting could stand alongside geometry and astronomy as one of the "liberal arts."

Today, Humanities has been challenged with the technology of the digital world including Remote Sensing. This is not to infer that the Humanities have been removed from the developments in this approach. In fact, the Humanities have accepted and made use of digital visualization technology in a variety of ways. (Figure 1).

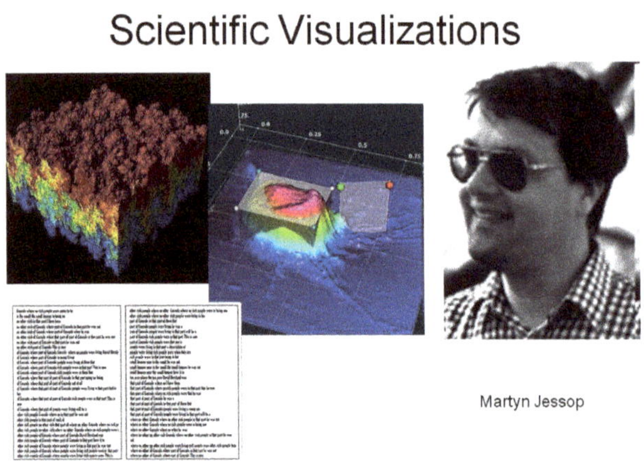

Fig. 1. Examples of scientific visualizations described by Martin Jessop.

Martyn Jessop calls our attention to six: i.e. numbers (quantitative analyses), text, narrative (diagrammed relationships), space (GIS), time, and 3-D (built environments) . He goes on to raise the question of whether Scientific Visualization has become "not just a methodology but a discipline in its own right." Humanities come to this research activity later than some other scientific fields. This is for good reasons; Scientific Visualization has been made possible only through the use of super computers and researchers at well funded national laboratories where the agendas do not normally include qualitative information. While scholars in the Humanities come late to the process, their needs for visualizations based on analysis of large data sets are increasing at the same speed as the growth of available digital information.

The organization of this year's conference on Remote Sensing which includes representatives from the Humanities is an important milestone. We are brought together to ask questions of one another about the feasibility and effectiveness of including Remote Sensing in the approaches used to study culture, texts, art, and all the varieties of the Human Record. The name Remote Sensing has only been in our vocabulary since the 1950s when Evelyn Pruitt at the U.S. Office of Naval Research used it to describe the images being obtained from satellites. Since then, the definition of the term has expanded and is now used to cover a wide range of techniques and data creation, as exemplified by the variety of approaches and topics discussed in our sessions.

As we see from the presentations at this meeting, Remote Sensing is still generally based on spatially organized data. When the Electronic Cultural Atlas Initiative (ECAI) was organized in 1996 at Berkeley, one of the first attempts of the community of scholars was to bring geo-registration of latitude and longitude to the markup of cultural data. (Figure 2) At first, it was difficult for some of our affiliates to truly appreciate the value of having metadata that referred to a spatial domain. Very shortly, the community adopting GIS realized the great step forward that spatial visualization represented. The use of dynamic maps built from our own data using Geographic Information Systems (GIS) gave us a powerful new tool for display, analysis, and even archiving. It has become an accepted part of much of the research done by Humanists and GIS is now included as a regular part of proposals to the National Endowment for Humanities in the U.S. As a further move, ECAI teams began to work with ways of marking up data by latitude and longitude but with the additional element of temporality.

Without adding the dimension of time to our markup, we realized that data association with mapped points and polygons constructed from latitude and longitude could contain too much information for our ease of use and analysis. We had to add the ability to take a slice of time for any given spot on earth in order to have better control of massive data. Still, there are skeptics among the disciplines of the Humanities.

For example, those who study philosophy continue to question me about what help the spatial could be for understanding Plato or Aristotle or in Buddhist Studies, Nagarjuna. They concede that cultural heritage sites are geographic in terms of location and can be profitably studied as spatial and temporal, but a number of them reject the idea that "thought" can be treated in a similar fashion. There are some who see no advantage to understanding Plato or Nagarjuna as a person who lived at a certain time in a specific location on the surface of the earth. (Figure 3)

However, we have seen a growing awareness in Greek studies that time and the place are crucial elements in philosophical development and the study of it. The *Perseus* website provides a wealth of information for scholars in the Classics and that includes maps and timelines.

As we meet here in Tiruchirappalli a new challenge is addressed, one that is even more complex and demanding than the issues associated with GIS and Time. We have yet to decide if Remote Sensing technology can have a place in the Humanities equal to the one now occupied by GIS. Some of these same questions have been asked with regard to the Social Sciences. As far back as 1996, the National Research Council in the U.S. organized a large workshop for social scientists who use satellite data. The members of that workshop were mainly involved with the study of human-environmental interactions. This involved such interests as how ecological change affects human health.

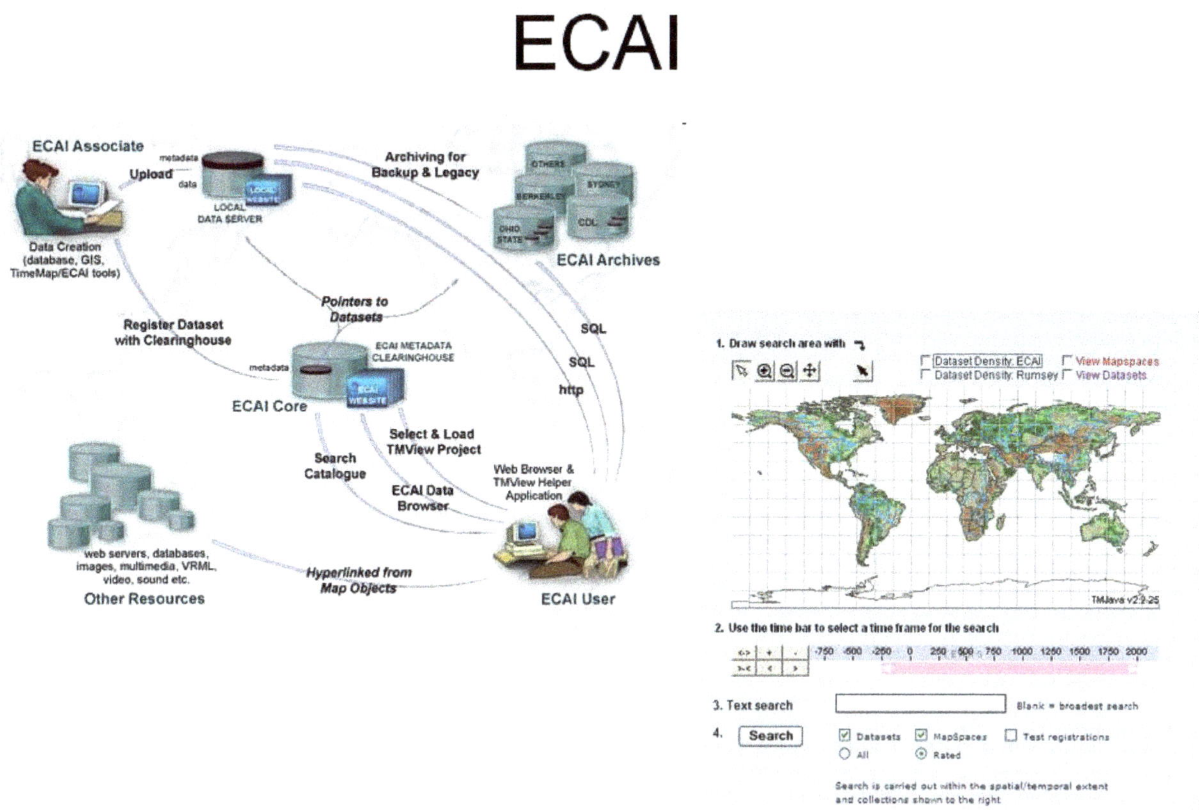

Fig. 2. Structure and software of the Electronic Cultural Atlas Initiative of the University of California, Berkeley.

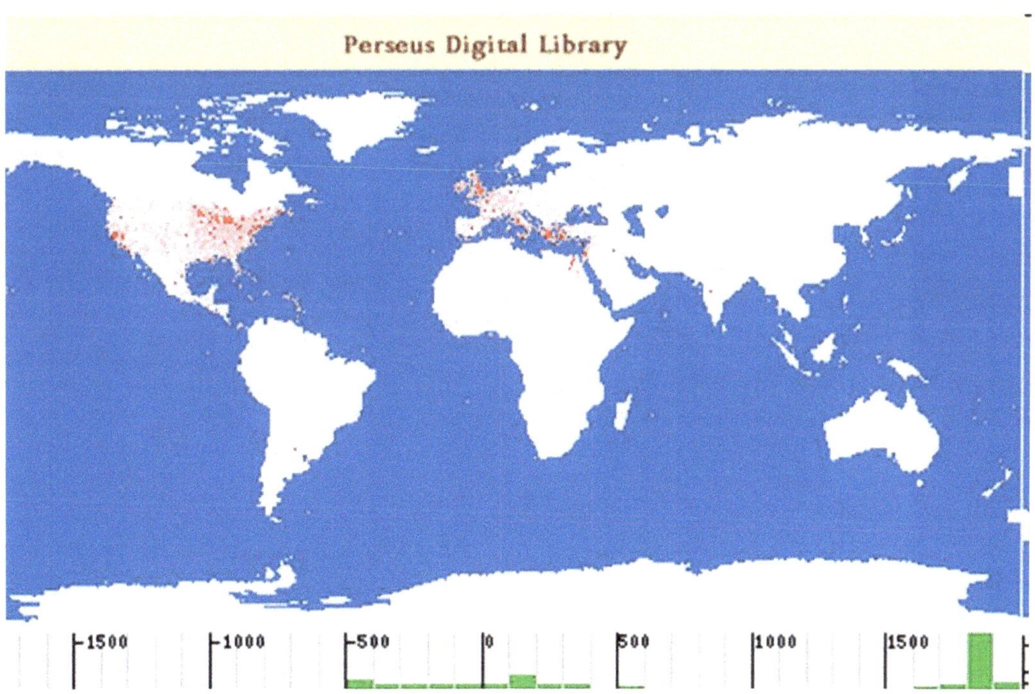

Fig. 3. Map and timeline interface of the Perseus Digital Library project at Tufts University.

NASA has given recognition to this type of data and analysis by setting up SEDAC (Socio-economic Data and Applications Center). The Humanities have not been included in the discussions that have marked the interaction between Remote Sensing and Social Sciences. There are still daunting hurdles for the Humanist to overcome. What can be the value for the Humanities of organizing pixels in such a way that we have an image that corresponds to features of the landscape or of other objects? The issues of Remote Sensing data are many and complex in terms of the cost of transforming raw data into some form that can be useful. Beyond the first interpretations of the images, we have the matter of cataloging, storing, and maintaining the data. Remote Sensing interpretation aims at classifying the features in the image into categories and these can become thematic maps that are familiar to us in cultural studies. The Humanities scholars can certainly contribute information for supervised classifications of the Remote Sensing data. That is, insights based on previous knowledge of the site can be of immense help when doing interpretation of the imagery generated for a site. On the other side of the equation, we have great interest in having unsupervised classification when the map is made up only based on spectral properties and computational analysis without input from prior knowledge and explanations. The unsupervised can give us concepts and options that have never been considered

The question still remains, can techniques that measure and record evidence collected at a distance, using sensors for subtle aspects such as electromagnetic radiation, acoustic energy, and force fields ever become a major strategy for the study of culture? It has been much easier to introduce this technology into medicine. There we find CATscans and Magnetic Resonance Imaging readily available for our use, as doctors access data about sub-surface tissues and organs without resorting to surgery. The value of making the invisible available for viewing is quickly recognized when it results in physical issues for health. It is much more difficult to see how such technology can ever have a major impact on how we read and interpret ancient texts.

Ironically, it is just in the application to ancient texts that some of the early successes of integrating Humanities and Remote Sensing have been achieved. (Figure 4) The famous Archimedes Palimpsest is an example of the value of the technology. Written on vellum, the document was erased in order to reuse the surface for copying a prayer book. Scholars could still see the dim shadows of the original ink but only portions could be recovered. When image processing of the vellum pages was done and pseudo-color techniques applied, it was possible to easily read the whole of the erased document. Scholars in the field were delighted to have a clear example of Archimedes' Method Proposition 14 which has given them a fuller understanding of his concept of infinity. Therefore, we have an example of where it is crucial, even in the Humanities, to make the invisible available through one form of Remote Sensing.

Fig. 4. The Archimedes Palimpsest showing the reconstruction of erased text after overwriting by a later scribe.

An even more difficult task has been undertaken by Professor Brent Seales. (Figure 5)

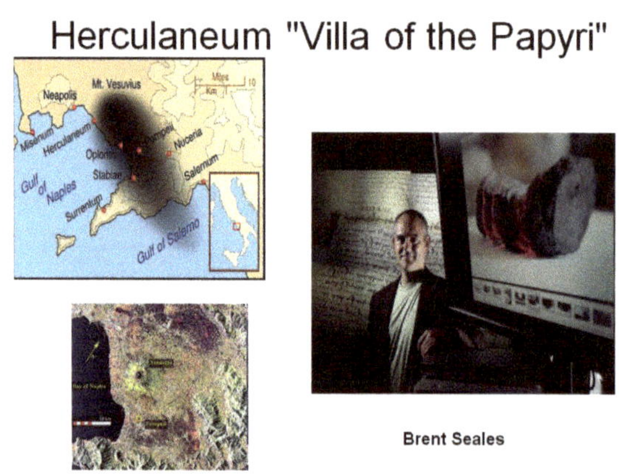

Fig. 5. Brent Seales project at University of Kentucky using MRI to read carbonized papyri.

He has turned his attention to the hundreds of papyrus scrolls that were buried when Mount Vesuvius erupted in 79 C.E. The heat of the ashes turned them into carbon and the blacken bundles simply disintegrated when scholars attempted to unroll them. The team working with Professor Seales at the University of Kentucky is using X-ray CT scan on two of these objects to make images of the layers of the scrolls. They are seeking to collect interior images, i.e. slices of the bundle that will show the writing at each layer. In this way, they will digitally provide images in a form that will appear as if one is unrolling the charred mass. There will be no objection on the part of Humanities scholars to have Remote Sensing used in this fashion on the physical objects.

As we see from the papers at this conference, Remote Sensing has been well accepted by archaeologists and in some cases has changed the way in which students of culture think about a site. (Figure 6)

Angkor Wat

Fig 6. Remote sensing data at Angkor Wat, Siem Reap Cambodia by the University of Sydney.

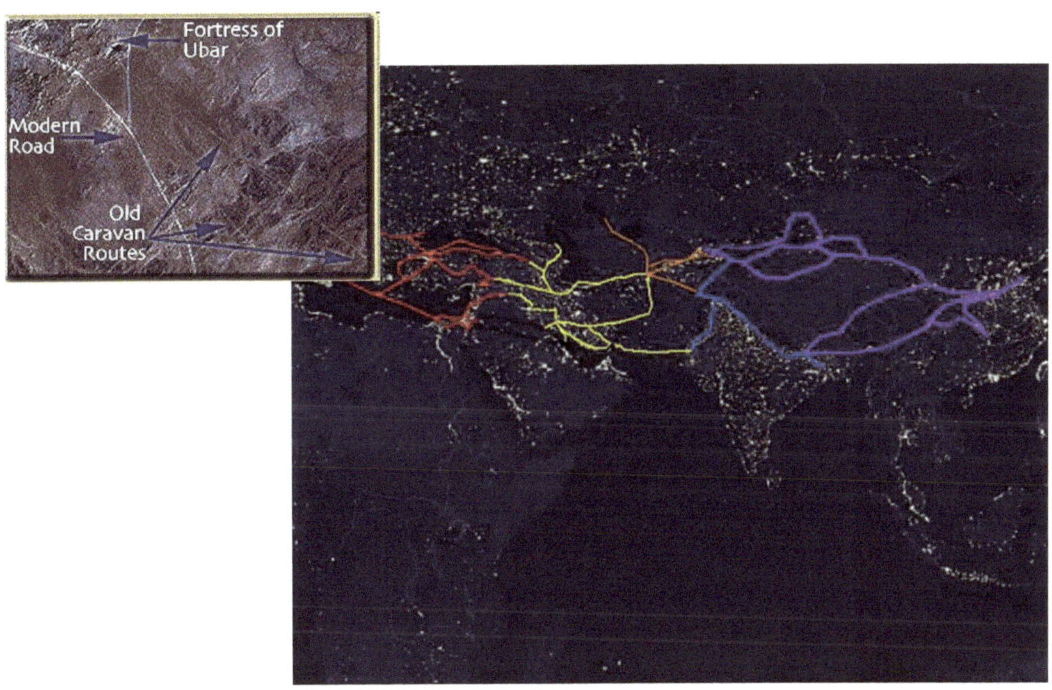

Fig. 7. Example of the use of GIS and satellite imagery for the lights of the earth and ancient trade routes.

The work at Angkor Wat stands out as an important model for this shift of interpretation based on Remote Sensing. The size of Angkor Wat and the complexity of the urban environment can only be seen in the images that provide data from the buried field patterns, roads, and irrigation works. We are not able to see many of these Remote Sensing features when physically walking through the forest and fields that overlay the ancient site.

There are still many problems for the study of the surface of the earth and culture. (Figure 7)

The so called "Silk Road" that is a name for the multiple trade routes crossing Eurasia is often just a line on a map. In one example, the ECAI team took the Lights of the Earth at Night imagery and attempted to place the routes alongside them. In remote places, we can see that the travel and human habitation still occurs in the same locations over the centuries. However, we do not know the exact route for all of the mercantile roads. It is Remote Sensing that will help us construct these more accurate maps based on longitude and latitude imagery from space. This technology has made it possible to locate the "lost" city of Ubar in Oman. Unlike the

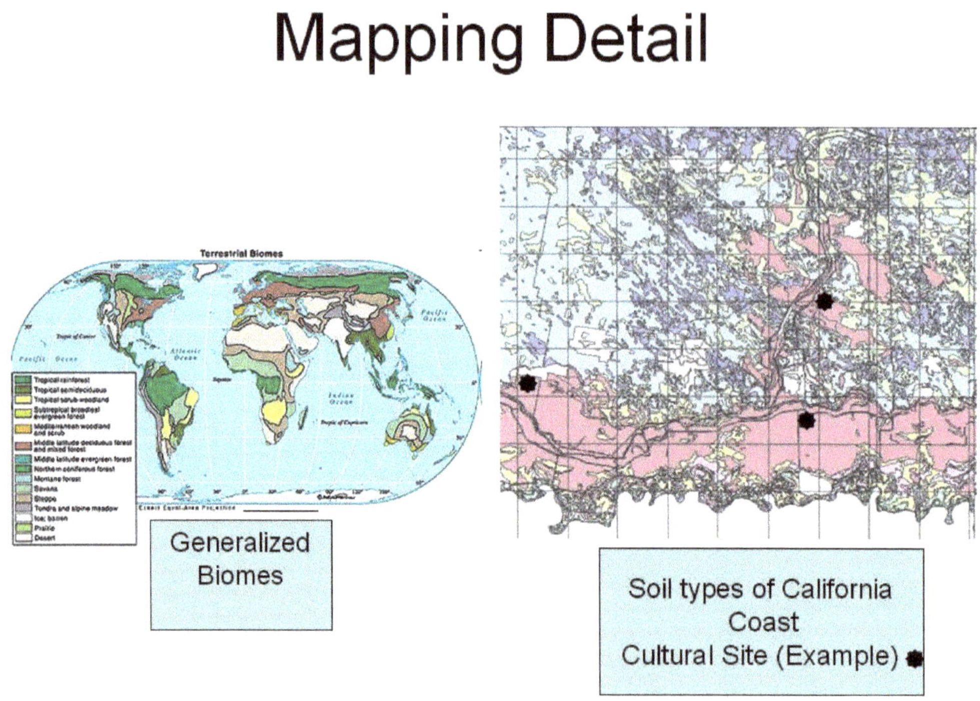

Fig. 8. Example of increased granularity of data using remote sensing of soil type compared to a generalized biomes.

previous methods of mapping by ground survey or imaging through aerial photography, we now have Lansat images from space shots that cover an area of 13,000 square miles in one frame. Not only do we have this enlarged view of the surface of the earth, we have new ways of doing analysis through the use of computers and applications such as GIS. There is much to be learned from the interpretations of the images of Lansat. For example, we can see the relationship between rock types and vegetation. Some plants thrive in certain soil types and when we study cultural sites, it is now possible to see if using geobotanical data collected through Remote Sensing can help scholars study the correlation between rocks and human activity. There is a great difference between the solid colored generalizations of biomes and the detail of soil types that can appear on closely studied Remote Sensing. (Figure 8) The images made from reflected light often offer us the finest granularity available for spatial study.

Our work with cultural mapping has often been based on aggregation of data by political or geographical units. The Historical Gazeteers show us political boundaries over time for provinces, cities, and towns. However, we are aware that cultural patterns lack clear geographical boundaries. We are in need of other layers of information that depict a wider and more varied array of boundaries. If our sites of interest show up with similar soil types, it alerts us to a possible explanation for the location of these places.

We have yet to fully research the patterns of the surface of the earth that can result from Remote Sensing and the placement and activities of cultural sites. It is still not apparent as to whether different ways of presenting the imagery will open the doors of analysis and interpretation in the Humanities. Can we find information for the Humanities in Remote Sensing that goes beyond True Color? Will Electronic Absorption show us something that is different from Vibrational Absorption? Can the depiction of something so subtle as the bonds in a lattice or a molecular compound be used as part of the aggregated cultural data for a site? These are questions that need to be addressed and I appreciate the fact that many of the participants in this workshop are at the forefront of exploring some of the queries. The level of training necessary to handle research of this magnitude shows us that our research must be multi-disciplinary. I use this term "multi-disciplinary" rather than "inter-disciplinary" because the work will require teams of highly trained researchers representing a number of disciplines.

In order to give an example of how this multi-disciplinary teamwork can function, I would like to describe one effort. In ECAI, we have a major research project to construct an Atlas of Chinese Religions. In order to do this, we have collected geo-referencing for thousands of religious sites in China and Vietnam. Our "team" includes scholars who work with religious history, comparative cultural features, geography, gazetteer construction, doctrinal and philosophical developments in the various traditions, social history, art history, Chinese, Vietnamese, Japanese, English, French, and German language materials, GIS mapping, interface design, software development for search and retrieval, digital library science and others. At the moment, working with the GIS Center at Academia Sinica, we can create maps for these sites and then lay down a series of map layers to look at elevation, water ways, historical roads, modern demographic density.

We need to add one more element to this mix. That is Remote Sensing data as additional map layers.

Once we start to use the data from Remote Sensing based on spatial aspects, the range of "what can be considered" increases in type as well as volume. (Figure 9) For example, in the Buddhist texts we often read passages that mention earthquakes. When the Buddha was born the earth shook in six ways, i.e. north, south, east, west, up and down. I had not thought of such information as being of spatial importance until I saw the map of epicenters for quakes on the earth during the last three decades. It is obvious that there are bands of epicenters and there are large areas where none have been observed. Looking at India, I can see that the original homeland of Buddhism was located in or near a very dense band of epicenters. Further south in India, the seismic activity is much lower. This suggests that the references to earthquakes would probably arise in a region that is covered by the band of epicenters. Such information is just an example of "what can be considered" when I read a Buddhist *sutra* in an era of Remote Sensing.

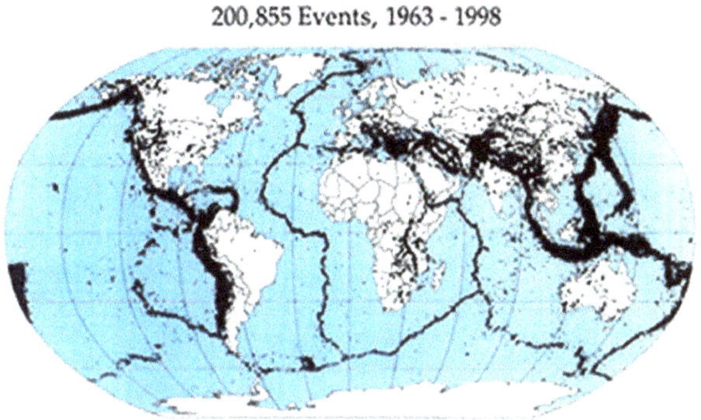

Fig. 9. Earthquake epicenter belts as one way of mapping cultural data.

Another ECAI project has been the creation of a digital publication dealing with the archaeological discoveries at the ancient capital of the Kushan Empire, Begram in present day Afghanistan. We have become more acquainted with this name because of the presence of a large military airport. Sanjyot Mehendale made a detailed study of the ivories contained in the palace structure that was unearthed by French archaeologists. She used the published drawings of the palace in the website with GIS markup. When we constructed the website, we started with a very distant view and continue to move into the data until we finally arrive at an image of one of the small pieces of the ivory carvings. With Remote Sensing, "what can be considered" for Begram expands. Our former approach which relied on the elevation mapping from satellites gives us an overview of what I term "generalizations" of terrain. However, as we move into the more precise and granulated Remote Sensing data, the solid map starts to fragment into distinct units that are not political nor easily identified surface features. With a greater degree of granularity, we see elevation, soil type, vegetation in a way that provides the context for the location of Begram that is more understandable that the "generalizations" of mapping that is often used.

We have seen in these few examples, a wealth of possibilities for the application of Remote Sensing to the studies conducted in the Humanities. The main conclusion is that Remote Sensing can provide scholars with a greatly expanded list of "what can be considered." This operates in both directions. As cultural information is made available for a region that has been explored with Remote Sensing, it provides the scientists with an expanded method of understanding that particular surface of the earth. Finding the remains of an ancient city in Oman or buried field patterns in Cambodia can give us information about the past. This past which the Humanities has studied as cultural and historical can now be explored with a host of new approaches. For example, we can now compare cultural mapping with a host of interpretations of Remote Sensing.

At this moment in time, it is still very difficult for us to define and map cultural influence by spatial and temporal methods. We do not have accurate data on the networks that relate to cultural heritage sites. For example, the famous stupa at Sanchi has been physically reconstructed and the "ontology" of the site, i.e. the physical elements of it, has been extensively studied by art historians and archaeologists. In the field of religious studies, the questions that arise have to do with the "catchment" area of those who supported and revered this site. Fortunately, Dr. Dayalan has shown us a method for drawing a map of the "catchment" area of Sanchi. Donor inscriptions often give the names of the donors and where they lived. If we geo-register this data and show a map of the places where donors lived, Sanchi is shown to have a "catchment" area that covers hundreds of miles. The same approach can be used at the Fang Shan caves in China where thousands of stones slabs were inscribed with the Buddhist canon. Many of the stones have donor inscriptions that include information about the home of those who provided the support for the carving. This can give us a wealth of information when it is put into our maps. We know which Buddhist text was chosen by the donor and we know where that person resided and we even have the date of the carving in the Tang dynasty. Such mapping will give us a suggestion for the type of teaching that was prized in certain locations based on the choice of text to be carved. We can also note what other corollaries exist between the donors and Remote Sensing data. The "catchment' area of Sanchi and Fang Shan can then be displayed with scores of map layers. With such expanded information, we will have data that indicates the spatial dimension of relic veneration as well as merit making from copying of certain religious texts.

I am very pleased to have this opportunity of speaking to issues which are crucial for the future of Humanities study. (Figure 10) We find much to emulate in the approach and the methodology of archaeologists who have so successfully incorporated both GIS and Remote Sensing. In the future, I hope that Humanities will be able to expand the use of Remote Sensing and contribute to the activities that seek to preserve and display the human record through digital technology.

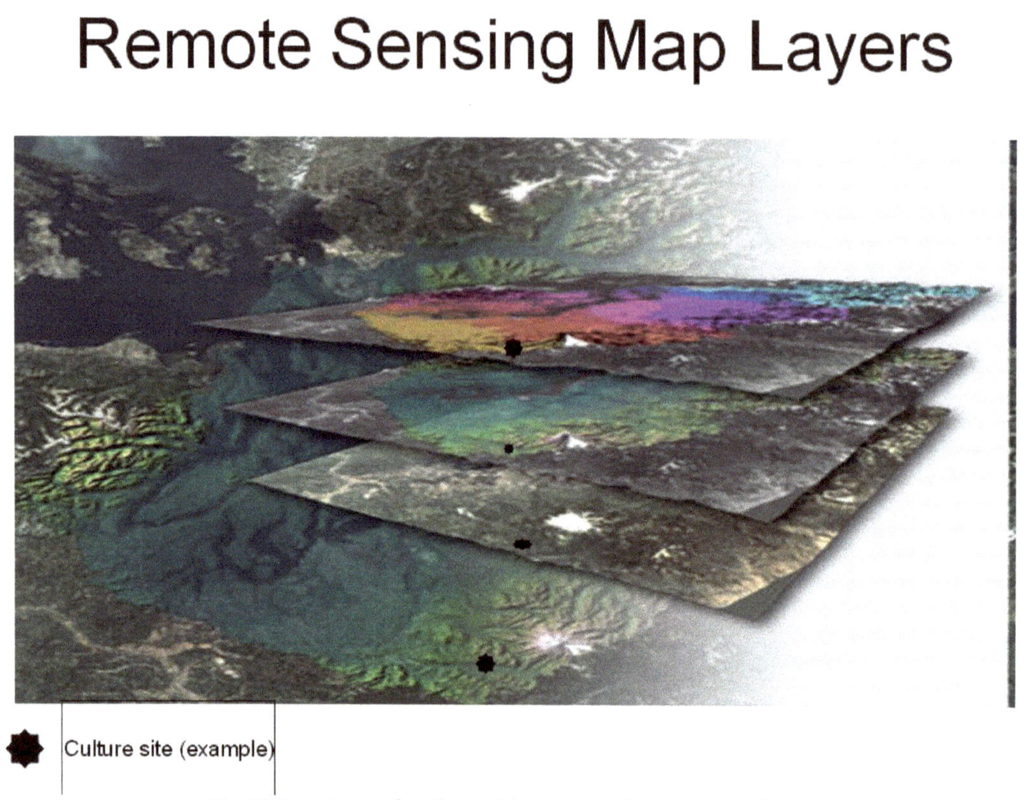

Fig. 10. Locating a cultural site within a series of remote sensed layers.

Identifying, mapping and managing the unmanageable: the implications of long term multi-sensor research into the archaeology of the Vale of Pickering, Yorkshire, England

Professor Dominic Powlesland FSA

Director, The Landscape Research Centre, Visiting Professor, The institute of Medieval Studies, University of Leeds

Introduction

This paper is intended to question why we do remote sensing; what, why and how do we discover things; where do our responsibilities lie with regard to our discoveries; our interpretation and ground truthing of those discoveries and what are our long term responsibilities with regard to securing a relevant future for those discoveries.

Remote sensing in archaeology, the identification of buried features without physical intervention owing to distinctive characteristics of the soils or crops reflecting past human activity, has been a characteristic of archaeological observation for more than four centuries with the observation of crop-marks documented for instance, by John Leyland at Silchester in 1541 (Fagan 1959). Leyland and a number of other later antiquaries observed features at many sites in Britain in the succeeding centuries and, although we tend to see remote sensing as a space age science, dependent upon multi-million dollar satellites, we should remind ourselves that the most significant interpretive tool, the eyes, have been around as long as humanity. A plan of Silchester incorporating the plot of the crop-marks was published in 1748 still two centuries before the space age.

Remote sensing can expose the 'invisible' and thus has a widespread appeal, an appeal that relates to the excitement of discovery. I well remember the exhilaration more than 40 years ago when I saw my first crop-marks from a light aircraft, an emotion that continues to play a part every time I fly over the Vale of Pickering during the crop-mark season. Whilst the excitement generated by seeing crop marks from a plane is not quite matched by the thrill of discovery during the interactive analysis of satellite images or geophysical survey data one can experience, for instance, through a virtual flight of discovery using Google Earth of Microsoft Virtual Earth in 3D, a similar excitement, viewing areas of landscape with high resolution air photography taken at a time when crop marks are showing.

Excitement can of course be distracting and we need be aware that the mere fact that one gets a buzz out of the process of discovery does not mean that we are doing a good job. If we view remote sensing as a method of finding great sites or good stuff then we are no better than treasure hunters and should be disbarred from practicing our archaeological craft. The past has a finite and unique evidence base that documents both human development and the interference and/or adaptation/utilisation by humanity with the planet upon which we live. Global warming, industrialisation and global economies are eroding and destroying the evidence base for the past at an ever increasing rate, and archaeologists, as interpreters of that evidence base, have a responsibility to the past and a responsibility towards securing a future for the evidence of the past. The weight of responsibility is heaviest upon those engaged in remote sensing as it enhances our ability to see the unseen. The use of satellite imagery to re-discover a large monument known to the local population is of limited interest, but the deployment of remote sensing to identify a whole landscape of features reflecting chronological and social depth is something else.

In the Vale of Pickering, North Yorkshire, England The Landscape Research Centre has been engaged in remote sensing since the mid 1970's, focused upon a single landscape. This work, primarily funded over the life of the project from many different English Heritage grants represents the largest project of its kind undertaken in Britain. We have employed conventional air photography, airborne multi-spectral and hyper-spectral scanners, Lidar and ground based geophysics and other ground based surveys in addition to about 30Ha of open area excavation. The results of this collective and long term research are formidable and challenge our comprehension of the landscape and its population for much of British prehistory through to the medieval period. The concept of site is likewise challenged; however, as a voyage of discovery this work, based in a single landscape over more than 30 years, has exposed substantial problems through observation and re-observation combining to form an anecdotal record of archaeological loss with which we should all be concerned. The mapped evidence is astonishing but, at a time when the management of archaeological sites is increasingly presented as a concept in the literature, this work exposes inadequate arrangements for the long term sustainability of archaeological landscapes.

Whilst to some, particularly geographers who have been pivotal in developing approaches to the application of space-borne sensors for observing the earth, the meaning of remote sensing is limited to space borne applications, our interpretation is broader. In the view of this author, a view which is reflected by the papers presented at conferences focussed upon remote sensing in archaeology, any non invasive approaches to viewing the buried and nominally 'invisible' evidence of past activity should be included within remote sensing for archaeology. These approaches include space and air borne sensor platforms - these might be traditional air photographic sensors as well as technology based multi-spectral or hyper spectral scanners, as well as ground based geophysics, under sea remote sensing and other non-invasive techniques such as surface collection; in essence any method that facilitates observation of the buried evidence without impacting the surviving stratigraphy is encompassed within the remit of remote sensing.

A casual observer wandering through a library, particularly in the United States, might be forgiven for believing that satellite remote sensing is the prime application for archaeological remote sensing. This is of course not the case; for every 'site' identified from space thousands have been identified through air-photography. In Europe air photography has been used to observe and document archaeological landscapes for more than a century, and this method remains by far the most significant contributor to the ever expanding archaeological record; we can see for instance, from the papers presented at the annual Aerial Archaeology Research Group (AARG) conferences, the tremendous impact air photography is now having in those European countries where flight restrictions had in the past made air photography from light aircraft impossible. In a sense we need to be very clear to separate the 'remote' from the 'sensing' as, regardless of the sensor platform employed, it is something on or, more importantly beneath the ground that we are trying to identify. Viewed at a global scale there are significant differences in the types of return that we might expect from open agrarian landscapes as opposed to deserts or dense forests, and the associated differences on the ground will influence the fitness for purpose of the many different remote sensing methods in each setting. LiDar for instance is a relatively new technology but, in contrast with other technologies, it is absolute and relies at the most basic level on the simple reflection of a laser beam from the ground to recover high precision relief information that might be obscured from view, both from the air and the ground, by dense vegetation. The results of high density LiDar surveys carried out over dense forest both in Europe and Central America have been nothing short of astonishing, revealing prehistoric barrow cemeteries and field systems in Europe and entire pre-Columbian cites in the jungles of Central America.

We are only now learning of the great potential for the recovery of additional information from the reflective intensity of the LiDar beam which can indicate degrees of wetness or the level of organic component in the soil facilitating, for instance, the mapping of palaeochannels. Whilst LiDar can be applied to produce highly accurate large area and high resolution topographic maps we need to be aware that what we are mapping is the present day land surface and that in former wetlands for instance the present landform may not accurately reflect the prehistoric or later land surface which has been transformed, in the case of the Vale of Pickering, by desiccation of buried peats as a consequence of both deliberate drainage and climate change. In contrast, most other forms of remote sensing rely on complex combinations of variables which, to produce good returns, are heavily dependent upon particular prevailing conditions.

Resolving the Past

The fitness for purpose of any remote sensing technique is heavily dependent on the scale and nature of the features we wish to discover and thus on the resolution of the data collected. Similarly, the returns from any remote sensing application are determined by a multitude of variables which combine to enable us to visualise contrast between archaeological features and the background within which they exist; this is most obvious in the case of crop mark features whether recorded from the visible or invisible parts of the spectrum. The relationship between imaging resolution and feature size is not absolute and thus we can, for instance, resolve the white lines on a major road in relatively low resolution (1.5m) multi-spectral data even though they are very much smaller than the pixel size; the skid marks left by traffic are frequently far larger but have a very low contrast relative to the dark grey tarmacadam road surface rendering them 'invisible'.

Air photography and the recording of crop-mark evidence at high resolutions of a few centimetres per pixel has made by far the greatest contribution to the identification of archaeological features through remote sensing. The returns from air photography are wholly determined not by the technology employed to see them but by the combined factors of crop, soils and the weather both over the long growing period and short term stress at stages during crop growth. The temporal factors influencing crop mark visibility are such that the returns are heavily dependent upon being in the right place at the right time and, although the broad relationship between soil moisture deficit, crop type and weather patterns are appreciated, the presence or absence of visible crop marks remains exceptionally difficult to predict. The underlying conditions which determine whether crop-mark evidence can be recorded using conventional film, digital cameras or multi and hyper spectral scanners are essentially the same and thus we need to remain aware at all times that the absence of crop mark evidence by no means implies evidence of absence.

In the Vale of Pickering where we have been engaged in remote sensing for more than 30 years, the returns from airborne remote sensing have been extraordinary. The overall density of evidence returned, primarily reflects frequently repeated observation of the same landscape over a very long time. It has become startlingly clear that continued re-examination of the same landscape with multiple methodologies is necessary if we are to even consider approaching our objective of identifying the total extent of the archaeological resource. If this cannot be identified then we need to be honest about the limitations of our visible identified sample. Frequent observation from light aircraft employing both film based and digital air photography has demonstrated a fall in the returns of 'new sites' over time; however, the sudden appearance of previously unobserved 'sites' indicates that the visibility of archaeological features is being greatly influenced by changes in agricultural practices on the ground. More worrying is the fact that through continuous repeated flying over the same landscape what amounts to anecdotal knowledge has been revealed, the importance of which has only now become clear. During the long period of observation we have witnessed significant changes which reflect changes in the crops grown, the ravages of increasing plough damage and the effects of climate change, the accumulated significance of which is now very clear. The formerly well preserved archaeology of our chosen research area has been seriously compromised by gradual change. Changes in the crops grown and the influence of changing climate have led to a point where the nature and timing of the crop-marks themselves have changed so that established 'rules' regarding the most likely time to see the best crop marks are no longer valid. New strains of wheat and barley reflecting the reduced need for straw and the need to produce crops that are more resistant to intense weather events, particularly flattening by high winds, means that the average height of the crops themselves has been reduced by more than half. Although air photography now has a less important role in our work in the Vale of Pickering than it has in the past, air reconnaissance over the last few years indicates that we need to be more aware of the returns from vertical photography undertaken earlier in the growing season, in May rather than June or July, when clear crop-marks have been observed in the lower vegetation of cereal crops which remain largely invisible when viewed obliquely.

The objectives of our work in the Vale of Pickering remain focussed upon trying to recover a complete picture of the archaeological landscape through time, rather than simply identifying 'good' sites. Gaps in the crop-mark record which tend to emphasise a picture of a landscape containing disconnected 'sites' were identified as a key problem more than 20 years ago, and much of our work over the last decade has been directed towards testing these areas. Multi- and hyper-spectral imaging recording images from visible and invisible wavelengths including near Infra-Red and thermal wavelengths was seen as a potential method to enhance the returns from conventional air photography. An initial survey supported by a data grant from the Natural Environment Research Council (NERC) and processing supported by a grant from English Heritage, was undertaken in 1992. A second NERC supported survey was conducted in 2006, when LiDar and CASI data were also collected covering much of our primary research area; the second experiment included an additional flight, undertaken at night, intended to secure data that would allow us to investigate whether features could be identified on the basis of thermal decay; both of the NERC surveys incorporated the recovery of high resolution vertical photography. The 1992 survey revealed a wealth of crop-mark evidence; the flights were undertaken at exactly the right time when crop-mark conditions and contrasts were exceptional and, although some additional returns from the infra-red and thermal wavelengths could be isolated, the most important results gained were from the high resolution colour vertical photographs, with a scanned ground resolution of 8-10cm per pixel. The pixel size of the multi-spectral data with a processed ground resolution of 1.8-2m per pixel, gathered with a Deadalus 12 band scanner, was insufficient to show detail with the clarity observed in the air-photographs. Had it not been for the fact that the 2006 NERC survey included the gathering of LiDar data, the returns would have been at best minimal. Crop marks in one or two fields confirmed the situation observed in the first survey, that whilst there can be a very slight gain in the return from the extended wavelength response of the Multi- and Hyper- spectral instruments the low pixel resolution (2m) of the airborne instruments deployed, and the fact that the response is directly related to the presence or absence of conventionally visible crop marks, mean that at current resolutions this technology has at present limited viability. Once again, where we could see results in the multi-spectral imagery, these were also visible in the vertical photographs. A second problem arose since the original high resolution film recorder had been replaced by a digital camera; when flown at 800m in order to get the highest multi-spectral resolution, the digital photographs covered a smaller area than those gathered by the film camera such that in places there was insufficient overlap to allow image stitching of the photographic cover. In our case the costs of the data gathering process, which amounted to several thousand pounds, were covered through NERC research grants. The LiDar data was also collected at a relatively low resolution giving 2m. pixels; since the landscape under investigation has very limited tree cover the limited density of the LiDar coverage caused little impact.

More importantly, it has become clear that the mapped surface derived from the LiDar data, which articulates the present landform very well, does not give an accurate representation of the landscape that was here in the past as much of the area has been subject to changes resulting from desiccation of buried peats and loss of surface peats which would have existed during the past. Novel techniques have had to be developed to try and model the past landform based on estimates of the peat loss and subsequent reduction of the ground level over the former wetlands. These experiments in modelling have revealed 'islands', the limits of which have to some extent been confirmed by the distribution of features recorded both from the air and through geophysical survey. It is absolutely clear that if we are to effectively employ Multi- and Hyper- spectral instruments for survey over intensively farmed arable landscapes then we need far higher resolution instruments returning data at 25cm per pixel resolutions or better. The lessons learned in the Vale of Pickering relate both to the resolution of the data we wish to map and to the type of landscape being investigated; we are well aware that what we would consider low resolution survey over desert areas for instance can produce significant returns, but the limitations of pixel size against feature size will still impact the results. We can however, look to the future with some confidence as improving resolutions from satellite imagery are giving significant returns, particularly from half metre resolution imagery and, in time, we shall see even higher resolution data recorded from space. More importantly, much of this new data is freely available through such internet based platforms as Google Earth.

Whilst still advocating continued aerial survey it became clear that an alternative approach would be necessary if we were to approach any form of absolute record of the archaeology of the landscape. During the early 1980's we had undertaken a number of limited tests employing ground based geophysical survey techniques; both magnetic and resistance methods had produced poor results, admittedly from very small areas. Following the production of some astonishing results by English Heritage, using gradiometery as part of a very large excavation program in West Heslerton during the 1990's, we tried gradiometery again. Initially our objectives were to see if we could use high resolution surveys undertaken during excavation to assist in the management of multi-hectare open area excavations. The results were magnificent and, by re-surveying areas after removal of the plough soil, we were able to see detail that greatly contributed towards the success of the excavation program. A series of new tests were undertaken covering areas that had not produced good crop-mark evidence; these confirmed that our original samples had been both too small and in the 'wrong' place. During the last decade a series of very large scale surveys funded by English Heritage, both from Archaeology Commissions programme and the Aggregates Levy Sustainability Fund, have produced a combined survey covering a contiguous area of nearly 1200Ha. This has been an enormous undertaking but the results reveal a landscape completely covered with archaeological features from the Neolithic to Medieval periods, evidence which challenges almost everything we thought we knew about the scale and density of past activity, not only in the Vale of Pickering but in lowland valleys throughout much of the East and South-East of England. Sands and gravels have not traditionally been seen as the best sub-soils for returning good gradiometer results. In the Vale of Pickering however, the magnetic contrasts between cut features such as ditches and pits, and the natural sub-soils, tend to be very high with good response from features even when they are buried beneath more than a metre of blown sands. As time has progressed changes in the gradiometers used and, in particular, the collection interval have changed such that we now collect data using a wheeled Foerster DLG Kartograph, employing an RTK GPS to secure precise data positioning with a point density of 0.1m x 0.5m or 0.1m x 0.25m allowing us to clearly resolve small features including on occasion post holes. The Foerster gradiometers are theoretically slightly less sensitive than the Bartington sensors employed previously but we can see no difference in the response within our working environment. Fundamental flaws in the Foerster software and problems with the default setup of the instrument were resolved by producing our own software following an extensive series of tests using many alternative settings; the results now speak for themselves and reveal the most significant limitation - simply that the kit has to be operated manually, restricting the coverage that can be achieved in a day to around three hectares. The area covered could be greatly enhanced by constructing a new cart or trolley that could be pulled by a quad bike; this has been done in France by Michel Dabbas of Geocarta, who are able to cover well over 10Ha per day.

When we combine the results of the many aerial campaigns with the geophysical data we can see that neither approach gives the same returns, confirming that if we are to reach our target to identify the archaeological capacity of the landscape then we must employ a multi sensor approach. It is likely that we could further enhance the returns in some areas by employing resistance surveys but, when conducted manually, these are very time consuming and, in contrast with magnetic techniques, resistance surveys like crop-mark observation are affected by multiple variables and long term environmental conditions which make the effectiveness of the technique on any particular soil or at any particular time very difficult to predict.

In contrast with the landscapes surrounding the Mediterranean for instance, the archaeology of the Vale of Pickering does not respond well to surface collection surveys. The majority of archaeological structures were of timber, and the quantity of ceramics and other material culture evidence is relatively low; even heavily ploughed areas do not necessarily reveal much on the field surfaces with the exception of times following new plough damage events when material is brought to the surface as a result of the underlying archaeological surface having been truncated. An alternative and effective, if labour intensive, approach is to sieve the plough soil in metre squares on a regular grid which allows us to investigate the full range and condition of material within the plough soil as a whole. This approach allows us not only to quantify the material in the plough soil and secure dating evidence, but also secure evidence on site condition as indicated by the relative size and abrasion of the ceramics held in the plough-soil in particular.

Why Bother

One might ask why it is worth spending so long investigating a single landscape in so many ways, especially since true insight into the nature of the buried archaeology can realistically only be gained through excavation. In just the same way as we need to make sure that our remote sensing projects are fit for purpose at the 'site' level, we also need to identify precisely what we want out of these projects in general. If one is happy producing dot density maps reflecting emptiness interspersed with dots denoting 'sites' deemed to be important, or simply those places noticed when looked for, the contribution to the understanding of the past will remain minimal. If we take the example of West Heslerton, the archaeological record in 1977 for an 8x10km area included a handful of 'sites' comprising a long barrow, a few round barrows, many surviving if severely truncated as visible bumps in the landscape which had been 'excavated' or perhaps more correctly 'robbed' in the 18th or 19th century, a number of major linear earthworks again severely truncated, a Late Bronze Age/Early Iron Age palisaded enclosure at Devils Hill, excavated in the 1960's and again in the 1980's (Brewster, 1963,. 1981; Powlesland, 1987), a set of monastic earthworks with a partially surviving monastic structure incorporated in a post dissolution farmhouse, and some spreads of Iron Age and Roman pottery recovered by school children in the 1960's. As is almost invariably the case, our attention or focus on this area did not derive from a proactive research programme but as a consequence of chance discovery in 1977, when an employee working at a sand quarry and walking his dog at lunchtime observed Early Anglo-Saxon burials eroding out of the sand after the topsoil had been removed prior to sand extraction. Most people would have failed to notice the fragments of human bone eroding on the surface. However, by another element of chance, Jim Carter, the quarry worker, had been taught at school by the passionate and I believe eminent, Yorkshire archaeologist T. C. M Brewster at West Heslerton primary school, where he had been one of the children collecting material from the fields in the 1960's. He recognised the potential importance of the site and alerted the authorities. A salvage excavation was mounted, and after the rapid excavation of 34 burials the site was declared cleared. The following summer I was invited to examine another area of the quarry where fragments of non-diagnostic pottery and a few post holes were visible on the stripped surface; these were initially considered to be Early Anglo-Saxon in date, although we quickly learned that they related to a Late Bronze Age and Early Iron Age settlement (Powlesland et al. 1986). Yet another chance discovery, an Early Anglo-Saxon brooch on the surface beyond the area covered in the salvage excavation, led us to examine the southern margin of the quarry and ultimately to the excavation of a further 169 Anglo-Saxon and 12 prehistoric burials from the cemetery (Haughton & Powlesland 2001).

Had it not been for those multiple chance events or serendipity none of the work in the succeeding 32 years would have come about; work could have stopped after the salvage excavation or even after the excavations at Cook's Quarry in 1986; in either case the result would have been articulated through small dots on maps and perpetuated a story of the past, in the apparently archaeologically barren landscape of the Vale of Pickering, that was fundamentally flawed. Since the first air photographic sorties associated with the excavations in 1978 the fundamental basis of our remote sensing program has been to address context - to connect the excavations and therefore the past that they revealed in a landscape context defined in terms of both space and time. At the simplest level, the discovery of the various cemeteries, whether Late Neolithic and Early Bronze Age or Anglo-Saxon, implies a population, and therefore settlement, and thus an economy based in the landscape. Whilst there has been a tremendous emphasis on archaeological theory in academic institutions over the last 30 years, much of the theorising has been data free and, whilst an interesting intellectual exercise, has not necessarily moved us forward in terms of understanding the past. Alternatively, when data has been used it has invariably been data discovered only as a consequence of serendipity, and therefore lacking context and flawed from the outset. We can, for instance, develop useful theories of settlement development only when we know where those settlements were.

The Bigness of Tractors

Whilst we have been heavily focussed upon identifying what we might term the '*Real Archaeology*', or identifying the archaeological volume of a segment of landscape both through large scale excavations covering more than 30Ha in total, or combined and intensive multi-sensor remote sensing, we have failed to take sufficient notice of the eroding fragility of the resource. This is not simply a reflection of our own lack of attention with respect to the Vale of Pickering, but a more widespread or national problem in failing to fully appreciate the combined impact that changing agricultural practices have made upon the buried resource on the one hand and the impact of climate change and changing land use on the other.

The word '*unique*' is one of the most misused terms with reference to archaeological 'sites'; every 'site' is unique, but to stress this uniqueness frequently gives a false impression of the importance of a particular 'site'; some 'sites' such as Stonehenge do of course deserve the term unique, but we should reserve the use of the term for the truly exceptional. In the case of our work in the Vale of Pickering, a combination of conditions make the archaeology exceptional; perhaps the most exceptional aspect of the archaeology is the long term investment of time and individual energy applied to the discovery of the buried resource. The most important physical aspect relating to the archaeology of the our research area, is the presence of a broad band of blown sand which has served to preserve the archaeological stratigraphy in a way rarely encountered in Britain. Despite the fact that the conditions that prevail here are exceptional, we should not preclude the survival of similar archaeology, similarly preserved, elsewhere. The blown sands that are a characteristic of the landscape to the east and west of West Heslerton were identified as an important characteristic of the landscape as early as 1978, and it was quickly realised that the gradual build-up of blown sand in the area, which was present and mobile as early as the Mesolithic, had buried old land surfaces over large areas. The original form of these deposits was probably as slight sand dunes which were mobile throughout prehistory such that land surfaces could be buried for very long periods and then re-exposed before later being re-buried. In some areas, perhaps as early as the Bronze Age, the blown sands started to accumulate so that the ground level began to rise, preserving the floor layers associated with structures in a way rarely found in rural lowland England. This remarkable sequence which led, for instance, to the complete burial of an Early Bronze Age barrow or burial mound beneath an apparently flat field, is now in reverse; the blown sands continue to move, becoming trapped in the hedges that bound the fields and occasionally blocking roads across the blown sand zone. In the early 1980's the fields over the blown sand were classified as of very poor agricultural value; the iron-rich blown sand is well bonded and relatively sterile. Changes in the types of crops grown and the increased use of irrigation have made this land more productive, but associated deep ploughing is releasing the blown sand, cutting through these deposits into the underlying archaeology which is much softer than the blown sand itself. The impact of increasing irrigation is leading to desiccation of former peats and other wet deposits in the centre of the Vale of Pickering, with a consequent loss of organic and environmental evidence, whilst increasingly deep ploughing is having a catastrophic effect.

As new larger and more powerful tractors are introduced, so are larger ploughs. This, combined with an emphasis on the production of potatoes in particular, is leading to the wholesale destruction of the very deposits that make the archaeology of the area so important. With hindsight we can see the progressive destruction of the buried archaeology is directly linked to tractor size and it appears that whilst we have been busy trying to identify the buried resource it has been progressively truncated. Plough damage is the single most progressive threat to the archaeology of lowland Britain, and probably most agrarian regions in Western Europe. As the levels of plough damage increase, the contrast between the undisturbed subsoil and the fills of archaeological features such as ditches increases and the crop-mark response 'improves'; we therefore need to be careful not to overemphasise the importance of 'good' as opposed to 'poor' crop-mark sites. We now realise that a large number of the crop-mark sites which were afforded some degree of protection through being designated as scheduled monuments in Britain in the 1970's were in fact already compromised as a consequence of plough damage, and we should have been focussing our attention on the less clear crop-marks in adjacent fields which in many cases have subsequently become similarly compromised. At a landscape level the subsurface truncation of enclosure ditches or track-way ditches is perhaps less significant; we can still map these features and excavation may yet reveal both dating and cultural evidence. However, it is the truncation of occupation deposits, floor layers and structural component's most of which remain invisible to most forms of remote sensing, that have the real potential to change our understanding of the past that is most critical. Some years ago we investigated through excavation part of a Late Roman and Iron Age linear settlement in the Vale of Pickering. The site had been preserved by aeolian sands just a few centimetres thick; floor layers and structural elements survived and, although banks associated with the ditched settlement enclosures did not survive, the distribution of snail shells revealed on which side of the ditches hedges had formerly been positioned.

Twenty years of increasingly intensive ploughing have since badly damaged this settlement complex, which now not only shows regularly from the air but is visible in the winter as soil marks. With online tools such as Google Earth (GE) and Microsoft Virtual Earth (VE) (it is important to use both systems as they do not necessarily use the same imagery) we are now able, in some areas, to view high resolution air-photographic snap-shots of whole landscapes, revealing both astonishing crop-mark complexes but also significant levels of damage. If for instance you search for 'Yorkshire Wolds' in VE, the focus will zoom to a location just outside the village of Tibthorpe; zoom right in and you will see crop-marks of a trackway and associated enclosures clearly visible in the field immediately to the west of the VE place-marker; the same fields viewed in GE show no crop-marks - in both cases the data source is high-resolution air photography, but the time of year at which they were recorded is different. The inclusion of a time based control in GE version 5.0, which allows one to see other earlier image coverage where it exists, has tremendous potential, especially as the control can be used to manipulate the display of overlain data over time. Sadly, the extraordinary imagery visible in VE covering much of the Yorkshire Wolds does confirm the anecdotal knowledge, identified by reflection over 30 years of air photography, that plough damage has been occurring at an increasing rate and intensity; the clarity of features in crop and soil marks (with characteristic zig-zag patterns made as the plough drags the soil to either side of the feature) reflects both the presence of the features themselves and the substantial damage that has occurred.

Towards a broader perspective

If, as we believe, the buried evidence of the past in the Vale of Pickering has become compromised through truncation whilst we have been discovering it, there are lessons to be learned from this work. In an increasingly globalised world with a burgeoning population with consequent pressures on food production at a time of devastating climate change, large parts of the world are undergoing the transition from subsistence and essentially un-mechanised agriculture to industrial farming. What we might term High Density Remote Sensing (HDRS), employing multiple techniques, and repeat observation of large areas of landscape does offer the potential to identify the archaeological resource in a way and at a scale that could not only transform understanding, but provide the basis for securing the future of the archaeological resource, or at least substantial landscape scale samples. No-one has the resources to undertake the sort of work we have undertaken in the Vale of Pickering at a national scale, even with the tremendous improvements, in areas that can be covered, using mechanised gradiometer or resistance arrays; however, there is no excuse for not investing in sample areas and selectively targeting appropriate techniques to recover a landscape based rather than 'site' driven picture.

The archaeological remote sensing community has a clear role here in trying to encourage governments and organisations like Google and Microsoft to invest in high resolution imagery, ideally taken at an appropriate time, which can not only serve as a primary data resource but also a map base for targeted research areas. Poor map coverage in many parts of the world has hampered research for generations; the global provision of satellite and air-photographic cover through internet based platforms such as GE, which are regularly updated with new imagery often at greatly improved ground resolution, is a tremendous resource which provides a good and free starting point for research in almost any part of the world. That we can also integrate our own data in systems such as GE should not be overlooked, as this provides a mechanism for rapid dissemination of research results. We have seen rapid changes over the last decade in Europe with regard to the possibilities for domestic air photography in countries where formerly it was almost impossible for archaeologists to take to the air, and we can look forward to more countries freeing up flying restrictions.

Ultimately, whilst discovery is exciting, the results of HDRS surveys raise uncomfortable issues with regard to long term management that need to be addressed. How do we facilitate the long term management of a few thousand hectares of archaeology which is being intensively farmed under an arable regime? How do we even secure a sufficient understanding of the landscape identified beneath our feet?

In our work in the Vale of Pickering, to excavate to an appropriate standard all the features identified through our remote sensing programme would take literally hundreds of years; this is neither achievable nor within our objectives. If we are to preserve samples of landscape there is little purpose in preserving something we do not understand, and thus we need to develop strategies that combine large area preservation with ongoing examination so that we improve our understanding of the resource and both draw upon and develop new methods to enhance our understanding of the buried landscape. Whilst there is some hope that we can secure the future of some landscape samples it is clear that the majority of the evidence of our past will be either lost altogether or compromised by gradual erosion of the buried deposits. The gradual accumulation of accidental knowledge that is increasingly used in Sites and Monuments databases throughout the world to manage development can be greatly enhanced through pro-active long term examination of sample landscapes which provide comparative data-sets for the less intensively studied areas. The studies originally focussed around West Heslerton in the Vale of Pickering, covering c.100 sq km, have a far reaching impact on our understanding of the archaeology of the whole of the Vale which extends over c. 500 sq km, allowing us to draw upon information from the detailed study area to develop risk management strategies for other areas which have not benefitted from the same level of attention. The anomalous nature of the record for the Heslerton area is not a function of the levels of activity here in the past, but of the level and long term nature of the research. Crop-mark and other evidence in other valleys throughout lowland England suggest that similar levels of activity would be identified in these areas given the same investment of research resources.

Although the work in the Vale of Pickering raises difficult issues regarding long term management and sustainability and the need to understand the resource, we need to adopt a positive attitude to the results. The model of landscape research combining rescue excavation and HDRS approaches is applicable in most environments around the world; the technical details as to which method of remote sensing returns the most or best data will of course vary tremendously from one landscape to another, but the principles of multi-sensor, multi-resolution and multi-temporal data collection are universally appropriate. Difficulties relating to the future of the discovered or observed past should be considered in the context of the undiscovered landscape which might simply otherwise be represented by individual unconnected 'sites' or spectacular individual crop-marks. If the interpretations we place upon ancient landscapes are to have any validity at all we need to be clear that we have sufficient evidence upon which to build our interpretations, and that the evidence we use is representative. One has to remind oneself constantly that the returns from one technique may work in some environments brilliantly and not at all in another; in remote sensing the lack of evidence can never be viewed as evidence of absence. Whereas in the Vale of Pickering surface collection has very rarely proved a successful indicator of buried archaeology, surface collection has produced excellent results in the intensively ploughed Mediterranean environment. Geomagnetic surveys have been particularly successful in our research but the response tends to be poor, for instance, for Neolithic features; the ditched barrows of the Late Neolithic and Early Bronze Age sometimes occur as geomagnetic anomalies but in some cases are only visible in the airborne dataset. The ditched trackways and enclosures that dominate the Late Iron Age and Roman landscape show particularly well in geomagnetic survey's but internal details such as post-hole or stone structures are very rarely visible.

We are currently engaged in work to bring all this research into the public domain, work that not only combines the detailed feature by feature analysis and interpretation but also addresses the strengths and weaknesses of different remote sensing methods. It is absolutely clear, for instance, that the returns from satellite images and multi- and hyper spectral imaging are constrained both by the image resolution as well as by the same sort of ground conditions that influence crop mark formation in agrarian landscapes. It is equally clear that mapping the resource is technically very simple, especially in our working environment where we have had our own dedicated GIS operating for more than 25 years; what is more problematic is what happens when we make our data available on Google Earth or some similar platform. Where does our responsibility lie when it comes to revealing the details of past landscapes which can form the basis for treasure hunting and the consequent loss primarily of dating evidence, and how do we secure long-tem sustainability of the resource? Rather than see these questions as a negative result of such a comprehensive research programme, they reflect the depth of knowledge which is transforming our view both of the investigated landscape and the comparable landscapes beyond, a view that has forced us to change our views of population and continuity in a landscape that since the Bronze Age has been alive with activity, rather than in any way empty and devoid.

BREWSTER, T. C. M. 1963 *The Excavation of Staple Howe,* E. Riding Arch. Res. Committee

BREWSTER, T. C. M. 1981 'The Devil's Hill', *Current Archaeol.,* 76 (1981), 140-41

FAGAN, B.M. 1959, 'Cropmarks in Antiquity', *Antiquity* 33, 132, 279-281

HAUGHTON, C.A. And POWLESLAND, D.J. 1999 *West Heslerton - The Anglian Cemetery,* Landscape Research Centre Monograph 1,2 vols. Yedingham

POWLESLAND, D.J., HAUGHTON, C.A. and HANSON, J.H. 1986 'Excavations at Heslerton, North Yorkshire 1978-82', *Archaeol. J.* 143, 53-173

Session 1

AERIAL ARCHAEOLOGY: SATELLITE

Pushing the envelope for satellite archaeology in Egypt: Quickbird feature detection, predictive site modeling, and thermal site signatures

Sarah Parcak

Director, Laboratory for Global Health Observation
Department of History and Anthropology
The University of Alabama at Birmingham
Birmingham AL 35294 USA
sparcak@uab.edu

Abstract

This paper will present a variety of techniques for the detection of archaeological features in Egypt, in order to understand greater anthropological issues, including population densities, ancient settlement locations, and modeling past human-environment interactions. This includes the utilization of Quickbird high resolution satellite imagery in the detection of subsurface architectural features, and advanced interpolation techniques. In addition, ASTER imagery is used to show how thermal responses (bands 13-12-11) might aid in the detection of archaeological features.

Keywords Satellite archaeology; Egypt; Quickbird; Predictive site modeling; ASTER; kriging; population studies

Background

Archaeologists have stressed the bias of Egyptian archaeology towards monumental and funerary sites and virtual neglect of ancient settlements in the Nile Valley and the Delta. In addition, how Egyptologists conduct landscape surveys (to the extent they have been conducted in the past) has been at odds with other regional survey techniques in the Ancient Near East. Regional survey work in Egyptology had been largely neglected due to the mostly non-settlement bias and has consequently missed thousands of ancient settlement sites and related landscapes, all of which are currently being destroyed by modernization. Using 30 years of satellite imagery, the author (see Parcak 2009a, 2009b, 2008, 2007a, 2007b, 2005, 2004, 2003) has noted modern cities on top of or beside archaeological sites in the East Delta and Middle Egypt growing between 50-200%. A technical manual for future ground survey work in across Egypt is desperately needed for Egyptologists and members of the SCA, so future survey work can be conducted in a scientifically rigorous fashion.

At no other time in our human history has the protection of past ancient remains been so critical: Given escalating development, urbanization, population growth, and ancient site looting (which is funding the insurgency in Iraq and Afghanistan), world governments have a limited time window in which to detect and protect ancient archaeological sites before they are gone forever. The key to this protection and detection is training, and the implementation of this research would train future generations of scientists in cutting-edge technological applications to global problems. In addition., the United States has just (2008) ratified the 1954 Hague Agreement, which states that, in times of war, invading countries have an obligation to protect and preserve cultural resources. At present, there is no directed government funding operation to support this effort. This funding would help to fill the gap. Hundreds of thousands of archaeological sites remain undiscovered and protected by governments not just in Egypt, but around the world.

Fig. 1. Deir el-Medina, Luxor, Egypt, imagery courtesy of Google Earth Pro and NASA

In addition, landscape modeling with satellite imagery can aid in understanding past people's relationships with their landscapes, and can illuminate patterns not otherwise evident using satellite imagery alone. Draping Quickbird satellite imagery over NASA SRTM data aids in relating human modified landscapes to their natural environments, seen in Figure 1 showing Deir el-Medina (ca. 1500-1300 BC). This town, which housed the main workers responsible for constructing royal tombs in the Valley of the Kings, Luxor, is often not appreciated from the perspective of its distance from the Nile and floodplain area. Its "hidden" location is better understood from satellite imagery, and the ancient

Egyptian's subtle placement of the town in an area not viewable from external locations.

On understanding settlement patterns in ancient Egypt

A great deal of literature exists on specific settlement sites, and a great deal of additional attention has been drawn to the excavation of settlement sites in the past 25 or so years. Karl Butzer's seminal work on *Early Hydraulic Civilization of Ancient Egypt*, published in 1976, described settlements along the Nile Valley and in the Delta, attempting to examine changes in settlements through time. For its time, it represented a breakthrough work. It-unfortunately-remains the most comprehensive study of its type in the field of Egyptian archaeology, and is over 30 years out of date. As an example: the book only lists 6 sites for the entirety of the Egyptian Delta. Well over 700 have been documented to-date (see www.ees.ac.uk/delta).

To what extent should archaeologists care about evaluating settlement patterns in ancient Egypt in the face of numerous new "headline" discoveries? Archaeologists should care about reconstructing past patterns of occupation because the field of settlement pattern archaeology in Egyptology is in its nascent stage. Given how much it has revealed about broader patterns in ancient near eastern archaeology, it promises to enrich the field of Egyptology just as much. We simply do not know where the majority of cities and villages were in ancient Egypt, their relationships to one another, and how many people lived in them.

The other reason for advocating the subfield of settlement pattern archaeology is the rapid urbanization of Egypt, which is systematically destroying hundreds, if not thousands, of ancient sites remaining to be discovered. Examining a Quickbird satellite image of Tell Tebilla, a multi period site dating to ca. 600 BC located in Egypt's East Delta , in Figure 2, one can see how much of the site has been cut into, especially over the past 40 years, and some cropmarks are still evident to the north and east of the site.

Thus, how can one test the very nature of predictive modeling, and what techniques work the best? Typically, archaeologists predict the location of likely archaeological sites, and do surveys to determine their accuracy with on-ground surveys. What happens when we have the data to attempt this in reverse? Using data collected from the Middle Egypt Survey project (directed by the author, and active since 2004, see Parcak 2004 and 2007a) tests the very nature of archaeological site predictive modeling. Starting in 2004, the MESP has used a variety of satellite images to detect a number of archaeological sites, combined with survey. Of the 70 sites found using the remote sensing analysis, 17 had been known from previous surveys. A predictive modeling technique known as "kriging," otherwise connected to optimal prediction, aided in the predictive modeling.

Fig. 2. Tell Tebilla, East Delta, Egypt, imagery courtesy of Google Earth Pro.

Predictive site modeling for Egyptian archaeology

A major question in general archaeological survey has been, to what extent does archaeological site predictive modeling work? If archaeologists know the location of archaeological sites X, Y and Z, how can they use that information to find sites A-W? This is an important question, and one that has not been well-tested with survey data. If archaeologists can answer this question using remotely sensed data, then perhaps they can start to fill in the gaps of archaeological site locations over much broader landscapes.

Kriging is a method of interpolation, or a technique that creates new data points from a set of known points, or a set of linear regression steps. It allows for predicting "new" data points from known data sets or points, and shows relationships between random and deterministic variables. The project tested several krigs to verify the results on the on-ground project work, one of which is shown in Figure 3. It does not reveal much information about the potential location of other sites. Larger dots represent known sites, and the smaller black dots represent sites found during the MESP survey.

Typically, data is kriged, and the likely location of sites is estimated based on overall size area. Instead of following the standard kriging steps, the project examined the overall archaeological site areas and location information from the 17 previously known archaeological sites in the survey region and ran a series of krigs to test the feasibility of the technique. 53 sites, already known in terms of their size and extent, could be cross referenced to determine to overall accuracy of the krig models.

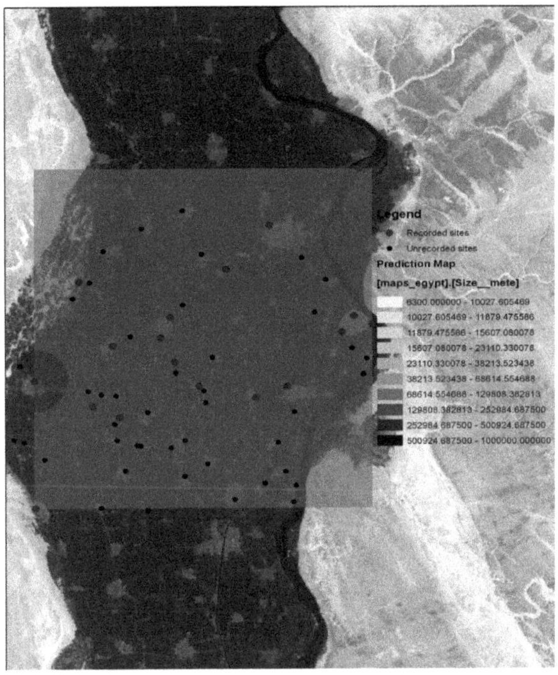

Fig. 3. Simple krig draped over NASA Landsat satellite imagery, imagery courtesy of NASA.

Like any statistical model, the various krig models vary in their overall accuracy. Krig models attempted include ordinary kriging, universal kriging, Inverse Distance Weighted (IDW), and simple kriging, some of which produced smoother results than others. Ordinary Kriging, Universal Kriging, and Inverse Distance Weighted (IDW) predicted the total area and general location of a number of archaeological sites, most of which represent settlements dating to ca. 400-800 AD.

Kriging did show a high likelihood for sites larger in area to appear in relationship to the largest site found in the survey, over 1km in size and 40m in height, which may be skewing the data slightly. In fact, the Ordinary Krig, seen in Figure 4, shows how size site and occurrence should increase the closer one gets to the Bahr Josef and the desert, not the Nile, showing the importance of sites connected to possible trade routes along these routes. This is the opposite of what traditional Egyptologists might hypothesize about archaeological settlement location. As with any archaeological data, this is representative of the particular surface data and time period found (ie, the late Roman period), and may not be an accurate reflection of site locations from earlier time periods.

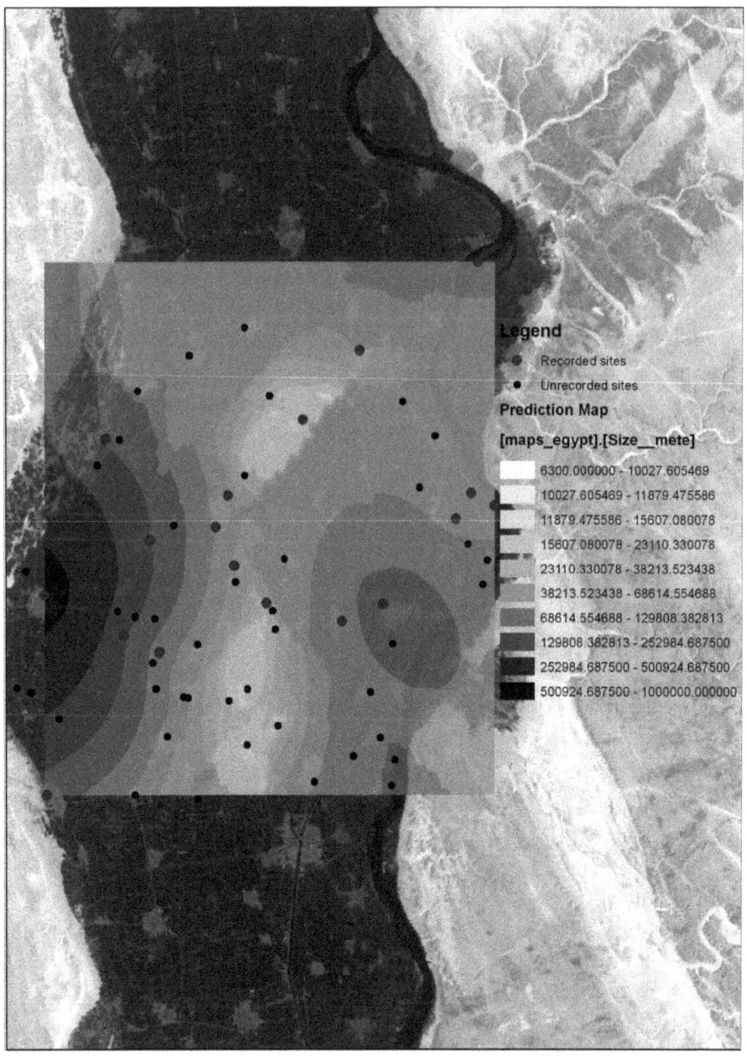

Fig. 4. Ordinary Krig, draped over a NASA Landsat satellite image, imagery courtesy of NASA.

Population reconstruction

In order to answer questions surrounding why sites might be located in particular locations, it is important to examine how we might reconstruct ancient population sizes based on archaeological sites densities indicated by the kriging, remote sensing and survey work. Population reconstruction was achieved through several means: calculating the total amount of land in the Hermopolite nome (which 40% of the MESP area covered) and the maximum number of people it could have fed, examining town size and reconstructing populations based on the average numbers of people living in each category of town (city, village, small hamlet). A hypothetical population of the rest of Egypt can be created, assuming a similar population density. Examining the spectral signatures of potential sites north and south of the survey area suggest that population density remained consistent throughout the Nile Valley, while survey results from the Delta suggest the same.

What do the "experts" say about the population of Egypt in the Late Antique period, ca. 300-800 AD? Diodorus Siclus says that 3 million people lived in Egypt at the end of the 1st century BC, while Josephus stipulated that 7.5 million people lived there based on a poll tax. The number could be potentially as high as 8-9 million (Bagnall and Frier, 1994) This latter population figure is one that Egypt would not have until the late 19th century, in the post-perennial irrigation and industrial era. Census figures under Mohammed Ali in 1882 record a total population of 6.8 million, while Lane stated the land capacity, if all the land were sown, could support a population of 8 million (Lane, 1981).

Using figures from the satellite remote sensing analysis and subsequent survey, shown in Figure 5, there are seventy ancient sites out of which there are ten sites we can call large villages based on remains and site size. Rural population numbers are estimated, with small villages having between 2,000-3,000 people, or average populations between 1,000-4,500 people. These numbers are the same for Egypt's towns in the pre-19th century (Bagnall and Frier, 1994). Large villages measured 4,000 hectares, and could have more than 5,000 inhabitants, whereas small village measured between 250 and 300 hectares and had populations of a few hundred people (Bagnall, 1993). With ten "cities" in the survey area (measuring around 5000 people), and sixty smaller villages (each with an average population of 1,500 people), we get a total of 140,000 people for the survey area. Considering it only covers 40% of the likely total area in the Hermopolite nome, we have a more realistic figure based on actual town density of 430,000 people. This is close to the maximum number of people the land could have supported with its agriculture. We can give an overall total population number of 5.3 million for Late Roman Egypt, which is based on specific sites located, mapped and recorded during ground survey.

Subsurface archaeological feature detection

How can archaeologists create ways to plan entire sites using cost-efficient forms of technology, given that so many of these sites are under threat of destruction? While archaeological geophysics had had a long history of success in diverse environments, and continues to be a key tool in selecting promising areas for archaeological investigation across the globe, it remains a high-cost device, and is not a tool one can use in broadscale sites surveys. Decreasing funds from national and international scientific bodies make it necessary to consider potential time and cost-saving methods for detecting subsurface archaeological remains.

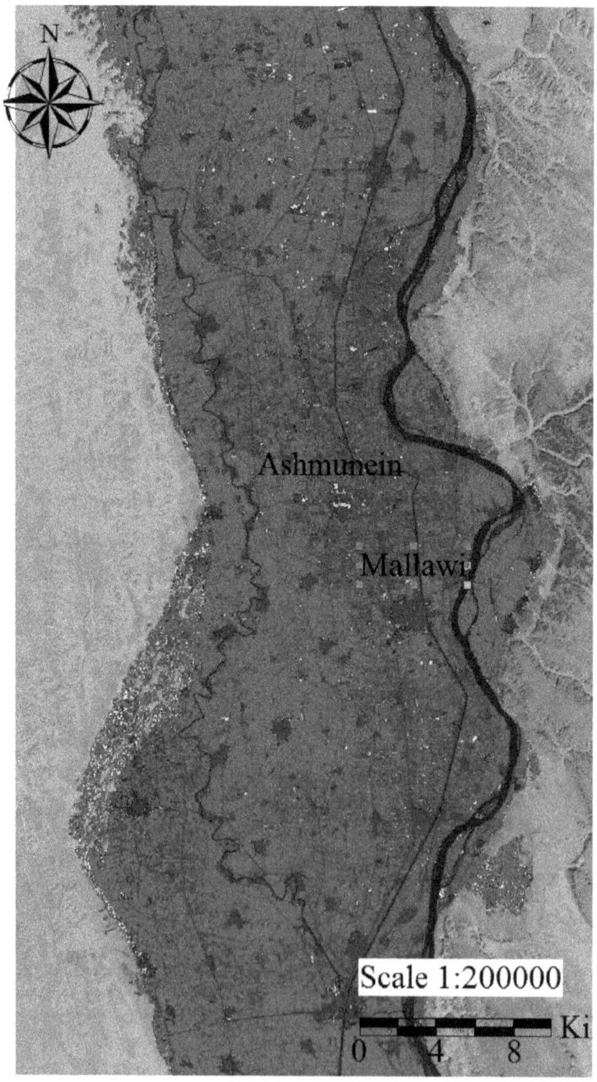

Fig. 5. 432 RGB maximum likelihood classification, showing the Middle Egypt Survey Area (measuring 15km x 30km, from Mallawi to the north edge of the scale bar) and areas to the north and south of the survey. Archaeological sites are represented by the white pixels and lighter grey pixels abutting modern settlements.

Using mechanisms on Quickbird high-resolution satellite remote sensing datasets (with a resolution of .61meters in the panchromatic) to detect potential buried subsurface architectural remains may be the way to detect buried remains, and, over time, map significant numbers of buried structures. Quickbird imagery has only begun to be tested in diverse archaeological areas, with Elizabeth Stone's 2008 paper illustrating how it can be used to monitor site destruction rates in areas under conflict. Using Quickbird's multispectral capabilities may also shed light on how archaeologists can map sites in advance of seasons, or in areas such as Iraq or Afghanistan where sites cannot be visited and mapped at present.

Buried ancient architecture and other features such as roads have properties which can be easily detected with magnetometers, resistivity and ground-penetrating radar, and those same properties make it possible to detect buried features from space. Mudbrick architecture, which represents the majority of ancient architecture in the ancient Near East, tends to retain more moisture than the soil surrounding it. This makes it easiest to see in the morning, when the raking light allows archaeologists to discern the architecture from the surrounding soil. Fortunately, Quickbird imagery is taken in the morning, when it is still early enough to differentiate between soil and mudbrick.

A Quickbird March 2006 satellite image of the city of Tell el-Amarna, seen in Figure 6, was used to test a variety of techniques for detecting potential subsurface remains. The image was first pan-sharpened, a remote sensing technique that merges multispectral remote sensing data with panchromatic data at a higher resolution. The image was pansharpened prior to attempting different filtering techniques. Spatial filtering is a common remote sensing method used to highlight differences in tonal image variations, and has many applications for archaeological contexts. Subtle variations between neighboring pixels can be emphasized or deemphasized depending on the filters applied. This is done via the image process of convolution, where kernels or coefficients can be applied to weight the digital numbers. Filtering can remove image noise, or can stress differences in edge values in either low or high frequency change areas, which is particularly useful when attempting to detect potential buried remains.

This work tested 1000 possible combinations using over 100 filter types in the program ER Mapper and 9 contrast enhancement techniques. While this work is in progress, early results are promising: using a high pass edge sharpening filter with an autoclip transform revealed a great deal of possible subsurface architecture, seen in Figure 7. Buried architecture appears to be quite similar to magnetometry survey results. Future publications will detail work testing buried and already excavated results with known maps to determine percent accuracy, as well as testing similar techniques on tell sites with published magnetometry results.

Fig. 6. A Quickbird March 2006 321-RGB satellite image, Tell el Amarna, Middle Egypt.

Thermal analysis in satellite imagery

Although maximum likelihood supervised classification and Principal Components Analysis had high success rates in detecting previously unknown archaeological sites in the Delta and Middle Egypt (Parcak 2007a), the project tested the thermal part of the spectrum, to date not used for the detection of settlement mounds in the Middle East. Bands 13 (8.925-9.275, 90m), 12 (8.475-8.825, 90m), and 11 (8.125-8.475, 90m) were combined to test the thermal response rates of archaeological debris from multi-period settlement sites ("tells") in the East Delta. Combining bands 15-14-13 and bands 14-13-12 did not distinguish the archaeological sites well from the surrounding landscape types, suggesting that it is the slightly lower part of the thermal IR that is most useful for this type of analysis, shown in Figures 8 and 9. Given the preliminary results, it appears that archaeological sites have a higher heat emission than modern towns and surrounding agricultural fields. This may be connected to the high amount of organic debris which a majority of archaeological sites contain. This technique holds great promise for the future detection of known and unknown archaeological sites in Egypt.

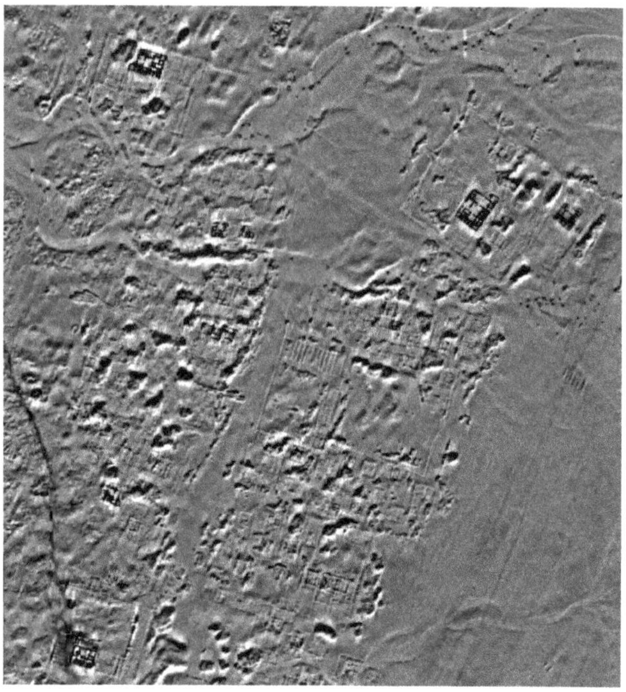

Fig. 7. Pansharpened IR+ panchromatic Quickbird satellite image, showing a high pass edge sharpening filter with an autoclip transform. The black architecture is above ground, and all other architecture is buried by desert sands. Imagery courtesy of Digital Globe.

Conclusions

This paper has touched on a number of areas within satellite archaeology and Egyptian archaeology. It is clear that a multifaceted remote sensing and ground-based approach is needed for the detection, protection, and preservation of archaeological sites in Egypt. All remote sensing analysis will depend on temporal, spectral, spatial, and weather-related characteristics of the remotely sensed and ground data. What we can say is that the field of settlement pattern archaeology in Egypt is ripe for analysis when combined with data mining from satellite datasets and accompanying ground survey. Archaeologists are not even at a point where they can begin to answer the questions they might pose, as they are only beginning to ask the right sorts of questions. Work over the next few years with advanced satellite remote sensing analysis in Egypt may begin, at last, to provide some answers.

Acknowledgements

The author wishes to thank Dr. Greg Mumford for his support during research and fieldwork, the University of Alabama at Birmingham NSF ADVANCE program and NASA for funding and satellite imagery, and the Supreme Council for Egyptian Antiquities for their permission to conduct fieldwork.

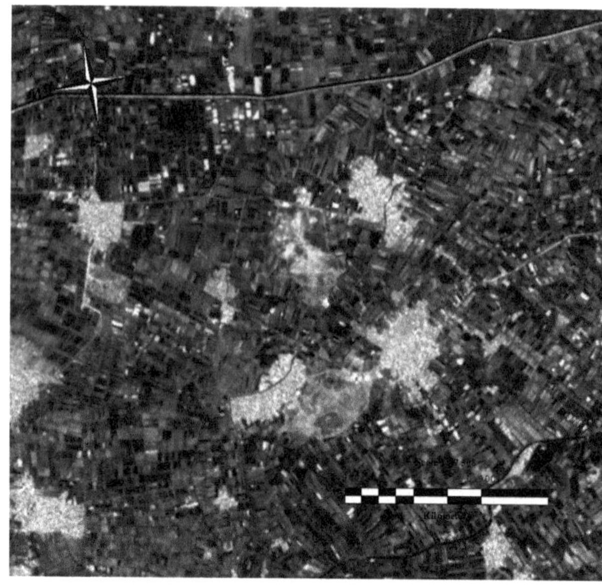

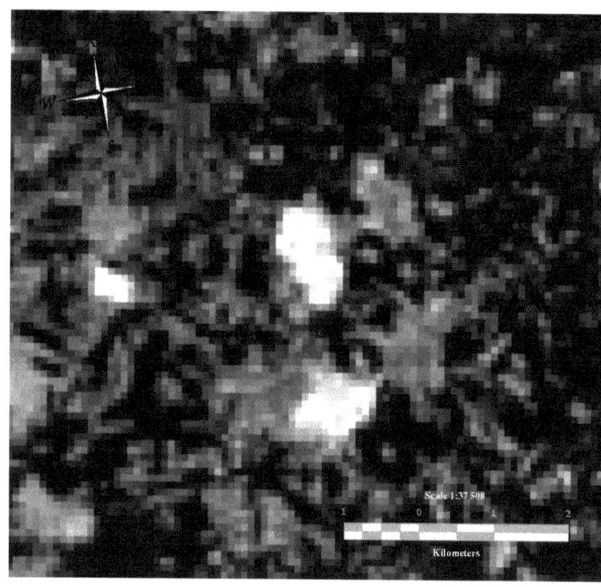

Figures 8 and 9. Image 1.8a is a 321-RGB ASTER image, showing the archaeological sites of Mendes and Tell Timai, seen as the two oval features in the central part of the image. Image 1.8b is a 13-12-11 RGB ASTER, showing the thermal reflectance of the two archaeological sites as well as a previously unknown site to the west, as bright white in color. Imagery courtesy of NASA.

References

BAGNALL, R. 1993. *Egypt in late Antiquity*. Princeton, Princeton University Press

BAGNALL, R. AND B. FRIER. 1994. *The Demography of Roman Egypt*. Cambridge, Cambridge University Press

Lane, E. 2000. *Description of Egypt*. J. Thompson, ed. Cairo, The American University in Cairo Press

PARCAK, S. 2009A. *Satellite Remote Sensing for Archaeology*. London, Routledge.

- 2009b. The skeptical remote senser. In S. Ikram and A. Dodson (eds.) *Beyond the* Horizon: Studies in Egyptian Art, Archaeology and History in Honour of Barry J. Kemp, Cairo: American University in Cairo Press.

-2008. Survey in Egyptology. In R. Wilkinson (ed.) *Egyptology Today*, Cambridge: Cambridge University Press.

- 2007a. Going, going, gone: towards a satellite remote sensing methodology for monitoring archaeological tell sites under threat in the Middle East. *Journal of Field Archaeology*, 42, 61–83.

- 2007b. The Middle Egypt Survey Project: 2005/6 season report. *Journal of Egyptian Archaeology*, 92, 3–8.

- 2005. Satellites and Survey in Middle Egypt. *Egyptian Archaeology*, 29, 28–32.

-2004. Finding new archaeological sites in Egypt using satellite remote sensing: case studies from Middle Egypt and the Delta. In C. Wang (ed.) *International Conference on Remote Sensing Archaeology*. Chinese Center for Remote Sensing Archaeology, Beijing.

-2003. New methods for archaeological site detection via satellite image analysis:case studies from Sinai and the Delta. *Archaeologia Polonia*, 41, 243–5.

STONE, E. 2008. Patterns of looting in southern Iraq. *Antiquity,* 82 (315), 125–38.

High-resolution, multi-spectral satellite imagery and extensive archaeological prospection: case studies from Apulia, Italy and Kazanluk, Bulgaria

Adela Sobotkova[1] and Shawn A. Ross[2]

[1]University of Michigan, Ann Arbor, Michigan, USA
[2]University of New South Wales, Sydney, New South Wales, Australia

Abstract

High-resolution, multi-spectral satellite imagery can be deployed in combination with ground control, archaeological surface survey, and environmental research to produce a richer understanding of the archaeological landscape. While intensive surface survey remains the gold standard for site detection, remote sensing allows larger areas to be evaluated quickly and efficiently. During 2007 and 2008, investigators analysed approximately 70 sq km surrounding the site of l'Amastuola, in Apulia, Italy, supplemented in 2009 by another 85 sq km in the environs of the ancient city of Seuthopolis, near Kazanluk, Bulgaria. Previous and concurrent surface survey provided site definitions and a large sample of sites discovered through an independent process. Integration and comparison of remote sensing and surface survey data indicated that although remote sensing dramatically increased the area evaluated in both regions, only certain sites could be detected and patterns of discovery varied by region. In Italy, remote sensing primarily revealed sites associated with geological conditions amenable to past human habitation. In Bulgaria detection depended upon a combination of cultural and environmental factors. Integrating the two methods allows remote sensing to extend the reach of surface survey, while surface survey indicates the detection rates and patterns provided by remote sensing.

Keywords remote sensing, satellite image analysis, reconnaissance, prospection, surface survey, site discovery rates.

This paper presents the results of a three-year research program in Italy and Bulgaria, during which high-resolution multi-spectral satellite image analysis was combined with archaeological surface survey for regional archaeological prospection.

In addition to archaeological reconnaissance our project's objectives included the improvement of remote sensing methodology for the archaeological investigation of large areas. Our ultimate goal was to gain a better understanding of the strengths and limitations of satellite image analysis by comparing the results of remote sensing, especially rates and patterns of site discovery, with those of surface survey.

Study Areas

The two study areas differ considerably with regard to environment, type of archaeological remains, and degree of previous archaeological investigation. It was hoped that developing a remote sensing methodology across two divergent regions would improve its general application.

The l'Amastuola region in Apulia, south Italy, is a well explored archaeological landscape, offering an existing pool of sites recorded through the long-term efforts of the Dutch investigators from the Free University of Amsterdam, with whom we collaborated (Burgers and Crielaard 2007). Its archaeological remains include few structures, but many "flat" sites consisting of subsurface remains that represent permanent settlements built from durable materials, dating from the Bronze Age through the Roman period. Environmentally, the l'Amastuola region includes the transitional zone between the karstic Murge tableland and rolling landscape cut by deep ravines (*gravina*) opening into coastal plain. Its climate is of a moisture-deprived Mediterranean type, in which water is the limiting resource for agriculture and habitation. Thin soils support perennial agriculture (especially vineyards and orchards) while fallow land is overgrown with Mediterranean scrub (*macchia*), or tall grass limiting access and surface visibility (van Joolen 2003, 5-7).

Archaeological landscape in the Kazanluk region is dominated by countless burial mounds dating from the Bronze Age through the Roman period, with comparatively few (known) "flat" settlement sites – and these mostly of short duration and built of perishable materials. Environmentally, Kazanluk region lies in a narrow valley enclosed by two parallel mountain ranges north and south. The valley's flat bottom is formed by colluvial deposits from the mountain slopes and alluvial sediments of the Tundzha River. The region is well-watered with a mild continental climate, further moderated by winds from the Aegean (Archibald 1998, 11-21). Most of the valley features open land divided between agriculture and herding. Land cover features fields of perennials such as rose, lavender and fruit orchards, with sunflower and cereal crops grown on better alluvial soils. Agricultural practices are still relatively low-intensity.

Methodology

Image Analysis

In Italy over 70 sq km of archival Quickbird imagery (0.6m panchromatic, 2.4m multi-spectral resolution), centred over the site of L'Amastuola, were analysed during summers 2007 and 2008. In Bulgaria, some 85 sq km of archival IKONOS imagery (1m pan, 4m multi-spectral resolution) from the GeoEye Foundation were investigated in March and April 2009 during an ongoing surface survey campaign. The Quickbird image arrived as ortho-ready product and was georeferenced and rectified in the field, while the IKONOS image arrived as orthorectified GeoProfessional product ready for use.

Both pieces of imagery were manipulated to reveal suspect features and patterns, such as soil and crop marks or other indicators of buried archaeological remains. The panchromatic and multi-spectral images were first overlaid or fused to combine the high spatial resolution with spectral information. Four spectral bands in each image were recombined to reveal latent patterns (a process where one of the four bands – red, green, blue, or near-infrared – is arbitrarily loaded into one of the three display channels – red, green, or blue). The red and NIR bands were processed to yield a Normalised Difference Vegetation Index (NDVI).

When an anomaly (a "remote sensing feature") was found during visual inspection of a particular image, its visibility across different layers was examined and recorded. In order to assess our methodology later, other information about each anomaly was recorded, including: (1) the suspected cause for the feature (earthworks, cut, natural phenomenon, etc.), (2) the presumed period (3) the layer or recombination in which best visible. The feature was then digitized in the GIS and passed on to the ground control team who went into the field with PDAs for verification.

Ground Control

Ground control was carried simultaneously with image analysis, so as to provide immediate feedback on the accuracy of feature identification and to improve feature classification. A team of two to five people intensively surveyed any large, interesting, or problematic features using the same methodology as the related pedestrian survey. The location, extent and nature of any surface remains (artefacts or architecture) associated with features were recorded in the field and compared to the assumptions made during image analysis. Ultimately the features were placed into one of five categories: sites (displaying significant surface remains associated with the feature), off-site scatters (displaying sparse surface remains associated with the feature), ambiguous (significant image anomaly with no surface material), false positives (marginal image anomaly with no surface materials), or indeterminate (inaccessible or no surface visibility). Strong anomalies with no surface material were not immediately eliminated since the presence of surface material, even over known sites, is not stable and varies with time ("stoplight effect" discussed in Barker 1984). The distinction between sites and off-site scatters was made using criteria of the associated surface survey. Sites and off-site scatters discovered during surface survey or ground control but not identified in the image previously were categorised as "false negatives".

Information from ground control and surface survey greatly refined ongoing feature identification in the image, as the difference between features representing natural phenomena, modern activity and those associated with ancient surface materials became clearer.

Results

L'Amastuola, Italy

In Italy, 114 features were visited for ground control. Fourteen met the Murge Tableland Survey project's criteria for a site, while another 15 were classified as off-site scatters (Burgers et al. 2003, 3-4). Thirteen features remain ambiguous or unassessed (distinct pattern but no surface material, or inaccessible). Characteristics of the features varied, but most comprised rectilinear or circular patterns or areas of high NIR intensity or high contrast in brightness that did not align with modern field divisions. In contrast to Bulgaria, all sites and off-site scatters were flat sites confirmed through the presence of surface artefact scatters.

Seventy-two of the suspect features were classified as false positives. They fell in two categories: modern features (such as recent irrigation ditches or pipelines) or natural (usually geological) phenomena. The rectilinear faulting of the local karst was particularly deceptive, and accounted for over a half of all false positives. A small number of false positives were associated with modern debris especially the remnants of *masserie,* early-modern estate houses, common in the area. Compared with the results of the Murge Tablelands Survey, remote sensing detected 12 and missed 52 known sites or off-site scatters (23%). The false negatives fell in the category of small surface scatters with median area ca. 0.1ha, while the remotely sensed features encompassed on average an area of 0.65ha (see Figure 1).

Kazanluk, Bulgaria

In Bulgaria, out of 166 features investigated, 114 were sites (107 mounds, including 29 that were defunct or destroyed, and seven shard scatters) and two were off-site scatters. Forty three were false positives (13 modern, 30 natural) and seven ambiguous features. There were 131 false negatives, 90% of which were mounds. (117mounds, 11 flat sites, and three masonry remains) (see Figure 1). Not surprisingly, the vast majority of all sites identified in the Kazanluk region were burial mounds. The few flat sites that were identified through remote sensing appeared as crop marks.

	Italy	Bulgaria
Study area (sq km)	70	85
Surface survey (sq km)	10	10
RS features investigated	114	166
Sites	14 (flat sites)	114 (107 mounds, 7 flat sites)
Offsite scatters	15	2
Ambiguous	13	7
False positives	72	43
False negatives	52 (small flat sites)	131 (117 mounds, 11 flat sites, 3 walls)

Fig. 1. Remote Sensing Summary.

Twenty-five of the remotely sensed features lay in the area examined by the simultaneous Tundzha Regional Archaeological Project (TRAP) surface survey. Of these, 23 sites were correctly identified (22 mounds and one flat site), while two were false positives. There were, however, 71 false negatives, again mostly mounds, (58 mounds, 10 shard scatters, and three masonry remains), which escaped detection due to their concealment in areas of rugged topography and forest or scrub land cover. The ratio of remotely discovered sites to those recovered by surface survey is 25%.

Comparison and Interpretation

Overall, remote sensing in Bulgaria yielded fewer false positives, and a higher ratio of sites to false positives, than Italy. The higher ratio of sites to false positives in the Kazanluk region is attributable to the fact that the most common type of archaeological site – the burial mound – had excellent visibility, at least in open landscape.

Two other factors contributed to the lower overall number of false positives in Bulgaria: (1) less deceptive geology than in Italy and, (2) more conservative analysis on our part. In Italy, the single most common cause of false positives was geological; protruding bedrock fractured in rectilinear patterns that looked manmade but did not align with field divisions. In Bulgaria most of the false positives turned out to be small islands of bushes and trees in agricultural fields.

Thousands of such clusters dot the Kazanluk landscape, sometimes concealing a mound or other surface remains and sometimes representing only a depression filled with field stones and overgrown. Early stages of ground control yielded many false positives, and this low accuracy led to more conservative assessments. Only features identified with high confidence were flagged, which lowered the number of false positives – but raised the number of false negatives. The accuracy of mound identification, moreover, decreased with the increasing density of surface vegetation, particularly trees and scrub. Identification rates also decreased along with the size of the mounds. Researchers in the Near East claim near 90% recovery of larger surface earthworks or mounds (Wilkinson et al. 2004, 196), while our rate of discovery varied widely from 75 to 23% depending on terrain, land cover and site type and size. Some 89% of the overlooked mounds in Kazanluk belonged to the category of small, often defunct, mounds whose height was below 0.5m and radius under 10m. Given the geomorphological diversity of the region, these were extremely hard to resolve against clusters of field stones, colluvial deposits, and under dense vegetation cover.

A comparison between results from Italy and Bulgaria also underscores how geology and land cover matter. The semi-arid Italian landscape proved to be an advantage in flat site detection due to site association with near-surface water sources, which are easily detectable in satellite imagery. We have not yet discovered such an important and detectable single factor governing habitation patterns in the Kazanluk region, and thus site detection through remote sensing has been limited to easily discernible surface remains or crop marks.

The topographic manifestation as mounds and high density of cultural remains in Bulgaria brought the number of detected image features to 166 as opposed to 114 in Italy, suggesting the recovery of a higher percentage of the site universe. Ground control demonstrated that the success rate of remote sensing was high in Bulgaria (70%) and relatively low in Italy (25%). Success rate represents the ratio of suspect image features that yielded sites to those that produced nothing.Recovery rate, on the other hand, tells us what percentage of known sites on the surface was remotely detected. Despite different absolute numbers and success rates across the two study areas, remote sensing on average recovered 25% of all archaeological features found by pedestrian survey in both regions. In Bulgaria as well as in Italy, the majority of overlooked sites were either significantly smaller or obscured by natural features.

Conclusions

Labor cost summary will, furthermore, help contextualize the results. Computer work for this project took three or four man/days during each campaign. Ground control in Italy, where it was a stand-alone operation (comparison was done "blind"), lasted four weeks. In Bulgaria, where an ongoing surface survey served in part as ground control, dedicated ground truthing took only one week.

Results from the analysis demonstrate that with limited time and resources, relatively low-cost archival imagery can be successfully employed to locate new sites and enhance the overall understanding of an archaeological landscape.

Enthusiasm for remote sensing, however, needs to be tempered by the realisation that its effectiveness – the rates and patterns of site recovery – require not only a thorough understanding of the satellite image characteristics and local environment, but also ground control or comparison with archaeological surface survey data. Especially in temperate regions of Europe, the seasonal land cover, variability of geology and local archaeological residues compound to challenge the satellite image analyst. Wide availability of satellite imagery invites researchers to be carried away with the fast and easy site detection from the comfort of their distant offices (Madry 2007). Such euphoria, however, is premature before we firmly establish both the success rates and recovery rates of remote sensing for the region under study. To achieve this, further comparisons of discovery rates across different landscapes and datasets are needed to provide some control over remote sensing accuracy and completeness.

References

ARCHIBALD, Z. H., 1998. *The Odrysian Kingdom of Thrace. Orpheus unmasked,* Oxford: Clarendon Press.

BARKER, G., 1984. The Montarrenti Survey, 1982-3. *Archaeologica Medievale*, 278-89.

BURGERS, G.-J. AND J.-P. CRIELAARD, 2007. Greek Colonists and indigenous populations at L'Amastuola, southern Italy. *BABesch,* 82, 77-114.

BURGERS, G.-J., P. A. J. ATTEMA AND M. VAN LEUSEN, 2003. Walking the Murge. Interim report of the Ostuni Field Survey (Apulia, Southern Italy). *Studi di Antichità,* 11, 1-26.

LASAPONARA, R. AND N. MASINI, 2006. Identification of archaeological buried remains based on the Normalized Difference Vegetation Index (NDVI) From Quickbird satellite data. *IEEE Transactions on Geoscience and Remote Sensing Letters,* 3(3), 325-8.

MADRY, S. L. H. 2007. An Evaluation of Google Earth for Archaeological Aerial Prospection and Site Survey. *Digital discovery: exploring new frontiers in human heritage: CAA 2006 : computer applications and quantitative methods in archaeology: proceedings of the 34th conference, Fargo, United States, April 2006*, 303-11.

VAN JOOLEN, E. 2003. *Archaeological land evaluation. A reconstruction of the suitability of ancient landscapes for various land uses in Italy focused on the first millenium BC.* Groningen: Rijksuniversiteit Groningen.

WILKINSON, T. J., J. UR AND J. CASANA, (2004). From Nucleation to Dispersal: trends in settlement pattern in the northern fertile crescent, in S. E. Alcock & J. F. Cherry (eds.) *Side-by-side survey: comparative regional studies in the Mediterranean world*, 189-205. Oxford: Oxbow Books.

Archaeological remote sensing approach in Honduras. A project for cultural heritage and human habitats protection

Rejas, J. G.,[1] Pineda, M.C.,[2] Véliz, S.V.,[2] Euraque, D.,[3] Martínez, E.,[3] Rodríguez J.R.[3] y Farjas, M.[4]

[1]National Institute for Aerospace Technology (INTA), Spain
[2]*Central American Astronomical Observatory of Suyapa*, Universidad Nacional Autónoma de Honduras, (OACS/UNAH), Honduras
[3]Honduran Institute of Anthropology and History (IHAH), Honduras
[4]*E.T.S.I. in Topography, Geodesy, and Cartography,* Polytechnic University of Madrid (UPM*), Spain*

Abstract

There are natural risk factors affecting the protection of the cultural heritage in Honduras, caused directly by floods and fires, or by seasonal effects of the climate change, producing significant damage to archaeological sites and, for that matter, the communities that live in their environments. In recognition of this, the research team have started a project with the overall aim of developing a methodology for evaluating, recording, surveying and monitoring cultural heritage sites on a national scale, focusing on some study cases.
Remote sensing, as a non-destructive technique of archaeological exploration, and new technologies, enable the measurement of spectral anomalies and biophysical parameters, such as reflectance and temperature, that may be indicative of the presence of anthropomorphic structures, or diseases that affect them. The satellite images evaluated (Landsat ETM+, ASTER, ALI and Hyperion) have been used to make a preliminary sites cartography, to extract current land cover/use and to test models for detecting and/or confirming possible buried archaeological structures. We describe in this paper the objetives of the project, the data processing and the first results and conclusions.

Key words archaeological remains, cultural heritage, reflection and emission spectrum, protection.

1 Introduction

Currently, there are risk factors affecting the protection of the material or immaterial cultural heritage in Honduras, and in general in Central America. Natural disasters, bush fires, deforestation caused directly by phenomena associated with the seasons, or indirectly by the effects of climate change, all of which account for huge, and sometimes irreversible damage to cultural and archaeological heritage, and consequently to the communities living in the surrounding areas.

Archaeological prospection by means of non-destructive techniques and the new technologies for detection and records in cultural heritage, out of which outstands remote sensing through active and passive sensor, different facets (space, airborne and short range), all represent an important tool (Rejas *et al.*, 2008) for the improvement, protection and prevention in the face of negative and destructive elements acting on the national heritage.

In the last few decades the development of instrumentation undertaken via Earth observation satellites or via ground-level platforms have widened the possibility of the creation of methodologies particularly in the representation and study of biophysical characteristics of the Earth, and from the outset, in the area of archaeologicaland patrimonial research in general. Hyperspectral sensors that record images in a wide range of electromagnetic spectrum bands (0.4 µm to 12 0.4 µm wavelength) visible (VIS), near infrared (NIR) shortwave infrared (SWIR) and thermal infrared (TIR), produce data with high spatial and spectral resolution of wide regions in continuous time sequences. We can measure biospheres and calculate spectral and thermal anomalies that could be indicative of the presence of anthropomorphic structures in Honduras or pathologies that affect these structures.

At the same time they provide us with information on the changes the use of the land undergoes, geomorphologic changes resulting from natural causes which affect inland and coastal Honduras seasonally, or from human action.

On the other hand, new spatial technologies in the microwave region, among which is found synthetic aperture radar (SAR) endowed with polarimetric and interferometric capacities, permit the recording of information below the forest and cloudy covering that characterizes Honduras in particular, and the various Central American ecosystems.

In the present project, we analyse the factors that affect areas susceptible to natural risk as well as the territorial distribution, with a view to enabling compatibility between the protection and conservation of places of interest to the national heritage through sustainable development and the preservation of the immaterial and human heritage of the communities that inhabit them and their surrounding areas. Under this premise, the objectives being put forward include the use of data coming from the Earth Observations Systems and its integration in GIS for the design and development of non-intrusive methodologies meant to improve the cultural heritage protection in Honduras, including human habitats in the face of situations of risk and natural disasters.

We are stating the trial areas identified for study, the characteristics of the data used and the exploratory analyses of remote sensing imagining carried out which has permitted the extraction of the biophysical parameters that will be used in the modules generated in GIS.

2 Study Cases

Three different type areas have been defined as test zones for the investigation. These areas are in different ecosystems, different geomorphological landscapes, and inhabited by different ethnic groups. The result of these area definitions has been intentional (Véliz 1983), trying to encompass as varied a spectrum as possible in order to be able to extrapolate and validate the results obtained through teledetection to other areas of the country in a second phase of the project.

2.1. Jesus de Otoro Valley

Spread along the Rio Grande de Jesus de Otoro, in a center-west position within the country, next and southwest of a big interior lake called Yojoa, in a county which gives the Valley its name. We are dealing with a not too extensive zone, with an abundance of archaeological sites. The zone is within the Lenca people area, one of the principal ethnic groups in Honduras. There are abundant ceremonial stone plazas in the valley, with an invariable ceremonial pyramid at the center and surrounded by six or more mounds, smaller pyramids, with a general orientation to the cardinal points. It was during the 1990s when the first archaeological reports appeared. Archaeologists from the Honduran Institute of Anthropology and History made the first visits and maps of the archaeological sites in the valley. North American archaeologists began afterward a more continuous project which has been obtaining better maps and putting test pits so as to have a better idea of the cultural development in the zone.

2.2. Southern Region: Choluteca

This region is basically the coastal Pacific plain in the southern part of the country, centered in the city from which the region takes its name, and through which runs the Choluteca River, which drains into the Pacific, in the Fonseca Gulf, running east and south from the Capital of the Republic, Tegucigalpa. It is an area little known archaeologically, but with great archaeological potential since it is considered an obliged passage for the early movements of Amerind populations (Stone 1957). Only one archaeological project has been carried out in the zone, in early 1960s. The basic result was a type ceramic sequence, based on surface and test pit materials.

2.3. Rio Aner, Mosquitia

This is a wide, little inhabited area and spreading practically over one third of the northeast region of Honduras. This is the area occupied and influenced by the Tawahka, Pech (Lázaro and Griffin, 1991), and Misquito people, three of the principal ethnic Honduran groups. The study of a radar image detected remains of what could well be the mythical White City, a non maya pre-Columbian city, mentioned in various occasions as one of the principal cities in the zone, but with an unknown location at present. In spite of several expeditions to the zone, archaeological literature is almost non-existent. The most recent expeditions by archaeologists have not yet been published.

3 Data and pre-processing

For the first phase of the project, data from passive (multi and hyperspectral) satellite sensors have been obtained for the first two areas of study, and for the third, data from SAR.
We have processed two images from the ALI (Advanced Land Imager) sensor corresponding to the dates 10-02-2004 and 17-01-2007 for the Jesús de Otoro Valley and the Choluteca Valley respectively. The ALI images were obtained from USGS on the L1b process level. It is a phusbroom sensor type which records data in 10 electromagnetic spectrum bands, between 433 nm and 2350 nm with spatial resolution of 30m in the multispectral band and 10m in the panchromatic band.

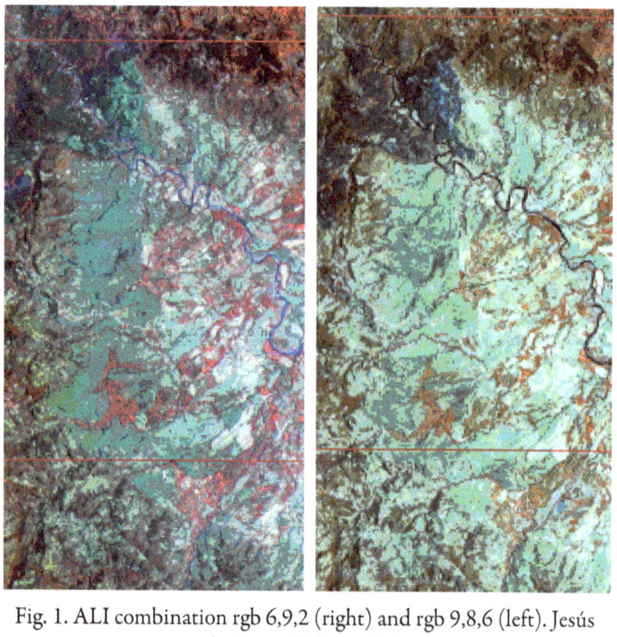

Fig. 1. ALI combination rgb 6,9,2 (right) and rgb 9,8,6 (left). Jesús de Otoro Valley, 2004.

Similarly, two images from the Hyperion sensor corresponding to 2003 and 2004 for the Valle de Jesús de Otoro and the neighbouring areas have been processed. These images were obtained on the L1R to USGS process level. It is a hyperspectral satellite sensor of the whiskbroom type, which records data in 242 electromagnetic spectrum bands, in 300nm to 2500nm wavelength, with a spectral width of 20 nm per band and special resolution of 30m.

4 Exploratory Analysis

We have developed, for these areas and with these images, typical remote sensing techniques which are constantly being contrasted with field projects, archaeological supervision-prospection at ground level and spectroradiometry, principally. The three important topics we are going to study concerning remote sensing are geology and geomorphorlogy, vegetation and spectral and thermal prospection applied to the detection of anthropomorphic elements or archaeological remains. These parameters will permit us to establish connections and time-space correlations between different pre and post-Hispanic archaeological sites.

To carry out the supervision in the selected training areas, areas have been visited while identifying the type of material in the coverings of interest and collecting samples of materials present in archaeological sites to be analyzed in laboratories later on.

Fig. 2. Field radiometry with USB4000, San Marcos site, Jesús de Otoro Valley.

We have done a spectral laboratory analysis with the aim of determining the spectral features characterizing each one of the selected training areas. For this purpose, we have used the spectra measured in situ, as well as spectra of artificial surfaces obtained from spectral libraries. Laboratory spectra and those measured in situ have been compared directly with reflectance data appearing in images processed with the aim of designing the pure pixels present in the training areas (Malpica et al., 2008). It has therefore been necessary to calibrate the image from radiance to reflectance, and for which ENVI 'Emperical Line' algorithm has been used.

We have begun to study the geology of the three study areas. We have applied a called Crosta Technique (Loughlin 1991), widely used and contrasted for the detection of materials that have been transformed by hydrothermal processes in different geological sceneries (Bragado et al. 2008). They are based on two analyses of principal components (PCA); in one of them, four bands are selected which permit the engraving on clays - related to the absorption interval of A1 –OH of clays in SWIR – and in the other, action is carried out on the four bands that exhibit the spectral behaviour of iron oxides and the jarosite – associated with electronics that are typical of iron in the VIS.

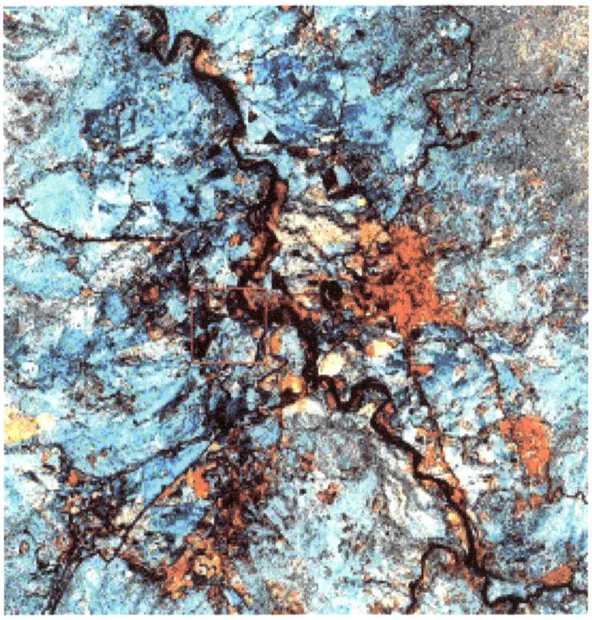

Fig. 3. Detail of ETM+ Crosta combination image rgb H,(H+F),F. Jesús Otoro Valley.

The theory consists of finding, on the one hand, the principal component that exhibits the greatest spectral contrast of the type: reflection in 1,650 nm and absorption in 2,215 nm (image H associated with clays). and on the other hand, the component with the greatest spectral contrast of the type : absorption in 485 nm and reflection in 660 (Image F related to iron oxides) . The entry bands in PCA clays are those whose wavelengths correspond to 485 nm, 830 nm, 1.650 nm and 2.215 nm in the case of ETM + images. Completely analogous is the PCA of iron oxides. Here again, four entry bands are used which are determined by the wavelengths 485 nm, 660 nm, 830 nm y 1.650 nm.

Similarly, we have generated a cartographic base for the entire country using ETM+ images, which we will also be using as an entry layer in a GIS and in strictly research and modeling tasks.

The classification based on the image, one of the principal techniques applied in the satellite image, requires a prior knowledge of the different types of surfaces present, although not necessarily of its spectral feature, which can extracted from the calibrated image. Ten area of supervision have been recorded and measurements with USB400 of discrete spectra have been taken in several of the areas that were accessible.

4 Conclusions

The research presented constitutes an approximation in the development of remote sensing techniques and methodologies for the study and characterization of archaeological sites and other areas of heritage value in Honduras. The characteristics and strategies for acquisition of data taken in areas of reflective and emissive spectrum, and in the microwave region, for three cases of study defined in the research.

The results show the great potential of the hyperspectral and multispectral thermal image for the detection of alignments and structures buried in the valleys of Jesús de Otoro and Choluteca. We have studied the capacities of interferometry and SAR polarimetric differential for the calculation of evidences and remains of anthropic structures in the area of Aner, in a region characterized by its extensive forest cover.

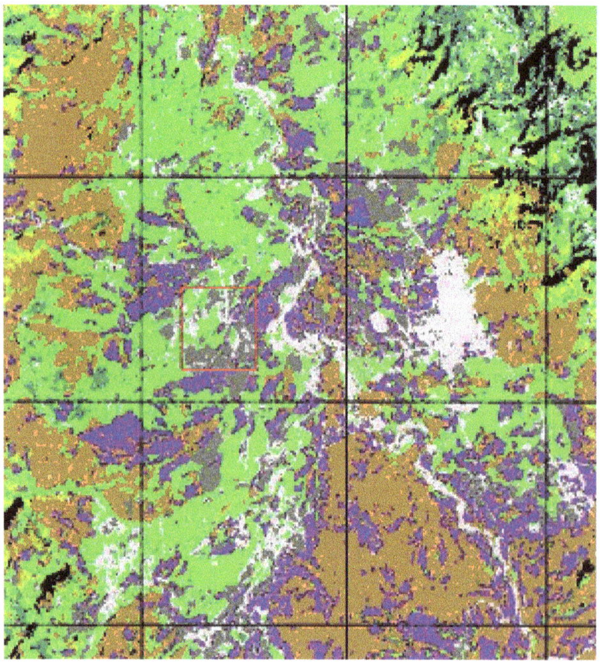

Fig. 4. Maximun Likelihood classification from ETM+ data, Jesus Otoro Valley landcover.

The integration of these multisource data with Geographic Information System permits the geospatial analysis of vast regions of the territory, the interconnection between different processes, both natural and historical. Similarly, they permit the generation of models capable of preventing and anticipating any effects in the face of possible situations of natural risk.

Acknowledgements

We are grateful to the Honduran Institute of Anthropology and History (IHAH) for providing us with the geographical and archaeological information needed for carrying out this project.

This project has been developed within the framework of the PCI project: A/019450/08 financed by the AECID, Spanish Agency for International Cooperation and Development.

References

BRAGADO, E., REJAS, J.G., MARCHAMALO M. AND MARTÍNEZ, R., 2008. Characterization of hydrothermally altered materials in the Central Volcanic Range, Costa Rica, using TM data. *Remote Sensing and Photogrammetry Society Conference 2008 "Measuring Change In The Earth System". University of Exeter (UK), 15-17 September 2008.*

MALPICA, J.A., REJAS, J.G. AND ALONSO, M.C., 2008. A projection pursuit algorithm for anomaly detection in hyperspectral imagery. *Pattern Recognition* 41 (11) pp: 3313-3327, 2008. ISSN 0031-3203; DOI: 10.1016/j.patcog.2008.04.014.

LÁZARO, F. AND GRIFFIN, W., 1991. *Los Pech (Payas): Una Cultura Olvidada. Departamento de Ciencias Sociales, ESPFM (Escuela Superior del Profesorado Francisco Morazán).* Editorial Guaymuras, Tegucigalpa.

REJAS, J. G., FARJAS, M., BURILLO, F., LÓPEZ, R., CANO, M.A., SÁIZ, M.E., MOSTAZA, T. AND ZANCAJO J.J., 2008. Comparative archaeometric analysis through 3D laser, short range photogrammetry, and hyperspectral remote sensing applied to the celtiberian city-state of Segeda. *37th International Symposium on Archaeometry, Siena (Italy), 2-6 May 2008.*

STONE, DORIS Z., 1957. The archaeology of central and southern Honduras. *Papers of the Peabody Museum of Archaeology and Ethnology*, Vol. 49, No. 3.

VÉLIZ, VITO, 1983. Síntesis histórica de la arqueología en Honduras. *Yaxkin, Órgano de divulgación del Instituo Hondureño de Antropología e Historia*, ISSN 0254-7627, Vol. VI, 1-9 pags.

WELLER, ERRIN T., 2006. Satellites, survey and settlement: the Late Classic Maya Utilization of Bajos (Seasonal Swamps) at Tikal and Yaxha, Guatemala. *From Space to Place, 2nd International Conference on Remote Sensing Archaeology, Rome (Italy) 4-7 December 2006*, BAR International Series 1568.

Integrating SAR data and hyperspectral analysis for the archaeological survey of the Segeda city, Spain

Rejas, J. G.,[1] Burillo, F.,[2] López, R.,[2] Cano, M.A.,[2] Sáiz, M.E.,[2] Farjas, M.[3]

[1]National Institute for Aerospace Technology (INTA), Spain
[2]Facultad de Humanidades y Ciencias Sociales de Teruel, Universidad de Zaragoza (UZ)
[3] *E.T.S.I. in Topography, Geodesy, and Cartography,* Polytechnic University of Madrid (UPM*), Spain*

Abtract

The archaeological site of Segeda located in Mara (Zaragoza), corresponds to one of the major Celtiberian state cities of the Iberican peninsula. The extent, specificity and heterogeneity of the landcovers in Segeda, coupled with a state of excavation, have enabled the implementation of active and passive technologies to facilitate the detection and recording of archaeological structures buried or appearing in the excavations.

In this article is the review of the development of remote sensing techniques and methodologies applied to the study of the archaeological site of Segeda. The project has utilized a series of different non-destructive methods, including close range photogrammetry, 3D laser scanner, hyperspectral remote sensing, and field radiometry. We have studied the capabilities of SAR differential interferometry for the calculation of subsidence as evidence of cavities or pathologies remaining at the site and its surroundings. We set out the results obtained from hyperspectral and multispectral thermal data to detect buried structures and alignments, and data integration in Geographical Information System (GIS), thus allowing the geospatial analysis for large areas of the territory and the interconnection between natural and historical processes.

Keywords thermal, hyperspectral, SAR, spectral anomaly, archaeological remain.

1 Introduction

In the last few years active and passive sensors have been adopted as non-intrusive instrumentations that have a great application potential in archaeological prospection, as indicated by the corresponding scientific publications and events ((Belvedere et al., 2001; Emmolo et al., 2004; Weller 2006). At this stage of the study the response from the surfaces in terms of the appropriate wavelengths of the reflective (VIS-SWIR) and emitting (TR) spectrum on the one hand, and of the region of microwaves on the other, may represent relevant information in that it supports the exploration and detection of man-made structural signs.

In this sense, the hyperspectral data provide global information about a wide area of study, presumably, a medium or large scale archaeological site. However, an initial problem subject to investigation is that of the dimensional scope of the data, that is to say, to what potential in terms of wavelengths and bands should information be raised? That is, the information necessary to determine most accurately the extent to which separation can be carried out between materials present in either low or compound minerals associated with anthropic structures (Malpica *et al.*, 2008). To address this difficulty, research has centred principally on three ways: improvement on the extraction or selection of features so as to reduce the dimensional scope of the information, regulation of the covariance matrix of the samples and structuring of a true covariance matrix determined by a small number of parameters.

Following a multi-scale and multi-source work strategy, a research-based approximation has been carried out since 2005 with a view to applying active and passive remote sensing techniques in the archaeological site of Segeda (Burillo 2006), composed of Segeda I (Celtiberian) and Segeda II (Roman). An initial experience in hyperspectral information was presented in the 2nd International Conference of Remote sensing in Archaelogy, held in Rome (Rejas *et al*, 2006). The continual study carried out since then has focused on thermal bands, given that, in principle and on the basis of earlier experiences (Belvedere *et al.,* 2001; Farjas *et al.,* 2003) this is where remote sensing can help to distinguish phenomena that affect buried archaeological structures.

The project being presented serves as a review of the progress made during the geospectral study of the Celtiberian town of Segeda and its surroundings, using hyperspectral data integrated with SAR data.

2 Pre-processing data

A multi, micro, medium and macro scale work strategy has been followed that affects the findings and artefacts, the site and the region respectively. The data with which work has been carried out correspond to this approach. In this way, a set of images from different sensors acquired from various relating platforms have been made available (Figure 1).

It became necessary to transform the parameters of the

images into an absolute scale, both spatial and spectral, to enable us to refer the measurements extracted to others originating from different sensors or taken on different dates. The objective is to prepare the set of data for later analysis that may permit the spatial correlation of elements of archaeological interest (walls, surfaces, structures, remains of organic origin etc) with a temperature superficial to the ones that were recorded and other parameters extracted through image-treating techniques, such as reflectance or the deformation of fields by means of SAR techniques.

Data	Sensor	Platform	Year
Multispectral	ETM+	Landsat 7	2002
Hyperspectral	AHS	Airborne	2005
SAR	ASAR	ENVISAT	2004 2006
SAR	Prototipe-INTA	Airborne (*)	2006
Thermal	Thermacam	Terrestrial	2006
Laser 3D	Trimble GX 3D	Tripod	2007
Photografy	Nikkon D70	Terrestrial	2007
Photografy	Digital camera	UAV (*)	2008

Fig.1. Data used during research (* currently not processed).

For the hyperspectral images, work began from a process level L1b. The images are therefore transformed through reflectance and superficial temperature (for the AHS thermal channels), and georeferenced by means of a direct method based on GPS/IMU with a special resolution of 3m.

Two ascending passes ASAR (ENVISAT) corresponding to the dates, 31-08-2004 and 05-09 -2006 have been acquired. The sceneries cover an area of 100 x 100 km, situated approximately in the town of Calatayud. They were processed with the programme RAT (http://srv-43-200.bv.tu-berlin.de/rat/), transforming them into Simple Look Complex and later into images of amplitude. The 2004 scene has been georeferenced through polinomic method using 55 GCP's (Ground Control Points), coregistering to its result the one corresponding to the year 2006.

3 Image analysis and results

Various techniques for image treatment have been used for characterisation in the mineralogical composition of the grounds, the use and vegetation, as well as the detection of thermal anomalies.

Secondly, anomalies were calculated by applying a combolution using a median filter to later establish a thermal index (Rejas *et al.*, 2006) based on the possibility of separation of the types presented by AHS channels 76 and 80. The result is an image in which the detected pixels outstand as possible anomalies.

Four areas have been detected in Segeda I, Segeda II and its surroundings characterized by the presence of thermal anomalies. One of them presents a concentric circular shape, and could well be some kind of buried structure. In 2008, through archaeological prospection at ground base, it was confirmed that it corresponded to a deposit belonging to the Roman Empire era, presumably a villa or a factory for ceramic production. Within Segeda I another one of the anomalies detected (3 Figure 2) correspond to the Titos' settlement town, which became part of the Celtiberian city at the outside of their walls. The anomaly detected, numbered 2, (Figure 2) is found within the confines of Segada II, corresponding to the marked alignment in the direction SE-NW of 350m long. In 1 (Figure 2) we observe a structure with a characteristic geometrical pattern in the form of a 5-tipped star, which could be the remains of some kind of defensive structure or the remains of some form of irrigation.

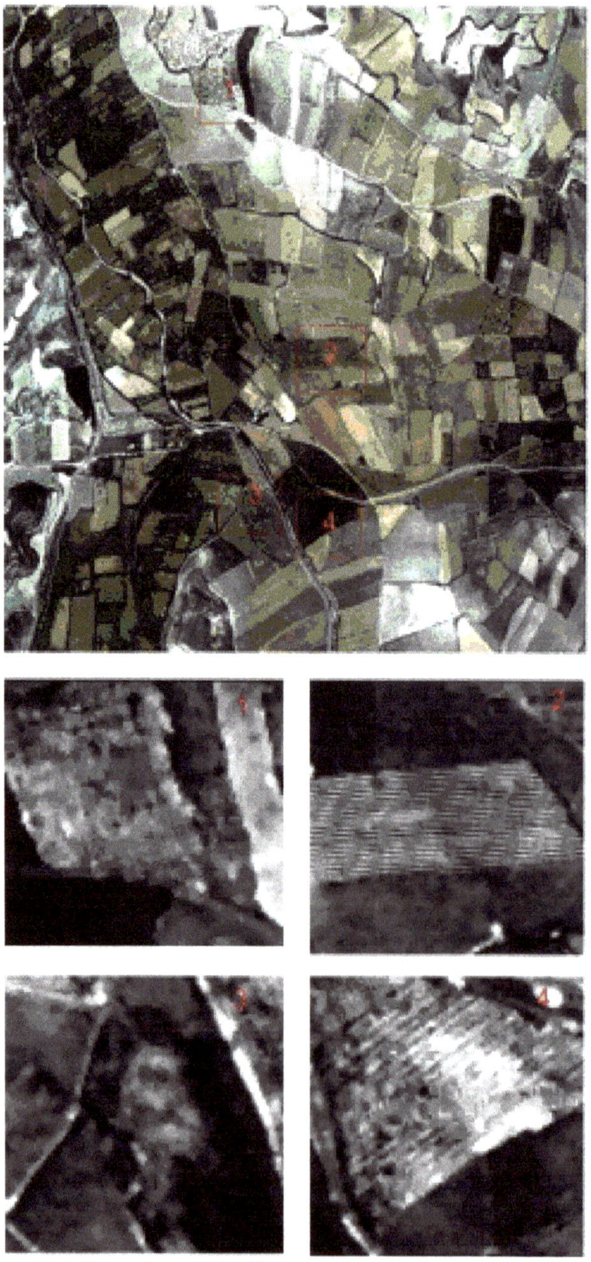

Fig.2. Areas characterized by anomalies calculated from AHS thermal chanels.

One of them is called Crosta Technique (Crosta *et al.*, 2003), widely used and contrasted for the detection of materials that have been transformed by hydrothermal processes in different geological sceneries (Bragado *et al.*, 2008). We have used AHS bands corresponding to the 463nm, 805nm, 1,670nm, and 2,210nm wavelengths to detect clays and AHS bands corresponding to 463nm, 659nm, 805nm and 1,670nm to detect iron oxide concentrations.

The result is very influenced by the vegetation cover, which acts as a source of error. Nevertheless, we observe the scarce presence of materials of clay-phylic alteration, contrasting with the abundance of iron oxide components. The initial analysis is the correlation existing between this type of materials and the presence of anthropic activity, enabling the verification that results cannot be extrapolated due to the fact that hydrothermal materials of alteration significant for Segeda I and II have not been detected, unlike the earlier or later archaeological deposits of its surrounding areas.

Fig. 3. AHS Crosta combination image.

Similarly, a PCA has been carried out focusing on the AHS thermal channels fused with aerial photograph. Earlier experiences (Traviglia 2006) have shown satisfactory PCA results in hyperspectral data for archaeology.

The analysis of the principal components for the thermal channels aim to calculate a new variable that might display the highest thermal contrast, mainly between 10,16 μm and the 12,89 μm wavelength. 16 CP's have been calculated by introducing channels AHS 64 to 80 as entry variables. The CP 14, registers the highest concentration of humidity, and it is observed that this corresponds to a large and well demarcated region at the extreme ends of the walled area of the Segeda II.

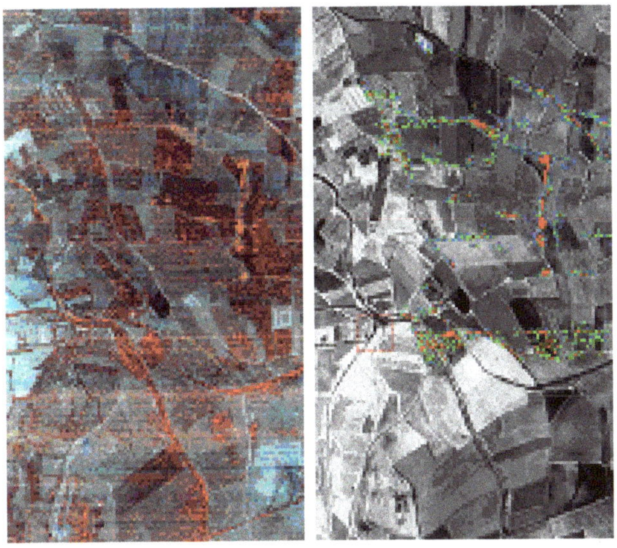

Fig. 4. Combination of rgb PC14, PC3 colour and PC14+PC3 (left) and AHS 8 channel with strutures extraction (right).

An interferometric par has been developed from ASAR images of 2004 and 2006, giving rise to a low coherence. Currently, SAR data are being analysed and integrated into SIG, together with the remaining pieces of information that has been generated, with the aim of exploring the possibilities for its interconnection with possible Celtiberian paths in the route between Segeda and Numantia, a city that linked up to Segeda against Rome.

The results obtained from the hyperspectral analysis have been integrated with the preprocessed SAR data. The responses from the surfaces with humidity concentration in ASAR images correspond to those detected in the AHS images and those that can be deduced from the NDVI vegetation index calculated for Segeda II. One of such areas is the old course of one of the aqueducts carrying water to Segeda II.

The main pre and post Celtiberian archaeological sites found between Segeda and Numantia have been recorded in the ASAR interferometric par. Under current study is the possible spatial correlation of these sites with optimal ways calculated in a GIS for the two most likely alternatives of the usual route between both Celtiberian towns. This analysis has not been completed yet.

4 Conclusions

This present project constitutes a review of the successive approximations in the application of active and passive remote sensing techniques and methodologies which have recently been used in the archaeological site of the Celtiberian town of Segeda.

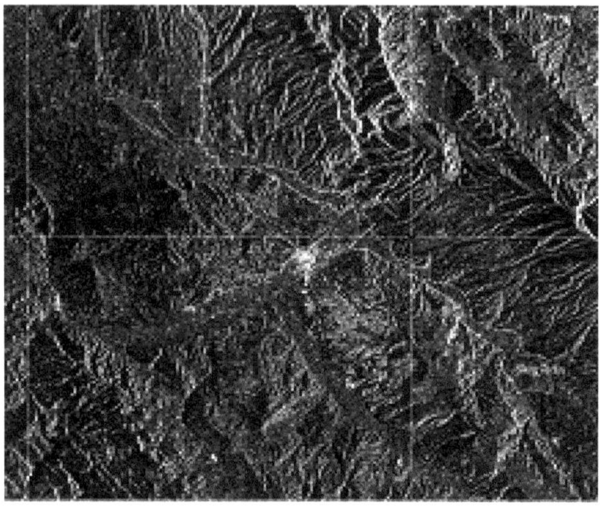

Fig.5. Multitemporal ASAR image (2004-2006) and possible ways (yellow and green) to Numantia city from Segeda (cyan).

The results confirm the ability of the hyperspectral data of the AHS sensor to detect and quantify the effective distribution of various types of anomalies and coverings on the deposit and on a large area of its surroundings. The hyperspectral images improve the distinction of these surfaces in standard numerical classification processes.

It has been observed that the findings of the thermal surfaces can help to characterize the types of anomalies associated with buried archaeological structures, bodies which, in other areas of the spectrum behave in a similar way and which can be distinguished by its response in MIR and LWIR wavelengths.

Multi-source files have been generated in a way that an accurate spatial coregistration between the different pieces of information becomes crucial for the subsequent data analysis. The integration of these results with SAR data has allowed the establishment of relationships between ground deformations and thermal variability.

Acknowledgements

We are grateful to the Remote Sensing Department of INTA for providing us with the thermal and hyperspectral information needed for carrying out this project.

This project has been developed within the framework of the R&D project: HUM2005-03369/HIST, financed by MEC and FEDER funds.

References

BELVEDERE, O., BURGIO, A., CIRAOLO, G., LA LOGGIA G., MALTESE, 2001. *Hyperespectral MIVIS data analyses for archaeological applications.* Fifth International Airborne Remote Sensing conference and Exhibition, San Francisco, 17-20 Sept. 2001.

BURILLO MOZOTA, F., 2006, "Segeda and Rome. The historical development of a Celtiberian city-state", en L. Abad; S. Ramallo & S. Keay (eds.), Early Roman Town in Hispania Tarraconensis. *Journal of Roman Archaeology. Supplementary Sereis* Number 62, Portsmouth Rhode Island, pp. 159-170.

BRAGADO, E., REJAS, J.G., MARCHAMALO M. AND MARTÍNEZ, R., 2008. *Characterization of hydrothermally altered materials in the Central Volcanic Range, Costa Rica, using TM data.* Remote Sensing And Photogrammetry Society Conference 2008 "Measuring Change In The Earth System". University of Exeter (UK), 15-17 September 2008.

CROSTA, A. P., FILHO, C. R. DE SOUZA, AZEVEDO, F. AND BRODIE, C., 2003. Targeting key alteration minerals in epithermal deposit in Patagonia, Argentina, using ASTER imagery and principal component analysis. *International Journal of Remote Sensing*, 10, pp. 4233-4240.

EMMOLO, D., FRANCO, V., LO BRUTTO, M., ORLANDO P. AND VILLA, B., 2004. *Hyperspectral Techniques and GIS for Archaeological Investigation.* Geo-Imagery Bridging Continents XXth ISPRS Congress, 12-23 July 2004 Istanbul, Turkey Commission 7.

FARJAS M., REJAS J.G., GÓMEZ J. A., DE MIGUEL E., FERNÁNDEZ-RENAU A., 2003. "Airborne Multispectral Remote Sensing Application In *Archaeological Areas". CAA 2003, The E-way into the four Dimensions of Cultural Heritage*, Vienna, Austria, April 2003.

MALPICA, J.A., REJAS, J.G. AND ALONSO, M.C., 2008. A projection pursuit algorithm for anomaly detection in hyperspectral imagery. *Pattern Recognition* 41 (11) pp: 3313-3327, 2008. ISSN 0031-3203; DOI: 10.1016/j.patcog.2008.04.014.

REJAS AYUGA, J. G., BURILLO MOZOTA, F., LÓPEZ, R. Y FARJAS ABADÍA, M., 2006. "Hyperspectral remote sensing application in the celtiberian city of Segeda". *From Space to Place, 2nd International Conference on Remote Sensing Archaeology, Rome (Italy) 4-7 December 2006*, BAR International Series 1568.

TRAVIGLIA, ARIANNA, 2006. MIVIS Hyperspectral Sensor for Detection of Subsoil Archaeological Sites and Interpretation Support GIS: an Italian Case Study. *CAA 2006*, Fargo, ND-USA.

WELLER, ERRIN T., 2006. Satellites, survey and settlement: the Late Classic Maya Utilization of Bajos (Seasonal Swamps) at Tikal and Yaxha, Guatemala. *From Space to Place, 2nd International Conference on Remote Sensing Archaeology, Rome (Italy) 4-7 December 2006*, BAR International Series 1568.

Wavelet based feature extraction and classification using differential evolution

R.A. Alagu Raja,* K.Vignesh, P. Dorothy Lizy, V. Abhai Kumar

Thiagarajar College of Engineering, Madurai-15
*Corresponding author, Email: alaguraja@tce.edu

Abstract

This paper presents a wavelet based texture classification technique that is applied to a Multi Spectral Scanner (MSS) image of Madurai City, South India. The process consists of wavelet based texture feature extraction and texture classification. The feature extraction stage of the technique uses coiflet wavelets to extract a texture feature set and this set is input to Differential Evolution (DE) based classification, an unsupervised classification technique. Four indices (user accuracy, producer accuracy, overall accuracy and Kappa co-efficient) are calculated to assess the accuracy of the classified data. The experiment results show that the performance of the proposed technique is better than the traditional unsupervised classification techniques.

Keywords Texture Feature Extraction, Wavelet Transform, Unsupervised Classification, Differential Evolution, Accuracy Assessment.

1 Introduction

Remote-sensing research focusing on image classification has long attracted the attention of the remote-sensing community because classification results are the basis for many environmental and socioeconomic applications. However, classifying remotely sensed data into a thematic map remains a challenge because many factors, such as the complexity of the landscape in a study area, selected remotely sensed data, and image-processing and classification approaches, may affect the success of a classification. In general, Image classification approaches can be grouped as Per-pixel, Sub pixel, and per field classification. Per-pixel classification is still most commonly used in practice. However, the accuracy may not meet the requirement of research because of the impact of the mixed pixel problem (Fisher 1997). Sub pixel algorithms have the potential to deal with the mixed pixel problem, and achieve higher accuracy for medium and coarse spatial resolution images (Foody and cox, 1994). For fine spatial resolution data, although mixed pixels are reduced, the spectral variation within land classes may decrease the classification accuracy. Per field classification approaches are most suitable for fine spatial resolution data.

In order to have accurate classification of remotely sensed data Spectral pattern recognition based classification technique is required. Texture is one of the efficient pattern recognition methods and texture is seen in all satellite images. So texture based classification will discriminate the features in remotely sensed data accurately. Texture classification is a process to extract features from a set of texture classes. In order to have an effective classification, features with good discriminatory details have to be obtained. Most of the existing approaches for texture feature extraction make use of statistical techniques in which processing the texture image data requires large storage space and computational capability.

In this paper, a Coiflet wavelet based texture feature extraction technique followed by differential evolution based classification is presented. The background on traditional unsupervised classification techniques, the proposed methodology and experimental results are the different sections of the paper discussed.

2 Unsupervised Classification Techniques

Classification includes a broad range of decision-theoretic approaches to the identification of images. All classification techniques are based on the assumption that the image in question depicts one or more features and that each of these features belongs to one of several distinct and exclusive classes. The classes may be specified a priori by an analyst (as in supervised classification) or automatically clustered (*i.e.* as in unsupervised classification) into sets of prototype classes, where the analyst merely specifies the number of desired categories. The user should be an expert and require some prior knowledge for the selection of good training data. This is not possible in all situations. So, unsupervised classification is becoming increasingly popular in many areas. The concept of traditional classification techniques are discussed below.

2.1 k-means Clustering

K-means clustering is an algorithm which, for K clusters $\{C_1, C_2, \dots\dots\dots C_k\}$ each with n_k patterns aims to find cluster centers m_k to minimise the cost function E_K^2 where,

$$E_k^2 = \sum_{k=1}^{k} \sum_{x \in C_k} \left\| x - m_k \right\|^2$$

$$m_k = \frac{1}{n_k}\sum_{x \in C_k} x,$$

The initial cluster centers are chosen randomly and the algorithm is applied repeatedly until a steady state is reached. While k-means classification will return the same number of classes that were chosen by the user at the beginning of the process, ISODATA will merge empty or insignificant classes.

2.2Fuzzy c- means clustering

Fuzzy c-means (FCM) is a method of clustering (Bezdek 1981) which allows one piece of data to belong to two or more clusters. This method is frequently used in pattern recognition. It is based on minimization of the following objective function:

$$J_m = \sum_{i=1}^{N}\sum_{j=1}^{C} u_{ij}^{m} \left\| x_i - c_j \right\|^2 ,$$

$$1 \le m < \infty$$

where m is any real number greater than 1, u_{ij} is the degree of membership of x_i in the cluster j, x_i is the ith of d-dimensional measured data, c_j is the d-dimension center of the cluster, and ||*|| is any norm expressing the similarity between any measured data and the center. Fuzzy partitioning is carried out through an iterative optimization of the objective function shown above, with the update of membership u_{ij} and the cluster centers c_j by

$$u_{ij} = \frac{1}{\sum_{k=1}^{c}\left(\frac{\left\| x_i - c_j \right\|}{\left\| x_i - c_k \right\|}\right)^{\frac{2}{m-1}}} ,$$

This iteration will stop when $\max_{ij}\left\{\left| u_{ij}^{(k+1)} - u_{ij}^{(k)} \right|\right\} < \varepsilon$, where ε is a termination criterion between 0 and 1, whereas k is the iteration steps.

3 Study Area and Data Used

The multi temporal satellite sensor image used in this study is of Madurai city, South India, shown in Figure.1.
Madurai is the second largest city in the state of Tamil Nadu, after Chennai and is one of the oldest cities in India, over 2,500 years old. It is identified as one of the 12 heritage cities of India and it is situated between longitude 78^0 04' 47" E to 78^0 11' 23" E and latitude 9^0 50' 59" N to 9^0 57' 36" N.
The topography of Madurai is approximately 101 meters above sea level. The land cover features of this study area include urban, vegetation, water body, waste land and hilly region.

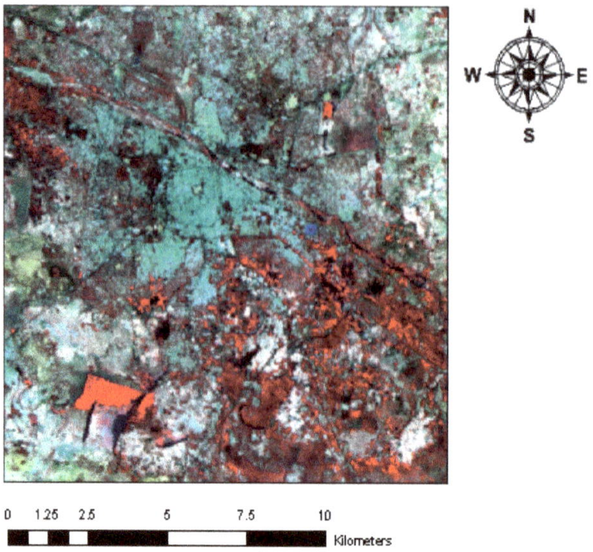

Fig. 1. Madurai City LISS III Image – March 2004.

The scene details of the area are,

Date of acquisition : 19th March 2004

Satellite/Senor : IRS P6/ LISS III

Resolution : 23.50 meter

The scene was geometrically registered using Survey of India topographic sheets 58 K/1 in 1:50,000 scale and four 58 K/1 SW, SE, NW, NE in 1:25,000 scale were used as reference data for GCPs (ground control points) collection and validation . The standard errors were below 0.5 pixels along the X and Y directions. A sub area of 407x411 pixels was extracted from the original scale.

4 Methodology

In this paper, we analyze the performance of the proposed feature extraction approach against the classical feature extraction approaches which are used for classification of remote sensing images with different types of landscapes. Figure. 2 shows the steps of the methodology for the proposed wavelet frame based feature extraction approach.

4.1 Wavelet transform

The wavelet transform is a bound alterable window method. A more accurate low frequency information can be produced using a longer interval and high frequency information using shorter interval. The advantage of the wavelet transform is information maintenance, which provides the space frequency information. Wavelet plays a central role in "Knowing what is where?".

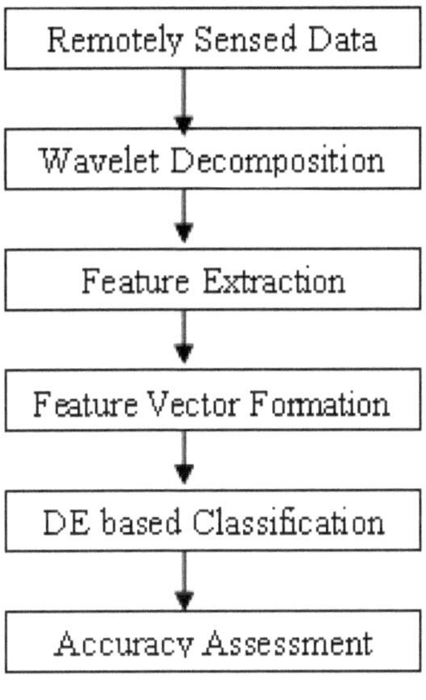

Fig. 2. Methodology for Texture Classification.

The frequency content provides the information content about "what" and spatial co-ordinates provide the information about "where" which is important for texture analysis. When used as a foundation for a texture measure, the wavelet transform enjoys several advantages such as spatial discrimination and multi-scale representation over other methods (Coggins and Jain, 1985).
The basis function is,

$$\psi_{a,b}(X) = \frac{1}{\sqrt{|a|}}\psi\left(\frac{x-b}{a}\right)$$

Here a, b are real. Each wavelet function in the decomposition is formed from a 'mother' wavelet function ψ (x) which is scaled (*a*) and translated (*b*) and the result being localized in both the frequency and spatial domains. The continuous time wavelet transform of a function *f*(*x*) is defined as,

$$CTWT(a,b) = \int_{-\alpha}^{\alpha} f(x)\frac{1}{\sqrt{|a|}}\psi\left(\frac{x-b}{a}\right)dx$$

Most previous work on texture classification has focused on the discrete wavelet transform with decimation by which some of the information is lost due to the size reduction of input image. This may be applicable for processing classical images. But for images like remote sensing data, minor information is also important to obtain higher classification rate. To achieve this, an algorithm making use of the Discrete Wavelet Frame Transform (DWFT) for satellite data classification is proposed in this paper.

4.2 Texture Feature Extraction using Wavelets

In order to apply a wavelet to image texture analysis, the two-dimensional discrete wavelet transform is generated via the tensor product of 1-D wavelets along the vertical and horizontal directions (Mallat, 1989). In Quadrature Mirror Filters (QMF), this means the L and H filters are applied to the image in both the horizontal and vertical directions. The outputs are sub sampled by a factor of two to produce three high pass sub bands, HH, LH, HL and one low pass sub band LL. The process is repeated on the LL sub band to generate the next level of the decomposition using the same algorithm (DeBrunner and Kadiyala, 1999). If we do not downsample the low pass and high pass sub band images, we will get the DWFT solution for that image. DWFT has been investigated and applied to many fields. In DWFT, the down sampling operation is removed after the signal has passed through L and H filters. Thus, the size of the image remains unchanged in every processing stage. As a first step, DWFT is applied to the input image and we get the LL subband image that contains only low frequency information, but has the same image size as the original image plot (Chang and Kuo,1993). When applied to images the detail extracted is finest 'i.e.' of highest frequency from the first level of the filter-bank and becomes progressively more coarse; it is this progression that allows the wavelet transform to provide a multi-scale representation of texture plot (Hiremath and Shivashankar ,2006).

4.3 Texture Classification based on Differential Evolution

Texture classification is a major task in the process of satellite sensor image analysis which involves a method of identification of the different texture regions based on the corresponding features. The output from the texture feature extraction stage is a set of images. These images represent different features of the raw image. They are used to form feature vectors, where each feature image corresponds to one element in the feature vectors. For the purpose of texture classification, a feature vector is needed to successfully characterize a textural region. When texture feature vector is used for classification there is heterogeneity within a texture in one class that causes spectral overlap with texture in other classes. So the traditional K-means clustering algorithm cannot handle data points that are close to more than one cluster. To overcome this we go for differential evolution based classification. Differential evolution (Omran 2005) is a population-based search strategy very similar to standard evolutionary algorithms. The main difference is in the reproduction step where offspring is created from three parents using an arithmetic cross-over operator. DE is defined for floating-point representations of individuals. Differential evolution does not make use of a mutation operator that depends on some probability distribution function, but introduces a new arithmetic operator which depends on the differences between randomly selected pairs of individuals. For each parent, $x_i(t)$, of generation *t*, an offspring, $x_i'(t)$, is created in the following way: Randomly select three individuals from the current population, namely $x_{i1}(t), x_{i2}(t)$ and $x_{i3}(t)$, with $i1 \neq i2 \neq i3 \neq i$ and $i1, i2, i3 \sim U(1, \ldots, s)$, where s is the population size. Then, for all parameters $j = 1, \ldots, N_d$, if $U(0,1) < P_r$, or if $j = r$, let

$$x_{i,j}'(t) = x_{i3,j}(t) + \gamma(x_{i1,j}(t) - x_{i2,j}(t))$$

Otherwise, let

$$x_{i,j}'(t) = x_{i,j}(t)$$

In the above, P_r is the probability of reproduction (with $P_r \in [0, 1]$), γ is a scaling factor with $\gamma \in (0, \infty)$, and $x_{i,j}'(t)$ and $x_{i,j}(t)$ indicate respectively the j-th parameter of the offspring and the parent. Thus, each offspring consists of a linear combination of three randomly chosen individuals when $U(0, 1) < Pr$; otherwise the offspring inherits directly from the parent. Even when $Pr = 0$, at least one of the parameters of the offspring will differ from the parent (forced by the condition $j = r$).

The mutation process above requires that the population consists of more than three individuals. After completion of the mutation process, the next step is to select the new generation. For each parent of the current population, the parent is replaced with its offspring if the fitness of the offspring is better, otherwise the parent is carried over to the next generation. DE is easy to implement, requires little parameter tunning, can find the global optimum regardless of the initial parameter values and exhibits fast convergence.

4.4 Accuracy Assessment

Analysis of confusion matrix has been one of the most common means of expressing classification accuracy. Error matrices compare, on a category by category basis, the relationship between known reference data (ground truth) and the corresponding results of an automated classification. Various accuracy indices such as overall accuracy, producer accuracy, user accuracy and kappa coefficient (Lillesand *et al.*, 2004) have been calculated using the field visit data in determining the accuracy of classification. Commonly, most of the researchers recommended the use of the Kappa coefficient of agreement as the standard measure for accuracy (Tung Fung, 1990).

5 Experimental Results

To demonstrate the potential of wavelets in extracting features for texture classification, we have applied the algorithm to remotely sensed data of Madurai City, South India. Also we have implemented the unsupervised classification algorithms like Kmeans, Isodata Fuzzy-c- means for comparative analysis. Figures 3 - 6 depict the classification results for the unsupervised methods and the DE based results respectively. Even by a visual analysis of the Figure 3 and Figure 6, it is clear that the features were significantly misclassified with respect to the classification results obtained with the proposed approach. Figure 7 summarizes the quantitative results obtained by comparing the reference map (through field visit) with the classification map yielded applying the proposed and traditional classification methods. The traditional Kmeans, Isodata classifier and Fuzzy C-Means produced the classification accuracies (in terms of Kappa coefficient) about 0.60 and 0.67 and 0.70 respectively. Whereas, the Differential evolution based Classification method produced acceptably high accuracy (0.78 Kappa). This is possible due to the spatial-frequency nature of wavelet frames which maximizes the simultaneous localization of energy in both the spatial and frequency domains. Also the proposed method extracts texture features by focusing on the dominant frequency regions whereas the traditional approaches fail to do so.

6 Conclusion

This paper provides a wavelet based feature extraction technique combined with DE clustering for texture information extraction from multispectral satellite data and the comparison is made with classical supervised classification techniques. The results found from the experiment demonstrate the higher performance of the proposed approach. When processing aperiodic function like remote sensing images, the efficiency of wavelet frame transform is higher. This can be attributed to the fact that the wavelets analyze the textures at multiple resolutions of the image. Since many textures in remotely sensed data have dominant frequency in middle frequency channels, the wavelets which are capable of focusing in the dominant frequency region provides good space-frequency information which gives unique texture signature that leads to higher classification accuracy. Further research is underway to analyze the potential of this method for texture analysis of high spatial resolution multispectral data obtained after merging with panchromatic data.

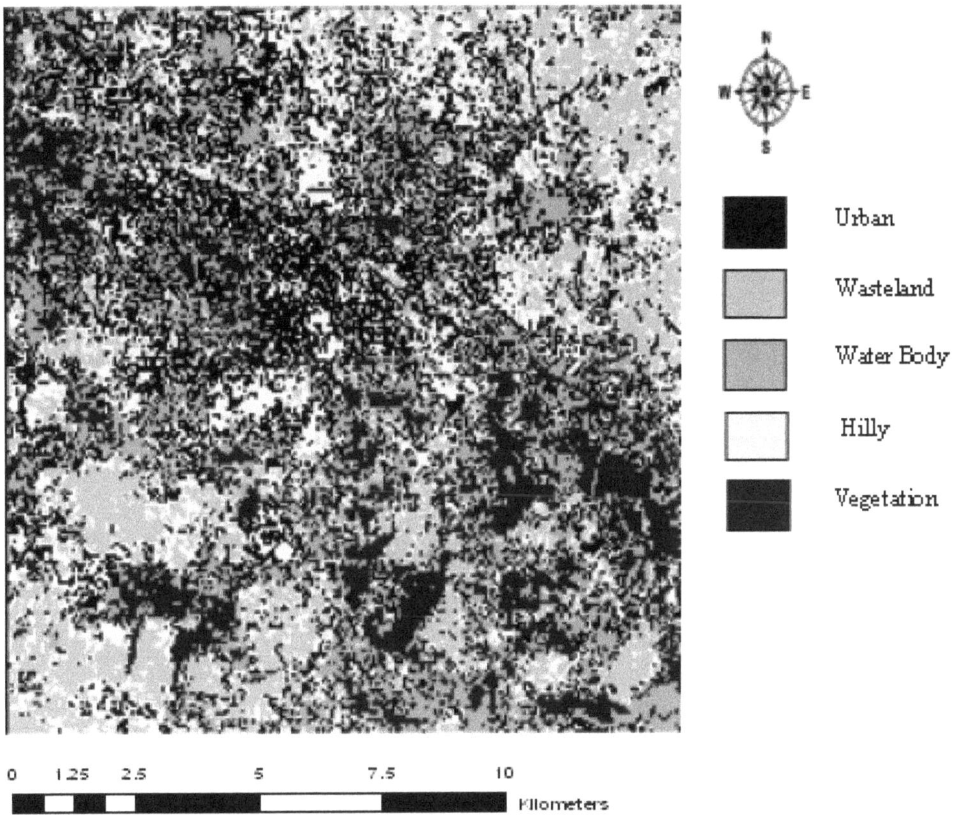

Fig. 3. Kmeans Classified Output.

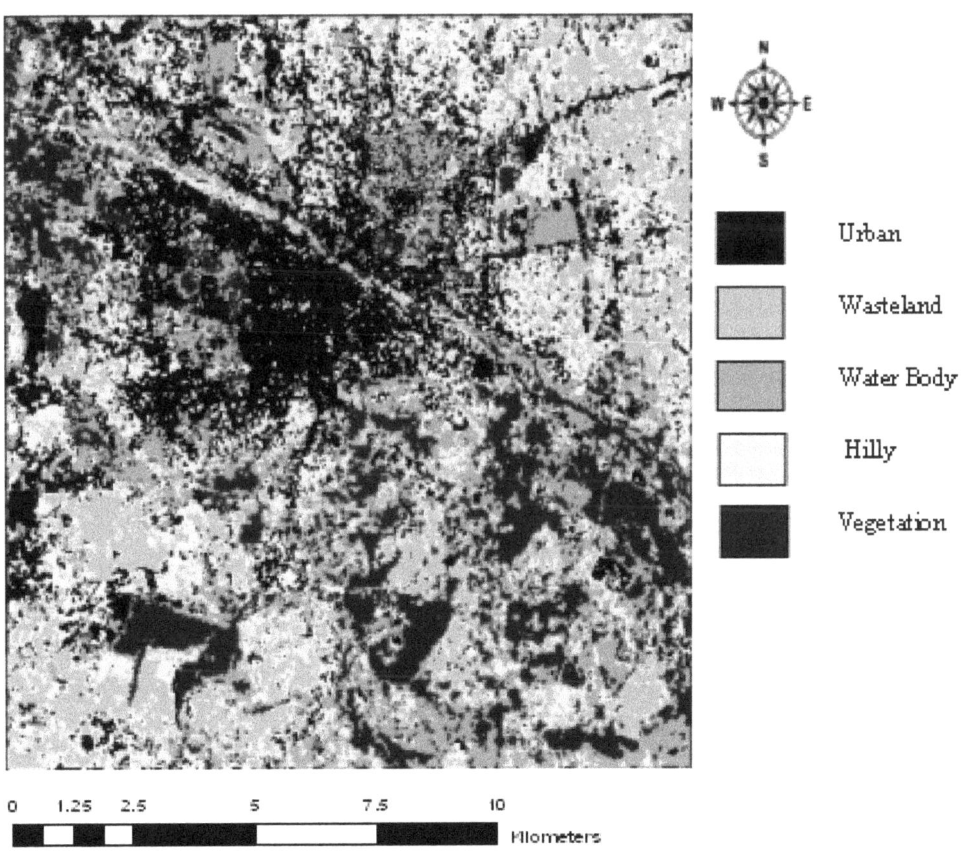

Fig. 4. Isodata Classification output.

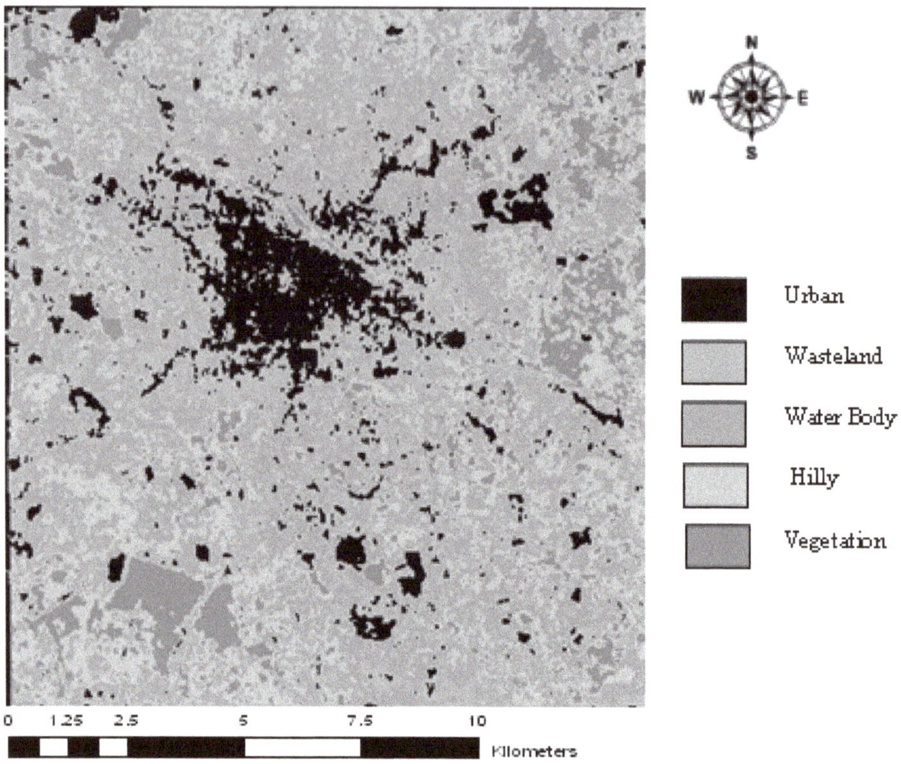

Fig. 5. Fuzzy C Means Classification output.

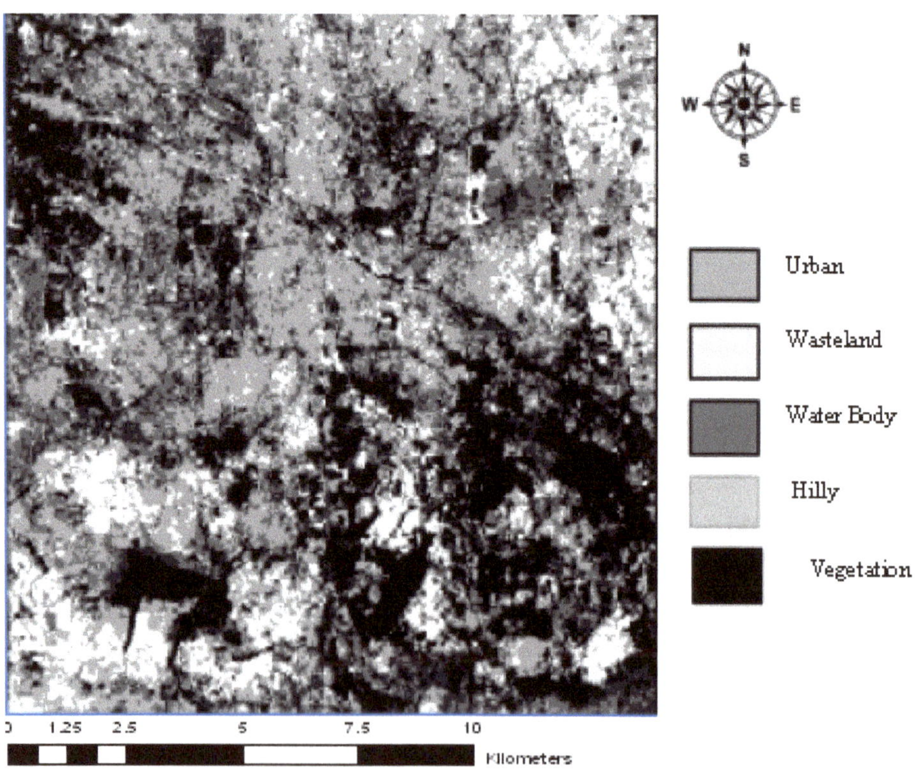

Fig. 6. DE based Classification output.

Techniques	Accuracy indices			
	User (%)	Producer (%)	Overall (%)	Kappa
K-means	38.33	47.60	53.88	0.6023
Isodata	40.12	45.23	50.45	0.6745
Fuzzy - *c*- Means	46.13	54.93	51.27	0.6184
Differential Evolution	58.27	60.46	65.64	0.7884

Fig. 7. Assessment of Classification Accuracy.

Reference

BEZDEK J.C. (1981) 'Pattern Recognition with Fuzzy Objective Function Algorithm' Plenum Press, New York.

CHANG, T. AND JAY KUO, C.C., 1993, Texture analysis and classification with tree structured wavelet transform. *IEEE Transactions on Image Processing.*, 2(4) :429-441.

COGGINS, J.M. AND JAIN, A.K., 1985, A spatial filtering approach to texture analysis. *Pattern Recognition Letters*, vol. 3, no. 3, pp. 195–203.

FISHER, P., 1997, The pixel: a snare and a delusion. *International Journal of Remote Sensing*, 18, pp. 679–685.

FOODY, G.M. AND COX, D.P., 1994, Sub-pixel land cover composition estimation using a linear mixture model and fuzzy membership functions. *International Journal of Remote Sensing*, 15, pp. 619–631.

HIREMATH, P.S. AND SHIVASHANKAR, S, 2006, Texture classification using wavelet packet decomposition. *International Journal on Graphics, Vision and Image processing*, Volume 6,Issue 2.

MALLAT, S.G., 1989, A Theory for multiresolution signal decomposition: The wavelet representation. *IEEE Transactions on Pattern Analysis and Machine Intelligence*, vol. 11, pp. 674–693.

THOMAS LILLESAND, M., RALPH KIEFER, W., AND JONATHAN CHIPMAN, W., 2004, *Remote Sensing and Image Interpretation.*, 5th edition, Wiley International edition. Tung Fung., 1990, An assessment of TM imagery for land cover change detection . *IEEE Transactions on Geoscience and Remote sensing*,Vol.28,No.4.

VICTOR DEBRUNNER AND MADHAVI KADIYALA., 1999, Effect of wavelet bases in texture classification using a tree structured wavelet transform.. *Proceeding of the IEEE*: 0-7803-5700-0/99.

OMRAN, M.G.H. ENGELBRECHT, A.P. SALMAN, A , 2005, "Differential Evolution Methods for Unsupervised Image Classification" *IEEE Transactions on Evolutionary Computing*(966- 973) Vol. 2.

Remote sensing of ancient Maya land use features at Caracol, Belize related to tropical rainforest structure

John F. Weishampel,[1] Arlen F. Chase,[2] Diane Z. Chase,[2] Jason B. Drake,[3] Ramesh L. Shrestha,[4] K. Clint Slatton,[4] Jaime J. Awe,[5] Jessica Hightower,[1] James Angelo[1]

1 Department of Biology, University of Central Florida, Orlando, Florida, USA
2 Department of Anthropology, University of Central Florida, Orlando, USA
3 United States Department of Agriculture, Forest Service, Tallahassee, USA
4 National Center for Airborne Laser Mapping, University of Florida, Gainesville, USA
5 Institute of Archaeology, National Institute of Culture and History, Belmopan, Belize

Abstract

To understand the social, political, economic, and ecological implications of historic land use patterns of Caracol, one of the largest (~200 sq km) Maya archaeological sites, we are conducting a regional survey using satellite and airborne sensors. About one-fifth of this polity has been mapped with traditional ground-based methods. In this NASA-funded project, we acquired Landsat TM/ETM, IKONOS, and small-footprint LiDAR data to detect archaeological features and measure rainforest properties based on canopy spectral signatures, three-dimensional canopy structure, and below-canopy topography. Using pattern analysis techniques, we will delineate causeways, agricultural terraces, reservoirs, plazas, and buildings to determine the extent and spatial organization of the urban and suburban settlements around Caracol.

Although the Maya typically clearcut ~75% of the land they inhabited, the vegetation rebounded to yield the most vast natural reforestation event in recent history (*c.* 1000 BP). From the imagery and point-cloud data, we will derive aboveground biomass and habitat structure (e.g., canopy top height, canopy openness, crown diameter of dominant trees, canopy texture) and assess forest recovery patterns. Thus, this study will link ancient land use legacies to current ecological condition and represents the most ambitious application of LiDAR for below-canopy archaeological prospecting to date.

Keywords Below-canopy, LiDAR, Maya, MesoAmerica, Rainforest, Settlement patterns.

1 Introduction

The social, political, and economic interpretation of ancient Maya settlement patterns requires broad-scale, regional surveys that extend beyond the urban epicenter (Chase and Chase 2003). Traditional ground-based survey methods are time-consuming and labor intensive. Recent advances in remote sensing and digital image processing show tremendous promise to assist with this process.

Here, we describe an effort, funded by NASA's Space Archaeology program that combines passive and active hyperspatial data to simultaneously record three-dimensional forest structural and below-canopy archaeological features from a largely unexcavated Maya site in Belize. This will permit the identification of previously unknown historic formations while determining the relationships between millennial-old land use practices and contemporary forest composition and structure.

2 Project Objectives

1) Use satellite remote sensing to detect Maya archaeological features at Caracol based on spectral signatures of the rainforest vegetation. The spatial extent and distribution of structures that comprise the urban metropolis, suburban hamlets, and trade routes will be determined.

2) Use airborne light detection and ranging (LiDAR) to map below-canopy topography in areas encompassed by known and detected Maya archaeological features. New causeways, reservoirs, agricultural terraces, plazas, and buildings around Caracol will be identified.

3) Relate contemporary rainforest canopy structure to historic Maya land use patterns. From satellite imagery and airborne LiDAR, we will derive forest biomass and habitat quality estimates (i.e., from measures of canopy top height, vertical distribution of return, degree of canopy openness, crown diameter of dominant trees) at Caracol.

3 Study Area

Caracol is the largest Maya archaeological site in Belize. Located near the Guatemala border on the Vaca Plateau of the Cayo District (Figure 1), it encompasses roughly 180km^2. At its cultural peak (c. AD 750), the metropolitan population is estimated to be ~150,000, which makes it one of the most populous cities in the Pre-Columbian world (Chase and Chase 1996). Caracol was, in essence, the capital city that maintained control of a 12,000 sq km region for several centuries (Chase and Chase 2003). It differs from other Maya lowland city states in terms of its large population size (e.g., it was nearly twice the size of the competing Maya city-states of Tikal and Calakmul) and its high population density (e.g., suburbs 6-8km outside of the Caracol urban center consisted of 1000 inhabitants per sq km). Because of the size and density of archaeological features found at Caracol, its outer limits have not been clearly demarcated.

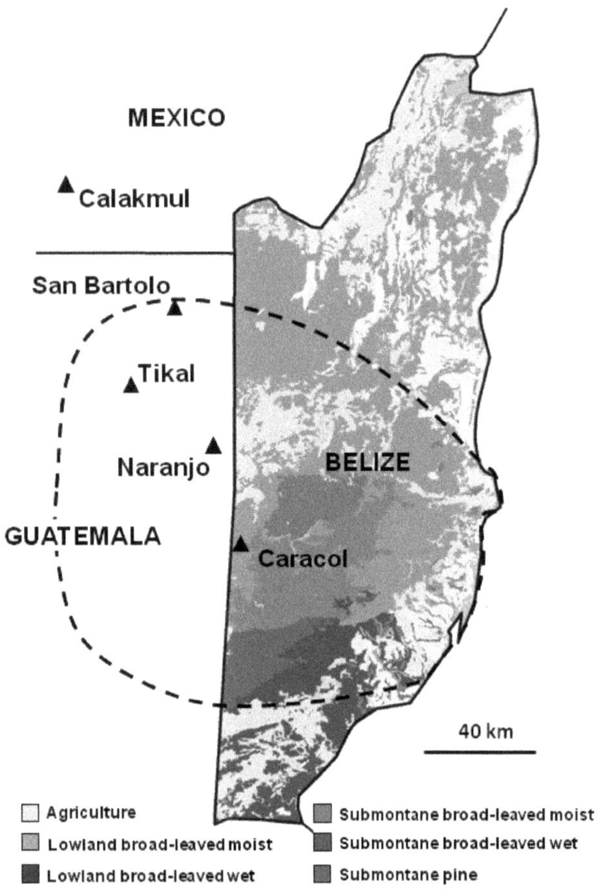

Fig. 1. Vegetation map of Belize (Meerman 2006) depicting location of mentioned Maya sites. The dashed line indicates the region controlled by Caracol, *c.* AD 750 (Chase and Chase 1996).

Unlike other Maya sites, the physical organization of Caracol is laid out in a radial plan like Washington, DC or Paris (Chase and Chase 1996). At its hub is the administrative and ceremonial plaza that contains the Caana pyramid monument. The 42m Caana is the tallest building in ancient as well as modern-day Belize. An estimated 75km of intra-site roads extend outward to surrounding hamlets that were absorbed into the city-state and to new residential settlements that arose as the prominence of Caracol increased (Chase and Chase 2001). The areal extent of metropolitan Caracol is estimated from the distribution of the termini of the causeways. These consist of a series of suburban complexes which are thought to have served as secondary administrative plazas, pre-existing centers, and large residential clusters. The termini form two rings around Caracol. The inner ring is approximately 3km from the urban epicenter and is comprised of low lying non-residential buildings. The second ring, located 4.6 to 9.6km way from the epicenter, is comprised of hamlets that were annexed. Portions of inter-site causeways totaling more than 75km have been detected in Landsat imagery (Figure 2). These extend ~40km northwest of Caracol towards another annexed Maya city-state, Naranjo, and ~24km to the south and southeast. However, these inter-site roads have not been verified at ground level.

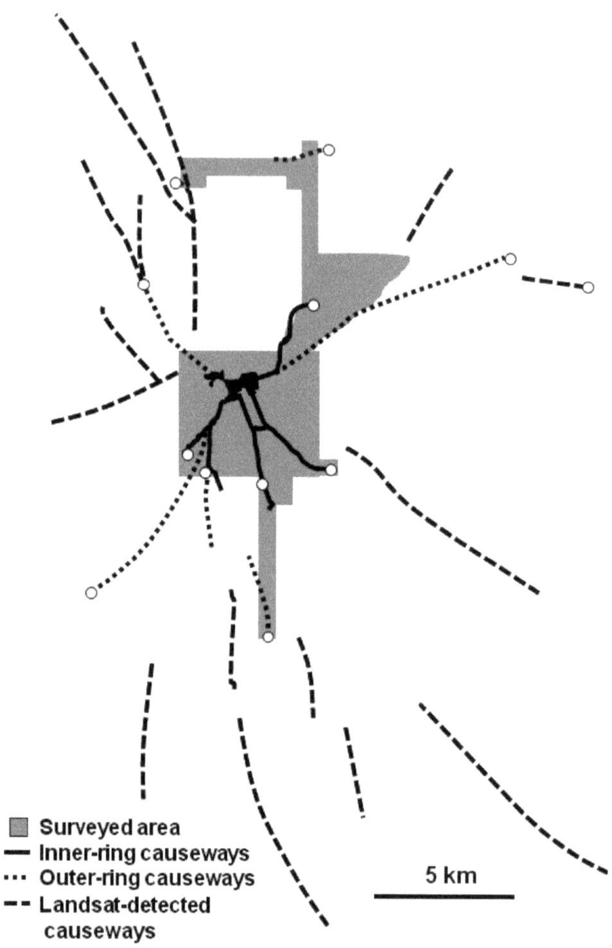

Fig. 2. Causeway map of Caracol (www.caracol.org). The inner blackened polygon corresponds to the urban epicenter from which the forest has been cleared. White circles designate termini.

Though discovered by archaeologist A. H. Anderson in 1938, only about a twelfth (~22 sq km) of this important site, i.e., the epicenter, has been intensively surveyed (Figure 3). Most of Caracol remains hidden below the canopy of lowland broad-leaved moist forest, designated by UNESCO as tropical evergreen seasonal broad-leaved lowland hill forest located on steep karstic terrain (Meerman 2006). As of 1998, the locations and dimensions of over 106 plaza groups of 4400 building structures have been mapped. Thus, if the density across the landscape remains constant, thousands of structures (on the order of 30,000; Chase and Chase 1998) and perhaps new termini plazas await discovery.

4 Remotely Sensed Data Acquisition and Analysis

In support of Objective 1

We have acquired hyperspatial IKONOS imagery (1m resolution panchromatic and four, 4m resolution multispectral bands) to explore the 200 sq km area immediately surrounding the Caracol urban epicenter and coarser scale Landsat TM (30m resolution) to examine a broader (2000 sq km) region.

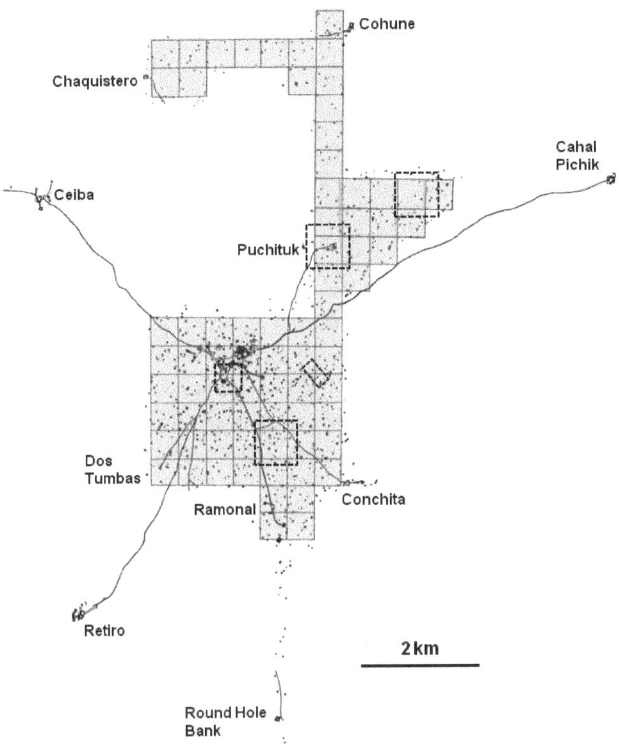

Fig.3. Settlement map of ~21 km^2 of Caracol (Chase and Chase 2001). Additional areas have since been mapped. Names correspond to causeway termini. Black specks represent plaza groups of buildings that have been mapped. Dashed rectangles are mapped terraced areas (Chase and Chase 1998).

Though hyperspatial imagery has been used to observe Maya archaeological features readily in the absence of forest canopy (e.g., Figure 4), only recently have methodologies been developed to reliably detect subcanopy features. To map the presence of archaeological structures in Caracol, we will employ the spectral signature analysis techniques that proved successful in identifying the presence of Maya structures below the forest canopy in the Petén in Guatemala (Sever and Parry 2006).

This methodology has not been shown to identify Maya archaeological features consistently, and may depend on vegetation type and/or phonological state (see Garrison *et al.*, 2008). The IKONOS and TM imagery were acquired from both wet and dry seasons and will be used to evaluate the robustness of these procedures in the broad-leaved moist forests around Caracol, as compared to the semi-deciduous moist forests, savannas, and wetlands in the Petén. We will use automated feature extraction techniques to identify forest patches that exhibit unique spectral signatures associated with underlying Maya structures. Through these procedures, we should be able to generate a more accurate measure of the extent of the greater metropolitan Caracol area and its regional influences.

Fig. 4. A panchromatic IKONOS image of the excavated epicenter of Caracol surrounded by rainforest canopy. The Caana pyramid is located in the upper right-hand corner. The Temple of the Wooden Lintel (A group) is on the left.

In support of Objective 2

Passive satellite sensors (e.g., IKONOS, Landsat) have been used to locate historic Maya features (Chase and Chase 2001); however, they generally only discern archaeological structures that: (1) extend above the canopy; (2) are found in areas devoid of vegetation; or (3) disrupt existing vegetation patterns in some way such as altering the spectral signature. Alternatively, an active microwave sensor, STAR-3i, was used to map canopy topography in the Petén (Saturno *et al.*, 2007). Elevated canopy heights corresponded to the presence of underlying Maya structures which tended to be situated on higher, drier ground.

Using the first laser return from the upper portions of the trees LiDAR, like STAR-3i, can be used to measure canopy topography (Weishampel *et al.*, 2000). As light penetrates the forest canopy and is reflected back, it records information about the three-dimensional distribution (x,y,z) of reflective surfaces. Because canopies are porous, some photons penetrate into the lower canopy and some reach the ground. Returned signals simultaneously provide measures of forest architecture and ground elevation (Figure 5). The horizontal and vertical resolution of these surface height measures are on the order of centimeters yielding very accurate topographic maps. Thus, LiDAR sensors can penetrate the forest canopy to measure obscured archaeological features.

Fig. 5. LiDAR point cloud data for a theoretical forest profile on a relic agricultural terrace. Returns would be filtered out to identify ground returns versus those from understory vegetation. From the point clouds, digital surface (DSM) and digital terrain (DTM) models, would be produced. The right panel represents a pseudo-waveform created by summing the returns across vertical bins.

Though results from early applications of single-track, profiling LiDAR sensors were not especially encouraging (McKee and Sever 1994; P. Sheets, pers. comm.), recent applications of significantly more advanced airborne laser swath mapping system (ALSM) technology shows considerable promise for archaeological prospecting (see Devereux *et al.,* 2005; Risbøl *et al.,* 2006; Sittler and Schellberg 2006). A pilot study, directly related to our proposed research, showed the remarkable accuracy of LiDAR for detecting below-canopy Maya ruins in Copán, Honduras when compared to a total station survey (Gibeaut *et al.,* 2003).

LiDAR sensors have been shown to provide accurate, detailed digital elevation models (DEMs) below dense, closed-canopy wet tropical rainforests at La Selva, Costa Rica (Hofton *et al.,* 2002) of sufficient quality to map hydrologic features (i.e., drainage patterns), which provide insight into human settlement patterns, water management, and storage capability (Saturno *et al.,* 2007). The rainforests at Caracol are seasonal and more open than those in Costa Rica, which should provide ample ground returns especially when flown at the end of the dry season, (i.e., March to May). Within a ~200 sq km region defined by the known causeway termini, state-of-the-art airborne LiDAR data were acquired in April 2009. Detailed topographic maps will be produced from these data, revealing natural (e.g., ground terrain) and anthropogenic (e.g., buildings, roads, terraces) features.

Previous and ongoing field investigations/excavations will provide comparisons of surface topography, allowing correlation of the digital topographic map with height measures of decomposed human made features. These GIS and hand-drawn maps include >22 sq km of Caracol roadways, public buildings, and residential groups, as well as approximately 4 sq km of Caracol's dense agricultural terracing (Chase and Chase 1998; 2001). It is expected that previously unmapped features will be detected by extrapolating linear ground features across the landscape and by using pattern recognition algorithms and image enhancement techniques such as edge detection and textural analysis (Lee *et al.,* 2005). The LiDAR data will also be used to delineate the intra-site network of causeways. Since they facilitated transportation and communication, these causeways can be thought of as political and economic statements that provided administration and integration of this Maya metropolis.

Caracol is known as the 'Garden City' because its dense settlements were embedded in a complex matrix of agricultural terraces (Chase and Chase 1998). However, there is no naturally running water for crop irrigation near the urban center. To adapt sufficiently to these conditions, the Maya constructed a network of thousands of kilometers of stone terraces considered to be the most elaborate in the Maya lowlands (Dunning 1996) and hundreds of reservoirs (Healy *et al.,* 1983).

We will map the 3-dimensional surface topography (i.e., representing Maya buildings, landuse patterns, and forest canopies) using the remote sensing/data processing facilities available through the National Center for Airborne Laser Mapping (NCALM; Carter *et al.,* 2001). The NCALM system records four returns per pulse as it scans across a ~0.5km swath. When a pulse is transmitted, part of the pulse reflects off an object close to the aircraft (e.g., the canopy top), while the remainder of the pulse continues through the trees and reflects off the hard surface (e.g., the ground or a building). At a nominal above-ground altitude of 600-1000m, the system yields 10-15 laser returns per square meter with a height accuracy of 5-10cm and a horizontal accuracy of 25-40cm. This system also records the relative intensity of each return. By crisscrossing flight lines over the central Caracol area, a higher density of photons and more accurate DEM will be obtained. A comprehensive set of the ALSM elevation and intensity features will be used to characterize the canopy and to identify and segment anthropogenic structures. A robust probabilistic method to isolate the optimal ALSM feature set for building segmentation features (Luzum *et al.,* 2005) as well as information-theoretic filters for separating vegetation structure from bare surface topography will be used (Kampa and Slatton 2004). These have been applied to detect small walking trails (Lee *et al.,* 2005) underneath forest canopies.

In support of Objective 3

The idea that ecology needs archaeology and vice versa focuses on human causes of environmental change and how, in turn, environmental change affects human society (Briggs *et al.,* 2006; Foster *et al.,* 2003). Because present-day rainforests (and other "wild" areas) bear the fingerprints of historic human disturbance (Heckenberger *et al.,* 2003; Willis *et al.,* 2004), archaeologists are necessary to provide ecologists with a view of past human land use. Such environmental impacts may require several centuries for the disturbed system to return to its pre-human-disturbed state, if, indeed, it does recover as most systems are in a state of continual disturbance. The system, once disturbed, may undergo an entirely different successional trajectory. Recovery of forest composition and structure tends to be relatively rapid when the disturbance primarily impacts the forest canopy and considerably slower when soils (e.g., fertility and texture) are also heavily impacted (Chazdon 2003). For example, secondary, 400 year-old central Amazon forests growing on pre-Columbian archaeological landscapes detected with remote sensing have not returned to "high" forest and still exhibit signs of acute alteration (Heckenberger *et al.*, 2003).

The collapse of the Maya civilization (*c.* AD 950) is the most recent example of a vast naturally reforested landscape (Sterpa 2005). Evidence consistently supports the hypothesis that there was a massive transformation of the central Maya lowlands from seasonally dry tropical forests to an open, cultivated landscape (Figure 6). The contribution that environmental degradation (e.g., deforestation and/or soil erosion) played in this collapse is subject to debate (Abrams *et al.*, 1996; Turner *et al.*, 2003). The ancient Maya landscape between the ceremonial centers and dense rural/agricultural settlements typically consisted of less-managed forests, complex agroforestry systems, and abandoned agricultural lands with secondary growth. Nearly 75% of the Maya environment was anthropogenically cleared for settlement and agriculture at its peak; thus, much of the present 'old-growth' or 'virgin' Central American rainforests are 600-1000 years old and have grown on what was a largely human-dominated, agrarian landscape (Redman 1999). As such, this example affords insights to other regions that are presently undergoing reforestation or woody encroachment (Chazdon 2003; Gómez-Pompa and Kaus 1999).

Caracol is located in the Chiquibul Forest Reserve, which is one of the largest remaining continuous tracts of tropical rainforest in Central America. Some of the forest stands surrounding Caracol were virtually undisturbed by humans over the last millennium. Studies of post-agricultural recovery of tropical forests show that the recovery of forest structure, soil nutrients, and species richness is far more rapid than recovery of species composition (Chazdon 2003). However, some ecological processes in tropical rainforests (e.g., plant growth rates) rebound more quickly than certain measures of ecosystem structure (e.g., vertical canopy structure) after a disturbance. The recovery of species composition is dependent on the landscape matrix surrounding the disturbance, (e.g., the presence of seed sources). Thus, the land use patterns around an abandoned field may dramatically influence recovery. Moreover, species abundance patterns may also reflect past Maya land use practices in at least two ways. First, the presence of agriculturally/economically important tree species may reflect the presence of Maya orchard-gardens (called *pet kotoob*; Gómez-Pompa *et al.*, 1987) and other forestry practices or second, the Maya-disturbed, often limestone-rich soils may promote tree

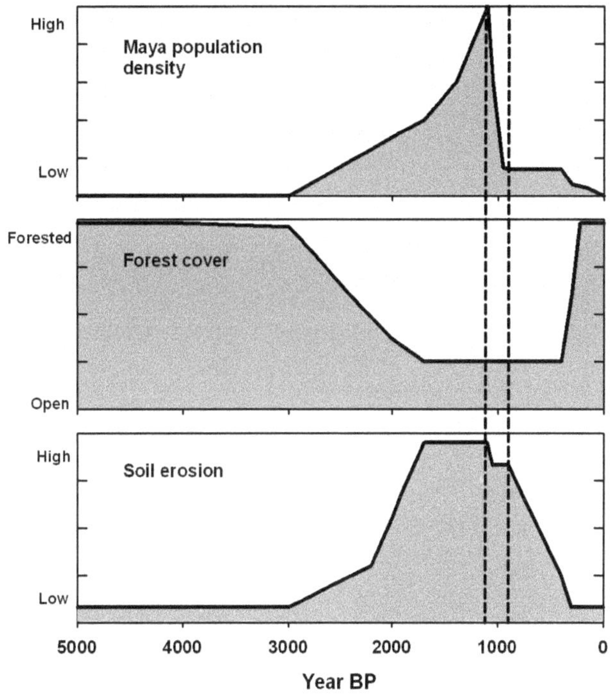

Fig. 6. impacts to tropical forests in the southern Yucatán adapted from Binford *et al.* (1987). The first and second dashed lines correspond respectively to the last date recorded on a Caracol stela and when the site was totally abandoned.

species, such as the ramón (*Brosimum alicastrum*), that are adapted to these conditions or are exhibiting stress due to them (Foster and Turner 2004; Lambert and Arnason 1982; Turner *et al.*, 2001). These latter micro-environmental conditions may explain the differences in spectral signatures from canopies above the archaeological ruins (Saturno *et al.*, 2007). As noted by Healy *et al.* (1983), the tree species composition on the agricultural terraces, which were designed to reduce soil erosion, does not match the composition found on ruins in the plazas. Because species composition influences canopy structure, structural differences may exist, reflecting past land use legacies (Weishampel *et al.*, 2007). However, the physiognomy of *pet kotoob* sites has been reported as similar to other tropical forest stands in the vicinity (Gómez-Pompa *et al.*, 1987).

Canopy structure largely dictates the biophysical environment of a forest. Its organization influences tree physiology, atmospheric exchange, and arboreal and understory habitats. Discrete LiDAR and hyperspatial imagery are well-suited to characterize canopy architecture (Lim *et al.*, 2003; Clark *et al.*, 2004a). Tropical forests, such as those engulfing the Maya city of Caracol, contain large stores of labile carbon (i.e., wood) that are not easily quantified by passive remote sensing (e.g., Landsat) methods. The estimation of forest canopy height (or digital canopy models, DCMs) using small-footprint LiDAR is accomplished by subtracting the digital surface model (DSM, the canopy topography) from the digital terrain model

(DTM) and has been shown to be very accurate in tropical rainforest environments (Clark *et al.,* 2004b). In addition to height, small-footprint LiDAR DCMs permit the detection and segmentation of individual, canopy dominant tree crowns (Popescu and Zhao 2008). Measurements of crown area of trees in closed-canopy tropical rainforests with IKONOS imagery has shown mixed results (Asner *et al.,* 2002; Clark *et al.,* 2004a). Given height and crown area, Brown *et al.* (2005) developed allometric relationships for functional groups of Belizean tree species and were able to discern aboveground carbon stocks (biomass) in forests with a high degree of accuracy.

Analysis of LiDAR waveforms from a closed-canopy, tropical forest in Costa Rica revealed the three-dimensional structure of the forest (Weishampel *et al.,* 2000) and the relationship between the waveforms properties and aboveground biomass (Drake *et al.,* 2002). The robustness of these LiDAR/biomass relationships will be determined in conjunction with previous flights over tropical rainforest sites in Costa Rica. These relationships will be extrapolated to Caracol to produce maps of tropical forest structure that includes data on height, biomass, and internal canopy organization across this previously intensively occupied landscape. The hyperspatial imagery coupled with LiDAR will readily permit the segmentation of emergent trees which often account for the majority of biomass in rainforests (Clark and Clark 1996). Using existing and newly generated maps of prehistoric agricultural and other land use features (Chase and Chase 1998; 2001; 2003), contemporary forest structure will be compared with the past constructed landscape to examine possible correlations with Maya land use and calculated ages since abandonment. Patterns of forest recovery on causeways, agricultural terraces, urban plazas, and other land use classes will be identified and quantified.

Acknowledgments

This research was supported by NASA Grant #NNX08AM11G through the Space Archaeology program. We appreciate the discussions at NASA Marshall Space Flight Center with Tom Sever, Bill Saturno, and Burgess Howell, and over the telephone with Payson Sheets.

References

ABRAMS, E. M., FRETER, A., RUE, D. J., AND WINGARD, J. D. 1996. The role of deforestation in the collapse of the Late Classic Copán Maya State. In: L. E. Sponsel, T. N. Headland, and R. C. Bailey (eds.), *Tropical Deforestation: The Human Dimension*, 55-75. New York, Columbia University Press.

ASNER, G. P., PALACE, M., KELLER, M., PEREIRA JR., R., SILVA, J. N. M., AND ZWEEDE, J. C. 2002. Estimating canopy structure in an Amazon forest from laser range finder and IKONOS satellite observations. *Biotropica* 34, 483-492.

BINFORD, M. W., BRENNER, M., WHITMORE, T. J. HIGUERA-GUNDY, A., DEEVEY, E. S., AND LEYDEN, B. 1987. Ecosystems, paleoecology and human disturbance in subtropical and tropical America. *Quaternary Science Reviews* 6, 115-128.

BRIGGS, J. M., SPIELMANN, K. A., SCHAAFSMA, H., KINTIGH, K. W., KRUSE, M., MOREHOUSE, K., AND SCHOLLMEYER, K. 2006. Why ecology needs archaeologists and archaeology needs ecologists. *Frontiers in Ecology and the Environment* 4, 180-188.

BROWN, S., PEARSON, T., SLAYMAKER, D., AMBAGIS, S., MOORE, N., NOVELO, D. AND SABIDO, W. 2005. Creating a virtual tropical forest from three-dimensional aerial imagery to estimate carbon stocks. *Ecological Applications* 15, 1083-1095.

CARTER, W., SHRESTHA, R., TUELL, G., BLOOMQUIST, D., AND SARTORI, M. 2001. Airborne laser swath mapping shines new light on Earth's topography. *Eos, Transactions of the American Geophysical Union* 82, 549-555.

CHASE, A. F. AND CHASE, D. Z. 1996. A mighty Maya nation: how Caracol built an empire by cultivating its 'middle class.' *Archaeology* 49, 66-72.

CHASE, A. F. AND CHASE, D. Z. 1998. Scale and intensity in classic period Maya agriculture: terracing and settlement at the "Garden City" of Caracol, Belize. *Culture and Agriculture* 20, 60-77.

CHASE, A. F. AND CHASE, D. Z. 2001. Ancient Maya causeways and site organization at Caracol,Belize. *Ancient Mesoamerica* 12, 273-281.

CHASE, A. F. AND CHASE, D. Z. 2003. Minor Centers, Complexity, and Scale in Lowland Maya Settlement Archaeology, in G. Iannone and S.V. Connell (eds.), *Perspectives on Ancient Maya Rural Complexity* Monograph 49, 101-118, Los Angeles, Cotsen Insitute of Archaeology at UCLA.

CHAZDON, R. L. 2003. Tropical forest recovery: legacies of human impact and natural disturbances. *Perspectives in Plant Ecology, Evolution and Systematics* 6, 51-71.

CLARK, D. B. AND CLARK, D. A. 1996. Abundance, growth and mortality of very large trees in neotropical lowland rain forest. *Forest Ecology and Management* 80, 235-244.

CLARK, D. B., READ, J. M., CLARK, M. L., CRUZ, A. M., DOTTI, M. F., AND CLARK, D. A. 2004a. Application of 1-m and 4-m resolution satellite data to ecological studies of tropical rain forests. *Ecological Applications* 14, 61-74.

CLARK M. L., CLARK D. B., & ROBERTS D. A. 2004b. Small-footprint lidar estimation of sub-canopy elevation and tree height in a tropical rain forest landscape. *Remote Sensing of Environment* 92, 68-89.

DEVEREUX, B. J., AMABLE, G. S., CROW, P., AND CLIFF, A. D. 2005. The potential of airborne lidar for detection of archaeological features under woodland canopies. *Antiquity* 79, 648-660.

DRAKE, J. B., DUBAYAH, R., KNOX, R., AND CLARK, D. 2002. Relationship between vertical canopy profiles and biomass in a Neotropical rainforest. *Remote Sensing of Environment* 81, 378-392.

DUNNING, N. P. 1996. A reexamination of regional variability in the pre-Hispanic agricultural landscape, in S. L. Fedick (ed.) *The Managed Mosaic: Ancient Maya Agriculture and Resource Use*, 53-68, Salt Lake City, University of Utah Press.

FOSTER, D. R., SWANSON, F., ABER, J., BURKE, I., BROKAW, N., TILMAN, D., AND KNAPP, A. 2003. The importance of land-use legacies to ecology and conservation. *BioScience* 53,77-88.

FOSTER, D. R. AND TURNER, B. L. II. 2004. THE LONG VIEW: HUMAN-ENVIRONMENT RELATIONSHIPS IN THE REGION, 1000 BC–AD 1900, in B. L. Turner II, J. Geoghegan, and D. R. Foster (eds.) *Integrated Land-Change Science and Tropical Deforestation in the Southern Yucatán*, 23-37, Oxford, Oxford University Press.

GARRISON, T. G., HOUSTON, S. D., GOLDEN, C., INOMATA, T., NELSON, Z., AND MUNSON, J. 2008. Evaluating the use of IKONOS satellite imagery in lowland Maya settlement archaeology. *Journal of Archaeological Science* 35, 2770-2777.

GIBEAUT, J. C., SMYTH, R. C., GUTIERREZ, R., AND HEPNER, T. 2003. LiDAR point classification. Presented at LiDAR Surveys for Coastal Hazards and Resource Mapping workshop. Bureau of Economic Geology, University of Texas, Austin.

GÓMEZ-POMPA, A., FLORES, J. S., AND SOSA, V. 1987. The "pet kot": a man-made tropical forest of the Maya. *Interciencia* 12, 10-15.

GÓMEZ-POMPA, A. AND KAUS, A. 1999. From pre-Hispanic to future conservation alternatives: lessons from Mexico. *Proceedings of the National Academy of Sciences* 96, 5982-5986.

HEALY, P. F., LAMBERT, J. D. H., ARNASON, J. T., AND HEBDA, R. J. 1983. Caracol, Belize: evidence of ancient Maya agricultural terraces. *Journal of Field Archaeology* 10, 397-410.

HECKENBERGER, M. J., KUIKURO, A., KUIKURO, U. T., RUSSELL, J. C., SCHMIDT, M., FAUSTO, C., AND FRANCHETTO, B. 2003. Amazonia 1492: pristine forest of cultural parkland. *Science* 301, 1710-1714.

HOFTON, M. A., ROCCHIO, L. E., BLAIR, J. B., AND DUBAYAH, R. 2002. Validation of vegetation canopy lidar sub-canopy topography measurements for a dense tropical forest. *Journal of Geodynamics* 34, 491-502.

KAMPA, K. AND SLATTON, K. C. 2004. An adaptive multiscale filter for segmenting vegetation in ALSM data. *IEEE Transactions in Geoscience and Remote Sensing Symposium (IGARSS)* 6, 3837-3840.

LAMBERT, J. D. H. AND ARNASON, J. T. 1982. Rámon and Maya ruins: an ecological, not an economic, relation. *Science* 216, 298-299.

LEE, H., SLATTON, K. C., AND JHEE, H. 2005. Detecting forest trails occluded by dense canopies using ALSM data. *Proc. IEEE International Geoscience and Remote Sensing Symposium (IGARSS)* 5, 3587-3590.

LIM, K., TREITZ, P., WULDER, M., ST-ONGE, B, AND FLOOD, M. 2003. LiDAR remote sensing of forest structure. *Progress in Physical Geography* 27, 88-106.

LUZUM, B. J., SLATTON, K. C. AND SHRESTHA, R. L. 2005. Analysis of spatial and temporal stability of airborne laser swath mapping data in feature space. *IEEE Transactions in Geoscience and Remote Sensing* 43, 1403-1420.

MCKEE, B. R. AND SEVER, T. L. 1994. Remote sensing in the Arenal region, in P. D. Sheets and B. R. McKee (eds.) *Archaeology, Volcanism, and Remote Sensing in the Arenal Region, Costa Rica*, 135-141, University of Texas Press, Austin.

MEERMAN, J. 2006. Belize Ecosystems Map version 2004. Biodiversity and Environmental Resource Data System of Belize.

POPESCU, S. C. AND ZHAO, K. 2008. A voxel-based lidar method for estimating crown base height for deciduous and pine trees. *Remote Sensing of Environment* 112, 767-781.

REDMAN, C. L. 1999. *Human Impacts on Ancient Environments*. Tuscon, University of Arizona Press.

RISBØL, O., GJERTSEN, A.K., AND SKARE, K. 2006. Airborne laser scanning of cultural remains in a Norwegian forest. *From Space to Place: Proceedings of the 2nd International Conference on Remote Sensing in Archaeology*, 107-112, Oxford, BAR Publishing.

SATURNO, W., SEVER, T. L., IRWIN, D. E., HOWELL, B. F., AND GARRISON, T. G. 2007. Putting us on the map: remote sensing investigation of the ancient Maya landscape, in J. Wiseman and F. El-Baz (eds.) *Remote Sensing in Archaeology*, 137-160, New York, Springer.

SEVER, T. AND PARRY, J. T. 2006. Archaeological remote sensing of human settlements, in M. K. Ridd and J. D. Hipple (eds.) *Remote Sensing of Human Settlements*, 431- 516, Bethesda, ASPRS.

SITTLER, B. AND SCHELLBERG, S. 2006. The potential of LiDAR in assessing elements of cultural heritage hidden under woodland canopies: Possibilities and limits in detecting microrelief structures for archaeological surveys. *From Space to Place: Proceedings of the 2nd International Conference on Remote Sensing in Archaeology*, 117-122, Oxford, BAR Publishing.

STERPA, J. P. (2005, MAY 5). Using local wood, harvested by 'sustainable' methods, will help planet, they say; biggest rebound in 1,000 Years. *Wall Street Journal,* p. A.1.

TURNER, B .L., II AND HARRISON, P. D. 1981. Prehistoric raised-field agriculture in the Maya lowlands. *Science* 213, 399-405.

TURNER, B. L., II, KLEPEIS, P., AND SCHNEIDER, L. C. 2003. Three millennia in the Southern Yucatán peninsula: implications for occupancy, use, and carrying capacity, in A. Gómez-Pompa, M. F. Allen, S. L. Fedick, J. J. Jiménez-Osornio (eds.) *The Lowland Maya Area: Three Millennia at the Human-Wildland Interface*, 361-387, New York, Food Products Press.

TURNER, B. L., II, VILLAR, S. C., FOSTER, D., GEOGHEGAN, J., KEYS, E., KLEPEIS, P, LAWRENCE, D., MENDOZA, P. M., MANSON, S., OGNEVA-HIMMELBERGER, Y., PLOTKIN, A. B., SALICRUP, D. P., CHOWDHURY, R. R., SAVITSKY, B., SCHNEIDER, L., SCHMOOK, B., AND VANCE, C. 2001. Deforestation in the southern Yucatán peninsular region: an integrative approach. *Forest Ecology and Management* 154, 353-370.

WEISHAMPEL, J. F., BLAIR, J. B., DUBAYAH, R., CLARK, D. B., AND KNOX, R. G. 2000. Canopy topography of an old-growth tropical rainforest landscape. *Selbyana* 21, 79-87.

WEISHAMPEL, J. F., BLAIR, J. B., KNOX, R. G., DUBAYAH, R., AND CLARK, D. B. 2000. Volumetric lidar return patterns from an old-growth tropical rainforest canopy. *International Journal of Remote Sensing* 21, 409-415.

WEISHAMPEL, J. F., DRAKE, J. B., COOPER, A., BLAIR, J. B., AND HOFTON, M. 2007. Forest canopy recovery from the 1938 hurricane and subsequent salvage damage measured with airborne LiDAR. *Remote Sensing of Environment* 190, 142-153.

WILLIS, K. J., GILLSON, L., AND BRNCIC, T. M. 2004. How "virgin" is virgin rainforest. *Science* 304, 402-403.

A study for analyzing the impact of Wavelet scaling on satellite data classification

R.A. Alagu Raja, V. Anand, V. Abhai Kumar

Thiagarajar college of Engineering., Madurai-625015, India

Abstract

Texture classification has recently attracted a world wide attention and is mainly used in classification of remotely sensed data since textures can be seen in all images from multispectral scanner images. For classifying remotely sensed data, texture is considered as an important parameter. Though the classical texture feature extraction techniques such as Laws mask, Edge filters, Ring & wedge filters and DCT can perform well in satellite data classification they fail to extract texture information in middle frequency channels. To overcome this, wavelets which are capable of focusing in the dominant frequency region can be used .This paper describes an approach for wavelet based texture feature extraction and classification remotely sensed data .Different types of wavelets such as Daubechies , Lemaire Battle , Coiflet, Symlet and Biorthogonal wavelets at different scales are used for texture feature extraction. The feature vector formed using the wavelet coefficients are subjected to fuzzy C-means unsupervised classification algorithm. Also this paper analyzes and compares the effect of wavelet scaling on classification accuracy .It is found that the wavelet decomposition with combined scales is giving better results

1 Introduction

Texture analysis is a fundamental methodology to many applications in areas like remote sensing, digital imaging, quality inspection, medical imaging, etc. Textures are replications, symmetries and combinations of various basic patterns usually with some random variation. Although texture analysis has a long history, its applications to satellite image data have been limited. Classification of textures in digital images has received considerable attention in literature and a large number of approaches have been suggested. Texture classification is a process to extract features from a set of texture classes. In order to have an effective classification, features with good discriminatory details have to be obtained. Most of the existing approaches for texture feature extraction make use of statistical techniques in which processing the texture image data requires large storage space and computational capability. The scalar features calculated from feature matrix is not efficient to represent the characteristics of image content. However space-frequency domain methods can be attempted for extracting texture features. Recently, multiscale filtering methods have shown significant potential for texture description, where advantage is taken of the spatial-frequency concept to maximize the simultaneous localization of energy in both spatial and frequency domains. Recent developments in the wavelet transforms provide a good multiresolution analytical tool for texture analysis with high classification accuracy. In this paper, different types of wavelets such as Daubechies, Lemaire Battle, Coiflet, Symlet and Biorthogonal wavelets at different scales are used for texture feature extraction. The feature vector formed using the wavelet coefficients are subjected to fuzzy C-means unsupervised classification algorithm and the classification accuracy for various wavelets at different scales is compared.

2 Wavelet transform

The wavelet transform is similar to a Fourier transform in the sense that it decomposes a signal into a set of basis functions. Wavelet transform is a bound alterable window method. It can get more accurate low frequency information using longer interval and get high frequency information using shorter interval. The advantage of the wavelet transform is information maintenance, which provides the space frequency information. The continuous wavelet transform of a function $f(x)$ is defined as

$$CTWT(a,b) = \int_{-\infty}^{\infty} f(x)\frac{1}{\sqrt{|a|}}\psi\left(\frac{x-b}{a}\right)dx$$

Unlike the short time Fourier transform, the windowing of the input signal is contained in the basis function, giving better spatial resolution at high frequencies and allowing the transform to adjust to the characteristics of the signal.

In practical applications the wavelet transform must be applied to discrete signals of finite length. For this purpose a discrete form of the wavelet transform has been developed. For the discrete wavelet transform (DWT) we make use of the fact that a signal may be represented in two parts—a smoothed signal and a detail signal. The smoothed signal is decomposed over a set of basis functions f(x) the scaling function while the detail signal is over the wavelet basis $\psi(x)$. Furthermore, it is possible to represent these basis functions using a pair of sequences, h(n) and g(n).

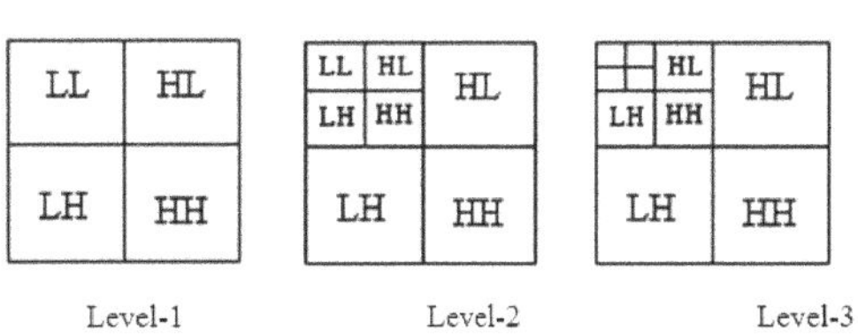

Fig. 1. Wavelet decomposition at different levels.

3 Study area and data used

The multi temporal satellite image used in this study is of Madurai city, South India. Madurai is the second largest city in Tamil Nadu, after Chennai and is one of the oldest cities in India, over 2,500 years old. Madurai is identified as one of the 12 heritage cities of India. Madurai is situated between the longitude 78^0 04' 47" E to 78^0 11' 23" E and the latitude 9^0 50' 59" N to 9^0 57' 36" N. The topography of Madurai is approximately 101 meters above the sea level. The land cover features of this study area includes urban, vegetation, water body, waste land and hilly region. The scene details are,

Date of acquisition : 19th March,2004

Satellite/Senor : IRS P6/ LISS III

Resolution : 23.50 meter

A sub area of 407x411 pixels were extracted from the original scale.

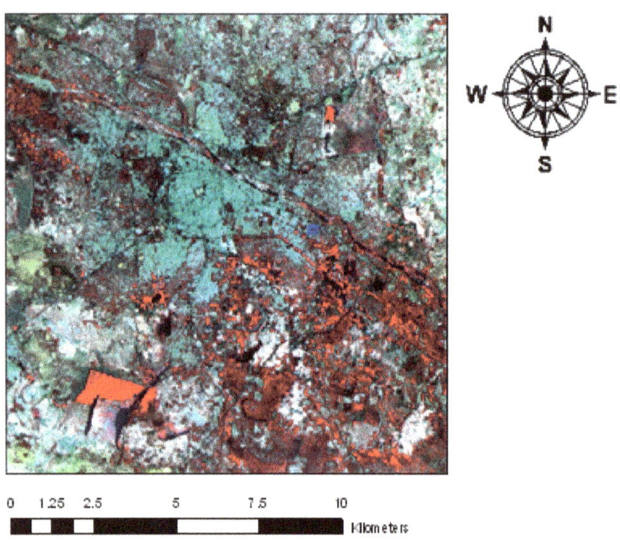

Fig. 2. Madurai City LISS III Image – March 2004.

4. Methodology

In this study, we analyze several texture methods applied to the classification of remote sensing images with different types of landscapes. Fig.3 shows the steps of the overall methodology for wavelet based texture classification.

4.1 Preprocessing

The Multispectral satellite data of Madurai taken in the year 2004 is geometrically registered using Survey of India topographic sheets 58 K/1 in 1:50,000 scale and four 58 K/1 SW, SE, NW, NE in 1:25,000 scale were used as reference data for GCPs collection and verification. The standard errors were below 0.5 pixel along the X and Y direction.

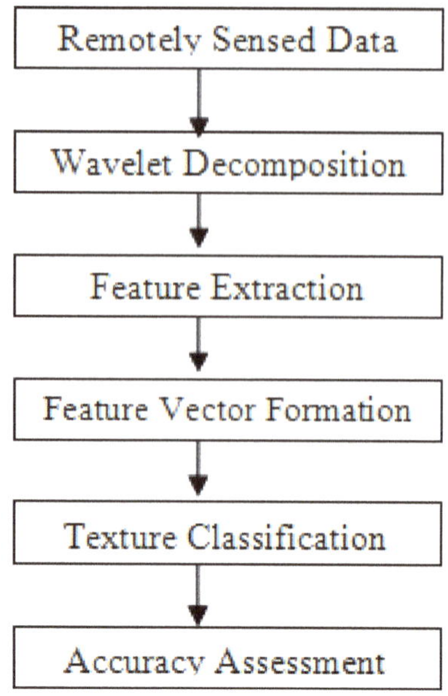

Fig. 3. Methodology for Texture Classification.

4.2 Texture Feature Extraction

Transforming the input data into useful details is called features extraction .The goal of feature extraction is to obtain a set of measures that can be used to discriminate different textures. The extraction of texture features from remotely sensed imagery provides a source of data for identification or classification of spectrally heterogeneous landscape units. However, there is a wide range of texture analysis techniques that are used with different criteria for feature extraction: statistical methods and filter techniques. The basic assumption for most filtering approaches is that the energy distribution in the frequency domain identifies a texture.

4.3 Texture Classification using different wavelets

The output from the texture feature extraction is a set of images. These images are the feature images that will form the basis for the classification. They are used to form feature vectors, where each feature image corresponds to one element in the feature vectors. Classification is performed based on the feature vectors in this project for feature extraction, we have used different types of wavelets at different scales. They are as follows.

4.3.1 Daubechies wavelet

The Daubechies wavelets are family of orthogonal wavelets defining a discrete wavelet transform and characterized by a maximal number of vanishing moments. Daubechies orthogonal wavelets D2-D20(even index numbers only) are commonly used. Each wavelet has a number of zero moments or vanishing moments equal to half the number of coefficients. A vanishing moment limits the wavelet's ability to represent polynomial behaviors or information in a signal.

For example, D2-with one moment easily encodes polynomials of one coefficient or constant signal components.D4-encodes polynomials with two coefficients or constant and linear signal components. D6-encodes polynomials with three coefficients or constant linear and quadratic signal components. Hence Daubechies wavelets are widely used in solving a broad range of problems. Daubechies D4 transform has four wavelet and scaling function coefficients.

The scaling and wavelet functions are calculated by taking the inner product of the coefficients and four data values. The scaling function and wavelet functions at scale 1 are shown below.

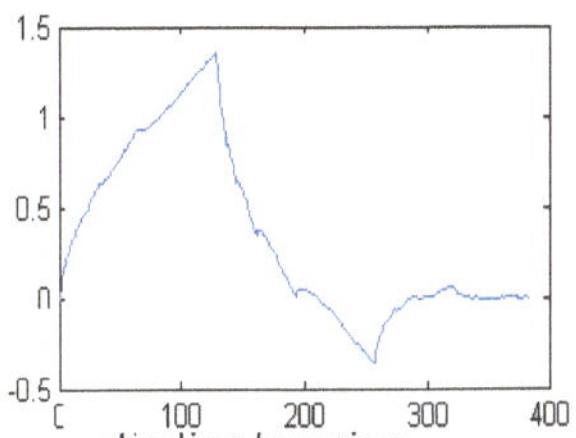

Scaling function

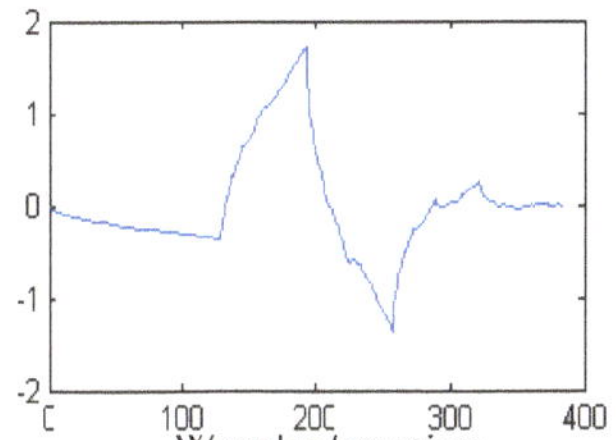

Wavelet function

Scaling coefficients = [0.4830 0.8365 0.2241 -0.1294]

Wavelet Coefficients = [-0.1294 -0.2241 0.8365 -0.4830]

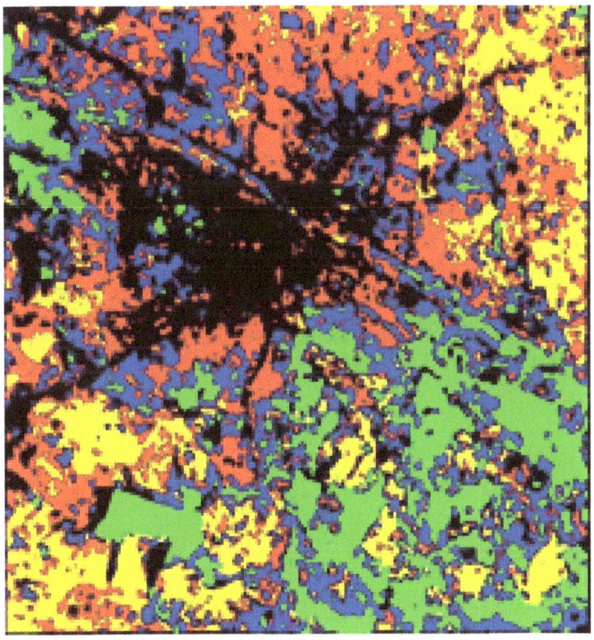

Fig. 4. Classified output- Scale 1

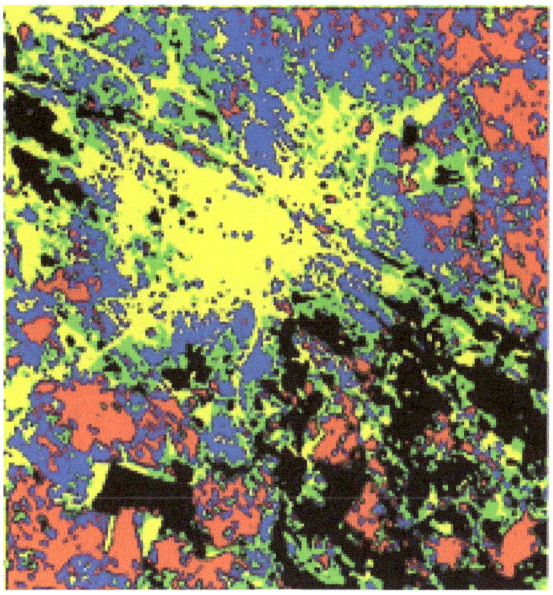

Fig. 5. Classified output-Scale 4.

4.3.2. Lemaire-Battle Wavelet

Polynomial spline wavelets introduced by Lemaire and Battle are computed from spline multiresolution approximations. The wavelet has an exponential decay. Since it is a polynomial spline of degree m, it is m-1 times continuously differentiable. Polynomial spline wavelets are less regular than meyer wavelets but have faster asymptotic delay. For m is odd, it is symmetric about ½. For m is even, it is anti-symmetric about ½.The scaling function and wavelet functions at scale 1 are shown below.

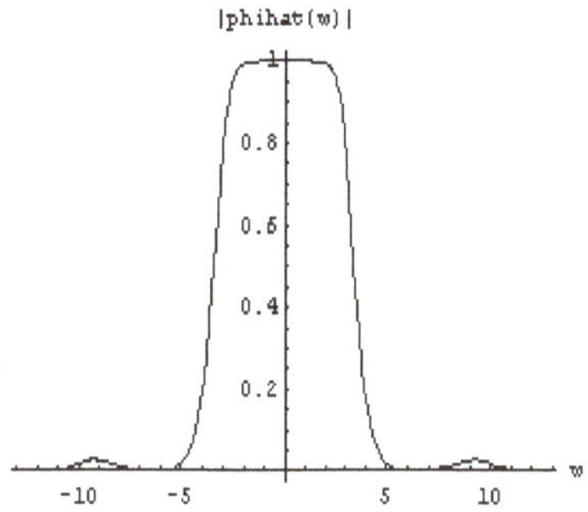

Scaling function

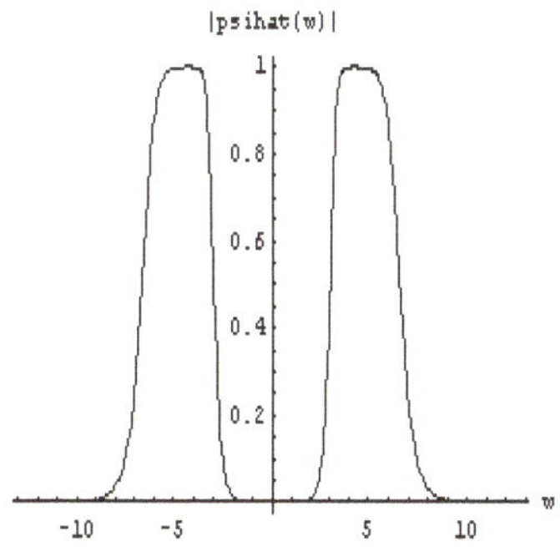

Wavelet function

Scaling coefficients = [0.4607 0.5339 0.0393 -0.0339]

Wavelet Coefficients = [-0.0339 -0.0393 0.5339 -0.4607]

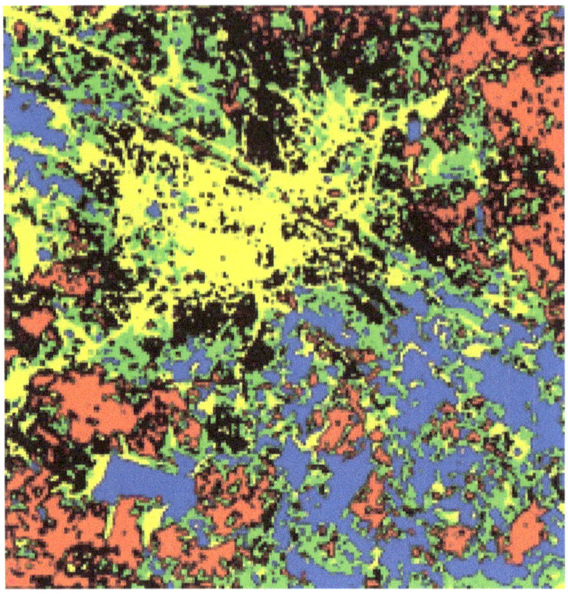

Fig. 6. Classified output-Scale 1.

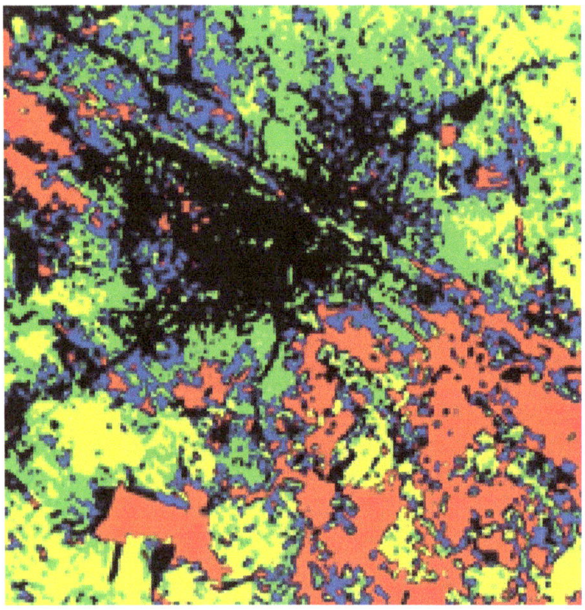

Fig. 7. Classified output-Scale 4.

4.3.3 Symlet Wavelet

Daubechies proposes modifications of his wavelets that increases their symmetry while retaining great simplicity. Daubechies wavelets are very asymmetric because they are constructed by selecting minimum phase square root Q(e-jw).One can show that filters corresponding to a minimum phase square root have their energy optimally concentrated near the starting point of their support. Symmlet filters of daubechies are obtained by optimizing the choice of the square root R(e-jw) of Q(e-jw) to obtain almost linear phase.The resulting wavelets still have a minimum support [-p+1,p] with p vanishing momemts but they are more symmetric. The scaling function and wavelet functions at scale 1 are shown below.

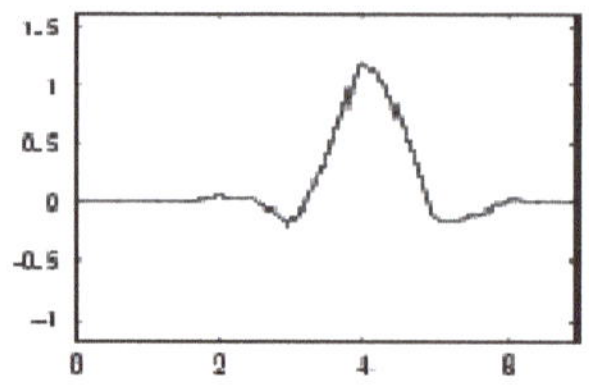

Scaling function

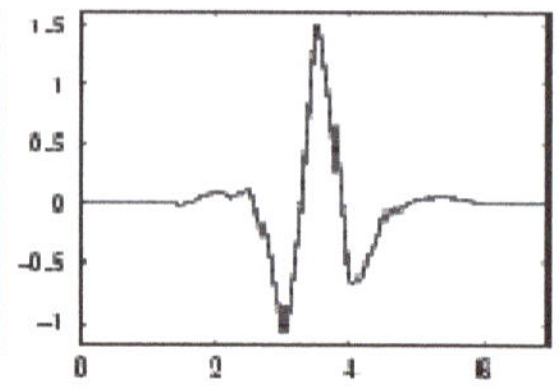

Wavelet function

Scaling coefficients = [0.2352 0.5706 0.3252 -0.0955 -0.0604 0.0249]

Wavelet Coefficients = [0.0249 0.0604 0.0955 -0.3252 0.5706 -0.2352]

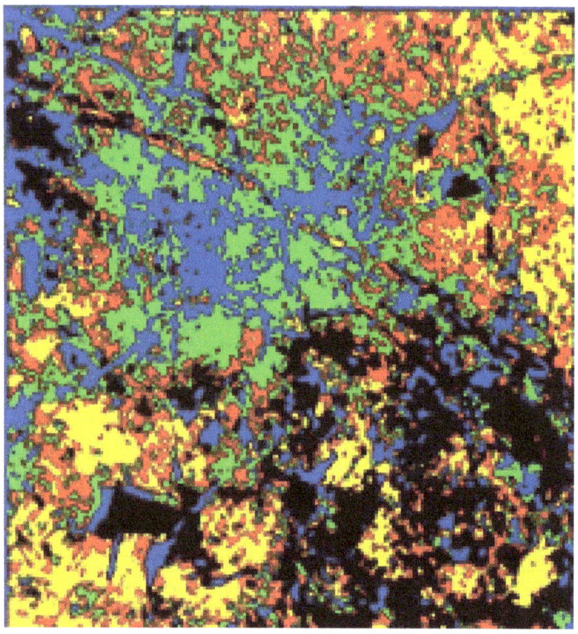

Fig. 8. Classified output-Scale 1.

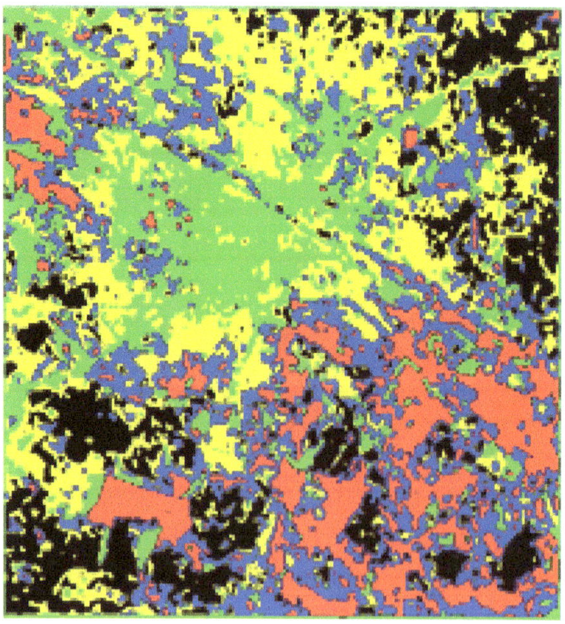

Fig. 9. Classified output-Scale 4.

4.3.4 Coiflet Wavelet

Coiflet is a discrete wavelet designed by Ingrid Daubechies to be more symmetrical than the Daubechies Wavelet whereas Daubechies wavelets have $N / 2 - 1$ vanishing moments, Coiflet scaling functions have $N / 3 - 1$ zero moments and their wavelet functions have $N / 3$. Both the scaling function (low-pass filter) and the wavelet function (High-Pass Filter) must be normalised by a factor ½

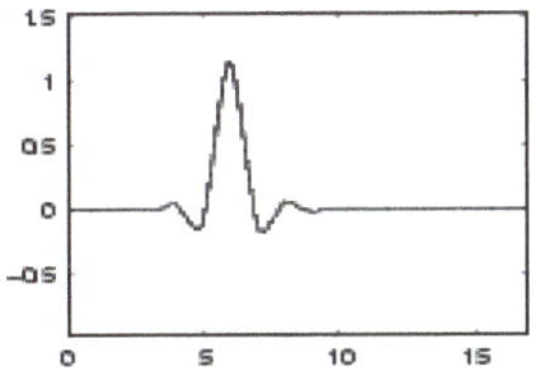

Scaling function

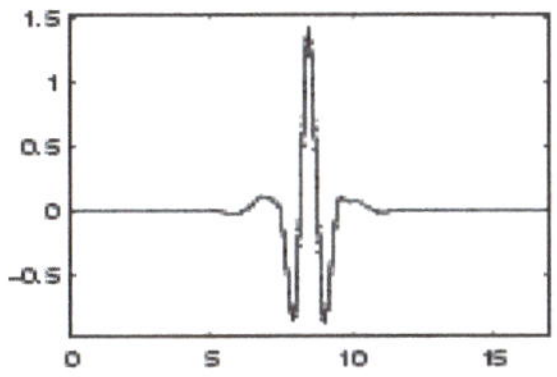

Wavelet function

Scaling coefficients = [-0.0514 0.2389 0.6029 0.2721 -0.0514 -0.0111]

Wavelet Coefficients = [-0.0111 0.0514 0.2721 -0.6029 0.2389 0.0514]

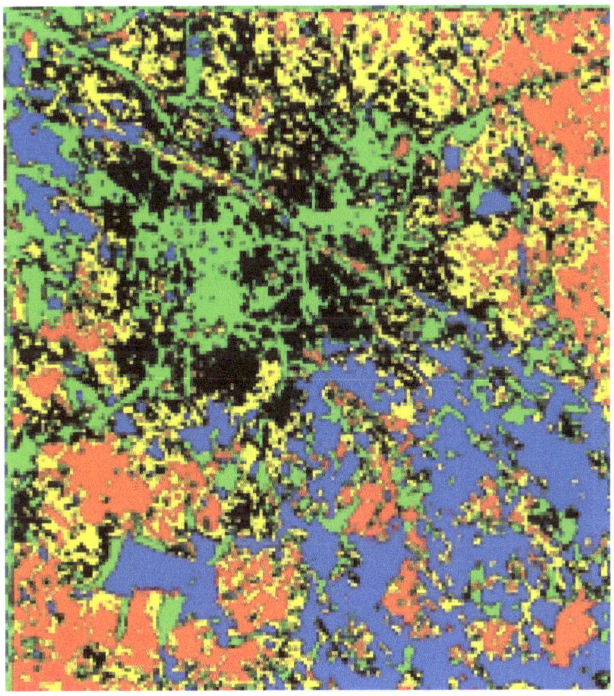

Fig. 10. Classified output-Scale 1.

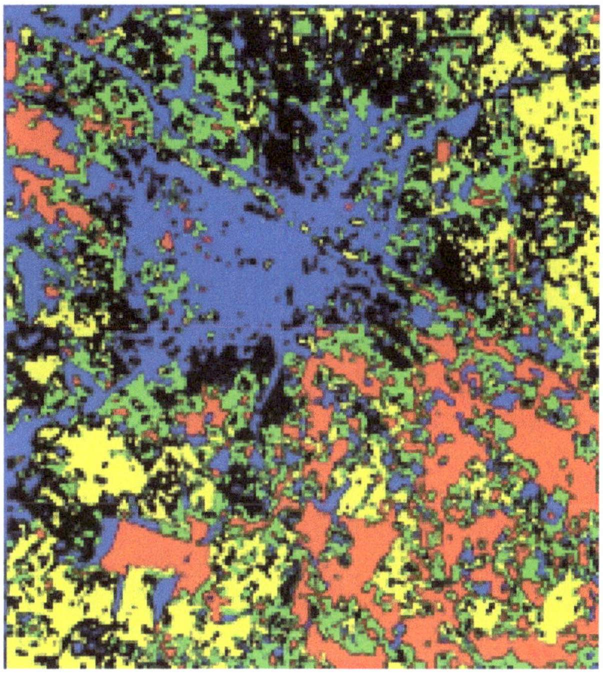

Fig. 11. Classified output-Scale 4.

4.3.5 Biorthogonal Wavelet

It is well known in the sub band filtering community that if the same FIR filters are used for reconstruction and decomposition, then symmetry and exact reconstruction are incompatible (except with the Haar wavelet). Therefore, with biorthogonal filters, two wavelets are introduced instead of just one.

One wavelet is used in the analysis of the coefficients of a signal. The other wavelet is used in the synthesis. The two wavelets can have very different regularity properties, one being more regular than the other.

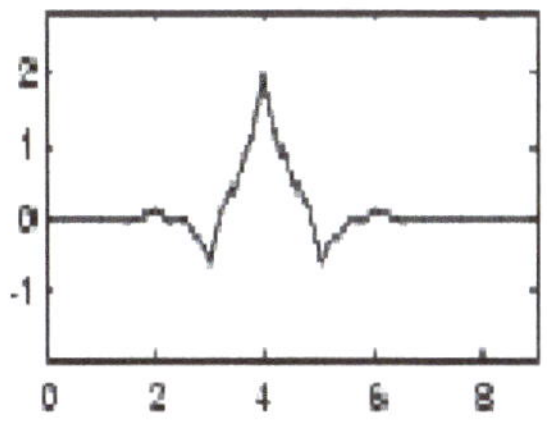

Scaling function

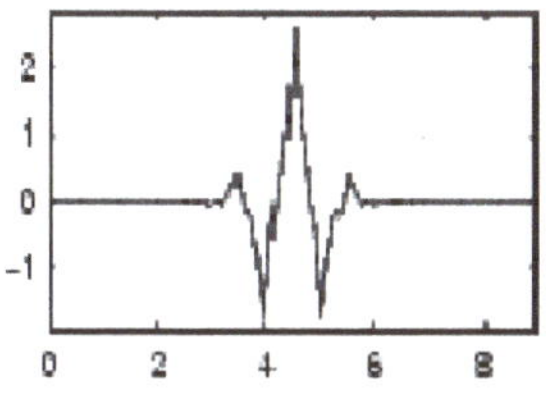

Wavelet function

Scaling coefficients = [0.4607 0.5339 0.0393 -0.0339]

Wavelet Coefficients = [-0.0339 -0.0393 0.5339 -0.4607]

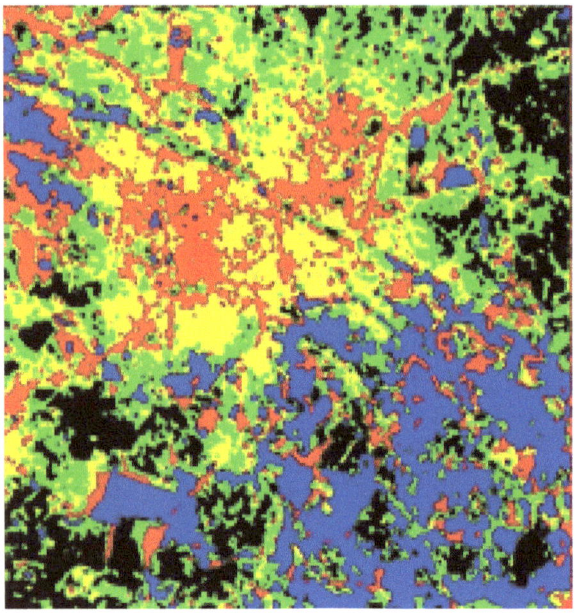

Fig. 12. Classified output-Scale 1.

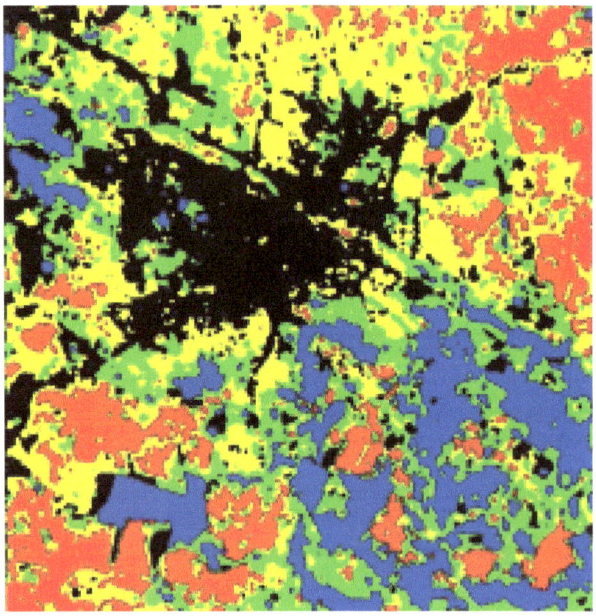

Fig. 13. Classified output-Scale 4.

5 Results and discussion

Classification Error Matrix is one of the most common means of expressing classification accuracy is the preparation of a classification error matrix also called as a confusion matrix or a contingency table. Error matrices compare, on a category-by-category basis, the relationship between known reference data(ground truth) and the corresponding results of an automated classification. Various accuracy indices such as overall accuracy, producer accuracy, user accuracy and Kappa coefficient have been discussed in determining the accuracy of classification.. Overall accuracy is obtained by dividing the total number of correctly classified pixels by the total number of pixels. Producer accuracy is obtained by dividing the correctly classified pixels in each category by the number of training set pixels used for that category. User accuracy is computed by dividing the number of correctly classified pixels in each category by the total number of pixels that were classified in that category. Kappa coefficient is a measure of difference between the actual agreement and the chance agreement. It is denoted as K. The error matrix for scale 1of Daubechies is shown in the table below

class \ ref	Urban	Vegetation	Waste land	Hilly	Water body
Urban	384	14	80	3	13
Vegetation	0	240	53	0	7
Wasteland	22	2	70	4	2
Hilly	11	1	7	26	5
Water body	18	0	11	1	20

User Accuracy:68.07 Overall Accuracy:76.52

Producer Accuracy:70.38 Kappa Coefficient:0.8254

Fig. 14. Error Matrix of fuzzy C-means clustered output (Scale 1).

Like this for all the scale's error matrix is formed and performance can be analyzed by calculating the kappa coefficient and the results are tabulated as in table.

Wavelets Used	Kappa Coefficient			
	Scale1	Scale2	Scale3	Scale4
Daubachies	0.8254	0.8164	0.8472	0.8642
Coiflet	0.7675	0.7845	0.8392	0.8844
Symlet	0.7831	0.7675	0.8037	0.8342
Lemaire	0.8152	0.8472	0.8644	0.8731
Bi-Orthogonal	0.7509	0.7290	0.8162	0.8312

Fig. 15. Performance Analysis of Wavelets at Different Scales.

From the above table it is clear that Coiflet at higher scale (4) is giving better results than the other wavelets. For coiflet and Lemaire battle wavelet the accuracy of classification is increased when the scale is increased but not for other wavelets. In general all wavelets give higher classification accuracy at higher scale. From the results it is inferred that there is no wavelet to classify all the features of the image accurately at a particular scale. So in order to achieve maximum classification accuracy, a feature vector with distinct features is required. To obtain such a feature vector the features are extracted using a wavelet at different scales.

5.1 Performance analysis of wavelets with combined scales

From the following table the Coiflet wavelet with combined scales is found to be superior. The Daubechies and Lemaire wavelets are also found to be comparatively good for texture classification. The Image for that Coiflet is shown in figure below.

Wavelets used	Fuzzy C-means Classification			
	User Accuracy (%)	Producer Accuracy (%)	Overall Accuracy (%)	Kappa Accuracy (%)
Daubachies	77.35	83.39	89.40	91.72
Coiflet	80.11	88.39	91.32	93.16
Symlet	75.69	79.24	85.04	88.54
Lemaire Battle	79.75	87.36	91.01	92.92
Bi-orthogonal	75.20	77.11	83.40	87.44

Fig. 16. Performance Analysis of Wavelets with Combined Scales.

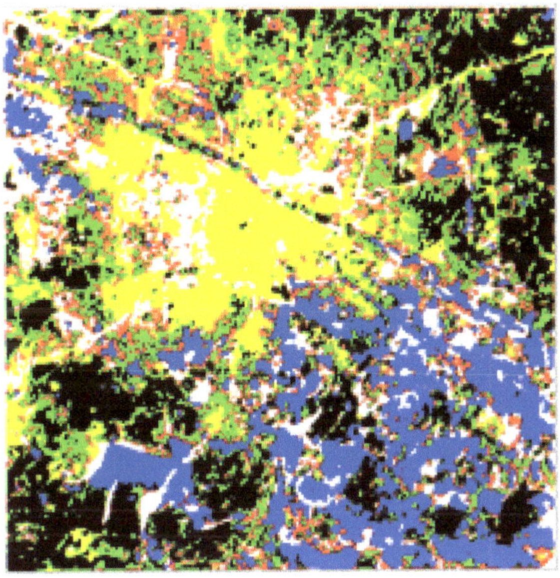

Fig. 17 Classified output for Combined Scale of coiflet.

6. Conclusion

A new method for classification of remotely sensed data through wavelet based texture feature extraction algorithm has been proposed. The results are compared with the Laws mask, Edge filters, Ring and wedge filters and DCT based feature extraction techniques and the wavelet based technique is found to be superior. The texture features are extracted using different wavelets such as Daubachies, Symlet, Coiflet, Lemaire Battle, Biorthogonal wavelet at different scales and combined scales. The results are compared and the coiflet wavelet with combined scale is found to be superior.

References

JOAN S.WESZKA, CHARLES R.DYER AND AZRIEL ROSENFELD, A Comparative Study of Texture Measures for Terrain classification, *IEEE Transactions on Systems, Man and Cybernetics*, vol. smc-6, No.4,April 1976.

JONG KOOK KIM AND HYUN WOOK PARK, Statistical Textural Features for Detection of Microcalcifications in Digitized Mammograms, *IEEE Transactions on Medical Imaging*, Vol. 18, No. 3, March 1999

M.UNSER, Texture Classification and segmentation using wavelet frames, *IEEE Transactions on Image Processing*, 1995.

THOMAS M.LILLESAND, RALPH W.KEIFER, *Remote Sensing and Image Interpretation*, Fourth edition, published by John Wiley & sons.Inc.

M.SINGH, N.SINGH, *Spatial Texture Analysis: A comparative Study*

DONYUN KIM, DONGSUB CHO, Texture Segmentation Using Walsh Spectrum and Modified ART. An Overview of different Clustering Algorithms, http:// www.comp.lancs.ac. uk / ~kristof/research/notes/clustr/

MICHAEL UNSER AND MURRAY EDEN, Multiresolution Feature Extraction and Selection for Texture Segmentation, *IEEE Transactions on Pattern Analysis and Machine Intelligence*, Vol.11, No.7, July1989.

LUCIA BALLERINI, LEONARDO BOCCHIO, *Segmentation of Liver images by texture and genetic snakes.*

Western Han landscape and remote sensing applications at Xi'an (China)

Maurizio Forte

University of California, Merced

Abstract

In 2008 the University of California, Merced, School of Social Sciences, Humanities and Arts started a multidisciplinary project in the area of Xi'an (China) aimed to the study of the Western Han Dynasty in the light of the new archaeological discoveries in the area of Xi'an and on the use of 3D integrated technologies for the analysis and the reconstruction of sites and landscapes. The final goal of this project will be the creation of the Virtual Museum of the Western Han Dynasty but, in the meantime, the production of so many data and applications is addressed to different output also for geospatial analyses and interpretations. In this frame the work on the landscape's reconstruction contributes to the comprehension of the ancient entire territory, peopling and morphology. The landscape has suffered relevant agricultural and environmental transformations during the Cultural Revolution in the '60s and, recently, because of the rapid urban expansion of the city of Xi'an which today counts 8 millions of inhabitants. The study of the transformation of landscape is necessary for interpreting the cultural asset of the territory at the time of the Western Han Dynasty where different "mindscapes" were coexisting: the sacred-power landscape of the imperial mausoleums and tombs, the construction and plan of the city of Chang Han, the use of Feng Shui as systematic methodology of space organization. In the North-Western area of Xi'an the imperial mausoleum were constructed according to specific principles of dominance based on different morphological and spatial features. The calculation of the elevation of the mounds by DGPS (with an accuracy of 15 cm for the vertical and 11 for the horizontal position after post-processing) has allowed understanding the sense of power attributed to each single tomb and mausoleum according to the shape, the size and the elevation. All these elements and the construction of the city of Chang Han show that the Western Han Period was one of the first remarkable example of landscape transformation and organization of the territory of the history of China. This is still a work in progress but the following geospatial analyses and reconstruction were applied for the landscape interpretation:

- Multitemporal analysis and spatial comparison of Corona , Landsat ETM+ and aerial photos;
- DGPS survey and mapping in the areas of the imperial mausoleums and in the city of Chang Han;
- laser scanning in the area of the Mausoleum of Maoling;
- 3D reconstruction of the ancient landscape of the Western Han Period.

Introduction

A multilateral project between the University of California, Merced, the Jiaotong University of Xi'an, the Xi'an Municipal Cultural Relics Conservation and Archaeological Research Institute and the Institute of Technologies Applied to Cultural Heritage (CNR, Rome) has started in 2008 with the goal to create a Virtual Museum of the Western Han Dynasty, using an integrated approach of different technologies: laser scanning, photogrammetry, remote sensing, computer vision, GIS, DGPS, virtual reality. The final goal of the entire project is the Virtual Museum, based on virtual immersive collaborative systems (with installations at Xi'an and in California), but the ongoing scientific research has allowed to reach interesting results in other multidisciplinary fields, such as remote sensing, 3D laser scanning and close range applications (Forte, 2003;2003a). In our case the Virtual Museum will represent not only a final interactive 3D installation, but it will be also the outcome of a complex research interdisciplinary methodology; in some ways it will constitute a virtual musealization of archaeological and historical spatial datasets.

The project is carried on by archaeological fieldwork (summers 2008-2009), data recording and digital post-processing. The UC Merced team works on laser scanning, data capturing and 3D reconstructions; the CNR team on the study and reconstruction of the archaeological landscape; the Chinese institutions on the scientific interpretation and validation of the cultural content and on the management of the fieldwork activities. Of course there are many overlapping in all the activities.

UCM and Chinese students are involved in the fieldwork, digital training and post-processing activities. The general impact on the project is either on the technological aspect or on the use of sustainable cultural management models for documenting and saving the local disappearing heritage. In fact the rapid growing and the urban development of the city of Xi'an in conjunction with the radical transformation of the landscape in the countryside during the Cultural Revolution, constituted a serious danger for the archaeological remains, in particular given the high concentration of multilayered sites in the territory. It quite difficult to estimate the density index of the cultural sites, but we know, for example, that an amount of 200-300 tombs (mainly dating in the Western Han Dynasty) are excavated every year due to the expansion of the city.

This paper is specifically focused on the remote sensing applications and long range scanning for the comprehension of the archaeological landscape and of the ancient topography in the Western Han Period (Forte, 2003). We present here just preliminary data and results, but they can give a correct overview of the methodological and technological problems we are facing in the first part of the project.

During the fieldwork 2008 and 2009 we have recorded by laser scanners two monumental and wall painting tombs of nobles of the Western Han Period, 17 sculptures in the Maoling Mausoleum, around 70 objects (jade, pottery, iron, bronze, terracotta) from the archaeological museum of Xi'an and, finally, the monumental East Gate of the ancient city of Chang Han.

It is interesting to note that while Xi'an is famous all over the world for the terracotta's army and the Chin Mausoleum, visited from millions of people, not enough attentions is given to the other archaeological sites and monuments, in particular in relation with new archaeological discoveries of the Western Han Period. An additional problem is that most part of the archaeological scientific publications is in Chinese and very few in English and a relevant amount of archaeological excavations are still unpublished, so that the international knowledge of the state of the art in this field is very limited worldwide.

Cultural Background

Western Han Dynasty sets its capital in Chang'an (at North-West of the modern Xi'an) from 206 BC to 8 AD. The society, culture and civilization of the Western Han Empire are further consolidated and developed, with the rise of the nationality-Han, from which the main identity of China is generated. The most important archaeological discoveries in China, dating back to the Qin (221-206 BC), Han Dynasties (206 BC-220 A.D.) and subsequent periods have greatly widened our field of vision and knowledge of ancient Chinese history. The study of imperial tombs of past dynasties has always been a central aspect in Chinese archaeology. Around 200.000 tombs of different types in different dynasties have been surveyed or excavated by archaeologists so far. Unfortunately, in the case of the Western Han Dynasty, a few of them were scientifically reconstructed in 3D and they are not digitally accessible. Therefore, the entire process of cultural re-contextualization from the archaeological excavations to the museum display is always very difficult. Even other archaeological contexts (mausoleums, settlements, temples, tombs) need an adequate 3D digital documentation accessible for a wide audience. The virtual museum of the Western Han Dynasty would like to fill this gap.

Close range and long range 3D laser scanning

The activity of laser scanning has involved different areas using a multiscale approach, from the close range to the long range 3D data recording. We have used a long range time of flight scanner (Riegel LMS Z390i) for recording datasets of monuments, sites and portions of landscape, and an optical-triangulation scanner (Nextengine, HD) for capturing small size archaeological objects and finds (typically funeral goods and vessels). The strategy of 3D data recording is to acquire the most representative and significant elements of the Western Han Dynasty, so that it is possible to re-contextualize them in the Virtual Museum, but also for enhancing the relevance of the new archaeological discoveries in China and their relations with the territory.

The archaeological fieldwork in 2008 was aimed principally to the documentation of a wall painting tomb (M27), found within the campus of the Jiaotong University (fig.1).

Fig.1 Satellite imagery (by Google Earth©) of the tomb M1

In 2009 we have divided the work in two different kinds of activities. The first team has worked outdoor with DGPS and time of flight laser scanners; the second one has recorded objects and artifacts in the Archaeological Museum of Xi'an. In this way we were able to collect a relevant amount of archaeological datasets: two monumental wall paintings tombs, 17 monumental sculptures (TOF laser scanners), 12 tombs and mausoleums (by DGPS) and 70 museum objects (by optical scanner).

Remote Sensing Applications

The landscape has suffered relevant agricultural and environmental transformations during the Cultural Revolution in the '60s and, recently, because of the rapid urban expansion of the city of Xi'an which today counts around 8 millions of inhabitants (fig.2).

The study of the transformation of landscape is necessary for interpreting the cultural asset of the territory at the time of the Western Han Dynasty where different "mindscapes" (Forte, 2003; Forte, Kay, 2003a; Forte et alii, 2005; Ingold, 2000) were coexisting: the sacred-power landscape of the imperial mausoleums and tombs, the construction and plan of the city of Chang Han and the use of Feng Shui as systematic methodology of spatial organization.

Fig. 2 Overlay of a satellite imagery Landsat ETM+ (2007), a Corona (1953) and the DTM-SRTM (fusion of the bands 3-2-1 + Pancromatic of Corona)

The correct understanding of the landscape and environment and its anthropomorphic relations are a crucial problem in our research project. We can say that the organization, the planning and cultural transformations of the landscape in the Western Han Dynasty are probably one of the most remarkable in the history of ancient China and have marked and influenced all the following dynamics of territorial peopling and settling.

In the North-Western area of Xi'an the imperial mausoleum were constructed according to specific principles of dominance based on different morphological and spatial features.

Approaching the classification of the landscape by remote sensing we have collected multitemporal datasets in different scales and format: Landsat ETM and ETM+ (1999-2008), Corona (1952), Google Earth (Quickbird) and aerial photos (from the archives of the Jiaotong University, fig.2). The analysis was focused on the spatial organization of the Western Han imperial Mausoleums,[1] in the North-Western area of Xi'an and on the topography of the ancient city of Chang Han. Below an itemized selected list of the principal applications and goals.

- Remote Sensing and DGPS interpretations.
- Multitemporal analysis and spatial comparison of Corona, Landsat ETM+ and aerial photos.
- DGPS survey and mapping in the areas of the imperial mausoleums and in the city of Chang Han.
- Laser scanning in the area of the Mausoleum of Maoling.
- 3D reconstruction of the ancient landscape of the Western Han Period (preliminary simulation).

For getting these goals we are studying and comparing several spatial datasets such as, in this first part of work (fig.2):

- Satellite imagery Corona, 1 band, Panchromatic (1952)
- Satellite imagery Landsat ETM+, 8 bands (2008-05-29T17:28:34Z)
- Satellite imagery Landsat ETM+, 8 bands (2004-02-12T17:36:41Z).
- Satellite imagery Landsat TM, 7 bands (2002-02-12T17:36:41Z).
- Satellite imagery Landsat ETM+ (2008-11-02T23:46:42Z)
- DEM SRTM, 30 mt
- Aerial photos of Xian (declassified, archives of the Jiaotong Unversity, courtesy of Prof. Li).

The multispectral analysis of the city of Chang Han (fig.3) has highlighted very well the boundaries of the urban area and the defense walls.

In fig.3 sun shading algorithm and edge detection filters were applied on the Landsat imagery and the result is quite satisfying: the outlines defined by remote sensing coincide with the historic reconstruction of the topography and with the archaeological excavations.

In particular the state of conservation of the defense walls, made by clay and raw bricks, is much more evident and visible in the Corona imagery and in its multispectral fusion with the multispectral Landsat ETM+ (fig.2). The combination of bands 4-3-2, 1-2-3, 3-2-1 was processed integrating the panchromatic band (n.8) in pseudo-color. The reconstruction of the gate by laser scanner (fig.4) can show the critical state of conservation of the gate.

Unfortunately today the conservation of the defense walls is at high risk because of erosion's phenomena (pollution, acid raining, and artificial works) and of the urban development of the city of Xi'an (in particular around the East Gate).

Additional very interesting results were obtained in 2008 multispectral imagery (fig.5) in the area of the Pingling Mausoleum.

[1] There is no plan of excavating the imperial mausoleums in the following years. This decision of the Chinese government is justified by the lack of adequate technologies of excavation and preservation of the cultural relics.

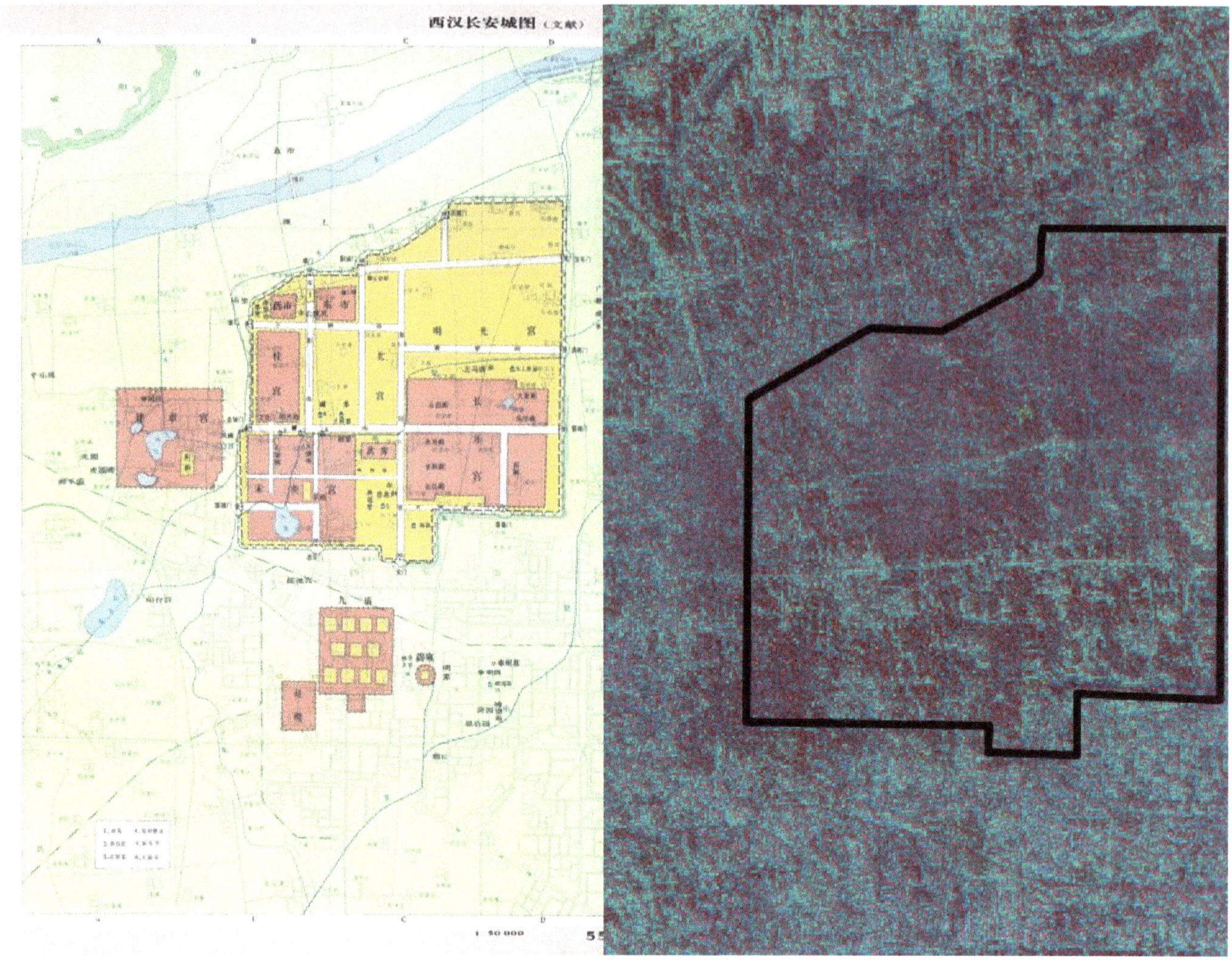

Fig.3 Sun-shading visualization of the Landsat ETM+ (2008, on the right) compares with the historical reconstruction of the city.

Fig. 4. Chang Han: East Gate recorded by laser scanner.

The processing in gray levels, after equalization and pixels' smoothing, has highlighted the entire perimeter of the monumentalized area of the structure (fig.5). We can still see the perfect orientation of external walls (fig.5, n.1-2), the South Gate (fig.5, n.3), and the burial pits to the South of the city walls and out of the South Gate (fig.5, 4). The entire area of the mausoleum surrounded by the walls is about 2915 sqm and today is visible just from a multispectral satellite, since there are no visible traces on the ground, apart the big mound. Immediately to the Southern side of the trunk-pyramid, it is possible to see a very consistent anomaly, likely identifiable as burial pit (fig. 5, 5).

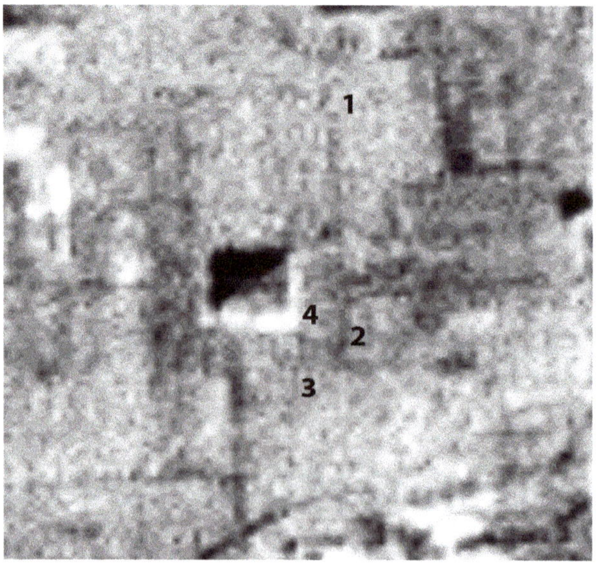

Fig.5 Landsat ETM+ 2008:

The evidence of these linear and well marked traces in a 2008 multispectral satellite imagery, shows how much persistent the Western Han Landscape is, but also how much relevant was the monumentalization of the landscape across the 1st cent B.C. and the 1st A.D. The sense of power and the monumental architectural features of the imperial mausoleums, marked by walls, gates and burial pits, characterized all the Western Han landscape.

The DGPS survey with accuracy in post-processing of 10-11 cm, was concentrated specifically in the areas of the 11 mausoleums, recording 10 imperial mausoleums and related tombs of the imperial relatives or nobles. For this goal we used a very portable instrument: the Trimble GeoXH Handheld. The construction of the imperial mausoleums, with the inclusion of relatives, favorite concubines, generals and nobles, is one of the most terrific demonstrations of power of the Western Han Dynasty and it marked the landscape for 2000 years. The rigorous application of the Feng-Shui rules has culturally contaminated entire generations of people living in the countryside and in the surroundings of the imperial tombs. For example even in the following ages many other smaller tombs (including modern ones) were constructed in a radius of a few hundred meters from the imperial mausoleums. We counted many of them during our archaeological survey and it is evident a sort of "constellation" of tombs around the agricultural fields and terraces. The tradition of Feng Shui has influenced territory and people until the last generations of Han, mainly before the Cultural Revolution (also many tombs were re-used as houses in the last centuries).

The mausoleums are a unique example of the imperial Han power, a grandiose demonstration of the control on the territory and they show clearly the hierarchical-spatial relations between sites and landscape.

They were planned and constructed over the alluvial terraces dominating the basin of the Weihe river, (fig.6) and the city of Chang-Han.

This is particularly evident in the reconstruction process of overlapping of the Landsat satellite imagery on the DTM (fig.6).
The use of the DGPS was fundamental also for the geo-spatial recording and mapping of the mausoleums and for a first analysis of the archaeological landscape. The original dominant positions of the mausoleums over flat alluvial terraces, is today less evident because of the landscape transformation operated during the Cultural Revolution. Most part of these lands were cut and re-shaped for getting more space for agricultural intensive cultivations, the canalizations of water and for increasing the use of aratro ploughing with mechanical instruments.

However, even with so many artificial deep changes, the landscape still shows today all its power and untouched sense of place. The systematic artificial construction of burial pits, mounds, canals, gates, fences, walls, roads, all of them perfectly oriented N/S, has deeply influenced all the landscape for over 2000 years.

Fig.6. 3D Texture mapping of the city of Chang-Han over the DTM. The sequence of the 5 points on the North-Western side of Chang-Han represent a sample of the imperial mausoleums.

The use of DGPS by sample (in 2009 fieldwork) has well highlighted the orientation of all the structures and the regular boundaries and shapes of the trunk-pyramidal structures.

By GPS some basic measurements (area, volumes, and height) were calculated in order to analyze the hierarchical spatial relations between each "familiar" chronological group (fig.7). The result is quite impressive and shows clearly that any relation of power is expressed in terms of height, volumes and area. The construction of imperial mausoleums is the most remarkable expression of the Western Han society, its "meme".

For example, in the case of the Martial's Mausoleum, the hierarchy is formulated in the following terms: the greatest tomb is that of the emperor (h.490.626 m; area 5919 sqm; perimeter 987.433 m), the second is the tomb of Madame Lee (h. 470.412; area 1444 sqm; perimeter 487 m), the third one is the tomb of the uncle of Martial, his general (h 465.609 m; area 1031; perimeter 472.582 m). Similar situation for the other family groups where the size of the tomb represents the level of importance of its owner.

The 3D visualization of the landscape generated by the SRTM DEM allows perceiving the power of the imperial mausoleums standing up over the river valley and dominating the ancient city, the settlements and the tombs (fig.6). It is also important to note that there no archaeological excavations in these areas, neither for the mausoleums either for the burial pits surroundings the main tombs, apart the case of the Yanling Mausoleum.

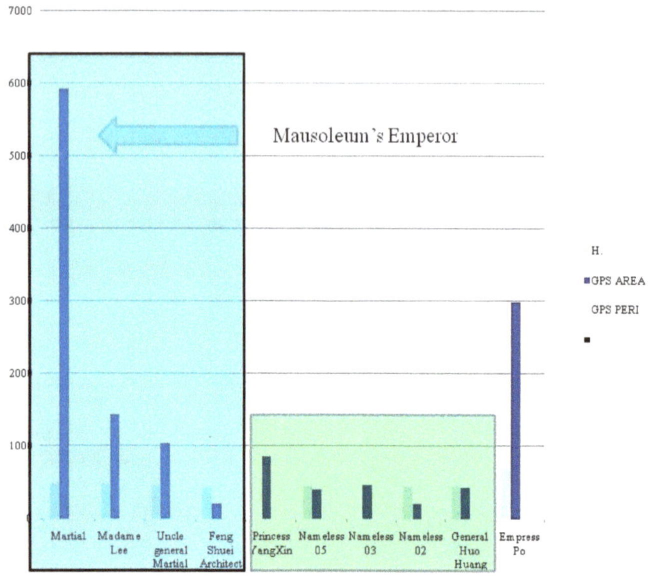

Fig.7. Statistical histograms of the dimensions (height, area and perimeter) of the tombs recorded by DGPS

Chang Han

Han Chang'an was the capital city of the Western Han Dynasty, located at North-West of the modern city of Xi'an (this last one built during the Tang Dynasty). With a population of hundreds of thousands of people and covering an area of 35 square kilometers, Chang'an, was the world's largest metropolis, even three times larger than ancient Rome.

Today the ancient city is still visible and identifiable from satellites, such as Landsat, Quickbird (Google Earth) and Corona (fig.2).

The greatest problem of the ancient city of Chang Han, is that, given the enormous dimensions, was not systematically excavated but, nevertheless, it is well known also from historical texts; so its topography is derived by the ancient historiography and just in part from archaeological layers. Some archaeological trenches have been recently preserved and now are become open air museums.

In our project we have analyzed principally the boundaries and the defense walls, focusing the first reliefs and data recording on the East gate (fig.8).

The defense walls were built in clay and raw bricks, with foundations in stones. This kind of precaria deperibile fragile architecture rends very difficult the conservation, and in particular the construction of a new big canal in front of the East gate is seriously treating what remains of the entire structure. If we compare the line of the defense wall visible on the Corona dating in 1952 (fig.9) along the East side, we can distinguish all the perimeter and the state of conservation appears still good. When we overlap on the same imagery the DGPS survey recorded in 2009 (fig.9) and concerning the gate and the related walls, we see that more than 80% of the structure now is destroyed or not identifiable. A 3D laser scanning of all the East gate with part of the wall has been carried out (June 2009) for saving a very detailed (accuracy 0.8 mm) model of the structure (fig.4).

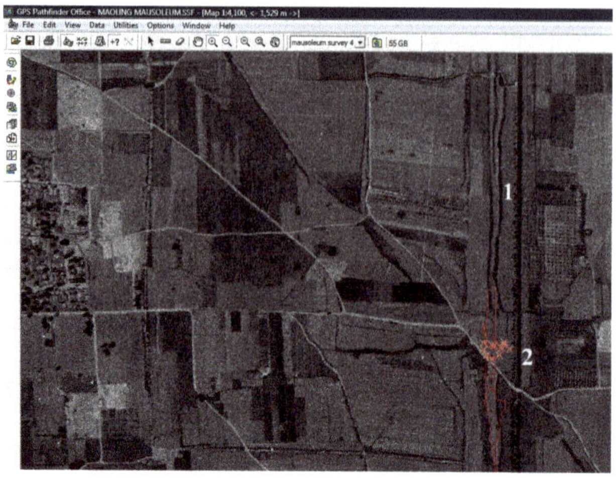

Fig.8. Corona imagery, 1952. It is very evident the good state of conservation of the Eastern defense wall (1). In overlay the relief by DGPS showing what remains in 2009 near the East gate (less than 20% of the defense wall).

The correlation of the DGPS survey with 3D scanning represents today the last possible documentation of this part of the city. It is likely that in a few months or in couple of years there won't be any trace of these monumental archaeological remains.

Conclusions

This project is a work in progress but it shows the importance to use an integrated approach in the study and reconstruction of the micro and macro scale landscape of the Western Han period. It is really impressive the persistence of the ancient landscape on the modern one and the good visibility we have obtained using different digital datasets: Landsat, Quickbird, Corona. The use of the panchromatic band (n.8) of Landsat (2007-08) has been particularly fruitful and surprising: it has clearly put in light the entire structure of the Pingling mausoleum and the burial pits, but it was also very useful in the comprehension of the surrounding landscape.

The combination of the panchromatic Corona with the multispectral data has increased the capacity of interpreting this kind of image, but it has been also very useful in the study of the archaeological landscape, much better preserved in the 50s than today. The agricultural landscape, in fact, was deeply transformed during the Cultural Revolution, with intensive cultivations and the creation of artificial terraces for extending the areas of cultivations and plantations. Nevertheless, in the last decades this landscape transformation did not touch directly the mausoleums, and the consistent presence of many recent tombs in this area demonstrates a high sense of respect for the imperial monuments.

In conclusion, the combination of all these imagery constitutes a very important multitemporal overview of the Western Han landscape and the spatial relations between sites and landscapes. The 3D visualization of the valley and its terraces, the dominant position of the mausoleums and the spatial and dimensional hierarchy between mausoleums and tombs have well contextualized the relations between tombs, mausoleums and the ancient city. This analysis is able to interpret the ancient landscape as very complex cultural system, specifically made for communicating a strong sense of power and the hierarchy of the spatial organization.

The final goal of the creation of a Virtual Museum is a stimulus for specific analyses on the landscape sites and monuments; but also for the re-contextualization of artefacts and data. The multitemporal use of Landsat imagery (1999-2004), Corona (1957), Google Earth and aerial photos (just for a little portion of the territory) has given a strong contribution to the interpretation of the landscape and the spatial organization of the territory.

The cooperation project will create a wide platform for Xi'an Jiaotong University and University of California, Merced to commit to preserving cultural heritage. In addition, it will open up an approach in which the virtualization technologies will help people worldwide to better understand the history of the West Han Dynasty. And the complete documentation and virtual reality system will help to establish important archives for archaeological restoration and conservation. In fact it is important to underline that the Chinese cultural institutions are currently in need of international help in order to document and save on time their own heritage. The rapid urban development of the city of Xi'an and its suburbs constitutes a high risk for the territorial rescue archaeology and put new urgent issues of cultural resource management. Therefore, in the meantime, the virtual museum is a great opportunity for saving, documenting, interpreting and communicating a so wide archaeological context.

The integrated approach of digital technologies applied to this pilot project opens new and interesting perspectives in the field of world heritage, particularly in training, research and management. It will also be a significant development of collaborations that provide a clear description and visual image of a cultural monument that is going to disappear. The goal of the process we are dealing with is a "multi-layered" reconstruction of an archaeological context. With "multi-layer" we intend the visualisation and interaction of different categories of complex data inside the same geo-referenced space. With Virtual Reality applications we can manage, in the same metric environment *acquired data*, such as survey results, geological and geodetic prospection, GPS data, Scanner Laser point clouds, air photography, satellite imagery, archives, meta-data or other textual resources, but also *processed data,* such as remotely sensed data, or *interpreted data*, such as thematic layer elaborated with GIS analysis, etc. Another important advantage is that it's possible to integrate data at different scales of accuracy (metric - GPS, centimetric - DGPS, millimetric – DGPS, scanner laser).

Acknowledgements

The author wishes to thank all the Chinese institutions involved in the project and in particular Cheng Linquan from the Institute of Cultural Relics Preservation and archaeology of Xi 'and Li Jiajun from the Art Museum, Xi'an Jiaotong University. A very special aknowledgment to Ms Abby' Lizhi Zhang, Assistant of the Art Museum in the School of Humanities and Social Sciences of the Xi'an Jiaotong University. The project is co-supported by Pacific Rim Research Program and GRC-UC Merced grant.

References

FORTE M., 2003, *Mindscape: ecological thinking, cyber-anthropology, and virtual archaeological landscapes*, in "The reconstruction of Archaeological Landscapes through Digital Technologies", Proceedings of the 1st Italy-United States Workshop, Boston, Massachussets, USA, November 1-3, 2001, BAR International Series 1151, Oxford, 2002, 95-108.

FORTE M., KAY S., 2003A, *Remote Sensing, GIS and Virtual Reconstruction of Archaeological Landscapes*, in "The reconstruction of Archaeological Landscapes through Digital Technologies", Proceedings of the 1st Italy-United States Workshop, Boston, Massachussets, USA, November 1-3, 2001, BAR International Series 1151, Oxford, 2002, 109-116.

FORTE M., PESCARIN S., PIETRONI E., DELL'UNTO N., 2005, *An integrated approach to archaeology: from the fieldwork to virtual reality systems*, in "Beyond the Artifact", Proceedings of the Conference CAA2005, Prato, Italy, BAR International Series (in press).

INGOLD, T., 2000, *The perception of the Environment. Essays in livelihood, dwelling and skill*, London and New York, Routledge.

Session 2
AERIAL ARCHAEOLOGY:
AERIAL PHOTOGRAPHY AND AIRBORNE SCANNING

Terrestrial remote sensing in archaeology

Munish Kumar Tiwari

Keywords GROUND-BASED REMOTE SENSING, Magnetometry, Electrical resistivity / electromagnetic conductivity, Ground-penetrating radar.

Archaeological remote sensing

Remote sensing in archaeology is an evolving enterprise, largely owing to rapid advances in technology, but also to the changing needs and goals of the discipline. As the planet's exploding human population results in massive developments and changes to the landscape, there is a consequent need for efficient and cost-effective methods to locate, map, and acquire information from sites of our cultural heritage before they are forever lost. At the same time, traditional archaeological excavations and pedestrian surveys are becoming increasingly expensive, and commonly examine only trivial areas. Archaeological remote sensing allows large regions to be rapidly investigated for archaeological features, at relatively low cost; it can detect features unseen on the surface, precisely map them, and offer interpretations based on their form, distribution, and context. In short, archaeological remote sensing may offer the only pragmatic means to locate, map, and inventory much of the world's archaeological resources.

Research Method:

The research design will be descriptive in nature. The method of data collection will be mainly based on secondary data. However, an effort will be made to collect primary data by taking information from competent persons off and on.

The nature of archaeological sites

An archaeological site is a three-dimensional matrix of materials existing in the present, containing artifacts, human-generated constructions, and the deposits in which they lie (see Schiffer 1976). Our collected knowledge of this matrix suggests features that might be remotely detected and sensor requirements (see Scollar *et al.* 1990:4-7). Artifacts_are material objects modified by people, both smaller portable artifacts (spear points, pots, and knives) and non-portable artifacts that include such larger items as cut posts, building timbers, bricks, or shaped stones used in architectural constructions. Constructions_include the many types of structures, buildings, and other facilities, including places of occupancy (dwellings, public structures), non-occupancy (storage pits, wells, burial mounds, fortification ditches), and transportation facilities (roads). Sediments_and_soils_are the deposits within which artifacts and constructions lie. Most result from natural processes, but many are anthropogenic or are altered by human activity.

Properties of archaeological sites: what can be remotely sensed?

In general, large non-portable artifacts such as building foundation or pavement stones frequently can be remotely detected, but small portable artifacts cannot (except metallic artifacts). Large masses of individual portable artifacts and other anthropogenic materials might_occasionally be revealed by remote sensing as when a concentration of ceramic sherds subtly raises the local magnetic field or a large amount of bone from an animal kill changes subsurface electrical or thermal properties or surface vegetation patterns. Human constructions receive focus in remote sensing because their larger sizes and relative masses of contrasting physical properties sufficiently increase detection probabilities and their regular geometric shapes make them more easily recognized as cultural anomalies.

Sensor mixes and detection probabilities

The likelihood of a buried archaeological feature being detected depends on a complex interaction between at least five factors: (1) matching the contrasting physical properties of the feature with sensors capable of detecting those properties, (2) the amount of contrast between the physical properties of the feature and surrounding deposits with respect to sensor sensitivity, (3) the size of the feature relative to the spatial resolution of the measurements, (4) the depth of the feature with respect to signal attenuation and levels of confounding noise, and (5) the degree of regular pattern the feature exhibits.

Ground-based remote sensing

Ground-based remote sensing employs instruments on or near the surface. Most methods are geophysical, measuring physical and chemical properties of near-surface deposits—usually in the uppermost 1-2 m. Four methods are principally employed: magnetometry, electrical resistivity, electromagnetic conductivity (EM), and ground-penetrating radar (GPR). All but the first are active methods. Resistivity and conductivity measure the same earth property, while the others generally assess different and mostly independent dimensions (Weymouth 1986:371).

Magnetometry

Magnetometry surveys are generally the most productive of ground-based methods owing to very fast instrumentation (8-10 measurements/s) that allows coverage of large areas in short amounts of time (.5-1 ha/day), high sampling rates (8-10/m), and the many characteristics of archaeological deposits that yield magnetic variations. Eight principal phenomena contribute to the formation of magnetic anomalies within archaeological sites (Clark 2000; Scollar *et al.* 1990; Weymouth 1986):

1. Firing of the soil beyond the Curie point (about 600° C, depending on the material), whether purposeful (hearths, kilns) or accidental (a burned house), can intensify the local magnetic field owing to "thermoremanent magnetism."

2. Accumulations of fired artifacts, such as ceramics or bricks, can intensify the local magnetic field.

3. Variations in soils and sediments can yield different magnetic susceptibilities, the ease with which a material becomes magnetized when subjected to a magnetic field such as the Earth's. Magnetic susceptibility is determined by the composition and concentration (mineralogy, size, and shape) of magnetizable materials.

4. Surface soil layers become magnetically enhanced by several processes that include burning, weathering, and chemical reactions that change certain iron compounds to more magnetic forms (Dearing *et al.* 1996) and magnetotactic and other bacteria that concentrate magnetic compounds (Fassbinder *et al.* 1990). Extended human occupations exacerbate this phenomenon through the introduction of organic and fired materials to the topsoil. Paleosols tend to retain this effect.

5. Removal of magnetically enriched topsoil during the construction of ditches, house pits, or other excavations, lowers the local magnetic field over these features.

6. Accumulations of topsoil, in mound or sod constructions, or when storage or other pit features become filled, increase the magnetic field locally.

7. Imported stone employed in the construction of buildings or pavements might be more (igneous rocks) or less magnetic (certain limestones) than surrounding soils.

8. Iron-bearing artifacts markedly alter the Earth's magnetic field, producing large dipolar measurements consisting of paired positive and negative extremes.

Electrical resistivity / electromagnetic conductivity

Resistivity and conductivity methods quantify identical aspects of deposits in very different ways. Soil resistivity to an electrical current depends on material type, moisture, dissolved ion content, density, pore space and other factors (Weymouth 1986:319). Soil conductivity is simply the inverse of resistivity: a highly resistant material is a poor conductor. These methods are sensitive both to subsurface contrasts caused by highly resistant stone or brick (e.g. foundations or floors) and to more subtle changes in soils and sediments. Moist sediments filling a ditch might exhibit low resistivity (high conductivity), but low compaction and high porosity within a midden could yield high resistivity (low conductivity) measurements.

Electromagnetic (EM) induction instruments utilize a radio transmitter to induce an electromagnetic field in the ground. If conductive materials are present eddy currents are generated that produce a secondary electromagnetic field, proportional to soil conductivity, which is measured by a surface receiver. Conductivity is quantified in millisiemens (mS) per meter (10^{-3} Siemens), with a theoretical relationship with resistivity given by: mS/m = 1000/ohm/m (Bevan 1983). Without probes to insert in the ground, EM instruments are very fast, but typically lack the resistivity meter's ready ability to target specific depths through simple adjustments in probe separation. By changing the transmitting frequency or the separation between transmitting and receiving antennas responses from different depths can be measured, but most instruments are manufactured with fixed settings. The most popular averages soil conductivity through a depth of about 1.5 m with peak sensitivity at .4 m below the instrument in a "vertical dipole mode"; others can prospect much more deeply. All instruments can also operate in a "horizontal dipole mode" simply by turning them on their side, and approximately halving prospecting depth (McNeill 1980).

Ground-penetrating radar

A complete GPR system is typically composed of a control unit, a display unit, a transducer, a number of cables, and a power source. Initial results may be viewed on a video display in real time as data are collected, making instant field interpretations possible. The transducer contains transmitting and receiving antennas in the 100-1500 MHz range, with 300-500 MHz most popular in archaeology. As radar pulses are transmitted into the earth their velocity changes with the electrical properties of the materials they are traveling through. Each velocity change returns a reflection recorded at the surface.

Other geophysical methods

In seismic refraction surveys artificial shockwaves are created by a sledgehammer striking a rubber pad or through explosive detonation caps. A geophone sensor at intervals along transects record the time (in milliseconds) and amplitude of the waves as they reach each one. Metal detectors are electromagnetic induction instruments optimized to detect metallic objects of any kind in the near-surface, usually the uppermost .5 m (Connor & Scott 1998). Other methods far less explored in archaeology include induced polarization (Ovenden & Aspinall 1991), self potential (Wynn & Sherwood 1984), and gravity techniques (Linington 1966).

Surface elevation data

Many archaeological sites possess subtle, or even pronounced, terrain variations caused by what lies immediately beneath. In aerial archaeology, low light angles have been exploited to emphasize small altitudinal changes through shadowing, allowing poorly visible settlement and architectural details to be mapped. This can be digitally replicated with high-density surface elevations recorded across an area by on-the-ground survey teams. Vertical relief exaggeration and analytical surface shading algorithms (common GIS methods) highlight archaeological signatures in the terrain.

Aerial remote sensing

Remote sensing from the air or space utilizes electromagnetic radiation reflected or emitted from the Earth's surfaces, offering a cost-efficient means for archaeological reconnaissance over wide areas (Bewley 2000:3). Aerial remote sensing uses airborne platforms; space remote sensing orbiting satellites, the topic of a later section. Both employ similar sensors; their principal difference lies in grossly divergent distances from the Earth's surface which, until recently, has profoundly affected the spatial detail of what could be resolved from space. Aerial photography is the oldest and most-used domain of archaeological remote sensing and receives greatest focus, but other sensing devices have been placed in the air in recent decades including passive multispectral and thermal sensors, and active radar and laser altimeter systems, making aerial remote sensing truly multidimensional.

Theory and sensors

The electromagnetic spectrum

The sun, source of most of the electromagnetic radiation (EMR) impinging on the earth, gives a spectrum ranging from biologically lethal gamma rays (short wavelength, high frequency) to passive radio waves (long wavelength, low frequency). For convenience, the spectrum is divided into regions possessing similar characteristics, "spectral bands." In order of increasing wavelength, the bands used in remote sensing include ultraviolet (UV), visible, infrared (IR), and microwave. Although EMR travels through empty space unimpeded, the earth's atmosphere allows certain wavelengths to pass freely (atmospheric windows or transmission bands) while others are blocked (atmospheric blinds or absorption bands), including short wavelength gamma, x-ray, and most UV radiation.

Spectral signatures

Any surface reflects or emits EMR over a range of wavelengths in a unique manner depending on its physical composition and state. The average amount of incident radiation reflected by a surface in some wavelength interval is its spectral reflectance or albedo, while emitted radiation yields information about temperature properties. This distinctive spectral response pattern, diagnostic of particular surfaces and their states, forms a spectral signature, allowing surfaces to be recognized from their spectrum. Water tends to reflect blue wavelengths and almost no NIR; most healthy vegetation reflects some green light but is enormously reflective in NIR; and many distinctive properties of soils and geological deposits are discriminated in NIR and MIR bands (Avery & Berlin 1992:14-15).

Sensor types

Sensor type determines the regions of the spectrum utilized. Some sensors provide photographs, pictorial representations recorded directly onto photographic film; others generate an "image," a more general term referring to all pictorial representations of remotely sensed data. All photographs are images, but not all images are photographs. Photographic cameras react to reflected radiation from a ground scene by recording variations in the intensity of light through a photochemical reaction on a frame in a roll of film. The region a camera views is determined by its angular field of view (AFOV), a geometric function of the camera's focal length relative to the edge of the film plane (format size) and platform altitude. Mapping or metric cameras are specially designed for vertical aerial photography at high resolution with minimal geometric distortion, allowing precise measurements and accurate maps to be generated. Films contain a thin emulsion of light-sensitive silver halide grains (ranging from .1-5 μm in diameter) that undergo a photochemical reaction proportional the number of photons hitting them during an exposure. In Microwave systems, a third class of sensors, radar sends pulses of microwave energy to the ground; depending on the physical characteristics of the surfaces and angles of incidence, and varying amounts of energy are echoed or backscattered to the sensor, making it an active method employable day or night.

Platform issues

Platforms employed for airborne imagery range from

balloons, to kites, and aircraft of various kinds, including helicopters, small airplanes, and high-speed jets. All sensors possess a fixed AFOV, so the ground area captured in an image is solely a function of altitude: low-altitude platforms image smaller areas but provide greater spatial resolution; high-altitude or space imagery captures larger areas, commonly with less spatial detail, and problems with clouds and atmospheric haze are exacerbated. Platform velocity also affects spatial resolution. In aerial photography an aircraft's forward motion during the brief exposure can blur the image. The orientation of the sensor relative to the Earth determines whether an image is vertical or oblique, vertical when the sensor is within 3° of a perpendicular to the surface, oblique when its axis is tilted away.

Aerial photography

Aerial photography was pioneered from balloons in the 19th century, with the earliest aerial photograph of an archaeological site, Stonehenge, taken in 1906. World War I (1914-18) saw vast improvements with a move to maneuverable aircraft and trained personnel, of whom several focused on archaeological sites during the war. Outstanding aerial photographic surveys of Roman occupations were carried out in Syria during the 1920s-30s and in North Africa during the 1940s. O.G. S. Crawford, an observer in the British Royal Flying Corps, was primarily responsible for the early development and systematization of aerial archaeology.

Air photo interpretation

Air photo interpretation is the process of identifying objects or conditions in aerial photographs and determining their meaning and significance. It is an art and a science, with a large subjective element allied to a basis of scientific principles. Good interpretations are based on a large store of experience and knowledge of geography, geology, forestry, biology, other landscape sciences, and—in the archaeological case—familiarity with the forms and characteristics of past site and settlement types. It is important not to just recognize archaeological occurrences; other features—type of crop, ground cover, access, terrain form, aspects of the local environment—may be important. The basic question asked is, "What is going on here?" Positive identifications are frequent, but many are qualified as "probable" or "possible" (Lyons & Avery 1977).

Advances in and sources of aerial photography

Recent advances in aerial photography include digital cameras more amenable to computer processing operations and in-flight aids like global positioning systems (GPS) that help locate photographs and the sites and features within them. Walker & DeVore (1995) have introduced "Low Altitude Large Scale Reconnaissance," with cameras in remotely controlled model airplanes designed to fly at very low altitudes and slow speeds. Their results yield spatial resolutions of a few centimeters with greatly reduced image blurring.

Aerial multispectral methods

Aerial multispectral scanning (MSS) potentially offers more than conventional aerial photography, because the photographic spectrum is only a small part of the electromagnetic spectrum; sensing more means seeing more. *et al.* This was demonstrated in a pioneering study by Morain (1981) that employed a Bendix 11-channel (from .38-12.25 µm) MSS with 1.25 m resolution over Bandelier National Monument, New Mexico. Supervised image classification methods showed a high degree of separability between several archaeological and natural land cover classes, confirming that human disturbances could be detected through spectral analysis. Hemans *et al.* (1987) used a Xybion multispectral video camera (.4-1.1 µm in 6 bands) from a tethered blimp at Corinth, Greece, gaining data particularly suited to defining the pattern of standing structures.

Aerial thermography

Radiation is emitted by all objects with spectral properties and magnitudes determined by temperature and the material's emissivity, its efficiency as an absorber and emitter. As temperature decreases, emitted energy shifts to longer wavelengths: the sun's surface temperature of about 6000° C emits peak radiation at a wavelength of .48 µm, seen as light, while the earth's ambient temperature of about 27° C yields a maximum emissivity at 9.7 µm, felt as heat. The latter wavelength conveniently falls within one of two transmission windows of thermal infrared radiation (TIR) in the lower atmosphere (Avery & Berlin 1992:120).

Airborne radar

Aerial applications of radar became available with Airborne Synthetic Aperture Radar (AIRSAR), built by NASA's Jet Propulsion Laboratory as a radar test platform aboard a jet aircraft. While early versions offered a nominal swath width of 10 km, an across-track resolution of about 20 m, and only the L-band (25 cm), recent flights also provide P- (67 cm) and C-band (5.7 cm) channels, and a resolution of 5-10 m. Adams *et al.* (1981) explored L-band (25 cm) data acquired in 1977-80 over Guatemala and Belize that yielded bright returns from the favorable geometry of sloping pyramid faces, and indicated a canal system of previously unrealized extent (covering 12,400 km^2), made visible by slight differences in the elevation of vegetation.

Laser altimetry

Recent advances in laser altimetry, the measurement of height using a laser rangefinder, make acquisition of high resolution elevation data trivial. Frequently called Lidar (LIght Detection And Ranging), it is the optical equivalent of radar and consequently an active method, capable of rapidly generating accurate and dense digital models of topography as well as the vertical structure of other surfaces (buildings, trees) from the air. Lidar is a fast evolving technology for wide-area mapping that provides remarkable surface detail, with absolute vertical accuracy currently as good as 15 cm (relative accuracy between adjacent measurements on a uniform surface is considerably higher) and horizontal sampling densities well below a meter in conventional applications (Flood 2001).

While relatively new and currently expensive, Lidar has already seen archaeological results (Barnes 2003; Holden *et al.* 2001). With considerable surface expression in many archaeological sites around the globe (e.g. Figure 1g), from standing ruins in the Middle East and American Southwest, to abandoned medieval towns in Europe or earthlodge depressions and mound groups left by Native Americans, Lidar represents a huge breakthrough in the mapping and understanding of archaeological landscapes, and quantification of topographic effects that might affect other remote sensing data (e.g. electrical resistivity, GPR, and other methods with a terrain response).

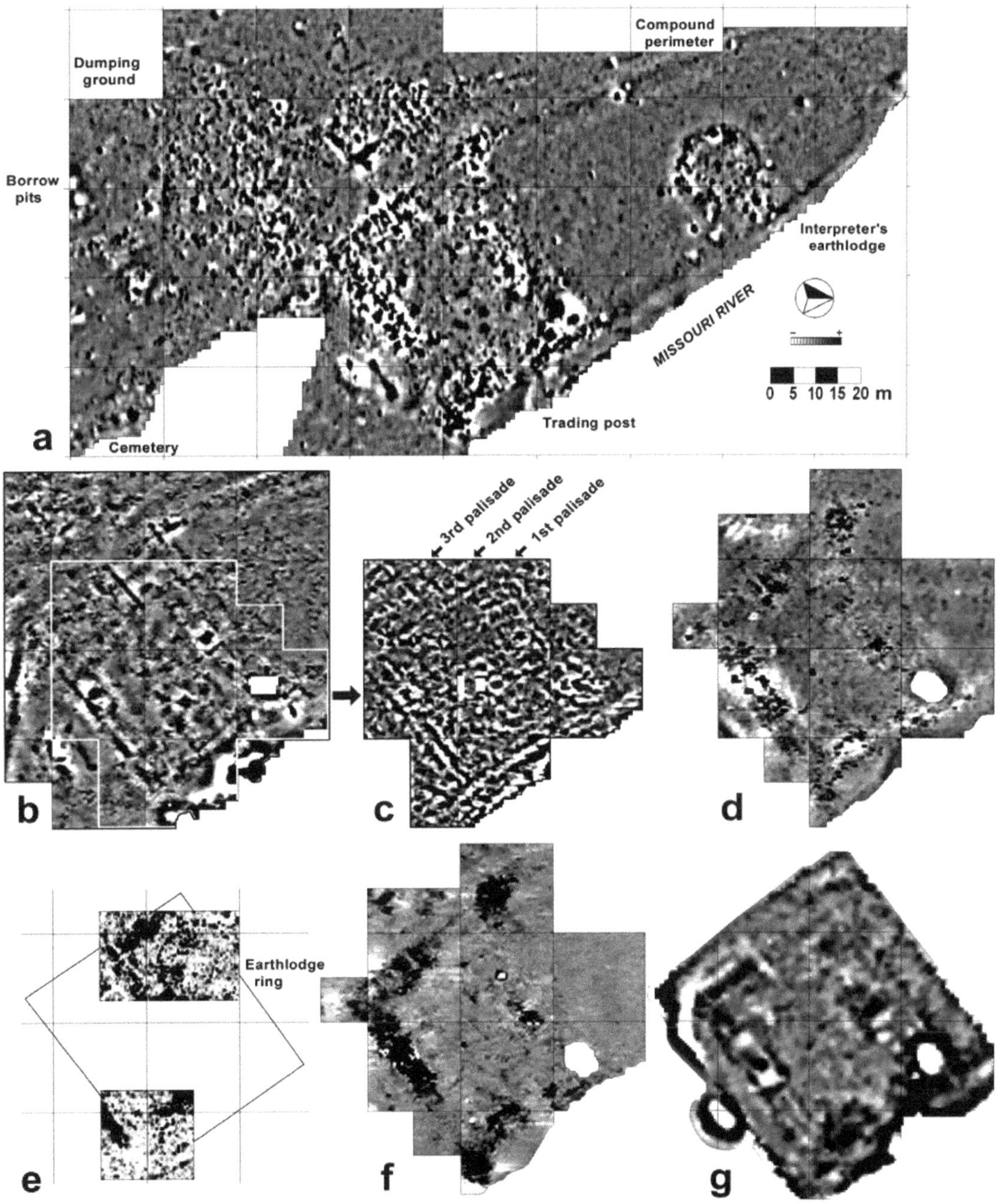

Fig. 1. Ground-base remote sensing results at the ca 1831-61 Fort Clark Trading Post, North Dakota: a) magnetometry, b) resistivity (50 cm probe separation), c) resistivity (1.5 m probe separation), d) EM conductivity, e) GPR, f) magnetic susceptibility, g) exaggerated and enhanced surface elevation data.

satellite system	bands*	wavelength	pixel size	swath	launch date
LANDSAT 1-3	Multispectral (G, R, NIR1, NIR2)	.5-1.1 μm	79 m	185 x 185 km	1972-78
LANDSAT 4,5	Multispectral (B, G, R, NIR, MIR1, MIR2, FIR)	.45-12.5 μm	30 m, FIR 120 m	185 x 185 km	1982, 1984
LANDSAT 7	Multispectral (B, G, R, NIR, MIR1, MIR2, FIR) Panchromatic	.45-12.5 μm .52-.9 μm	30 m, FIR 60 m 15 m	185 x 185 km	1999
SPOT 1-3	Multispectral (G, R, NIR) Panchromatic	.5-.89 μm .5-.73 μm	20 m 10 m	60 x 60 km	1986-93
SPOT 4-5	Multispectral (G, R, NIR, MIR) Panchromatic	.5-1.75 μm .48-.71 μm	20, 10 m 10, 5 m	60 x 60 km	1998, 2002
CORONA KH-4B	Panchromatic – film camera	.5-.7 μm	2-10 m	14 x 188 km	1967
KVR-1000	Panchromatic – film camera	.51-.76 μm	1-2 m	40 x 180 km	1987
IKONOS	Multispectral (B,G, R, NIR) Panchromatic	.45-.9 μm	4 m 1 m	13 x 13 km	1999
QUICKBIRD	Multispectral (B, G, R, NIR) Panchromatic	.45-.9 μm	2.5 m .61 m	17 x 17 km	2001
SIR-C / X-SAR	Microwave (X, C, L)	3-24 cm	10-50 m	30-60 km	1994
RADARSAT	Microwave (C)	5.7 cm	8-100 m	50 -500 km	1995

*B=blue; G=green; R=red; NIR=near infrared; MIR=mid-infrared; FIR=far infrared; X, C, L=microwave bands discussed in text

Table 1. Satellite systems that have been employed in archeological remote sensing.

Satellite remote sensing

A host of satellite programs are sponsored by government agencies and commercial enterprises; several have played a more central role in archaeology. Landsat (for "land satellite") was the first satellite program for collecting repetitive, synoptic, multi-spectral imagery for monitoring and analyzing earth resources and environment; Landsat 1 was launched in 1972, with Landsat 2-5 and 7 following in subsequent years. Several sensor systems have been deployed, including the Multispectral Scanner (MSS), the Thematic Mapper (TM), and the Enhanced Thematic Mapper (ETM+). SPOT, a French program, is the 'Systéme Pour l'Observation de la Terre'; five SPOT satellites have been launched since 1986, each carrying High Resolution Viewers (HVR). Both programs were important in the early development of archaeological space remote sensing, and continue to be utilized (Limp 1989; Sever 2000). Details of these and other satellites are summarized in Table 1.

Recent advances: high resolution satellite programs

High spatial resolution satellite imagery (< 4 m) sufficient to detect and map individual archaeological features is very recent; we are now at the dawn of true archaeological feature mapping from space-based platforms (Fowler 2002). High resolution satellite data were first made publicly available in the mid 1990s in the form of Russian KVR and US Corona panchromatic photographs, both from spy satellite programs (the latter acquired in the 1960s-70s and recently decommissioned), with spatial resolutions as high as about 2 m (Table 1).

Space radar

In 1981 the US Space Shuttle Imaging Radar (SIR-A) captured the attention of archaeologists by revealing ancient canyons and paleochannels buried meters below the Sahara desert. The microwave energy easily penetrated the subsurface in this region of dry sands with low dielectric constant (El Baz 1997; McCauley et al. 1982). Conventional excavations later found prehistoric sites, dating as far back as the Acheulian, along these "radar rivers" (Wendorf et al. 1987), but this work did not directly detect archaeological sites. In the 1994 SIR-C flight, X-band SAR was acquired over the ancient city of Angkor Wat, Cambodia (Figure 3b), revealing temples, new structures, mounds, roads, and other features (Moore & Freeman 1998; see also page 000 above). Sever (2000:36) shows L- and C-band SIR-C images of the

Great Wall of China, while Holcomb (1998) overviews theory, method, and additional archaeological applications, including discovery of the "lost city of Ubar" buried beneath the sands of Oman but located in SIR-A, SIR-B, and other imagery at the convergence of tracks through the desert.

Computer methods: image processing & GIS

Computer manipulation and processing of data is now essential to nearly all remote sensing, an activity nearly as important as collecting the actual data because noise, defects, and distortions in raw imagery obscures important information. Figure 4a illustrates how defects in raw resistivity data, including data spikes (erroneous extreme measurements), a broad geological trend, survey grid imbalances, and a plow zone response, are effectively removed to produce a sharp, clear, and more interpretable image of the 14th-century earthlodge village of Whistling Elk, South Dakota (Figure 4b). Frequently, general purpose image processing, photogrammetry, or geographical information systems (GIS) software (which often includes elements of the first two) contains all necessary tools; special-purpose software is sometimes also required.

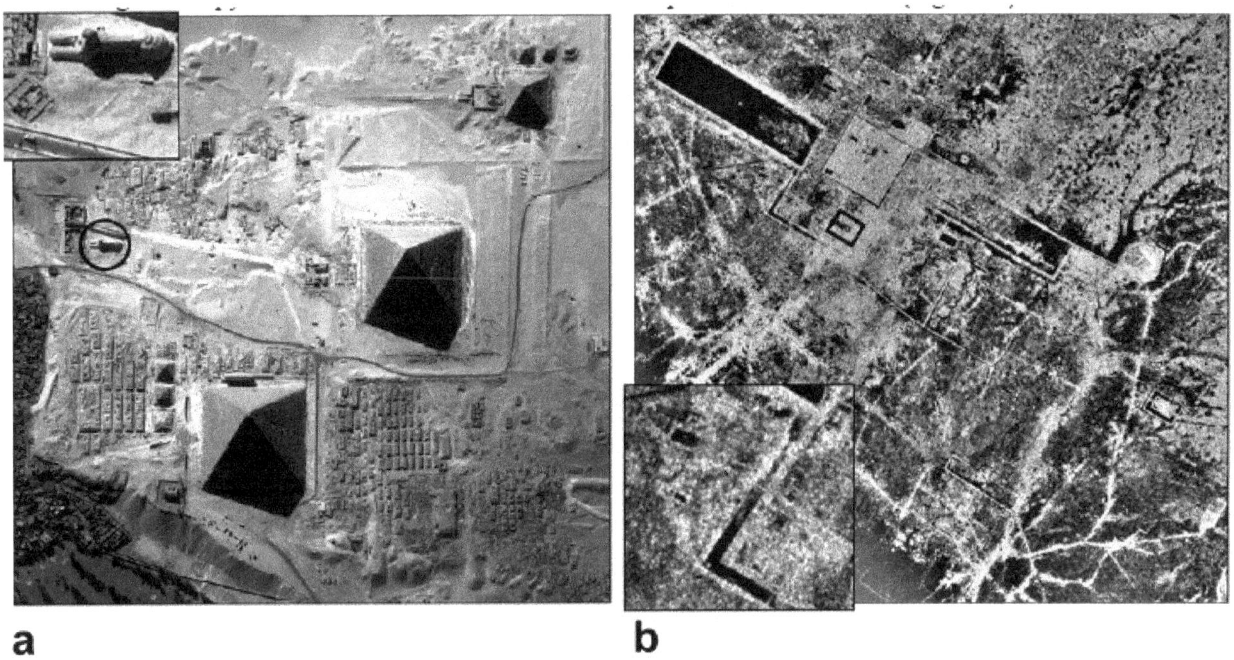

Fig.3. Satellite remote sensing results. a) Ikonos 1m resolution panchromatic image of Giza, Egypt, featuring the Great Pyramids and Sphinx, the latter shown in detail in the inset (credit: Space Imaging, collected November 17, 1999). b) The 1994 SIR-C flight acquired this X-band SAR image of Angkor Wat, Cambodia, and environs (inset shows detail; credit: NASA/JPL image p-45156).

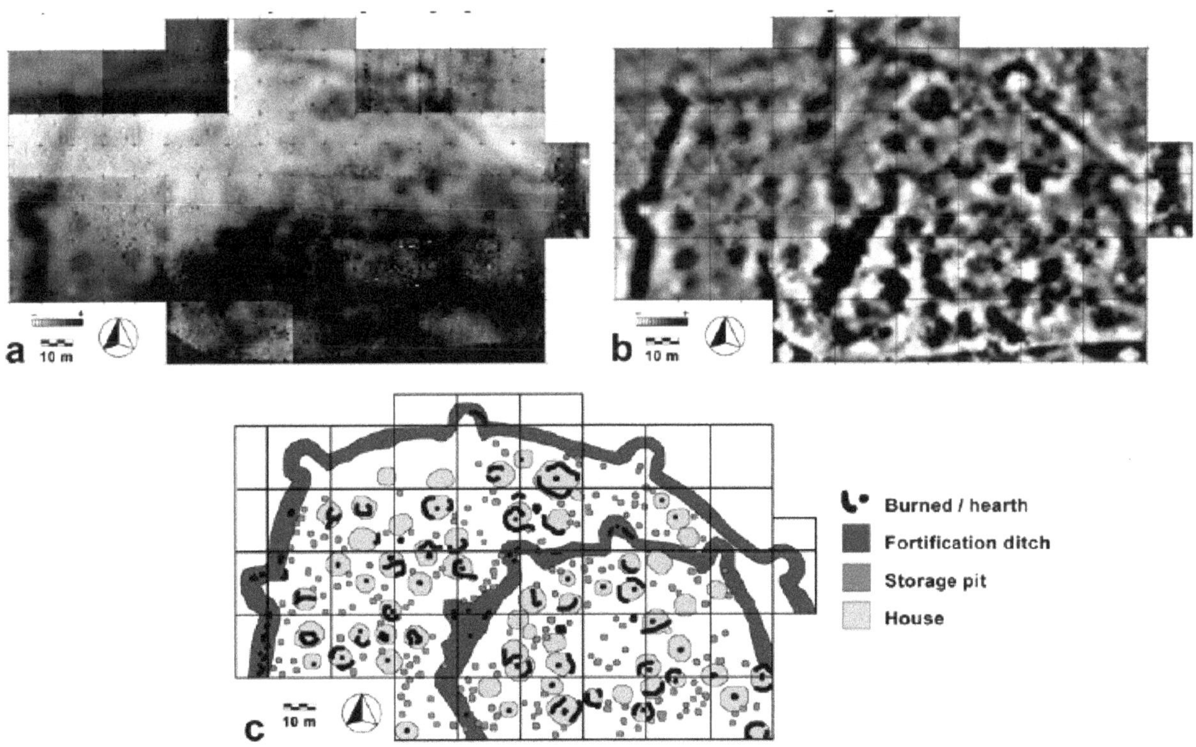

Fig.4. The importance of digital image processing is evident in comparing a) the row resistivity image with b) the final processed outcome at the 14th-century Whistling Elk village, South Dakota. c) Interpretations of aerial (Figure 2a), resistivity (b), magnetometry and conductivity data (not illustrated) from Whistling Elk village allow creation of a series of vector polygons transcribed onto a base map.

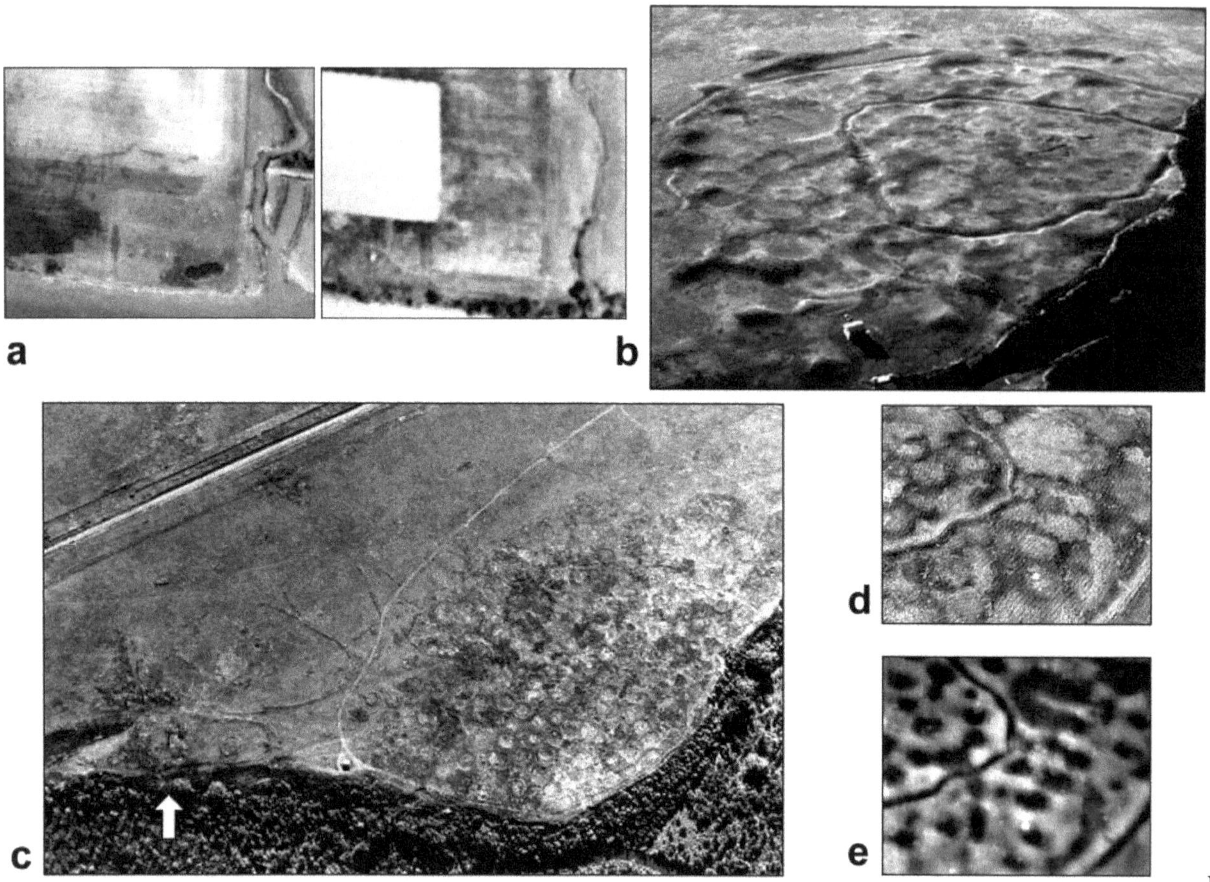

Fig. 2. Aerial remote sensing results. a) Vegetation marks from 1968 (left) and 1946 (right) showing different portions of the 14th-century fortification system at Whistling Elk village, South Dakota. b) Shadow marks reveal numerous circular depressions signifying former earthlodges, fortification ditches, and midden mounds at the 15-18th century Double Ditch site, North Dakota (photo credit: State Historical Society of North Dakota). c) Features of the 1831-61 Fort Clark Trading Post (indicated by arrow) and its environs, including a nearby Mandan-Arikara settlement (ca 1822-61), in North Dakota, are revealed by a combination of vegetation, soil, and shadow marks (photo credit: State Historical Society of North Dakota). d) Fortification ditches and depressions in the surface at the loci of former earthlodges are represented by vegetation differences in this enhanced normal-color view at Double Ditch. e) Cooler temperatures occur in the ditch and house depressions compared to nearby high points in this thermal infrared view of the same region depicted in d).

Benefits, goals, future directions

Remote sensing surveys provide a rapid means of data collection that can provide 100 percent coverage of large areas, with costs far less than excavation. They can guide expensive excavation and testing programs to features of potential archaeological interest. Indications can be given about the depth, areal extent, and distribution of cultural features, as well as of soil types and other subsurface conditions. The volume of excavations can be reduced, with smaller artifact collections recovered, and lowered curation costs. Traditional invasive methods generally damage or destroy the very resources they were designed to investigate; remote sensing methods, non-destructive, leave the resource intact—an important consideration for culturally sensitive burial, sacred, or ceremonial sites. Finally, when conditions and sensor mixes are correct, remote sensing can produce detailed maps of subsurface features over large areas, revealing arrangements of dwellings or other structures, rooms within them, internal room features, streets, lanes, trails, public spaces, dumping areas, middens, mounds, gardens, storage features, fortification systems, and other elements.

Cultural resources inventory: interpreted and transcribed maps

The production of a final "end-image" that clearly portrays subsurface features of possible cultural origin through extensive data processing and enhancement is not necessarily the ultimate goal of remote sensing: the utility of results is greatly benefited by an analytical phase that interprets and transcribes features accurately onto maps. Transforming information in remotely sensed imagery into a record of the archaeology is complicated, but important. It requires knowledge of the kinds of features that might occur, experience in their remote sensing signatures, which can vary regionally, the amount of regular or geometric pattern expressed, and technical knowledge of cartography and mapping.

Future directions

The goal of digital archiving is to preserve information in any type of medium while maintaining its content and structure so that records can be transferred from one format and medium to others that might evolve. The issues are (1) media degradation (5-10 years for magnetic; perhaps 30 for optical), (2) hardware evolution (changes in disks, tapes, CDs, DVDs), and (3) changes in data storage formats. Digital archiving requires reliable backups (duplicate copies), data refreshment (copying information to new media as the original approaches its lifespan, or to new forms as technology evolves), data migration (copying from outmoded data formats to contemporary ones supported by current software), and good documentation. The last is essential because the structure and content of the information must be fully known to migrate from one set of data structures or formats to another without information loss (see Bewley *et al.* 1999; Schmidt 2001).

Conclusion

Satellite sensors will continue to improve, with greater spatial, spectral, and temporal resolutions, lower costs, and greater ease of access. Yet, orbital constraints limit frequencies and times of data collection, cloud cover obscures many scenes, spatial resolutions are generally lower than airborne sensors, and most perspectives are vertical or near-vertical; imagery from small airplanes can be acquired at any time, from any angle to optimize lighting and view, and at much higher spatial resolutions. For this reason, the 'old technology' of oblique aerial photography taken by archaeologists will remain important. The coverage of satellite scenes offers some potential for wide-area prospecting, should automatic image classification and other detection algorithms become sufficiently developed to identify archaeological features (e.g. see Methias-Lemmens et al. 1993).

It is only through remote sensing that archaeologists will be able to map, analyze, and interpret large archaeological regions and cultural landscapes in much of the globe. Its central role in a planetary inventory of cultural resources is obvious. With prohibitive costs, labor requirements, and ethical issues raised in excavating large regions—the only other means to realize what lies beneath the surface—only small exposures can ever be made. Remote sensing is sure to occupy a more central stage in the future of archaeological research and practice.

References

ADAMS, R.E.W., W.E. BROWN, JR. & T. PATRICK CULBERT. 1981. Radar mapping, archaeology, and ancient Maya land use, *Science* 213:1457-1463.

AL CHALABI, M.M. & A.I. REES. 1962. An experiment on the effect of rainfall on electrical resistivity anomalies in the near surface, *Bonner Jahrbücher* 162:226-271.

ALFODI, TOM, PAULINE CATT & PETER STEPHENS. 1993. Definitions of "remote sensing," *Photogrammetric Engineering and Remote Sensing* 59:611-612.

ASPINALL, A. & J.G. CRUMMETT. 1997. The electrical pseudo-section, *Archaeological Prospection* 4:37-47.

AVERY, THOMAS E. & GRAYDON L. BERLIN. 1992. *Fundamentals of remote sensing and air photo interpretation*, 5th ed. New York: Macmillan.

BARNES, IAN. 2003. Aerial remote-sensing techniques used in the management of archaeological monuments on the British Army's Salisbury Plain training area, Wiltshire, UK, *Archaeological Prospection* 10:83-90.

BECKER, H. 1995. From nanotesla to picotesla - a new window for magnetic prospecting in archaeology, *Archaeological Prospection* 2:217-228.

BERLIN, G.L., J.R. AMBLER, R.H. HEVLY & G.G. SCHABER. 1977. Identification of a Sinagua agricultural field by aerial thermography, soil chemistry, pollen/plant analysis, and archaeology, *American Antiquity* 42:588-600.

BEVAN, BRUCE W. 1983. Electromagnetics for mapping earth features, *Journal of Field Archaeology* 10:47-54.

BEVAN, BRUCE W. 1998. *Geophysical exploration for archaeology: an introduction to geophysical exploration.* Midwest Archeological Center Special Report 1. Lincoln, Nebraska: National Park Service.

BEVAN, BRUCE W. 2000. An early geophysical survey at Williamsburg, USA, *Archaeological Prospection* 7:51-58.

BEWLEY, ROBERT H. 2000. Aerial photography for archaeology, in Linda Ellis (ed.), *Archaeological method and theory: an encyclopedia:* 3-10. New York: Garland Publishing.

BEWLEY, ROBERT, DANNY DONAGHUE, VINCE GAFFNEY, MARTIJN VAN LEUSEN & ALICIA WISE. 1999. *Archiving aerial photography and remote sensing data.* Oxford: Oxbow Books.

CHALLAND, ADRIAN. 1992. Field magnetic susceptibility measurement for prospection and excavation, in P. Budd, B. Chapman, C. Jackson, R. Janaway, & B. Ottaway (ed.), *Geoprospection in archaeology*: 33-41. Oxbow Monographs 18. Oxford: Oxbow Books.

CIMINALE, M. & M. LODDO. 2001. Aspects of magnetic data processing, *Archaeological Prospection* 8:239-246.

CLARK, ANTHONY. 2000. *Seeing beneath the soil: prospection methods in archaeology.* London: Routledge.

CLAY, R.BERLE. 2001. Complementary geophysical survey techniques: why two ways are always better than one, *Southeastern Archaeology* 20:31-43.

COMFORT, ANTHONY. 1997. Satellite remote sensing and archaeological survey on the Euphrates, *Archaeological Computing Newsletter* 48:1-8.

CONNOR, MELLISA & DOUGLAS D. SCOTT. 1998. Metal detector use in archaeology: an introduction, *Historical Archaeology* 32:76-85.

CONYERS, LAWRENCE B. & CHRISTOPHER GAFFNEY. 2000. Some reflections on the Society for American Archaeology annual meeting, *Archaeological Prospection* 7:145-146.

CONYERS, LAWRENCE B. & DEAN GOODMAN. 1997. *Ground-penetrating radar: an introduction for archaeologists*. Walnut Creek, California: Alta Mira Press.

CUSTER, JAY F., TIMOTHY EVELEIGH, VYTAUTAS KLEMAS & IAN WELLS. 1986. Application of Landsat data and synoptic remote sensing to predictive models for prehistoric archaeological sites: an example from the Delaware coastal plain, *American Antiquity* 51:572.

Ground-based icing condition remote sensing system definition

Munish Kumar Tiwari

Keywords: NASA, Profiling Microwave Radiometer, Polarized Ka-Band Radar, Dual Band Radar, Multi-Band Radar.

Background

The NASA In-flight Icing Remote Sensing activity started with the findings of 1997 White House Commission on Aviation Safety and Security, which directed NASA to significantly increase the level of safety for aircraft, including all-weather operations. NASA then initiated the Aviation Safety Investment Strategy Team (ASIST), which prioritized aviation safety activities required to meet the White House goals. The ASIST Weather team identified In-flight Icing as one of its top 3 priorities to improve flight safety. Simultaneous to this activity, the NASA Advanced General Aviation Transport Experiment (AGATE) was defining technologies required to enhance General Aviation (GA) aircraft safety and operation. Within AGATE, the Ice Protection Systems Work package was defining the Avoid and Exit strategy as the key to improving flight safety in the icing environment.

Key to success of the Avoid and Exit strategy was the ability to remotely measure the icing environment. In 1997, NASA Glenn Research Center (then Lewis Research Center), the U.S. Army Cold Regions Research and Engineering Laboratory (CRREL), and the FAA sponsored the In-flight Remote Sensing Icing Avoidance Workshop. The outcome of this workshop was the formulation of the NASA Icing Remote Sensing activity. The NASA Icing Remote Sensing Activity was designed around the findings of the 1997 workshop. Tasks were defined that address the three major areas of concern: Meteorological Issues, Operational Issues, and Technological Issues. The Meteorological and Operational Issues are being addressed to define the requirements of future remote sensing systems that will adequately measure the environment and pass the appropriate information to the user in the most appropriate manner and format.

Research Method

The research design will be descriptive in nature. The method of data collection will be mainly based on secondary data. However, an effort will be made to collect primary data by taking information from competent persons off and on.

Technology Assessment

Basic Requirements

The primary requirement for the ground based icing condition remote sensing system is that it be capable of measuring environmental conditions sufficient to identify areas of icing hazard. However, in-flight icing potential is not directly measurable. It does not exist until an aircraft passes through the environment. To determine icing hazard, one would ideally measure the liquid water content (LWC) of the cloud, the ambient temperature, the droplet size distribution of the cloud, and the horizontal and vertical extents of the icing conditions. Icing becomes a hazard to aircraft when the ice accretion rate exceeds the capacity of the aircraft's ice protection system or when excessive ice accretes on unprotected areas. The accretion rate is a function of many factors, but the most significant is the cloud LWC (assuming below freezing temperatures).

Technology Options

Three basic technologies were selected for assessment based upon the findings of the 1997 In-flight Remote Sensing Icing Avoidance Workshop. The three technologies are radar, lidar, and radiometry. For the radar option, several variations were identified for individual evaluation. Four radar measurement techniques were identified: Ka-Band Radar, Polarized Ka-Band Radar, Dual (X/Ka) Band Radar, and Multi-(X/Ka/W) Band Radar. The radiometry technology assessed is the Profiling Microwave Radiometer. And the lidar technology examined is the Multiple Field of View (MFOV) Lidar.

The individual technologies and their scores are listed in Table 2, Technology Assessment Table, and the two hybrid technologies and their scores are listed in Table 3, Hybrid Technology Assessment Table.

Profiling Microwave Radiometer

System description

The Radiometrics Corp. TP/WVP-3000 Water Vapor, Temperature, and Cloud Liquid Water Profiling Radiometer (Figure 1) were assessed for use as an icing condition remote sensing system. The profiling radiometer utilizes discrete measurements in the 22 to 60 GHz range3. This range of frequencies provides the observation of several spectral features in the atmosphere. Broadening around the 60 GHz oxygen band provides information on temperature profiles. Similarly water vapor profile information can be obtained by observing the broadening around the 22 GHz water vapor line. Liquid water emits over the entire range and increases approximately with the second power of the frequency, so by examining the temperature and water vapor measurements, liquid water profile information can be obtained.

	Score = 0	Score = 5
Current Readiness	Technology only a concept	All hardware and software available for purchase and fielding
Strength	No strengths	Technology capable of accurately measuring all parameters required for the determination of icing conditions aloft
Weakness	Technology incapable of producing any information that would help detect icing conditions aloft	No weaknesses
Cost	Very expensive (well over $1 million)	Very inexpensive (complete system for under $100,000)
Current Operational Constraints	Must be constantly attended and unable to operate in many meteorological conditions	Capable of running unattended for long periods of time with 100% accurate data capture

Table 1. Meaning of technology scoring.

Technology	Current Readiness	Strength	Weakness	Cost	Current Operational Constraints *	Total **
Profiling Microwave Radiometer	4 (off the shelf)	4 (temp and int. LWC)	3 (LWC profile inaccuracy, no dropsize)	4	4 (radome contam.)	19
Ka Band Radar	4	1 (indication of cloud)	1 (no LWC, no dropsize, no temp)	3	4 (radome contam.)	13
Polarized Ka Band Radar	3 (build planned)	2 (indication of liquid water, indication of large drops)	2 (no LWC, no temp)	2	3 (attended)	12
Dual Band Radar	2	2 (LWC and indication of large drops in Rayleigh regime)	2 (no LWC and no dropsize in Mie regime, no temp)	2	3 (attended)	11
Multi-Band Radar	2	3 (LWC, indication of large drops)	3 (no temp)	2	3 (attended)	13
MFOV Lidar	3	2 (cloud edge LWC, dropsize)	1 (poor cloud penetration, no temp)	3	2 (attended, window contam., eye safety)	11

Scoring is on a scale of 1 (poor) to 5 (excellent)
* "Attended" indicates the need for attended operation and/or analysis
** Total out of a possible of 25

Table 2. Technology Assessment Table.

Technology	Current Readiness	Strength	Weakness	Cost	Current Operational Constraints *	Total **
Profiling Microwave Radiometer with Ka Band Radar	4	4 (temp and improved LWC profile)	4 (no dropsize)	3	4 (radome contam.)	19
Profiling Microwave Radiometer with Polarized Ka Band Radar	3 (build planned)	5 (temp, improved LWC profile, indication of large drops)	5	2	3 (radome contam., attended)	18

Scoring is on a scale of 1 (poor) to 5 (excellent)
* "Attended" indicates the need for attended operation and/or analysis
** Total out of a possible of 25

Table 3. Hybrid Technology Assesment Table.

Fig. 1. Radiometrics Inc. Prolifing Radiometer.

The radiometer is controlled at a low level by a self-contained microprocessor which in turn is controlled by a data acquisition computer running FORTRAN software. The data acquisition computer takes the raw radiometer measurements at the discrete frequencies and along with several meteorological measurements and calculates temperature, water vapor, and liquid water profiles with a neural network. The neural network is trained using historical radiosonde data from the intended operational region over the various seasons. Work is currently underway to eliminate the reliance on the neural net and the need for location and season specific programming and to include some physics of cloud formation into the calculations.

Scoring

1) Current readiness scored 4 out of 5:

The profiling radiometer examined in this activity is a commercially available product. Radiometrics, Inc. manufactures and markets their MP3000 radiometer, and it is available for purchase with relatively short lead times.

During the assessment of this system several shortcomings have been identified, but their correction should not significantly alter the cost or availability of the system.

2) Strengths scored **4** out of 5:

The profiling radiometer examined in this effort typically does a good job measuring temperature profiles and does nearly as well measuring water vapor (humidity) profiles4. Temperature and humidity profiles together

typically can provide reasonable indication of cloud location. The microwave radiometer technology has been shown to provide accurate integrated (over the view path) liquid water measurements.

3) Weaknesses scored **3** out of 5:

The profiling radiometer currently doesn't do very well with liquid water profiles5. Also, the technology as currently configured cannot provide any indication of large droplets in a cloud. Work is currently underway to improve the liquid water profiling measurement. While typically strong, the temperature and humidity profile accuracy can be limited in inversion situations or whenever the parameter changes very rapidly with altitude. Since the radiometer uses the temperature profile along with the cloud base temperature measurement to define the cloud base, additional work is required to improve cloud base identification in inversion conditions.

4) Cost of system, scored **4** out of 5:

Due to the fact that a profiling radiometer is currently commercially available, the cost of this technology is quite reasonable. Complete profiling radiometer system costs (including instrument and data acquisition computer) are approximately one quarter of a million dollars.

5) Current operational constraints scored **4** out of 5:

The profiling radiometer examined has been run unattended in numerous filed tests. As with all microwave instruments, this device has problems with water contamination on the radome. This instrument has a blower and heater to minimize the length of time of contaminated measurements. However, the measurements from the instrument are invalid for a fairly significant amount of time during liquid precipitation events. This is considered to be a problem that can be solved with adequate engineering effort.

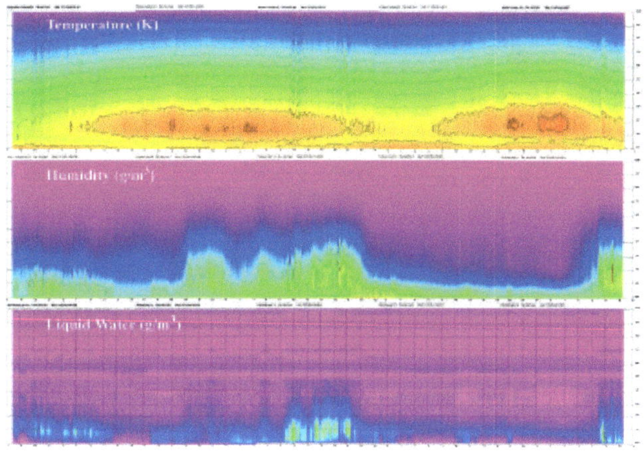

Fig. 2. Example of Profiling Radiometer Data.

Ka-Band Radar
System description

While not a subject of the active NASA assessment activities for icing remote sensing systems, the Ka-Band cloud radar has the potential for significant contribution to a hybrid system (see below). Therefore, its attributes have been examined and scored here. The system described here is the NOAA/ETL Millimeter-wave Cloud Radar (MMCR)6 (Figure 3). It functions at 34.86 GHz with a peak transmit power of 100W. It has a 1.8 m diameter antenna that produces a 0.3 deg wide beam. It is designed to stare at the zenith. Its range is approximately 20 km with a resolution of 45 m.

Scoring

1) Current readiness scored **4** out of 5:

Ka-Band radars are currently available for purchase from radar manufacturing firms. The radars currently available may not be the optimal instrument for use with an icing remote sensing system, so some development effort is likely to be required.

Fig. 3. NOAAA Environmental Technology Laboratory (ETL) Millimeter-Wavelenght Cloud Radar (MMCR)

2) Strengths scored **1** out of 5:

As a stand-alone device, the cloud radar can provide a good indication of the boundary of the clouds and an indication of the density of hydrometeors in the cloud; however this information by itself is of little value.

3) Weaknesses scored **1** out of 5:

By itself, the cloud radar is weak for icing condition remote detection. The cloud radar can provide no information on the liquid state of the cloud, the amount of moisture in the cloud (even if the liquid state is already known), temperature of the environment, or cloud droplet size.

4) Cost of system, scored **3** out of 5:

Currently available cloud radars are quite expensive since they are custom made instruments for each research application. However, there is a good potential that the costs of such devices required to define the boundaries of icing clouds can be significantly lower.

5) Current operational constraints scored **4** out of 5:

Cloud radar systems have been designed to run unattended. However, as with all of the technologies examined, the problem with radome contamination has not been adequately addressed for this form of remote sensing system to ensure 100% availability for the winter operations environment.

Polarized Ka-Band Radar
System description

While NASA has not funded examination of Polarized Ka-Band Radar for use as an icing condition remote sensing technology, NOAA ETL, with funding from the FAA, has completed an extensive examination. At the MWISP field test, NOAA ETL operated their Polarized Ka-Band Radar (NOAA/K)8 along with their X-Band Radar and dual frequency microwave radiometer . The NOAA/K radar was originally built in the early 1980's and has been continuously upgraded since then. It has a very flexible configuration including multiple polarization methods, scanning techniques, signal processing methods, and even antenna selection. The system configuration that resulted in the best results for super-cooled large droplet (SLD) identification at MWISP is the one described here9.

The system was configured to perform horizon-to-horizon range-height indicator (RHI) scans. A slant linear polarization was obtained by transmitting a quasi-linear polarization state at a 45 deg slant and receiving the corresponding co- and cross-polarization signals. This configuration has yielded detection methodologies that appears capable of determining cloud particle type, and if spherical, also size range.

Scoring

1) Current readiness scored **3** out of 5:

The technology required for designing and constructing polarized Ka-band radar are well understood. Currently no radar exists that has been designed specifically for icing detection. NOAA ETL has a project underway to build such radar, and this radar should be available in several years.

2) Strengths scored **3** out of 5:

This form of radar is capable of detecting spherical particles and then providing information on their size. The detection of SLD conditions is an important piece of an aviation icing hazard identification system, since for full coverage, it is necessary to identify the exceedance of Appendix C conditions in the droplet size direction.

3) Weaknesses scored **2** out of 5:

While this technology can detect spherical cloud particles and identify if they are in the SLD size range, it has neither way of providing neither the liquid water content nor the temperature of the icing environment. Therefore this technology can only answer a piece of the icing severity question.

4) Cost of system, scored **2** out of 5:

Development of radars that can provide the sensitivity required for the cloud particle classification is still quite expensive (over $1million). However, as advances are made in radar hardware and particle identification signal processing, the cost of such a system should come down dramatically.

5) Current operational constraints scored **3** out of 5:

Although some unattended operation of radars like the NOAA/K is possible, they are typically operated with an engineer or scientist present. Future systems underdevelopment will hopefully eliminate this operation constraint. Also, as with the other instruments operating in the microwave waveband, this form of radar will be susceptible to radome water contamination.

Dual Band Radar
System description

Weather radar users have long been aware of the frequency dependence of attenuation on atmospheric liquid water. In the early 1990s NOAA attempted to utilize this attenuation variation at X- and Ka-Bands (known as the dual band attenuation technique) as a means of remotely measuring liquid water content of a cloud10. They concluded that the technique looked promising, but that among other things, the sample volume of the two radars needed to be matched to obtain valid results. Since that time, several attempts have been made to test the concept in the field. NOAA ETL brought their X- and Ka- Band radars to MWISP and gathered data for the assessment of the technique11 and Stratton Park Engineering Corporation (SPEC) operated X- and Ka-Band radars in Colorado in 1997 (under a U.S. Army Small Business Innovative Research, SBIR, contact) and at AIRS in 2000 (under a NASA Small business Technology TRansfer, STTR, contract)12.

Scoring

1) Current readiness scored **2** out of 5:

The primary concern for the readiness of this technology is the design and manufacture of the required antenna system. The antennas for the two frequencies are required to be perfectly aligned and produce beams with matching widths. While possible with current engineering practice, further development is required to ensure a low cost solution.

2) Strengths scored **2** out of 5:

The dual band radar technique can measure the cloud liquid water content and provide an indication of the relative size of cloud particles. However, this capability is limited to the Rayleigh regime.

3) Weaknesses scored **2** out of 5:

The measurement of liquid water content and indication of droplet size is not valid when a significant number of large particles (typically ice crystals) are present in the cloud. Recent research is indicating that most icing clouds exist in a mixed phase or combined liquid/ice crystal state13. Also the technique can't provide any information about the temperature of the cloud environment.

4) Cost of system, scored **2** out of 5:

Current radars that can provide a quality measurement utilizing the dual band technique are quite expensive. To date, no low cost radar manufactured to utilize this measurement technique has been of sufficient quality to ensure valid results.

5) Current operational constraints scored **3** out of 5:

No multi-frequency radar is known that has been configured to run in extended unattended mode, but there is no particular limitation to the technology that would prevent it. As with all of the other technologies examined here, the contamination of the radar radome by water is an issue that would need to be resolved for extended, uninterrupted operations.

Multi-Band Radar
System description

Quadrant Engineering, Inc. under a US Army contract funded by the FAA started assessment of Multi-Band Radar technology in 199714. They performed a series of computer simulations of the radar response to icing conditions and determined that an X- /Ka-/W-Band configuration was the optimal system. They determined that this configuration would be capable of measuring cloud liquid water content and also provide an indication of cloud droplet size. Quadrant utilized a neural net analysis technique for their simulations. With a neural net, they were able to examine the response of different radar configurations to thousands of various environmental states. Quadrant tested the different configuration with varying levels of signal noise. Interestingly, they found that the Multi-Band Radar configured in an X- /Ka-/W-Band system could also measure cloud temperature if no noise was introduced, however, when any noise was introduced to the simulation the measurement failed. They theorized that the noisy-signal simulation was the best comparison to how the system would function in the real world.

Scoring

1) Current readiness scored **2** out of 5:

Research radars exist that can be used for this type of measurement. However, development would be required to produce a good Multi-Band Radar that has matching beam widths and is optimized for icing cloud measurement.

2) Strengths scored **3** out of 5:

The Multi-Band Radar has been shown to produce measurements of liquid water content and indication of large droplets. These measurements do not appear to be limited to the Rayleigh regime of only relatively small cloud particles as is the dual-band attenuation technique.

3) Weaknesses scored **3** out of 5:

Analysis has shown that radar systems with realistic noise values will not be capable of measuring cloud temperature.

4) Cost of system, scored **2** out of 5:

As with the Dual-Band Radar technology, current radars that can provide a quality measurement utilizing a multi-band technique are quite expensive

5) Current operational constraints scored **3** out of 5:

Again, like the Dual-Band Radar, no multi-frequency radar is known that has been configured to run in extended unattended mode, but there is no particular limitation to the technology that would prevent it. And, as with all of the other technologies examined here, the contamination of the radar radome by water is an issue that would need to be resolved for extended, uninterrupted operations.

Multiple Field of View (MFOV) Lidar
System description

The system described here is the Defense Research Establishment Val artier (DREG) Multiple-Field-Of-View (MFOV) Nd: Yag lidar16 (Figure 13). The lidar output is 70 mJ with a pulse repetition rate of 100 Hz, a pulse width of 12 ns, a beam diameter of 12 mm and a beam divergence ranging from 0.1 to 12 mrad. The system can be operated in the RHI mode at a single FOV or in the staring mode using all FOVs. The lidar transmits a linear polarized beam and receives both the parallel and perpendicular backscatter. The MFOV provides a capability to obtain both single and multiple backscatter information.

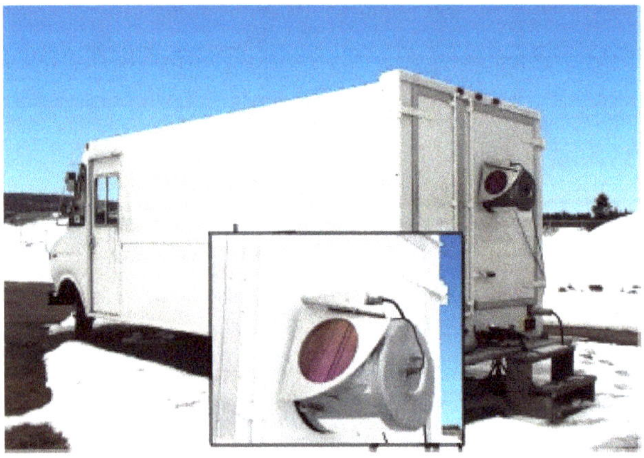

Fig. 13. Canadian Defense Research Establishment, Valcartier (DREV) MFOV Lidar.

Data example

An example of a system RHI output for a fog and liquid cloud environment is shown in Figure 14. Evident in this example is the high spatial resolution of the measurements.

Scoring

1.) Current readiness scored **3** out of 5:

The MFOV Lidar is a research tool. Some of the components of the system are available from commercial manufactures, but a development effort would be required for this to function as a dedicated aviation icing remote sensing system.

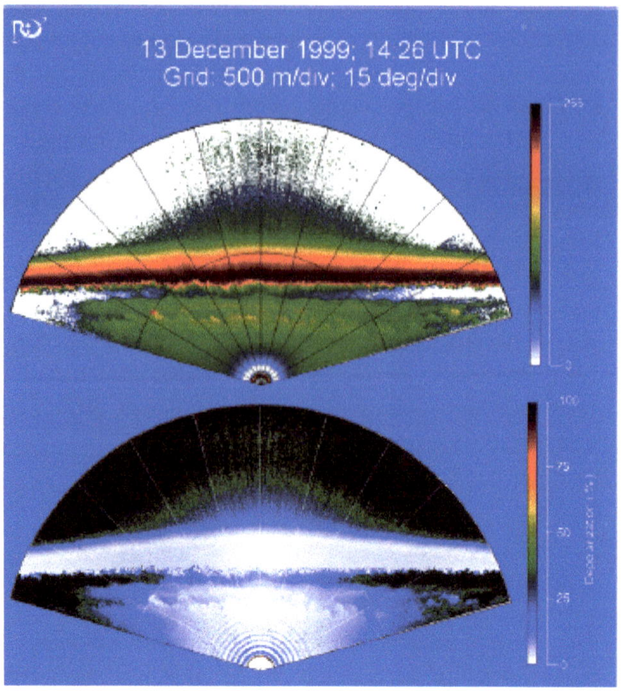

Fig. 14. Example of Data from DREV FOV Lidar.

2.) Strengths scored **2** out of 5:

MFOV Lidar can provided most of the parameters required to remotely infer cloud icing potential including phase, liquid water content, effective drop size, and information about the presence of large drops. Of all the systems reviewed, lidar can provide the greatest spatial resolution.

3.) Weaknesses scored **1** out of 5:

The optical depth associated with the cloud limits the lidar cloud penetration depth. For both the MWISP and AIRS experiments, the effective penetration depth of the MFOV Lidar was on the order of 150 to 200 meters.

4.) Cost of system, scored **3** out of 5:

Lasers are commercially available and should be fairly inexpensive. But, the MFOV Lidar contains unique components that would drive up the cost. Additional expense would be required to optimize the system output for icing severity indication.

5.) Current operational constraints scored **2** out of 5:

In the present configuration the lidar cannot run unattended. By using a lidar operating at an eye safe wavelength it may be possible to develop an unattended system.

Hybrid Profiling Microwave Radiometer and Ka-Band Radar
System description

The concept behind this hybrid system is to address the shortcomings of the Profiling Microwave Radiometer by adding more accurate ranged data. By combining a Ka-Band Radar and a Profiling Microwave Radiometer into one integrated system, one becomes able to accurately identify where the clouds are, accurately determine the temperature in that cloud, and to accurately determine the amount of liquid water in that cloud. With these three pieces of information, this combined system will satisfy the most of the information needs for aircraft icing hazard identification. The combined system will require the Profiling Microwave Radiometer to define the temperature profile and the path line integrated liquid water content.

Scoring

1.) Current readiness scored **4** out of 5:

Since both the radiometer and the radar individually were ranked at 4 out of 5 for readiness, this combined system is also ranked at that level. There will be some developmental issues in combining the two technologies; however, they are not seen as being too severe.

2.) Strengths scored **4** out of 5:

The combined use of the two technologies will result in a more accurate measurement of the liquid water content profile. The integrated liquid water content from the radiometer will initially be evenly distributed between the cloud boundaries. This will result in NASA/TM 2001-211102 27 somewhat inaccurate liquid water content magnitudes, but the location of the liquid water should be very accurate.

3.) Weaknesses scored **4** out of 5:

As with its component technologies, this hybrid system will not be capable of providing any cloud droplet size information.

4.) Cost of system, scored **3** out of 5:

While this hybrid system is made up of two relatively lower costing systems, its combined cost is considered to be moderate. If higher volume production is achieved, the cost of this system could become competitive with the current lower cost remote sensing technologies.
5.) Current operational constraints scored **3** out of 5:

While this system is expected to operate in an unattended mode, it will still be susceptible to radome contamination, which is still an issue with all of the microwave frequency instruments.

Hybrid Profiling Microwave Radiometer and Polarized Ka-Band Radar
System description

The combination of a Polarized Ka-Band Radar and a Profiling Microwave Radiometer is even more powerful than the combination of a cloud profiling Ka-Band Radar and a Profiling Microwave Radiometer. As discussed above, by adding the polarization to the Ka-Band Radar, one gains the ability to detect super-cooled large droplet (SLD) conditions within an icing cloud. With the SLD detection, the icing condition remote sensing system then becomes an even more powerful tool to aid the aviation community keep aircraft out of severe icing conditions. In their presentation at the 2000 In-Flight Icing Remote Sensing Workshop18, NOAA ETL described such a system that they call the Ground-based Remote Icing Detection System (GRIDS). Figure 15 shows a concept drawing of such a system.

Scoring

1.) Current readiness scored **3** out of 5:

The radiometer portion of this system is currently available for commercial purchase, but the Polarized Ka-Band Radar is not. However, the radar is currently being developed by NOAA ETL and if funding continues should be available in several years.

2.) Strengths scored **5** out of 5:

The combination of a Polarized Ka-Band Radar and a Profiling Microwave Radiometer provides all the information required for a vertically profiling icing severity remote sensing system. This system would be able to provide an accurate measurement of the temperature profile, location of super-cooled liquid water, liquid water content within the cloud boundaries, and an indication of SLD conditions.

3.) Weaknesses scored **5** out of 5:

The only weakness of this system is the constraint of not scanning (volume scans). It will be limited to vertical profiles. However, the restriction to vertical profiles is an assumed constraint of the overall activity at this time. Sometime in the future the technologies may progress to the point where a volume scan type operation will be possible.
4.) Cost of system, scored **2** out of 5:

The addition of polarization to the Ka-Band Radar will make this system significantly more expensive than the one that does not include polarization.

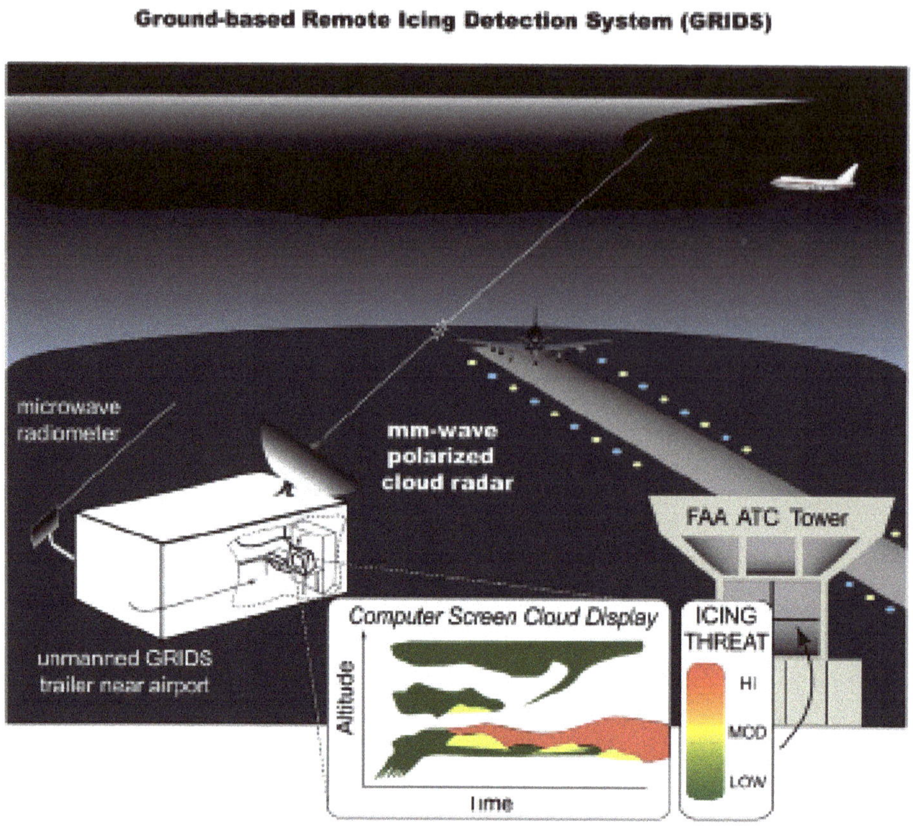

Fig. 15. Concept Drawing of NOAA ETL GRIDS (combination Polarized Ka-Band Radar and RAdiomeer System).

5.) Current operational constraints scored **3** out of 5:

Radome contamination is the primary constraint to operation of such a system. No Polarized Ka-Band Radar has yet operated in an extended unattended mode however NOAA ETL is planning to operate GRIDS in this manner.

Proposed Ground Based Icing Condition Remote Sensing System

Proposed system description

Since the future funding potential for implementing icing condition remote sensing is unknown, three versions of ground based icing potential remote sensing system are defined. The difference between these stand-alone systems is capability, accuracy, and cost. As would be expected, the lowest cost system is the least capable, with the next more expensive system adding an additional level of capability, and the most expensive system being the most capable and accurate. These three systems will be described as Level 1, Level 2, and Level 3, with Level 1 the least costly, least capable and Level 3 the most costly and most capable. The Level 1 system consists of a Profiling Microwave Radiometer. This system utilizes technology that has been shown to produce accurate temperature profiles and liquid water integrated values.

Status of hardware for proposed system

One benefit of the selection of this set of technologies is that all are already separately in some level of development, and within 2 to 3 years, hardware should exist within the government for even the Level 3 system. NASA Glenn Research Center has been examining the Radiometrics Corp. Profiling Microwave Radiometer that can be the basis of the Level 1 system. NASA Glenn Research Center has also managed an SBIR contract with Technology Service Corp., which will deliver an X- and Ka-band radar system. This radar should be sufficient for at least the initial development of the Level 2 system. And NOAA ETL, with funding from the NWS and FAA, are developing their Ground-based Remote Icing Detection System (GRIDS) which should satisfy the longer term radar requirements for Level 2 and the polarized radar requirements for Level 3.

Recommended Development Path

While much of the basic hardware required for the proposed system either already exists or is in the process of being procured, there are still developmental activities that must take place to successfully field the three levels of system proposed. Level 1 requires additional work to improve the all-weather capability of existing radiometers. Current radiometers have difficulty operating in precipitation that wets the instrument radome. This issue impacts the accuracy of temperature profiles, liquid water profiles, and integrated liquid water, and thus is seen as a critical issue. Also, the accuracy of the liquid water profiles can very likely be improved with further software and/or hardware development.

Additional Development Opportunities

In addition to the development of the proposed ground based icing condition remote sensing system it is important that NASA continue to foster the development of Multi- Band Radar and Airborne Radiometry. Multi-Band Radar holds the greatest promise of accurate measurement of LWC and indication of large drops in a scanned volume out to a reasonable range. This form of scanning system is the long term vision of ground based icing condition remote sensing. However, this form of system will require advancements in radar technology that may be 10 years in the future. Therefore the Multi-Band Radar does not warrant the primary effort of this activity, but should continue to be developed for future application.

References

RYERSON, CHARLES C., *Remote Sensing of In-Flight Icing Conditions: Operational, Meteorological, and Technical Considerations*, NASA/CR 2000-209938, ERDC–CRREL–M–00–1, March 2000.

PRESENTATIONS FROM 2000 *In-Flight Icing Remote Sensing Workshop*, NASA Glenn Research Center CD-ROM M–0738–1, January 2001.

SOLHEIM, F., GODWIN, J., WESTWATER, E., HAN, Y., KEIHM, S., MARSH, K., AND WARE R., Radiometric Profiling of Temperature, Water Vapor, and Cloud Liquid Water using Various Inversion Methods, *Radio Science* 33, 393–404, March–April 1998.

WARE, R., AND SOLHEIM, F., Microwave Profiling of Atmospheric Temperature, Humidity, and Cloud Liquid Water, *Microwave Remote Sensing of the Atmosphere and Environment* (SPIE Conference Proceedings), October, 2000.

REEHORST, ANDREW L., *Comparison of Profiling Microwave Radiometer, Aircraft, and Radiosonde Measurements from AIRS*, NASA/TM 2001-211101.

MARTNER, BROOKS E., HAZEN, D., MORAN, K., UTTAL, T., POST, M., AND MADSEN, W., *NOAA/ETL's Vertical-Profiling Cloud Radar and Radiometer Package*, NOAA ETL web site: http://www6.etl.noaa.gov/instruments/mmcr.3pager.pdf

U.S. DEPARTMENT OF ENERGY'S ATMOSPHERIC RADIATION MEASUREMENT (ARM) program web site: http://www.arm.gov/

MARTNER, BROOKS E., BARTRAM, B., GIBSON, J., CAMPBELL, W., REINKING, R., AND MATROSOV, S.,"An Overview of NOAA/ETL's Scanning Ka-Band Cloud Radar," NOAA ETL web site: http://www6.etl.noaa.gov/instruments/noaak.3pager.pdf

REINKING, ROGER F., MATROSOV, SERGEY Y., RYERSON, CHARLES C., KROPFLI, ROBERT A., AND BARTRAM, BRUCE W., Verified Detection of Super cooled Large Droplets with Dual- Polarized, Millimeter-Wave Radar, *Proc. Ninth Conference on Aviation, Range, and Aerospace Meteorology,* 8.10, pages 537–542, September 2000.

MARTNER, B.E., KROPFLI, R.A., ASH, L.E., AND SNIDER, J.B., *Cloud Liquid Water Content Measurement Tests Using Dual-Wavelength Radar,* NOAA Tech. Mem. ERL ETL–235, 1993. NASA/TM 2001-211102 37

VIVEKANANDAN, J., ZHANG, G., AND POLITOVICH, M.K., Multi-frequency and Polarization Radar-Based Detection of Liquid Droplets, *Proc. Ninth Conference on Aviation, Range, and Aerospace Meteorology*, 8.14, pages 557–562, September 2000.

LAWSON, R.P., *Improved Measurements and Detection of In-flight Icing and Freezing Drizzle, Final Report* NASA Contract NAS3–98007, October 2000.

POLITOVICH, M.K., AND RILEY, J. T., Mixed-Phase Icing Conditions, *Proc. Ninth Conference on Aviation, Range, and Aerospace Meteorology*, 4.11, pages 274–279, September 2000.

MEAD, J., PAZMANY, A., GOODBERLET, M., *Evaluation of technologies for the design of a prototype in-flight remote aircraft icing potential detection system,* DOT/FAA/AR- 98/72.

The role of aerial photographs in the interpretation of Iron Age communities in eastern Scotland

Amanda Brend

Orkney College Geophysics Unit/University of Durham

Abstract

Aerial photographs are a valuable means for exploring Iron Age settlements and landscapes across Scotland, but have been largely overlooked as an interpretive media. Cropmarks in eastern Scotland enable examination of Iron Age life at different scales, complementing excavated information. Detailed analysis of these sites provides alternative ways of looking at the past, but the aerial resource remains underexploited. Using a case study from the Lothians and Fife, this paper argues that the aerial material for eastern Scotland can contribute directly to our understandings of the later prehistoric period in eastern Scotland.

Keywords Cropmarks; Iron Age; Regionality; Eastern Scotland

Introduction

Aerial survey has vastly increased known distributions of prehistoric sites through the discovery of cropmarks in lowland areas across Britain throughout the twentieth century. Air photographs are a valuable resource for the exploring Iron Age settlements and landscapes across Scotland as they provide core settlement data for the eastern lowlands. Although aerial photographs are important for locating later prehistoric landscapes in the plough-zone, they are largely overlooked as an interpretive media. This paper will examine the current role aerial photography plays in interpreting past landscapes in eastern Scotland (Figure 1) by providing an overview of its aerial resource, from the Late Bronze Age (1300-800BC) to the Roman Iron Age (50BC-AD400). It will use a case study from the Lothians and Fife to illustrate the contribution air photographic material can make to interpretations of later prehistoric landscapes.

Aerial photographs are used mainly as a tool for site discovery and as illustrative devices to accompany text in publications. Part of their current marginalisation stems from the two-dimensional nature of cropmark sites. Whereas excavation provides three-dimensional information and dating evidence, cropmarks lack stratigraphic depth and tight chronological control. Archaeologists rely on morphological characteristics and typological comparison to provide chronological and cultural meaning for cropmarks when there has been no excavation. In cropmark-rich areas where programmes of commercial- and research-based excavations provide them with a chronological and social context, aerial and non-aerial archaeologists remain reluctant to give them broader social/historical contexts. Chronological uncertainty still seems to be a significant hindrance to interpretation. Some archaeologists (e.g. Cowley forthcoming) argue cropmark sites must be systematically mapped and interpreted before we can create robust regional settlement frameworks. However, as there are financial and time limits to the number of cropmarks that can be excavated, and as they are being steadily destroyed, we cannot continue to sideline this important resource.

Aerial photographs are a very different media to traditional excavated material with which to explore past landscapes. They should not be viewed as *subsidiary* to excavation, but *complementary* to it. Cropmarks are often viewed as less glamorous when compared to extant monuments found in other parts of Scotland. Conventionally, they have been omitted from accounts of the Scottish Iron Age in favour of regions abundant in upstanding sites, or have been incorporated within generalised interpretations. Part of the current stasis in which aerial archaeology finds itself derives from difficulties in making an 'interpretive leap' between what we see on an aerial photograph and what it means in terms of past actions and activities. If we think about cropmarks as an outcome of *past social practices*, we immediately bridge this gap; these practices provide a historical context for the emergence of particular traditions and kinds of site recorded as cropmarks. We can move away from the 'detached' perspective the aerial view apparently promotes, and for which it has received criticism from interpretative archaeologists (e.g. Thomas 2001).

Large-scale transcription and mapping programme of cropmark sites in England through English Heritage's National Mapping Programme (NMP) is vital for contextualising and disseminating aerial material over large areas. In Scotland, thousands of photographs of cropmark sites have accumulated through half a century of air reconnaissance, conducted mainly by the Royal Commission on the Ancient and Historical Monuments of Scotland (RCAHMS). Mapping of cropmark sites and their incorporation with GIS is under way in Scotland, but the majority remain unmapped. The publication of RCAHMS's regional surveys (e.g. southeast Perth volume (1994)) is invaluable for disseminating this information more widely, but much of the RCAHMS information and that produced via the NMP, awaits explicit interpretation and integration within period-specific volumes.

Current understandings of Iron Age landscapes have benefited from recent synthetic regional accounts produced by period specialists (e.g. Davies 2007; Giles 2007). Through detailed analysis of cropmarks, archaeologists can address key themes in contemporary British Iron Age studies, such as regionality and the role of enclosure (e.g. Davies 2007; Giles 2007). We need to build methodologies for cropmark interpretations that derive from period specialists' input and bring a balance to interpretations focusing largely on taphonomic aspects. These methodologies should not have to rely upon the results of extensive excavation in any particular region; this approach would leave hundreds of cropmark sites with little contextual information or interpretation beyond basic morphological classification. This situation exists in many Scottish regions, such as Fife, where interpretation has stagnated due to the relatively small handful of excavated sites.

This paper will address some of the issues relating to cropmarks in eastern Scotland. A case study, focusing on two regions either side of the Firth of Forth estuary, the Lothians and Fife, will explore potential avenues for dealing with this underexploited resource. The aerial photographs for these regions provide unique broad-brush views of the landscape and can contribute directly to concerns within British Iron Age studies, whilst bringing cropmarks back from the sidelines in archaeological interpretation.

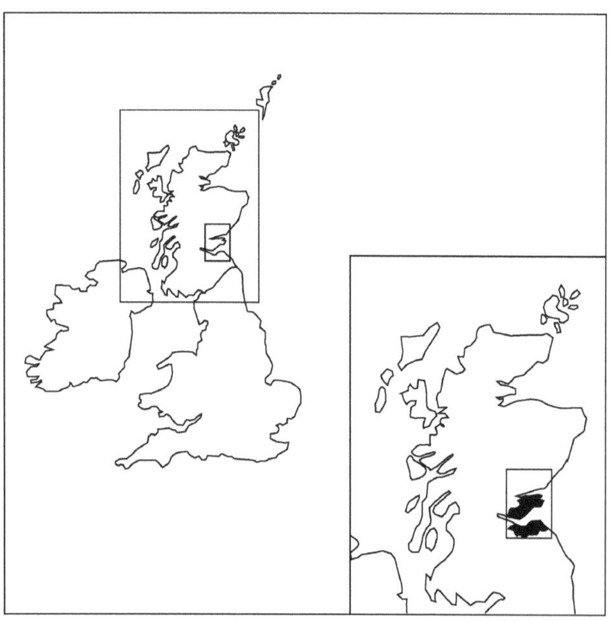

Fig.1. General location map of the United Kingdom with eastern Scotland, the Lothians and Fife demarcated

Eastern Scotland: an overview

Eastern Scotland (Figure 1) encompasses the fertile lowlands of the Lothians, Fife, Angus and Perthshire. The area is divided by two major estuaries, the Firth of Forth and the Tay Estuary. The majority of Iron Age sites in eastern Scotland survive as cropmarks, characterised through the appearance of a variety of ditched enclosures, spreads of roundhouses, and large-scale land divisions, supplemented by numerous earthwork sites. We can examine of Iron Age life at a number of different scales, which both complement and enhance information gathered from excavation.

Dating Cropmarks

Cropmarks are notoriously difficult to date from air photos alone. Programmes of developer-funded, and research-based, excavation (Driscoll 1997; Finlayson *et al* 1999; Haselgrove. forthcoming; Lelong & MacGregor 2007; Watkins 1988) have been carried out in parts of east and northeast Scotland since the 1970s. Although there has been less sustained developer-funded work in Fife, Angus and Perthshire than in the Lothians, the information we do have provides a huge 'back-catalogue' of comparative data and robust chronological framework with which to contextualise and inform interpretations of cropmark data.

Extrapolating information from excavated sites for interpreting cropmarks is not a one-way street. The perspective provided by aerial photography provides a broader context for information we retrieve on the ground. Through integrated methods we can build a detailed, intimate picture of Iron Age life in eastern Scotland. Weaving together critical re-evaluations of traditional morphological analyses of aerial photographic information with excavated evidence and embodied approaches (e.g. Giles 2007) allows exploration of the *practices* behind the features recorded as cropmarks. This approach also highlights the array of different tempos and scales of human life that existed during the Iron Age, from large-scale settlement patterns to small-scale acts of deposition.

Enclosures/Enclosed Settlements

Eastern Scotland is rich in cropmark and earthwork ditched enclosures, across upland and lowland areas, which represent the remains of later prehistoric enclosed settlements. Cropmark and earthwork sites display a variety of forms: curvilinear, rectilinear and irregular. Their distinctive geometric forms generally make them easy to spot from the air. They dominate the aerial record south of the Forth, where they are visible in almost all the main topographic zones. In Lothian, a large number of excavations yielded radiocarbon dates placing the construction of some of these enclosures in the Late Bronze Age, with a further *floruit* during the mid-late Iron Age. North of the Forth, excavations have dated enclosure construction from the early first millennium BC (e.g. Brown Caterthun, Angus (Dunwell & Ralston 2008), to the mid first millennium AD (e.g. Clatchard Craig, Fife (Close-Brooks 1986)). While numerous cropmark and earthwork enclosures are present in Angus, Perthshire and Fife (Davies 2007; Dunwell & Ralston 2008), their distributions are less widespread as those to the south. North of the Forth, enclosed settlement is restricted to distinctive topographic locations. Its absence in areas where open settlements abound suggests their distribution is both culturally and socially significant, not only matter of differential preservation and visibility. The majority of enclosures south of the Forth are modest in overall scale, with relatively wide ditches compared to internal area. North of the Forth, lowland enclosures are smaller in overall scale, defined by much narrower ditches; they are relatively modest in terms of internal area (e.g. at North Straiton, Fife (Figure 4)). The bulk of lowland enclosures in eastern Scotland are univallate, although the presence of multivallate enclosures and palisades at a number of sites demonstrates the variety of ways in which communities defined their settlements.

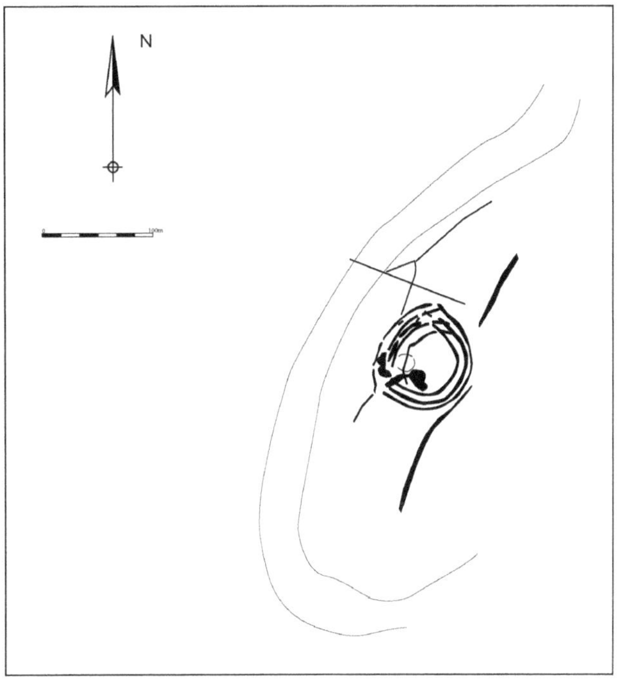

Fig.2. Transcription of the substantial multivallate curvilinear settlement at Bolton, East Lothian (after RCAHMS).

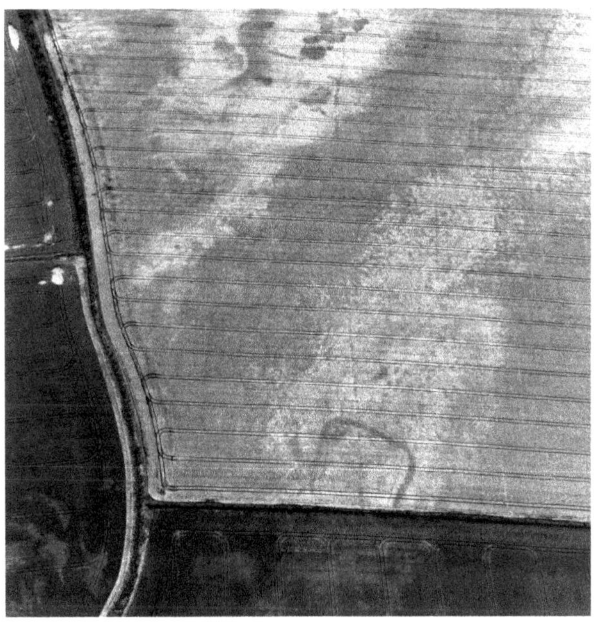

Fig. 3. Rectilinear enclosure at Congalton, East Lothian, with the enclosure. At the top of the photograph is a scooped settlement of possible Roman Iron Age date. Crown Copyright: RCAHMS. Licensor www.rcahms.gov.uk

Curvilinear sites are the most common enclosure form south of the Forth, ranging from circular to oval and 'D'-shaped enclosures. These vary in size, but are generally modest in internal area. Curvilinear sites also vary in terms of vallation; the majority are univallate, but they can have as many as five or six circuits, as at Bolton, East Lothian (Figure 2). A number of curvilinear enclosures south of the Forth contain traces of internal features and excavations have confirmed many of these sites contained one or more houses. Excavation has also shown (e.g. Haselgrove & McCullagh 2000) that activities, such as industrial/agricultural processes, were carried out within the broader enclosure space; the same can perhaps be extended to rectilinear and irregular enclosures.

A high number of palisaded enclosures have been recorded north and south of the Forth. Dates from excavations suggest their construction spans the Late Bronze Age to Roman Iron Age periods. The settlement at Dryburn Bridge, East Lothian, was constructed during the early to mid-first millennium BC (Dunwell 2007; Triscott 1982). Large, centrally-placed roundhouses are visible in several of these enclosures across eastern Scotland, while a number of ditched enclosures in the Lothians also incorporated palisades inside the main ditch. Excavations at cropmark sites, such as Fishers Road West (Haselgrove & McCullagh 2000) suggest palisaded elements often preceded the construction of more substantial ditches. This supports a trend across the region towards increasingly heavy enclosure over time, a development perhaps visible in the juxtaposed cropmarks of palisade and ditch at Congalton, East Lothian (Figure 3).

Rectilinear enclosures are also found across eastern Scotland in rectangular, square and trapezoidal forms. Their physical form, location and purpose on either side of the Forth seem quite different. Like the curvilinear sites, a number contain traces of internal buildings, such as the sub-square enclosure at Morham Mains, East Lothian, which contains several scooped structures. Excavation at rectilinear settlements in the Lothians, such as Brixwold, Midlothian (Crone & O'Sullivan 1997), Knowes, East Lothian (Haselgrove forthcoming) and Fishers Road West, East Lothian (Haselgrove & McCullagh 2000) showed that they were built during the same period as curvilinear sites. North of the Forth, rectilinear enclosures do not appear to have enclosed houses, but are often closely associated with them, situated amidst or nearby groups of unenclosed roundhouses (e.g. at Ironshill, Angus). As with curvilinear enclosures north of the estuary, rectilinear examples were modest in overall area.

Irregular enclosures are less numerous in cropmark form across eastern Scotland, tending to be restricted to prominent topographic locations, such as promontories, summits and knolls. A high percentage of them have not been subjected to the degree of plough-damage as their lowland counterparts; consequently, as with some curvilinear and rectilinear forms, the level of boundary preservation is good, and we can extrapolate something of the former physicality of cropmark enclosures. Their irregular form results from close correlation between the course of the circuits and the contours of the topography feature on which they were built; accordingly, these sites do not conform to neat classificatory schema that place primacy on morphological analysis. The intimate relationship between enclosure form and location emphasises the significance of particular locales in the landscape, a phenomenon apparent throughout eastern Scotland. Several enclosures in the Lammermuir Hills, along the southern edge of the Lothians, have a triangular form, dictated by the shape of the spurs they were constructed upon. In Angus, Fife and Lothian, promontories also formed a focus for enclosure, such as juxtaposed palisade trench and pit-alignment running through the coastal promontory at West Mains of Edzell, Angus (Davies 2007). Enclosure form appears to be less of a concern than the *act* of enclosing the location. When dealing with extant examples, it is often difficult to unpick 'natural' features from those that are human-made. The boundaries of the substantial cropmark enclosure at Kilmurdie are tightly wedded to the form of the knoll on which they were constructed, giving the site a distinctive pear-shape. Terrace-edge sites similarly incorporate distinct topography, in which enclosures are hinged onto dramatic changes from level ground to precipitous drops to ravines or watercourses.

Substantial Enclosures

A large number of enclosures across eastern Scotland can be classified as 'substantial' in scale, in terms of internal area and depth of enclosure. Either side of the estuary these enclosures share the common feature of: being situated in prominent topographic locations. Located on summits, terrace-edges, knolls and promontories, they share many features with irregular enclosures. The sites in eastern Scotland demonstrate a broad chronological range; some excavations place hillfort construction in the uplands during the Late Bronze Age (Rideout 1992), while excavations in East Lothian (Haselgrove forthcoming) revealed that substantial enclosures were also built in *lowland* areas during the same period. The latter excavations also produced evidence for activity at these sites during the later Iron Age/Roman Iron Age, demonstrating a sustained history of occupation. Excavations at other multivallate and substantial enclosures across eastern Scotland suggest substantial enclosures were in use from the early-late Iron Age, such as the terrace-edge enclosure at North Mains, Perthshire (Barclay 1983), and summit enclosure at the Brown Caterthun in Angus (Dunwell & Ralston 2008).

North and south of the Forth, the majority of large, multivallate sites enclosed, and incorporated, prominent natural features, but in the Lothians, enclosures defined by more than one ditch were constructed across a wider suite of topographic zones, including the flat coastal plain. In Fife, Angus and Perthshire, substantial enclosures were largely restricted to distinctive topographic locations, such as terrace-edges, summits and ridges (Davies 2007; Dunwell & Ralston 2008). Multiple ditches and ramparts defining some enclosures on either side of the Forth may be contemporary features; they may also represent long-term accumulation of boundaries. The superimposed boundary features at Bolton, East Lothian (see Figure 2) exemplify the possible longevity of settlement boundaries in low-lying areas.

Open Settlements

Dense distributions of open settlement are recorded throughout Fife, Angus and Tayside. These settlements are characterised by maculae (dark 'blob'-shaped cropmarks), ring-grooves and ring-ditches, occurring singly and in spreads, with no discernible enclosing boundary. Excavations either side of the Forth (e.g. at Douglasmuir, Angus (Kendrick 1995) and Easter Kinnear, Fife (Driscoll 1997) have demonstrated that open settlements have a long currency of use, spanning the Early/Middle Iron Age to the Early Historic period. As with most features on aerial photographs, it is difficult to establish the contemporaneity of open settlement features, but their spatial relationships between these features, however, suggest some chronological possibilities (see below). The open settlements north of the Forth generally take three spatial arrangements: clustered, dispersed and linear (Figures 4 and 7). In Angus and Fife, they were commonly grouped on localised knolls or ridges.

South of the Forth, fewer open settlements are recorded as discrete cropmarks. Ring-ditches are visible individually and in groups, but they are different in character to those Fife and Angus. A few examples of closely-set maculae are evident, for example at Congalton, East Lothian, (Figure 3); these probably represent the remains of intercutting scooped buildings similar to those excavated at Knowes, East Lothian, which were of Roman Iron Age date (Cowley forthcoming; Haselgrove forthcoming). Excavations have identified open settlements across the Lothians in forms less conducive to air survey, for example, the stone-built settlement at Phantassie, East Lothian (Lelong & MacGregor 2007). Open settlements were often constructed before and after periods of enclosure on the same site, creating complexity that is difficult to discern from the air, and which may conceal the presence of more open settlements than are currently known.

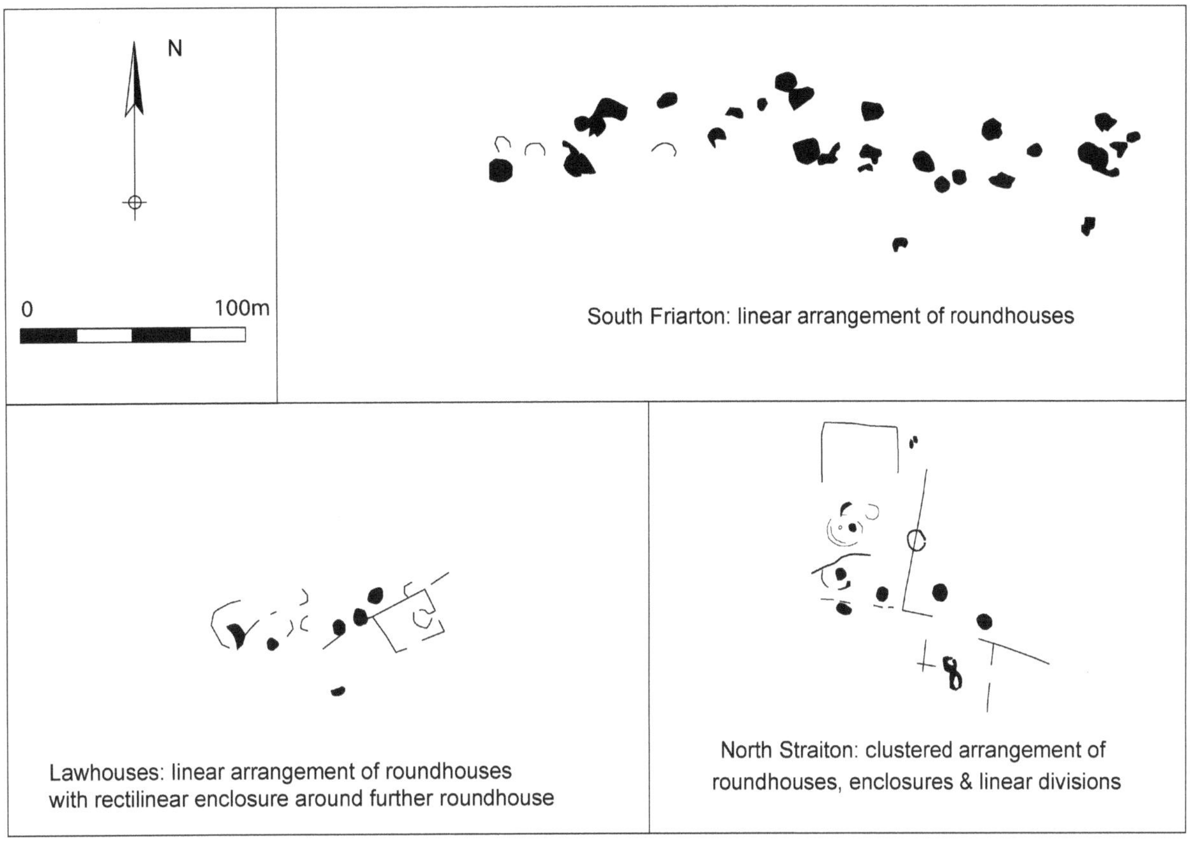

Fig.4. Transcriptions of linear and clustered open settlements in northeast Fife (after RCHAMS)

Linear Boundaries

There is a variety of evidence for past land-use across eastern Scotland. There are approximately 300 pit-alignments throughout the Lothians, formed by linear series' of pits varying in length from 80m to 350m. The majority of these alignments appear to be associated with settlement features. A high percentage is found in close proximity to the substantial enclosures of Kaeheughs, Barney Mains and The Chesters, Drem. Excavations at pit-alignments across the Lothians (Barber 1985; Rees 2002) suggest that these monuments date to the late centuries BC. Very few pit-alignments are recorded as cropmarks north of the Forth; where they occur, their meandering, irregular character is markedly different to the Lothians alignments. Excavation of a short alignment at North Straiton, Fife, produced a Neolithic date (Carter 1996). Morphologically-similar alignments in Perthshire and Angus occur in close proximity to Neolithic/Bronze Age monuments (RCAHMS 1994), suggesting an earlier construction for these sites.

Land-divisions throughout Fife and Angus comprise linear ditches, running for relatively short distances in proximity to, and among, spreads of open settlement (e.g. Ironshill, Angus (Pollock 1997) and Leuchars, Fife). The nature of these features suggests land division and agricultural activities were organised at a more intimate scale than that south of the Forth. Linear ditches may also have acted as boundaries between houses where there was no enclosing boundary on the scale of those to the south. Many settlements in Fife and Angus are situated on localised knolls and ridges and it is possible that areas of wetland formed 'natural' boundaries between groups of open settlement. Subsidiary watercourses in the Lunan Valley, Angus, may have formed both physical and conceptual boundaries between open settlements (Dunwell & Ralston 2008).

Case study: interpreting the Iron Age of the Lothians and life

The Forth estuary has traditionally been viewed as a significant physical and cultural barrier during certain periods of the Iron Age and less so during others (e.g. Macinnes 1982). The settlement patterns to either side have been summarised as enclosed settlement dominating to the south and open settlement to the north (e.g. Hingley 1992). The later prehistory of the Lothians is significantly well-developed, whereas the Iron Age of Fife, despite its rich aerial record, has been defined, socially and chronologically, by developments in the Lothians and Angus. Closer scrutiny of aerial photographs from the Lothians and Fife shows that the nature of enclosure is markedly different in both regions. The tendency to broadly characterise the settlement evidence as open or enclosed at particular periods in time denies the complexity evident in the aerial record for both regions. This variability highlights the necessity of picking out the nuances in the material and the recognition of the significance of particular topographic locations during later prehistory.

Fig.5. Transcriptions of linear pit-alignments and enclosed settlements at East Field, Inveresk in Midlothian (after Armit 1997)

Figure 6 Curvilinear settlement at Doon Hill, East Lothian, with maculae (roundhouses) possibly overlying and extending beyond its boundaries. © Crown Copyright: RCAHMS. Licensor www.rcahms.gov.uk

Intensive aerial survey of the Lothians and Fife allows us to consider broad-brush views of the later prehistoric landscape and is a powerful way to examine large-scale spatial relationships. Cropmark distributions fill in apparently 'blank' areas between earthwork sites; we now have an abundance of probable Iron Age sites across the Lothians and Fife, suggesting densely inhabited landscapes, with long histories of settlement and activity. The large-scale perspective also allows us to consider specific cropmark forms within a broader context and to think about how they relate to other kinds of monument.

In the Lothians, several clusters of settlement and land-division features are visible amongst the density of sites, which may represent large-scale groupings of communities beyond the level of house and enclosure. These clusters may reflect differential preservation and visibility from the air, with blank areas a product of areas less favourable for cropmark production. A recent motorway upgrade through East Lothian (Lelong & MacGregor 2007) demonstrated the presence of prehistoric features in 'quieter' cropmark areas. Amongst the relatively dense and uniform distribution of cropmarks across the region, the clusters still stand out; as has been argued for similar settlement clusters in the Central Tweed Valley to the south (Wise 2000), they may reflect archaeological reality rather than survey bias.

The Lothians is a densely occupied landscape of enclosed settlement with occasional discrete open settlement, interspersed with traces of large-scale land-divisions. The spread of features at East Field, East Lothian, is a stark example of this intensive activity (Figure 5). The cropmarks present a regular layout of sites, suggestive of a planned, ordered landscape. While excavations demonstrate some of these monuments were in use during the late centuries BC, they also show many of these features were produced over a more protracted period. There is considerably more time-depth to these landscapes than the aerial view presents.

During the later Iron Age in Lothian, there was a shift towards larger-scale agricultural endeavour, formalised through the construction of large-scale land boundaries. Formerly open settlements were enclosed with often substantial boundaries, entrenching communities within the landscape. This complex settlement history may exist at East Field, although we are not able to see it from the cropmarks alone. Reuse of sites can mask traces of open settlement, as they are destroyed by subsequent boundary construction and other activities. The relationship between these different settlement forms is occasionally clear from the air, for example at Doon Hill, East Lothian (Figure 6), where a series of roundhouses overlie the enclosure boundaries, continuing beyond its limits in a linear arrangement.

Compared to the Lothians, a very different landscape organisation existed in Fife, from the level of the house to the broader agricultural landscape; this difference seems particularly stark during the late centuries BC. Unenclosed settlements were constructed over many centuries across the peninsula, with concentrations in the northeast. Enclosure was generally small-scale, interspersed among, or associated with, these open settlements (e.g. at Lawhouses (Figure 4)). These enclosures may have defined discrete areas for agricultural/industrial activities away from the main

Figure 7 Open settlement at Ravenshall, Fife, with ring-ditch, ring-groove and scooped buildings clustered together. © Crown Copyright: RCAHMS. Licensor www.rcahms.gov.uk

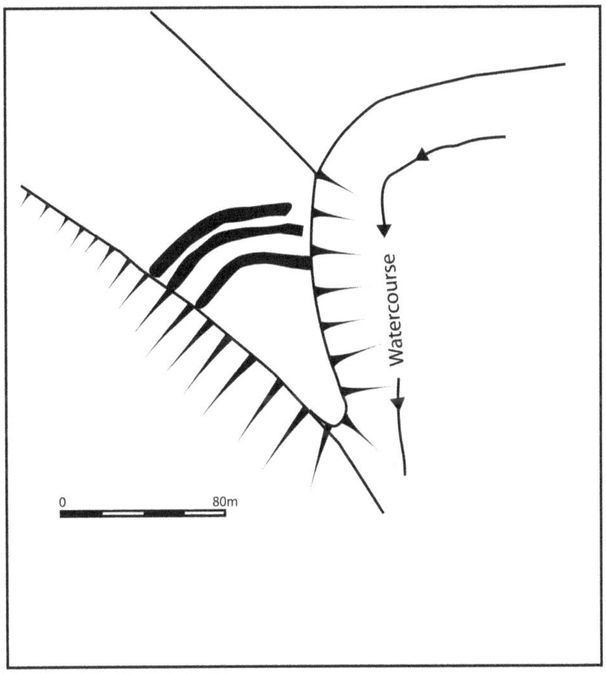

Fig.8. Promontory enclosure at Waukmill, Fife. Transcribed by the authors from RCAHMS aerial photograph.

dwellings. Groups of maculae, ring-ditches and ring-grooves forming open settlements, as at Ravenshall (Figure 7), may represent contemporary buildings. It is possible these structures were chronologically distinct: settlements such as South Friarton (Figure 4) may represent several centuries of inhabitation, with footings of individual dwellings shifting over time and space. In Fife, as in Angus and Perthshire, substantial enclosures were confined to distinctive topographic locations. At Waukmill, Fife (Figure 8), cropmarks revealed a narrow promontory closed off by three substantial ditches drawn across its neck.

The demarcation of locales through boundary construction in the Lothians is a long-standing process with roots in the Late Bronze Age (Haselgrove forthcoming). These early enclosures were an integral part of the Iron Age landscape and the social lives of Lothians' communities during the 4th to 3rd centuries BC. Their presence provides a material, historical and social context for the construction of later Iron Age enclosures across the same landscape. The re-use of these early sites during the later centuries BC underlines the continuing importance of particular places and the architectural features defining them, that is, banks and ditches. Drawing on aspects of power and place, tenure and perhaps continuity, the use of these features was extended across a fuller variety of topographic zones in the Lothians than in Fife.

North of the Forth, specific topographic locales were the focus for substantial, long-term enclosure and activity. A lack of excavation in Fife has led to poor chronological control for when these enclosures were built; they may have similarly early roots in the Late Bronze Age/Early Iron Age as seen elsewhere in eastern Scotland. The paucity of substantial enclosure in low-lying areas, where open settlements proliferate, and its restriction to prominent locales, suggests the general trend towards more heavily enclosed settlements during the late centuries BC, was a tradition specific to the Lothians. The intentions and motivations of the Fife and Lothians' communities who were involved in building their settlements were different, tied into local landscapes forms, attached histories and the social networks and relationships built up through different scales of agricultural endeavour. These themes are manifest in the diverse articulation of architectural space visible in the cropmarks and earthworks in both regions.

Conclusions

The air photographic archives held within RCAHMS are a fascinating resource through which to explore many aspects of the Iron Age in eastern Scotland. There are chronological problems when using cropmarks, but we need to provide a historical and social context for these features. Consideration of the possible relationships and scenarios that account for what we see as cropmarks can inform research aims and priorities for their future investigation. Detailed examination of aerial material and its classification using other criteria (e.g. topography), in addition to morphology, can highlight particular trends that confirm and add to current understandings of later prehistory around the Firth of Forth, despite the paucity of ground-based archaeological investigation undertaken in Fife.

Note: Figure 5 is after an illustration in Armit, I. 1997 *Celtic Scotland*, 59. London: Batsford. Historic Scotland,

References

BARBER, J. *ET AL.* 1985. The pit alignment at Eskbank Nurseries [Midlothian]. *Proceedings of the Prehistoric Society* 51, 149-66.

BARCLAY, G. J. 1983. Sites of the third millennium bc to the first millennium ad at North Mains, Strathallan, Perthshire, *Proceedings of the Society of Antiquaries of Scotland* 113, 122-281. Edinburgh: Society of Antiquaries of Scotland.

CARTER, S. 1996. A radiocarbon dated pit alignment at North Straiton, near Leuchars, Fife; in *Tayside and Fife Archaeological Journal* 2.

CLOSE-BROOKS, J. *ET AL.* 1986. Excavations at Clatchard Craig, Fife. *Proceedings of the Society of Antiquaries of Scotland* 116, 117-84.

COWLEY, D. FORTHCOMING. The Traprain environs in a regional perspective, in Haselgrove, C. (ed.), forthcoming.

CRONE, A & O'SULLIVAN, J. 1997. Excavation of a cropmark enclosure at Brixwold, Dalhousie Mains, Bonnyrigg, Midlothian. *Proceedings of the Society of Antiquaries of Scotland* 127, 387-405. Edinburgh: Society of Antiquaries of Scotland.

DAVIES, M. H. 2007. Dominated by unenclosed settlement? The Later Iron Age in eastern Scotland north of the Forth, in Haselgrove, C. & Moore, T. (eds.) *The Later Iron Age in Britain and Beyond*, Oxbow Books, Oxford, 266-285.

DRISCOLL, S. 1997. A Pictish settlement in north-east Fife: the Scottish Field School of Archaeology excavations at Easter Kinnear, in *Tayside and Fife Archaeological Journal* 3, 74-118. Glasgow: Cruithne Press.

DUNWELL, A. 2007. *Cist burials and an Iron Age settlement at Dryburn Bridge, Innerwick, East Lothian.* Scottish Archaeological Internet Report 24.

DUNWELL, A. & RALSTON, I. 2008. *Archaeology and Early History of Angus.* Stroud: Tempus Publishing.

FINLAYSON, B., COLES, G., DUNWELL, A. & RALSTON, I. 1999. The Angus and South Aberdeenshire Field School of the Department of Archaeology, University of Edinburgh, in *Tayside and Fife Archaeological Journal* 5. Glasgow: Cruithne Press.

GILES, M. 2007. Good fences make good neighbours? Exploring the ladder enclosures of Late Iron Age East Yorkshire, in Haselgrove, C. & Moore, T. (eds.) *The Later Iron Age in Britain and Beyond*, 235-249. Oxford: Oxbow Books.

HASELGROVE, C. *ET.AL* 2001. *Understanding the British Iron Age: an agenda for action.* Wessex Archaeology. Cromwell Press.

HASELGROVE, C. FORTHCOMING *THE TRAPRAIN ENVIRONS PROJECT: FIELDWORK AND EXCAVATIONS 2000-2004.* Edinburgh: Society of Antiquaries of Scotland Monograph Series.

HASELGROVE, C. & MCCULLAGH, R. 2000. *An Iron Age Coastal Community in East Lothian: The Excavation of Two Later Prehistoric Enclosure Complexes at Fishers Road, Port Seton, 1994-5.* Midlothian: STAR Monograph 6.

HINGLEY, R. 1992. Society in Scotland from 700 BC to AD 200, in *Proceedings of the Society of Antiquaries of Scotland 122*, 7-53. Edinburgh: Society of Antiquaries of Scotland.

KENDRICK, J. 1995. Excavation of a Neolithic enclosure and an Iron Age settlement at Douglasmuir, Angus, in *Proceedings of the Society of Antiquaries of Scotland 125*, 29-67. Edinburgh: Society of Antiquaries of Scotland.

LELONG, O. & MACGREGOR, G. 2007. *The Lands of Ancient Lothian: Interpreting the Archaeology of the A1.* Edinburgh: Society of Antiquaries of Scotland.

MACINNES, L. 1982. Pattern and Purpose: the settlement evidence, in Harding, D.W., *Later Prehistoric Settlement in South-East Scotland*, Edinburgh: Department of Archaeology, University of Edinburgh Occasional Paper 8.

POLLOCK, D. 1997. The excavation of Iron Age buildings at Ironshill, Inverkeilor, Angus', in *Proceedings of the Society of Antiquaries of Scotland* 127, 339-58. Edinburgh: Society of Antiquaries of Scotland.

RCAHMS 1994. *South-east Perth: an archaeological landscape.* Edinburgh.

REES, A. 2002. A first millennium ad cemetery, rectangular Bronze Age structure and late prehistoric settlement at Thornybank, Midlothian, in *Proceedings of the Society of Antiquaries of Scotland* 132, 313-55. Edinburgh: Society of Antiquaries of Scotland.

RIDEOUT, J.S. 1992. *Hillforts of Southern Scotland.* Edinburgh: AOC (Scotland) Limited.

THOMAS, J. 2001. Archaeologies of Place and Landscape, in Hodder, I. (ed.), *Archaeological Theory Today*, 165-86. Cambridge: Polity Press.

TRISCOTT, J. 1982. Excavations at Dryburn Bridge, East Lothian, in Harding, D. W. *Later Prehistoric Settlement in South-East Scotland*, 117-124. University of Edinburgh, Department of Archaeology, Occasional Paper No. 8.

WATKINS, T. 1988. North Straiton, Balmullo, NE Fife, *University of Edinburgh Department of Archaeology Annual Report*, vol.34 No.9.11

WISE, A. 2000. Late prehistoric settlement and society: recent research in the central Tweed valley, in Harding, J. & Johnston, R. (eds.), *Northern Pasts: Interpretations of the Later Prehistory of Northern England and Southern Scotland*, 93-100. BAR British Series 302. Oxford: BAR Publishing.

Historic aerial photographs for archaeology and heritage management

Dave Cowley and Lesley Ferguson

Royal Commission on the Ancient and Historical Monuments of Scotland, Edinburgh, UK

Abstract

Historic aerial photographs are an invaluable resource for archaeological studies and heritage management. Against this background of change, the view of the past from the 1940s and earlier is increasingly valuable, both as a resource to explore areas that through urban expansion or vegetation change have since been destroyed or obscured, and as a means of assessing changes in condition and context. The value of these images as reservoirs of archaeological, economic and social knowledge, and as records of events and changing landscapes, will be illustrated through case studies, drawing on imagery held in The Aerial Reconnaissance Archives at RCAHMS, Scotland.

Keywords Historic aerial photographs; landscape change; aerial photograph archives.

Introduction

Around the globe, the second half of the twentieth century has witnessed some great changes to environment and landscape: the extent of urban areas has increased dramatically, vegetation cover has been altered and agricultural regimes have often become increasingly extensive and mechanised. These changes have directly impacted on the archaeological resource and on the built heritage in general. Urban redevelopment has destroyed historic centres, sprawling towns and cities have covered ever greater areas of ground and changes in land use regimes have destroyed ancient monuments or otherwise covered them up, for example in the spread of scrub vegetation. Against this background of change, any record of both urban and rural landscapes from the mid twentieth century has become increasingly valuable. Aerial imagery in particular has an established track record as a major source of archaeological, economic, historical and social knowledge, and as a powerful medium for recording events and changing landscapes. This paper explores some of the issues around the use of historic aerial photographs, illustrated with examples drawn from the collection of the Royal Commission on the Ancient and Historical Monuments of Scotland (RCAHMS), which includes material from around the world.

Historic aerial photography – some terms of reference

By definition reference to 'historic' aerial photography is a relative statement, and within the scope of this paper a broad view of what constitutes historic imagery will be taken to encompass material collected since the early twentieth century and extending up to the relatively recent past. For some countries aerial imagery is available from the 1910s and 1920s, stimulated by the First World War (Bourgeois and Stichelbaut 2009), and it is more routinely and widely accessible from the 1930s and 1940s onwards. However, the picture is not uniform and in many parts of the globe access to aerial photography may still be highly restricted. Implicit in these terms of reference is that the majority of aerial imagery has not been taken for archaeological or heritage purposes, though archaeologists did appreciate its value from the early years, foremost amongst them OGS Crawford (Crawford and Keiller 1928). The principal sources of aerial imagery are from sorties flown for military intelligence, both within the context of warfare and in 'peace-time', and photography commissioned to aid map making. The majority of this photography is block-coverage near vertical imagery, rather than oblique photographs which are generally targeted at specific subjects. In the United Kingdom, for example, this type of photography numbers many millions of frames and outnumbers imagery collected for archaeological or other heritage reasons many fold. This likely holds true for most countries, in particular those areas where there is no long-standing tradition of archaeological airborne reconnaissance.

Why use historic aerial photography?

In some countries aerial photography is one of the principal means of discovering and recording archaeological sites (e.g. British Academy 2001). It is certainly a highly effective means of prospecting widely and can provide a broad view of the overall distribution of sites and their general character. However, the value of aerial photography and the added dimension that imagery dating back through the twentieth century can provide should extend well beyond the simple discovery and recording of sites. Aerial imagery is key to understanding landscapes beyond the tight focus of traditional sites and monuments, allowing a broad consideration of topography, context and land-use (see Palmer and Cowley this volume). It provides a unique perspective and a textured rendering of landscapes that cannot be matched by maps. Such imagery is a source of high quality geospatial data, which can underpin accurate mapping. Indeed, such a requirement may be inbuilt to imagery captured for military intelligence and to assist in map making. Change is readily recorded – be it gradual, ongoing development or rapid, wholesale change, that comes from wartime destruction, for example. Such imagery also has a proven record in recording moments in time, and in these cases historic imagery can be the canvas on which world events are recorded. These are unique advantages not available in other data sources, such as maps, which may not always be available, may not be regularly revised and anyway are limited in what they show. However, it is vital to stress that aerial imagery is only one tool, albeit a very powerful one, in the armoury of approaches and sources that the archaeologist or landscape historian should draw on.

Successful applications of historic aerial photography have often concentrated on military remains (e.g. Hegarty and Newsome 2007; papers in Stichelbaut *et al.* 2009) drawing on imagery often captured during the dying days of conflict. There is a long-standing recognition of the archaeological value of military aerial imagery, including both conventional photography and satellite imagery (e.g. Fowler 2003; Fowler and Fowler 2005; Going 2002; Rączkowski 2004) but accessibility has remained a block on the fuller exploitation of this type of imagery. More recently, however there has been a wider recognition of the archaeological applications of historic aerial imagery with some of these results going forward to publication (e.g. Hanson and Oltean forthcoming; Cowley *et al.* forthcoming;).

Understanding the landscape – Understanding change

Approaches to understanding past landscapes are underpinned by what might be referred to as 'historical geography' or 'landscape archaeology'. That is, in very general terms, the unravelling of the processes and stages that landscapes have gone through over time, from the changing patterns of field-systems to the shifting courses of rivers and drainage systems. To undertake this type of landscape study successfully a wide range of sources must be drawn into consideration. These may include historic maps and documentary sources, but foremost amongst them are aerial photographs.

Historic aerial photography may record an enormous amount of information, from the landscape in the split second the photograph was taken, to material remains many thousands of years old. It is argued elsewhere (see Palmer and Cowley this volume) that the totality of this information requires assessment and understanding, even if the archaeologist is only interested in a small portion of it. This is vital if basic mistakes of interpretation are not to be made, for example in identifying modern agricultural marks as archaeological features. It is suggested here that the routine of working with aerial imagery that provides snapshots dispersed across periods of time, from the earlier decades of the twentieth century to the recent past, obliges the user to confront these issues, and arguably contributes to making more critical use of the observed information.

This 'time-lapsed' aerial imagery may not be available everywhere. However, by way of example, the whole of the United Kingdom was photographed in 1944-50 when the entire country was flown by the Royal Air Force (RAF) to aid the Ordnance Survey in revising mapping. Most areas of the country have also been photographed in the 1950s/1960s, with rather more intermittent coverage in the 1970s/1980s, much of which was taken specifically to aid map-making. In Scotland, the entire landmass was recorded in 1988/9 to inform land-use mapping, and since then photographs continue to be captured for a variety of purposes (Figure 1). Taken together, these 'time-lapsed' photographs provide invaluable insights into a changing landscape, but may also record complementary information that is a product of serendipitous timing in appropriate conditions for registration of archaeological features. Seen as a whole, they oblige the user to understand the changing contemporary landscape in order to make fuller sense of the past that they may be seeking to extract from the imagery. In lowland Scotland for example, this 'time-lapsed' imagery records the change of the landscape as increasing mechanisation recast field patterns and provides varying snapshots of the more distant past as coincidences of archaeologically advantageous dates of photography fortuitously record otherwise unknown sites.

Chance is always a significant factor in the utility of aerial photographs taken for other purposes to the archaeologist. A mapmaker may prefer bland lighting without too many deep shadows and highlights, exactly the characteristics that might make the imagery more useful to an archaeologist, throwing into relief the variations in topography and texture that might reveal archaeological sites. However, an appreciation by the archaeologist of the reasons for photographs being taken and of the environmental and other factors that may bear on their utility will help to identify potentially useful imagery, or conversely, to help understand why particular photographs may be less informative. In the experience of the authors, there are large numbers of photographs, including, for example, those taken by the Ordnance Survey to aid map revision, that routinely record otherwise unknown sites and monuments. Such images have been stored in the collection of RCAHMS for many years but only a small proportion have so far been examined for archaeological purposes, when their potential may at last be realised.

Moving beyond patterns of archaeological sites, usually presented as simple dots on maps, to site foci in a landscape context is one challenge. Beyond this is the difficulty of understanding dynamics in the past – how settlement and land-use systems have developed or how human activity has interacted with palaeo-geography and vegetation patterns. In the present and recent past, the understanding of rapidly changing landuse may be vital to understanding the biases in the aerial information. In Europe, for example, the last 60 years have seen widespread changes in the extent of arable

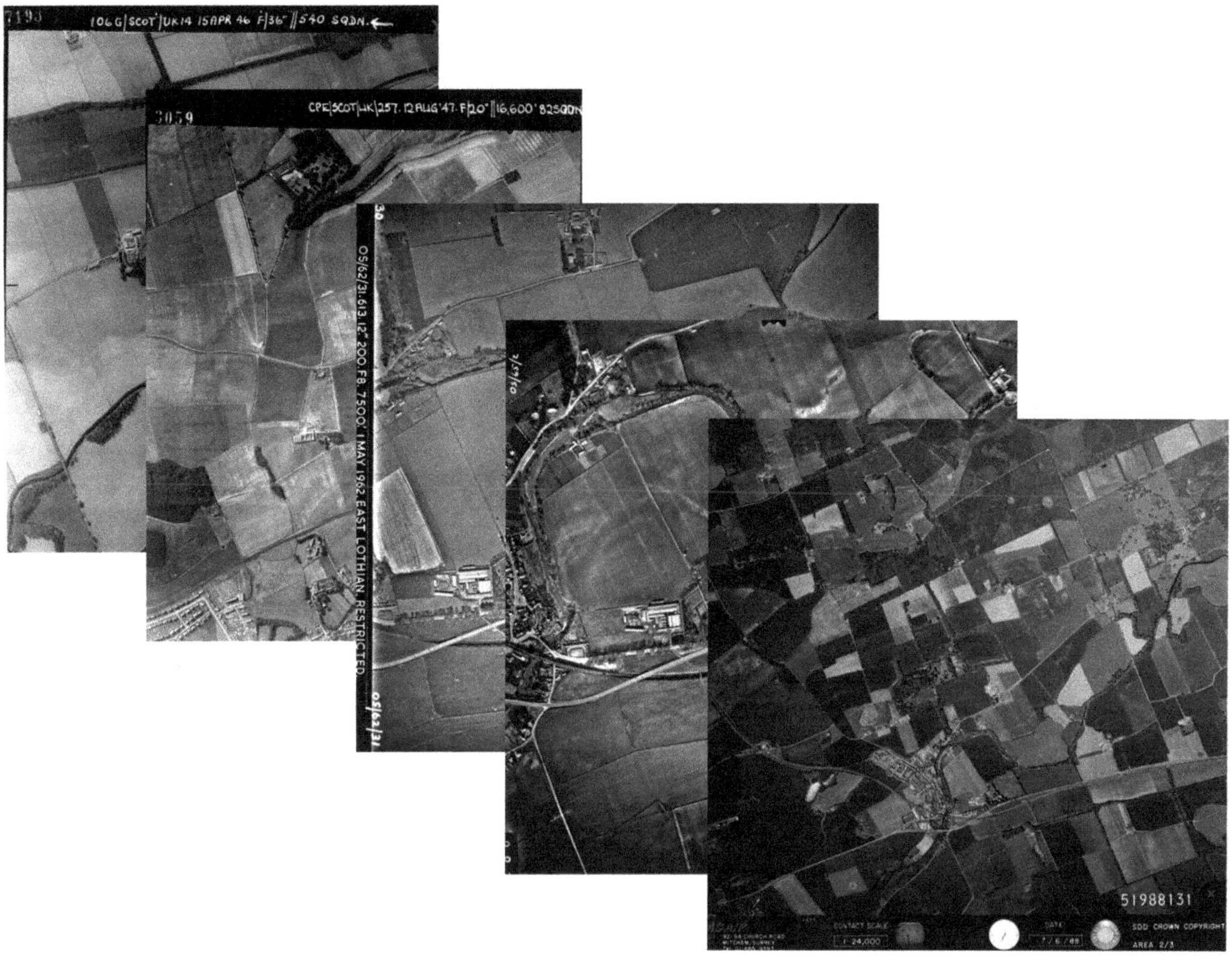

Fig. 1. Sequences of photographs spanning the second half of the 20th century are a very powerful means of understanding change in the landscape and may also record archaeological sites and monuments that have been swept away by subsequent developments, including agricultural intensification, forestation and urban sprawl. Crown Copyright © RCAHMS.

landuse and in the degree to which it has become mechanised. These are processes that impact directly on the archaeologist's ability to see particular types of site, especially buried plough-levelled ones whose presence may only be revealed by differential crop growth. Key to this is the importance of understanding the biases inherent in our data sources (e.g. Cowley 2002) if we are to use them in a robust way. This can extend from biases such as time of year and its impact on the manner in which vegetation may betray buried features (e.g. cropmarks) to understanding why a particular suite of photographs were taken – to assess land-use, to aid map making or for military intelligence.

Beyond the narrow confines of archaeological interests, the value of aerial imagery across a wide range of disciplines, from earth sciences to historical studies has been repeatedly demonstrated (e.g. St Joseph 1977; Lillesand and Kiefer 2000) and common to many of these applications has been the study and measurement of change. This is one of the major strengths of aerial imagery that spans the twentieth century and is especially evident in studies where 'time-lapsed' photography records change, for example in studying the declining extent of Antarctic glaciers (Fox and Cziferszky 2008) to patterns of landslides through time (Walstra *et al.* 2007). Also with a specific landscape emphasis, is the use of integrated suites of geospatial data, including aerial imagery, to characterise the contemporary landscape and its historic components – for example in Historic Landscape Characterisation/Assessment (Herring 2009; Rippon 2007 and papers in same volume). These approaches illustrate the power of aerial imagery to record the life of the landscape and its historic components.

Separated by over fifty years, two photographs of Silchar in southern Assam, India, amply demonstrate the power of such imagery to study change (Figure 2). Sitting on the Barak River near the border with Bangladesh, the town was photographed in September 1944 by RAF 681 Squadron, who undertook reconnaissance missions over Burma and Siam and monitored enemy ports and railways in south-east Asia. Visible are distinct patterns of land-use, areas of settlement, agriculture, a network of roads and tracks, and parkland. Now a major trading and processing centre for tea, rice and other agricultural products, there has been a

Fig. 2. Urban sprawl is a global phenomenon and this example of Silchar in India shows. In these cases historic photographs can be the only record of sites and landscapes absorbed by towns and cities. Left RCAHMS TARA JARIC 681 168 1002; right © Google earth 2009.

significant influx of people with corresponding infrastructural developments, including the construction of the Sadarghat bridge to replace the ferry. Such imagery provides a unique insight into the character of the landscape and the documentation of change. This is a global phenomenon, and is amply illustrated in Scotland, where city planners have recast many cities over the last 50 years, with vast areas of slum housing demolished, huge road networks slicing through historic urban cores and experiments in urban planning played out in new high-density, often high-rise housing schemes (see for example Cowley and Crawford 2009).

Recovering the lost

The massive rate of change in many landscapes during the twentieth century has already been commented on. The aerial photographs captured over the last 60 or more years will, in many cases, be the only record of sites and monuments that have been completely destroyed or at least covered up and effectively put beyond the reach of the archaeologist. The extent of areas no longer accessible to the archaeologist or landscape historian in this way can be quite dramatic. While this will vary from region to region, in Scotland, as in many parts of Europe, major areas of land have been impacted over the last 60 and more years. Since the 1940s urban areas have expanded dramatically, for example as 'New Towns' were built on former 'green field' sites, while residential and commercial buildings have spread out from historic town cores, in many cases increasing the urban footprint many fold. In rural areas, a drive to increase self sufficiency in timber supplies initiated in the 1910s and 1920s saw large areas of upland ground blanketed with commercial conifer plantations, many planted without any archaeological assessment or survey. Encouraged by government tax incentives in many areas the rate of planting increased in the 1970s and 1980s. Also in upland areas, during the 1970s and 1980s a government supported drive to improve the quality of grazing saw the ploughing of large areas of previously rough grazing. In these areas the earthworks of archaeological monuments routinely survived, and this programme of upland improvement undoubtedly contributed to the destruction of parts of this resource. In all these instances it is the historic aerial photography alone that provides at least some means of recording what has been lost, helping to fill significant gaps in distribution maps and our understanding of the past.

One such example comes from aerial photography undertaken by the Ordnance Survey for mapping purposes and shows the fort at Pirn, Innerleithen in 1963. With multiple ditches and the remains of at least twenty houses visible, the fort itself was surveyed by RCAHMS in 1960 but without the benefit of the aerial imagery, the associated field banks and remains were not noted. The hillside was planted with forestry in the early 1970s and the opportunity to record the cultivation remains was lost (Figure 3).

Documentation of what has been lost can also extend to large houses and their estates, often with significant designed landscapes, which have been the victims of changing economic and social conditions during the twentieth century. Designed by important architects, these houses are often of historical significance in their own right but when placed in a wider context they have influenced the development of the surrounding landscapes. In Scotland, as elsewhere, large

Fig. 3. While the Iron Age fort was recorded in some detail in 1960, the ephemeral remains of the adjacent field system (A) were not recognised at the time and have since been forested, so this photograph (1963) is the principal record of a rare, probably prehistoric, field system. Crown Copyright © Ordnance Survey.

numbers have become derelict and then destroyed, while others have been redeveloped for other uses such as hospitals or school. One example from the west of Scotland was designed in 1857 by one of Scotland's foremost architects, David Bryce. The decline and loss of Craigends House is clearly illustrated through military, mapping and modern aerial photography. Visible in 1947 RAF photographs with associated market garden, home farm and landscaped garden, some signs of decline are noticeable on 1963 Ordnance Survey photographs by which time the house was empty and the contents auctioned off. The house was demolished in 1971 and by 1988 housing has encroached on the estate.

Changing requirements for civic and domestic buildings, for retail outlets and leisure complexes have all had a marked impact on the survival of the fabric and contexts of historic buildings and areas. Urban cores have been destroyed – in wartime and by peacetime development – and recast in contemporary frameworks, often to the extent that it is difficult to envisage what was there one or even two generations ago. Here too, aerial photography provides what other sources, such as maps, cannot, with the broad view of street layout and the detail of individual buildings, all within a geospatial framework that allows accurate mapping. The value of 'time-lapsed' imagery is enhanced by the fact that urban areas are also some of the world's most dynamic locations, where wholesale change can occur very rapidly, driven, for example, by booming population and changing requirements for housing and recreation (see above).

Windows on the past

The ability of historic imagery to freeze moments of time from 50, 60 or more years ago also makes them very appealing to the general public. This adds considerably to their value as historical documents, capable of recording change in fabric, but also as records of economic and social change that may be of interest to historians, economists and geographers, amongst others. That historical events may be frozen in time certainly adds to the impact of this type of imagery. For example, images taken during the D-day landings in 1944 remain incredibly evocative, powerful documents in a population that still has many who remember the Second World War. At a more everyday scale, the RAF photograph of prefabricated buildings being delivered to a new housing estate in Arbroath, in north-east Scotland (Figure 4), is a vivid demonstration of the economic and social challenges facing the country in the years immediately after the Second World War. Images such as these appeal directly to cultural memory in contemporary populations, and can be a very immediate means of engaging communities with aerial photography and their past (e.g. Brockman and Barton

Fig. 4. Historic photographs can be powerful records of events and processes. On the left Omaha Beach on the Normandy coast is recorded in vivid detail on 6 June 1944 – D-day, a major milestone in the history of the Second World War. On the right, the event is mundane by comparison – the delivery in 1947 of prefabricated sections of emergency housing to Cairnie, Arbroath – but it illustrates Scotland's struggle to house a growing population in the grim economic period immediately after the Second World War, when many estates of prefabricated units were built. Left RCAHMS TARA ACIU 27 1744 4014; right RCAHMS RAF National Survey CPE/SCOT/UK 233 FFO, 0157.

2009). These may include images that record changes or events that an individual or a community have lived through, or that span the memories of recent generations in a population, and as such are powerful economic, historical and social documents, illustrating what had gone before, but also showing the way to what was to come.

Unlocking archives

The problems of making vast collections of aerial imagery available to potential users have been long recognised. The sheer scale of collections, which may number many millions of images, creates issues of curation, storage and delivery amongst others. At a very basic level, older film and print stocks carry their own concerns with physical survival and the curation of unstable, and potentially dangerous, media. Beyond this, all of the historic aerial photographs discussed here were not created for archaeological or heritage concerns – indeed the majority of it has a military origin. This highlights one of the key issues in its survival, in particular in countries where access to aerial photographs is circumscribed, and that is that much such imagery may not be destined for the public domain. In Britain, RAF photographs are deposited in public archives when they are declassified, but this is not the case globally – a situation where ensuring long-term preservation for the future, let alone access in the present is a major concern. Even the identification that such imagery exists has historically been a problem in many countries, for example in the former Eastern European Block of Soviet states, and may remain so in other countries.

Maximising accessibility is undoubtedly necessary if fuller use of valuable archives is to be encouraged. Working against this aim are a number of factors, chief among them the sheer scale of some collections. Beyond this, are the nature of finding aids which underpin all points of access to collections. Some potentially very valuable imagery may not have finding aids or they may be constructed for a specific purpose that has limited broader applicability. Luftwaffe photographs captured by the American's and British at the end of the Second World War have few indices with the result that it can be very difficult to even identify frames to particular countries. In the case of the Aerofilms Collection, the index is a hand-written ledger with only the most generic place name information. In this case, Aerofilms employees knew where photographs were and how to find them – they had their own system specific to their own use of the photographs. As might be expected this system does not easily translate into other means of accessing the collection, and is anyway not digital, which also limits its broader utility. Equally RAF and Ordnance Survey photographs have different indices, specific to the purpose they were designed for (Figure 5).

These examples underline the importance of understanding why a collection was created to enable access to its content. Beyond this is the massive challenge of mediating or modifying, or even creating from scratch, finding aids that will allow a broad spectrum of users to access the material. This is partly a problem of funding, to invest in creating finding aids, but it is also a challenge to collaborative working and recognition of the importance of these unique and irreplaceable records as a part of a shared heritage.

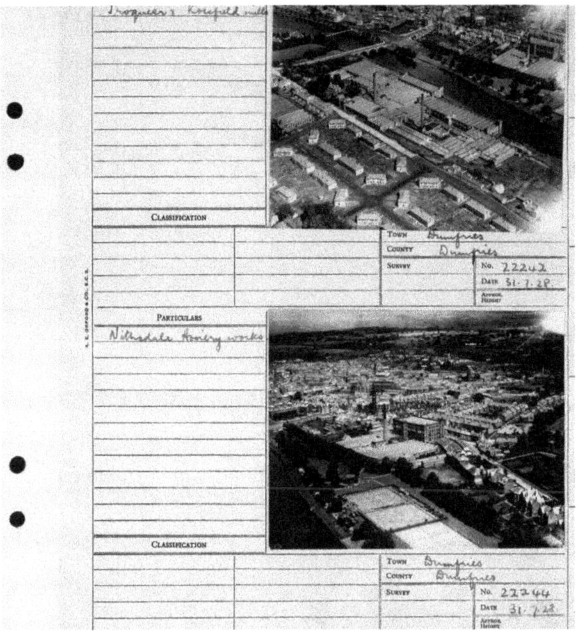

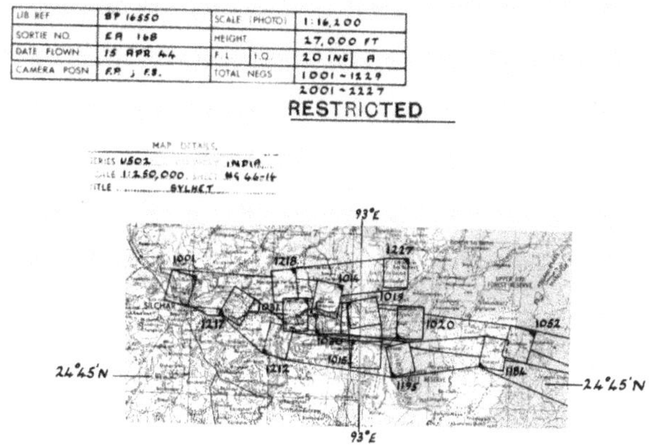

Fig. 5. Indices and other finding aids to collections invariably reflect the origin of the photographs and the purpose for which they were captured. While they may have been fit for purpose when created their value in facilitating other applications may not be so straight-forward. On the left, a page from the principle index of the Aerofilms collections records limited location information and was probably designed for retrieval by people who knew that particular system. On the right a RAF sortie plot for a part of India has limited geo-spatial information. In both cases the creation of a digital index that maximises access to the photographs can be laborious. Crown Copyright © RCAHMS.

The Aerial Reconnaissance Archives (TARA)

One of the largest aerial collections in Europe was recently transferred to the RCAHMS in Edinburgh. Formerly held at Keele University it comprises imagery from around the world dating from 1938 until relatively recently. At present the main use of the archive has been by European bomb-disposal experts as a means to locate unexploded ordnance with some limited use by historians, environmentalists and archaeologists. RCAHMS has recognised the potential within the archive, and aims to open up access to much wider audiences such as schools education and local history research where aerial imagery can contribute to the cultural memory of the recent past. Within Scotland, RCAHMS has been in the fortunate position of having had regular access to historic archives since the immediate post-war period and this has considerably enhanced the collective knowledge base. With this experience and recognising how historic aerial imagery has affected survey methodologies and research, archaeologists and historians are one of the target audiences for the archive.

Within TARA, there are four separate and distinct collections, all with huge potential. The most accessible with finding aids, comprises prints of the areas of Western Europe affected by World War II and taken by the Allies. Similar to this is the Mediterranean Allied Photo Reconnaissance Wing archive which comprises prints of southern and central Europe, including many classical sites and communities. Luftwaffe imagery discovered by the British and Americans at different locations across Germany illustrates whole areas of Eastern Europe c. 1940s and marks a crucial time in history. With incredible potential for both site discovery and analysis of landscape change, this imagery will also provide the earliest aerial views in some of the countries. The final component within TARA is the archive from the Joint Air Reconnaissance Intelligence Centre, the main provider of imagery intelligence to the UK Ministry of Defence. Dating from 1938 to 1974 this can offer different snapshots in time in areas across the world where there has been or might have been military conflict.

With limited or no finding aids for much of TARA, opening up access presents a significant challenge and the key will be to build partnerships across the globe to further research into the content and to explore funding opportunities. Enabling access to everything will take decades and may ultimately never be complete but with collaborative projects there is the opportunity to frame research questions that can inform future directions.

Conclusion

Historic aerial photography is an immensely rich source of information for the archaeologist, historian and many other related disciplines. Its use should extend beyond the simple identification of sites and monuments to informing the analysis and understanding of the past and present across landscapes from city centres to fields and wastes. Its wider application and accessibility present very real challenges to curators and archaeologists that are best answered through collaboration and recognition of the broad-based importance of these sources in recording a global landscape which has often been changed out of all recognition during our lifetimes, and those of our parents and grandparents.

References

BOURGEOIS, J. AND STICHELBAUT B., 2009. Images of Conflict: An Introduction. In Stichelbaut Bourgeois, J., Saunders, N. and Chielens, P., (eds) *Images of Conflict – Military Aerial Photography & Archaeology*. 1-12. Cambridge Scholars Publishing

BRITISH ACADEMY, 2001. *Aerial Survey for Archaeology. Report of a British Academy Working Party 1999.* Compiled by Robert Bewley. British Academy

BROCKMAN, A. AND BARTON, K., 2009. Air Photographs, Military Archaeology and Human Memory: a case study from South East London. In Stichelbaut B., Bourgeois, J., Saunders, N. and Chielens, P., (eds) *Images of Conflict – Military Aerial Photography & Archaeology*. 221-240. Cambridge Scholars Publishing

COWLEY, D. C., 2002. A case-study in the analysis of patterns of aerial reconnaissance in a lowland area of southwest Scotland. *Archaeological Prospection*, volume 9, 255–265

COWLEY, D. C., STANDRING, R. A. AND ABICHT, M., (eds) forthcoming. *From Archives to Applications: historic aerial photographs for archaeology and understanding landscape change*. Oxbow

COWLEY, D. AND CRAWFORD, J. 2009. *Above Scotland – The National Collection of Aerial Photography*. RCAHMS, Edinburgh

CRAWFORD, O. G. S. AND KEILLER I., 1928. *Wessex from the air*. Oxford

FOWLER, M. J. F., 2003. The archaeological potential of declassified KH-7 and KH-9 intelligence satellite photographs. *AARGnews* 26 (September 2003), 11-16

FOWLER, M. J. F. AND FOWLER, Y. M., 2005. Detection of archaeological crop marks on declassified CORONA KH-4B intelligence satellite photography of Southern England. *Archaeological Prospection*, volume 12, issue 4, 257-64

FOX, A. J. AND CZIFERSZKY A., 2008. Unlocking the time capsule of historic aerial photography to measure changes in Antarctic peninsula glaciers. *The Photogrammetric Record*, volume 23, issue 121, 51 – 68, Published Online: 27 Mar 2008

GOING, C. J., 2002. A Neglected Asset. German Aerial Photography of the Second World War Period. In Bewley, R. H. and Rączkowski, W., (eds) 2002. *Aerial Archaeology Developing Future Practice*. 23-30. Nato Science Series. Amsterdam

HANSON, W. S. AND OLTEAN I. A., (eds) forthcoming. *Spy in the Sky: Archaeology from Aerial Archives.* Cambridge University Press

HEGARTY, C. AND NEWSOME, S., 2007. *Suffolk's Defended Shore – Coastal Fortifications from the Air.* English Heritage

HERRING, P. C., 2009. Framing Perceptions of the Historic Landscape: Historic Landscape Characterisation (HLC) and Historic Land-Use Assessment (HLA). *Scottish Geographical Journal*, volume 125, no. 1, 61-77, March 2009. Routledge

LILLESAND, T. M. AND KIEFER, R. W., 2000. *Remote Sensing and Image Interpretation*. John Wiley & Sons

RĄCZKOWSKI, W., 2004. Dusty treasure: thoughts on a visit to The Aerial Reconnaissance Archives at Keele University (UK). *AARGnews* 29 (September 2004), 9-11

RIPPON, S., 2007. Historic Landscape Characterisation: Its role in Contemporary British Archaeology and Landscape History. *Landscapes*, volume 8, no. 2, 1-14, Oxbow

STICHELBAUT B., 2009. The Interpretation of Great War Air Photographs for Conflict Archaeology & Overview of the Belgian Royal Army Museum Collection, in in Stichelbaut B., Bourgeois, J., Saunders, N. and Chielens, P., (eds) *Images of Conflict – Military Aerial Photography & Archaeology*. 185-202. Cambridge Scholars Publishing

ST JOSEPH, J. K. S., (ED.) 1977. *The Uses of Air Photography*. John Baker, London

WALSTRA, J., DIXON, N. AND CHANDLER, J. H., 2007. Historical aerial photographs for landslide assessment: two case histories. *Quarterly Journal of Engineering Geology and Hydrogeology*, 2007, volume 40, issue 4, 315-332

Towards an improved archaeological record through the use of airborne laser scanning

Ole Risbøl

Norwegian Institute for Cultural Heritage Research

Abstract

The archaeological survey situation in Norway is rather uneven because it is largely settled farmland areas that have been the object of systematic survey campaigns. The archaeological records for forests and other outfield areas are very flawed, which makes proper cultural heritage management of these areas difficult. This paper outlines how airborne laser scanning (ALS) can contribute to an improvement in the situation. The advantages and challenges of using ALS are discussed and analysis of statistics from two completed ALS projects shows how large a percentage of the cultural remains it is possible to detect, to what degree we have succeeded with interpretation of the digital elevation models and the accuracy of remote sensing based documentation of the remains. The conclusion is that ALS can successfully be used to conduct remote sensing based surveys and documentation of cultural monuments and remains with competitive accuracy in all kinds of landscape. Forest and outfield areas which so far have been given little attention can now be surveyed efficiently and improve research and cultural heritage management.

Keywords Airborne laser scanning, archaeological survey, cultural remains, forest, outfield, slag heaps, charcoal pits, pitfalls for reindeer

Introduction

The best possible record of archaeological monuments and remains situated in the landscape is a crucial premise for reliable archaeological research and cultural heritage management. To what extent areas are surveyed and the archaeological record known varies from country to country. Coverage within the various countries is also uneven. In Norway, which is the subject of this paper, we have a fairly good – but by no means perfect – overview of visible monuments and remains in some parts of the landscape. This is primarily because of a national survey campaign initiated in 1963 and carried out more or less systematically for the following three decades covering approximately 80 % of the 430 Norwegian municipalities (Guribye and Holme 2001, 58). The result of the surveys is accessible in the national 'Norwegian National Cultural Heritage Database' which is regularly updated when new surveys are carried out.

It is important to emphasise that the national surveys to a very large extent were restricted to settled farmland areas while forest and other outfield areas were almost completely left out. This prioritising was partly the result of economic limitations but also because of lack of knowledge concerning the extent and the age of the versatile outfield exploitation. According to the Norwegian Cultural Heritage Act, all man-made monuments and remains predating the reformation in 1537 are automatically protected by law. Cultural remains like charcoal pits, pitfalls for reindeer, iron extraction sites, etc., which are common features in forest and outfield areas, were not protected by law before 1978 (Larsen and Bøe Sollund 1993). It was not until some large archaeological projects were carried out in outfield areas in the 1980s and 1990s that the extent and significance of the prehistoric resource utilisation in outfields were properly realised. Considering that only 3 % of Norwegian land is cultivated, 38 % is covered with forest and 44 % is mountainous, the disparity between surveys carried out systematically and the size of the land areas is obvious. This point can be emphasised by using the Gråfjell project as an example. In the Gråfjell survey project, which was carried out in a vast forest and sub-alpine area covering 230 km², a total of 2191 cultural features was recorded from 1999 to 2003 (Risbøl 2005, 5). Of these, only 17 were recorded within the national survey programme and another 127 in the course of smaller projects carried out prior to the main survey project (Risbøl 2000, 13). This shows clearly the size of the problem. The lack of a good record in outfield areas has been proven to result in a high degree of damage to cultural remains and their erasion especially owing to forestry activity (Risbøl 2006).

Thus the need for a better archaeological record is evident, especially where forest and outfield areas are concerned. Attempts have been made to improve the situation by use of unemployed manpower in organised survey campaigns (Bårdseng 1996) or to integrate cultural feature surveys as part of regular forest inventories (Harby 2003). Neither of these attempts, however, has succeeded in implementing archaeological surveys on a larger scale and they have now been stopped mainly owing to changes in policies and economic circumstances.

The emergence of airborne laser scanning (ALS) and its increasing application for archaeological purposes in recent years offers a remedy that makes it possible to improve the survey situation dramatically. ALS has been successfully used, mainly in Europe, to recognise and document monuments and to detect hitherto unknown ploughed-out remains on farmland (Barnes 2003; Bewley et al., 2005; Challis 2006; Crutchley 2006; Holden et al., 2002; Shell and Roughley 2004) as well as to detect cultural monuments and remains in forested areas (Devereux et al. 2005; Doneus and Briese 2006a; Gallagher and Josephs 2008; Risbøl et al. 2006; Sittler 2004). ALS is essentially a ranging device which has proved to be an efficient technique for mapping with high accuracy three-dimensional topographic features and other objects appearing on or near the ground surface over large areas. When scanning is performed, a dense cloud of points with x, y, z coordinates near and on the ground is produced. When the ground is obstructed by man-made objects like buildings or by vegetation, an emitted laser pulse may completely or partly be reflected back to the instrument from such objects. A DEM derived from the ground returns is a typical product of ALS, and owing to their high accuracy and detailed spatial resolution such DEMs may uncover structures on the ground surface, i.e. non-natural terrain features. For this method to be applicable, the remains must be visible on the surface. Thus, ALS data collected over large areas can be used for visualisation of anomalies that may indicate cultural remains. As opposed to optical satellite remote sensing and ordinary aerial archaeology using vertical or oblique photos, airborne scanning offers an opportunity to detect remains below vegetation (Devereux et al. 2005). Even though this is an obvious advantage, vegetation is not completely unproblematic and still represents a challenge to laser scanning.

In order to optimise the interpretation possibilities the quality of the bare-earth DEM is of the utmost importance. The use of proper filtering processes in order to eliminate as much vegetation as possible is essential. The compilation of better filtering techniques adapted to archaeological purposes is a way of improving ALS data instead of using customary filtering algorithms (Crow et al. 2007).

As shown by Doneus et al. (2007, 2008) another step in the direction of achieving a better interpretational platform is to bring into use full-wave laser techniques where more information is collected from the returned signals. The full-wave scanning offers the possibility to do a more thorough filtering of the data and thereby exclude a higher rate of returns from vegetation. A new scanning system called Fli-map has been developed recently and offers a high vegetation penetrating ability and a collection of higher resolution data (Corns and Shaw 2008).

Work is constantly going on to improve ALS, to adopt new techniques and to analyse the significance of the different scanning parameters. Scan frequency and angle are parameters of importance that have not been given proper attention in ALS projects studying the method in relation to the detection of cultural monuments.

In the Norwegian ALS projects carried out so far only conventional laser scanning has been used. It is the results from two of these projects that are presented here with a focus on using ALS as a tool to detect and document cultural remains in forest and outfield areas. Thus the objective of this paper is to study how ALS can contribute to an improvement in the archaeological survey situation in Norway. To what extent can ALS be used to collect quantitative survey data and ensure qualitatively good documentation of cultural monuments and remains situated in forest and outfield areas?

Methods

The study is based on the results of two ALS projects carried out respectively in Elverum, Hedmark county municipality in south-east Norway, and Tana, Finnmark county municipality in north Norway (Figure 1). The areas were chosen because of the presence of many cultural remains, the majority of which were not recorded earlier. In addition, the two areas represent two different landscape categories exploited in various ways by two different ethnic groups. The Elverum study area is a forest area dominated by conifers with some deciduous trees and is situated 500 to 560 m above sea level. The cultural remains found in this area are mainly charcoal pits and slag heaps connected to a fairly large scale bog ore based iron production in late Viking age and early Middle Ages (tenth to thirteenth century AD) (Figure 2). The scanned area in Tana is located 60 to 300 m above sea level in a sub-alpine landscape dominated by alpine birch and willow. In this landscape a large catching system for reindeer was in use by the Sami people between the fourteenth and sixteenth centuries AD. The system consists of almost 600 pitfalls lying in a row nearly 9 km long (Figure 3). Near the catching system a dwelling-site with 14 house-grounds is situated where the Sami people lived temporarily during the hunting season.

Based on the laser data bare-earth digital elevation models (DEM) were generated that constituted the base for analysis and interpretations carried out mainly by use of Quick Terrain Modeler (QTM). QTM is a 3D-modelling software developed to facilitate real-time manipulation of large amounts of 3D-data (http://www.appliedimagery.com/). QTM models allow surfing and the software allows shift of light angle and a 360° change of light direction in real-time in analysis of the models. This was the function used most often along with exaggerating the elevation (the z-value) and making digital height profiles (cross-sections) of anomalies.

The Elverum data were analysed in order to find out how large a percentage of the total stock of cultural remains in a certain area it was possible to detect by the use of ALS. Further, the same data were used to create statistical evidence of the validity of the interpretations of the detected anomalies. Finally, a selection of the cultural remains from both Elverum and Tana was measured digitally to sort out the accuracy of in-office documentation of such remains. The test followed standard procedures used for documenting pits and heaps where outer diameter, inner diameter and depth were measured on pits and length, width and height were measured on heaps.

All three steps in the computer-based analysis were followed up by fieldwork where the test results were controlled by use of conventional archaeological surveys and documentation methods.

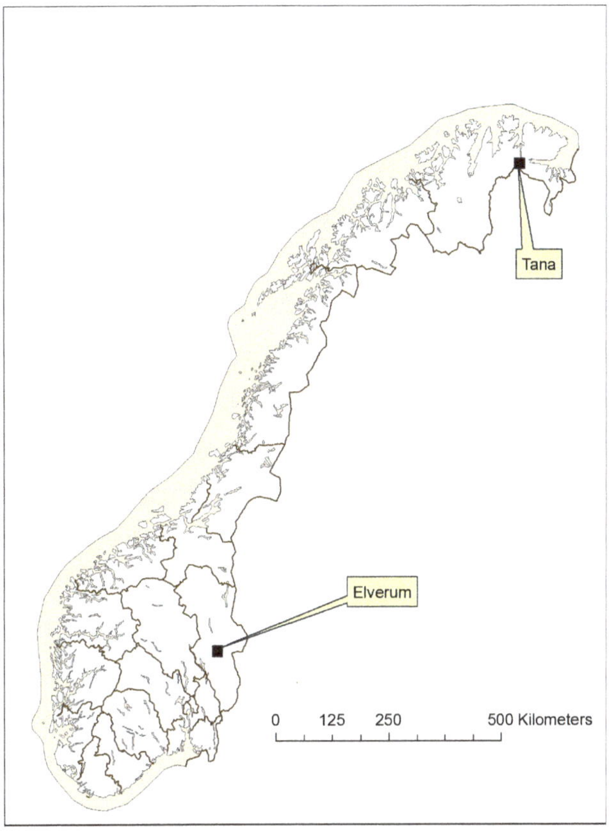

Fig. 1. The location of the two scan areas discussed in this paper.

These two areas were scanned with the parameters given in Table 1.

	Elverum	Tana
Altitude	1000 m	1200 m
PRF	70.000 Hz	100.000 Hz
Scanner frequency	47 Hz	55 Hz
Half angle survey	17°	18°
Aircraft speed	75 m/s	75 m/s
Points per square m.	4 pt.	3 pt.
Area	10 km^2	30 km^2
Degree of latitude	App. 61°	App. 70°
Date	28.06.2005	17.09.2007

Table 1. Technical specifications of the laser scanning in Elverum and Tana.

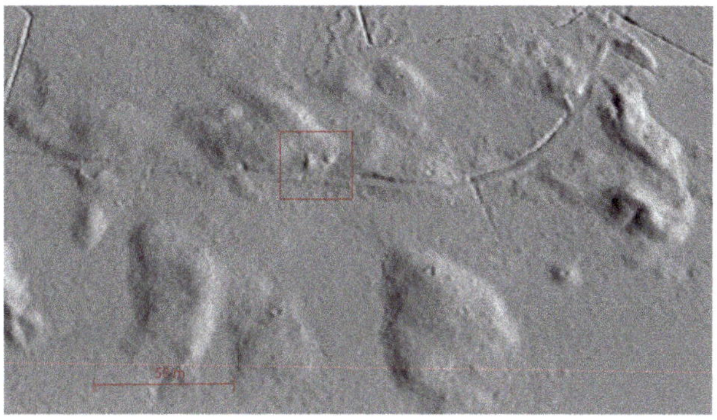

Fig. 2. A section of the Elverum DEM showing an iron production site with two slag heaps and adjacent charcoal pits scattered in the area around.

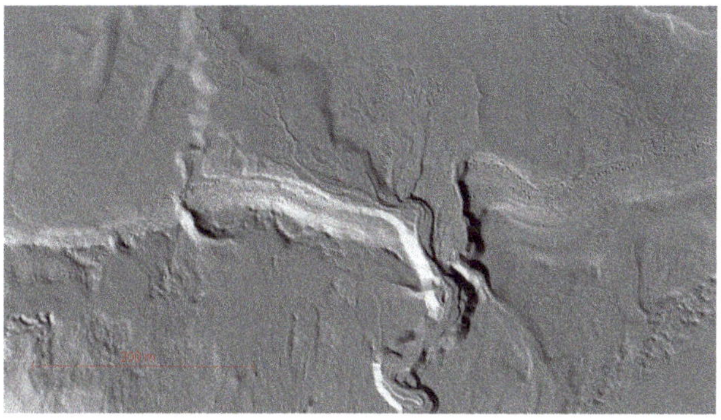

Fig 3. A section of the large system of pitfalls for reindeer in Tana.

Results

In Elverum, two smaller areas were studied with different approaches. In the first area (1.5 km^2) a plain interpretation of the DEM was carried out in ArcGIS without use of QTM. Later on, the area was systematically surveyed in the field by a search party and 62 charcoal pits and six iron extraction sites were found. As shown in Table 2, none of the six iron production sites found during fieldwork were disclosed on the DEM as opposed to 74 % of the charcoal pits. In the interpretation of the other area (0.25 km^2), QTM was taken into use in the interpretation, resulting in an increase of detection of charcoal pits to 83 %. One of two iron production sites situated in this area was found as well.

Another approach was to analyse to what degree the interpretations of anomalies were correct. In analysis of anomalies on the DEMs the interpretations made are more or less reliable. What you expect to be a cultural remain sometimes turns out to be a natural feature instead.

	Type	Number	Detected	Undetected	Percentage found
Area 1	Charcoal pits	62	46	16	74
	Iron production sites	6	0	6	0
Area 2	Charcoal pits	20	17	3	83
	Iron production sites	2	1	1	50

Table 2. The percentage of detected cultural remains in the two areas studied in Elverum.

		Number	Cultural feature	Natural feature	Percentage correct
Area 1	Charcoal pits	51	46	5	90
	Iron production sites	2	0	2	0
Area 2	Charcoal pits	17	16	1	94
	Iron production sites	1	1	0	100

Table 3. The percentage of correct interpretation of anomalies in the two areas studied in Elverum.

Area	Cultural remains	Number	Measurements	Average variance in metres	Accuracy
Elverum	Charcoal pits	20	Outer diameter	1,5	85 %
			Inner diameter	0,7	88 %
			Depth	0,1	92 %
	Slag heaps	10	Length	1,6	97 %
			Width	1,7	70 %
			Height	0,2	80 %
Tana	Pitfalls	47	Outer diameter	1,5	88 %
			Inner diameter	0,7	91 %
			Depth	0,3	72 %

Table 4. Comparison of digitally-based measurements of cultural remains with those from field work.

As shown in Table 3, this turned out to be the case in 10 % of the cases concerning charcoal pits in area 1 and for both the anomalies interpreted as possible iron production sites. In area 2, the results were somewhat better, as the one iron production site and 94 % of the charcoal pits were given correct interpretations.

Besides detecting and mapping cultural remains a documentation normally also include a verbal description plus measurements. By using a suitable soft-ware it is possible to carry out a quite accurate digital based documentation of anomalies on the DEM. This step of the analyze involved 20 charcoal pits and 10 slag heaps from Elverum as well as 47 pitfalls from Tana. All these features were measured both on the DEM and in field and the measurements were compared (Figures 4 and 5). The field-work measurements were set as the right result with which the QTM-based measurements were compared (Table 4).

An average charcoal pit measures 8 m in outer diameter, 4 m in inner diameter and is 0.7 m deep in the Elverum area. The results show that depth and inner diameter of charcoal pits were most successfully measured while outer diameter was a little less accurate but still within 85 % of the fieldwork measurements.

An average slag heap in the Elverum area is approximately 6 to 7 m long, the width is 4 m and the height 0.7 m. The accuracy of measurements of slag heaps varies from a very good result, 97 % (length), and a poorer result of 70 % (width) while the accuracy of the height is somewhere in between, with 80 %.

The average pitfall measurements in Tana are 6.4 m for outer diameter, 3 m for inner diameter and 1 m for the depth. The results resemble the accuracy of the charcoal pit measurements, except for the depth, which was significantly poorer.

The figures show that the remote sensing based documentation of cultural monuments does not deviate substantially from field work based documentation. The average of all measurements is within 85 % accuracy. The inequalities probably do not differ much from the result obtained by two different survey teams documenting the same features.

These analyses show that digital screen based documentation using data from ALS can be carried out in a manner that to a large degree satisfies the standards required for mapping cultural monuments and remains.

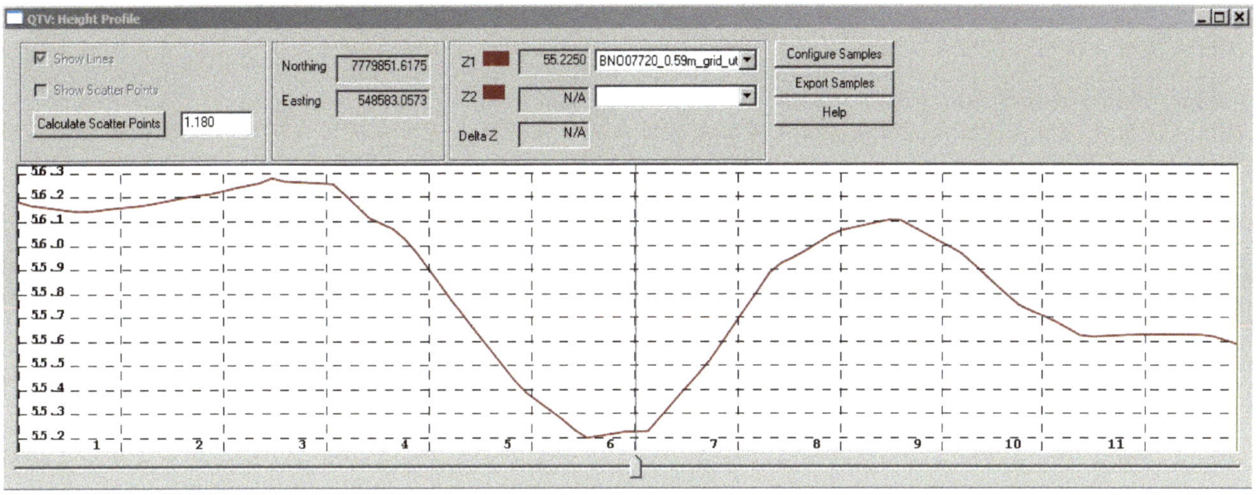

Fig. 4. A cross-section of one of the pitfalls from Tana. The image gives a good impression of the shape and makes it possible to make horizontal as well as vertical measurements.

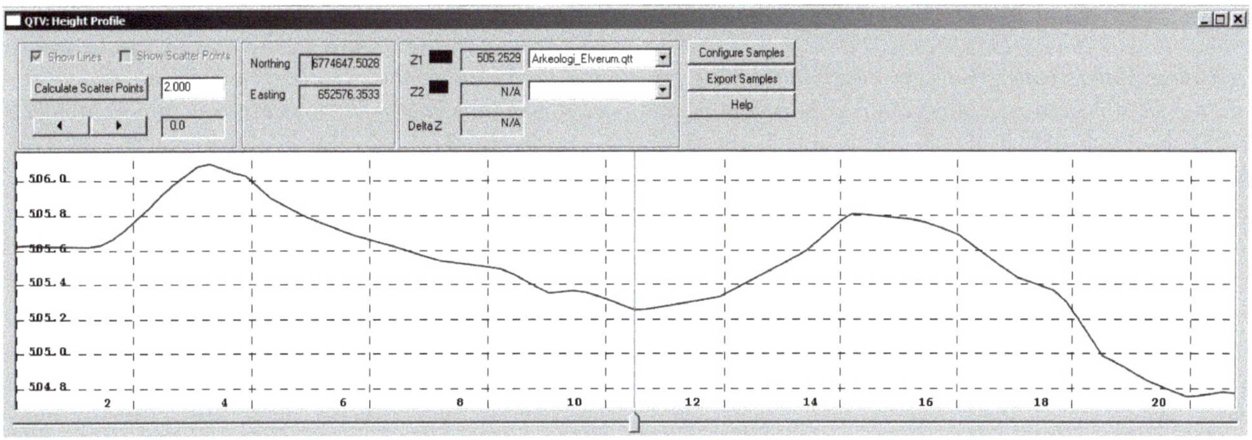

Fig. 5. A cross-section of two parallel slag heaps at an iron production site from Elverum.

Discussion

Analysis of the results from two case study areas has proved that quite a large percentage of cultural remains potentially found in an area can be detected with ALS. Even though this statement is based on a rather small-scale study the tendency is clear. The ability to detect remains and make reliable interpretations by the use of additional software developed to facilitate the handling of large laserscan data sets is proven. As many as 18 out of 22 cultural remains were found in our test area 2. Gallagher and Josephs (2008) carried out a somewhat similar project in a forested environment in Michigan, USA, where 25 of 32 anomalies (78 %) turned out to be cultural remains.

Usually a large number of anomalies are seen on the DEM and it is a sizeable challenge to sort out if they are owed to the presence of cultural or natural features. It is our experience that the size but also the shape of the remains is very decisive for the ability to detect and interpret the anomalies. Cultural remains with a regular geometric form are more likely to be detected and interpreted correctly than those with a less distinct shape that are more easily mistaken for natural features. Further, it can be mentioned that experiences from other ALS projects have proven that aligned and linear structures are the easiest to disclose (Sittler and Schellberg 2006; Crow et al. 2007). Another aspect to take into consideration is the fact that personal skills in analysing DEMs for cultural remains will surely improve with increased experience in handling such data.

When anomalies are detected and interpreted as cultural remains the archaeologist often wants to make some kind of documentation with measurements. In our study, a selection of charcoal pits and slag heaps from Elverum were measured along with pitfalls from Tana. As far as the charcoal pits were concerned, the accuracy of the inner diameter and the depth measurements were somewhat better than those of the outer diameter. This is clearly because the former measurements are of a more objective kind while it is harder to define a clear limit between the outer parts of the pit and the natural ground around it. This line is in many cases quite diffuse. Still, the result is quite good with an accuracy of 85 % for the outer diameter.

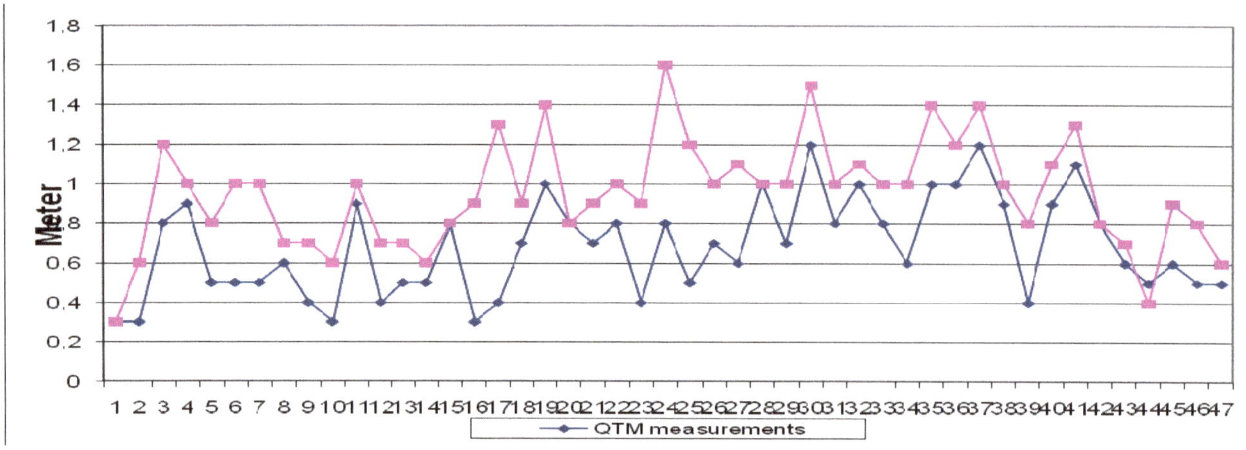

Fig. 6. The graph shows the divergence of the analysed depth measurements of the 47 pitfalls for reindeer.

The accuracy of pitfall measurements is almost the same as for the charcoal pits except for the depth measurements, which are quite poor – 72 % versus 92 %. This can probably be explained by the angle of the scan. To obtain the full depth of a pit, one is dependent on ground hits in the bottom of the pit. Owing to a combination of the scanning angle, the depth of the pit and the steepness of its sides this will not always be the case. When the DEM is created by an interpolation of laser points the pit will be given a shallower form than the reality. That this is probably the right conclusion is supported by the fact that pitfalls generally are somewhat deeper than charcoal pits and almost 100 % of the digital depth measurements are shallower than those gathered by field work (Figure 6).

The differences between field measurements and digital measurements were also quite large as regards the height and width of slag heaps. An explanation of the relatively poor result could be related to the less homogeneous form of slag heaps compared with the more regularly shaped ones like pitfalls and charcoal pits. Also, the height and width of slag heaps are more uneven than the length, which is easier to define in these long-oval shaped heaps. This means that the result of the comparison is dependent on where exactly on the object the measurements are taken. In our case, the measurements were taken in a rather random fashion.

To map the cultural remains detected is also an important part of the documentation required in a survey. The accuracy of ALS is of such a quality (approximately 15 to 20 cm vertically and 25 to 50 cm horizontally) that it would satisfy most survey standards. In any case it will outdo mapping done with ordinary GPS equipment that can only guarantee accuracy within a few metres. This also means that ALS data are suitable for quality assurance of previously mapped cultural monuments and remains, as Bewley et al. observed (2005).

As mentioned above, the ability to penetrate vegetation is one of the greatest advantages of ALS. On the other hand, the vegetational situation in areas that are scanned is of the utmost importance for the quality of the DEM. There are two main challenges connected with vegetation: its density and its height. Usually the main part of the laser returns stemming from the top of vegetation can be removed from the point cloud by use of filtering algorithms in such a way that the remaining ground returns can be used to generate DEMs based on the resisting ground hits. Sometimes, however, the density of the forest is so compact that it does not allow any laser beams to pass through to the ground. In deciduous forest the problem can be coped with by scanning at times of the year when foliage is at its lowest. Thus in Norway and other geographical areas at the same latitude the best time for scanning is early spring or late autumn, the time frame in which vegetation obstacles are at their lowest level and the ground is free from snow. The understory represents a large challenge as it is usually very dense but also because it is more difficult to separate ground points from points returned from the top of low close-to-ground vegetation when filtering the data (Crow et al. 2007). This is the case for vegetation measuring 1.5 m or less (Doneus and Briese 2006b). This represents a great challenge for those working with ALS data because many cultural remains are low, only slightly visible and easily at risk of being obscured by bushes, grass, etc.

The ALS technique is being continuously improved. The use of full-wave scanners has been tested with good results as regards disclosing cultural remains in forested areas (Doneus et al. 2007). With full-wave scanning more information is collected from the returned signals, which makes it possible to make a better distinction between low understory and bare earth than is the case with conventional scanning. This means that laser hits from scrub and logging waste can be moved from the data, resulting in a better and 'cleaner' DEM (Doneus et al. 2008). Fli-map is a new and improved scanning technique that has been developed recently, as mentioned above. The initial projects using Fli-map are promising, as it offers a high vegetation-penetrating ability and a collection of high resolution data (Corns and Shaw 2008). So far, neither full-wave scanning nor Fli-map has been tested in Norway, where the projects carried out have used the conventional technique.

The results obtained through the Norwegian ALS projects accomplished up to now have encouraged the application of this method in more projects. As regards the Norwegian forest areas, a project has been initiated to implement ALS on a national scale. The project is being carried out by the Norwegian Forest and Landscape Institute in cooperation with the Norwegian Institute for Cultural Heritage Research and the intention is to implement the ALS-based survey of cultural remains as part of the national forest inventory campaign. For almost a century the extent and state of Norwegian forest resources have been surveyed and mapped every ten to fifteen years in order to obtain statistical data for ensuring sustainable management of forest resources. The results are used for establishment of silvicultural plans for the forest properties. This mapping was initially carried out by appraisers working in the field and estimating the quantity and quality of the forest but the task has now been replaced by the collection of ALS data for the same purposes. The ongoing project is meant to provide the forest laser-scanning projects with the right parameters so that the scanning carried out for silvicultural purposes can also be used for detecting cultural remains on a national scale.

Conclusions

ALS can successfully be used to conduct remote sensing based surveys and documentation of cultural monuments and remains with competitive accuracy. This applies to all kinds of landscape, and scanning can be executed in an efficient manner. Like other techniques, ALS has its limitations and does not replace conventional archaeological survey methods. The method offers, however, an extremely valuable supplement that archaeology will benefit from in the future to an increasing degree. The parts of the landscape which so far have been given little attention can now be surveyed efficiently and important information added to the archaeological record as a base for improved research and management.

Acknowledgements

The ALS projects mentioned in this paper have been carried out with financial support from the Norwegian Directorate of Cultural Heritage.

References

BARNES, I. 2003. Aerial remote-sensing techniques used in the management of archaeological monuments on the British Army's Salisbury Plain training area, Wiltshire, UK. *Archaeological Prospection* 10, 83-90.

BEWLEY, R. H., CRUTCHLEY, S. P. and Shell, C. A. 2005. New light on an ancient landscape: lidar survey in the Stonehenge World Heritage Site. *Antiquity* 79, 636-647.

BÅRDSENG, L. 1996. *Utmarksregistrering i Hedmark.* Rapport. Utarbeidet på oppdrag fra Statens Kulturminneråd. Oslo, Statens Kulturminneråd.

CHALLIS, K. 2006. Airborne laser altimetry in alluviated landscapes. *Archaeological Prospection* 13, 103-127.

CORNS, A. AND SHAW, R. 2008. Short report: Fli-map 400 LiDAR system. *AARGnews* 36, 42-47.

CROW, P., BENHAM, S., DEVEREUX, B. J. AND AMABLE, G. S. 2007. Woodland vegetation and its implications for archaeological survey using LiDAR. *Forestry* 80,(3), 241-252.

CRUTCHLEY, S. 2006. Light Detection and Ranging (lidar) in the Witham Valley, Lincolnshire: an Assessment of New Remote Sensing Techniques. *Archaeological Prospection* 13, 251-257.

DEVEREUX, B. J., AMABLE, G. S., CROW, P. AND CLIFF, A. D. 2005. The potential of airborne lidar for detection of archaeological features under woodland canopies. *Antiquity* 79, 648-660.

DONEUS, M. AND BRIESE, C. 2006A. Digital terrain modelling for archaeological interpretation within forested areas using full-waveform laser scanning. M. Ioannides, D. Arnold, F. Niccolucci and K. Mania (eds.), *The 7th International Symposium on Virtual Reality, Archaeology and Cultural Heritage VAST, Aire-La-Ville 2006*, 155-162.

DONEUS, M. AND BRIESE, C. 2006B. Full-waveform airborne laser scanning as a tool for archaeological reconnaissance. S. Campana and M. Forte (eds): From space to place: 2nd international conference on remote sensing in archaeology (*Proceedings of the 2nd international workshop, CNR, Rome, Italy, December 2-4, 2006*). British Archaeological Review, International Series 1568. Oxford, BAR Publishing, 99-107.

DONEUS, M., BRIESE, C., FERA, M., FORNWAGNER, U., GRIEBL, M., JANNER, M. AND ZINGERLE, M.-C. 2007. Documentation and analysis of archaeological sites using aerial reconnaissance and airborne laser scanning. *XXIth International CIPA Symposium (1-6 October, Athens Greece, 2007), Athens, Greece*, 275-280.

DONEUS, M., BRIESE, C., FERA, M. AND JANNER, M. 2008. Archaeological prospection of forested areas using full-waveform airborne laser scanning. *Journal of Archaeological Science* 35/4, 882-893.

GALLAGHER, J. M. AND JOSEPHS, R. L. 2008. Using LiDAR to detect cultural resources in a forested environment: an example from Isle Royale National Park, Michigan, USA. *Archaeological Prospection* 15, 187-206.

GURIBYE, R. AND HOLME, J. 2001. Kapittel II - Automatisk fredete kulturminner. J. Holme (ed.): *Kulturminnevern. Lov, forvaltning, håndhevelse.* Oslo, Økokrim.

HARBY, S. 2003. Kulturminneregistreringer under lupen. *Fortidsminneforeningens årbok* 2003, 57-72.

HOLDEN, N., HORNE, P. AND BEWLEY, R. 2002. High-resolution digital airborne mapping and archaeology. In: R. Bewley and W. Raczkowski (eds.) *Aerial Archaeology: Developing Future Practice*. Amsterdam, Elsevier, 173-180.

LARSEN, J. H. AND BØE SOLLUND, M.-L. 1993. Fornminneregisteret ved Oldsaksamlingen. *Universitetets Oldsaksamlings Årbok* 1991/1992, 37-50.

RISBØL, O. 2006. Bevaring og forvaltning av kulturminner i skog. In: I. M. Egenberg, B. Skar and G. Swensen (eds): *Kultur - Minner og Miljøer. Strategisk instituttprogrammer 2001-2005*, NIKU - Norsk institutt for kulturminneforskning. 18, 211-221.

RISBØL, O. 2005. Registrering av kulturminner i skytefelt - kulturhistoriske resultater. K. Stene, T. Amundsen, O. Risbøl and K. Skare (eds): 'Utmarkens grøde' Mellom registrering og utgravning i Gråfjellområdet, Østerdalen. *Varia*, vol. 59, KHM - Kulturhistoriske Museum, Universitetet i Oslo. 59, 5-26.

RISBØL, O. 2000. Registrering av automatisk fredete kulturminner innenfor Regionfelt Østlandet, Åmot kommune, Hedmark fylke. Fase 2. Oslo, NIKU: 1-23.

RISBØL, O., GJERTSEN, A. K. AND SKARE, K. 2006. Airborne laser scanning of cultural remains in forests: some preliminary results from a Norwegian project. S. Campana and M. Forte (eds.): From space to place: 2nd international conference on remote sensing in archaeology (*Proceedings of the 2nd international workshop, CNR, Rome, Italy, December 2-4, 2006*). British Archaeological Review, International Series 1568. Oxford, BAR Publishing, 107-112.

SHELL, C. AND ROUGHLEY, C. 2004. Exploring the Loughcrew landscape: a new approach with airborne lidar. *Archaeology Ireland* 18(2), 22-25.

SITTLER, B. 2004. Revealing historical landscapes by using airborne laser scanning. A 3-D model of ridge and furrow in forests near Rastatt (Germany). In: Laser-Scanners for Forest and Landscape Assessment (*Proceedings of the ISPRS Working Group 8/2, Freiburg, Germany, October 3-6, 2004; International Archives of Photogrammetry, Remote Sensing and Spatial Information Sciences Volume 36, Part 8/W2*), 258-261, Freiburg, Germany.

SITTLER, B. AND SCHELLBERG, S. 2006. The potential of LIDAR in assessing elements of cultural heritage hidden under forest canopies or overgrown by vegetation: Possibilities and limits in detecting microrelief structures for archaeological surveys. S. Campana and M. Forte (eds): From space to place: 2nd international conference on remote sensing in archaeology (*Proceedings of the 2nd international workshop, CNR, Rome, Italy, December 2-4, 2006*). British Archaeological Review, International Series 1568. Oxford, BAR Publishing, 117-122.

LiDAR surveys of ancient landscapes in SW Germany: assessment of archaeological features under forests and attempts for automatic pattern recognition

Johannes Heinzel[1] and Benoit Sittler[2]

[1]University of Freiburg, Department for Remote Sensing and Landscape Information Systems, 79106 Freiburg, Germany.
[2]University of Freiburg, Institute for Landscape Management, 79106 Freiburg, Germany

Abstract

Through its capacity for obtaining high resolution measurements of surface elevations and as a remote sensing tool preserving site integrity, airborne laser scanning has the potential to spark unprecedented developments in the assessment of our cultural heritage. Until now, applications have been primarily limited to the visual detection of structures appearing in the derived digital terrain models; their capacities for automatic pattern recognition and object measurements are rarely assessed.

Based on an ongoing project documenting medieval fields, it is shown that the microrelief patterns of preserved ridge and furrow can be detected clearly and assessed metrically, even when fossilized under forests.

Presented here is the automatic assessment of the corrugated fields and other elements. This approach includes algorithms for automatic pattern and feature extraction. The procedure adopted consists of 3 general steps, including transformations of the laser derived terrain model through the extraction and filtering of line features up to the parametrization of the single fields.

A verification with manual structure delineation based on raster DTMs by using GIS has shown that under favorable conditions 80 % of the corrugated fields could be automatically extracted. Noteworthy is that even height differences of just 15 cm could be detected.

Keywords LiDAR, pattern recognition, archaeology, ridge and furrow field systems, line detection

Introduction

LiDAR altimetric applications are rapidly emerging as a highly efficient method of data collection for a variety of high-resolution topographic mapping surveys. Especially its capacity of 3D modelling of landscape features at resolutions not previously possible is causing it to become a great asset for the detection and documentation of natural or manmade structures. This potential also offers new prospects for archaeological investigations, as exemplified in recent studies of ancient field systems (Sittler 2004; Sittler et al. 2006) and other archaeological sites (Deveureux et al. 2005; Doneus et al. 2008).

As a growing number of land survey agencies and other users interested in fine scale topographic data opt for this technology to upgrade existing altimetric data, enormous datasets are becoming available for users of many disciplines, including heritage agencies. Besides improving documentations of known sites, the access to such state-wide laser data sets also promises to contribute to the detection of additional archaeological resources in new sites, especially those fossilised under forests.

For such a purpose, automatic pattern recognition could prove quite helpful in locating particular structures in LiDAR data sets. Thus far, only a very small number of studies have dealt with pattern recognition in archaeological remote sensing. One successful approach is reported by de Boer et al. (2008) who describes template matching techniques to search burial mounds on LiDAR derived DTMs. Based on the LiDAR data now available for Baden-Württemberg, such tests have now been conducted to systematically detect ridge and furrow fields systems, the legacies of medieval agricultural practices.

Test sites

The two sites considered here are located in the upper Rhine Valley, in Baden, some 30 km south of the city of Karlsruhe (see fig 1.) One site includes an exceptionally well preserved pattern of a medieval ridge and furrow field system covered by forests and revealed a great extent and good state of preservation of the microtopography in earlier terrestrial surveys (Hauger et al., 2001). The area is located on a sandy and dry flat terrace from the River Rhine near Rastatt. It includes various stands with older mixed forests (beeches, pine trees and Norway spruces), as well as areas with denser young stands. These corrugated fields preserved under woodland display height differences of 30 to 50 cm, often hardly visible from the ground due to obscuring underwood vegetation.

In contrast to these ancient fields that returned to forest more than 3 centuries ago, the second site near Illingen (north of Rastatt) is also part of the floodplain and decoupled from the Rhine river by levees that have remained under cultivation, with modern practices having largely leveled these former system of ridge and furrow. Besides arable fields, it also includes some orchards and grassland.

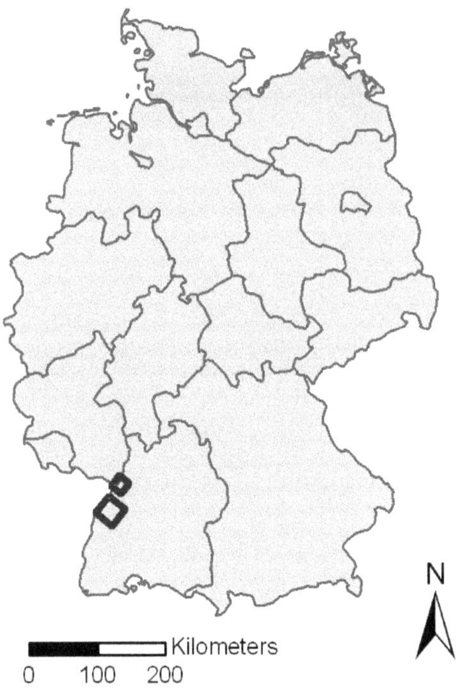

Fig.1. Location of the two test sites in SW Germany.

Data

1. The Lidar data sets provided by the Land Survey Agency

The data was captured during a long term flight campaign by the Land and Survey bureau of the state of Baden-Württemberg (Landesvermessungsamt). It has adopted the airborne laser scanning technique for the purpose of upgrading existing contour maps of the entire state (Hoss, 1997; Gültlinger et al., 2001). In this respect, it was the first state in Germany to collect and comprehensively process high-resolution LiDAR elevation data for the entire state (encompassing an area of approximately 35,000 km^2). The goal was to provide comprehensive altimetry data at a resolution of approximately one meter mesh width in space and less than 50 cm in height.

The company Topscan was commissioned with the flight campaigns and delivery of data sets. The project started in 2000 and was completed by the end of 2005. Technical details of the flight missions are listed in Figure 2 Data processing scheme: The data sets provided by Topscan were first preprocessed by the services of the Land Surveying Agency. This step included the automatic filtering of ground and height points, as well as georeferencing the data into Gauss Krüger coordinate system. A further step consisted of an interactive postprocessing of this first filtering result to fully remove all objects from the bare ground. Once this was done, a digital terrain model (DTM) of the ground surface could be obtained. Data are available in tile form (square of 1km * 1km). The LiDAR points are distributed in the binary LAS format. Each point is attributed for intensity, classification, etc., according to the LAS standard specifications. The metadata distributed with the LAS file provides details on the classification.

2. Use and manipulation of these data for the purpose of pattern recognition

Because the initial terrestrial surveys of the sites failed to assess many of the ridge and furrow due to obscuring underwood vegetation, laser data from the main site were acquired from the Land Survey Agency as DSM and DTM for visualisation. In addition, some preliminary tests were conducted to manually assess geometric characteristics of the corrugated fields, which also included some cross sections (Sittler 2004).

Complementing these preliminary tests, the present approach included LiDAR raw data that cover the test sites in the form of point data of first and last pulses. From the raw data, digital terrain models (DTM) are calculated by using the active surface algorithm from Weinacker *et al.* (2004). It can be described roughly as straining a cloth from underneath up to the point cloud.

Each point has a certain magnetic force which attracts the cloth. From this the cloth builds a 2.5 dimensional surface which represents the terrain without buildings and vegetation.

Automatic assessment

Besides the visual interpretation of image data in remote sensing, there is a strong need to automate data analysis. To handle large datasets in the shortest possible time increased research is done in developing algorithm to extract relevant information. For this purpose, existing methods from pattern

Altitude	Scan frequency	Scan angle	Wave length	Strip width	Pulse repetition	Point spacing	Flight velocity
1000 m	25 Hz	+/- 20°	1,55 μm	400 m	25.000 Hz	1,5 m	80 m/s

Fig. 2. Topscan laser scanner performance parameters and specifications.

recognition are transferred to remote sensing applications, with newly developed and adapted solutions. LiDAR data is going to have an enormous impact for archaeological applications since it is possible to filter most kind of forest and therefore allow wide range surveys in areas which were inaccessible using other techniques. This let archaeological remote sensing proceed from taking pictures of of visible pattern with handheld cameras from the air. With the new technique, the areas to be photographed are not decided upon in advance, but rather large suspicious areas are scanned and the data evaluated afterwards. This uncovers objects and features not visible from above or any other perspective, some of which were not even meant to be found.

1. Method

The method to detect ridge and furrow field systems covered under forest and in open areas uses elementary operations from pattern recognition and uses these to build a new algorithm. The main procedure is subdivided into three major steps, as shown in Figure 3. First, an applicable raster image that represents the field ridges in a capable way. has to be derived and transformed from the laser raw data. Second, all line features are extracted from the image and filtered. Third and lastly, boundaries of the single fields and parameters are derived.

The program is written in HDevelop which is a programming language integrated into an own development environment from MVTec® and uses the Halcon image processing library. For pattern recognition tasks it has the important advantage that all operation results can be visualized and controlled in real time while writing the code.

1.1. Calculation of the input raster

Each of the three main steps consists of numerous single operations. Those which are relevant for the detection method and not only for technical programming reasons are explained in the following and visualized as a flowchart after ISO 5807 in Figure 5.The only input is the DTM, which was calculated from the LiDAR point cloud as described above. Our approach is to extract the single ridges as line features. To get the best possible results it is necessary to threshold the ridge region in advance. Since the height of the ridges is not constant and the terrain might be inclined, it is not possible to find a unique threshold directly from the DTM. Instead, the second derivative of the grey values in the raster is calculated, from which the minima describe the local maxima of the original DTM and the zeroes the inflexion points. So when setting thresholds to the part between zero and below on the second derivative raster, the region of the ridges are obtained (see equations 1 and 2).

$$\min(f'') \leq f''(x, y_0) \leq 0 \tag{1}$$

$$\min(f'') \leq f''(x_0, y) \leq 0 \tag{2}$$

Since the orientation of the ridges might be more in the direction of the x-axis or the y-axis of the raster, we calculate both partial derivative images, one in x-direction and one in y-direction (see Figure 4). All further calculations are done separately for each direction image until a certain point, upon which the extracted features of both directions are combined.

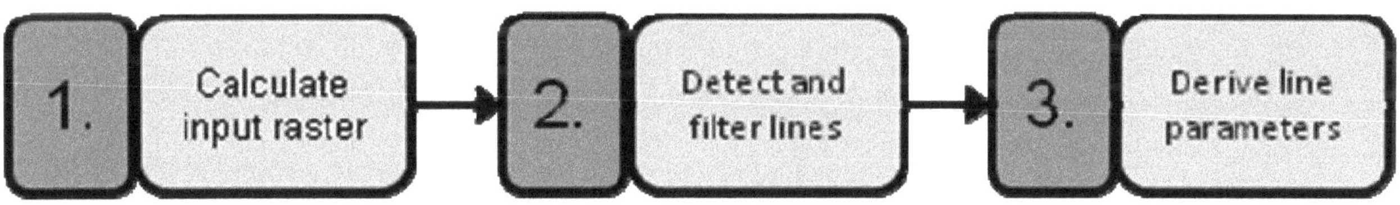

Fig. 3. Main steps for automated detection of field systems.

Fig. 4. Left: subset from DTM of reference plot 1, center: second derivative in X direction, right: second derivative in Y direction.

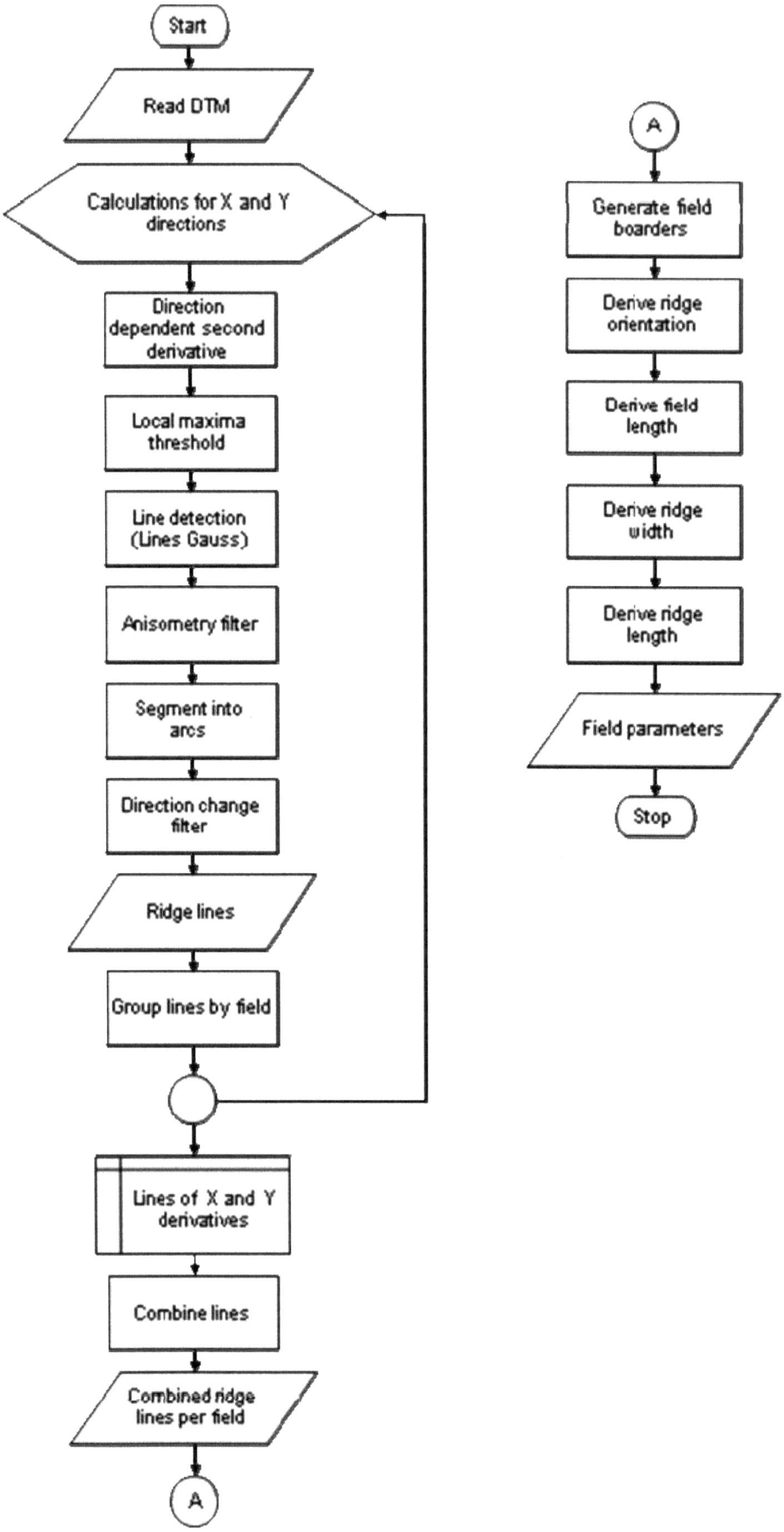

Fig. 5. Flowchart for automated ridge and furrow field system detection (after ISO 5807).

Fig. 6. Series of line filters for the x-directional derivative image. Left: all line segments obtained from lines gauss operation, center: lines filtered by anisometry, right: final lines after directional change filter.

1.2. Detection and filtering of lines

From the threshold regions all line features are extracted with the Lines Gauss operator (MVTec Software GmbH 2009). This operator is based on differential geometric properties of the raster. It calculates the second order polynomial for each pixel by using directional derivatives.

For each point of the line the lot is computed, which must be characterized by a vanishing gradient and a high curvature (Steger 1996a). A detailed description of the procedure can be found in Steger (1996b). The operator finds the sub pixel precise center line of the ridge regions. Since all ridge shaped regions are taken into account and not only the field system stripes that were searched for, a complex filtering of the extracted lines has to be done.

The two major implemented filters are concerning, first, the anisometry of the single lines and, second, the kind of direction change within a line. The anisometry of a polygon is described by the proportion of the axis of the smallest enclosing ellipse. When R_a is the length of the major axis and R_b the length of the minor axis, then the anisomtery A = Ra/Rb. Since in this case lines are used instead of polygons, the contour points of the lines are extracted in advance and treated as a point cloud for further processing. By setting an appropriate threshold to the anisometry it is possible to filter long straight lines from those that are more crimped (see figure 6).

The required ridge lines of the field system are characterized by their relatively long and straight linear form. Since the linear objects which are received as raw output are often interconnected with smaller lines they appear S-shaped. The task is to delete the shorter connecting line segments that show a significant change in direction. For this purpose, all contours are first segmented into line segments. The algorithm after Ramer (1972) is used to approximate polygons to the contours, whereas the obtained vertices represent the line segments. Ramer (1972) subdivides the contour recursively until the distance of the resulting line segments to the contour falls below a user defined threshold (Steger et al. 2008). When n is the number of contour points $p_i = (r_i, c_i)$ for i = 1, ..., n with r and c as rows and columns of the image function, then the polygon is the subset of m control points of the contour p_{i_j} for j = 1, ..., m with $m \leq n$ (Steger et al. 2008). For each line segment, the orientation within a half circle from 0 to π radians (rad) is calculated and adjacent segments are determined together with their length. From these parameters, the rule is defined that from two adjacent segments that have a difference in orientation of more than a certain threshold, the shorter one is deleted. The remaining long lines represent the ridge lines of the field system (see figure 6.).

After the extraction of the ridge lines, they are grouped by orientation intervals and neighborhood relation. Defining each group of lines as one object, the convex hull was calculated which represents the boarder of the group and approximates one parcel as well as it is automatically detectable. The last step is the combination of both the lines and the border regions of each directional derivative image into one dataset. Intersections are allocated to that region which has the higher cumulated line length within the intersection.

1.3. Derivation of line parameters

From the extracted and grouped ridge lines the following parameters are derived automatically for each parcel: Mean field length, mean field orientation, mean ridge height and mean ridge width.

The mean field length is defined as the mean length of all ridge lines within the convex hull of a parcel. If a shorter ridge line does not intersect the convex hull and is therefore overtopped in length by the neighbored lines, it is assumed that the ridge was not detected in its full length. Those lines are elongated until they intersect the convex hull in two points. With this technique, failures due to partly detected ridges or ridges with overlength are compensated.

The ridge parcels were grouped by orientation intervals of constant size. To calculate the exact mean orientation within the concerning interval, the orientation of all ridge lines within the parcels are averaged.

To receive the mean ridge height within a parcel, a cross-section through the center and orthogonal to the mean ridge orientation is computed. For this cross-section a function of the DTM heights is derived. The local extremes of this function match with the position of the ridges and furrows from the field system. Taking the mean of the subtracted neighbored minima and maxima the mean ridge height can be received. From the position of the maxima and the known resolution of the image it is also possible to calculate the mean distance between the ridges.

Finally, some corrections are performed to the ridge parcels. First, those parcels with three or less parallel lines are deleted because in most cases they relate to other linear structures, such as roads or rivers. Second, all parcels with a mean height above a certain threshold are removed. The threshold is chosen by the user and pertains to the maximal estimated height of the ridges. From practical experience with the inspected data sets, a value of 100 cm produces good results.

Results and Verification

The algorithm was run on 20 raster images with each covering 1 km^2 of area. For three raster DTMs representing different situations, a verification was conducted. The output of the program are two raster images, one with the grouped ridge lines and one with the field parcels and four ASCII tables for the concerning parameters. Figure 8 displays the results of the three reference plots, which were prepared as map in a GIS system.

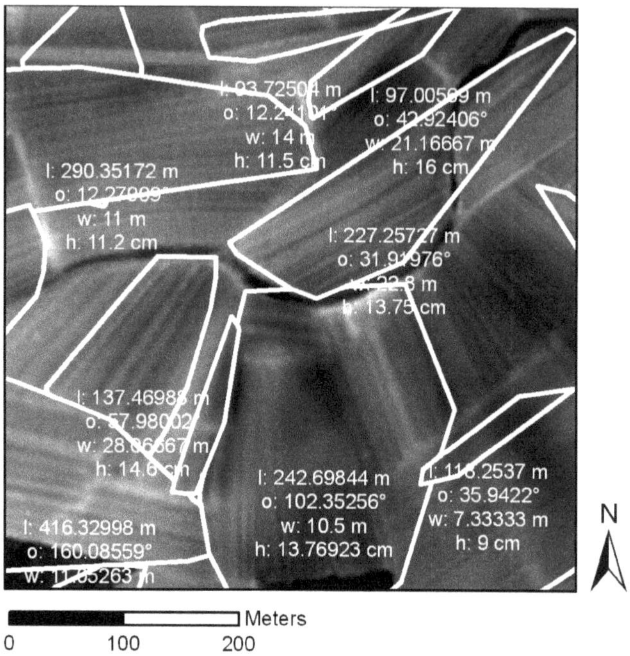

Fig. 7. Subset of reference plot 1 with each parcel labeled exemplary with four parameters: mean field length (l), mean ridge orientation (o), mean ridge width (w) and mean ridge height (h).

Figure 9. shows the four parameters derived as mean value for each field parcel. To reduce the complexity of the table, the parameters are averaged for each reference plot.

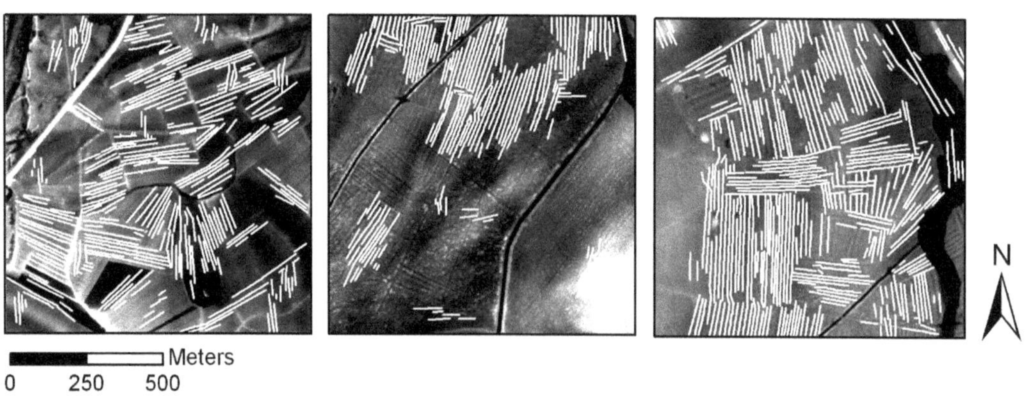

Fig. 8. From left to right: reference plot one to three with automatic derived ridge lines of the medieval field system. The borders of field parcels are not displayed for illustrative reasons.

	Plot 1	Plot 2	Plot 3
Mean field length (m)	150.91	114.49	194.97
Mean ridge orientation (°)	67.7	67.7	81.51
Mean ridge width (m)	10.32	11.51	9.08
Mean ridge height (cm)	13.32	23.4	24.7

Fig. 9. Field parcel parameters averaged for each reference plot.

	Producer's accuracy	User's accuracy
Total area (m^2)	709939.99	446960.92
Overlapping area (398169.86 m^2)	56.09 %	89,08 %
Overlapping area with correct orientation of stripes (377249.58 m^2)	53.14 %	84.4 %

Fig. 10. Verification statistics for plot 1

	Producer's accuracy	User's accuracy
Total area (m^2)	306918.21	324622.64
Overlapping area (227893.37 m^2)	74.25 %	70.2 %
Overlapping area with correct orientation of stripes (223974.30 m^2)	72.98 %	69,0 %

Fig. 11. Verification statistics for plot 2

	Producer's accuracy	User's accuracy
Total area (m^2)	656310.66	654305.14
Overlapping area (555248.82 m^2)	84.6 %	84.86 %
Overlapping area with correct orientation of stripes (550500.87 m^2)	83.88 %	84.14 %

Fig. 12. Verification statistics for plot 3

Verification of the results was conducted with help of ArcGIS from ESRI®. Based on the DTM, a manual delineation of the field system parcels with uniform ridge and furrow orientation was conducted. Then, the automatically derived parcels were clipped with the reference parcels. The orientation of the parcels, which were classified with intervals of 0.3 rad, was visually verified and wrong parcels marked in the attribute table. A visual inspection of the orientation was sufficient in this case, as it is easy to decide if the automatic extracted ridges cover exactly the ridges identifiable from the DTM. A field measurement of the ridges was not possible because they are not clearly recognizable in most cases as described above. Due to definition ambiguities of a parcel, for example if cells of same orientation which are separated by a recent forest road still belong to one parcel, only the correctly detected overall area as square metres was validated. This was done first for the whole overlapping area from the reference and the automatic result and secondly only for the overlap with correct orientation of the stripes.

For each situation, the percentage relative to the reference area, called producer's accuracy, and the percentage relative to the total area of automatic detected fields, called user's accuracy, are calculated. The results for each reference plot can be seen in the tables of Figure 10 to Figure 12.

Reference plot 1 which belongs to the test site north of Rastatt is characterized by a large total area of ridge and furrow field systems. The orientation varies strongly within a small region and the boarders of the overall field area is very irregular. The surroundings contain several other linear features, such as roads and drainage lines. This makes it the most complex situation. Another major difference compared to the other plots is the fact that this area represents a medieval field system not covered by forest. The ridge and furrow structure is therefore less preserved, which explains the low mean ridge height of 13,32 cm (see figure 9.). The difficulties are expressed in the low producer's accuracy of 53,13 % for correctly detected field area and orientation.

Plot 2, which is situated within the test site close to Rastatt, represents a more patchy distribution of medieval parcels. They are only clearly identifiable in some areas and are surrounded by unstructured forest floor. The recognition rate is quite high with 72.98 % correctly detected area.

On plot 3, which is also located close to Rastatt, most of the image is covered with field structures. The ridges and furrows emerge very clearly and the borders to the surroundings are very distinct. These conditions are represented by the best accuracy result of 83.88 % correctly detected area.

Discussion

The algorithm developed here to extract the geometric pattern of the corrugated microrelief of medieval fields has, as a whole, demonstrated its feasibility and high performance.

The verification results show that the accuracy depends on the complexity of the geographic situation, such as the landscape settings including the field system. Under good conditions more than 80 % of the ridge and furrow field systems can be automatically extracted but under worse conditions this number can fall to 50 % and maybe less. In the majority of the cases, visual inspection showed that the boundary of the automatic parcels does not exactly fit with the manually digitized parcels. This is caused on the one hands side by interruptions of parcels by other structures like roads or smaller paths, which are a boundary for human interpreters but not detected by the program. On the other hand, several parcels are only partially detected by the algorithm, so that the boundaries are smaller than the reference plot. Additionally, it must be remarked that there are still problems in distinguishing other linear structures with a high number of line segments from those lines that belong to the ancient field system. From the current state of the algorithm, the major advantage of the automation comes from the derived parameters, which are highly precise if the extracted lines match with the structure from the DTM. The high conformity of this matching can be seen from the very small difference between the given accuracies of overlapping area and overlapping area with correct orientation. This difference is less than 5 % in most cases.

An amazing result is that those very small height differences of only 13 cm in the first reference plot can be detected with the described accuracy. Taking a width of 10 m for each field stripe into account it is obvious that these low height differences can hardly been recognized in the field. The higher ridges of plots two and three (see Figure 9.), which are located under forest, prove the rule that the field systems are better conserved under forest than in open areas.

Among the few other attempts of pattern recognition of archeological features, De Boer (2005) and De Boer et al. (2008) have opted for a template matching to detect hemispherical shapes of earth mounds. As this approach did not include any verification of the quality of the results, these may however hardly be compared to ours.

Adapting the algorithm to other linear structures (irrigation systems, roads, headland ridges etc.) is also encouraged to widen its applicability to further heritage resources.

References

DE BOER, A. 2005. Using pattern recognition to search LIDAR data for archeological sites, in *The world is in your eyes – Proceedings of the XXXIII Computer Applications in Archaeology Conference*, am 21.-24. März 2005, Tomar/Portugal: 95-96.

DE BOER, A. G., LAAN W. N. H., WALDUS, W. AND VAN ZIJVERDEN, W. K. 2008. LIDAR-based surface height measurements: applications in archaeology, in Frischer, B. and Dakouri-Hild, A. (eds), *Beyond Illustration: 2D and 3D Digital Technologies as Tool for Discovery in Archaeology*, 76-84, Oxford, BAR Publishing.

DEVEREUX B.J, AMABLE, G.S, CROW,P. AND CLIFF, A.D. 2005. The potential of airborne lidar for detection of archeological features under woodland canopies. *Antiquity* 79: 648-660.

GÜLTLINGER M., SCHLEYER A. AND SPOHRER M. 2001. Flächendeckendes, hochgenaues DGM von Baden-Württemberg. *Mitteilungen des Vereins für Vermessungswesen* 48-2. 63-77.

DONEUS, M.; BRIESE, C.; FERA, M. U. JANNER, M. 2008. Archaeological prospection of forested areas using full-waveform airborne laser scanning. *Journal of Archaeological Science* 35: 882-893.

HAUGER K., RIEDINGER R., SITTLER B. 2000. Wölbäcker bei Rastatt – eine Dokumentation zur Analyse und Erhaltung überkommener Altackerkomplexe. *Kulturlandschaft – Zeitschrift für Angewandte Historische Geographie*. 10 (2) 113-118.

HOSS, H. 1997. Einsatz des Laserscanner-Verfahren beim Aufbau des digitalen Geländehöhenmodells (DGM) von Baden-Württemberg. *DVW Mitteilungen*, 44-1. 5-26

RAMER, U. 1972. An iterative procedure for the polygonal approximation of plane curves. *Computer Graphics and Image Processing*, 1, 244–256

SITTLER, B. 2004. Revealing historical landscapes by using airborne laser scanning – A 3-D model of ridge and furrow in forests near Rastatt (Germany), in *International Archives of Photogrammetry* Vol. 36 (8/W2) . 258-261.

SITTLER, B., WEINACKER, H., GÜLTLINGER M. & L. KOUPALIANTZ 2007. The potential of Lidar in assessing elements of cultural heritage hidden under forests.. In Z. Bochenek (ed.), *New Developments and challenges in remote sensing*, 539-550. Warsaw, Millpress.

KATZENBEISSER, R. U. KURZ, S. 2004. Airborne Laser-Scanning, ein Vergleich mit terrestrischer Vermessung und Photogrammetrie. *Photogrammetrie, Fernerkundung, Geoinformation*, 3: 179-187.

MVTEC SOFTWARE GMBH 2009. *Halcon/HDevelop Reference Manual*. München, MVTec Software GmbH.

STEGER, C. 1996A. Extraction of curved lines in images, in M. E. Kavanaugh and B. Werner (eds), *Proceedings of the 13th International Conference on Pattern*

Recognition, Volume II, 251 – 255. Los Alamitos, IEEE Computer Society Press.

STEGER, C. 1996B. An unbiased detector of curvilinear structures. *IEEE Transactions on Pattern Analysis and Machine Intelligenc,* 20, 113 – 125.

STEGER, C., ULRICH, M. AND WIEDEMANN, C. 2008. *Machine Vision Algorithms and Applications.* Weinheim. Wiley-VCH.

WEINACKER, H., KOCH, B. AND WEINACKER, R. 2004. TRESSVIS – A softwaresystem for simultanious 3D-real-time visualisation of DTM, DSM, laser raw data, multispectral data, simple tree and building models. In Thies, M., Koch, B., Spieker, H. and Weinacker, H. (eds) *Proceedings of the ISPRS working group VIII/2, laser-scanner for forest and landscape assessment*, 36, 90-95 Freiburg..

Application of multispectral remote sensing imagery in detection of enclosure walls of ancient settlements in South India

M.B.Rajani * and S.Settar

National Institute of Advanced Studies, Indian Institute of Science campus, Bangalore – 560012. Email: *rajanihf@gmail.com

Abstract

Organized communities for long have developed a system of protecting cities and town by means of defensive walls or fortifications, many times surrounded by moats. These features got destroyed or obliterated in course of time due to natural disasters, expansion of urban or agricultural activities, or sometimes just got decayed with time. The sub-surface composition of ruined ancient sites over time effect surface cover and traces of this phenomenon can be identified on remote sensing images taken from satellites. This paper explores the potential of multispectral imagery in identifying dried and buried moats, and possibly the adjacent fortifications. The variation in reflected/emitted spectral radiation is used to distinguish the various land surface features. A combination that includes infrared band is particularly useful because most archaeological anomalies, so far identified, present as vegetation marks, and infrared wavelengths are very sensitive to subtle variation in vegetation. The present work discusses three sites (Beluru, Halebidu and Somanathapura) in south India. They belong to Hoysala period, when a systematic town planning was followed. Traces of specific configuration of moats can be detected around each of them. This kind of multispectral remote sensing data analysis can be used to detect moats of other archaeological sites.

Key words Multispectral satellite image, remote sensing, wall, fortification, moat, Beluru, Halebidu, Somanathapura.

Introduction

A defensive wall or a fortification was used to protect a city or settlement from any kind of encroachment. Generally, these are referred to as city walls or forts. Traditionally, moats were excavated around outer wall and filled with water. They provided a preliminary barrier against attacks and made access to the walls difficult. With the decline of the ruling power and changing times, the defensive wall got into disuse. Over the time, either they collapse or the material was appropriated for fresh constructions. Some walls remain almost intact, some are in ruins and some others are completely obliterated. But the underground remnants affect the land surface. These can be recognized when the area is viewed synoptically from air or space. The effect would be more conspicuous if the surface has vegetation, which can be easily picked up by multispectral satellite data.

The energy or light that is reflected from the earth surface and detected by sensors aboard satellites comprise of wavelengths beyond the visible range. Sensors can be designed to sense visible wavelengths and wavelengths that are not visible to human eye. It is difficult to get information from continuous spectral region; also some regions may be blocked by the intervening atmosphere. In most cases observation in continuous spectral region is not essential. For this purpose sensors are designed to sense reflected/emitted spectrum by making measurements at a few selected wavelengths (Joseph 2003). The wavelength region of observation called spectral band is defined in terms of the band width. Each band can be visualized as separate layers and/or as colour composites. The variations in reflected/emitted spectral radiation between the bands are used to distinguish different features of the land surface. A combination that includes NIR (near infrared) band is particularly useful because most archaeological anomalies, so far identified, have presented themselves as vegetation marks, and NIR wavelengths are very sensitive to variations in vegetation.

Study Area

The present study has identified features on multispectral satellite imagery, indicating fortification and adjoining moat in three archaeological sites Beluru, Haledbidu and Somanathpura in Karnataka, South India (Figure-1). All these belong to Hoysala time when a series of powerful kings ruled from the 11th and 14th century AD. The architecture of Hoysala settlements followed a pattern in which a temple formed the hub of a new township, both built simultaneously. On occasions, a tank either preceded or followed establishment of temple and village (Settar, 1992, 125). The layout of the Hoysala township was influenced by the faith followed by the religious community that formed its core. All the three townships selected for study here were not under the monopoly of a single religious community. A settlement established to serve the exclusive interest of a specific community – either *Sri Vaisanava* or *Saiva* – was dominated by the temple of that sect. A village with dominant *Sri Vaishnava* community most often had the *Vishnu* temple located in its middle with *Saiva* temples elsewhere. When temples were built in an existing village, a site was chosen either in the heart of the village or on the bank of a tank or river or on the summit of a mound or hill. Generally temples of Siva, Vishnu and Jaina stand in localities

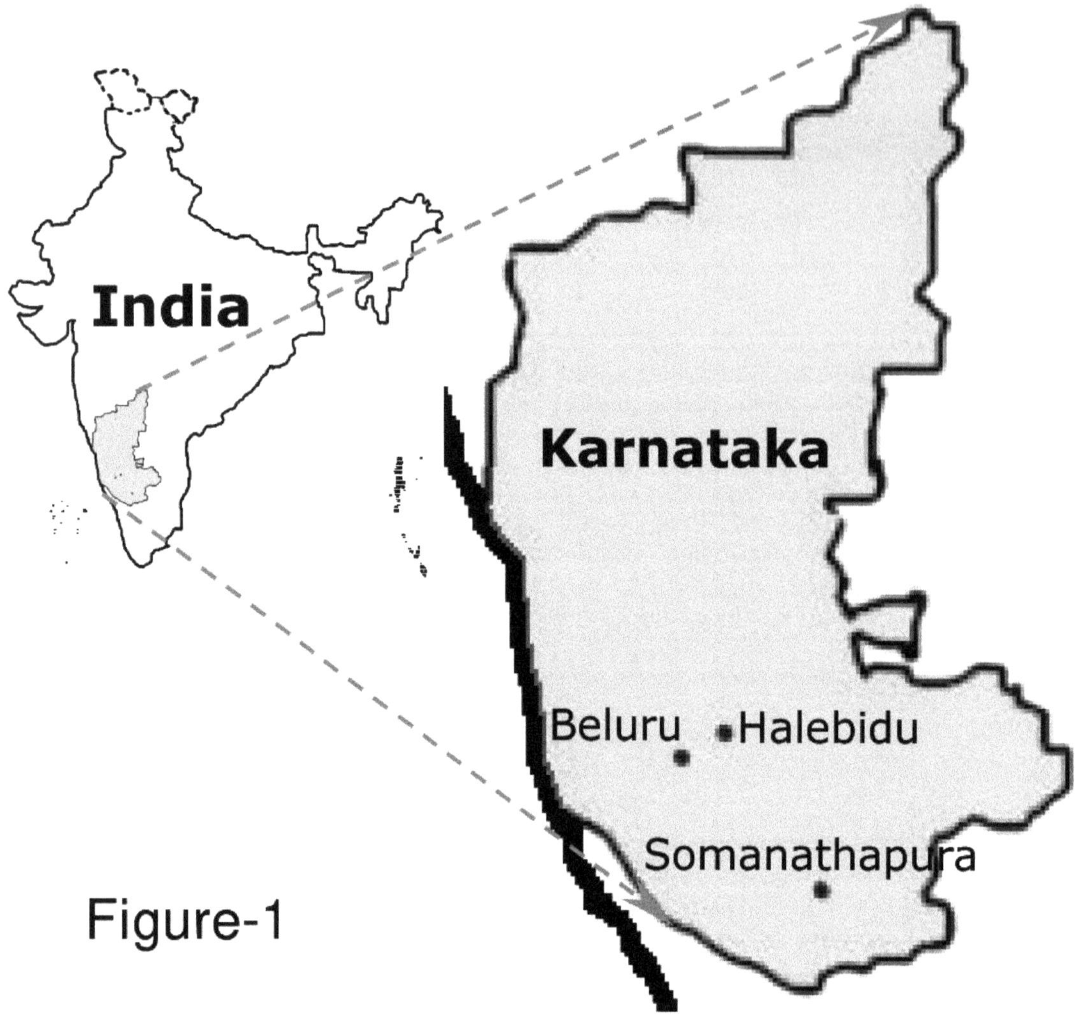

where the followers of these respective gods were concentrated, but on as many occasions the foundation of the temple provided an incentive for the followers to settle around it (Settar, 1992, 125).

Remote sensing data, analysis and interpretation

Beluru was dominated by the *Srivaishnavas* and all their temples are concentrated within an enclosure wall, which originally formed the heart of its settlement. Beluru is one of the earliest townships created under the Hoysalas and represents the characteristics of their early town planning. It was established around the Chenna Kesava temple which stood in its middle. The original plan has been considerably altered during the subsequent expansion of this township (Settar, 1992, 128). However the multispectral remote sensing images endorse the original layout. The RS imagery shows a circular signature of positive vegetation mark surrounding the temple. This indicates existence of a moat surrounding an enclosure wall.

The buried structures affect the surface vegetation by causing variation in their growth. These variations form patterns following the lines of buried features. Remote sensing images reveal these patterns which show the plan and layout of the site. These are caused by disturbance of subsoil in the past such as ditches, drainage, foundations, tanks, wells and pits. If the effect of underground archaeological features on the vegetation is favourable, then the vegetation grows taller and more copious called positive vegetation mark. This is caused by archaeological features that caused subtraction of subsoil, for example ditches for drainage, isolation or defense and also tanks, wells and pits. These become silted over the years and the greater depth of hard soil encourages roots to penetrate further to utilize moisture and nutrients that are exhausted in the upper levels (Rajani 2007). If the subsurface features effect vegetation unfavourably it results in restrained vegetation growth, where tightly packed features (such as stonewalled foundations, buried streets and solid floors) obstruct the roots causing negative vegetation mark. In either case they form patterns over the ground. These patterns are subtle and are not readily visible on ground. A synoptic view is vital to see these patterns over a large area in totality. Image processing techniques enhances features for human recognition, and analysis of satellite images leads to recognising features of significance. In the case of Beluru, the

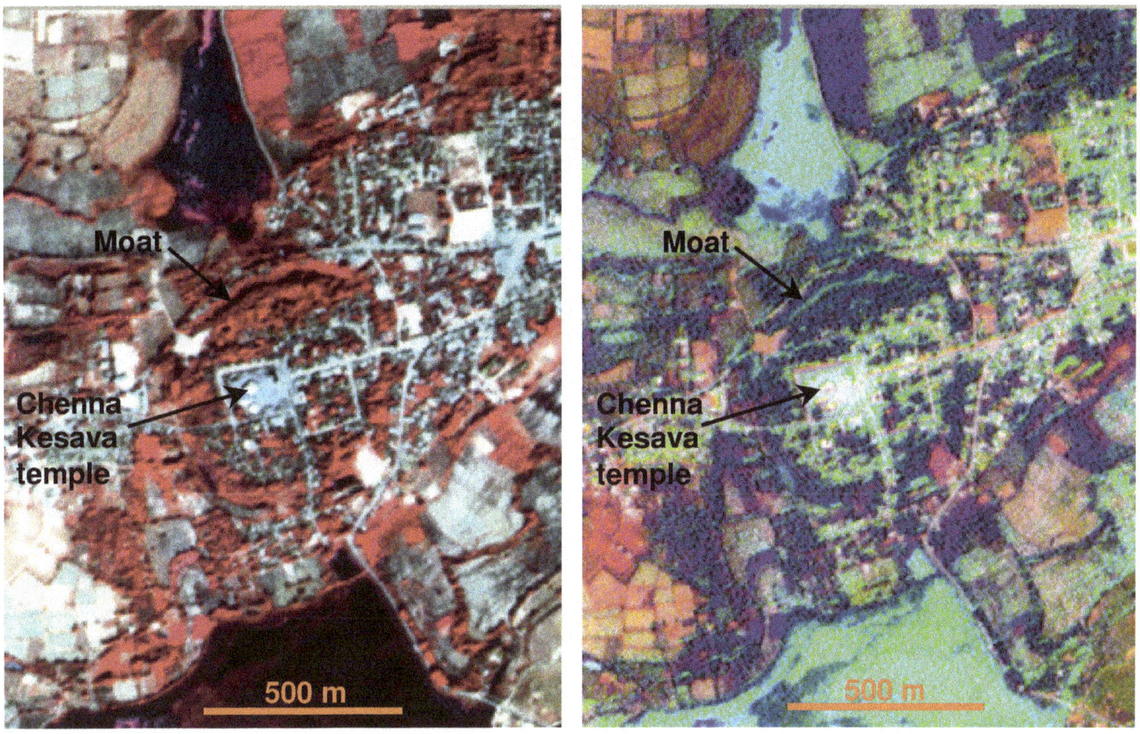

Fig. 2. Left: IRS-P6 Liss-4 image of Beluru; right: PC1,2,3 projected as RGB

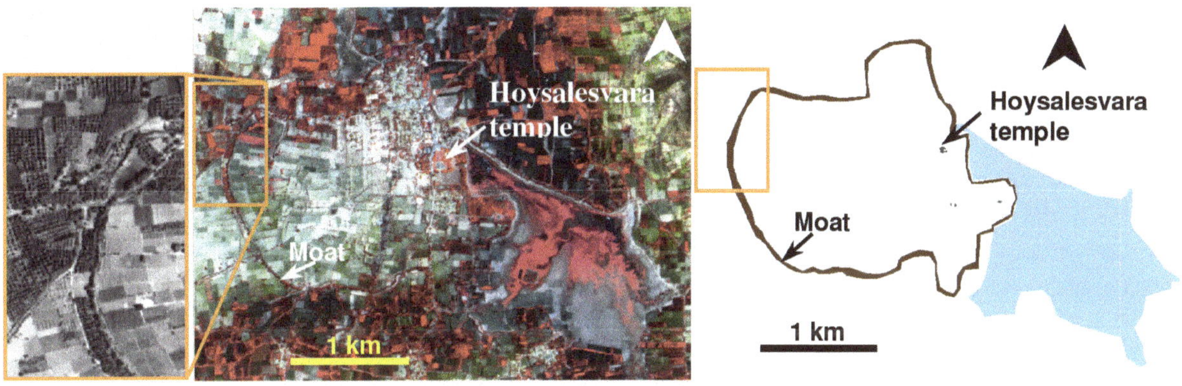

Fig. 3. Halebidu: *Left* Cartosat-1 image of a portion of the moat; *middle* IRS-P6 LISS-IV multispectral image; *right* Map showing the outline of the moat.

image clearly shows the main temple (Chenna Kesava) located in the centre surrounded by a circular positive vegetation mark indicating that there was a moat earlier. Selected image processing techniques that enhance multispectral character of images like ratioing, intensity hue saturation and principal component (PC) analysis can be applied to improve specific vegetation signatures. Some of these methods have enhanced the circular vegetation mark indicating the area of the moat. Figure-2 shows the original image of Beluru, and PC 1,2,3 projected as RGB (Red, Green, Blue) colour composite.

Halebidu (also known as Dorasamudra) was dominated by Saivas, but Jains also held a position of importance here. Dorasamudra, with a fortified area, became the capital of Hoysalas during the last decade of 11th century AD. Hoysalesvara, the dominant (Saiva) temple of the city occupied the eastern section of the city (Settar, 1992, 128). Halebidu had an asymmetric shaped fort covering a large area and attached with a deep moat. The archaeological report for the year 1930 (52) records that this moat was 50 feet deep. It describes that the fort wall had bastions at interval of every 100 yards, and several gateways. This report also published a "rough sketch map" indicating the fort, temples and tanks. An accurate map can now be prepared using multispectral imagery.

The outline of the fort can be identified following the traces of the moat which can be detected on multispectral imagery. Halebidu is a case in point to demonstrate that multispectral imagery is better than higher resolution panchromatic (PAN) imagery for detection of moat through vegetation signature. Once the moat is detected through difference in vegetation colour, higher resolution PAN image can be used for high definition of spatial characteristics of the boundaries. With this, one can map the exact shape of the moat. Figure-3 shows the whole area produced by multispectral LISS-IV sensor and a

section of the moat on the image produced by panchromatic camera aboard Cartosat1.

Somanathapura is located 30 km from Mysore, Karnataka, India. This town is famous for Kesava temple built in the second half of 13th century during the reign of Hoysala king Narasimha III. The Kesava temple is a good example of Hoysala architecture built in schist; it is fairly well preserved. Historical records mention that the temple was once surrounded by a protective wall and a moat (Settar 2009). At present it is difficult if not impossible to find vestiges of this wall or moat. But a remotely sensed image taken from space reveals the temple and its alignments with the surrounding land cover features displaying a pattern that befits a bounding wall. Figure-4 is a multispectral FCC image taken by IRSP-6 LISS-IV on 24th March 2005; offers an overall perspective of the landscape including the main temple, the river course, canal, roads and the subtle signature of the buried defensive wall. The features that are almost black are water bodies, with river Kaveri on the left and a lake on the right. A canal skirts Somanathapura and the land between canal and river formed the command area and is showing a high NIR reflectance because of intense cultivation there. A small grey rectangular feature in the centre of the image is the Kesava temple. Synoptic view reveals a much larger rectangle, the proportions of which are nearly similar to the temple. The temple is in a position that suggests some symmetry and harmonious arrangement of the whole layout. Most of the outer rectangle is indicated by positive vegetation-mark, which is probably due to the moisture retained in the buried moat. The rest of the rectangle is connected with boundaries of the fields and settlement. Probably the hard remnants of the stone fort hindered vegetation and obstructed fresh construction. These lines become separators between two fields or two types of land cover. These lines together with vegetation marks collectively reveal the shape of a near complete rectangle. This pattern also suggests that the planning of the outer rectangle was in harmony with the plan of the temple or vice versa, which suggests that both belong to the same time phase.

Conclusions

The ancient city walls and fortifications were generally surrounded by moats, which constituted the defense system of the settlement. Even if the fort/wall has gone into ruins, the buried moat affects the surface vegetation, which can be detected on remote sensing imagery. These effects can be better detected on multispectral imagery as the signature is determined by varying vegetation colour and texture. This paper has demonstrated detection of moats in three historical sites of the Hoysala period and has also illustrated the extent of identifiability of such signature on higher resolution PAN compared to medium resolution multispectral image. The paper has also discussed the geometric pattern and layout of the sites based on cultural and community factors. This combination of cultural information and associated geometrical patterns can be used for detecting moats and other buried historical sites.

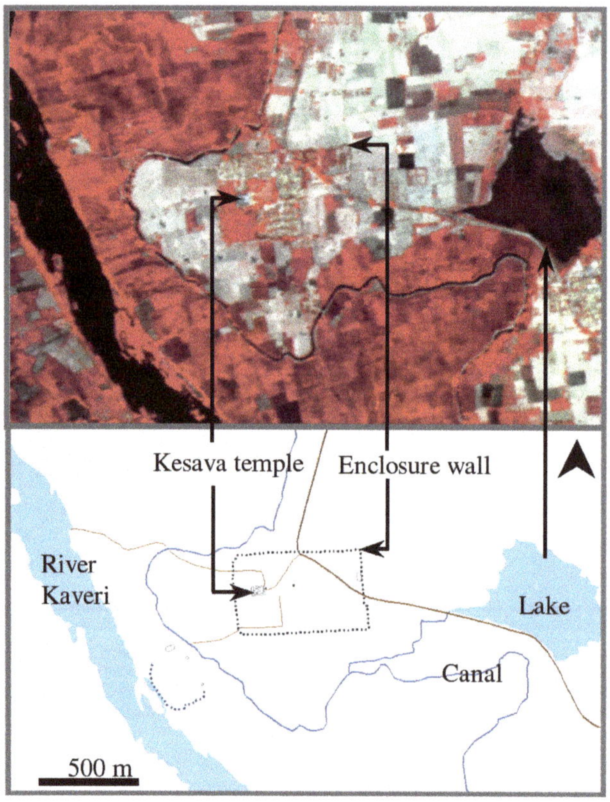

Fig. 4. IRSP-6 LISS-IV (24 March 2005) showing Somanathapura and environs.

Acknowledgements

We thank Dr. K. Kasturirangan, Director, NIAS, and Dr. R.R. Navalgund, Director, SAC, for their overall guidance, Institutional support and encouragement in carrying out this study. We also thank Prof. B.V. Sreekantan at NIAS for useful discussions. The financial support for this project has been provided by Department of Space, Government of India.

References

1934. *ANNUAL REPORT OF THE MYSORE ARCHAEOLOGICAL DEPARTMENT* for the year 1930, The Government Press, Bangalore, India.

JOSEPH, G. 2003. *Fundamentals of remote sensing*, Universities Press, Hyderabad, India.

RAJANI, M.B. 2007., Bangalore from Above: An Archaeological Review. *Current Science* 93(10), 1352-53.

SETTAR, S. 1992. *The Hoysala Temples*, Karnatak University and Kala Yatra Publications, Bangalore, India.

SETTAR, S. 2008. *Somanathapura*, Ruvari, Bangalore, India

Interpreting aerial images – developing best practice

Rog Palmer[1] and Dave Cowley[2]

[1]Air Photo Services, Cambridge, UK
[2] Royal Commission on the Ancient and Historical Monuments of Scotland, Edinburgh, UK

Abstract

This paper explores the complex issue of the archaeological interpretation of aerial data, including satellite images, multispectral/hyperspectral and traditional photographs. In addressing how archaeological interpretation works, the role of experience, intuition and 'pattern recognition' will be discussed, suggesting methodologies appropriate to particular contexts. It is argued that interpretation of aerial images is a specialist skill, improved by experience and that methods of auto-extraction, often applied to unsuitable images, are a poor substitute for this.

Keywords Aerial images, experience, image interpretation, auto-extraction.

Give half a dozen people the same information and you get half a dozen different interpretations, all equally valid and probably not even mutually contradictory in any significant way, but each of them will bear the style of the individual interpreter.
(Reginald Hill, 2008. The Roar of the Butterflies)

Introduction

This paper will explore the complex issue of the archaeological interpretation of aerial data, including satellite images, multispectral/hyperspectral and traditional photographs – which, together, will be referred to as Aerial Images. In addressing how archaeological interpretation works, the role of experience, intuition and 'pattern recognition' will be discussed, suggesting methodologies appropriate to particular contexts.

Illustrated with examples drawn from the authors' experience, this paper will advocate a 'best-practice' approach to the rigorous interpretation of aerial images. This will provide an appropriate context for the development and assessment of auto-extraction/auto recognition approaches, suggesting some principles for its application and warning against the uncritical development and application of the technique.

Why do we examine Aerial Images?

Within the range of optical wavelength sensors there is little difference between aerial photographs and satellite images, although the publicly-available resolution of satellite data is limited by the US government. One may have been taken on film (e.g. during the 20th century), and both are likely to be captured digitally in the early 21st century. In essence, we are dealing with information recorded by sensors that are above, and detached from, the ground and for ease of discussion, we will refer to all of them as aerial images unless specific cases require clarification. There is every possibility that a single aerial image will record information that spans time from the fraction of a second it was taken to millennia in the past. As archaeologists we are interested in only parts of that spectrum of information, which may relate, for example, to period-specific research interests. However, the understanding of this specific information requires an awareness of other aspects, such as modern and recent land use and agriculture, so that filtering of the mass of information on the images is based on a broad appreciation of what has been recorded. This is vital to avoid incorrect interpretations such as, for example, suggesting modern agricultural marks to be archaeological features (Wilson 2000, 163-209). Consideration of geological and geomorphological processes that took place before, during and since the anthropogenic events of the past are key to creating robust and reliable interpretations from aerial images. Invariably, aerial images record a palimpsest of information, in layer upon layer, some heavily truncated, some transformed by vegetation or post-depositional processes, that require a broadly-based understanding if, as archaeologists, we are to interpret and extract the data that are relevant to our particular interest. It will be argued that to do this effectively requires specialist skills (figure 1).

Specialist skills in archaeological practice

A first or second degree (i.e. BA and MA/M.Phil) in archaeology from almost any university will provide students with broad and general knowledge about theories, methods, practice and the results of applying these to our knowledge of past communities. Some specialist knowledge may have been acquired through writing a dissertation but, in general, this will come from work at doctoral level or from working in the profession. Excavation reports are full of contributions about pottery, stone, bones, plants, seeds, mud and other bits and

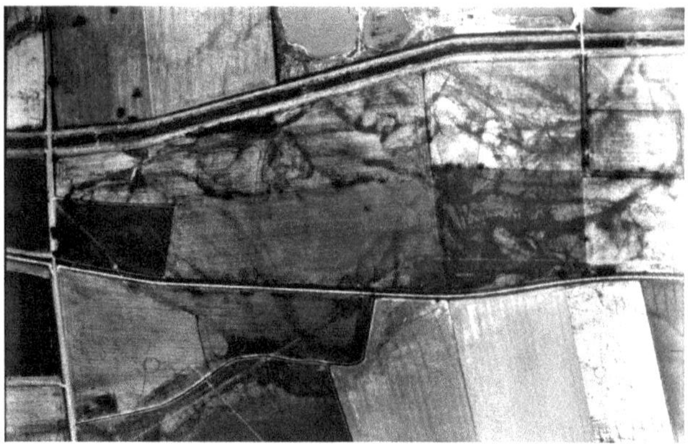

Fig. 1. Part of a single vertical photograph that shows good detail of natural, archaeological and recent features. Palaeochannels, known to have been active during Neolithic and Bronze Age show the relations of contemporary monuments to the water. By the Iron Age the channels had silted up and were sufficiently dry to allow local communities to construct tracks and ditched field systems over them. Modern pipelines cut through the area and quarrying (top) is actively destroying archaeological sites. (Photograph: RAF/39-3163T: 5460: courtesy Peterborough Museum and Art Gallery).

pieces that were dug up and need thorough analysis to make sense to the excavator. These are often called 'specialist reports' and (usually) have been produced by someone who specialises in that particular aspect of the past. The authors assert that interpretation of aerial images is similarly a specialist skill and – just as most people would not attempt to analyse and interpret pollen that had been sampled from a section – is something that needs to be effectively taught, practised and learned by anyone wanting to extract information from these sources.

Is interpretation of aerial images a specialist skill?

There is no one type of aerial image interpretation and one of the authors (RP) remembers a day spent at RAF Chicksands (the school where the Royal Air Force trains photo interpreters from all over the world). During one of the talks a slide was projected showing what he, as an archaeologist, mentally classified as 'factory', which at a basic level was correct (Figure 2). The instructor then asked one of his fellow instructors to tell us what was on the picture and he responded with a five minute analysis telling us what type of factory it was, what it was making, an estimate of the daily production, where the raw material came into the factory, where the product went out, and so on... This was an excellent demonstration that there are specialist skills within the overall specialism. The speaker was an expert in analysis of factories and production. Perhaps he would not have recognised an archaeological site and, if so, perhaps he would not have been able to say as much about it as an experienced archaeological interpreter. Specialist knowledge comes from a specialist background and is able to produce specialist output.

The interpretation of aerial images is a skill that results from a certain amount of teaching and accumulated experience; and experience is something that will continue to accumulate throughout an interpreter's life. We will argue that there are some fundamental knowledge and skills required for the interpretation of aerial images for archaeological purposes. Among these basic necessities we would include knowledge of extant sites and post-depositional processes, including

Fig. 2. In simple terms – and to most archaeologists – this is a factory, but an interpreter who specialises in factories and production can deduce far more information from this single photograph. There is no single correct interpretation of any photograph. (Photograph: Rog Palmer: 98.123/22).

contemporary landuse as are relevant to the area being studied. From these can come the ability to create mental pictures of how sites may now be visible from above. Knowledge of individual types of site and of how systems of sites may have functioned in past landscapes can suggest what elements of past society may be missing from the current record and which of those may be detectable from the air. For example, many past societies had particular burial traditions. Some of these – cemeteries or monumental mounds, for example – may be identified from the air, but isolated burials, perhaps marked by a single stone or small mound, may not be identified as such although it is possible that they will be seen by an airborne observer or noticed on existing aerial images. Past traditions and modern landuse can sometimes make effective interpretation impossible – as in Armenia where it is virtually impossible to distinguish a Bronze Age kurgan (a heap of stones) from a modern field clearance cairn (a heap of stones) on aerial photographs. To sort one from the other requires field investigation (Figure 3).

This paper begins from the premise that experience in a particular speciality improves output, and the reliability of that output, from a person engaged in that work. It is argued that image interpretation is not an inbuilt ability in every archaeologist and, like artefact identification or faunal analysis, is something that needs to be properly taught, learned, practised and developed as a skill. Therefore any archaeologist – however capable in a particular area, such as excavation or theory – who says, "I looked at the aerial photos and saw nothing", should substantiate their observation with answers to a series of questions. These may include noting the date of photography and its appropriateness for what they hoped to see, the scale of the photographs, what experience has the viewer of reading and interpreting aerial images, how were the images examined (monoscopically, stereoscopically, on screen, at what orientation and at what magnification(s)), whether the viewer's mind was open to perceive a range of features, whether elements of the natural background were visible and helped guide the search for archaeological features... and so on. "I saw nothing..." is the response of an unskilled and unaware observer because there is always something, however apparently insignificant, that will add new knowledge to a study area and perhaps help its understanding.

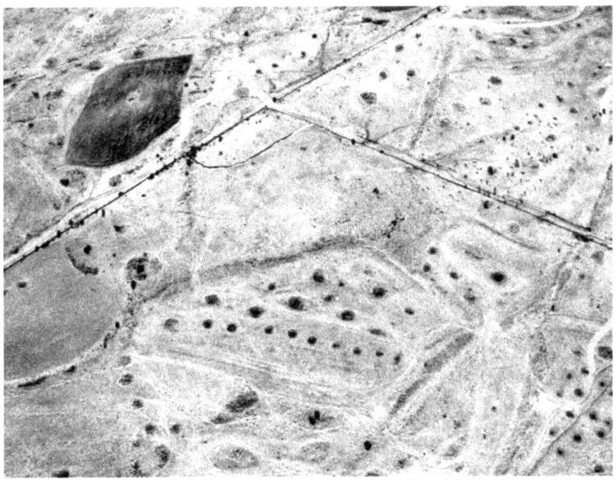

Fig. 3. Heaps of stones in Armenia. Most are likely to result from field clearance, but may include kurgans. Distinction of one from the other requires field investigation. (Photograph: Karen Martirosyan/Wings over Armenia: 20040815_103).

What information has come from interpretation of aerial images?

There is a tendency amongst some archaeologists to view the application of aerial images as little more than a mechanism to discover sites. Part of the fault for this view lies in teaching which, in many universities, does not extend much beyond the notion that 'aerial photography discovers sites', perhaps illustrated with a few dramatic aerial images. Some of us would argue that even at undergraduate level this is inadequate teaching about one of archaeology's major prospection methods and that more effective teaching may come from discussion of examples that have come from *uses* of aerial images. The authors argue that a narrow view – that the application of aerial survey for archaeology is largely limited to finding new sites – has the effect of limiting the exploration of the fuller potential of aerial images to extract information and understanding of the past – a problem that is compounded where expertise in aerial image interpretation is still not recognised as a specialism.

So, what kinds of information can come from the specialist interpretation of aerial images? The overhead views they provide are good for recording the results of earth moving in the past. These efforts can be divided broadly into civil engineering and farming and they have produced the majority of information that can now be recovered from aerial images. In different countries, in different parts of one country, these remains may be of upstanding features (e.g. walls, tells, earthen banks, cultivation traces) or those that were cut into the ground and now are backfilled and levelled (ditches, pits) and it is the evidence of such traces that the interpreter needs to be able to identify and understand regardless of the means by which they are visible on images.

Once we begin to piece together that information and combine it, if possible, with other sources of information, we may begin to see expanses of past occupation – field systems, tracks, settlements and other remains that were essential to the daily life and effective functioning of past communities. And often those same images provide a level of information about the environmental background of those societies and will indicate, for example, areas that were liable to flood, former watercourses, ground that is too steep for anything but goats – and also show evidence of later activities such as local quarrying that may have damaged or destroyed archaeological contexts. All this is just the kind of information that can be interpreted from aerial images once sufficient experience has been accumulated. Some results have been published, although these are a small percentage of the overall interpretative work that has been undertaken. Among these are landscape studies that analysed past settlement in an area of 1350 sq km of the Yorkshire Wolds (Stoertz 1997) and within the 3250 sq km of a whole English county (Deegan and Foard 2007). Other work has linked aerial photographic evidence with that from field walking (Hall and Palmer 1996), or with analytical field survey (Bowen 1990; RCAHMS 1994), or has examined coastal defences through time (Hegarty and Newcome 2007). Michael Doneus mapped the whole of the vast Roman town of Carnuntum (Austria) from aerial photographs and has used this to show how trivial can appear the work of digging archaeologists when we stand back and look at the greater whole. In a lecture he told us that he had mapped the 4.5 sq km area of the town in two years and that, after more than 100 years digging, excavators had examined 250-300 hectares. Similar comparisons of 'scale' have been apparent in the Danebury Environs Project in Britain, that followed from the 20 years' excavation within the hillfort of Danebury to examine a few of the 120 'sites' that surround it (Cunliffe and Poole 2000).

The point is that interpretation of aerial images can provide extensive information across the landscape that can then be analysed and studied in collaboration with other investigative techniques. For example, a landscape project in Italy, directed by Stefano Campana, combined information from four sources of aerial image – satellite, vertical and targeted

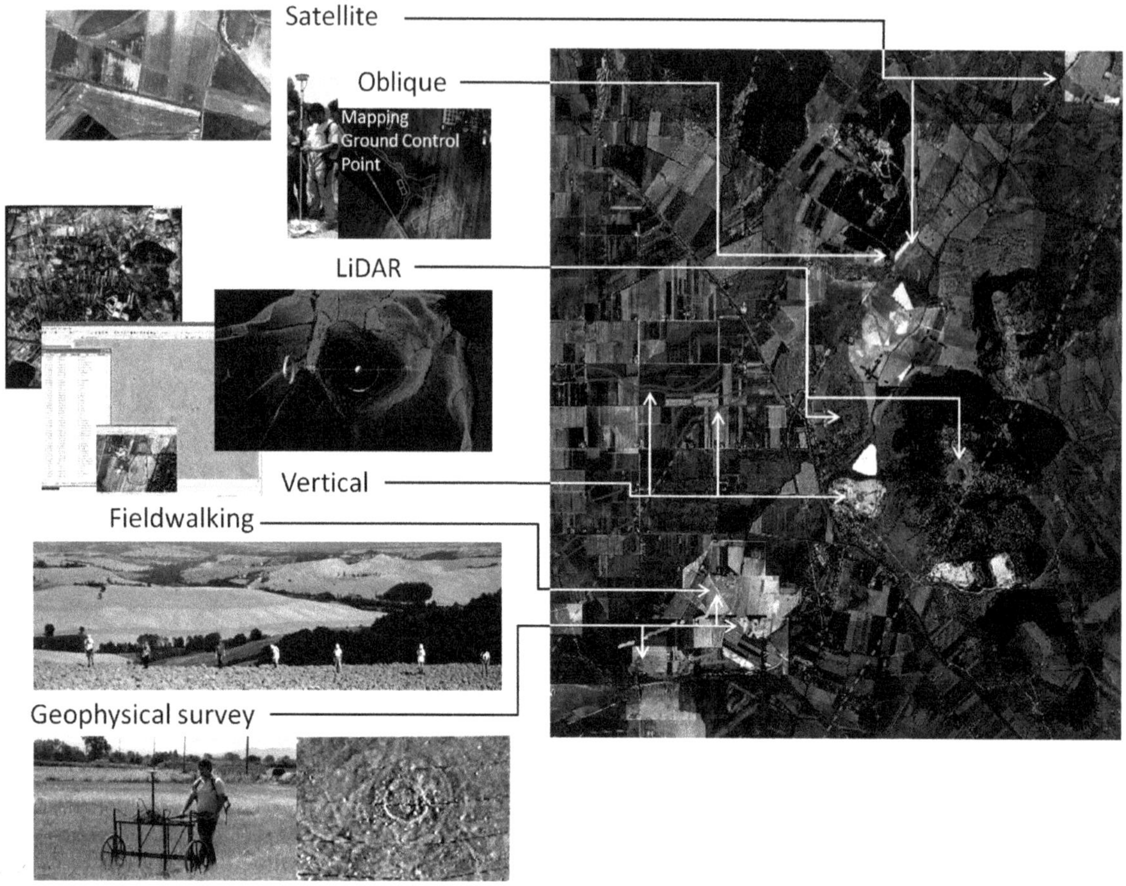

Fig. 4. Illustration of the range of non-destructive methods that were integrated to study features in a landscape project by the University of Siena, Italy. (Source: Stefano Campana).

oblique aerial photographs and Lidar – with ground data collected by field walking and geophysical survey (Figure 4). Such approaches represent a serious and sustained engagement with the wider landscape that is in contrast to some projects that have created so-called 'landscapes' by adding a few surrounding modern fields to a small excavated area (e.g. see the Avebury landscape – Pollard and Reynolds 2002). Once interest expands beyond that of a single modern field, skilled examination of aerial images is the only effective way to extract information at a considerable level of detail over an extensive area.

How do we look at aerial images?

Teaching for the Culture 2000 projects that were part of *European landscapes: past, present and future*, as well as much of that undertaken in Britain, always included an interactive session of *photo reading*. In this approach a single photograph stays on screen and students take it in turn to say one thing each about the image. We usually begin by asking them to tell us what time of year the picture was taken, then move on to other teacher-guided questions until we reach the archaeological content – if there is any. Proceeding in this way makes beginners (and others) aware of just how much information there is on a single image and good teaching will encourage class discussions that argue about particular things. All this is part of a useful learning process and it is difficult to think of a more relaxed way of introducing people to ways of seeing, reading and understanding aerial images.

While it is difficult, and probably unwise, to be prescriptive about a 'best-practice' approach to the rigorous interpretation of aerial images, it is vital that an interpreter is aware of the whole image. Anyone who tries to jump directly to the archaeological features is missing a lot of ancillary information that can provide context as well as the reasons those features are where they are (Figure 5) and help explain what we are (and may not be) seeing and why. Different views of the same photograph or image are likely to provide different types of information – on paper prints and on screen. Too close (or too zoomed in) and you see detail but not the whole picture; too far away and there may be some small or faint things that remain unseen.

An example from the archives of one of us, when working on material in Wessex, was at the site of an enclosure at South Wonston in Hampshire.

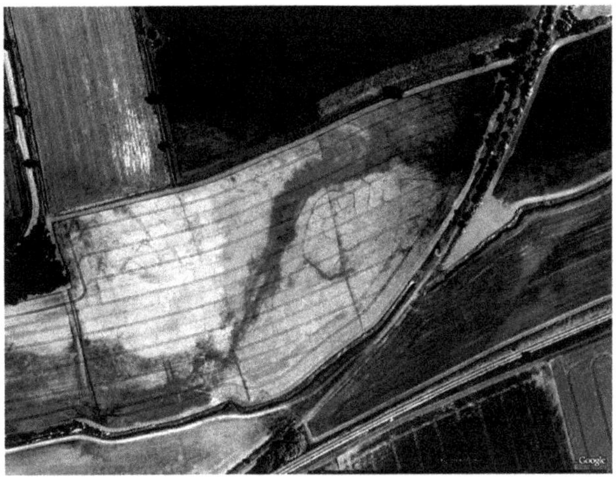

Fig. 5. Regardless of when, or if, there was water flowing in the dark-toned palaeochannel it now indicates the local topography and explains why the multi-enclosure system was sited there and it also explains its shape. Archaeological features can be placed in context using data from aerial images. (Source: Google Earth).

RP had seen and mapped a 'standard sized' curvilinear enclosure that was visible on photographs of several dates. One day he was shown a new colour print of the same site and immediately noticed a huge rectangular enclosure that surrounded the other site. 'This is the first time I've seen that', he said – but on checking back on the older photographs that large enclosure was there on all of them. The reason for not seeing it was almost certainly that all photographs were examined in close-up using a stereoscope and the larger feature was too big to be seen this way. The colour print held by someone else provided the more distant view from which the larger feature was obvious. Far from being embarrassed by this mistake RP is happy to use it as a teaching example because it makes just the point that is required – look at all photos in as many different ways as possible.

And in this mode, it is important to point to one of the awkward aspects of working on screen. Traditionally – and usually fixed so as to be unchangeable – north is to the top. This is fine for people working in the southern hemisphere but plays havoc with the brain for the rest of us. It is well known by those who study visual perception that the eye and brain read height and relief correctly when shadows fall towards the viewer and vertical aerial photographs and satellite images should be viewed in the same way (Figure 6).

Examples from the past of good lighting include hill shading on maps that lights topography from north-west of the map and exactly the same principle makes it so much easier to read vertical images. This means that GIS users are immediately disadvantaged unless they are working with prints (rotated) alongside them. This may sound petty, but image interpretation can be difficult at the best of times without making things harder for oneself. So if it is possible either to work only in the southern hemisphere or to rotate your images or GIS so that south is to the top life will be much easier for those who have to interpret only on screen. Unfortunately, one has only to look at the illustrations in a recent book on satellite archaeology (Parcak 2009) to realise that most GIS users and their teachers are blissfully unaware of this fact.

Added to this is a further issue directly arising from how aerial images are increasingly used. With the ready availability of digital images and the widespread application of GIS, there is a tendency for an aerial image to be loaded as a layer in the GIS, and used as 'wall-paper'. In many cases, this may be restricted to a single image that is easily available, and it may be the only aerial image that is used.

The authors' experience of using aerial images, principally in the UK, shows how important multiple views of the same sites, areas or features are – how important images from different years can be to assist detailed interpretation of what is visible – and this is at odds with the 'single snapshot' view that characterises many applications of aerial images. Admittedly, many parts of the world do not have the benefit of multiple suites of aerial images spread over many different years and taken under differing conditions, but where this is available we would argue that its use is vital.

The authors have outlined some of the issues of how aerial images are best interpreted and used appropriately to derive robust and reliable archaeological interpretation – stressing the importance of the specialist aerial image interpreter. The authors share a concern, widespread amongst many experienced photo/image interpreters, that many contemporary approaches using aerial images are undertaken with little consideration of these issues of interpretation. For example, GIS-based applications, while very powerful in combining data sources, may encourage the user to insert a single, often very recent, image as a layer among other data layers. The appropriateness of that image to the task in hand may not have been evaluated – i.e. the date, scale, resolution and spectral range may be totally inappropriate to the task in hand but it is likely to be accepted as serving the purpose of having 'done the aerial images'. By contrast, a thorough conventional photo-interpreter will begin by collecting together all available photographs of the study area. S/he will endeavour to examine these in appropriate ways that allow the extraction of information that will build up a picture based on the sum of the available evidence.

What information do we expect from auto-extraction or other mathematical methods?

We suggest that automated methods have been adopted by people who have no skills in interpreting aerial images in the hope that 'something' will come of it. They (wisely) do not trust themselves to interpret aerial data but expect that the

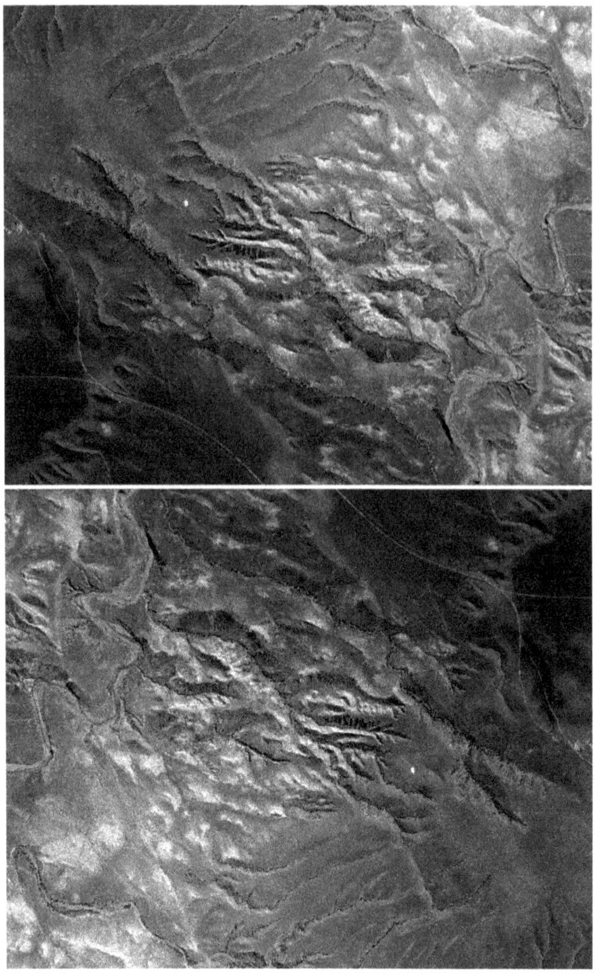

Fig. 6. This pair of images shows the effect of the direction of light on perception of topography. Top, is the image with the traditional north to the top. Note that rivers appear to run along the tops of ridges. Bottom, is the same image rotated 180 degrees so that the shadows fall towards the viewer and rivers now run in valley bottoms. This demonstrates the problems that GIS users will have if they keep north to the top. (Source: OFEK: June 2001).

machine can and will do it for them. To us this seems very poor science. We have seen examples of the use of edge detection in which results are much poorer than the information that can be seen by looking at the initial image – but the presenter has not been taught how to examine images so seems content with the result. Even with uninformative results there seems to be no willingness to state that the method is no use in that particular environment or by using that or those particular images.

Recent work in Norway (Trier *et al.* 2009) has resulted in an algorithm that has had some small success identifying circles on satellite images. This is a really basic level of pattern recognition and it could be suggested that if the algorithms have trouble identifying circles there is no hope in the foreseeable future that they will be able to detect the vast range of shapes and signals through which archaeological features are visible. The argument has been that this is a first stage of image reading that will save archaeologists time in visually examining the same area. Possibly this is true (although we have reservations about that) but it is difficult not to suggest that in the five years of that project it would have been possible to teach many Norwegian archaeologists a good level of interpretation skills *and* that we could have examined a large percentage of the image data during that training. Automated approaches also firmly embed what archaeologists think they know about the archaeological remains of an area in the prospection method: i.e. we 'know' what the sites in an area look like, therefore we will create algorithms that find these sites. This does not allow for one of the proven benefits of a skilled image interpreter that they may find the unexpected, and so change our view of the past.

Is this a good thing for the discipline?

In summary it seems that automatic detection is able to isolate certain anomalies in photographs and images *but* these then need to be interpreted by someone with archaeological knowledge. If users of automated methods do not have these skills they need then to find someone who does – and we would expect that person to ask to see the original material. Consider this in terms of other specialities: if an expert in flint artefacts was given the choice between making a decision based on a drawing or photograph of an artefact or the artefact itself which do you think they would chose?

Conclusion

We strongly argue that the interpretation of aerial images in archaeology is a specialism that is not as widely acknowledged at it should be. Routinely images are interpreted and used by archaeologists who have little experience or training and this produces unreliable information on the past. This has a direct bearing on the issue of auto-detection approaches. We fear that users of auto-detection may not know much about the images under examination, and that auto-detection is being developed *instead* of properly engaging with the information that is embedded in the aerial images. Is this not a case where a specialist subject may best be left to specialists in that subject?

It is useful, as we are sure that everyone will agree, to have a level of knowledge about the tools (in this case aerial images) that you are applying. Using a machine instead of an experienced brain is not an easy way to achieve useful or useable results and we would urge some reflection on the effectiveness of interpretation of aerial images, and the importance of taking the time to learn about the data you are using. It is encouraging that this is recognised amongst researchers in the field of image analysis (e.g. De Laet and Lambers 2009), who identify the value of collaboration and the appropriate application of new approaches and techniques.

References

BOWEN, H.C., 1990. *The Archaeology of Bokerley Dyke.* RCHME, London

CUNLIFFE, B. AND POOLE, C. (EDS), 2000. *The Danebury environs programme: the prehistory of a Wessex landscape.* Volume 2: parts 1-7. English Heritage/Oxford University Committee for Archaeology, Oxford

DEEGAN, A. AND FOARD, G., 2007. *Mapping Ancient Landscapes in Northamptonshire.* English Heritage, Swindon

DE LAET, V. AND LAMBERS, K., 2009. Archaeological Prospecting Using High-Resolution Digital Satellite Imagery: Recent Advances and Future Prospects - A Session Held at the Computer Applications and Quantitative Methods in Archaeology (CAA) Conference, Williamsburg, USA, March 2009. *AARGnews* 39 (September 2009)

HALL, D.N. AND PALMER, R., 1996. *Cambridgeshire Fenland Survey: the silt fens,* in D. Hall (ed.) *Fenland Landscapes, Cambridgeshire, Volume 3. East Anglian Archaeology*

HEGARTY, C. AND NEWSOME, S., 2007. *Suffolk's Defended Shore: costal fortifications from the air.* English Heritage, Swindon

PARCAK, S. H., 2009. *Satellite Remote Sensing for Archaeology.* Routledge, Abindgon/New York

POLLARD, J. AND REYNOLDS, A., 2002. *Avebury: the biography of a landscape.* Tempus, Stroud

RCAHMS, 1994. *South-east Perth: an archaeological landscape.* HMSO, Norwich

STOERTZ, C., 1997. *Ancient landscapes of the Yorkshire Wolds,* RCHME, Swindon

TRIER, Ø. D., LARSEN S. Ø. AND SOLBERG, R., 2009. Automatic detection of circular structures in high-resolution satellite images of agricultural land. *Archaeological Prospection* Volume 16 Issue 1, pp 1 - 15 (January/March 2009)

WILSON, D. R., 2000. *Air Photo Interpretation for Archaeologists.* Tempus, Stroud

Worldview: the importance of diversity in human (spatial) thinking

Judith van der Elst

University of New Mexico

Abstract

The use of remotely sensed data, notably satellite imagery, has allowed for understanding ecological processes in conceptually new ways: different scales and categories, temporally and spatially. Human groups are an integral part of ecological systems, however, since much of science is based on assumed universalities in human thinking, the role of culture(s) is often overlooked or marginalized in ecological studies. However, new findings in cognitive research and ethnogeography indicate that cultural differences in space –time concepts are significant between different linguistic groups, contrary to what was previously thought (Levinson 2003, Levinson and Wilkins 2006). This challenges cultural and digital heritage studies to adopt a broader conceptual framework for understanding human (landscape) experience. The raster model used in geospatial analytical approaches offers a way to explore diverse concepts of space-time that can be derived from landscape categories and placenames in relation to landscape processes, as it is not limited by preconceived (discrete) spatial categories used in vector models. This paper presents an exploratory case study focused on the ancestral Tewa communities of the Northern Rio Grande Valley region in New Mexico, USA.

Introduction

Archaeologists have interpreted ancestral Puebloan cultural material using a number of different frameworks; on a landscape scale 'sourcing' studies are a dominant approach (David and Thomas 2008). These research approaches often make use of economic and subsistence models for interpretation, while ecological systems approaches that incorporate or consider landscape/spatial knowledge of human groups, as mediated through language, are limited (Duvall 2008, Forte 2007). The possible existence of sophisticated spatial / landscape knowledge of indigenous groups is often overlooked and only considered for its practical application in resource management (Agrawal 2002), and, as argued by Duvall, many ethnoscience approaches focus on particular aspects of local knowledge, while few have analyzed "how knowledge of broad portions of the biophysical environment is structured in local knowledge systems (Duvall 2008: 327)." For archaeological studies this kind of information is seldom available, but also, the argument of cultural assimilation and cultural change has been used to only selectively include available indigenous knowledge in the interpretation of archaeological material. No doubt cultures undergo changes, especially European colonization has had a dramatic impact on Puebloan life and forced changes in indigenous societies in general. However, conceptual differences that still exist between different groups today only reinforce the idea that past landscape experience is based on a different conceptual framework(s) than those developed for current scientific studies. Data recorded in more recent times can therefore provide a starting point for exploring underlying spatial conceptualizations of past experience. The research framework that is explored herein considers the current scientific framework, the western knowledge system, too limited to understand the full dimension of knowledge represented and communicated in different knowledge systems (Casey 2008, Heen Wold, 1978 Wallerstein 2004).

In addition to the Indigenous philosophical discourse, in which the existence of metaphysical differences between western and non-western worldviews is discussed, this research draws on developments in two related disciplines to investigate these differences in the spatial domain. First, the developments in the field of Geographic Information Science (GIScience) over the last two decades, and secondly, recent research results in the field of psycholinguistics (Goodchild 1992, Levinson 2003, Mark, Turk and Stea 2007). The new direction in GIScience started out as Critical GIS (Kwan and Lee 2004, Schuurman 2002, based on the idea that space is not experienced by all people in the same way, such that general spatial (social) categories are not suitable in all approaches. Since then research has specifically focused on spatial ontologies; how humans experience space is gaining interest over understanding space purely geometrically.

The research program of the Max Planck Institute of Psycholinguistics focused on space has as its goal to understand the breadth of human spatial cognition through linguistic analysis. Several studies done under this program have indicated that significant differences in the spatial domain and subdomains exist between speakers of different language groups (Levinson 2003, Levinson and Wilkins,2006). These new insights challenge us to look at ancestral landscapes differently, traditional spatial analytical frameworks are too limited and environmentally deterministic, while our personal landscape experience, (a phenomenological approach) may not be suitable to serve as a model for past landscape experience due to our specific spatial cognitive development (Birnbaum 2008, Maijd, Enfield, and van Staden 2006).

The case study represents an exploration of the potential use of different Tewa landscape names and categories that were recorded and published as the Tewa ethnogeography, (Harrignton 1916). I integrate the placenames with other geospatial data to gain insight in the ecological relationships that underlie the cognitive landscape. The initial exploration exposes the complexity of the relationship between human spatial thinking and processes of the broader biophysical environment, which leads to new ideas regarding ways to address human experience. The general hypothesis addressed here is that frameworks founded on understanding of Indo-European spatial cognition are not valid and new frameworks or broader frameworks, theoretical as well as representational, are needed. Specifically, this paper addresses how differences in human spatial cognition can be integrated in current spatial analytical systems.

Research setting: The Tewa World, a brief introduction

Today Pueblo communities in New Mexico, descendant communities of people who have lived in this region for centuries, occupy 19 pueblos. Pueblo communities are autonomous and the different languages spoken are highly diverse, containing three distinctive language groups, Zunian, Keresan, and Tanoan, and dialectical differences within these groups (Dozier 1983). According to Dozier, the Tewa language, which is part of the Tanoan language group, does not seem to be influenced by Spanish to any great extent morphological or syntactically, although loanwords are present, especially for newly introduced items. Speakers of the Tewa language today live in six of the 19 New Mexican pueblos (Mithun 1999, Ortiz 1969). Ancestral cultural material of these communities is omnipresent and visible on the modern day landscape and is part of the ongoing Tewa world experience. The relationship between archaeologists and descendant Pueblo communities regarding research, interpretation, and management of this material however is contentious. This issue between western research and management approaches and local or indigenous viewpoints has been documented in other regions as well (Schmidt and Patterson 1995).

The conceptual world of the Tewa specifically has been discussed in several publications, notably by Ortiz and Dozier and general characteristics of American Indian metaphysics underlie this worldview (Cajete 1999 Dozier 1983 , Ortiz 1969, Waters 2004). Ortiz' study of the Tewa world is based on life in *Ohkay Owingeh* (previously known as San Juan Pueblo), one of the six modern Tewa villages, the largest, most conservative, and regarded as the 'mother village', according to Ortiz. Based on one variant of the Tewa origin myth Ortiz outlines a broad conceptual (space/time) framework loosely based on the dual organization of (Summer and Winter) moieties within Tewa society. "The Tewa classify all human and spiritual existence into a hierarchy of six categories, three human and three spiritual (linked in pairs), and the spiritual categories are further associated with specific geographical points in the Tewa world" (Ortiz, 1969: 9). Principal reference points for the Tewa world can be represented as a series of tetrads or groups of four (specifically for *Ohkay Owingeh*). This horizontal organization is complemented by a vertical structure. Specific points in the landscape are important in mediating between the different vertical levels of the Tewa universe (Ortiz 1969).

Moreover, Dozier states that in general the origin myths of all Pueblos are similar. In addition to a shared concept of origin, concepts regarding the nature of the universe (metaphysics), conceptual ordering and spatial arrangement of the pueblos is also shared. In this framework, the pueblo is the 'center of the universe'. Ortiz describes the Tewa world as existing of a number of zones with the Pueblo at its center; the surrounding physical landscape is tightly integrated with subsistence activities and the ritual calendar, thus maintaining a relationship with the larger world. This interconnectedness between the pueblo and the surrounding world, is for example reflected in the pueblo village layout, which exist of four parts and maintains open corners (Ortiz 1969:128)

The goal of puebloan design and of human existence in general is to maintain wholeness with the natural universe, people are not separate from nature; this is reflected clearly in puebloan architectural design and evident from the subtle or inconspicuous markers that reflect Tewa cosmological organization (Cajete 1999, Swentzell 1997). This is fundamentally different from western design principles. As described by Swentzell, directional forces, rather than being upward as in Christian ideology, are cyclical and move in and out of the earth. This flow of life or *Po-wa-ha*, (water-wind-breadth) symbolized by the movement of water and wind determines existence rather than a physical body or other physical manifestation. Therefore it seems inappropriate to base efforts toward understanding Pueblo culture, present as well as ancestral, on cultural material isolated from landscape processes. The use of discrete objects on a landscape scale seems similarly unsuitable as processes and relationships take precedence. The use of other types of categories that are based on relationships, introduced in the next section, seem to be more conducive for understanding a Pueblo worldview. For instance, one of the placenames recorded by Harrington, *Po-wa-wisi,* or 'water-wind-point' is described physiographical as a projecting point that forms a narrow gap through which the river passes. This gap is always windy, and "the Tewas also think of the water-air spirits *Po-wa-ha,* (water – wind-life/respiration) at this place". Another example is the terminology recorded for the Pueblo components, terms that refer to its relationship to the sun, such as *Pa'asi'*, translated as 'sunny side' (Harrington 1916). Furthermore, the rich set of cloud terms suggest that atmospheric processes are an important component of the Tewa world; placenames for locations on the earth's surface are likely more permanent than transient cloud patterns, however not necessarily more important. A few archaeologists have started to consider the idea that Puebloan people moved across the landscape in a cyclical pattern of settlements shifts as a long-term strategy for sustaining agricultural production, this instead of the more standard notion among archaeologists that people 'just' responded to climatological changes (Anschuetz 2007). This idea is supported by indigenous scholars; the characterization of life based on the directional forces moving in and out and similarly, Jojola describes the in- and outward migration pattern within the Tewa world, that is symbolized iconographically by a directional spiral (Jojola 2004, Swentzell, 1997).

Research methodology: human spatial cognition underlying landscape experience

Geographical categories

Spatial cognition is assumed to be similar cross-culturally even though variation due to environmental variability and culture may exist. The ongoing research program of the Max Planck Institute of psycholinguistics, loosely based on the linguistic relativity hypothesis and specifically focused on human cognition in the spatial domain, is testing this assumption. A few key components in the spatial domain that can be addressed with linguistic analyses include angular subdomains, e.g. spatial reference framework and non-angular subdomains, toponymy and topology (Levinson 2003). The combined results show that significant differences exist between language groups in the use of one or more main spatial reference frameworks, absolute, relative, and/or intrinsic, which indicate a different form of spatial reasoning. The second main line of inquiry is focused on placenames and the relationship between placenames and landscape categories that can provide information on the conceptual organization of a specific group.

For instance, Levinson found that landscape categories are not universal, based on a study of an isolated language Yéle Dnye, spoken on an island in Papua New Guinea, which is characterized by al rich set of seascape terms. Levinson's method identifies three main factors that influence the construction of landscape categories and placenames: *perceptual salience* (indicates the existence of universal categories), *affordances, and cultural ideas.* Within his study he found that landscape categories are most influenced by affordance and cultural ideas. An affordance can be defined as a quality of an object, or an environment that allows an individual to perform an action (Gibson, 1979).

Galton, in his work on spatial theory, discusses several object classes and the principles by which these can be organized. Objects that possess spatial attributes, he argues, must be dependent in some way on matter. This dependence forms one of the organizing principles to classify objects in three kinds of objects: material, parasitic, and perspectival. Material objects refer for instance to a piece of matter that exist in the same form over an extended period of time, this is not always straightforward as shown by his example of 'pebble' that changes form due to erosional processes. Parasitic objects, depend on existence of other objects. Examples are shadows and holes, the object of shadow for instance involves a four-place relation: x is a shadow cast by y on surface z in the light coming from w (Galton 2000). The Pueblo component terms can be thought of as belonging to this category of objects. A perspectival object depends for its existence on a particular point of view. A good example is the horizon, a category that involves a three-place relation: x is the silhouette of y as seen from point z. A clear example in Tewa is the 'great gap', *Wjio*, known to all Tewa, and at the Pueblo of Santa Clara the sun rises through this gap (Harrington 1916).

Representation of Space

Even though progress has been made in thinking about landscapes and focus has shifted in several disciplines to the importance of human experience of space (and place), the problem of representation looms large (Fisher and Unwin, 2005). In order to be able to represent the world around us, we somehow have to accept that there is a reality out there, a physical world that can be represented (Raper 2000). From research on the intersection of information systems theory, philosophy, and cognitive science innovative new ways of understanding space and spatial relationships emerge. Space and spatial entities are difficult to define and spatial relationships can be expressed in multiple ways. In archaeology, the 'discipline of things' our focus is often on discrete material objects and relationships between these objects, the landscape is the container, the empty space. In a theory of spatial representation spatial entities as well as relationships between entities need to be expressed. Several efforts toward a theory of spatial representation have been undertaken that move beyond the traditional geometrical framework (Casati and Varzi, 1999 Galton 2000, Raper 2000), discussion of which is beyond the scope of the present paper but underpins the exploratory case study.

Raster data model and ecological variables

The raster data model used in geographic information systems is argued to be better suited for representation of continuous surfaces, such as landscapes (Galton 2000, Schuurman 2002). Landscapes are not easily categorized and categories currently in use are not representative of the diversity in human thinking and therefore not useful for understanding the variation in human spatial experience. The raster data model therefore provides more flexibility for the exploration of other ways of conceptualizing and representing space. In addition data of many of the relevant environmental variables are available as raster data. Satellite imagery for instance, consists of data collected in different parts of the electromagnetic spectrum can be used to generate information on vegetation health. Sequential data can then be used to assess change, the pulsing of the landscape as responses to other environmental processes. For categories that are not discrete, statistical analyses has been used to a large extent to reveal changes and/or cycles in landsurface characteristics. This data model seems better suited for understanding the Tewa ecological relationships than other research models focused on cultural material as discrete, often isolated, material remains. Raster based representations are suited to represent kinds of terrain, also defined as landcover, for which the natural designation is in mass terms (Galton 2000: 139) Each raster cell has assigned values or attributes based on the characteristics of the earth's surface at that specific location, typically, but not limited to biophysical characteristics.

Case study: initial explorations of landscape categories, spatial models and representation

The research presented herein is anthropological, primarily interested in the paradigm (or worldview) within which the Tewa culture thrives. From the discussion above it follows that the construction of categories (ontologies of space) is not straightforward. The research into qualitative theory of space and geographic representation in addition to new insights in human spatial cognition can be used as a foundation to investigate differences in space/time concepts and to describe these differences.

The data used for this study is temporally diverse, however, since this study is exploratory and focused on how biophysical processes and ecological relationships can be integrated with human perception and conceptualization, precise chronological control to create discrete spatial units is not its objective. The spatial extent is based on the description of the Tewa world as bounded by sacred mountains and the importance of water related features; the Tewa world boundary is therefore tentatively defined by watershed boundaries, the Rio Grande being its primary stream. In this way, hydrological (cyclical) processes can be easily integrated into the analysis of landscape and landscape processes. Rainfall is a crucial element in this environment along with associated atmospheric processes, and as many factors are influencing the delicate ecological balance, cultural mediation is critical.

The data used are primarily 20th century data and information; for this timeframe a number of detailed data sets are available for the case study area that have sufficient overlap in order to make some inferences based on patterns that occur. For instance, instrument climate measurements have been used to calibrate a long record of dendroclimatic data in an effort to retrodict historic rainfall patterns in New Mexico in more detail (Ni et.al. 2002). Spatial rainfall patterns are influenced by orographic/ topographic effects and cyclically by ENSO, PSO and monsoon patterns (Allen, Lindesy and Parker, 1996). These coupled processes result in a complex hydrological pattern that can be difficult to model or predict. References to these changes are given within the descriptions of placenames; it is for instance mentioned that people moved downstream when dry farming at higher elevations became precarious in the past.

Instrument data were first recorded in New Mexico in 1896, and even though this is not long before, it does cover the time period during which the Tewa placenames, including many archaeological sites, were recorded (1910). Records show that one of the driest years and period measured occurred in 1904, followed by one of the wettest periods. Satellite derived data, only available for the last 30 years or so, such as a Normalized Difference Vegetation Index (NDVI) can be compared for dry years followed by wet year in modern times, to get an indication of changing vegetation patterns.

The Bureau of American Ethnology promoted anthropological research of indigenous groups in the United States in the early 20th century, resulting in in-depth studies of Native American culture. This interest was driven by the fact that indigenous cultures were expected to disappear, or at least assimilate into a general American culture as several policies were implemented to that effect (Haake 2007). One such body of anthropological work was the *The ethnogeography of the Tewa Indians*, a study primarily of placenames in the Tewa languages supplemented with related landscape categories (Harrington 1916) (Figure 1).

Exploring cognitive landscapes

The investigation of the use of landscape categories for the Tewa through toponomy is feasible due to linguistic continuity such that the recorded placenames can provide some information as to the use of landscape categories and potential differences with Indo-European based languages within the spatial domain. The current application represents an exploration of the potential use of different landscape categories and how these are part of a –logical- coherent paradigm, i.e. an indigenous worldview. I integrate placenames with other environmental data sets, such as satellite imagery within a geospatial system to gain insight in the ecological relationships that underlie the cognitive landscape.

The Tewa placenames are organized by region in 29 separate mapsheets, covering different size geographical areas based on the density of placenames and areas claimed by specific groups. The mapsheets/ or surveyed area/ cover the earth surface component of the Tewa world. Each placename is represented by a point location on a map and by a description, which includes information such as etymology, by whom this place is known, alternative names in different

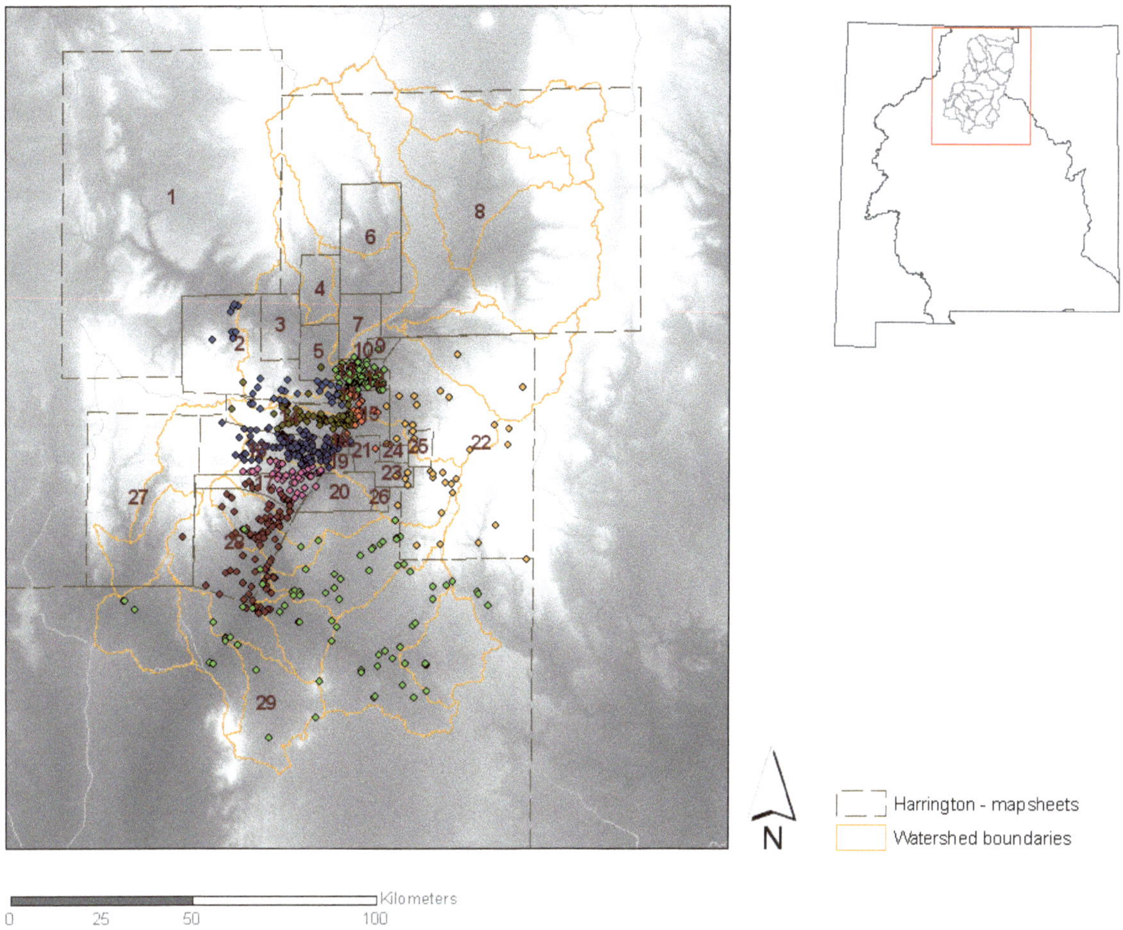

Fig. 1. Overview of Harrington's mapsheets

dialects and/or Spanish and English, the meaning of the name, the associated story, and/or description of the physical landscape feature.

The approximate locations of these names are entered in a Geographical Information System employing a digital elevation model (DEM) and 7.5' topoquads as basedata layers in order to match the 1916 description with the current landscape in an on-screen digitization process. The resulting shapefiles from the digitization process contain initially only basic information, location and original map number. In order to make it more usable other associated attribute tables need to be linked or joined, in doing this several issues come to the fore.

The original names and categories are transcribed in phonetic script; this is problematic in the sense that there is no easy way to enter this information in a database file. In addition, phonetic transcription often does not do justice to the variety and nuances in the spoken language and therefore may be misleading. More suitable would be to attach a recording of the spoken version of the name to a landscape representation as part of a learning and exploration tool in a multi-dimensional environment, for instance to retain or learn specific spatio-temporal reasoning skills based on oral language communication. In addition, information on to whom these places are known in integrated. This will provide insight in the area that is used or claimed by a specific group, overlapping regions, and corridors.

As was discussed by Jojola, community boundaries do not overlap, however regions outside the immediate community boundaries (or zone) are used by and are important for multiple groups, as well as the corridors or areas in between communities that facilitate travel and migration. The point data are usually not representative geometrically as most of these names indicate areas, region, streams etc, and are better represented as raster surface. Multiple raster datasets can thus be created each containing specific characteristics of landscape knowledge. The resulting raster files can be compared and analyzed in models based on a multiple variables, such as the correlation between names and occurrence of certain soil types, slope, aspect, as well as processes, e.g. we can begin to question the affordance of these places within a Tewa based paradigm. Figure 2 shows different raster surfaces representing places known to different or all groups, combined/summed to create a surface of shared knowledge. Analysis so far has been exploratory and is an iterative process considering that preconceived notions of landscape categories are abandoned and a new taxonomy is sought, cultural knowledge needs to be the foundation for coding these datasets.

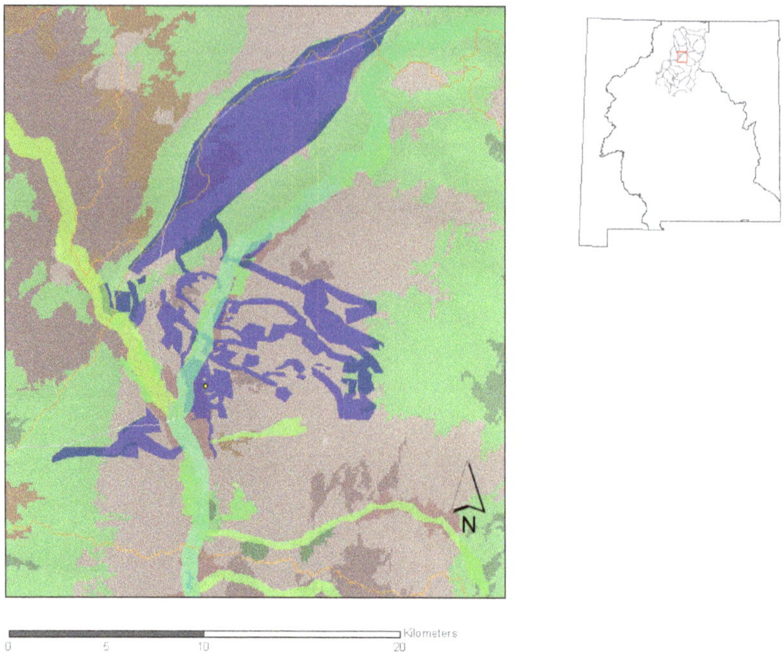

Fig. 2. "Knowledge surfaces" overlain on landcover map.

Fig. 3 . Exploring the Sun path.

Issues and problems:

The cognitive datalayer should not be used in isolation, but as part of a multidimensional analysis in understanding the role of spatial cognition. For instance, placenames referring to animal and plant habitats are not fixed locations, however can be used in change detection studies. In addition to the initial example, other placenames, particularly of the parasitic and perspectival kind, are of interest. Even though some parasitic objects can be explored, for instance the effect of sunlight on a geographic surface can be modeled for its effect, perspectival, as many placenames can be classified as such, seem more complex to explore and understand within the current analytical, GIS, framework. For instance, specific landmarks are of a different kind and less conspicuous. These are not mountaintops, high buildings, instead are – translated as- 'gaps' (Figure 3). This warrants a careful analysis based on a combination of understanding ecological relationships within the current analytical framework with extensive ground-truthing, to gain better insight in the multidimensional world of human spatial cognition.

Conclusion

In past research efforts, analyses of placenames were focused on identifying the pacename itself, its character, if possible comparing to names in other languages, and its tie to its location, but not necessarily the relationship to the larger landscape. Recent research initiatives are integrating toponymy as an important component of investigation spatial cognition and cultural diversity in that domain. For instance, why is a specific corner named, but not another similar one. This may be a result of the way places were named in the European tradition, the categories based on historic descriptions of these names and how they came to be. For this research, it is assumed that these characterizations may not be relevant for different cultures we are trying to understand. The categories that were explored showed that several categories and names are based on relationships between landscape and other cosmological elements. Categories that were discussed as part of the interpretive framework were divided into several kinds. Especially, so-called parasitic and perspectival categories that are based on a multi-place relationship are of interest. Placenames and landscape categories that potentially can be identified as such are the focus of this paper, however others that may not be obvious candidates can also be based on relationships that we so far have overlooked and are therefore not included in the initial exploration.

Acknowledgements

I would like to thank Chandra Bales at the Earth Data Analysis Center, University of New Mexico for providing data, and Dr. Beverly Singer for her continuous support of this research.

References

AGRAWAL, A. 2002. Indigenous Knowledge and the politics of classification, *International Social Science Journal,* 54 (3), 287-297.

ALLEN, B., LINDESAY, J., PARKER, D. 1996. *EL Nino Southern Oscillation Climate Variability,* Collingwood, CSIRO Publishing.

ANSCHUETZ, K. F. 2006, Tewa Fields, Tewa Traditions, in: V. B. Price and B. H. Morrow (eds), *Canyon Gardens: ancient Pueblo landscapes of the American Southwest,* Albuquerque, UNM Press.

BIRNBAUM, D. 2008. *The Hospitality of Presence, Problems of Otherness in Husserl's Phenomenology.* New York, Sternberg Press

CAJETE, G. (ED) 1999. *A Peoples's Ecology, Explorations in Sustainable Living,* Santa Fe, Clear Light Publishers.

CARROLL, J. B. (ED.) 1997 [1956]. *Language, Thought, and Reality: Selected Writings of Benjamin Lee Whorf.* Cambridge, Mass.: Technology Press of Massachusetts Institute of Technology.

CASATI AND VARZI 1999. *Part and Places, the structure of spatial representation,* Cambridge, MIT Press.

CASEY, E.S. 2008. Place in Landscape Archaeology: A Western Philosophical Prelude, in: B. David and J. Thomas (eds) *Handbook of Landscape Archaeology,* Walnut Creek CA, Left Coast Press.

DAVID, B, AND THOMAS, J. 2008. Landscape Archaeology: Introduction, in: B. David and J. Thomas (eds) *Handbook of Landscape Archaeology,* Walnut Creek CA, Left Coast Press.

DOZIER. E. P. 1983. *The Pueblo Indians of North America,* Illinois, Waveland Press.

DUVALL, C. S. 2008. Classifying Physical Geographic Features: the case of Maninka farmers in southwestern Mali. *Geografiska Annaler: Series B, Human Geography.* 90, (4), 327-348.

FORTE, M., 2007. Ecological Cybernetics, Virtual Reality, and Virtual Heritage, in F. Cameron & S. Kenderdine, (Eds.) *Theorizing Digital Cultural Heritage, A Critical Discourse,* (pp. 245-260), Cambridge, London: The MIT Press.

FISHER, P., AND UNWIN, D. (EDS) 2005. *Representing GIS,* West Sussex, John Wiley & Sons Ltd.

GALTON, A. 2000. *Qualitative Spatial Change,* Oxford, Oxford University Press.

GIBSON, J. J. 1979. *The Ecological Approach to Visual Perception,* Boston, Houghton Mifflin.

GOODCHILD, M. F. 1992. Geographical Information Science. *International Journal of Geographical Information Systems,* 6, 31-46.

HAAKE, C. B. 2007. *The State, Removal, ad Indigenous Peoples in the United States and Mexico, 1620- 2000,* New York, Routledge.

HARRINGTON, J.P. 1916. *The Ethnogeography of the Tewa Indians,* twenty-ninth annual report of the Bureau of American Ethnology, Washington, Government Printing Office

HEEN WOLD, A. 1978. *Decoding Oral Language,* London, Academic Press.

KWAN, M. AND LEE, J. 2004. Geovisualization of Human Activity Patterns Using 3D GIS: A Time Geographic Approach. in *Socially Integrated Social Science.* D. Janelle and M. Goodchild, (Eds) (pp. 48-66). New York, NY, Oxford University Press.

LEVINSON, S. C. 2003. *Space in Language and Cognition, explorations in cognitive diversity,* Cambridge, Cambridge University Press.

LEVINSON, S. C. 2008. Landscape, seascape and the ontology of places on Rossel Island, Papua New Guinea, *Language Sciences,* 30, 256-290.

LEVINSON, S. C., AND WILKINS, D (EDS). 2006. *Grammars of Space,* Cambridge, Cambridge University Press

JOJOLA, T. 2004. Notes on Identity, Time, Space, and Place, in: A. Waters (ed) *American Indian Thought,* 87-96,, Malden, Blackwell Publishing.

MAJID, A., ENFIELD, N.J., VAN STADEN, M. (EDS.) 2006. Cross-Linguistic Categorization of the Body, *Language Sciences* 28.

MARK, D., TURK, A. G., AND STEA, D. 2007. Progress on the Yindjibarndi Ethnophysiography, in *COSIT 2007,* Heidelberg, Springer.

MITHUN, M. 1999. *The Languages of Native North America,* Cambridge, Cambridge University Press.

NI, F, CAVAZO, T., HUGHES, M. K., COMRIE, A. C., AND FUNHOUSER, G. 2002. Cool-season precipitation in the southwestern USA since AD 1000; a comparison of linear and non-linear techniques for reconstruction, *International Journal of Climate,* 22, (13), 1645-1662.

ORTIZ, A. 1969. *The Tewa World, Space, Time, Being, & Becoming in a Pueblo Society,* Chicago, University of Chicago Press.

PEUQUET, D. J. 2002. *Representations of Space and Time,* new York, Guilford Press.

RAPER, J. 2000. *Multidimensional Geographic Information Science,* London, Taylor and Francis.

SCHMIDT, P.R., & PATTERSON, T.C. (EDS.). 1995. *Making Alternative Histories, the Practice of Archaeology and History in Non-Western Settings,* Santa Fe, School of American Research Press.

SCHUURMAN, N. 2002. Reconciling Social Constructivism and Realism in GIS, *ACME,* 1, 73-90.

SMITH, L. TUHIWAI, 1999, *Decolonizing Methodologies, Research and Indigenous Peoples,* London, New York, Zed Books,Ltd.

SWENTZELL, R. 1997. An Understated Sacredness, in: V. B. Price, and B. H. Morrow, *Anasazi Architecture and American Design,* Albuquerque, UNM Press.

WARE, C. 2000. *Information Visualization: Perception for Design,* San Francisco, CA, Morgan Kaufmann Publishers, Inc., Academic Press.

WALLERSTEIN, I, 2004. *The Uncertainties of Knowledge,* Philadelphia, Temple University Press.

WATERS, A, (ED.) 2004, *American Indian Thought,* Malden, Blackwell Publishing

To reconcile fire and water? Some discourse issues on the interpretation of aerial images (and not only)

Włodek Rączkowski

Institute of Prehistory, Adam Mickiewicz University, św. Marcin 78, 61-809 Poznań, Poland

Abstract

The use of remote sensing in archaeological heritage studies is becoming ever more common. Archaeologists are finding it harder to competently navigate the world of remote sensing and the images of the world they provide. Thanks to the range of remote sensing data available we are now more often able to acquire varied images of archaeological features and landscapes which are essential elements of our diverse activities from scientific studies to the popularisation of the past. However, we do not treat all of these images as equally valuable. Depending on the expected results, or the effectiveness of the image, we treat some as more useful or reliable reflections of the past and others less so. In practice the images that are acquired are the result of complex interpretation processes as they also are subject to interpretation. Among many archaeological specialists, and scientists from other disciplines involved in archaeology, there appears to be a widespread belief that the results they achieve through refined technical tools are an objectified representation of archaeological features or landscapes. On the other hand the less we (archaeologists) understand the complexity of the technical equipment and the procedures involved the more we may believe in the objectivity of the result. Contemporary philosophical thinking does not allow uncritical acceptance of this point of view. Hermeneutics and phenomenology steer our thoughts towards the complexity of the processes of building knowledge about the past (as well as understanding the past). The above mentioned phenomena have a significant impact on the process of interpretation and teaching using aerial photographs as well as other images of the past worlds.

Key words remote sensing archaeology, hermeneutics, phenomenology, cognitive scheme, aerial archaeology

Introduction

It is appropriate to begin discussion on the interpretation of aerial photographs (and other images) with the same quote used by Rog Palmer and Dave Cowley (in this volume) – 'Give half a dozen people the same information and you get half a dozen different interpretations, all equally valid and probably not even mutually contradictory in any significant way, but each of them will bear the style of the individual interpreter' (Hill 2008).

This means that the proposed discourse is carried out on a worldwide multicultural plain. But where does this statement lead? It is possible to conclude from this that everyone interprets the data he/she receives in his/her own specific way. Does this mean total tolerance for the opinions voiced? Perhaps some of these interpretations are treated as better, others worse? The answer to the latter question is most certainly 'YES' – some interpretations are unacceptable whilst others may be discussed. Such an answer directly indicates that regardless of our tolerance of worldwide polyphony there are certain images of this world which we regard as wrong. The result of such a position is the belief (hidden to some extent) that we know what is correct (if we know what is incorrect). I treat such thinking as an attempt to create solid points of reference and to maintain some certainty in traditional science in a world in which cultural relativism is slowly becoming widespread.

Belief in the cognitive possibilities of science is deeply embedded in our mentality, and the success and effectiveness of technology actually strengthens the conviction that contemporary science (especially technology/science) provides true knowledge about the world. I think it is possible to formulate the thesis that the less we understand the process of building knowledge the more we believe in the final result. This phenomenon refers to the whole field of remote sensing archaeology, starting with aerial photographs (superficially the least technologically complicated) through to multi- and hyperspectral images. The archaeologist using data/images created through such technology treats them as an objectified image of the world (archaeological features and past landscapes).

Objectivity vs interpretation

Remote sensing methods provide archaeologists with images of various aspects of the world including traces which can be interpreted as the remains of human activity in the past. These images are treated as objective representations of the world. How do these pictures arise? The simplified explanation may be as follows: 1) equipment with a built-in sensor registers the pre-determined electromagnetic radiation, 2) the relevant signal is transferred to the computer, 3) the computer collates the data obtained using the appropriate program, and 4) the computer then produces an image which is supposed to satisfy the archaeologist. It is clear that the human element is apparently eliminated from this process and this is meant to guarantee the objectivity of the result (most often in the form of a visualisation). However, if we take a closer look at this process (Figure 1) then we see that it is wholly subject to human decisions.

	1 Stages in the production, acquisition of images	Human involvement	Non-human involvement
1.	Construction of device, sensor, etc.	√	—
2.	Selection of device (by subject of interest, parameters, etc.)	√	—
3.	Software used in transfer of registered signal to computer	√	—
4.	Data processing (software, algorithms, filtering, edge detection, data fusion, etc.)	√	—
5.	Forms of data visualisation	√	—

Fig. 1. Subjectivity/objectivity in the production and/or acquisition of images

The above analysis clearly shows that all images created using remote sensing methods are the result of many human decisions and are therefore the result of interpretation processes. However, for the archaeologist, if he/she does indeed reach for such methods, they are an objective representation of the world. From this perspective we may have the impression that the archaeologist using these images treats engineers, computers technicians, and other experts as 'non-human creatures' who in their activities are not caught up in cultural contexts and use only culturally neutral language.

Aerial photographs in archaeology: a case presentation

The faith of archaeologists in the objectivity of remote sensing, in the image created through LiDAR or the visualisation of a geophysical survey, does not at all differ from the approach of the majority of archaeologists to aerial photographs. However, in contrast to other remote sensing methods the theoretical discussion on the cultural context of aerial archaeology has been on-going now for around 10 years (e.g. Brophy 2002; Brophy and Cowley 2005; Cowley 2002; Rączkowski 1999; 2002; 2005). Therefore, the approach to aerial photographs in archaeology may be treated as an example of a much wider phenomenon.

Photography first appeared in the first half of the 19th century when science was dominated by positivism and the belief that only empiricism was the basis of true knowledge about the world. From the very beginning, photography was treated as an objective representation of the world. In the same way, photography became part of the common imagination about the world and how we discover it. Do we not believe what we see in a photograph? Does a photograph not show the world in a neutral way? Belief in such objectivism of a photograph also comes from the conviction that visual perception is part of our natural in-born skills and any cultural experience has an impact on what and how we can see (Jenks 1995; Slater 1995). Such an approach is described as trivial realism encompassing: realism of representation, ontological realism and mechanical realism. From the perspective of this paper the most important point seems to be mechanical realism. Photography encompasses the means to present the world, the study of it and its manufacture and transformation within the framework of unified image technology. So it is possible then to formulate a thesis that as it is mechanical, it is also impersonal and as such, objective. Such thinking refers to all images used in archaeology – satellite images, aerial photographs, LiDAR images, magnetograms, resistivity maps, thermal photographs, etc.

Roland Barthes – phenomenology and hermeneutics of (aerial) photographs

In contemporary philosophical thinking the belief exists that what we see and the way we approach how to see something is a result of our cultural background. It is closely connected to the way in which forms of knowledge, power strategies and needs systems are shaped in each person's society. If this is true then the relation between seeing and reality is constructed socially. As a result looking at a photograph changes fundamentally. The process of using and working with photographs depends on the social and historical contexts of the observer, and is individualized. The conviction of its objectivity, neutrality and realism becomes scientific myth.

Roland Barthes, the French cultural philosopher, spent a lot of time analysing the nature of photographs and his study of them was two-fold. On the one hand he referred to Husserl's phenomenology, accentuating the experience of consciousness, confined to the analysis of impressions (Barthes 1996), and on the other – he treated photographs as cultural phenomena which are decoded within social discourse (Barthes 1995).

The phenomenological approach to aerial photographs encompasses many aspects of the archaeologist–aerial photograph relationship. It is not possible to discuss them all here in detail. The important point is that both in the process of taking aerial photographs (particularly oblique) and during the stage of looking at the photographs and their analysis, an emotional connection between the archaeologist and the subject of the photograph is created. An important element in the creation of this relationship is the deeply rooted desire to discover and the need to metaphorically appropriate fragments of the past (this is evident in the language used – i.e. 'my site') (Shanks 1992), and publications or conferences often turn into a 'vanity fair', where successful achievements such as new discoveries or modern documentary evidence are shown off.

Individual emotions allow an object in the world - its relation to perceived elements - to be defined (Heidegger's Being-in-the-World). But this means that in some way the understanding of these elements must be shaped in our minds. So here enters the question of hermeneutics, the art of reading (interpreting) a text. In the case of aerial photographs this is connected to the ability to decode the information present in the photograph. The whole process of 'reading' photographs is tied up in culture and is dependent on our abilities. Therefore, the teaching of the 'reading' of aerial photographs – if they are to serve research procedures – is important. So now we have a certain opposition to deal with – phenomenology highlights the role of individual emotions and how they affect perception, but hermeneutics, through the need to teach cultural codes used in social communication, steers us towards the unification of their understanding.

So then are phenomenology and hermeneutics in a mutually opposed relationship? To a certain extent they are, but this is an over-simplification. Rather, it is possible to say that they build a dialectical relationship. So the knowledge that we have up to this point influences how we perceive, and therefore also our emotions. However, emotions can lead to a deeper interest in the subject under study and as a result greater dedication, a more detailed analysis and understanding.

To teach or not to teach

Rog Palmer and Dave Cowley (in this volume) raised the problem of 'little consideration of issues of interpretation' of aerial photographs and other remotely sensed images in contemporary approaches. This 'bad practice' is to lead to many false results, misunderstandings etc. Of course I have no intention of questioning the need to teach how to work with aerial photographs. This question is beyond discussion. But we may take a look at this problem from various angles. Education is the foundation for the consolidation of cultural models, the creation of social identities and group continuity. This may refer to aerial archaeologists too. Within this particular group certain modes of functioning have been created, defining power relationships in the field of using aerial photographs (Rączkowski 1999), both in relation to archaeologists and the public. Not that there is anything wrong with this, it is a natural result of the specialisation process in every discipline.

There may be various strategies for the teaching of aerial photograph interpretation. One such strategy is founded on categorising features based on their characteristics (shape, colour, texture, etc.), on particular models, which allow for the identification and interpretation of features visible on the photographs. This practice is especially well developed in the geographical sciences, though a similar idea has appeared in archaeology (Wilson 1982). With today's technology this may lead to thoughts on the automatisation of the process of aerial photograph and satellite imagery interpretation (e.g. Redfern 1997; Trier et al. 2008). I have no doubts that such developments lead to a reduction of information about features, their specific characters, complexity and so on. Another strategy refers to the information we have to date about archaeological features (including formation processes) and the mental image they shape which makes their identification on aerial photographs possible. Both approaches accept (though to a different extent) the assumption that specialist knowledge has a fundamental influence on the interpretation result. Experience in the field of interpretation is a particularly important factor in the teaching process.

One of the conclusions which we seem to be coming to here is as follows: experience (knowledge) is possessed by a few (authority) and they have the essential competence which allows them to teach properly (power). Such a relationship leads towards a totalitarian approach. Of course this is common practice and it is practically impossible to imagine a teaching process without such a relationship (I myself work within it). So it is not aspects of totalitarianism in the process of interpretation of aerial photographs education which make me feel uncomfortable, but rather the results of the teaching.

Cognitive schemes vs perception

The consequence of the teaching process is not only knowledge, which may be utilised. To some extent cognitive schemes are shaped in the background. A cognitive scheme is a model of reality which has two inseparable and integrated functions. The first one is a description or representation of reality as well as a set of procedures for processing information about this reality. The second function guides our recognition of objects and the way we think about them. When our cognitive scheme is stimulated by incoming preliminary data, it autonomously, i.e. without us being aware of it, starts guiding our attention, perception and reasoning (Rączkowski 2006). Within several different actions the cognitive scheme activates our mind in: searching for pre-defined information, evaluating the incoming information, interpreting the information in a particular way (conditioned by our preliminary hypothesis), supplementing the incoming information (with our own information generated by our expectations) and even distorting the incoming information to make it match our expectations. Consequently, the cognitive scheme filters our search and perception making our mind notice pre-selected information, supplement it, as well as combine and categorize the information gathered (Figure 2).

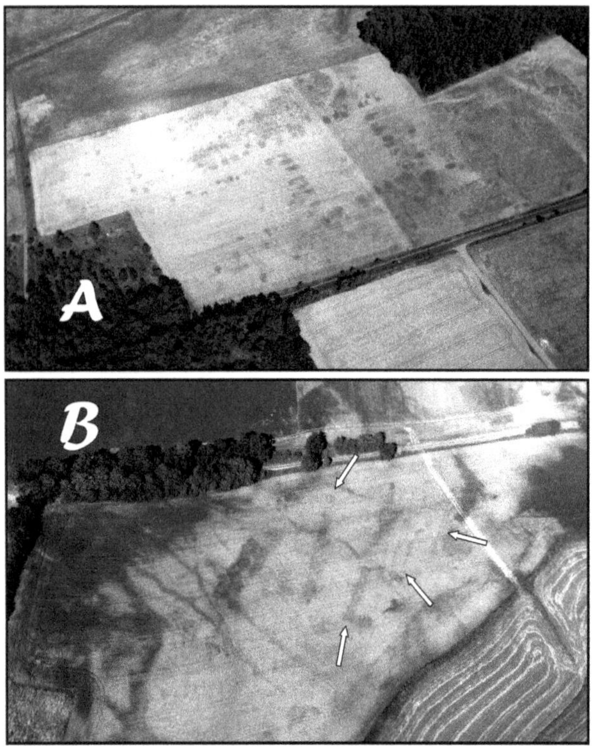

Fig. 2. Two aerial photographs showing archaeological features taken 20 minutes apart. Photograph A shows a plan of a medieval town visible thanks to the cropmarks. In B traces of leveled megalithic long barrows are just visible. Which of these archaeological sites would create a stronger emotional response during aerial reconnaissance? Perhaps surprisingly, it was photograph B which raised my adrenaline levels. The features visible in A were treated as a complex of pits (sunken houses) of no great significance. This approach (including perception and emotions) was affected by the cognitive scheme present in my consciousness which 'did not allow' me to identify, from the very beginning, the particularly valuable discovery of a medieval town, but which resulted in euphoria about a not-particularly convincing picture of damaged megalithic long barrows. It was not until I 'read' the photograph showing the town that I changed my opinion. I became quite excited on further analysis (reading) of it. How many such exceptional sites have been missed, gone unnoticed, 'blocked out' by our cognitive scheme?

The cognitive scheme shaped in the teaching process therefore has far-reaching consequences in aerial archaeology practice (and beyond). It will interfere in the phenomenology–hermeneutics relationship. On the one hand it will affect perception and the related individual emotions, and on the other it will shape the process of 'reading' photographs. This interference will lead not to a divergence but a unification of the image being shaped – we will see what our educated cognitive scheme wants to see.

Conclusions – can we reconcile fire and water?

No one questions the need for, the indispensability of, teaching, including the interpretation of aerial photographs. But the question of responsibility for the results of this teaching remains. How do we reconcile the need for cultural consensus within a group such as archaeologists interested in remote sensing methods, whilst at the same time retaining the right to individualised thought, to solve research problems? Is it possible to reconcile fire and water?

References

BARTHES, R. 1995. *Image–Music–Text.* New York, Hill & Wang.

BARTES, R. 1996. *Światło obrazu. Uwagi o fotografii.* Warszawa, Wydawnictwo KR.

BROPHY, K. 2002. Thinking and doing aerial photography. *AARGnews* 24, 33–39.

BROPHY, K. AND COWLEY, D. (EDS) 2005. *From the Air. Understanding Aerial Archaeology.* Stroud, Tempus.

COWLEY, D. 2002. A case-study in the analysis of patterns of aerial reconnaissance in a lowland area of southwest Scotland. *Archaeological Prospection* 9, 255–265.

HILL, R. 2008. *The Roar of the Butterflies.* London, HarperCollins.

JENKS, C. 1995. The Centrality of the Eye in Western Culture: an Introduction. In C. Jenks (ed.), *Visual Culture*, 1–25. London, Routledge.

RĄCZKOWSKI, W. 1999. Power of image: some ideas on post-processual aerial archaeology. *AARGnews* 19, 10–14.

RĄCZKOWSKI, W. 2002. Beyond Technology: do we need 'meta-aerial archaeology'? In R. H. Bewley and W. Rączkowski (eds), *Aerial Archaeology. Developing Future Practice*, 311–327. Amsterdam, IOS Press.

RĄCZKOWSKI, W. 2005. Theoretical dialogues – is there any theory in aerial archaeology? *AARGnews* Supplement 1, 12–22.

RĄCZKOWSKI, W. 2006. Towards integration: two prospection methods and some thoughts. In S. Campana and M. Forte (eds), *From Space to Place. 2nd International Conference on Remote Sensing in Archaeology*, 203–206. Oxford: BAR Publishing, British Archaeological Reports, International Series 1568.

REDFERN, S. 1997. Computer assisted classification from aerial photographs. *AARGnews* 14, 33–38.

SHANKS, M. 1992. *Experiencing the Past. On the Character of Archaeology.* London, Routledge.

SLATER, D. 1995. Photography and modern vision: the spectacle of "natural magic". In C. Jenks (ed.), *Visual Culture*, 218–237. London, Routledge.

TRIER, Ø., LOSKA, A., ØYEN LARSEN, S. AND SOLBERG, R. 2008. Detection of burial mounds in high-resolution satellite images of agricultural land. In R. Lasaponara and N. Masini (eds), *Advances on Remote Sensing for Archaeology and Cultural Management*, 17–21. Rome: Aracne.

WILSON, D. R. 1982. *Air Photo Interpretation for Archaeologists*. London, Batsford.

Geomatics techniques for the 3D documentation and visualization of archaeological building

Stefano Campana,[1] Matteo Sordini,[1] Alessandro Rizzi,[2] Fabio Remondino[2]

[1] Dept. Archaeology & History of Arts, Landscape Archaeology & Remote Sensing LAB - University of Siena, Italy, campana@unisi.it, sordini@lapetlab.it - http://www.lapetlab.it
[2] 3D Optical Metrology Group - Bruno Kessler Foundation, Trento, Italy
<rizziale><remondino>@fbk.eu - http://www.fbk.eu

Abstract

The work illustrates the surveying of a small but spectacular Romanesque church close to the castle of San Giovanni d'Asso (Siena, Italy). Using terrestrial laser scanning (TLS) and image-based techniques, a complete 3D model of the interior and exterior of the church was produced. . TLS was employed for the interior of the church, which consists of a main floor above a very small crypt, the recording of which was complicated by a plethora of columns and arches. The image data were acquired with consumer-grade digital cameras along with a model radio-controlled helicopter to capture details of the roof and the apse which were not visible from ground level. The 3D survey, representation and reconstruction of San Pietro in Villore should become a case study of scientific research addressed to test a simulation based open-space where it is possible in every moment to measure and compare historical evidence, stratigraphical relationship and other datasets but also where is possible to improve and upgrade the record in the future on the basis of new field work or post-processing.

Key Words 3D recording, archaeological needs, photogrammetry, laser scanning, visualization.

Introduction

The Heritage sites in the world (natural, cultural or mixed) suffer from wars, natural disasters, weather changes and human negligence. According to UNESCO, a heritage can be seen as an arch between what we inherit and what we leave behind. In the last years, great efforts focused on what we inherit as Cultural Heritage and on their documentation, in particular for visual man-made or natural heritages, which received a lot of attention and benefits from sensor and imaging advances. The importance of Cultural Heritage documentation is well recognized and there is an increasing pressure to document and preserve them also digitally. Therefore 3D data are nowadays a critical component to permanently record the shapes of important objects so that they might be passed down to future generations. This has produced firstly a large number of projects, mainly leaded by research groups, which have realized very good quality and complete digital models (Beraldin et al., 2002; Gruen et al., 2005; Remondino et al. 2009) and secondly has alerted the creation of guidelines describing standards for correct and complete documentations.

The actual geomatics technologies and methodologies for Cultural Heritage documentation (Ikeuchi and Miyazaki, 2008) allow the generation of very realistic 3D results (in terms of geometry and texture) used for many scopes like archaeological documentation, digital conservation, restoration purposes, VR/CG applications, 3D repositories and catalogs, web geographic systems, visualization purposes, etc. But despite all the possible applications and the constant pressure of international organizations, a systematic and well-judged use of 3D models in the Cultural Heritage field is still not yet employed as a default approach for different reasons: (i) the "high cost" of 3D; (ii) the difficulties in achieving good 3D models by everyone; (iii) the consideration that it is an optional process of interpretation (an additional "aesthetic" factor) and documentation (2D is enough); (iv) the difficulty to integrate 3D worlds with other more standard 2D material. But the availability and use of 3D computer models of heritages opens a wide spectrum of further applications and permits new analyses, studies, interpretations, conservation policies as well as digital preservation and restoration. Thus virtual heritages should be more and more frequently used due to the great advantages that the digital technologies are giving to the heritage world and to recognize the documentation needs stated in the numerous charters and resolutions.

The case study presented in this work is related to the church of San Pietro in Villore (Figure 1) mentioned in documents for the first time in 1278. Today the church is one of the most important Romanesque buildings in the Siena countryside for its rich range of structural and decorative features. The original roof, now only extant in the presbyterial area, is supported by blind arches set against the walls, as well as the large self-supporting cross-arch, and it is characterized by the presence of diagonal arches with large quadrilateral sections. Under the floor of the church there is a crypt, a feature that is relatively uncommon in the Romanesque churches of the Siena landscape. The crypt is divided into three naves with seven columns. This is a peculiarity not very common in the Romanesque churches of Siena countryside. Although this is a relatively small building, it represents an excellent case study to show why archaeologists should make use of 3D modeling. On this issue we should recognize that the returns in recent decades of archaeological research from the use of 3D digital documentation and representation, in terms of scientific investigation, are relatively disappointing, bearing in mind the possibilities offered nowadays by newly developed software and hardware.

Fig.1. Some views of the Romanesque church of San Pietro in Villore, Siena (Italy).

2 Surveying of a Romanesque churc

In archaeological research it is essential to apply geomatics techniques to correctly and metrically represent all peculiarities of the investigated object. Nowadays we can employ a wide amount of geomatics tools to obtain results usable in different scales of archaeological investigation (Campana et al., 2008). The most common used techniques for 3D documentation, reconstruction and interpretation process are image-based (Remondino and El Hakim, 2006) and range based data (Cignoni and Scopigno, 2008).

In order to obtain a virtual 3d model of San pietro in Villore at local scale, useful to be used in intra-site archaeological analysis, we employed a strategy based on the integration of multiple techniques: a multi-resolution and multi-sensor approach has been used choosing the most appropriate methods in relationship to the research scale, object size, location and required detail. The integration between different data sources was essential to overcome some limits of each method and to have as result the complete 3D geometric information of the heritage.

2.1 Data acquisition

The first step of reverse engineering process was the survey of the monument. We used a terrestrial laser scanner for massive data collection of the shape. TLS survey of the interior and exterior of the Church has been made using the TOF instrument Leica Scanstation 1 at the resolution of 5 millimeters, leading to 130 millions of points. Due to natural and unnatural occlusions, part of the roof and an half of the posterior apse could not be scanned. To overcome the limits of TLS survey, aerial and terrestrial images were acquired with consumer digital cameras (6 and 10 mega pixel). For the aerial images, a small remote-controlled model helicoper (drone realized by Microdrones GmbH, Germany and operated by Zenit s.r.l., www.zenit-sa.com) was used. The terrestrial images were acquired mainly for texture mapping purposes. Approximately 80 HDR textures (High Dynamic Range) of all surfaces were acquired to realize a very photo-realistic 3D model of the heritage monument.

2.2 Data processing and 3D model generation

Compared to the data collection stage, the data processing step is generally much more time consuming, saying at least 5 times more: each kind of dataset has to be processed with specific software to extract a clear dataset ready to be integrated with all the others.

TLS data has been cleaned from noise to remove unwanted features, registered (Figure 2), meshed and finally edited again to close holes and repair surface normal. On the other hand, the image data of the roof and apse have been processed with the typical photogrammetric pipeline consisting of calibration and orientation, 3D surface measurement and texture mapping. Both TLS and image based datasets have been merged using Innovmetric Polyworks. Finally the HDR images have been mapped onto the 3D geometry using TexCapture software. This software can texture a mesh by the selection of homologues points between the 3D model and the 2D image. This is still an interactive procedure as no automated and reliable approaches were proposed yet. Indeed the identification of homologues points between 2D and 3D data is a hard task much more complex than image to image or geometry to geometry registration.

Some views of the final textured 3D model of the church are shown in Figure 3 and 4.

Fig. 2. The range data of the exterior of the church (left) and the crypt (right).

Fig. 3. Different views of the modeled crypt, shown in shaded and textured mode.

3 Conclusions

The paper described how accurate, detailed and photo-realistic 3D results can be relatively easily derived from the integration of multiple geomatics tools and sensors. The 3D survey, representation and reconstruction of San Pietro in Villore should become a case study of scientific research addressed to test a simulation based "open-space" where it is possible in every moment to measure and compare historical evidence, stratigraphical relationship and other datasets but also where is possible to improve and upgrade the record in the future on the basis of new field work or post-processing. This approach should, therefore, allow us to achieve:

- better understanding of analytical study of the building - stratigraphy, structural, architectural, stylistic also assisted by automatic extraction of cross-sections, longitudinal sections and prospects;

Fig. 4. The 3D model of the church's exterior, shown in textured mode with artificial background.

- the interpretation of archaeological features in deep association with their physical and cultural context exploring relationships across a landscape;
- overcoming space and time sharing complex "objects" with other experts all around the world;
- the demonstration (firstly addressed to archaeologists) that digital reconstruction is a complex exercise and projective interpretation and also is much more than a graphic reconstruction, is a simulation, that allows exploration of three-dimensional model perspectives through endless investigations and non-invasive documentation;
- virtual reconstruction and simulation scenarios address to specialist as to the general public developing new tool to share but also preserve the memory of cultural heritage;
- development of 3D archaeological thinking.

Acknowledgements

The realized project would have been impossible without the support of the community administration of San Giovanni d'Asso, Italy. Our acknowledgements go to Stefano Girardi (FBK Trento, Italy) for his valuable aid and suggestions in data processing skills and to Daniele Sarazzi (Zenit company, Italy), for flying the Microdrones helicopter to capture aerial photographs. The final sincere thanks go to the PAVA Foundation with their founder and associate members: Lea Ricci, Nina Benders, Virginie Saverys, Raymond Benders, Michele Boscagli, Sebastiano Cameli, Sheppard Craige, Russell Wilkinson, Carol Van Wonterghem but also all the other friends of PAVA.

References

BERALDIN, J.-A., PICARD, M., EL-HAKIM, S.F., GODIN, G., VALZANO, V., BANDIERA, A. AND LATOUCHE, D., 2002. Virtualizing a Byzantine crypt by combining high resolution textures with laser scanner 3D data. *Proc. of VSMM* 2002, pp. 3-14

CAMPANA, S., REMONDINO, F., SORDINI, M., 2008: *Integration of geomatics techniques for the digital documentation of building and archaeological excavations, 1st Workshop Earth Observation: "Advances in Remote Sensing for Archaeology and Cultural Heritage Management",* Organised by EARSeL, CNR - National Research Council, Rome, Italy

CIGNONI, P., SCOPIGNO, R. 2008. Sampled 3D models for CH applications: a viable and enabling new medium or just a technological exercise? *ACM Journal on Computing and Cultural Heritage*, 1(1), article 2

GRUEN, A., REMONDINO, F., ZHANG, L., 2005. The Bamiyan project: multi-resolution image-based modeling. In *Recording, Modeling and Visualization of Cultural Heritage*, E.Baltsavias, A.Gruen, L.Van Gool, M.Pateraki (Eds), Taylor & Francis / Balkema, pp. 45-54.

GUIDI, G., REMONDINO, F., RUSSO, M., MENNA, F., RIZZI, A., ERCOLI, S., 2009a. A multi-resolution methodology for the 3D modeling of large and complex archaeological areas. *International Journal of Architectural Computing*, 7(1), 40-55

IKEUCHI, K., MIYAZAKI, D., 2008. *Digitally Archiving Cultural Objects.* Springer, 503 p.,

REMONDINO F. AND EL HAKIM S., 2006: Image based 3D modeling: a reviw, *Photogrammetric record*, Vol. 21(115), pp. 269-291

REMONDINO, F., EL-HAKIM, S., GIRARDI, S., RIZZI, A., BENEDETTI, S., GONZO, L., 2009A. 3D Virtual reconstruction and visualization of complex architectures - The 3D-ARCH project. *International Archives of the Photogrammetry, Remote Sensing and Spatial Information Sciences,* Vol. 38(5/W1), Trento, Italy (on CD-Rom)

Session 3

3D REMOTE SENSING: CYBER ARCHAEOLOGY AND LANDSCAPES VISUALIZATION THROUGH THE TIME

Digital videography: recording, preserving, and disseminating archaeological data

Munish Kumar Tiwari

Keywords Sony DCR-VV1000 Video Camera, Hard Drives, RAID Disk Configuration, Fire Wire (tm) (IEEE-1394), Adobe Premier 4.2.

Introduction

Developments in high-quality digital video cameras and consumer grade computer hardware and software have revolutionized the video industry, creating unparalleled opportunities for archaeologists, architectural historians, and historic preservationists. Developments in digital imaging will change the ways in which they collect and record data, and promote the dissemination of findings both within the historic preservation community and out to a public that has, in recent years, become enamored with archeology. This dynamic, flexible technology will take its place in the archaeological toolkit, providing a means for modeling data and testing hypotheses.

Research Method

The research design will be descriptive in nature. The method of data collection will be mainly based on secondary data. However, an effort will be made to collect primary data by taking information from competent persons off and on. *The Lost Towns of Anne Arundel Project,* in cooperation with the Anne Arundel County Trust for Preservation, uses digital imaging technologies to create three-dimensional (3D) graphics, edit digital video, and store graphical data.

Objectives

The principal goal of the project is to identify a suite of hardware and software tools that, taken together, can meet the educational and scholarly needs of historic preservationists and archaeologists in a cost-effective manner. While not the ideal system for commercial video production, the flexibility and cost effectiveness of this suite of tools make it an attractive option for archeologists and architectural historians. DV recording enables archaeologists to record daily field activities, adding to, but not supplanting, conventional field notation.

Specific goals of the project include:

1. Capture and store additional digital images of artifacts and archaeological features;
2. Create three—dimensional models of archaeological features;
3. Develop virtual historic standing structures from London and Providence archaeological data;
4. Assist the London Town Foundation in providing access to portions of London and Providence, in compliance with the Americans with Disabilities Act;
5. Provide video images to broadcast journalists and film makers; and
6. Disseminate research results and interpretations via the Internet.

To reach these goals the following products have been completed:

1. A digital video archive of excavations in both low and high intensity light;
2. A library of digital raster still images in both low and high intensity light;
3. Three-dimensional models of historic standing structures created from conventional and digital raster images;
4. Three-dimensional models of select artifacts created from conventional and digital raster images;
5. Development and broadcast of a television video, *Unearthing Lost Towns,* in cooperation with Anne Arundel County's cable television studio, incorporating animated 3D models;
6. Development of a 3D model of the Robert Burle House (1660s—1680s) for use in Maryland's Archeology Month poster;
7. An illustrated article on the digital Videography project published in the Maryland Archeology Month Calendar (Moser and Corder 1999);
8. A scholarly paper on the digital Videography project presented, with animated images, at the Annual Meeting of the Society for Historical Archaeology in Salt Lake City, Utah (Moser, et al., 1999); and
9. A digital Videography demonstration at the 1998 annual meeting of the Vernacular Architecture Forum in Annapolis, Maryland.

Mobilization

Acquiring state-of-the art technology requires considerable research, the prospective operators learning about the use and development of computer hardware, software, digital video production, storage, and file format compatibility, specialized vocabularies and concepts—just to name a few issues. Some members of our staff already were conversant in Geographic Information Systems (GIS), Computer Aided Drafting (CAD), Global Positioning Systems (GPS), and a variety of other software applications. The video includes some of the project's initial attempts at 3-D modeling of artifacts and buildings. The producer also provided access to the county's videographic facility, including an editing deck.

Methodology

The goal of this project is to create an affordable DTV (Desktop Video) editing system capable of capturing DV (Digital Video) for long term storage and dissemination. The products of this DTV included a digital database of sites on which *The Lost Towns of Anne Arundel Project* conducts ongoing archaeological research. Excavations are planned around filming (as they are around all forms of data recordation), and all significant features digitally recorded in the same fashion in which we had videotaped excavations. One significant difference involves the creation of three—dimensional control fields— placement of point provenienced, labeled pins within the excavation—to insure accurate three dimensional integration of images. Similarly, cleaning and conservation of poorly preserved and potentially significant artifacts are filmed, indexed with artifact images, and linked to computerized artifact catalogues.

Equipment

Creating a viable video editing system in this era of increasing technological innovation requires a firm grasp of many fundamental issues relating to the use and development of computer hardware, software, and video. Extensive research preceded equipment purchase. We purchased a Pentium II, 233 MHz computer with a Windows NT 4.0 operating system. This system contains a digital transfer technology referred to as FireWire. FireWire(tm) (IEEE-1394) technology is a high-speed cable that connects digital camcorders and DVCRs to computers.

Sony DCR-VV1000 Video Camera

The first purchase necessary to this project was a video camera. The technology and functionality of the camera then determined the DTV (Desktop Video) editing suite we could use. The first decision was whether to purchase a DV or a Hi8 camera. There are two differences between them. The first difference between the two is resolution. A Hi8 camcorder records images at 400 lines of resolution: DV camcorders record at 500 lines of resolution. Hi8 camcorders record to tape using an analog format: DV camcorders record in a digital format. To edit non-linear video on the desktop it must first be converted to a digital signal with a digitizing card (PCI). This digitizing process usually introduces defects or (generation loss) into the video.

Recording the images in DV and then transferring them to the DTV eliminates generation loss. Whether digitized analog video, or native DV format, once the video is loaded into the computer the resulting digital file can be accessed like any other computer file, or transferred with no loss of quality to another medium (an important issue given the curation issues confronting conventional magnetic tape). These two factors led us to purchase a Sony DCR-VX 1000 Video Camera (Figure 3-1). This camera is a 'high—end' consumer model, which possesses many of the features of professional equipment, including; 1 8nimlsecond speed with a 1/3" charge coupled device, l0X (5.9 to 59mm) lens, 1/4 to 1/10,000 shutter speed control for use in subdued light, AC power adapter, and a directional microphone (5—6°).

Fig. 3-1. Sony DCR-VX 1000 Digital Video Camera.

Hard Drives

Digital video editing requires more sophisticated hardware than typically used in most computer applications. NTSC(North American Television and Video Standard). Digital Video is captured at nearly 30 fps (frames per second). Every second of video is composed of thirty individual images, and each frame of video is approximately 1.2 MB in size (at 720 x 480 resolutions). This represents a tremendous quantity of data for the computer to process for every second of video playback. Two types of hard disks generally are used in computers: IDE (also called ATA) and SCSI. IDE hard disks generally are larger and more cost effective than SCSI drives; however, an IDE interface to the hard disk is not capable of high-speed data transfer.

RAID Disk Configuration

To further increase the performance, efficiency, and cost effectiveness of the SCSI hard disk system, we purchased two 4.4-Gigabyte SCSI Cheetah (10,000 rpm) drives. We then arranged these drives in a RAID configuration. This configuration provided 8.8 Gigabytes of storage space, operating faster than a comparably priced single 9 Gigabyte SCSI drive. Invented at the University of California at Berkeley in 1987, RAID is an acronym for Redundant Array of Inexpensive, or Independent, Disks. In this system multiple hard drives, usually SCSI drives, work together as a single disk, writing data across several hard disks, performing faster and with greater security *(PC Magazine,* November 4, 1997).

Initially, six varieties of RAID were designed, each having different characteristics and purposes; additional RAID variations, however, have been introduced. Until recently, RAID storage technology was used primarily in upper midrange to high end computing applications. This was due primarily to the higher cost and integration complexity of RAID technology and a lack of understanding of the benefits of RAID technology. Decreasing hardware costs, with increased performance and ease of use, have brought RAID into the realm of affordable video-editing applications. The RAID system consists of" an external desk side or rackmount subsystem that contains{s] a RAID controller, its supporting hardware, and a number of disk drives. The RAID controller is connected to a host computer with an interface cable using technologies such 'Measured in thousands of bytes per second.

Fire Wire (tm) (IEEE-1394)

Digital video cameras have introduced a new form of data transmission, creating images in a digital format (.AVI file) and negating the need for conversion. We selected DPS Spark as the means of affordably transferring, broadcast quality DV to our computer. This equipment permits movement of data from one storage device to another without alteration or degradation. The data then is transmitted from edited tape back to the DV camcorder for TV viewing or for recording directly to a VCR. FireWire eliminates the analog video digitization process of video capture hardware by transferring already digital DV data from DV tape to a hard drive. Despite repeated transfers, video and audio remain equal to the original DV source. DV technology provides near-Betacam SP quality on the desktop as well as a choice of display sizes; including, 90 x 60, 180 x 120, 360 x 240 or full DV resolution at 720 x 480. At the heart of this complete editing system is FireWire(tm) (IEEE-1394) technology, a high speed cable that connects digital camcorders and DVCRs to computers. DPS Spark (IEEE-l394) PCI bus adapter connects to a DV camcorder using the FireWire connector, thereby transferring digital audio and video to the hard drive in real time, without analog-to-digital conversions. Once loaded into the computer these video files can be manipulated using Adobe Premiere 4.2, a video editing software bundled with the DPS Spark card.

Software

Adobe Premier 4.2

It is designed for low and middle end budgets, and it is bundled with the DPS Spark FireWire interface card. Adobe Premiere stitches together multiple video clips, saving them as .AVI files. Each clip can be placed into proper sequence, allowing the creation of a continuous video file. In addition, Adobe Premier allows addition of effects to the video. These effects create smoother transitions between clips and add continuity to the video. Fades, wipes, scrolling, text formatting for credits and subtitles, and other effects and transitions help produce a compact, visually pleasing multimedia experience.

Photo Modeler Pro 3.0

Creating 3D images requires multiple programs. We capture images or scenes using either the Sony digital video camera, or a conventional 35mm camera with inserted fiducials. We transfer the images to the computer through the FireWire interface, or a Hewlett Packard Photo smart Photo Scanner. Photo Modeler Pro uses multiple photographs of an object or scene, using information about the camera and the lenses, and extracts measurements. Points in each are photograph marked with unique reference numbers and the common reference points between photographs are manually referenced to one another. Once completed, the computer can process the spatial relationships of the data and generate a 3D model of the object or scene (Figure 3-3).
Archaeologists can use Photo Modeler to:

- Model artifacts, particularly fragmentary objects and building reconstructions
- Perform morphological measurements
- Document excavations
- Record shipwrecks and other submerged sites
- Document standing historic structures for conservation and preservation

Fig. 3-3. Photomodeler image downloaded from the intenet.

Light Wave 3D

Light Wave 3D is a professional graphics-modeling program and, at approximately $2,000, and is less expensive than functionally comparable packages; e.g., 3D Studio MAX and Softimage. Fortunately, we were able to obtain access to Light Wave through the Arnie Arundel County video production studio. Light Wave 5.0 consists of two applications: Modeler and Layout. Modeler creates wire frame models, adds textures and lighting effects, and then transports the models to Layout for animation. Modeler offers four simultaneous viewports in which to design and construct objects. Although well-regarded by graphics artists, Light Wave is a difficult program to learn. Relatively simple tasks are convoluted in Light Wave: Unfortunately, neither Modeler nor Layout let you easily edit your curve tangents, which made fine-tuning our path difficult. Modeler has spline controls, but in Layout you can specify only the tension, continuity, and bias parameters.

Data and product storage

DV Tape

The best method for storing large video files, in terms of expense and space, remains the DV cassette tapes on which the images were first recorded. DV cassette tapes, a magnetic medium, are subject to the same failures as 1.44-MB floppy disks and hard disks. Magnetic media are particularly susceptible to temperature and humidity extremes, as well as strong magnetic variations. In addition video written to DV cassettes is compressed using a software code which our equipment compresses at a 5:1 ratio, thus, increasing the storage capacity of the tape by a factor of five.

Removable Drives

Removable storage drives also store digital data and they are relatively inexpensive; e.g., SparQ and Syjet drives manufactured by SyQuest, and the JAZ drive manufactured by Iomega. These drives store between one and two Gigabytes of data; the new Iomega JAZ drive has doubled its capacity to 2 GB, SyQuest will soon manufacture a 4.7 GB Quest drive more expensive than the disks offered by Iomega. Regardless, a portable drive is nearly essential for any desktop video editing solution, and greatly improves the performance of any computer system. Practical transporting of video files demands portable drives.

CD-R/R W

One of the newest and most exciting forms of storage media is the R/RW CD ROM. Unlike magnetic media, CD-ROM technology uses optical technology to store data. CD-R/RW technology uses a laser to write data to an optical compact disk (CD). This form of data storage is more stable than magnetic media. Temperature and humidity do not affect data stored on CD-ROM, nor is it subject to data degradation; however, it is susceptible to very high temperatures and physical destruction. Currently CD-R/RW recorders are available for about $200, with very inexpensive CDRIRW disks (about 1/6 cent per MB). These CDs can store about 650 MB of data. Current CDR/ RWs feature 2X to 4X write speed and 16X to 24X read speeds. Writing a CD on a 2X CD-R takes approximately thirty-six minutes and eighteen
minutes on a 4X CD-R.

DVD

Another technology under development is recordable DVD. Recordable DVD may be the single best answer to long term video storage. DVD technology is still in its infancy, with no consensus for standardizing the format. In fact, contention between standards organizations may lead to competing formats such as occurred between VHS and Beta during early VCR development. Right now several groups are vying for supremacy in writable DVD, including DVD-ROM, DVDRAM, DVD-R, DVD+RW, and DVD-RW. DVD-R is currently available with a capacity of storing approximately 3.9 Gbytes (Troop 1998).

Implementation

Testing

Following these calibrations, several digital still images of objects were captured as tests for the software.

Rumney's Tavern (18AN48)

One of the first significant tests for the system involved the creation of a three dimensional model of the Runmey's Tavern cellar. *The Lost Towns of Anne Arundel Project* began excavations at Rumney's tavern in 1996. The team began by excavating fifty-five 5—ft by 5-ft Squares, removing the plow zone and exposing features that intrude into the subsoil. Excavations exposed nine postholes with molds and a trash—filled cellar hole measuring approximately 18 by 16—ft at its surface and, as the excavations eventually revealed, 1 0—ft long at its base. Excavators divided the cellar into four quadrants and excavated stratigraphically to the cellar floor, a distance of about 5—ft below the current grade.

Prior to photographing and video-recording the excavation area was retrowled, cleaned, profiled, and mapped. The walls of the cellar were highlighted using 3" long steel pins with colored plastic heads. The pins were pushed into the subsoil to mark the location of major changes in cellar contours. The pins~ assist the Photo Modeler operator by enhancing the visibility of the cellar's irregular contours and referencing the same physical point in space among multiple photographs. These visual references greatly assist in marking the photographs (Figures 6-4 and 6-5).

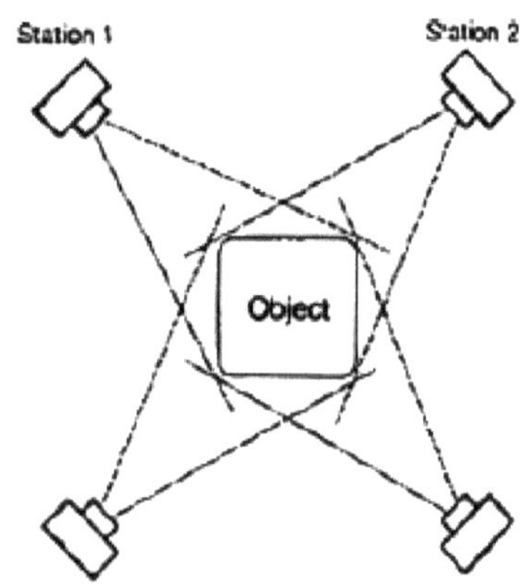

Fig. 6-4. Top view of suggested camera orientations for photographing objects for PhotoModeler (Eos System inc, 1997).

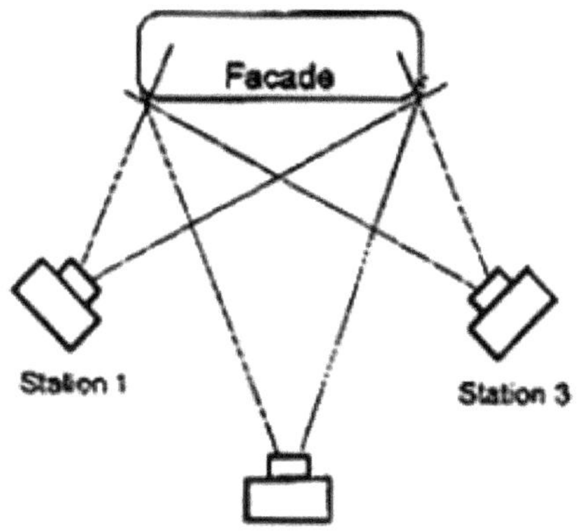

Fig. 6-5. Top view of suggested camera orientations for photographing facedes for PhotoModeler (Eos System inc, 1997).

The cellar was photographed and video-recorded in February and March of 1998. The Photo Modeler user manual suggests the following photography guidelines:

• shoot as close to right angles as possible, take at least three photographs;
• shoot all important points on at least three photographs;
• overlap adjacent photographs;
• take photos from above and below the object;
• take many photographs, but begin using only four until others are necessary; and
• Measure the distance between two clearly visible points.

To ensure that we captured a sufficient number of photographs with overlapping points, photographed the cellar within points all around its perimeter were captured (Figure 6-6).

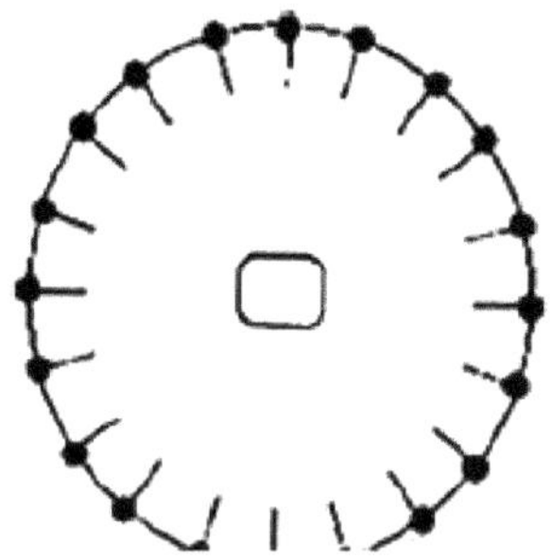

Fig. 6-6. Top viewof suggested camera positioning to photograph in 360 degree ring (Eos Systems Inc., 1997).

The software generates the object or scene by projecting straight lines from the camera position, through the point on the film or CCD (Charge Coupled Device), and out into space. The intersection of two rays defines the point position (Figure 6-7; Eos Systems Inc. 1997). Images were loaded into the computer and imported into a new Photo Modeler project. Photo Modeler saves the work in a project file, but it does not save each photo. Instead the software links the location of each photo with a reference to the project file.

Following importation, each photograph is marked with points designated by a unique reference number (Figure 6-8). The operator then common references photographs with common points. The computer processes the spatial relationships and generates a 3D model of the object or scene (Figure 6-9). We exported the wireframe model to AutoCAD via a 3D *.dxf (drawing exchange format) capability (Figure 6-10). The resulting wireframe was then imported into the Light Wave software (Figure 6-1 1).

Fig. 6-7. Method used by software for computing 3d points.

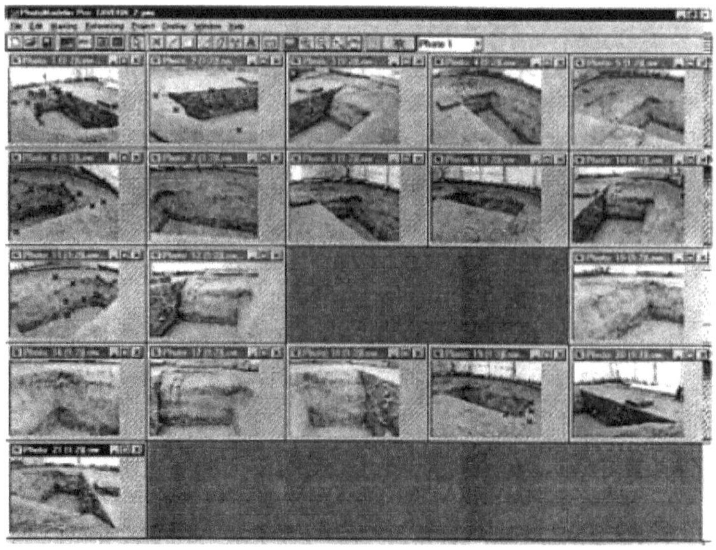

Fig. 6-8. PhoModeler screen illustrating point marking.

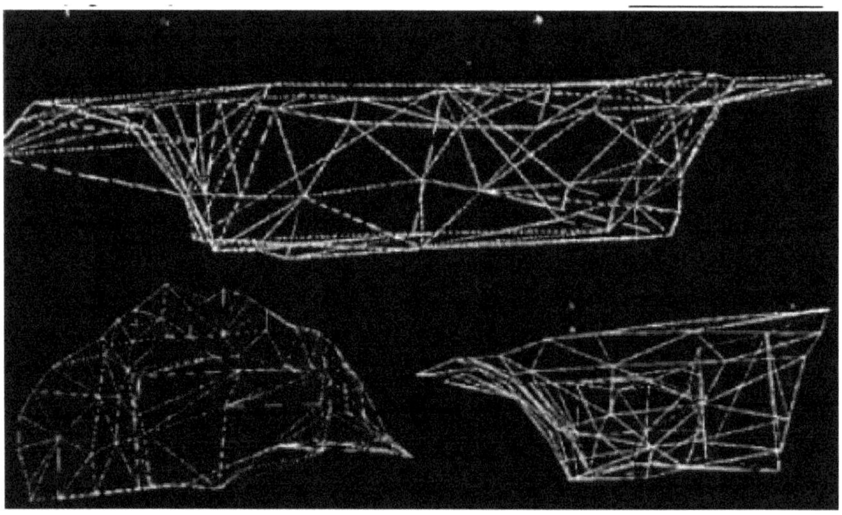

Fig. 6-9. Wireframe model of cellar from, top, and side in photomodeler.

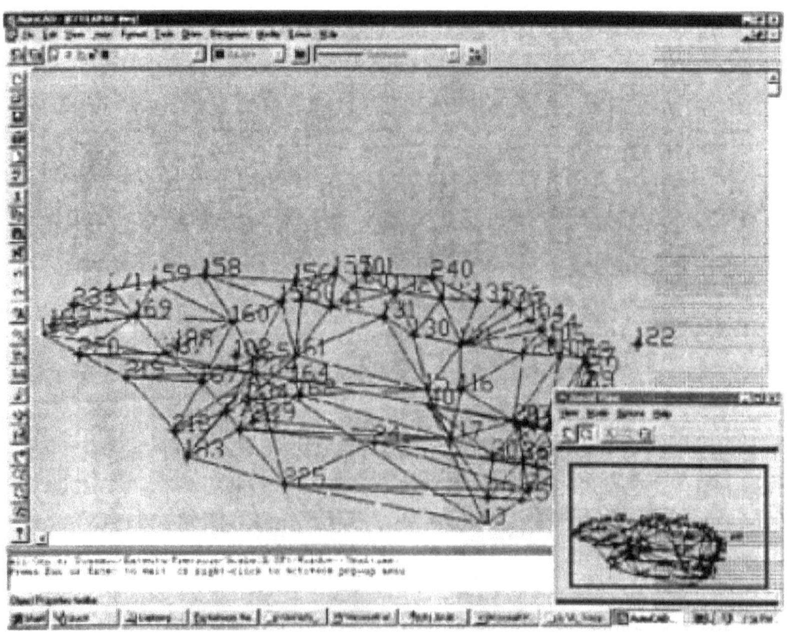

Fig. 6-10. Rumney's cellar model showmn in AutoCAD.

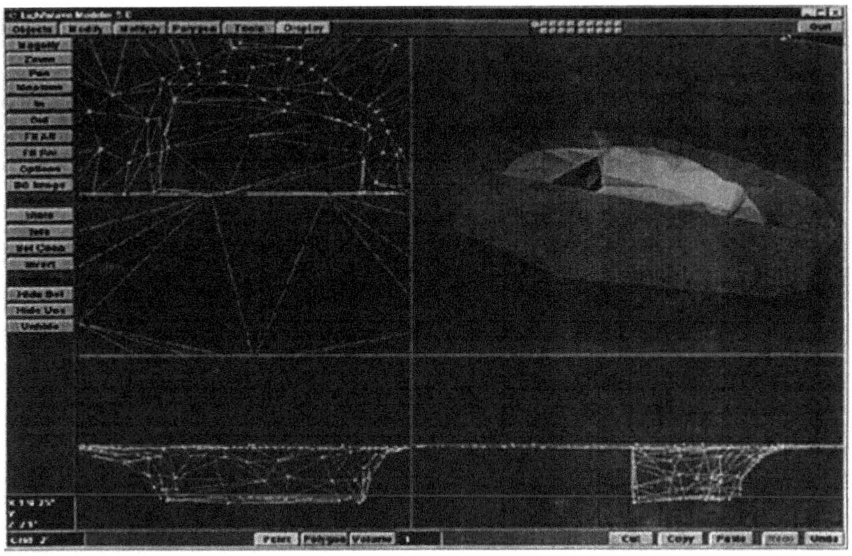

Fig. 6-11. LighWave Modeler construction window with east half of cellar.

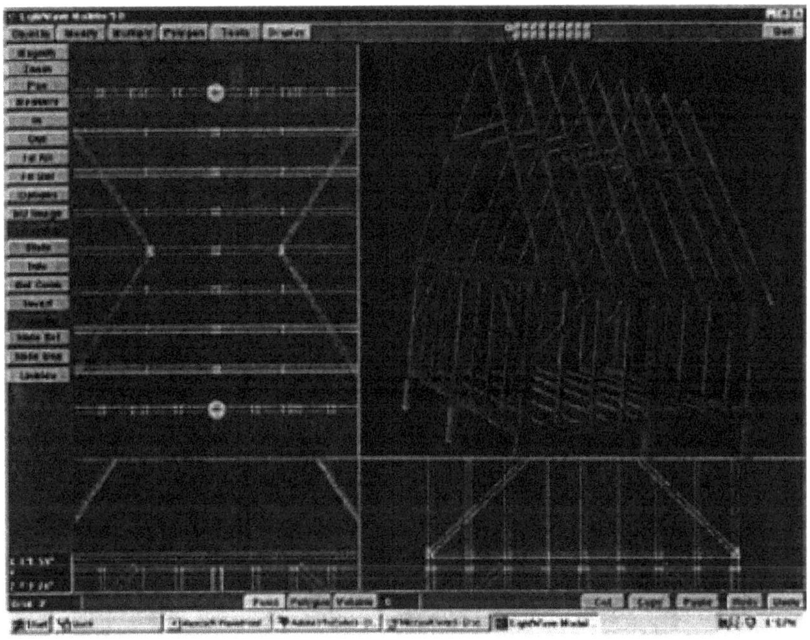

Fig. 6-12. LighWave Modeler construction window showing completed framing.

In the Light Wave program surfaces, texture, and color were added to the cellar model. These features added realism to the model. Following the completion of the cellar model, the project staff began reconstruction of the building above the cellar. Once the lower framing (the

sidewalls and up bracing) were completed, the upper framing was created. We created one roofing truss, then copied it at the appropriate spacing recorded for surviving early eighteenth-century structures in Maryland and Virginia (Figure 6-13).

Once the model is constructed such final touches as smoke from the chimney and flickering candles can be added. The framing model was then integrated with the cellar model creating a composite image of the cellar, framing, and

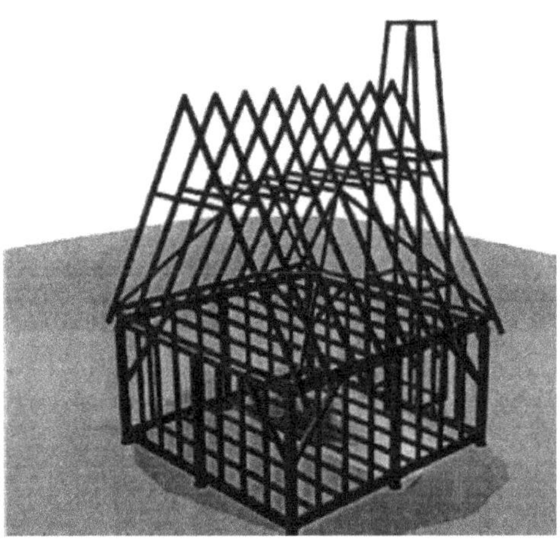

Fig. 6-13. Rendered version of Rumney's tavern with integrated cellar and framing.,

chimneys. The final process involves animations, Through Light Wave's Layout module

motion, lighting, and camera paths are added. For instance, the camera zooms from a distant point to a detailed "walkthrough" of the tavern. Once the camera positions and the lighting effects are added, the program begins the rendering process. Computer rendering of animation effects takes hours to create just a short movie sequence. Creating even a two-minute animation of the tavern required leaving the computer on over night. The animation is a raster graphics file that cannot be modified. To make modifications to the tavern animation, the changes have to be made in the model, which then has to be re exported to layout program. The rendering program then has to be run again.

Robert Burle's dwelling (18AN826)

The structure appears to have consisted of two sections (a "duplex"), each an inverted mirror image of the other. Each section had an interior wattle and daub chimney centrally located along either the west or east wall. The vertical support posts likely were raised in tied pairs using bent construction, and tied into an interrupted sill (Figure 6-15).

Virtual reconstruction of the Burles Town Land site used slightly different methods than those used in reconstructing Rumney's Tavern. Photo Modeler was not used at the Burle site. Instead, AutoCAD planviews were created from archaeological data to delineate the footprint. Like the Rumneys, Robert and Rebecca Burle built their house on earth fast posts, with smaller posts erected to create two wattle and daub chimneys. Spatial analysis of artifacts helped determine placement of doors and windows (Figure 6-16).

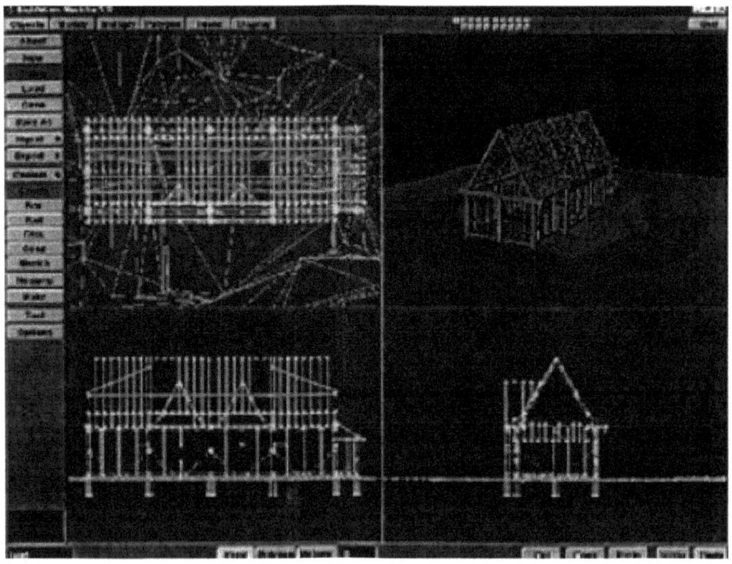

Fig. 6-15. LightWare model of the Burle house.

Fig. 6-16. Recostruction of Burle house superimposed on the archeological grid.

Rumney's Tavern Artifacts

One method used scanned line drawings in Targa (tga) format, imported into Light Wave, their profiles digitized. We then used Light Wave's lathe tool to extrude a cylindrical object from the profile. This technique produces an idealized version of the object, artificial in appearance, because the image is perfectly cylindrical and symmetrical: hand-blown glass vessels and wheel thrown ceramics are not perfectly shaped.

These vessels are not exact reproductions of objects from Rumney's tavern cellar hole, but they are close. Textures and surfaces were created from scanned or digital photographs. The photographs were then retouched to remove lines from breakage, or other imperfections. These textures were then appropriately scaled and mapped to the object. Figure 6—17 is an example of a line drawing and a photograph illustrating the materials used to create a three dimensional eighteenth-century wine bottle. Notice the asymmetry of the neck in the right hand version typical of many period bottles.

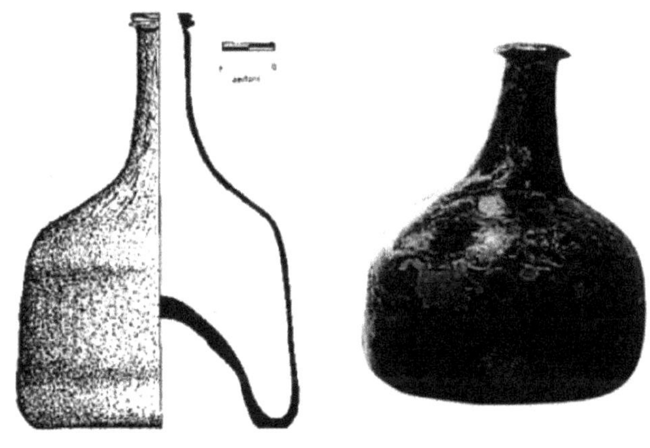

Fig. 6-17. Line drawing and photograph of two mallet bottles from Rumney's Tavern.

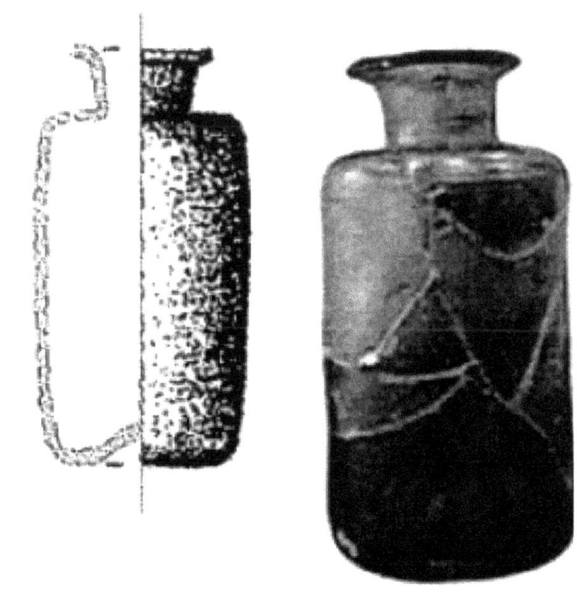

Fig. 6-18. Pharmaceutical vial recovered from Rumney's tavern.

The other method of creating three-dimensional models of artifacts uses scanned photographs of artifacts, digitized as 2D images within the Light Wave program, and then extruded as a 3D object. The digital photograph

of the pharmaceutical vial below (Figure 6—18) is an example of a 3D object created with this process (animated in digital format).

Conclusion

In conclusion, we see the growth of virtual reality, digital video, and the Internet as positive advances, offering American households a more expansive and more sophisticated view of the world. Archaeologists and historic preservations must become at least as technologically savvy as the average householder if we expect to grab and hold their attention. Just as important, multimedia environments can help scholars achieve more sophisticated and comprehensive views of the past. Space may not be the final frontier for PC graphics, but it certainly bears further exploration in the software arena. From static images to games to virtual worlds on the World Wide Web, 3D environments increasingly pervade our computing lives. The ability to circumnavigate objects adds a higher degree of realism and elegance to the ideas you convey, whether you're flying through an architectural design for a client or creating alien landscapes for entertainment (Grunin 1997). Complex digital technologies are not 'fix—alls,' and there are many issues that require extensive research and discussion; ethics, long—term preservation of data, protocols for hypothesis testing with animated 3D models, to name but three. But this is a powerful and exciting new technology that now lies within the reach of many historic preservation and archaeology groups.

Appendix

Appendix A: Camera calibration formula

The image was digitized by a Sony DigiCam DCR VX 1000 digital video camera, and then measured in Adobe

Photoshop. The total number of pixels in the whole image were determined to be 720 pixels wide by 480 pixels high (720 = Sx, 480 = Sy). The total number of pixels taken up by the 8.5 by 11 inch paper (11 = Px, 8.5 = Py) were calibrated by selecting the pixels with the "magic wand tool" to produce an image size of 422 pixels wide by 361 pixels high (422 = Nx, 361 = Ny). The distance from the piece of paper to the location of the camera's imaging chip (just behind the back of the lens) was measured to be 21 inches (D = 21). These factors were plugged into the following calibration formulas using a Focal lens size of 59mm (f= 59).

(W) idth = Px/D * f/Nx * Sx
(H) eight = Py / D * f Ny * Nx
W= 11"/21" * 59mm/422 * 720 52.728mm
1.) 11"/ 21" = .52380
2.) .52 * 59mm = 30.904mm
3.) 30.9mm / 422pixels = .07323 mm/pixel
4.) .07323 * 720 52.728mm

H = 8.5" / 21 * 59 / 361 * 480 = 31.753mm
1.) 8.5" / 21 = .4047
2.) .4047 * 59mm = 23.880mm
3.) 23.880 / 36lpixels = .0661 mm/pixel
4.) .066lmm/pixel * 480pixel = 3 1.753mm

Videographic Database — The Lost Towns of Anne Arundel Project

Tape	Site	Description	Time Code	Comments
Tape 1	Edmondo Site	Edmondo Site(I8AN)	0:00-20:05	Site overview and cellar excavation.
Tape 1	Grunwald Site	Site overview, plowzone, excavation	20:05-30:31	
Tape 2	Hancock's_Resolution (18AN)	Digital Stills (Standing Structure)	00.00-00.57	
Tape 2	William Brown House (18AN 48)	Digital Stills (Standing Structure)	00:00-1:18	
Tape 2	PhotoModeler	Test	0:00-0:30	
Tape 2	PhotoModeler	Test	0:00-0:06	
Tape 2	PhotoModeler	Test	0:06-0:30	
Tape 2	Allum_Works	Site Overview	0:01-7:11	
Tape2	Swan Cove (18AN934)	Pit Feature Overview	7:11-10:57	
Tape 2	Londontown Water Survey	Remote Sensing Survey	0:00-47:27	Views of the Shoreline.
Tape 3	PhotoModeler	Video Test	0:00-0:38	
Tape 3	PhotoModeler	Ceramic Chamberpot	0:00-0:39	
Tape 3	Steward Shipyard	Test Block A	0:00-1:01	
Tape 3	PhotoModeler	William Brown House-South Facade	0:00----	
Tape 3	PhotoModeler	William Brown House-East Façade		
Tape 3	PhotoModeler	Outbuilding		
Tape 3	PhotoModeler	Monument Stone	3:34	
Tape 3	Londontown (18AN48)	Public Dig Day	3:34-5:47	Screen Shots
Tape 3	Londontown (18AN48)	Brown House Basement	0:00-2:30	
Tape 3	Londontown (18AN48)	Well and Outbuilding	0:00-2:50	
Tape 3	Lost Towns Video	Composite Demo Clip	0:00-0:2:51	Jason and Tracy
Tape 3	Swan Cove (18AN934)	Remote Sensing Survey	0:00-6:38	Radar Images of cellar (very good)
Tape 3	Lost Towns_Video	Final Demo	0:00-2:28	
Tape 3	Larrimore Site (18AN	Larrimore Cellar	0:00-2:09	
Tape 4	Rumney's Tavern (18AN48)	Photomodeler Project	0:00-6:06	Unenhanced Control Points
Tape 4	Rumney's Tavern (18AN48)	Photomodeler Project	0:00-2:36	Enhanced control points
Tape 4	Rumney's Tavern (18AN48)	Pollen Analysis Explanation	0:00-9:25	
Tape 4	PhotoModeler	Tobacco Barn	0:00-1:11	
Tape 4	Rumney's Tavern (18AN48)	Cellar	0:00-4:11	SW Quad. May 12, 1998 3H_description_of 4C and 4B
Tape 4	Wye Plantation	Captains house	0:00-5:33	
Tape 5	Londontown (18AN48)	University of MD Held School	0:00-33:17	Field School Tour
Tape 6	Londontown (18AN48)	Rumney's Tavern	00:00-10:13	SW Quad. Strat. F.
Tape 6	Londontown		10:15-10:29	Digital Stills
Tape 6	Londontown (18AN48)	Rumney's Tavern	10:29-11:24	Video
Tape 6	Londontown		11:24-11:43	Digital Stills
Tape 6	Londontown (18AN48)		11:43-12:44	Video Discussion Fade Out

Videographic Database — The Lost Towns of Anne Arundel Project

Tape 6	Londontown (18AN48)		1 2:50-21:09	Pigskull Discussion
Tape 6	Londontown (18AN48)	Rumney's Tavern	21:13-23:40	SW Quad. Bed I Strat 1 H in process
Tape 6	Londontown (18AN48)		23:40-24:32	Artifacts
Tape 6	Londontown (18AN48)		24:32-25:10	Digital Stills
Tape 6	Londontown (18AN48)		25:10-29:28	
Tape 6	Londontown (18AN48)		29:28-30:32	Bernie Water Screening
Tape 6	Londontown (18AN48)		30:32-31:32	Rumney's Excavation
Tape 6	Londontown (18AN48)	Rumney's Tavern	21:32-34:29	SW Quad. March 13, 1998 end of Strat H
Tape 6	Londontown (18AN48)		34:29-35:21	
Tape 6	Londontown (18AN48)		35:21-36:32	Digital Stills
Tape 6	Londontown (18AN48)	Rumney's Tavern	36:32-42:09	Rumney's SW Quad. WR Stoneware Mug
Tape 6	Londontown (18AN48)		0:00-3:55	SW Quad. Bed 3A & 3B
Tape 6	Londontown (18AN48)		0:00-1:41	English Ladies Tour
Tape 6	Londontown (18AN48)		1:41-1:54	Digital Stills
Tape 6	Londontown (18AN48)	Rumney's Tavern	1:54-3:46	SW Quad. Strat E. April 21, 1998
Tape 6	Londontown (18AN48)		3:46-4:07	Digital Stills
Tape 6	Londontown (18AN48)	Rumney's Tavern	4:10-6:33	SW Quad Strats. 3E & 3F removed, base of 3E & 3F
Tape 6	Londontown (18AN48)		6:33-8:03	Digital Stills
Tape 6	Londontown (18AN48)	Rumney's Tavern	8:07-15:33	Rumney's Tavern
Tape 6	Londontown (18AN48)		00:00-00:37	Beginning of Univ. of MD Field School Lecture

References

BOLKAN, J.V. 1998 "Media Video RAIDPCI: Fast Video storage." *PC Graphics & Video* (September), 49-50.

BSALES, JAMIE M. 1998 Storage. *PC Magazine* July 1998. 156-161.

BRESS, STEVE 1997 "DVD and Computer Video: Myths and Facts about the Long Awaited Digital Versatile Disc." *Camcorder & Computer Video* (September), 19-22.

BUNZEL, TOM 1997 *Digital Video on the PC: Video Production on Your Multimedia PC.* Micro Publishing Press, (Torrence California).

BURKE-WEINER, LARRY 1998 "Keep Your Copies Clean and Pristine: Play the Generation Loss Game." *Video maker Magazine* (January), 74-79.

DAVIS, KAREN LEE 1997 Sights without Sites: Interpreting Closed Archaeological Sites. In *Presenting Archaeology to the Public: Digging for Truths,* edited by John H. Jameson, pp. 84-98. Altamira Press, Walnut Creek, California.

DOYLE, BOB, AND JEFF SAUER 1997 "Digital In, Digital Out." *Newmedia Magazine* 7:11 (September),46-5 5. Eos Systems, Inc.

1997 USER MANUAL. 12th Edition, Eos Systems Inc., Vancouver 1997. Feeley, Jim

1998 "DIGITAL DECKS FOR MEDIA STUDIOS" Digital Video Magazine (September), 24-26.

FEELEY, JIM, KELLY LYNCH, AND KIM REED 1998 "DV: State of the Format, Industry leaders discuss the successes, challenges, and future of DV-base video production." *Digital Video Magazine* (November), 22-38.

GIBB, JAMES G., AND THOMAS E. BEAMAN, JR. [Editors] In prep Seeking Lost Towns: Discovery and Exploration of Colonial Town Sites. (Special Issue) *Historical Archaeology.*

GIBB, JAMES G., AND AL LUCKENBACH 1997 Ceramic and Tobacco Pipe Seriations of Five 17th-Century Domestic Sites in Anne Arundel County, Maryland. *Maryland State Highway Administration Archeological Report Number 156.*

GRUNIN, LORI 1997 "Ideas Taking Shape" *PC Magazine* (July 22, 1997)

HONES, LUKE AND VERNON KATO 1998 "DV Spotlight: Six digital video cameras shoot it out, in both the field and the Studio." *Digital Video Magazine* (November 1998), 48-57.

JAMESON, JOHN H. [EDITOR] 1997 *Presenting Archaeology to the Public: Digging for Truths.* Altamira Press, Walnut Creek, California.

LUCKENBACH, AL 1995 *Providence, 1649: The History and Archaeology of Anne Arundel County, Maryland's, First European Settlement.* The Maryland State Archives and The Maryland Historical Trust.

LUCKENBACH, AL AND JAMES G. GIBB 1995 Dated Window Leads from Colonial Sites in Anne Arundel County, Maryland. *Maryland Archeology* 30(2):23—28.

MARKS, RICHARD 1997 "Build a Desktop Video System for Under $1000: Falling Prices and New Technology Make the Dream Come True." *Camcorder & Computer Video* (September), 26-32.

MERMELL, DEAN 1998 "RAIDs That Rock." *Digital Video Magazine.* January 1998, *46-65,* 68

MORALES, MANNY 1988 "Using 3D Studio MAX to Design A Walk-Through Virtual Museo de Arte," *PC Graphics & Video* (September), 24-30.

MOSER, JASON, JAMES G. GIBB, AND BONNIE PERSINGER 1997 Limited Site Examination of Lot 86 at the Colonial Port Town of London, Edgewater, Maryland. *Archeological Report No. 157,* Maryland State Highway Administration.

MOSER, JASON D., AND TRACY CORDER 1999 "The Lost Towns of Anne Arundel Project," *Maryland Archeology Month: Preserving the Past for the Future* (April 1999), 9.

MOSER, JASON D., TRACY CORDER, AND JAMES G. GIBB 1999 Digital Videography. Paper presented at the Annual Meeting of the Society for Historical Archaeology, Salt Lake City, Utah, January 1999.

OAKES, ROBERT 1997 "Good Camcorders Under $600: Prices Are Down; Quality is Up." *Camcorder & Computer Video* 13 (11) (November), 10-14.

PAAPE, KARMA 1999 Providence: A Case Study in Probate Manipulation, 1670-79. *Maryland Historical Magazine* 94(1): *65-87*

NOVITSKI, B.J. 1998 "Reconstructing Lost Architecture." *Computer Graphics World: Innovation in Visual Computing* (December): 24-30.

PLUMLEY, LISA E. 1999 Discovering Lost Towns: Seeking London in the Woods. *The Lost Towns of Anne Arundel Project.* Submitted to the Maryland Historical Trust, Crownsville, Maryland.

SEYMOUR, JIM 1997 "RAID Grows Up... Er, Down." PC Magazine (November 4, 1997).

SHAPIRO, MARK 1997 "What You Need to Know About Desktop Video Editing." *Camcorder &Computer Video* (September), 10-16.

SHAW, RICK 1998 "Living With Digital Video in an Analog World." *Digital Video Magazine,* 6:(2) (February), 28-33.

THOMAS, JOSEPH B., JR., AND ANTHONY D. LINDAUER 1999 Seeking Herrington: Settlement in a Very Early Maryland Town. *Maryland Archaeology* (in press).

TROOP, BILL 1998 "Removable Solutions." *PC Graphics & Video* (September), 18-22.

WAGGONER, BEN 1998 "Making a great first Compression: The New Delivery Codecs." *Digital Video Magazine* (June), 25-34.

WHITE, CHARLIE 1998 Mixed & Matched Video Editing Solutions." Digital Video Magazine (August), 22-26.

WILLIAMS, PAUL 1997 "DVD and Conquer." *Newmedia Magazine* 7:11(September), 64.

WODASKI, RON 1998 "MPEG-2 Encoders." *Digital Video Magazine* (February), *50-60.* An Application Discussion By: Stephen Ferrari, Product Marketing Manager, CMD Technology

Digital photo documentation of murals at Brihadisvara Temple, Tanjavur: a tool for art historians

P.S. Sriraman

Archaeological Survey of India
Fort St. George Museum, Chennai, India

Abstract

The Brihadisvara temple at Tanjavur is the magnificent creation of the Chola king Rajaraja I (985 AD – 1014 AD). The dark *pradakshinapatha* (the circumambulatory passage around the sanctum) was embellished by the king with extraordinary frescos, which are next best to the murals of Ajanta in the art of India. The murals are divided into huge panels of about fifteen feet high and ten feet wide and the clearance between the two walls of the passage is just seven feet. It was nightmare for all photographers while attempting to document the panels for art historians. The lack of natural light and the obvious restrictions on the use of artificial lights, the lack of clearance between the object and the camera were insurmountable in the past. As a result, the art historians were handicapped in the correct interpretation of the murals. A methodology was devised in documenting the murals in which a panel was divided into a grid of forty frames and the panels were documented using accurate measurements of light and distance. The digital files were then stitched using Photoshop software to create a single image of the panel without loss of colour and perspective. The art historian can now view these images in the system in any corner of the world by digitally enlarging the image.

Keywords Digital Documentation, Montaging, Chola Murals

Introduction

Periodic creation of records is the primary objective of any documentation. It may be an alternative to the original, particularly when the original itself is in very fragile condition. Among many forms of recording, like verbal, graphic and photographic, digital and the like, photographic documentation is a significant method, adopted by documentarians all over the world because of its advantages. An accurate visual reproduction is of the original is possible. Cameras being very sensitive can help us in detecting the change in the original when periodically documented. Photographic documentation of art objects is very fundamental to all students of art history, for the obvious reason that the researcher will not be able to travel and see all by themselves. Therefore, photo documentation brought an objective record of remains of art to the table of the researchers. Photo documentation of India's art remain is an imperative need as the assemblage is diverse and immense. It is impossible for any individual to visit them for his research purposes. Earlier, organisations like the Archaeological Survey of India, American Institute of Indian Studies, French Institute, Pondicerry have millions of photo negatives and prints dating back to the last decades of 19th century. These organisations have undertaken digitization of these records in their own way.

Colonial method of documentation

During colonial period, the British took a keen interest in understanding the artistic remains of Indian subcontinent and began to produce graphic documentation of the same. Initially, many British professional artists visited India, just to produce famous views of Indian monuments and sculptures. These are the first modern day recording. They produced the famous aquatints, through which the Western world took notice of Indian art. Among them, the Daniells – Thomas and William – stands out. In addition to the professional artists, the amateur artists like the government officials and the like also produced equally impressive collection of graphic record. The British Government also initiated many architectural surveys to produce fantastic collection of line drawings. The advent of photography added a new dimension to documentation. Photo documentation became concomitant aspect of documentation. Using photography accurate, fast, easy reproduction of the remains was prepared, helping the students of art history immensely. The colour photography coupled with modern printing technology made books on art more visually pleasing and undoubtedly it was not only a boon to a serious scholar but also attracted many connoisseurs to appreciate the greatness of Indian art. The digital technology is a wonderful tool. It helps us in various ways in our endeavour to take the art remains to a large section of people. One such effort was tried in the documentation of Chola murals.

Role of documentation in understanding the Ajanta Murals

The murals of Ajanta were immensely popular, enjoyed by all over the world and perhaps the best studied of all murals remains of India. The best reason for this state that it attracted the attention of the scholars for a long time since its discovery in 1819. There were periodic efforts to produce graphic documentation, including the modern day colour and digital photography (Ghosh 1967). The intense documentation made Ajanta murals the iconic representation of Indian art among the connoisseurs. But the Indian mural tradition did not begin and end with Ajanta. Even leaving out the sedate medieval and post-medieval mural traditions, the Classical murals tradition of ancient India was truly immense. In south India, such remains are found in bits and pieces. To be on record, the Badami caves were once had murals. Even few decades back, there were fairly a large area of painting was preserved. At present, sadly, few square feet of the murals remains. We had to depend on the documentation done in the past to understand and interpret them (Sivaramamurti 1959). Among the murals of South India, the Chola murals are relatively better preserved over a large area in the closed ambulatory path of the sanctum of the Brihadisvara temple, Tanjavur. Though murals formed as an integral part of the surface decoration of any temple, they were lost in many temples by 16th century itself. These temples were painted again, which survived till recently, only to vanish by the modern vandalism. The remains at Tanjavur are by far the best and stand next only, if not equal, to Ajanta in execution.

Chola Murals at Brihadisvara temple, Tanjavur

The Brihadisvara temple, built around 1008 AD by the Chola Emperor Rajaraja I. The *pradakshinapatha* (circumambulatory) around the sanctum of this temple is fully covered with murals, in an area measuring about 670m^2 approximately. Much of it were dilapidated by 16th century itself, as the Nayakas of Tanjavur, who ruled this region in 16th-17th century, repainted the whole area all over, without removing the then exiting Chola layer. The Chola layer was hidden for centuries when it was discovered in 1931(Govindaswami 1931, 1934). Efforts were taken to remove the Nayaka layer wherever possible to expose the Chola layer (Subbaraman 1991). About 112m^2 of Chola murals have been exposed so far, initially by scarifying permanently the Nayaka layer and later preserving them. Since then, they occupy a niche of their own in the history of Indian art, easily equalling their counterpart in bronze. It is a common fact that the temple was planned to the last details not only as a structure but also as an institution by its builder, the Chola king Rajaraja, the great (Balasubrahmanyam 1975, Venkataraman 1985, Srinivasan 1991). Why he chose the darkest alley for executing these fantastic murals, is still a mystery. Perhaps, he wanted to create a holographic world of gods or desired that they murals should survive the test of time.

The location of the murals

The murals depicting the themes from the Hindu mythology are executed in the narrow, dark circumambulatory path of the sanctum sanctorum. The panels are huge measuring 4.5 m x 3m, with barely 2m between the walls of the passage. The lack of distance between the walls, absolute lack of natural light, and massive size of the panels deprived the viewer an opportunity to have a comprehensive look at the canvas. The artificial lighting provided there in was restricted in quantity and duration of lighting for an observer to have a close look. As a result there were many incorrect interpretations of themes by those who happened to study them in situ (Sivaramamurti 1968, Champakalakshmi 1973). The location is a nightmare for any photographers, as there was less clearance between the walls and the natural light is absent. Therefore, these murals have never been photographed comprehensively. The photographers, taking the restrictions to their advantage, photographed only the best, choicest portion of the mural. Therefore, many mistakes are there in the present day interpretations and that being carried on by all scholars who had to rely on these descriptions. Further due to conservation measures undertaken by the Archaeological Survey of India, the custodian of the edifice, and due to religious prescriptions, entry of visitors is highly restricted. It was, therefore, decided to document the murals in such a way that the 'whole' is produced in a single frame so that the art historian and connoisseurs will get an opportunity to see the 'whole' and not the 'parts of the whole'. The large format reproductions from the documentation may be kept on view for the general visitors to see and appreciate. The aim is ambitious and is very difficult to achieve.

Overview of documentation

As said earlier, the location had no natural light and the space between the two walls is very short. Therefore, the photographer's task became all the more difficult due to the lack of two fundamental factors on which photography works. A similar documentation measure was done in 1950, when colour photography was at nascent stages. The results were published recently (Nachiappan 2004) but the quality was poor for reasons known to publishers only. Over the years the murals were photo documented by photographers in parts only. This work is significant on several accounts. It was done when the paintings were just exposed. The method was very similar to us but executed without the help of modern day tools like computers and the rest. It is indeed a commendable work. Even though it was executed for the Archaeological Survey of India, none in the department is aware of the records. It may be due to the fact that the copy submitted to the ASI was spoiled by the company entrusted with the exposing and joining (Nachiappn 2004). That is, the best preserved and easily accessible parts of the 4.5m high panel were photo documented. These were utilised and published by the art historians to illustrate the Chola School of paintings. Often, these images too were cropped according to the available space for publications. Therefore, the documentation produced so far were sketchy and never conveyed the full to the scholars and others. This was kept in background while formulating the documentation policy, which broadly included:

1. That the aim of the documentation should be to produce the 'whole' canvas in a single frame / digital file. The earlier photo documentations, due to conditions beyond the control of the individuals, were restricted to photographing the parts of canvas.
2. The documentation should to the maximum possible extant be true to the original in colour and perspective.

The location with total lack of light and space was the greatest impediment in the execution of project. The impeding factors were analysed and the following method of execution was devised and executed.

1. Each panel and the contiguous offsets should be imaginarily divided into small frames. Any marking on the surface or even suspending guide strings was not possible. After repeated experimental shoots it was found that each pane need to be divided into 40 frames (4 rows and 10 columns) each sufficiently overlapping with each other. That is each frame must overlap on all its four sides. Since boundaries of this frame is to be remembered because so markers are permissible.
2. In order to maintain the proper perspective, it is necessary to place the camera at a fixed length and fixed focal length. This parameter had to be maintained very accurately for shooting al the frames. Experimental shooting were done to fix the distance and range of the lens. In order to achieve this, a special platform was erected with simple scaffolding devices. Since every space was covered with painting of one period or other, extra precaution was taken while erection and dismantling the platform (Figure 1).

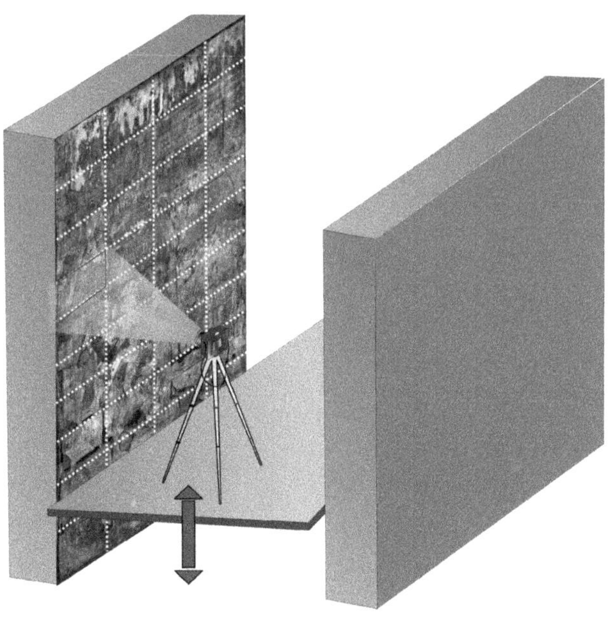

Fig. 1

3. The lighting need to be artificial and non-harming to the paintings. In order to achieve this, experimental shoots were conducted with various lighting and was found that normal 500w Halogen lights were required to produce the desired quality. Exposing the fragile layer of painting for long to harsh lights such as the halogen lamps of 500w intensity will be harming it. Therefore, it was decided to do all the groundwork for exposing every frame using a friendly 60w incandescent lamps and switch on the lights just few seconds before the actual clicking. The lighting was adjusted to be indirectly reflected light for practical considerations (Figure 2). Experimental shooting indicated that shooting forty frames will take about 15 to 18 hours.

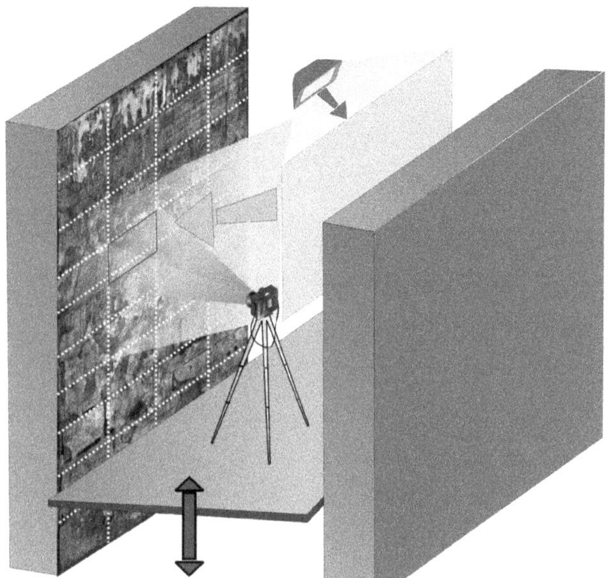

Fig. 2

4. The biggest and critical issue is the medium of exposure. The choice was between transparencies, negatives and digital. Exposing transparencies and developing them was a big challenge and required professionals were not readily available at that time of execution. The negative had disadvantages of losing quality in every step right from exposing. Digital documentation had several advantages, which will be discussed below.
5. Even though the panel had to be documented in forty or odd frames, the biggest task was joining them. At the time of execution of project Adobe Photoshop 7 (hereafter APs7) was the latest version available. This package allowed stitching of several photographs into one single image. This was used. Since the computer systems at that time (2005) were quite slow with a Pentium 4 processor and 256/512 MB RAM, 40 GB HDD, working on Photoshop with massive file size was really cumbersome. However, this package was accepted for stitching the frames. This method of digital stitching of the frames decided the medium of documentation. If transparencies were opted, in addition to less than satisfactory processing, the

conversion of images into digital file will end in large sized files. As they are being montaged in APs7, the resultant file size will be enormous as the frames increase. At that time when documentation was done, handling such large file size was very difficult. So transparency was not the medium of choice, even though it enjoyed relative advantage over negatives towards loss of quality. Negatives enjoyed relative advantage in conversion it digital files. But it suffered losses at every stage of conversion – exposing, developing, printing, and scanning. Therefore, negatives too were dropped in favour of direct digital files shot using the just introduced high end (at that time) SLR cameras. SLR cameras allowed several advantages. They provided direct digital images suiting stitching needs.

6. Then the high end SLR camera (Nikon D 100) was used in the documentation. The images were saved initially in .TIFF format and later in RAW and NEF formats in later schedules. The later formats produced bigger sized image at a lower file size.
7. The images were then stitched in APs 7 after due correction for colour and perspective (Figure 3 to 7).
8. The original camera images were saved in several storage devices like the system HDD, external isolated HDD and now in DVDs. So were the stitched output were saved in several formats. Large outputs of few panels were prepared by digital printing and exhibited for general visitors.

Fig. 3. One montaged row in the Tripurantaka panel.

Fig. 4. Camera images of the montaged row in the same panel.

Fig. 6. Camera images of the montaged row in the same panel.

Fig. 5. Camera images of the montaged row in the same panel.

Fig. 7. Camera images of the montaged row in the same panel.

Advantages of the documentation

Several distinct advantages had emerged from this kind of documentation. The whole principles lies in that an object covering a vast area, a mural in this case, was documented in contiguous frames and were stitched/montaged into a single frame. The resultant image is massive digital file of very high size and dimensions. This resultant image could be used in several ways.

1. An output of the same size of the original or of slightly smaller size could be prepared for exhibition purpose. It can effectively substitute the original, if the documentation is done perfectly. Such a reproduction will be very useful in satisfying the needs of general visitors when the access to the original is restricted.
2. The single frame view of the large canvas is indeed a boon to any art historian and connoisseur. Now the members of this tribe can glimpse at the whole canvas and research/appreciate the efforts and scheme of the artists. Hitherto, they were only seeing a fragment of the whole and not the whole itself.
3. These digital files are compact and made available to them quickly and a cheaper cost. This enables them to view and study the art at their own places reducing the need for travelling a long way. Hitherto available digital documentations were in parts, which are single frame views, which are definitely disadvantages for mural paintings.
4. The large size digital files can be enlarged in a suitable graphic software package like Adobe Photoshop and almost every square inch can be digitally enhanced and viewed closely. This exercise often throws out very interesting features about the painting, that often escapes the naked eye observation of the mural.
5. Very often, when the mural is located in disadvantageous situations like a place where there is no proper light or located at a distance, like the present place. In such case, many finer details of mural escape naked eye observation. The digital files are an excellent tool in such cases.
6. To satisfy the curiosity and often demanding public, particularly in India where the restoration ethics are differently perceived, the digital files may be used for limited digital restoration of the missing portions of the murals. One such attempt was made to restore a panel.

Disadvantages of the documentation

While the digital documentation and montaging of large murals into a single frame has several advantages as mentioned above, particularly that they could bring the art remains to the comfort area of the researcher / connoisseurs, it has some disadvantages.

1. These are digital files and hence there permanency is still debated. The best method of storing digital data is under strong discussion among the technology developers.
2. Presently the camera images are stored in .TIFF/ RAW/ NEF formats. These are compatible with presently available software packages. In case of a irreversible changes in the file formats or in the structuring of the packages, present formats will become obsolete. At that point of change the documentation holders must foresee the problems and find a solution to it.
3. The stitching per se is not absolutely free from subjectivism. At the areas of joining the use of burn tools to camouflage the differences in alignment of details. However accurate may the documentation, the stitching is laborious job. There bound to be some misalignments.
4. Though this kind of digital documentation, which is photo documenting a record in several frames and stitching it in the system was found to be immensely useful in case of Tanjavur, the same need not be the case at all places. It may turn out that this method may not be even useful at some places. Obviously, the documenter needs to take a studied decision.

Results of Digital Documentation of the murals of Brihadisvara temple, Tanjavur

A comprehensive analysis of the digital documentation had thrown ample light on the unstudied paintings, which was otherwise not feasible on site. Few of the significant discoveries may be given here. A panel which was identified as Dakshinamurthi, Siva as the teacher par excellence, now turned out to be a great royal teacher seated below a sacred banyan tree. The birds and animals that inhabited the tree were simply described as "many animals in the banyan tree" actually turned out to be a great moment there (Figure 8). The animals – birds, peacock, monkeys, squirrel and owls – were all perturbed by the entry of a cobra, which had already swallowed a prey. The four monkeys show different reactions, one running away in fear from the cobra, another too running in another direction, third is just watching and the fourth, with its back towards the snake, is yet to realise the threat. The birds are all attentive to the threat. The squirrel and the owls are not reacting to the threat. These emotions, wonderfully displayed on the faces of the animals, became apparent when the whole canvas is seen enhanced digitally.

Fig. 8. The banyan tree in the Dakshinamurthi Panel.

Another panel, which was passed over by every other scholar, happened to be a significant with a historical event being portrayed. The highly damaged panel was identified as the king Rajaraja worshiping the linga before its installation in the sanctum, an event witnessed by his courtiers. However it turned out to display two individual events – one the king worshipping the linga and the other the whole court listening to an exposition by a great saint. The significant feature is that the individuals were all seated in a strict protocol order.

The digital further threw out some interesting and minor features which could not be seen at the site with naked eyes. For example, the extraordinary coiffure of the Chola period women is very brilliantly portrayed by the artists. At several places, the bottom layer, where the artist had sketched the whole canvas in line is seen. However, while executing there were correction, particularly in the posture of the arm of the dancers and the like. Does this indicate that the technique of painting is not 'true fresco' as claimed by some?

Conclusion

The digital documentation of the Chola murals at the world heritage Site of Brihadisvara temple at Tanjavur, using montaging really opened a window for the art historians to study the paintings in depth without travelling to the site. These murals, out of bounds for persons other than those profess Hindu faith. Therefore, this kind of documentation will enable them to understand the canvas in full. Such detailed documentation will also serve the conservator and restorers to get an accurate understanding of the murals. However, monatging has its own limitations. Executing a documentation plan involving monataging require deft professional hands.

Acknowledgements

The present work was conceived by the author in 2003, without t any knowledge about the work cited above. The author was officially encouraged by T.Satyamurthy, Superintending Archaeologist, Archaeological Survey of India, Chennai Circle and the photography was handled by N.Thyagarajan, a freelance photographer. Without their contribution the conception would have remained a dream.

References

BALASUBRAHMANYAM, S. R. 1975. *Middle Chola Temples*. Faridabad, Thomson Press

CHAMPAKALAKSHMI, R. 1973. New Light on the Chola Frescoes of Tanjore. *Journal of India History* (Golden Jubilee Volume), 349-59

GHOSH, A. 1967. *Ajanta Murals*. New Delhi, Archaeological Survey of India

GOVINDASWAMI, S. K. 1931. On the discovery of Cola Paintings, The Hindu dated April, 11 1931

GOVINDASWAMI, S. K. 1934. Cola Paintings. *Journal of the Annamalai University*. Volume II

SIVARAMAMURTI, C. 1959. Western Chalukya Paintings at Badami. *Lalit Kala*, No. 5, 49-58

SIVARAMAMURTI, C. 1968. South India Paintings. New Delhi, National Museum.

SRINIVASAN, K. R. 1991. The Peruvudaiyar (Brihadisvara) Temple, Tanjavur: A Study. In C. Margabandhu et.al (ed), *Indian Archaeological Heritage* (K. V. Soundara Rajan Festschrift). Delhi, Agam Kala Prakashan , 525-538

SUBBARAMAN, S. 1991. The Mural Paintings in Brihadeeswara Temple, Thanjavur – a New Approach to their Study and Preservation. In C. Margabandhu et.al (ed), *Indian Archaeological Heritage* (K. V. Soundara Rajan Festschrift). Delhi, Agam Kala Prakashan , 745-752

VENKATARAMAN, B. 1985. *Rajarajesvaram, the Pinnacle of Chola Art*. Madras, Mudgola Trust

The importance of precision in georeferencing demonstrated in the case study of archaeological GIS of Chitradurga, Karnataka

N.S.Nalini and M.B.Rajani*

National Institute of Advanced Studies, Indian Institute of Science campus, Bangalore – 560012. Email: *rajanihf@gmail.com

Abstract

Chitradurga is a town located 202 kms to the northwest of Bangalore in central Karnataka. The town consists of a fortress on the rocky hill, which dates back to 16th century AD. There are many ancient temples and monuments within the fort and in the city. A project was taken up to generate an archaeological GIS that would include information about the monuments, temples and also 3D perspective views of the landscape. A variety of information got from satellite imagery, maps, history and ground truth is being used for producing this map. GPS survey was used to integrate this multi-layered information. For multi-layered analysis it is essential to bring all the layers to a common GIS format. In order to bring all the layers to a common GIS format a high precision georeferencing is vital. This paper will demonstrate three different methods to georeference imagery and discuss the efficacy of various methods, and their implications. This paper will also discuss the implication on 3D view, if the DEM and surface rendition image are not co registered accurately and demonstrate a method for correcting the difference.

Keywords Georeference, DEM, archaeology, GPS, Chitradurga, Google earth.

Introduction

Georeferencing is a system that links an image to a position on earth's surface. To georeference means, to define the existence of an object in physical space by assigning geographical coordinates. Georeferencing is the process of adding the geographic information to the image, which is scanned as a digital image. The scanned image may be a satellite image, an aerial photograph, topographic map. Scanned image is a matrix, consisting of rows and columns of pixels. In the process of georeferencing each pixel in the digital image is assigned geographic coordinates. When geographic information is added, the image relates itself to an appropriate position in the real world. This process allows the data to be viewed, queried and analyzed in comparison with other data.

Accurate georeferencing is fundamental to all GIS applications. Management and analysis of multilayered GIS data requires a very carefully referenced position on the earth's surface. Inaccuracy in referencing results in mispositioning of the data and may disrupt the data, making it invalid. Positional accuracy for a large extent can be got from using ground control points (GCPs) in order to generate an accurate transformation model. A GCP is a feature which can be uniquely identified in an image and whose geographic coordinates are known or can be determined. The chosen GCP should be prominent, stable and preferably a permanent feature that is not subject to seasonal changes so that its appearance is unambiguous. The coordinates of GCP can be found out from existing maps of GPS survey (Joseph 2003).

Georeferencing is carried out with the help of geographic coordinates. There are many different coordinate systems. Various layers of data needed for a GIS project are sometimes in two or three different reference systems. Converting or transforming these into one system should be done very carefully. Various coordinate systems are developed based on map projections. Map projection is a method where in curved surface of the earth is projected on a flat surface. While georeferencing, two more parameters to be considered along with projection are ellipsoid and datum. Even if the projection of two images is the same, if the ellipsoid and the datum are different, then also the registration will not match. Therefore ellipsoid and datum also have to be matched before georeferencing.

Objectives

There are many ways of doing georeferencing. In this paper three different methods are discussed and also a comparative study has been made to check the accuracy of georeferencing. The three methods are:-

- Taking GCPs from published topographical map (toposheet)
- Taking GCPs from Google earth
- Taking GCPs from the GPS reading on the site

Along with the study regarding georeferencing methods, DEM (Digital Elevation Model) generation and its relation to accurate georeferencing was also carried out. This was done as part of our ongoing project on Chitradurga. The project involved generating an archaeological GIS that would include information about the monuments, temples and also 3D perspective views of the landscape.

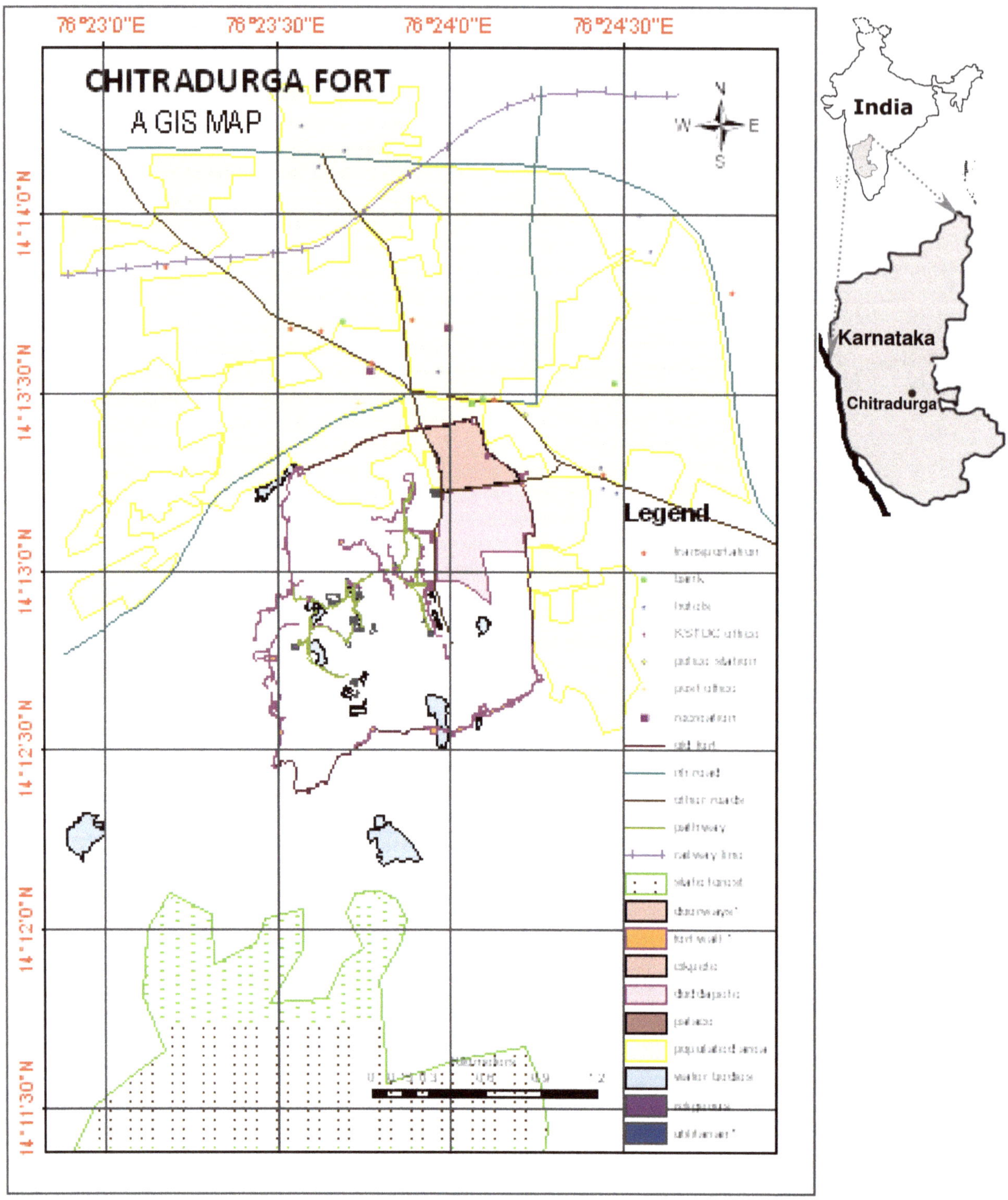

Fig. 1. GIS map of Chitradurga.

Study area

Chitradurga district extends from 13°34' - 15° 02' latitude and 75° 37' - 77° 01' longitude and is in the state of Karnataka in southern India. Entire district is in the valley of river *Vedavathi*. The rocky hill range stretches in southeast to northwest direction, consisting of rocky surfaces that are not covered by soil or vegetation. The altitude of the area is 2383feet from mean sea level. *Jogi maradi* is the highest point in the entire hill range with a height of 3863feet. Chitradurga was called by several names during the rule of different rulers like *Hidimbapattana, Chinmuladri, Bemmettana kallu, Soolagal, Perumalepura, Chitrakaldurga Mallarayana durga* . Tippu named it *Farukhabad*. Chitradurga Fort was built in parts by the *Palegar Nayakas*, Hyder Ali and Tipu Sultan, between 10th-18th century AD. It comprises a series of seven enclosure walls. Eighteen ancient temples can be found inside the fort. This seemingly impregnable fort has 19 gateways, 38 posterior entrances, a palace, a mosque, granaries, oil pits, 35 secret entrances, four secret tunnels and water tanks. The rulers of this fort built 2000 watch towers to guard and have a clear look at any potential intruders. Chitradurga has a total geographical area of 8440 Sq Kms and a road length of 1651 Kms. The fort has a total area of 300acres and a length of 8 kms (Nayak 1974, 401). The central coordinates of the site are Lat. 14°12'46"N, longitude 76°23'41"E.

Methodology

Satellite images were taken from Google earth. The area of interest was demarcated and several separate pictures were extracted with a closer zoom. Taking one picture of the whole area would not have produced a high definition image required for identifying the monuments. Therefore the zoom was adjusted so that the monuments could be clearly identified. These were mosaiced in Photoshop and geo referenced.
The three methods of georeferencing are:

1. ***Georeferencing using GCPs from toposheet*** (Geo-image A): This was done by using coordinates in the toposheet. Toposheet has polyconic projection and Everest datum. Georeferencing was first done with same parameters. Then reprojected into geographic lat/long projection and WGS 84 datum in order for it to be compatible with GPS survey data. The GCPs from toposheet was used to assign coordinates to the mosaiced image. A georeferenced image (Geo-image A) was generated with RMSE of less than one pixel.

2. ***Georeferencing using Ground control Points from Google earth*** (Geo-image B): Google earth image was taken as the reference and points identifiable in both the images like sharp corners of the monument and road intersections were marked. Ten well distributed points were chosen and their geographical coordinates were noted. These points were identified on the mosaiced image and their respective coordinates were assigned. Geo-image B was generated with RMSE less than one pixel. Here also the image was set to geographic lat/long projection and WGS84 datum.

3. ***Georeferencing using GPS coordinates*** (Geo-image C): GPS survey was done using – Garmin eTrex Vista HC x. This has an error of less than 10m.The coordinate readings were taken on the ground at locations that can be clearly identified on the image. These were used for georeferencing the mosaiced image. Geo-image C was generated with RMSE less than one pixel.

Results and discussion

For comparing the accuracy between three methods of georeferencing a GPS ground survey was conducted. In this ground survey GPS track logs/track points were recorded while navigating in the study area through footpaths and along fort walls, which can be clearly identified in the image. A GPS track log is a collection of track points along the course of a route. Track Logs are created by a GPS instrument as the user navigates a desired route or path. It's a record of the path that has been navigated. The instrument records one point in every 30 seconds and thus a series of points are tracked while navigating through a route.

The track log of the GPS survey done at Chitradurga was converted into a vector point layer in the GIS environment of Geomatica software. The vector layer of track points were separately overlaid on all the three georeferenced images (Geo-image A, B, and C). Geo-image C coincided with track points as can be seen Figure 2 (1) .The navigated path marked by the yellow track points fit on top of the fort wall and path distinctly seen on the image. This near perfect match is understandable because the source of both track log and GCPs used for georeferencing is same (GPS). Geo-image B showed slight linear shift of about 2m in the southeast direction from the track points. Since GPS itself has an error of 5-10m, an error of 2m is not so significant. But the track points were displaced considerably when they were overlaid on Geo-image A. The displacement was nearly 165m in the northwest direction. This could be due to the cartographic parameters (projection, datum and ellipsoid) followed when original maps were made, since they would be based on slightly different values of earth's circumference and radius calculations (Robinson *et al.*1995). Toposheets are in polyconic projection, Everest datum/ellipsoid. Whereas GPS follows WGS84, which is now the global standard. This is probably the reason why there is so small a difference between Geo-image B and C, since B is based on Google earth which follows WGS 84, same as GPS.

This study demonstrates that Google earth GCPs can be considered for georeferencing the images and a near accuracy can be achieved. This shows that if images of inaccessible areas have to be georeferenced, Google earth co-ordinates can be considered. This would mean that it is not necessary to go for field visit for the purpose of georeferencing. Even if the study area is accessible this method also enables us to georeference the image before going for field visit so that the georeferenced map can aid further investigations.

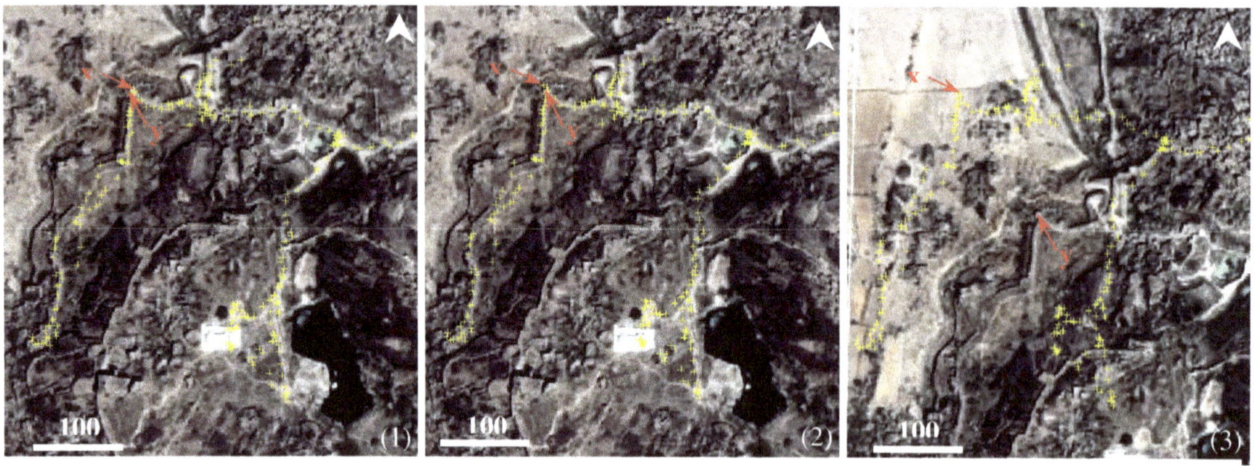

Fig. 2. The yellow color points are the GPS track log recorded during field survey. The track log is superimposed on (1) Geo-image C, (2) Geo-image B and (3) Geo-image A. 'x' is a point in the vector layer (GPS track log) marking the corner of the fort wall and 'y' is the same corner of the fort as seen on the raster image.

Digital Elevation Model (DEM):

DEM was generated to understand the overall landscape since the study area had hilly terrain. There are many different ways of generating DEM, some of which have been discussed in Rajani *et al.* 2008. For the present study DEM has been generated in a GIS environment from contours (given at 20 m interval), spot heights and bench marks provided in topographical Map (published by Survey of India) covering the study area at 1:50, 000 scale as well some of the data collected in field using GPS. Over this DEM Geo-image C was draped. When 3D perspective views were generated out of this combination and closely observed, linear shift between DEM and surface rendition image could be noticed. For example, Figure 3 is a section of the study area showing a slope of the hill and the adjacent lake. In Figure 3 (1) the lake seems to be draped over the slope. This is because of the miss match between the DEM and surface image. The displacement is same as what was demonstrated in Figure 2 (3). There fore the DEM had to be corrected for this displacement. The correction was carried out in the following way. The DEM was converted into shape-file of contours and this was overlaid on the Geo-image A. The source of DEM and the source of GCPs for Geo-image A is the same (toposheet) therefore they match perfectly. Since contour lines are not identifiable on raster surface image, some reference points identifiable on the image had to be marked, which will aid to perform linear shift. For this purpose a road intersection was chosen and arc lines were drawn on the contour vector layer marking the position of this intersection. The road intersection is clearly seen on the top portion of Figure 4(1). Apart from the intersection the rest of the vector lines are contours. This vector layer was then superimposed on Geo-image C (Figure 4(2)). The displacement between the two georeference is evident in the displacement of cross roads as seen on Figure 4 (2). There is a linear shift of 85m in the X direction, 106m in the Y direction and a total of 165m in the south east direction.

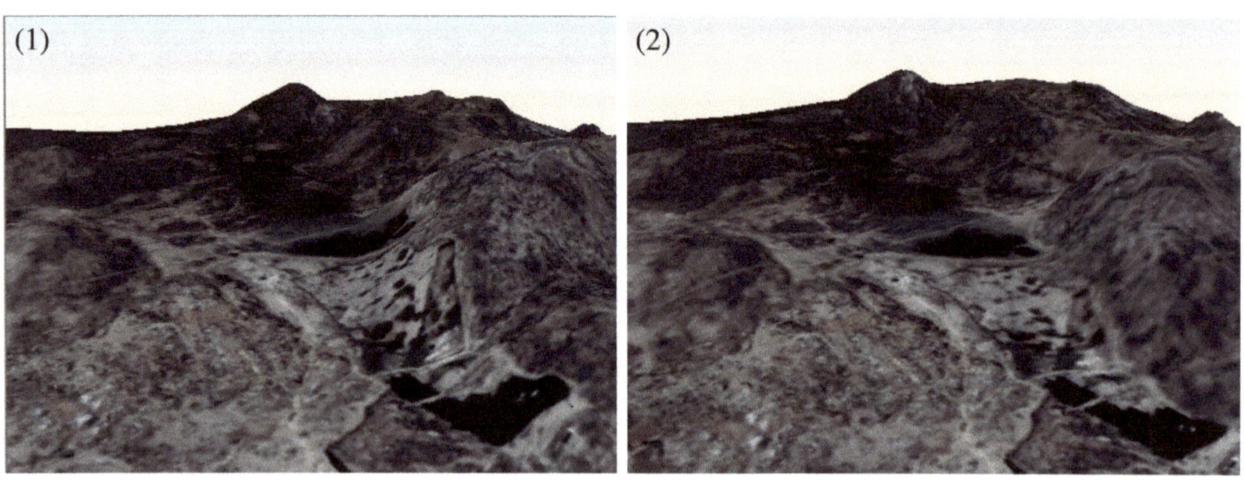

Fig. 3. 3D visualization of the study area (1) the georeference between DEM and surface image is displaced by 165m, (2) after correcting the DEM to match the georeference of Geo-Image C

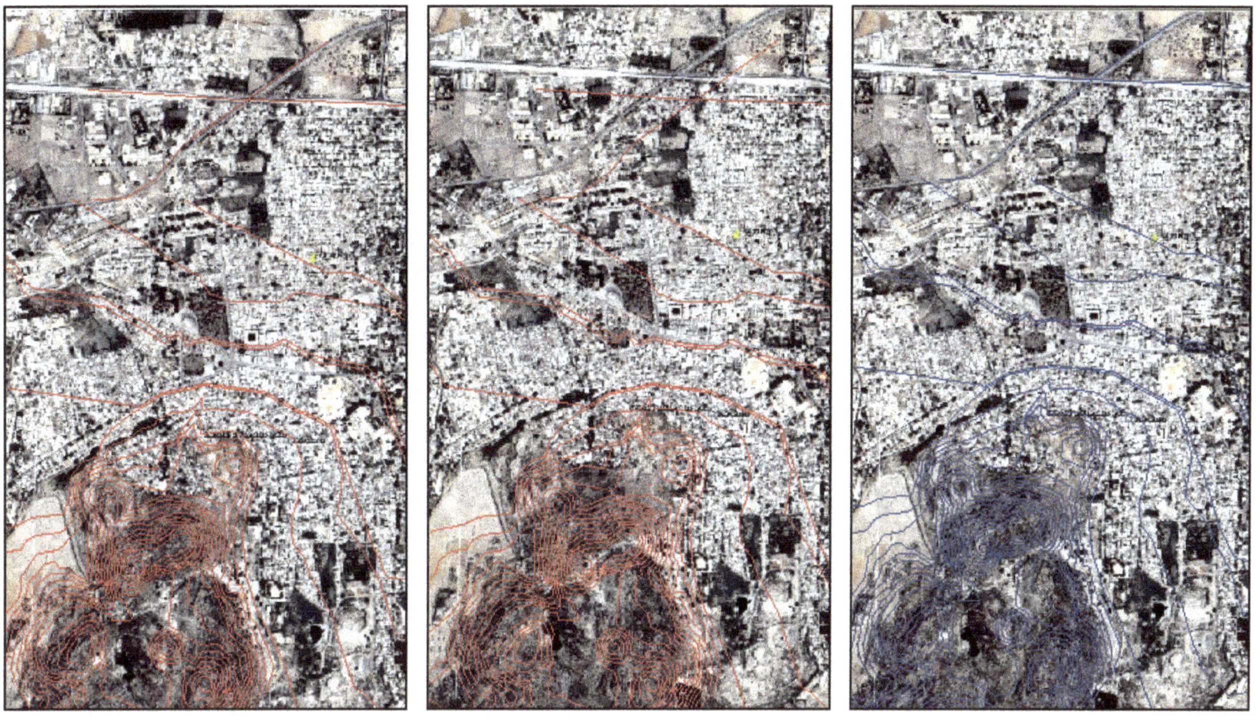

Fig. 4. (i) Contours (generated from original DEM) overlaid on Geo-image A. A cross road has been marked for reference (ii) contours overlaid on Geo-image C. The displacement between the two georeference is evident in the displacement of cross roads. (iii) Contours after shift to match the Geo-image.

The whole vector layer, including all the contours and cross road, was simultaneously moved in the direction reverse to the error until the cross roads on the vector layer matched perfectly the cross roads on raster Geo-image C (Figure 4(3)). After the shift, the arcs of the cross roads were deleted and the contours were converted to raster DEM. This DEM is now corrected and normalized with the GPS coordinates. This DEM was used along with Geo-image C for generating perspective views. Figure 3(2) is a view with the corrected DEM. In this version the lake is flat beside the hill unlike in Figure 3(1). The difference is clearly demonstrated as both views are of the same scene.

Conclusion

This study was carried out as part of an on going project on archaeological GIS of the site Chitradurga. This paper has discussed the importance of precision in georeferencing, especially while analyzing multi layer geospatial data. Three methods of georeferencing 1) acquiring GCPs from maps, 2) GCPs from Google earth and 3) from GPS survey have demonstrated and compared to evaluate the efficacy of each. It was found that GPS is the most accurate means and that use of toposheet can project a displacement of around 165m. The method in which Google earth is used for GCP is close to GPS accuracy and within the error range. Therefore near accurate georeferencing can be achieved without field visit. This will be very useful especially for areas that are inaccessible. This paper has also demonstrated a method to correct the mismatch between DEM and surface image by converting DEM to vector contours and shifting the vector to match the image and then converting the vector back to raster.

Acknowledgements

We thank Dr. K. Kasturirangan, Director, NIAS, for overall guidance, Institutional support and encouragement in carrying out the present study. We thank Prof. B.V. Sreekantan at NIAS for useful discussions.

References

JOSEPH, G. 2003. *Fundamentals of remote sensing*, Universities Press, Hyderabad, India.

NAYAK, H.M. (ED.) 1974. *Kannada Vishwakosa* (in Kannda), Vol.7,Institute of Kannada Studies, University of Mysore, Mysore, India.

RAJANI, M.B., PATRA, S.K., AND VERMA, M. 2008. 3D perspective views and their applications to Archaeology: A case study on the site of Badami. in Lasaponara, R. and Masini, N. (eds) Proceedings of Workshop on *Advances in Remote Sensing for Archaeology and Cultural Heritage Management*, Rome.

ROBINSON, ARTHUR H., MORRISON, JOEL L., MUEHRCKE, PHILLIP C., JON KIMERLING, A. AND GUPTILL, STEPHEN C. 2004. *Elements of cartography*, 6th edn., John Willey & sons, Inc., Singapore.

Semi-automated data capture and image processing: new routes to interactive 3D models

Katharina Galor,[1,2] Donald H. Sanders[3] and Andrew Willis[4]
with
David Cooper,[1] Benjamin Kimia,[1] and Gabriel Taubin[1] and Oren Tal[5]

[1]Brown University
[2]Rhode Island School of Design
[3]President, Institute for the Visualization of History, Inc., and Learning Sites, Inc.
[4]University of North Carolina, Charlotte
[5]Tel Aviv University

Abstract

Apollonia-Arsuf lies on the Mediterranean coast of Israel on a cliff overlooking a natural anchorage between Jaffa and Caesarea. It was continuously settled between the 6th century BC and 13th century AD, when the Mamluks captured and destroyed the site. The first systematic excavation was carried out in 1950, followed by numerous additional seasons through 2004. Architectural remains include a Roman villa, a Byzantine church, an Early Islamic Market Street, and a Crusader Castle. In 2006, the authors began a long-term collaboration with a team of computer scientists and engineers at Brown University that plan to implement a series of innovative digital archaeology applications, including a semi-automatic virtual reality reconstruction of the Crusader Castle, the semi-automatic virtual re-assemblage of amphorae and glass shards, and a new type of multi-camera-based excavation documentation system that will ultimately be made accessible as a series of tools for archaeologists (dramatically changing the methods whereby excavations are recorded, data are analyzed, and visualizations produced).

Keywords computer vision, castle reconstruction, automated 3D models, Apollonia-Arsuf, automated excavation documentation, multiple cameras, virtual reality

Introduction

In 2006, an interdisciplinary research group of engineers and archaeologists met to identify projects that would constitute problems of interest to be solved by merging the expertise of both disciplines. The biggest challenge was to create a system that would be highly innovative to scholars from both disciplines, not simply furnishing the engineers with some colorful data sets they would experiment with, nor using existing technological tools to facilitate the processing and analysis of data sets for the archaeologists.

What developed we have come to call *REVEAL* and it has four components:

1) an Internet-accessible archaeological space-time database containing both traditional information about finds as well as images, video, and extracted 3D representations;
2) an automatic and semiautomatic 3D fragment, object, and large architectural surface-capturing system based on geometry estimation from multiple still image and video cameras;
3) the automatic and semiautomatic re-assembly of glass and pottery vessels from precisely recorded 3D representations of their many fragments; and

4) interactive tools and real-time 3D immersion for searching a database and manipulating fragments, objects, and architecture with the goal of facilitating their virtual re-assembly, making precise metric measurements at an archaeological site from remote locations, enabling collaborative research, and providing for new types of hypothesis testing.

Initial tests will focus on the site of Apollonia-Arsuf, Israel (nominated by the World Monuments Fund as one of the 100 most endangered sites; Figure 1).

Apollonia-Arsuf

Our first test site of Apollonia-Arsuf is located on a coastal plain that spreads from Gaza in the south to the Mount Carmel range in the north. The ancient settlement rests on a sandstone ridge overlooking the Mediterranean Sea; the plateau upon which the site spread rises about 35m above sea level. Apollonia-Arsuf has been excavated by various teams since 1950.

The settlement developed into an urban community as early as the Persian period (Tal 1999: 205–14), in the 6th century BC when Apollonia-Arsuf, then a part of the province of Dor, was ruled by overlords from Sidon. As the principle city in the southern Sharon Plain, the settlement developed into the commercial and industrial center of the region.

Fig. 1. Aerial view of the castle at Apollonia-Arsuf (© Albatros Aerial Photography).

The first-century Roman historiographer Flavius Josephus is the first to mention the site. He refers to Apollonia, along with Joppa, as the "Jewish cities" that existed at the time of Alexander Jannaeus (*Ant.* 13.15.4 [395]). Pliny and Ptolemy write about Apollonia in the context of the coastal cities of Iudaea-Palaestina (Pliny, *Nat.* 5.13.69; Ptolemy, *Geo.* 5.15.2). On the Peutinger map (a medieval copy of an ancient Roman map), Apollonia appears on the coastal highway, located between Joppa and Caesarea. This suggests that it was an official stop on the country's Roman imperial road network. It continued to flourish through the Byzantine era.

In 1101, the town was conquered by the Crusaders and became a possession of the Crown (Roll 1999: 11–18). The new inhabitants repaired the city wall and a few adjacent buildings. John II of Ibelin who served as constable and baillif of the Kingdom of Jerusalem, began construction of the town's castle in 1241. The fortified structure (Area F) is the most important building complex erected at the site during Crusader rule (Roll 2008: 255–62; Roll and Arubas 2006). It is situated in the northwestern section of the fortified city (Figure 2).

After the city was captured by the Mamluks in 1265, the defenders of the castle were forced to destroy it themselves (Amitai 2005; Raphael and Tepper 2005). While the Mamluks ruled Palestine, they intentionally left the coast in ruins so as to prevent any potential incursions by the Christians. With Arsuf lying in ruins and far from the country's main communication lines, it was gradually forgotten.

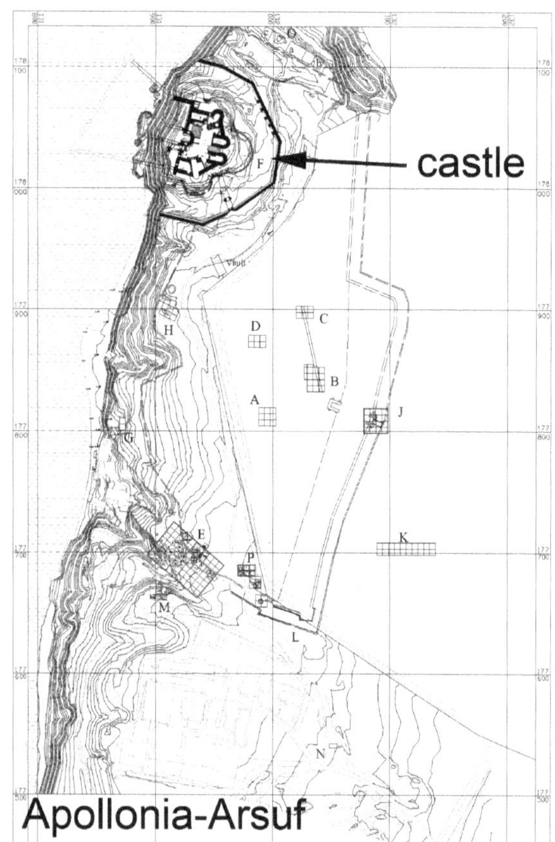

Fig. 2. Plan of Apollonia-Arsuf (at the close of the 2006 season; drawing and survey data by Benny Arubas; image courtesy of Oren Tal).

REVEAL 1 & 2

Numerous benefits will accrue to Apollonia's continuing excavation and architectural and historical interpretation as a result of REVEAL's initial tests, for example:

1. Automated 3D architectural reconstructions of the castle from drawings will provide increased accuracy and virtual reassembly speed, and enable new types of hypothesis testing that could not be accomplished pre-REVEAL.
2. Automated pottery and glass vessel reconstructions generated from multiple images will provide increased accuracy and virtual reassembly speed, and enable new types of hypothesis testing that could not be accomplished pre-REVEAL.
3. Intra- and inter-site analyses permitting more accurate and more efficient digital comparisons of fragments, artifacts, and architectural components.
4. 3D architectural reconstructions from aerial photos will allow for accurate and fast modeling of large architectural complexes.
5. 3D immersive virtual reality re-creations of archaeological excavations at any time during their operations.

Not all these features are ready yet, but this summer's tests along with previous field and lab work will merge to demonstrate the dramatic potential of these new digital and automated approaches to archaeological documentation, recording, image-gathering, data processing, querying, and publishing.

Castle Reconstruction

The 13th-century destruction of the castle left few building components in place. Enormous broken chunks consisting of partial vaults, ceilings, arches, floors, walls are still scattered in the surrounding moat and along the beach. Since a physical reconstruction of the complex is impossible, a virtual reconstruction remains the only viable means to visualize and understand the castle's original appearance. Thus, the general castle disarray provides an excellent platform for testing REVEAL's automated data capture and processing methods. Previous year's scanning results have become a prelude to further refinements of the REVEAL system, complemented with research of contemporary Crusader castles in England, France, Lebanon, Syria, and Jordan.

Part of the REVEAL system is software that will semi-automatically extract 3D model geometry from plan drawings. Apart from the application for modeling sites and buildings in archaeological contexts, such new tools provide useful ways to analyze and cross-validate varying representations as well as to evaluate sophisticated 3D spatial hypotheses. Our efforts represent a novel fusion of cultural heritage, automated data processing, and 3D modeling (see also: Muller et al. 2006a, 2006b; Yin et al. 2009).

We propose to fuse concepts from each discipline to generate domain-specific software that enables accurate and detailed analyses of this data in a digital context, alongside other relevant input such as survey or 3D scan data that would otherwise require a prohibitively large time and money investment from the researcher. Implementing this fusion has required new computational algorithms, shape models, and new types of visualizations (Figure 3).

Our approach consists of five steps:

1. digitize the documents of interest with a digital camera or scanner.
2. pre-process the digital image to remove text and extraneous lines.
3. vectorize a skeletonized plan into simple, non-intersecting curve fragments connected at a collection of special x,y locations or nodes.
4. using the vectorized image and hand-specified heuristics, each contiguous group of pixels in the image is assigned a semantic label designating the architectural element or space they represent.
5. based on the label for each curve fragment, the software automatically constructs a 3D model over the extent of the curve fragment.

Manual intervention is required only to record the data and in some cases modify the resulting image to prepare it for the automatic processing stages. The remaining steps are completely automatic, except for step 4 which requires a set of hand-specified heuristics to isolate semantically meaningful components within the image.

We are pleased with our results so far, which have reliably classified castle surroundings, outer walls, towers, tower rooms, and courtyards. As demonstrated by comparing the actual buildings to those reconstructed automatically, the semantic labels provide important visual information and a good estimate of the actual structure. Note that, at present, existing photographic information is not used to augment the 3D model; this is an area of interest for our future development.

We have devised software that processes archaeological plan drawings for purposes of identifying significant semantic sub-structures present within the drawings. Our approach differs from existing methods in both goal and methodology. Our goal is to process drawings for historic sites; and our approach generates visualizations form scanned documents using a process that is fully automatic in most tested cases and only in some cases requires modest user input. We are currently looking at several avenues of enhancement, including: interpreting other annotations occurring on typical plan drawings, using a procedural model to automatically generate local building details based on the semantic labels, integrating information from cross section drawings, and extracting

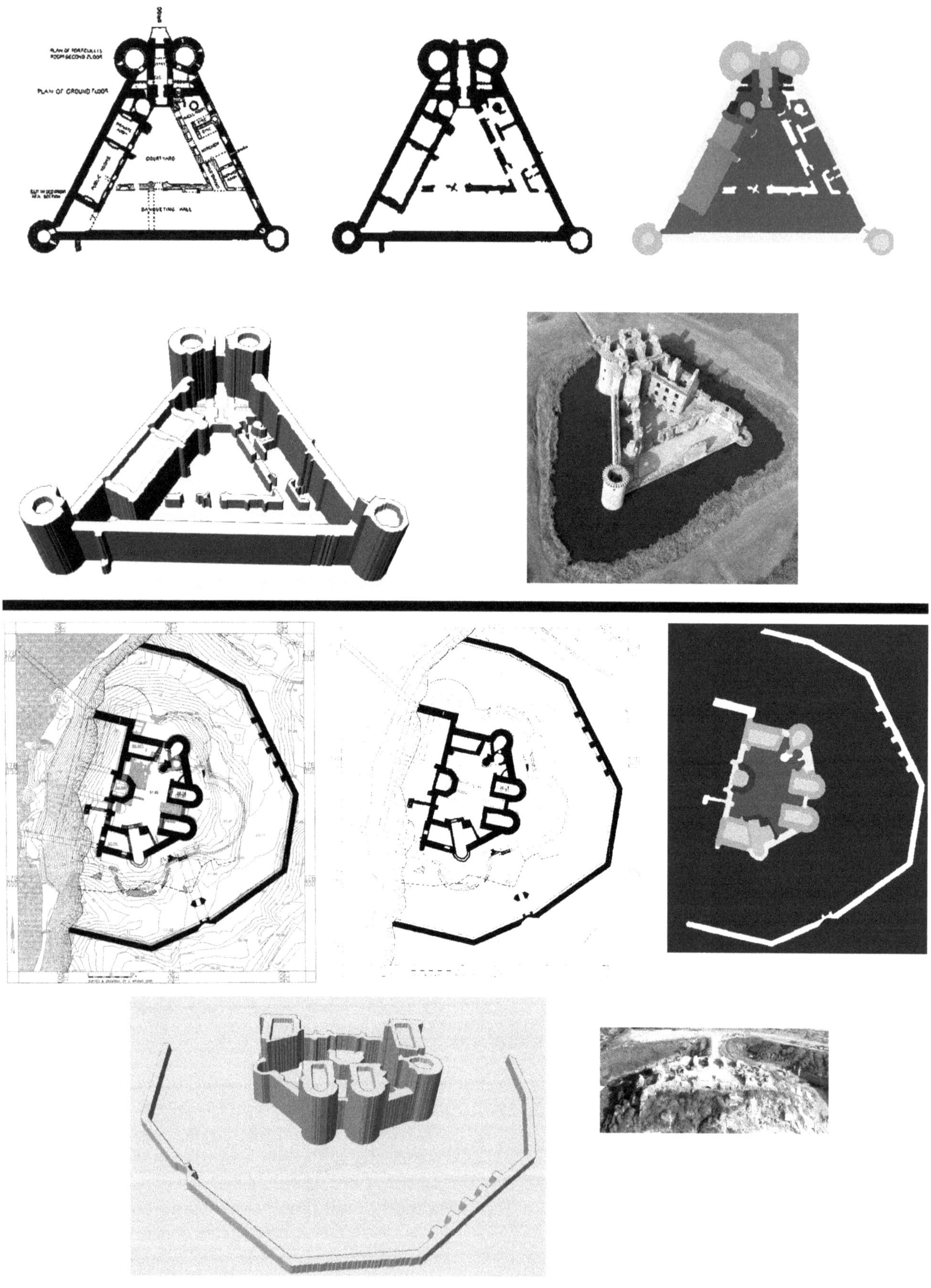

Fig. 3. Semi-automated castle reconstruction (top Caerlaverock Castle; bottom Apollonia-Arsuf) showing (a) original digitized image, (b) binary image with non-architectural features removed, (c) semantic labels assigned to portions of the image, (d) an estimated 3D model using the semantic labels, (e) an aerial view of the actual in-situ remains of the castle.

information from other types of archeological annotations that are currently removed as noise.

Pottery and glass vessel reconstruction

Pottery sherds are among the most ubiquitous finds in archaeological contexts, often reaching quantities that present a challenge for cataloging and storage. Archaeologists have joined with engineers over the past few years to develop mechanisms that attempt to cope with this problem, resulting in prototype systems capable of performing typological analysis (Gilboa *et al.* 2004) as well as virtual reconstruction of complete ceramic vessels (Willis and Cooper 2004; Kampel and Sablatnig 2003; and see Willis and Cooper [2008] for a survey of these techniques). Great challenges in the widespread use of such schemes are to: (1) make available a system that can quickly capture digital models of ceramic fragments; (2) produce mechanisms that are capable of storing and searching the digital models; and (3) create systems that can extract archaeologically significant information from these fragments. Ongoing research at the Apollonia-Arsuf site seeks to address each of these issues.

Another one of our goals at Apollonia-Arsuf is to virtually reassemble and reconstruct storage jars and transport amphorae (Galor in press). Apollonia's immediate location on the Mediterranean coast, the presence of its harbor, and its proximity to several land routes and highways were responsible for a flourishing trade during the Roman, Byzantine, Early Islamic, and Crusader periods. Our ability to understand its commercial ties with other Mediterranean cultures, the evolving and changing economies, technologies, and many other aspects of life in the region depends on our knowledge of pottery. Traditional methods of determining typological and chronological changes depend on minor variations in texture, size, and shape of the vessels. However, such methods of recording and interpretation that normally follow rim morphology hardly ever allows us to distinguish accurately between a jar that was produced in the 5th century AD from a jar type produced in the 6th century AD. Refined observations and measurements of 3D reconstructions would allow a much more accurate and precise understanding of Apollonia's international trade activities throughout the centuries, a factor that determined much of the region's history. Full dimensional reconstructions of certain local and imported groups of storage jars and transport amphorae will hopefully allow us to detect their capacity standard through different periods of time, as vessel standardization changed consistently throughout classical and medieval times.

Virtually reassembling glass vessels is potentially even more important than reassembling pottery vessels. Since glass is much more breakable than are clay vessels, manual reconstruction of the original objects is far more limited. Despite the fact that surface irregularities of broken glass are just as valuable as pottery sherds for the extraction of geometric characteristics, this method has never been explored. We will scan glass fragments and develop new techniques to handle those much more delicate pieces characterized by edges that have similar geometric features that can be digitally reproduced and virtually matched with other edges of broken glass sherds.

With these applications and procedures underway, we began to expand the functionality of the REVEAL prototype system, which will be tested this summer at Apollonia.

REVEAL 1 (Figure 4)

Modular packs of networked digital cameras and laptop computers with a central database are a main component of our collaborative digital archaeology system. These nodes will be used for: 1) archaeological site-excavation monitoring that is comprehensive, informative, and searchable from within the database; 2) site exploration including extracting 3D measurements and representations by remote users for purposes of 3D-immersive data analyses and virtual tourism; and 3) 3D fragment assembly and reconstruction of artifacts, facilitated among others by intra- and inter-database comparisons.

The images captured by the camera nodes will be displayed and manipulated in real time in a special interface.

1) Our site-excavation monitoring system consists of a camera array integrating both still and video digital cameras that take images of dig activities and transmit them to local laptops. Images can be taken automatically or interactively. Frame rate and image resolution can be kept at a minimum when monitoring general work progress. Frame speed and resolution can be increased when interesting soil changes are being noted or an artifact is being uncovered. Cameras will be mounted at a distance overhead that will not interfere with the activities in the trench. Computer vision algorithms enable the cameras automatically to eliminate views of moving individuals so that only the relevant material will be stored.

Several ten-megapixel cameras will be mounted on a geodesic-like dome scaffold made of rigid metal pipes (Figure 5). Cables carrying power and data to and from each camera node will be attached to the pipes, so as not to inconvenience the excavators. The technology for image-data collection involving small numbers of digital cameras is not new (Hampapur 2008; Vincent 2007), but has not been applied to archaeological contexts.

Therefore, existing software has to be adjusted to conditions of an archaeological site reality. Images have to be recorded in close to real time in order to capture situations, x-y-z data, structures, objects, fragments, and general site compositions and modifications throughout the excavation process, which

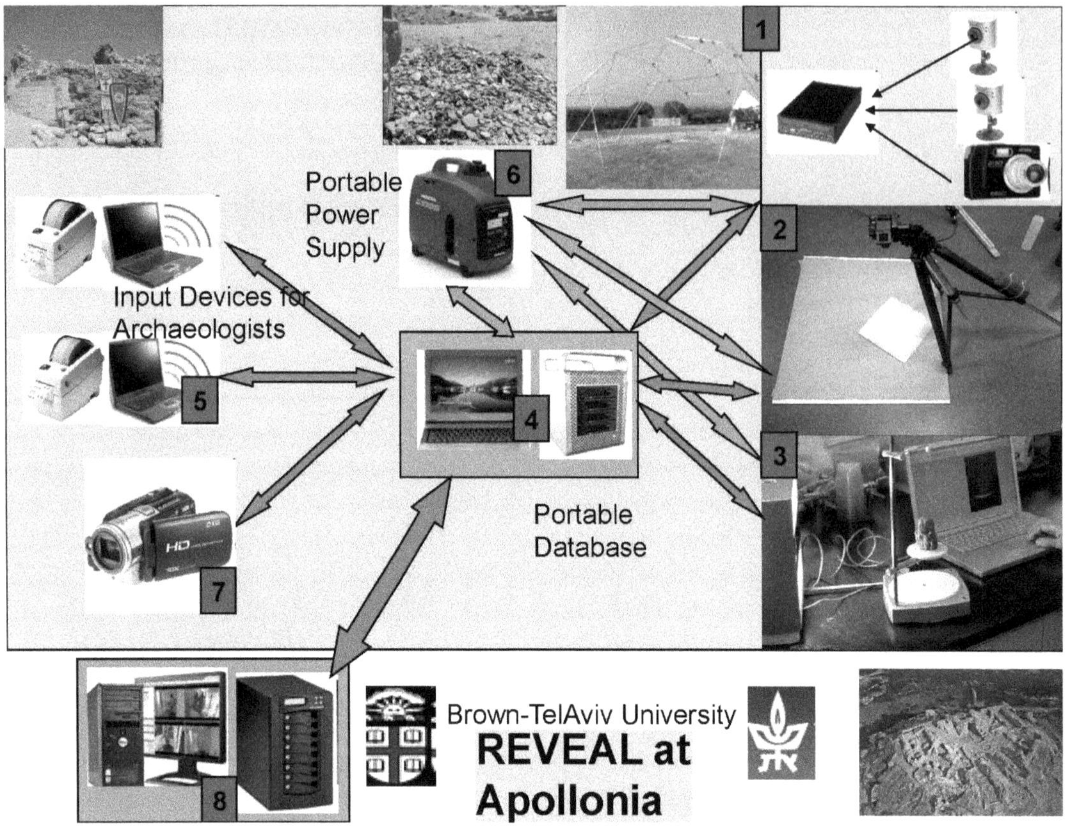

Fig. 4. Preliminary mock-up of the REVEAL 1.0 component relationships and data connection pointers.

Fig. 5. Sample mock-up of the camera nodes and the scaffolding to be used over trenches.

is destructive and under situations that are constantly changing. A ten megapixel overhead camera placed at a distance of four meters, with a ground resolution of roughly one millimeter has the ability to capture the ground level in a reasonably clear manner. We believe that four to six nodes will be required to monitor the activity in a typical trench. The location and orientation of fragments and small finds will be captured automatically in 3D. For this purpose, two or more "calibrated" cameras need to view the same site points used for the necessary calculation. By using two to three times the number of ten megapixel cameras and additional camera geometry as specified above, a distinct 3D point can, in theory, be automatically estimated to an accuracy of roughly two millimeters (Faugeras and Toscani 1987; Salvi, Pagès and Batlle 2004; Zhang 2000; Zhang and Huang 2006).

2) Once the images taken by calibrated cameras are stored in a database, they can be viewed remotely thus permitting researchers or lay people to review the excavation progress in almost real time. In addition to simply observing the process of excavation, the archaeologist can extract precise measurements of shapes, locations, and relative positioning of structures and objects. The technology used to estimate 3D points and their relative alignments, which are based on images taken by calibrated cameras is well established (Heuel and Forstner 2001; Liu *et al.* 2008; Lorenson and Cline 1987; Schmid and Zisserman 2000). Taking precise measurements and creating accurate representations of the existing physical data to create virtual 3D reconstructions of

fragmentary objects and structures is not exclusively meant to improve the visual appearance of the excavation results, but most importantly, to improve the ability of the archaeologists to answer complex questions and facilitate challenging inquiries. Our newly developed procedures will enable researchers to review the excavation process and uncovered data fully in 3D at any point in time. Furthermore, they will enable the general public, in any given location, to explore the site as if they were physically present. Finally, they will enable researchers to virtually reconstruct partially destroyed structures and objects.

3) To virtually reassemble a pottery or glass vessel, based on any number of fragments, computers need to estimate points, curves, and surfaces to an accuracy of about 1mm. This is not feasible with a camera network set up above an excavation square as described above. So, we will develop a mechanism to photograph these fragments with specially designed small camera arrays of anywhere between four and sixteen cameras, rigidly mounted together and equipped with a light source that projects the designed light pattern (Wu *et al.* 2006). Existing systems do not have the accuracy and processing speed required for our purposes (Lanman *et al.* 2008).

The image capture and display component can support images and 3D scenes (Figure 6). The information panel maintains a list of all objects, which can be an image or a 3D scene (polygon meshes, point clouds, indexed line sets, contained in a VRML scene graph). Each object is rendered on the desktop within its own bounding box. The location and size of each bounding box can be modified interactively with the mouse, or by typing parameters through menus. The objects are displayed as overlapping layers, like in Photoshop. The order of the objects in the list can be changed.

3D scenes can be rendered with transparent background. Eventually, we will implement more general transparency/blending by using the alpha channel, even for images. The cursor will have different behaviors in different locations in the display window. In particular, right clicking on a bounding box brings up an object-dependent popup menu. The application will display and play videos streamed from the capture computer servers, as well as be able to control the still cameras in the dome arrays. The videos can be assembled on the fly as requested.

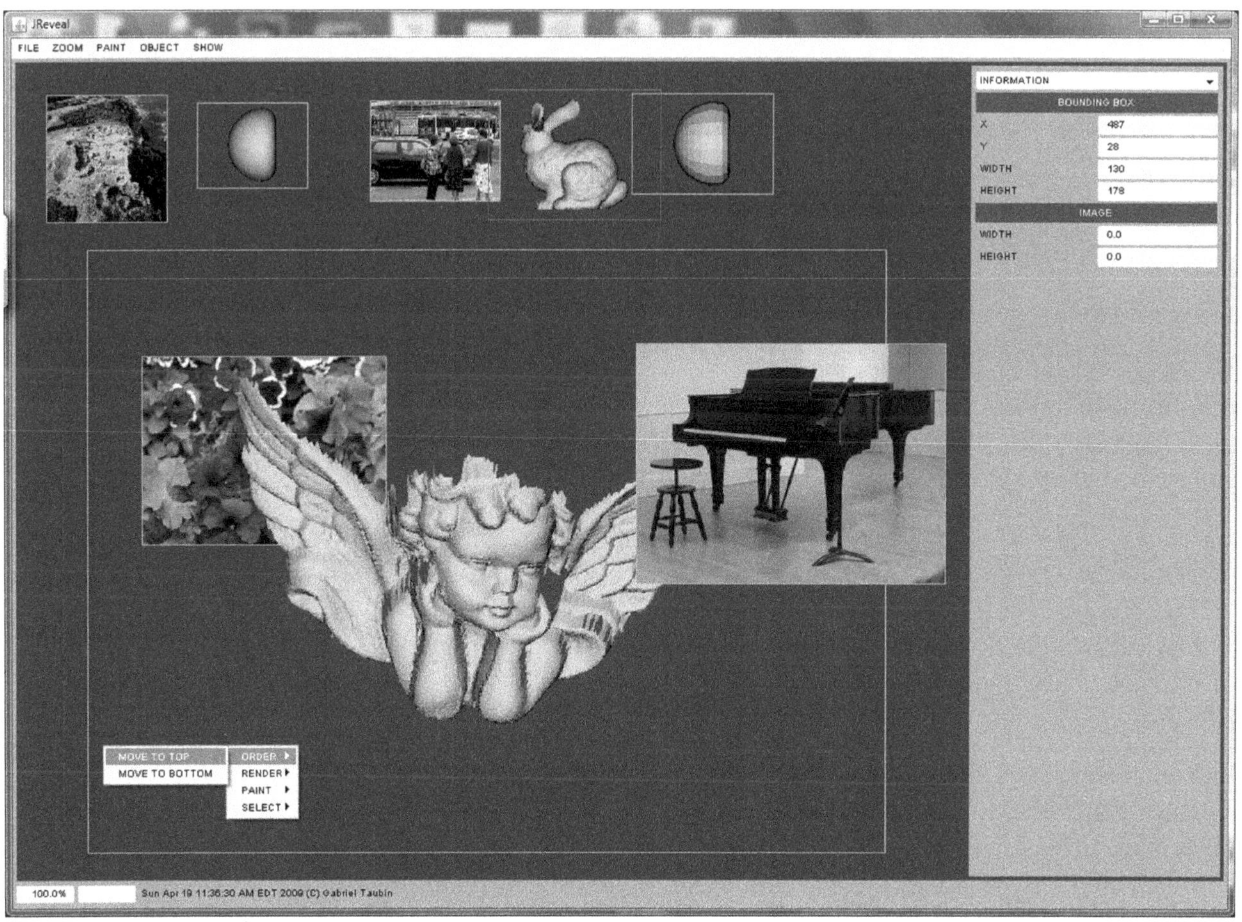

Fig. 6. Sample screen grab from the REVEAL 1.0 java-based image-capture manipulation interface, showing representative image types on the left, and one of the many interactive parameter menus on the right.

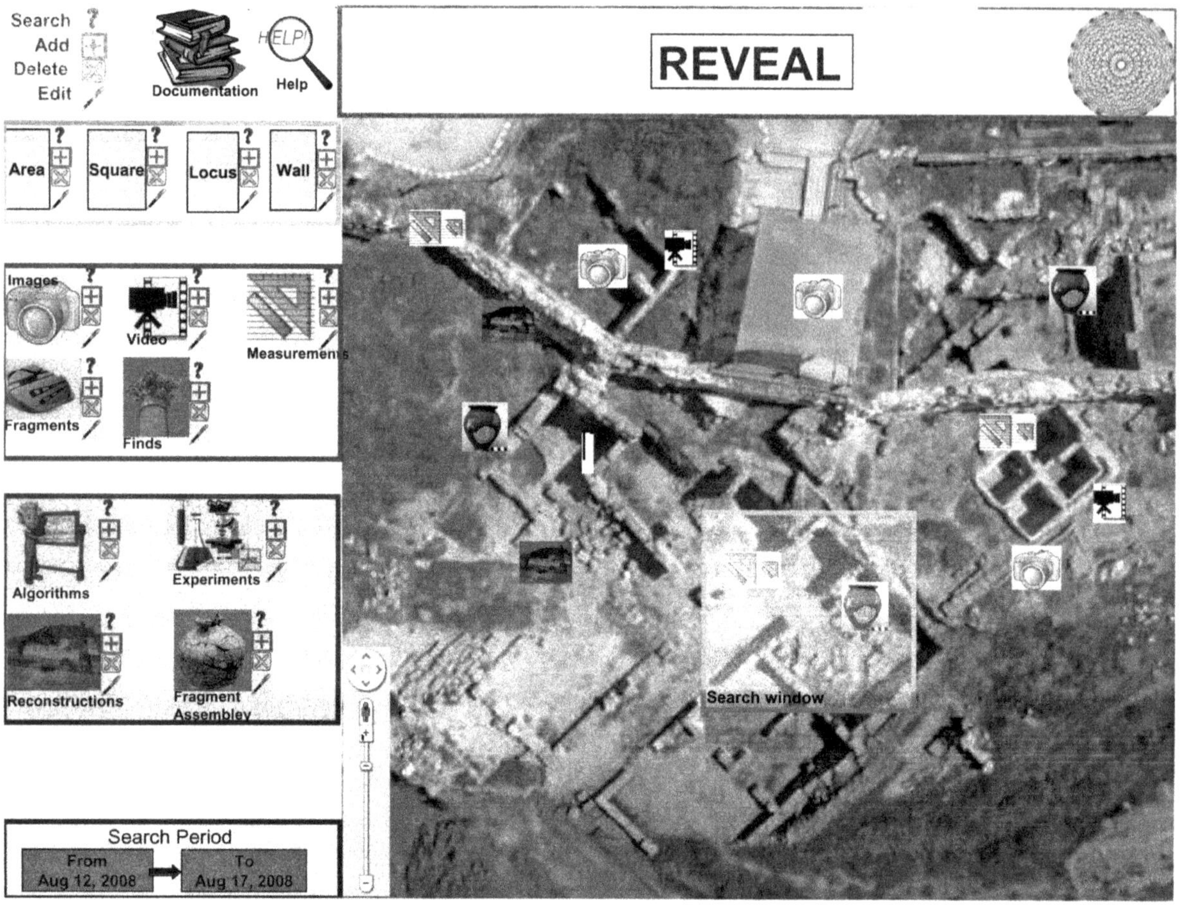

Fig. 7. Screen grab from the REVEAL 1.0 database query system.

Later on, we will also add support for unstructured tetrahedral meshes and isosurfaces extracted from them, which we can use to represent volumetric data such as soil layers. The internal scene graphs will allow users to insert other objects into the scene, and control their position, orientation, and relative sizes. Animation of 3D scenes and other exciting applications will follow in due course.

Essential to the retrieval of real-time 3D data is our visual archaeological database, which will be designed to dramatically transform the slow and tedious recording process of an excavation. The multiple camera nodes and scanning stations will maintain a continuous visual recording of excavation activity. The excavators will process artifacts immediately by inputting or selecting text, video, still images, dense-data 3D laser scans, and spatial position coordinate information into the database. The database will allow for immediate and efficient searches as well as retrieval of images and 3D scene reconstructions (Figure 7).

The procedure could ultimately replace the current manual recording and processing protocol with a less time-consuming and more accurate documentation process. The methodology should not only make it easier for archaeologists to review, analyze, document, synthesize, and publish excavation data, but should also provide the opportunity to review the excavation process and finds in a more accessible and organized presentation. During our 2009 summer tests, we intend to explore this system in 1 trench only.

Conclusions

As opposed to REVEAL 1 (the system we will be testing this summer), our future REVEAL systems will not only record visually all activity taking place in the excavation trench, but have the ability through human interface to select important points and key moments. Those points can be the top and bottom surfaces of a layer, the 3D characteristics of a wall, or the irregular shape of an installation. By placing colorful marbles or pointing arrows at specific points within the trench, the cameras will be given orders to record those 3D characteristics with special attention and separate them from the remaining data. The elimination of redundant data will thus allow the excavator to be more time efficient and focus primarily on shapes, objects or structures that are essential to the 3D reconstruction of all elements uncovered within the trench. As of 2010 the REVEAL system will record the activities and findings in multiple excavation squares, so that

the database too will have the ability not only to store the data, but to allow the archaeologist to search and establish differences, similarities and groupings.
Thus, REVEAL

- will document the excavation process more accurately and in a less biased manner.
- will limit the subjectivity of the area supervisor/excavation director.
- will allow full immersion into the excavation from remote locations

The goals are to automate many of the mundane, tedious, and time-consuming tasks of excavating, especially those often prone to inaccuracies. We intend to create a central database that can be queried in order to replay in realtime and in 3D any part of the dig. The database will acquire specific and detailed information about artifacts, architecture, and stratigraphy, thus leaving the archaeologist more time and flexibility to ask new questions about the data and test hypotheses about the site that could not have been posed before. Our expectation is that with our newly developed methodologies, we will be able to dismantle and reconstruct destroyed structures and fragmentary objects more effectively and accurately.

We hope that in a couple of years we will be able to deliver a set of Interent-based tools and excavation-tested products that will convince other archaeologists to venture into this new automated digital way of documenting and analyzing field activities. We look forward to nothing short of a paradigm shift in the way archaeological fieldwork is conducted, data are processed and analyzed, and interpretations disseminated as we look to a more accurate and precise view of the past.

References

AMITAI, R. 2005. The Conquest of Arsuf by Baybars: Political and Military Aspects. *Mamluk Studies Review* 9, 61–83.

FAUGERAS, O. D., AND TOSCANI, G. 1987. Camera Calibration for 3D Computer Vision. in *International Workshop on Machine Vision and Machine Intelligence, Seiken Symposium, Tokyo, Japan, February 2–5 198,* 240–47. New York: IEEE.

GALOR, K. in press. Hellenistic and Roman Apollonia-Arsuf: Digital Quantification and Reconstruction of Amphorae. *Bollettino di Archaeologia.*

GILBOA, A., KARASIK, I., SHARON, I., AND SMILANSKY, U. 2004. Towards Computerized Typology and Classification of Ceramics. *Journal of Archaeological Science* 31, 681–94.

HAMPAPUR, A. 2008. Smart Video Surveillance for Proactive Security. *Institute of Electrical and Electronics Engineers Signal Processing* 25, 132–34, 136.

HEUEL, S., AND FORSTNER, W. 2001. Matching, Reconstructing and Grouping 3D Lines from Multiple Views Using Uncertain Projective Geometry. in *CVPR 2001: Proceedings of the 2001 IEEE Computer Society Conference on Computer Vision and Pattern Recognition, Volume 2,* 517–24. Los Alamitos, Calif.: Institute of Electrical and Electronics Engineers Computer Society.

KAMPEL, M., AND SABLATNIG, R. 2003. Virtual Reconstruction of Broken and Unbroken Pottery. in *Proceedings of the Fourth International Conference on 3D Digital Imaging and Modeling (3DIM'03),* M. Rioux and G. Godin, eds. 318–25. Los Alamitos, CA: Institute of Electrical and Electronics Engineers Computer Society. Online: http://ieeexplore.ieee.org/xpl/freeabs_all.jsp?arnumber=1240265. Accessed April 10, 2009.

LANMAN, D., CRISPELL, D., AND TAUBIN, G. 2008. Shield Fields: Modeling and Capturing 3D Occluders. *Association for Computing Machinery Transactions on Graphics* 27: 131, 1–10.

LIU, S., KANG, K., TAREL, J.-P., AND COOPER, D. B. 2008. Free-Form Object Reconstruction from Silhouettes, Occluding Edges and Texture Edges: A Unified and Robust Operator Based on Duality. *Institute of Electrical and Electronics Engineers Transactions on Pattern Analysis and Machine Intelligence* 30, 131–46.

LORENSEN, W. E., AND CLINE, H. E. 1987. Marching Cubes: A High-Resolution 3D Surface Construction Algorithm. in *Proceedings of the 14th Annual Conference on Computer Graphics and Interactive Techniques,* 163–69. New York: Association for Computing Machinery.

MULLER, P., VEREENOOGHIE, T., WONKA, P., PAAP, I., AND VAN GOOL, L. 2006A. 3D Reconstruction of Puuc Buildings in Xkipche, in *The 7th International Symposium on Virtual Reality, Archaeology, and Cultural Heritage,* M. Ioannides, D. Arnold, F. Niccolucci, K. Mania, eds. 139-46.

MULLER, P., WONKA, P., HAEGLER, S., ULMER, A., AND VAN GOOL, L. 2006B. Procedural Modeling of Buildings, in *Proceedings of Association for Computing Machinery SIGGRAPH,* 614-23.

RAPHAEL, K., AND TEPPER, Y. 2005. The Archaeological Evidence from the Mamluk Siege of Arsuf. *Mamluk Studies Review* 9, 85–100.

ROLL, I. 2008. Der frühislamische Basar und die Kreuzfahrerburg in Apollonia-Arsuf.. in *Burgen und Städte der Kreuzzugszeit.* Studien zur internationalen Architektur- und Kunstgeschichte 65, M. Piana, ed., 252–62. Petersberg: Michael Imhof.

ROLL, I. 1999. Introduction: History of the Site, Its Research and Excavations. in *Apollonia-Arsuf: Final Report of the Excavations,* Vol. I: *The Persian and Hellenistic Periods (with Appendices on the Chalcolithic and Iron Age II Remains)*, I. Roll and O. Tal, eds., 1–62. Tel Aviv University, Sonia and Marco Nadler Institute of Archaeology Monograph Series 16. Tel Aviv: Emery and Claire Yass Publications in Archaeology of the Institute of Archaeology, Tel Aviv University.

ROLL, I., AND ARUBAS, B. 2006. Le château d'Arsur: Forteresse côtière pentagonale du type concentrique du milieu du XIIIe siècle. *Bulletin Monumental* 164, 67–81.

SALVI, J, PAGÈS, J., AND J. BATLLE, J. 2004. Pattern Codification Strategies in Structured Light Systems. *Pattern Recognition* 37, 827–49.

SCHMID, C., AND ZISSERMAN, A. 2000. The Geometry and Matching of Lines and Curves over Multiple Views. *International Journal of Computer Vision* 40, 199–233.

TAL, O. 1999. The Persian Period, in *Apollonia-Arsuf: Final Report of the Excavations,* Vol. I: *The Persian and Hellenistic Periods (with Appendices on the Chalcolithic and Iron Age II Remains)*, 83–222, I. Roll and O. Tal, eds. Tel Aviv University, Sonia and Marco Nadler Institute of Archaeology Monograph Series 16. Tel Aviv: Emery and Claire Yass Publications in Archaeology of the Institute of Archaeology, Tel Aviv University.

VINCENT, L. 2007. Taking Online Maps Down to Street Level. *Computer* 40, 118–20.

WILLIS, A., AND COOPER, D. B. 2008 From Ruins to Relics: Computational Reconstruction of Ancient Artifacts. *Institute of Electrical and Electronics Engineers Signal Processing Magazine* 25, 65–83.

WILLIS, A., AND COOPER, D. B. 2004. Bayesian Assembly of 3D Axially Symmetric Shapes from Fragments, in *Conference on Computer Vision and Pattern Recognition*, 82–89, R. Chellappa and A. Bobick, eds.. Los Alamitos, CA: Institute of Electrical and Electronics Engineers Computer Society. Online: http://ieeexplore.ieee.org/xpl/freeabs_all.jsp?arnumber=1315017. Accessed April 10, 2009.

WU, H. B., CHEN, Y., WU, M. Y., GUAN, C. R., AND YU, X. Y. 2006. 3D Measurement Technology by Structued Light Using Stripe-Edge-Based Gray Code. *Journal of Physics: Conference Series* 48, 537–41.

YIN, X., WONKA, P., RAZDAN, A. 2009. Generating 3D Building Models from Architectural Drawings: a survey, *Computer Graphics and Applications* 29.1, 20-30.

ZHANG, S., AND HUANG, P. S. 2006. Novel Method for Structured Light System Calibration. *Optical Engineering* 45.083601, 1–8.

ZHANG, Z. 2000. A Flexible New Technique for Camera Calibration. *Institute of Electrical and Electronics Engineer Transactions on Pattern Analysis and Machine Intelligence* 22, 1330–34.

CENOBIUM – putting together the Romanesque cloister capitals of the Mediterranean Region

M. Corsini,[1] M. Dellepiane,[1] U. Dercks,[2] F. Ponchio1, M. Callieri1, D. Keultjes,[2] A. Marinello,[2] R. Sigismondi,[2] R. Scopigno[1] and G. Wolf[2]

[1]Visual Computing Lab, ISTI-CNR Pisa, Italy
[2]Kunsthistorisches Institut in Florenz (Max-Planck-Institut), Florence, Italy

Abstract

CENOBIUM (Cultural Electronic Network Online: Binding up Interoperable Usable Multimedia) is a Cultural Heritage project regarding Romanesque cloister capitals from the Mediterranean region. The central aim of this project is to illustrate cultural exchange in the twelfth and thirteenth centuries through the example of architectural decoration. In this ambit, three Cloisters have been considered until now; the cloister of Monreale (Sicily), the cloister of Sant'Orso in Aosta and the cloister of Cefalù (Sicily). For the most important capitals of these cloisters several high resolution digital images and highly detailed coloured 3D models have been produced.
This big amount of data has been integrated in a web application where, textual descriptions, high resolution images and 3D models can be accessed from the same page for each capital. Moreover, a simple mechanism permits to compare images and models of capitals of any cloister: the user can select them during the exploration, and then a simple tool (called LightTable) can download and visualize them on the computer. In this way, it is possible to compare the selected elements, so that further similarities between them can be found and easily investigated.
In conclusion, the CENOBIUM website is a successful integration example of different types of data, in the context of an environment which is devoted not only to the wide public, but also to the community of experts and scholars of the Romanesque art.

Keywords Romanesque art, 3D scanning, integrated information, website, multimedia, Cultural Heritage.

Introduction

The use of new technologies in the context of Cultural Heritage has been an important issue since the 19th Century. The advent of new means to acquire and visualize information has brought to several revolutionary changes in the way Art and History are analyzed and presented.

More recently, 3D Scanning and digital photography provided ways to easily collect accurate information about the geometry and the appearance of objects of interest. The integration of these data with the other "usual" ways (like drawing or texts) to describe an artefact is still an open issue. In particular, the aim is to go beyond the mere archival and visualization of all the data: it is indeed critic for the experts and the restorers to be able to access and manipulate all the information in order to enhance the quality of work, both in terms of speed and of data analysis.

The CENOBIUM project aims at demonstrating the potential benefits of the integration of modern techniques of representation with cutting-edge web technology in the pursuit of new knowledge about artefacts of the past. The central theme of this project is the articulation of the dynamics of cultural exchange at work the twelfth and thirteenth centuries through the medium of architectural decoration. The specific case studies currently being considered for the assessment of our approach are a selected group of capital cycles, found in medieval cloisters throughout the Mediterranean.

CENOBIUM provides a web-based open-access research environment, which includes 3D models created by combining high-resolution scanning with on-site digital photographic campaigns of the highest professional quality. The technical tools dedicated to the integration and extension of available technologies include databases, image-viewers, 3D-viewers, content management, and the innovative *LightTable* tool (described in the next), as well as other digital tools and media.

Cenobium offers the user the opportunity to fully manipulate the 3D images of the capitals in order to allow for maximum visual clarity from any point of view, including high-resolution zoom capability. This allows convenient and interactive access to the formal, narrative, and material contents of the capitals that can exceed even on-site study of the cloister. Cenobium will provide for more comprehensive appreciation, documentation, and interpretation of the site by allowing the user to virtually move inside the cloister itself, to focus in on a particular capital, and to analyze and compare capitals from different cloisters at the same time on the same screen. By means of innovative technological solutions, all data (at the highest level of detail and resolution) will be accessible at a dedicated kiosk installation in the institute, locally on private computers, and from anywhere on the World Wide Web. One of the main components of this innovative solution is the *LightTable* tool. This works as a shopping cart in an e-commerce website; the users select items during the navigation in the website, but they can compare them instead of buying such items. Specifically, the high-res photos and the geometry of the capitals can be selected.

In this paper we will describe the steps which brought to the creation of the prototype of this integrated environment, which was made available online at the address http://cenobium.isti.cnr.it/. First, we will present the motivations which led to choice of the Romanesque cloisters around Europe as a large and promising test case. Then, we will shortly describe the technologies used to acquire and process the data which are present inside the collaborative environment. In particular, the pipeline and the tools used to produce the accurate coloured 3D models of the capitals will be presented. The Section about the website will show how all the data were integrated in the context of the developed web application, with particular attention to the mechanism which helps the user in comparing the data from different cloisters around Italy. Finally, the discussion and the conclusions will underline the achieved results and the future work.

The Romanesque cloisters around Europe: a complex and promising test case

The cloister of the Cathedral of Monreale in Sicily was the starting point of the Cenobium project. Its high quality Romanesque sculptural programme is extremely well suited to the technological analysis. Executed between 1174 and 1189 under the patronage of King William II, this cloister unites various artistic currents of Romanesque sculpture then circulating in the Mediterranean into an architecturally homogeneous ensemble. Along with Monreale, the cloisters of Sant'Orso in Aosta (c. 1150) and Cefalù (c. 1131/1166 o 1185/1189), belong to the few cloisters in Italy that display an extensive capital programme comprising both narrative and figural representations. The cloister in Aosta, for example, stands as an exemplary monument of the artistic development of medieval capital design in the Mediterranean and, stylistically, constitutes a bridgehead between the French and Italian cloister type.

The Italian Medieval cloisters (Monreale, Cefalú, and Aosta, shown in Figure 1) are only some of the examples of the Romanesque art, which includes several contemporary Spanish and French examples.

Shortly before the turn of the twelfth century a type of capital emerged in Romanesque sculpture that profoundly influenced the overall character of sacred spaces and monastic cloisters and became a primary element of medieval sculptural decoration of architecture.

Previously, capital decoration had confined itself mainly to ornamental, vegetative, zoomorphic, or anthropomorphic forms. Now, however, it broadened its spectrum to include narrative cycles, thus taking on the added function of depicting stories from the Old and New Testaments, historical events, *exempla*, satirical scenes, and allegories. These cycles of historiated capitals integrate narrative elements into their particular physical contexts. The nature of this type of cloister design – with free-standing columns and a moving viewer - is particularly well suited to narratives that must be scene from all sides. Therefore, they afford the possibility of telling stories through a series of relief compositions, while facilitating a dialogue among and between capitals and other decorative elements of the cloister.

Of additional importance is the spatial interrelation between the capital and the functional design of the cloister complex. The arrangement of themes and motifs on the capitals permits, in some cases, an aesthetic and physical interaction on the part of the viewer and gives us an idea about the liturgical role and ritual practices associated with the cloister. Moreover, the similarities between capitals around Europe can bring to the discovering of points of contact between different sculpting schools, which can be consequence of a common way of thinking or of contacts and contaminations between the artists.

Data collection and processing

The comprehensive acquisition campaigns performed in Monreale, Aosta and Cefalù brought to the creation of a large database of high quality 2D and 3D data. High-end devices were used for both image and geometry, as shown in the next subsections. Then the data were performed using various non-commercial tools, in order to produce extremely accurate colored 3D models.

High resolution digital imaging

A Sinar P3 camera was purchased by the Photo Library of the Kunsthistorisches Institut, providing for the integration of the digital multi-shot-mode Sinarback 54 H with a resolution of 22 million pixels (sensor resolution 5440×4080 pixels), as well as various Sinaron lenses. This is a very expensive but also very high quality device, which can produce impressive results if used by a professional photographer.

The high-resolution digital images are created in a two-step process. First, a digital image is produced with a color management tool by Gretagmacbeth, following the intent to save as much information as possible with the one-, four- or sixteen-shot (taken at different exposure levels). This master copy is used for producing further copies and for long-term preservation. Its size ranges from approximately 130 up to 520 Megabytes (TIFF format uncompressed, 16-bit colour depth and 300dpi). A working image copy is created from this master. This copy is digitally enhanced to allow improved quality on a low-dynamic range output device (screen or printing device). Its size is approximately 65 Megabytes (TIFF format uncompressed, 8-bit-per-channel colour depth, 4000 × 4000 pixels - approximately 33 cm on a 300 dpi printout): an example is shown in Figure 2, left.

Fig. 1. A view of the Monreale (top left), Aosta (top right) and Cefalù (bottom) Cloisters.

The photographic campaigns on the cloisters were performed by acquiring 8 images for the highly interesting capitals and 4 or less images for the low interest capitals. This brought to the acquisition of a total of 690 images for the Monreale cloister, 270 images for the Aosta cloister and 180 images for the Cefalù cloister.

Fig. 2. An example of an acquired image of a capital from Monreale (left) and a snapshot of a colored 3D codel of the Cloister of Aosta (right).

Geometry acquisition: scanning the capitals

For two of the acquisition campaigns (Monreale and Cefalù), a Konica Minolta VI 910 Laser Scanner (a device based on optical triangulation) was used. Triangulation scanning permits to acquire accurately geometry of an object with a sampling density of around 10 samples/sq.mm. and a sampling error lower than 0.05 mm. Since the scanner works at a distance between 50 and 100 cm from the objects, it was necessary to put it on scaffolding. It is well known that scanning any 3D object requires the acquisition of many shots of the artefact, taken from different viewpoints, to gather geometry information on all of its shape. Therefore, to perform a complete acquisition several range maps were needed; the number of range maps requested depends on the surface extent of the object and on its shape complexity. A number from 120 to 200 single scans (each scan samples around 0.3 Million points of the surface) was needed to cover the entire surface of each Monreale capital. In the case of Cefalù capitals, due to their smaller size and more simple geometry, a number from 50 to 80 was necessary. The Monreale scanning campaign (February 2006) brought to the acquisition of 20 out of the more than 100 capitals of the cloister (Baracchini *et al*, 2006), with a total of nearly 4000 range maps. The Cefalù scanning campaign produced ten 3D models, with a total of nearly 800 range maps.

The acquisition of the geometry of the Aosta cloister capitals was performed by ITABC-CNR of Rome using a Menci ZScan, which employs images taken from calibrated images to produce accurate range maps. This technology heavily relies on Computer Vision techniques for 3D reconstruction from images. The main difference with the Minolta device is the final accuracy of the geometry that is slightly lower. Moreover, the numbers of samples are hundreds of thousands instead of millions. Using this technology, eighteen 3D models were produced, starting from 24 range maps for each capital.

The range maps from both Minolta and Menci ZScan devices were processed to convert it into a single, complete and non-redundant 3D representation. The processing steps followed the usual 3D scanning pipeline (Bernardini and Rushmeier, 2002):

- range maps alignment;
- range maps merging (or fusion), to build a single, non redundant mesh out of the many, partially overlapping range maps;
- mesh editing, to improve (if possible) the quality of the reconstructed mesh;
- mesh simplification and conversion into a multiresolution representation;
- color mapping (see next Subsection).

This processing steps was obtained through the ISTI-CNR tools, which give the possibility to deal with large number of range maps and to produce the final model with the lowest possible human intervention. A complete overview of these tools is presented in (Callieri *et al.*, 2003).

The number of triangles of each model ranges from 4 to 6 millions, depending on the shape complexity and size of each capital.

Adding color to the 3D geometry

As already mentioned in the previous subsection, color mapping is an important step in the scanning pipeline. As a result of the acquisition campaigns we had high quality 2D and 3D information: the objective was to integrate them in a unique model, preserving the detail of both color and geometry. In order to produce a detailed colored model starting from the set of photos provided, two phases are necessary:

- each photo has to be "aligned" to the model: the extrinsic (position in the space) and intrinsic (focal length and lens distortion) parameters of the camera which took the photo were estimated with an appropriate tool (Franken *et al.*, 2005);
- due to the highly detailed geometry, we chose to represent color following a per-vertex approach: for each vertex, the color assigned is computed as a weighted sum of the contributions of every photo which framed that vertex (Callieri *et al.*, 2008a).

Following this approach, we produced a set of high detailed colored models (a snapshot of a model from the Aosta Cloister is shown in Figure 2, right). The union of 2D and 3D information can lead to a new way to archive and remotely represent Cultural Heritage objects.

In the next Section, which describes the CENOBIUM website, we describe how this data are integrated in the web application.

The CENOBIUM website

Nowadays, several digital instruments to archive and analyze the data about artefacts of interest in the ambit of Cultural Heritage are available. As previously stated, one of the main problem is the integration of all of such data in a useful manner. Obviously, the integrated information system can be designed for specific purposes, following the requirements of certain kind of users (e.g. hobbyists, tourists, experts or researchers).

Previous approaches in the presentation of multimodal data (Balzani *et al.*, 2004; Baracchini *et al*, 2004) were mainly

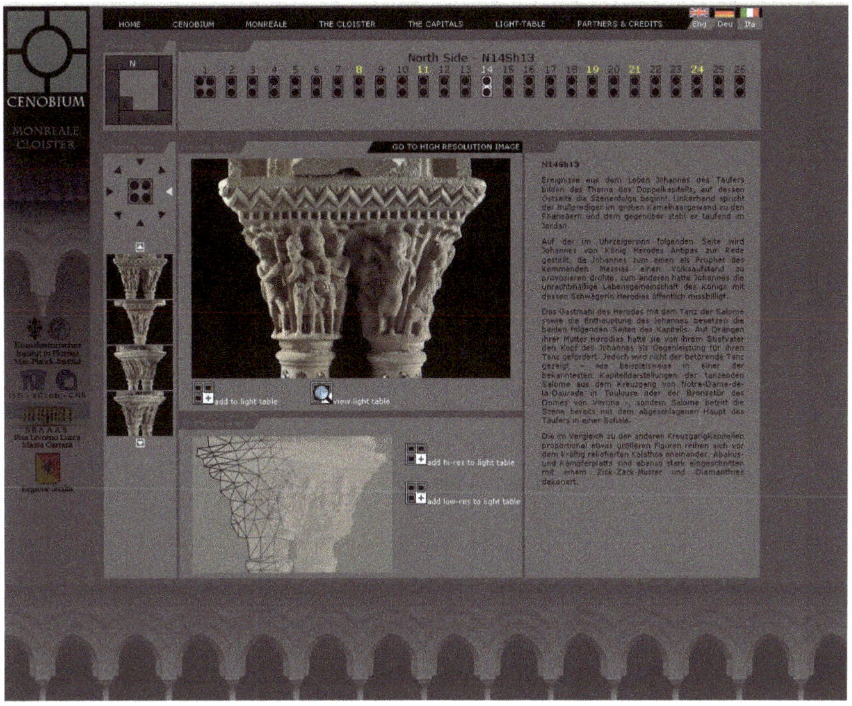

Fig. 3. Integrated information for the N14Sh13 capital of the Monreale Cloister.

devoted to the integration of 3D models with textual information. These environments were created to provide an easily navigable environment for naïve users.

The Cenobium web application tries to be an integrated environment for several categories of users. For example, naive users can simply read information about the capitals, analyse the high-resolution images and navigate the 3D models (Callieri *et al.*, 2008b). At the same time, experts and academic researchers can exploit all the potential of this environment in order to make further studies about the capitals.

The current web application includes two Cloisters (Monreale and Aosta) and a third one (Cefalù) will be soon available. For each Cloister, a brief description is provided. The most interesting section is the one about Capitals. In this section, the users can explore in an integrated manner all the multi-media information related to each capital. After a capital of the Cloister is chosen, all the information is shown in a single web page (see Figure 3). In order to be able to choose the capital of interest, in the upper part of the page a stylized version of the Cloister is shown; the capitals are arranged as in the real cloister, and their position is highlighted. The left part permits to navigate through all the images coming out from the photographic campaign: in particular the user can choose the side of the capital to visualize. By clicking on the selected view of the capital the full-resolution image can be visualized. Since the resolution of such images is very high, a dedicated image server, called *Digilib*, gives the possibility to navigate them interactively. A textual description containing historical information is given in the right part. If available, the description can be expanded to shown a more detailed one. The availability of a 3D model representing the chosen capital is shown in the bottom part.

Both the 3D model and the images can be selected as items into the LightTable tool, by clicking the appropriate option. It is important to underline that in order to enable such advanced functionalities the LightTable tool required to be previously installed on the PC where the website is explored. The tool can be downloaded for the installation directly from the website. When an item is selected, the LightTable takes care to download the item (high-resolution photos or 3D geometry) on the local PC (see Figure 4, left). Then, for whose items that are ready-to-use, the user has the possibility to visualize them simultaneously for comparison. In this way, further similarities between the items can be found and investigated easily (see Figure 4, right). The maximum number of items that can be visually compared is four. Even if the development of this tool is still in progress the LightTable could become a very useful resource for the work of art historians.

Preliminary presentation of the web application during the Annual meeting of the German Art Historians (25th – 29th March 2009 at Marburg) and as a part of the exhibition *Sicily. From Ulysses to Garibaldi* in the Art and Exhibition Hall of the Federal Republic of Germany (25th January – 25th May 2008 at Bonn) has shown that the objectives of the project are fulfilled. Both interested people and academic experts have found this way of present and access multimedia data about the capitals useful and with high potential to conducted in-depth studies.

Figure 4. The lighttable: the list of the items selected by the user (left) and an example of visual comparison of four elements (right)

Discussion and conclusions

As a comprehensive and long-term project, Cenobium not only serves as an instrument for academic research, but it also provides for the interests of several other branches of Art History.

For example, a possible starting point for research into the lively artistic exchanges of the 12th and 13th centuries, which precludes a monographic analysis, is the question concerning the relative coherence between programmes and capital cycles and the analysis of their specific design characteristics.

The Italian Medieval Cloisters (Monreale, Cefalú, and Aosta) form the first part of the project, which will expand to include contemporary Spanish and French cloisters. Cloisters that are no longer extant or that have undergone extensive rebuilding can now be usefully compared with capitals preserved in museums, while under certain circumstances destroyed structures can be reconstructed.

CENOBIUM combines both traditional and innovative methods of Art History with the latest technology in order to open up new horizons of art historical scholarship that have not, until now, been exploited to their full potential.

One of the main future developments of the Cenobium website will be the inclusion of other Cloisters and the development of a component to make it a collaborative environment where new contributions can be added and discussed in a "Wiki fashion".

References

BALZANI, M., CALLIERI, M., FABBRI, M., FASANO, A., MONTANI, C., PINGI, P., SANTOPUOLI, N., SCOPIGNO, R., UCCELLI, F. AND VARONE, A. 2004. Digital representation and multimodal presentation of archeological graffiti at Pompei, *Proceedings of VAST 2004*, 93-104, Bruxelles, Eurographics.

BARACCHINI, C., CALLIERI, M., CORSINI, M., DELLEPIANE, M., DERCKS, U., KEULTJES, D., MONTANI, C., SCOGNAMIGLIO, M., SCOPIGNO, R. AND SIGISMONDI, R. 2006 Starting the CENOBIUM project: the cloister of Monreale (Sicily) revealed, *Proceedings of VAST 2006*, 100-110, Nicosia, Eurographics.

BARACCHINI, C., BROGI, A., CALLIERI, M., CAPITANI, L., CIGNONI, P., FASANO, A., MONTANI, C., NENCI, C., NOVELLO, P., PINGI, P., PONCHIO, F. AND SCOPIGNO, R. 2004. Digital reconstruction of the Arrigo VII funerary complex, *Proceedings of VAST 2004*, 145-154, Bruxelles, Eurographics.

BERNARDINI, F. AND RUSHMEIER, H. 2002. The 3D Model Acquisition Pipeline. *Computer Graphics Forum* 21, 149–172, Blackwell.

CALLIERI, M., CIGNONI, P., GANOVELLI, F., MONTANI, C., PINGI, P. AND SCOPIGNO, R. 2003. VCLab's tools for 3D range data processing. *Proceedings of VAST 2003*, 13–22, Brighton, Eurographics.

CALLIERI, M., CIGNONI, P., CORSINI, M. AND SCOPIGNO, R. 2008 Masked Photo Blending: mapping dense photographic dataset on high-resolution 3D models. *Computer & Graphics*, 32, 464-473, Elsevier

CALLIERI, M., PONCHIO, F., CIGNONI, P. AND SCOPIGNO, R. 2008, Virtual Inspector: a flexible visualizer for dense 3D scanned models. *IEEE Computer Graphics and Applications*, 28, 44-55, IEEE

FRANKEN, T., DELLEPIANE, M., GANOVELLI, F,. CIGNONI, P., MONTANI, P. AND SCOPIGNO, R. 2005. Minimizing user intervention in registering 2D images to 3D models, *The Visual Computer*, 21, 619-628, Springer Verlach

The virtual museum of the Western Han Dynasty: 3D documentation and interpretation

Maurizio Forte,[1] Nicolo' Dell'Unto,[1] Paola Di Giuseppeantonio Di Franco,[1] Fabrizio Galeazzi,[1] Claudia Liuzza[2] and Sofia Pescarin[3]

[1]University of California, Merced
[2]Stanford University
[3]CNR-ITABC, Istituto per le Tecnologie Applicate ai Beni Culturali, Rome

Abstract

This project, aimed to the digital documentation of archaeological sites, artifacts and cultural relics of the Western Han Dynasty, started in 2008. The anticipated outcome will be the creation of a virtual museum, based on collaborative environments, dedicated to the Western Han Dynasty and able to integrate new archaeological datasets coming from fieldwork activities (most part of them unpublished), monuments, and famous collections of artifacts of the Xi'an archaeological museums. All the archaeological datasets will be virtually reconstructed in a very accurate way, keeping all the spatial information even in the cyber space: GIS, remote sensing and laser scanning data, DGPS surveys, 3D reconstructions (landscapes, sites, monuments, artifacts). The Virtual Museum of the Western Han Dynasty will be the first archaeological example of international collaboration between Chinese institutions and Western countries based on virtual heritage methodologies and real time outputs.

The archaeological fieldwork, mainly encompassing 3D documentation of new tombs in the center and in the surroundings of Xi'an, is managed integrating different technologies and methods: laser scanning, digital photogrammetry, photomodeling, remote sensing, and GIS. The final plan is to create at least two installations, one in Xi'an and one in California. At UCM, the Virtual Heritage Lab is working on the set up of two different virtual reality systems based on participatory learning: a stereo collaborative environment (virtual heritage room) and a simulation environment (Powerwall). The primary purpose is to visualize and display very high-resolution data from large scientific simulations performed via high-resolution imaging applications. This paper will give an overview on the principal technologies and methods used in the field for the documentation of monumental wall paintings' tombs, archaeological sites, landscapes and museum artifacts.

Key words Xi'an, Western Han, Archaeology, Virtual Museum, 3D datasets.

Introduction

The project "The Virtual Museum of the Western Han Dynasty started in 2008, thanks to the collaboration between the Xi'an Jiaotong University and the University of California, Merced, School of Social Sciences, Humanities and Arts. Later this collaboration was extended to the Xi'an Municipal Cultural Relics Conservation and Archaeological Research Institute (China), CNR-ITABC, Italian National Research Council (Italy). Scope of the project is the creation of different virtual museums (in China and in California) based on the 3D documentation and reconstruction of the most representative cultural relics in the Shanxi Province. Principally we have distinguished the following macro classes of cultural objects and sites: wall paintings tombs, mausoleums, archaeological landscapes, artifacts and the ancient city of Chang Han. Actually it is impossible to approach a so complex whole of datasets in exhaustive way, therefore the strategy was to proceed by samples, integrating fieldwork activities with processes of reconstruction. The project, still in progress, is also aimed to test different digital technologies: GIS, remote sensing, time of flight laser scanning, optical scanning, photomodelling. The combination of different 3D datasets in virtual reality environments is a great challenge for reinterpreting the past and saving, at least digitally, a meaninghful set of archaeological data and archives. It is important to highlight also that rescue archaeology in China has a fundamental role in the contemporary society, given the fast and destructive process of the urban development in the countryside. Hundreds of tombs and sites were excavated every year in the area of Xi'an with very few chances to record properly all the scientific data and to recompose this information for further digital processing and virtual visualizations/simulations.

Fieldwork and 3D data collection. The importance of integrated technologies

One of the aims of the *Western Han Dynasty virtual museum* project is the restitution of the socio-cultural aspects of the Dynasty in their whole, through the virtual reconstruction of death and life in the ancient city of Chang'an, partially corresponding to the modern city of Xi'an.

The virtual reconstruction has being realized starting from the 3D data collection of the material remains, obtained using integrated technologies. The work presented in this paper has been done during two fieldwork campaigns: summer 2008 and 2009.

Also if the final output of the project has a communication purpose, the use of integrated technologies has permitted the preservation of the archaeological remains relieved. The actual economical growth of Xi'an, in fact, has given a huge impulse to the urbanistic development of the city. Chinese archaeoogists constantly faced with the problems linked to the rescue archeology: first of all the lack of economical resources, which prevents both a good preservation approach, and the public fruition of the monuments discovered.

The same argument can be said on the archaelogical relics of the ancient city. A big part of them were made of adobe and, because of the climatic conditions of the city, and, again, of the lack of economic resources for a good preservation plan, they are desappearing.

For all these reasons the technological approach adopted for the data collection is of extreme importance, because it permits the preservation of the monuments through the digital documentation.

Unlike the monuments, the artifacts are movable and they can be preserved. Those relieved during the 2009 campaign, in fact, are all preserved in the Xi'an Municipal Institute of Cultural Relics Conservation and Archaeology Museum, and they have all been restored. The problem is that they are in a storage, and not inside a museum. That said, the value of 3D models for those artifacts is linked to the possibility of showing them to both scientific community and general pubblic. As a result it can be said that the virtual model make those artifacts more movable.

Many topics have already been examined for the creation of the virtual museum, such as the landscape, the ancient city, the imperial mausoleums, other burial monuments, the art and artifacts. The reconstruction of the landscape has been realized through DGPS (Differential Global Positioning System); the monuments data collection has been obtained using a time of flight scanner laser; a desktop triangulation 3D laser scanner has been used, instead, for the artifacts data capture.

Landscape and mausoleums

For a more detailed discussion of this subject we refer to the contribution of Maurizio Forte in the same proceedings. In this context it is just necessary to underline the importance of the ancient landscape reconstruction, for a full understanding of the city and its monuments. The ancient landscape can be recreated starting from the analysis of the contemporary one. In fact, the presence of imperial mausoleums mounds is still imposing in the actual environment of the Xi'an countryside.

Moreover, for a more accurate reconstruction of the landscape the DGPS technique has been integrated with the laser scanner one. It has been possible to acquire, in fact, part of the mausoleum inside the Mao Ling Museum. The acquisition has been made using a Riegl LMS Z390i. This kind of technology permits the acquisition of point clouds with an accuracy of 6 mm for a range between 1-400 m. The high geometry resolution of the final model depends on the point cloud density and the detail of the representation.

The data acquisition plan is the most important step during the fieldwork. It is fundamental to choose the level of detail in advance, while planning for the result of the final data output. Setting the resolution of the data capturing it is possible to decide in advance the quality of the final digital representation.

The same time of flight laser scanner was used to relief the amazing statues of the Maoling Museum. In this case the point clouds level of detail was between 6 and 8 millimeters. The high resolution point clouds obtained in the fieldwork have, in the specific case of those statues an important added value, which will be very useful in their musealization phase. The sculptures are, in fact, made using a local stone that makes sometimes difficult the perception of the subjects shapes and details, because of the color and consistence of the material (fig. 1). Looking at the point cloud, they are, instead, clearly perceptible (fig. 2). It becomes, in this context, an augmented perception level of the monuments, very useful for communication purposes. Through the use of digital technologies, it is, thus, possible to improve people learning processes, just working on their perceptions of material objects.

The city of Chang'an

For a more detailed discussion of this subject we refer again to the contribution of Maurizio Forte. Important information regarding the capital of the Western Han Dynasty have been found thanks to the study of digital images. Most part of the city structures were, as already said, made of adobe, and they are desappearing. In the last fieldwork it was just possible to scan the rests of the South Gate, with an accuracy of 1 cm.

Tombs

During the 2008 fieldwork it was possible to relief two tombs through laser scanner techniques. The more important, the Xi'an University of Technology Mural Tomb (M 27), is located on Leyou Plane, northwest of Yue Jiazhai Village in the south suburb of Xi'an. The model was realized the, and high resolution pictures have been used to texturize it. M 27 is one of the few Western Han Mural Tombs discovered in the city. Despite its importance, the preservation of this heritage is at risk. In particular paintings inside the tomb are disappearing, because of the critical conditions of plasters and colours. Since they are made over a very thin stratus of clay, directly applied on the bricks composing the structure, their removal from the walls and preservation inside a museum is impossible.

Fig. 1. Maoling museum. Picture of the horse statue.

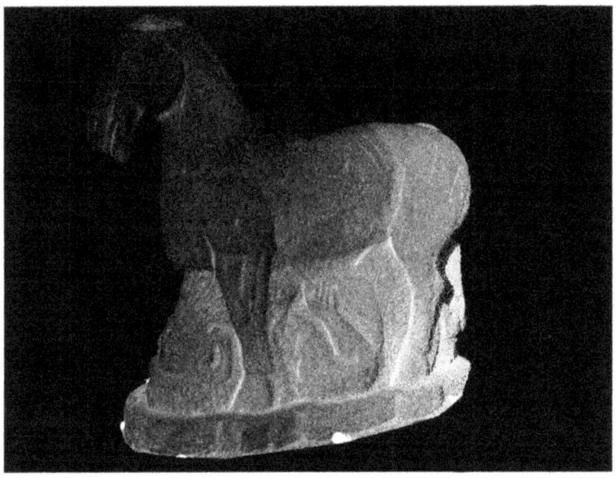

Fig. 2. Maoling museum. Point cloud of the horse statue.

In summer 2009 other two tombs were relieved, the most important one is the Green Bamboo Garden mural tomb(M1). Its 3D model was realized at a high level of detail (8 mm). The environmental and static conditions of the monument, supported by retaining structures, had considerably increased the scanning time. In order to fill all the twilight zones caused by the retaining structures, we had to make more scans. For this reason we couldn't fully utilize the potentiality of the Riegl, which usually permits a single acquisition of 360 degrees.

Moreover the registration of a so high number of point clouds increased the data weight, and complicated their managing during the post-processing.

The first post processing step was the filtering: in order to obtain reliable 3D models it is fundamental to apply different kinds of filtering that allow the point clouds smoothing. It is necessary to smooth point cloud data in reverse engineering or the inspection of free-form surfaces, because otherwise noisy points will have a negative influence on the post-processing phase.

The second step was the triangulation of the tomb point cloud. A very complex cleaning operation of the retaining structures meshes was necessary (fig. 3). The filtered and triangulated object was then exported in a 3D modelling software, 3dStudio Max, and mapped through the "unwrap modifier". This is a special tool that allows to apply the image directly on the mesh, and to link the plain of projection to the position of the digital camera[1] (fig. 4).

This project is still in its embryonic face, but we already started to think to the communication plan. For this reason a Cybermap of the Xi'an University of Technology Mural Tomb (M 27) was created (fig. 5)[2]. The M 27 cybermap is a 3D graphic representation of the relations between all the paintings realized in the main chamber of the monument, where the diseased was buried. It could be also defined as a mindmap, since it is the geometric translation of a mindscape[3].Through this map it was possible to show all the semantic relations of the scenes, making the contents very easy to be understood.

The idea is to use this tool also for the M1 Green Bamboo Garden tomb and for the other monuments representing the Western Han Dynasty. This will permit, in a second step of the work, to create a single 3D map underlining all the relations between structures, artifacts and paintings representing the whole Western Han Dynasty cultural background.

Fig.3. M1 Green Bamboo Garden. Point cloud after the filtering and the triangulation.

[1] Galeazzi *et al.* 2008.

[2] The map was presented at the TAG (theoretical archaeology group) conference(Stanford University, may 1st-3rd 2009): Galeazzi *et al.* in press.

[3] "The mindscape is a virtual landscape perceived and interpreted by mental maps"; Forte 2001: 95-108.

Fig.4. M1 Green Bamboo Garden. Mapped 3D model.

Fig. 5. Cybermap of the Xi'an University of Technology Mural Tomb (M27).

Artifacts

The most representative Western Han Dynasty artifacts have been chosen to be scanned, and in just a month, in the 2009 campaign, more than fifty objects were relieved. Some of those formed the Green Bamboo Garden Mural Tomb grave good. Their 3D acquisition has been foundamental for recreating the tomb context. The work was very challenging, because the engine permitted to scan objects manifactured using different kinds of material, especially pottery, bronze, and jade. The jade was the most difficult to be scanned, because of its transparency and brightness.

The Next Engine, a 3D desktop laser scanner, was used for making 3D relieves. This is a triangulation laser scanner, which permits a 3D scanning in high definition (HD technologies). Thanks to this engine it is possible to obtain an accuracy of 0.002 inches (about 50 microns), packing points finer than the accuracy spectrum (fig. 6).

The added value of this kind of hardware is the possibility to obtain the final 3D model with meshes and textures applied, without the need of post processing work. During the acquisition it consents, in fact, the automated alignment of scan sets, the scans merge into single mesh model, the reduction of edge and noise artifacts, and the maintenance of higher quality points through oversampling. Moreover, a seven channel color capture for high accuracy generates a mapping of the final model characterized by the high resolution of textures (fig. 7).

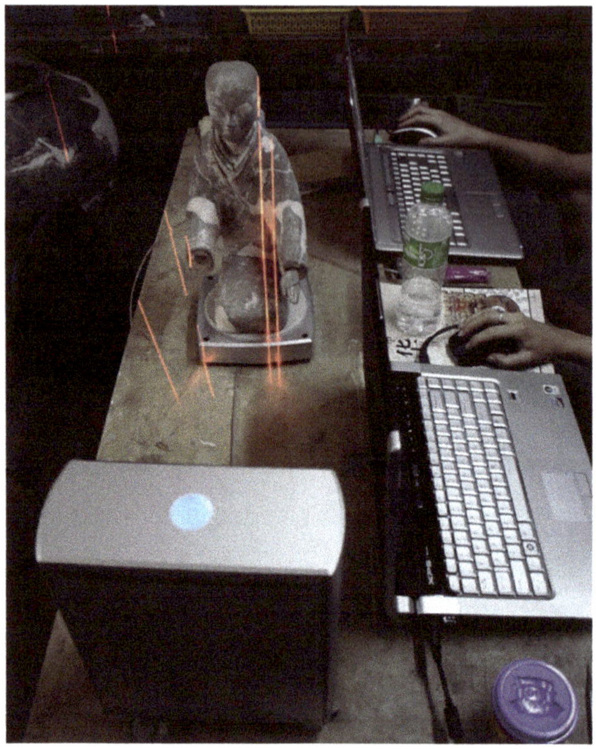

Fig. 6. 3D Desktop Laser Scanner during the relief.

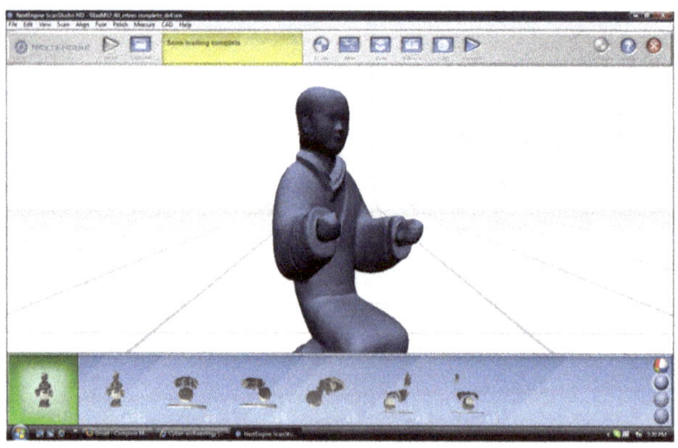

Fig. 7. Artifact 3D model after the relief.

The process of the 3D desktop Next Engine is automatic; a manual alignment is necessary only with objects characterized by a very complex shape; but also in this case the procedure is simple and fast.

Conclusions

The collaboration between the Jaotong University of Xi'an, the Italian CNR and the University of California Merced was originally aimed to the creation of a virtual museum. The integrated use of technology was decided in order to collect as more information as possible for a complete recostruction of the Western Han Dynasty, one of the most important archaeological period for the chinese civilization. We shouldn't forget, in fact, that the majority of Chineses today are Han.

Also if for the creation of the virtual museum, high resolution models weren't really necessary, an high level of detail was chosen in order to combine the communication purpose with the necessity to preserve this heritage at risk. We have collected a very big amount of data, so far, sufficient for the creation of a digital archive of 3D archaeological data.

The digital archive can be useful for different reasons: first of all it helps preserving all the information related to archeological remains. Moreover, it is a powerful instrument for scientific analysis related to that remains; that is mainly for the easy accessibility of the scientific data, that can be shared by a broad community of scholars, improving the interpretative process. Eventually, after the interpretation, there is always the possibility to elaborate the scientifc data for communication purposes. In this way also the general public would be able to take advantage of the information.

In conclusion the present contribution has shown that the integrated technologies have a huge pontential and versatility. They should be used for the creation of a new methodology applicable to all case-studies, and a code of interpretation broadly sharable. This is desirable and challenging for the Third Millennium Archeology, where the obsession for the collection of archeological remains gives way to a modern idea of preservation, interpretation and broad sharability of data.

Acknowledgements

Special thanks are due to all the Chinese institutions involved in the project and in particular to the two most important leaders of the research framework in China: Cheng Linquan, Institute of Cultural Relics Preservation and archaeology of Xi 'and Li Jiajun, Art Museum, Xi'an Jiaotong University. A very special aknowledgment to Ms "Abby' Lizhi Zhang, Assistant Art Museum, School of Humanities and Social Sciences, Xi'an Jiaotong University. The project is co-supported by Pacific Rim Research Program and GRC-UC Merced grant.

References

FORTE M. 2001

Forte, Maurizio. "Mindscape: ecological thinking, cyber-anthropology, and virtual archaeological landscapes". Forte, Maurizio – Williams, P.R., *The reconstruction of archaeological landscape through digital technologies*, proceedings of the 1st Italy-United States workshop (Boston, Massachussets, USA, november 1-3 2001), BAR International Series 1151 (2002): 95-108.

GALEAZZI *ET AL*. 2008

Galeazzi, Fabrizio – Moro, Alessia –Vico, Lola. "La modellazione architettonica. Il texturing". Forte M. (ed.) La *villa di Livia, un percorso di ricerca di archeologia virtuale*. Roma: L'Erma di Bretschneider 2008: 133-136.

GALEAZZI *ET AL*. IN PRESS

Galeazzi, Fabrizio – Di Giuseppantonio Di Franco, Paola – Dell'Unto Nicolo', 3D Cybermaps of Western Han Mural Tombs, *Proceedings of the TAG (theoretical archaeology group) conference* (Stanford University, May 1st-3rd 2009)

Session 4

CULTURAL AND NATURAL SITES RESOURCE MANAGEMENT: SUSTAINABILITY, UTILITY AND MANAGEMENT OF HERITAGE SITES

The 'Cultural Heritage Map of Apulia' project

A. Buglione,[2] A. Barbanente,[1] G. Volpe,[2] C. Annese,[2] A. Di Zanni,[2] R. Goffredo,[2] A.Valentino Romano[2]

[1] Regional Department of Land Management – Apulia Region, Italy
[2] Human Sciences Department- University of Foggia, Italy

Abstract

The '*Cultural Heritage Map of Apulia*', started in September 2007, is a mapping project carried out by the Regional Department of Land Management (Apulia, Italy) in collaboration with the Regional Department for Cultural Heritage, the four Regional Universities of Apulia and the Regional Central Department for Cultural and Environmental Heritage. The project aim is to create a thematic cartography in order to archive, map and describe in detail the Cultural Heritage of the territory of Apulia, overcoming the traditional way of cataloguing based on a concept of Cultural Heritage as spots on a map. The other main purpose of the project is to analyse and describe the stratified historical landscapes of the regional countryside, from prehistory up to now, as long-term evidence of the identity of people and places. These aims are realized through the integration of different sciences and methodologies and a geo-database which is part of the Apulia Region G.I.S. organized in different themes and informative layers. This is intended to be an innovative and dynamic instrument of Cultural Heritage preservation and the accordance with the Cultural Heritage Map of Apulia is expected to be the necessary requirement for approving every land-planning activity in order to preserve the local Cultural Heritage.

Keywords Italy, Apulia Region, Cultural Heritage, Landscape, Planning and Conservation, WebGIS.

Introduction

The editing of the new Regional Territorial Landscape Plan (*PPTR*), started by the Apulia Region at the end of 2007 with the coordination of Alberto Magnaghi (University of Florence), has opened a new season in the planning of the territory of Apulia and in the protection and enhancement of its landscapes. The Region Authority, actually, through the Regional Department of Land Management, has intended to promote a complex series of initiatives culminating in the draft of a new Plan, which will replace the existing Thematic Territorial Urban Plan for the Landscape (*PUTT/P*) (Deliberation of Regional Council 15 December 2000, No. 1748, BURP No. 8, suppl., 17 January 2002).

The process of drafting of the new Plan includes a specific activity in order to give a systematic aspect to the knowledge of the Regional Cultural Heritage, through the cataloguing and georeferenced localization of the Cultural Heritage of Apulia (Lucchesi, Carta, Di Zanni 2008, Volpe, Di Zanni, Laurenza 2008). This methodological route was born to develop «an idea of the landscape planning which, overcoming the constraining character applied to some preservation areas, would have the objective of enhancing actively territorial and landscape heritage, combining long duration identity and short term innovation, landscape and economy, the value of existence and the value of use in lasting and self-sustainable forms» (A. Magnaghi).

This activity is embodied in the draft of the Cultural Heritage Map of the Apulia Region, with which the four Universities of Apulia and the Regional Central Department for Cultural and Environmental Heritage, with the technical collaboration of InnovaPuglia have been entrusted.

The project has so far led to the census of almost 10.000 sites, but the number is expected to grow rapidly (Fig. 1).

Actually, the Map as well as the Plan, will continue to be updated by the activities of the Regional Observatory for the Quality of Cultural Heritage and Landscapes, in the implementation phase, and with the creation of which, as provided by the Code of Cultural Heritage and Landscape, Apulia becomes a leader in the Italian scenario, since for the first time in Italy it will be possible to make a constant and integrated monitoring of Cultural Heritage and landscape on the regional territory.

Since the Cultural Heritage Map «must form the basis for the identification of unbuildable areas, respectable areas, controlled use areas, to be determined, through a graduation of interest, in the planning phase» (Deliberation of Regional Council No. 1787, 28 November 2006), the subject of the census were all the buildings and the areas of cultural and landscape value located in suburban areas (in particular the buildings and the areas referred to in articles No. 136 and No. 142 of the Code of Cultural Heritage and Landscape), already published or whose data were present in the archives of Superintendence Offices, Universities or other Research Organizations that have been operating in the Apulia Region.

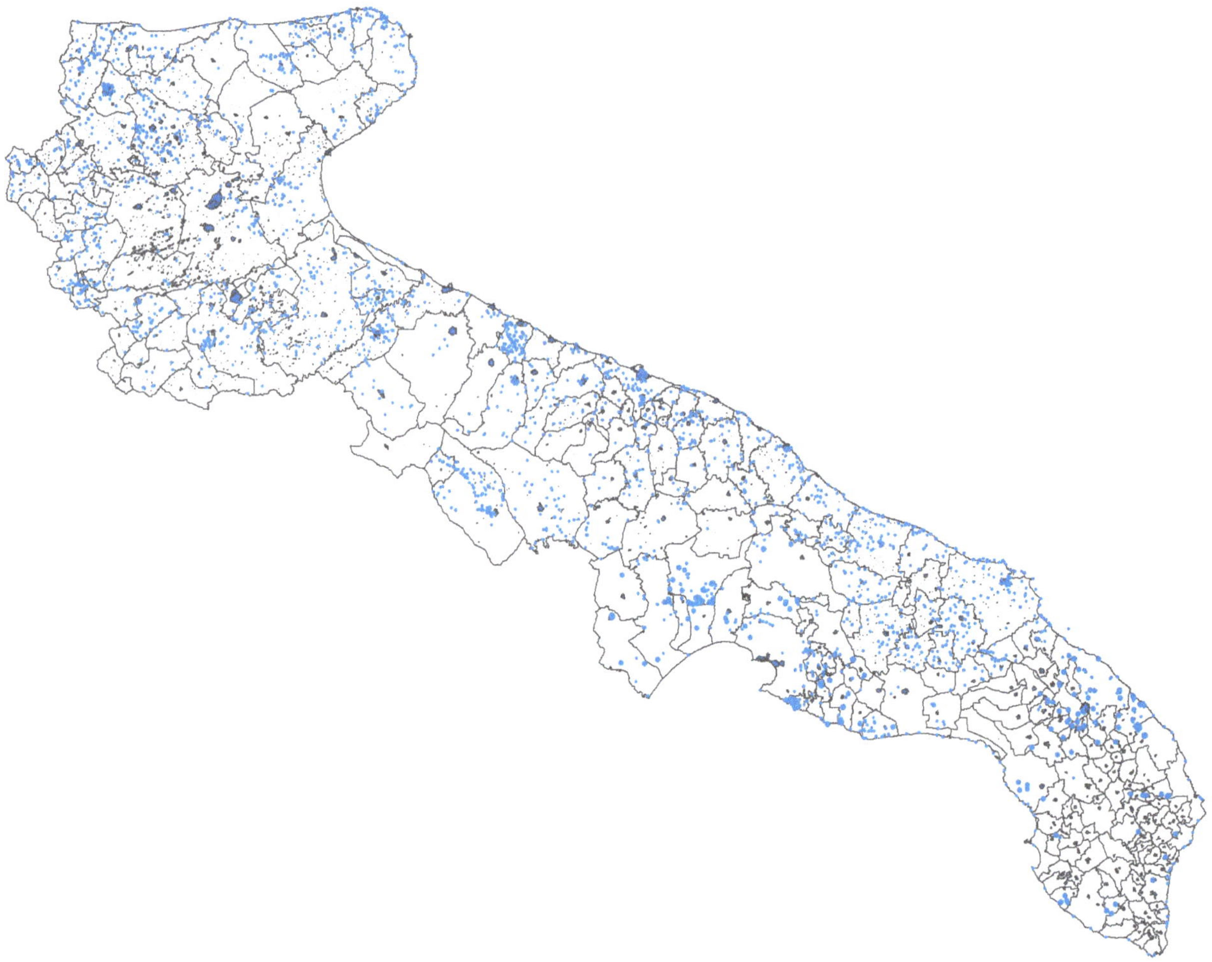

Fig. 1. The Apulian Cultural Heritage localized by the Project.

To these data all the other data gathered through previous instruments of regional, provincial and municipal planning have been added. It is therefore a *corpus* of data quantitatively and qualitatively significant that for the first time is collected and managed through a single computerized data management system, composed of a G.I.S. platform, and an alphanumeric archive associated with it, destined to become soon a webGIS.

The particular system of sources used ensures that the Map represents nowadays the mirror of the state of knowledge of the Cultural Heritage of Apulia. So, if for some territories, such as the Tavoliere Plain, the long tradition of systematic researches on the field, has revealed their extraordinary wealth in terms of Cultural Heritage, other territories now seem insufficient regarding the quantity of assets counted and the quality of descriptive data associated with them.

Moreover, regarding the type of data used, the working groups have chosen to examine and assess also reports provided by publications of local relevance or more ancient ones, sometimes the only evidence of no more visible or unknown assets. In addition to estates it has been decided to file even the rich Epigraphic Heritage, for the most of Roman Age, that the Region preserves. This heritage, of which sometimes only the memory of the site of origin is kept, often allows supposing the presence of sites otherwise unknown. Although characterized by the limitations as described above, the Map is to all intents and purposes a turning point in the state of knowledge of the Cultural Heritage of Apulia, as well as a valuable instrument thanks to which it is possible to plan a new season of research, also in accordance with the intentions of the Region, or as a «dynamic instrument and in progressive updating in accordance with the progress of knowledge» (see Deliberation above).

The acquisition of such huge heritage of knowledge will allow planning a sustainable development of the territory, able to strengthen the already existing infrastructure apparatus, promote the exploitation of available natural resources and the retrieval of new sources of energy, limiting the conflicts with the needs for protection and enhancement of the Cultural Heritage. Therefore, the Cultural Heritage Map is an instrument now more than ever necessary in a Region as Apulia, which has witnessed in the last years a rapid spread of wind plants, expansion of cities and industrial areas, research for places and modalities for the disposal of waste.

The objective of the first stage of the Map, which started on 16 September 2007 and which ended on 31 December 2008, has been to identify all the estates and areas of cultural and landscape value, located in suburban areas, already published or whose data are present in the archives of Superintendence Offices or Universities. In total about 10.000 sites (5683 polygon sites, 4017 point sites, 161 not located sites) have been filed including 346 ancient cities, and 319 modern cities, 1000 topographical units (743 polygon, 248 point, 10 not located); and about 1400 constrained areas (605 direct architectural constraints, 99 indirect architectural constraints, 457 direct archaeological constraints, 129 indirect archaeological constraints, 141 landscape constraints).

For this purpose a computerized system for data management composed of a G.I.S. and an alphanumeric archive has been realized. The activity of processing of the alphanumeric archive has revealed itself to be precious thanks to the support offered by InnovaPuglia, with the coordination of C. Caroppo, whose contribution to the work of the Map has been to ensure compatibility with the structure of the Territorial Information System (S.I.T). of the Apulia Region. A significant contribution to the creation of the models of the cards and the construction of the information architecture of the archive has been supplied by S. Laurenza.

A.BA., G.V.

The Structure of the Cultural Heritage Map

The themes, the conceptual pattern

The map has been divided into three essential themes: the Cultural Heritage Map, the Landscape Map and the Map of Constraints. It describes the cultural characteristics of the territory, with particular regard to estates and areas referred to in articles No. 136 and No. 142 of the Code of Cultural Heritage and Landscape. The Landscape Map includes civic uses, landscapes to be restored, anthropic landscapes and water landscapes. The Cultural Heritage Map includes all the Cultural Heritage scattered all over the landscape, the historical routes, the sheep-tracks, essentially all the buildings and the areas of cultural and landscape value located in the suburban areas. The historical cities, which are naturally part of the Cultural Heritage, have been described and represented as Cultural Heritage assets themselves. The Regional Central Department for Cultural and Environmental Heritage of Apulia has also instructed consultants to identify and report, with reference to the whole Cultural Heritage of Apulia, farmhouses, towers, castles, churches and anthropic landscapes of particular value in order to arrange appropriate forms of protection. In the end, the Map of Constraints has traced out the perimeter and described all the located archaeological, architectural and landscape constraints existing on the regional territory. Printed documentation in digital format of such constraints has been acquired.

The implementation of such an articulated and complex project has naturally required a long phase of study, working out and testing of appropriate instruments to collect and describe effectively different types of assets. The first phase has provided the analysis and study of the most important Italian and European experiences of cataloguing and application of information systems to the field of Cultural Heritage (Gelichi 2001, Buora, Santoro 2004, Forte 2002). However, none of these experiences could represent, alone, a reference model, since generally the projects for cataloguing that were realized concerned only one type of assets, i.e. in particular the archaeological ones. Regarding the Map of Apulia it has been necessary to extend this census also to the so-called architectural assets, cities, sheep-tracks and the network of historic roads, and constrained assets. The source used in the work of census has been first of all the very rich bibliography existing on the Cultural Heritage of Apulia, with special attention also to the publications of local significance or to the more ancient ones, sometimes the only evidence of tracks nowadays completely vanished.

Then the census deriving from the analysis of all the publications has been integrated with the data collected during the still unpublished researches conducted on the territory by the same Universities of Apulia or by other Italian and foreign Research Organizations. Previous regional projects such as the Thematic Territorial Urban Plan for the Landscape (*PUTT/P*) and its accomplishments/ adjustments, provincial ones such as the Coordination Territorial Plan of Foggia (*PTCP*) and municipal ones such as the General Urban Planning (PUG) have represented sources for the research of the Cultural Heritage reports as well.

The *corpus* of data, collected for the first time thanks to a single computerized management system composed of a G.I.S. platform and an alphanumeric archive associated with it, is based on the concepts of *UT* (Topographical Unit), *SITO* (Site), *SITO PLURISTRATIFICATO* (Pluristratified Site), *CTS* (Stratified Topographical Context) (Fig.2).

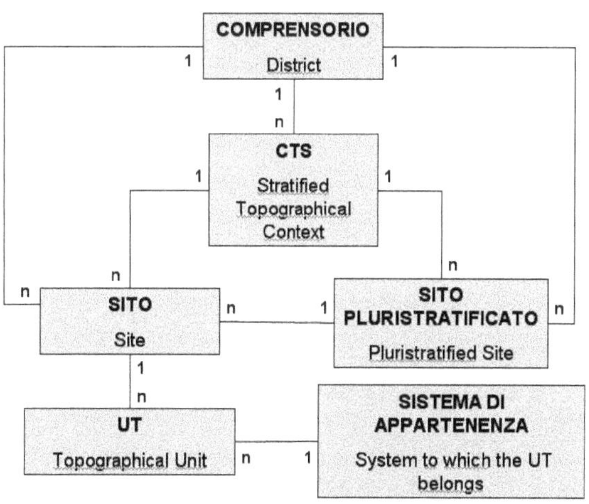

Fig. 2. The architecture of the DBMS.

The concept of *SITO* unites all the Cultural Heritage assets counted in the Map and allows overcoming the typical subject definitions referred to in the past (archaeological assets, architectural assets, etc.). Adopting the definition given by Manacorda, the site has been considered as «every place where history is deposited as a form of stratification» and as a «three-dimensional portion of space that bears within itself the signs of time, or the fourth dimension that has shaped it» (Manacorda 2007).

The minimum unit that helps to define a site is the *UT*. This definition within the context of the Map has a different meaning from the one that the concept points to in Landscape Archaeology. An example of *UT* with respect to the whole *SITO* may be one of the buildings that form a typical Apulian farmhouse called 'masseria', one of the huts that form a village or one of the tombs of a necropolis. This hierarchical type of breaking down allows, on the one hand, a more detailed reading of the settlements and, on the other hand, avoiding the risk of the multiplication of sites, also reported by Manacorda (Manacorda 2007, 12-14) and resulting from awarding any type of evidence with the rank of site.

The use of the concept of *SITO PLURISTRATIFICATO* has also allowed taking into account and representing the rather frequent cases of stratigraphic overlapping of more sites.

In consideration of the deeply contextual nature of the Cultural Heritage and, therefore, of the landscape too, the effort currently made is to overcome a model of a census that we could define 'philatelic' (or in other words a model which is based on a concept of the Cultural Heritage asset as an isolated point) and to represent instead, through an interpretative effort firmly based on scientific data at our disposal, a reconstruction of the stratified landscapes of Apulia, from Prehistory to nowadays, able to reveal the *continuum*, or the plot in which those assets have their origins and meaning, appearing to us as guardians of the identity-memory of the places and populations that lived in them (Settis 2002, Volpe 2007).

The considerations made above clearly show that the methodological reflection, which directed some of the critical decisions of the project, not only regarding terminology but above all nature and evidence of the Cultural Heritage assets to record, has been substantially affected by what we could define an 'archaeological perspective', or a scientific perspective which, by merging the principles of Landscape Archaeology (Cambi and Terrenato 1994) and those from Global Archaeology (Mannoni 1997, Manacorda 2004), considers the Landscape and Cultural Heritage with the approach of Global Landscape Archaeology (Volpe 2008).

A New Conceptual Model: the CTS (Stratified Topographical Context)

A further attempt to innovation in the construction of the knowledge and in the interpretation of Cultural Heritage is represented by the experimentation of new conceptual models and at the same time of cataloguing ones. Starting from the observation of the deeply contextual nature of Cultural Heritage and therefore of the landscape, the working groups are currently engaged in an analysis aimed at recognizing, on the entire regional territory, some regional contexts significantly characterized by the peculiarities of their territorial mass, which is considered as «...historical accumulation of territorializing acts of different nature (such as: buildings, monuments, cities, communication infrastructures, ports, bridges, terraces, divisions of lands into farms, reclamations, channels, hydro-geological and environmental solutions)» (Magnaghi 2000, 63). These areas have been defined *CTS*, or Stratified Topographical Contexts. Starting from the analysis of the wide basis of data now available, the attempt is to provide an integrated and diachronic reading of Cultural Heritage sedimented in a given territorial context, overcoming the model of census normally adopted, that we could define 'philatelic', which is based on a concept of asset as an isolated point (Volpe 2007, 20-32). One of the purposes of this interpretation level is to allow an integrated and diachronic reading of the relationship that has linked together some different Cultural Heritage assets and of the relationship between these and their attendant historical, environmental and landscape context (combining factors), in order to grasp the co-evolutionary relationship and to provide, thanks to integration with data collected from the analysis of the Technical Secretariat of the Project, a support for identifying the most appropriate forms and means of protection for all these territorial areas (Fig. 3). Moreover, through tracing out the perimeter of *CTS*, the potential use of landscape in certain areas has been reported highlighting the stratification of the Cultural Heritage and reporting any methods of use.

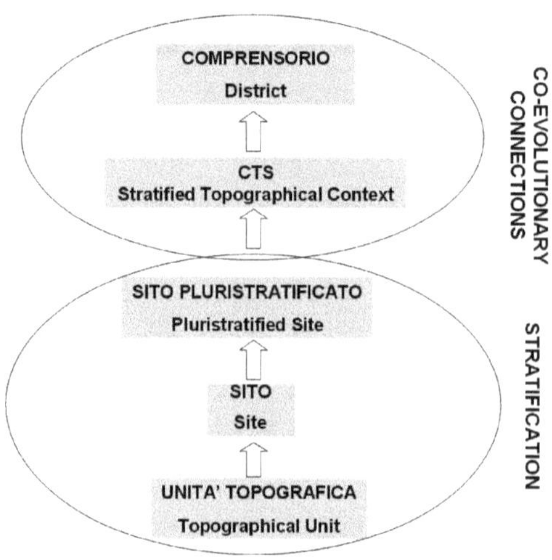

Fig.3. The conceptual scheme of the Cultural Heritage Map.

One of the first attempts made in this direction has led to the reading in terms of Stratified Topographical Context of a portion of land between the current city of Ascoli Satriano (Foggia) and the Carapelle River, where, along the definitely easier route rising from the river to the city, a concentration of settlements from the Early Age of Iron follow one another continuously until nowadays. The experimentation of the reading of cultural landscapes on a larger scale, or on the District, has been until now carried out on the Ofanto River, but expects to be improved through the overlap of the other descriptive components of the territory, in particular on the basis of environmental data. The detailed reading of this territorial sector was made possible thanks to data, in part still unpublished, collected during the research activities in the Carapelle River Valley conducted by the Laboratory of Landscape Archaeology of the University of Foggia (Goffredo, Volpe 2007, 219-246, Volpe, Goffredo, Di Zanni 2007, 109-124). In some cases the possibility to insert, in the context of *CTS*, urban centres already subjected to protection being historical cities, such as e.g. Ascoli Satriano (Foggia), has been considered. The proposal for the card for historical centres advanced in the context of the more general card for the construction of the Cultural Heritage Map has been inspired by the concept of historical city proposed by A. Magnaghi. The concept of historical city includes the ancient city (which starts roughly from the nucleus of the foundation and goes on to the various additions related to subsequent civilizations within the city walls or other natural or artificial boundaries) and the modern city (the city existing until the 1950s of the twentieth century, which has a variable date of foundation in different contexts, but, unlike the ancient city, concerns a stratification of two or three centuries). The group of the Cultural Heritage Map has been in charge of the historical city as part of the cataloguing, mainly with the purpose of preservation and enhancement; the Technical Secretariat of the Regional Territorial Landscape Plan (*PPTR*) has been in charge of contemporary urbanization (built after the second post-war), defining the main settlement morpho-types at a regional level, for which there is the need to define the critical points and, therefore, the types of landscape retraining and reconstruction interventions.

The proposed objective has been to combine in a unified regulatory framework, tracing out consistent perimeters, historic city (ancient and modern cities) and contemporary urbanization. The criterion for tracing out the perimeter of the *ancient city* is based on the systematic comparison between the buildings reported in IGM Cartography on a 1:25.000 scale of 1949 and the built elements reported in the historic IGM Cartography of 1870 available on a 1:50000 scale; the criterion for tracing out the perimeter of the modern city is based on the systematic comparison between the built elements in IGM Cartography on a 1:25000 scale of 1949 and the Technical Regional Cartography.

R.G.

The Structure of the Database

The implementation of a Management System Database, articulated in a series of cards called *UT, SITO, VINCOLI, BIBLIOGRAFIA* (Topographical Unit, Site, Constraint, References), has been preceded by a long and complex cooperative work (Fig. 4). In the first phase the cards produced by the Central Institute for the Catalogue and Documentation (*ICCD*) have been scrupulously analyzed: in particular the detection of fixed attributes according to the national regulations of Italian Ministry for Cultural Assets and activities has allowed assuring all the Institutions involved in the project of the respect of the basic parameters of the national catalographic standard.

It has thus been created a new model of sheet, focused on the basic concept of Site as Cultural Heritage, which allows cataloguing globally all types of territory data, representing, already at this stage, the different levels of vertical complexity (the stratification) and horizontal complexity (the co-evolutionary relationships) (Fig. 3).

The innovative element represented specifically by the *UT* sheets and the *SITO* ones is the fact that they do not refer to disciplinary and academic type distinctions among the assets, but they tend to bring back, and therefore to describe, the asset on the basis of its 'objective' characteristics.

By adopting a classification based on the concepts of *Tipo* (Type), Categoria (*Category*) and Funzione (*Function*) for which appropriate glossaries have been arranged, such sheets allow describing any type of Cultural Heritage asset provided by the Map without distinctions such as "archaeological asset", "architectural asset", "historical-artistic asset", which do not reveal anything about the real nature of the assets, but they rather concern the discipline that traditionally made it the subject of study. According to the classification of assets made on the basis of *Tipo*, *Categoria*, *Funzione*, for example, an asset such as "Masseria" *Tipo* belongs to the "Settlement" *Categoria* and has a Living/Residential and Production *Funzione*; an asset such as "Aqueduct" *Tipo* belongs to the "Infrastructure" *Categoria* and has the "Water" *Funzione*; an asset such as "Sanctuary" *Tipo* belongs to the "Area/Place" and has a "Religious/Cult" *Funzione*.

The need to define in detail the different levels of DBMS has brought to the realization of a synthesis of the several terminological lexicons produced by *ICCD*, for the attributes of primary significance, i.e. Type, Category and Function. Thus, on the basis of predetermined codes for each Category and Function, the system, according to the choice of the Type, automatically proceeds with the registration of the Category associated with it. For the Function field instead a list of multiple-choice values freely associated with each Type has been preferred.

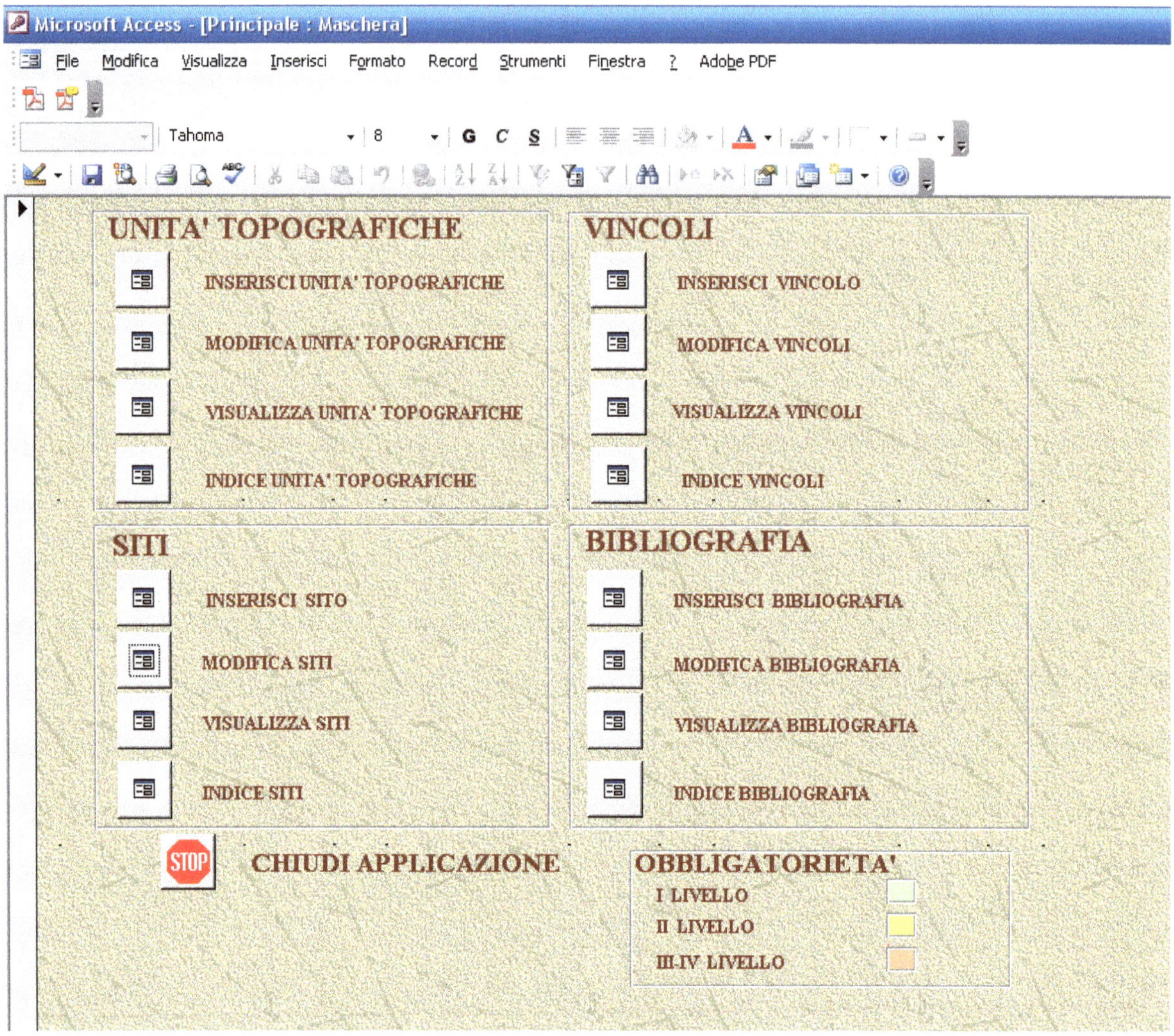

Fig. 4. The entry form of the DBMS.

The objective of this new approach given to cataloguing work has been to overcome the fragmentation in different sheets according to the categories of assets to describe, which is typical of models produced by *ICCD* and rather to articulate the descriptive text through data normalization.

This overcoming has been made necessary, moreover, by the clear inadequacy of similar concepts to define pluristratified and complex assets which are often those which stand on the territories of such a wide and uninterrupted anthropization as the Italian ones. Actually there are frequent cases of a Medieval or Modern building with walls in a better state of preservation and therefore suitable for architectural type analysis which are generally defined 'architectural asset'. This building may reuse part of masonry of a more ancient partly buried site and thus defined as archaeological.

The sheets related to the *VINCOLI* with all the information spaces useful for the registration of information regarding protection have been connected to the *SITO* sheet, which may include one or more *UT*. In addition, the *SITO* sheet can be part of a more complex stratified and topographical set, identified only on a representative level, such as the Pluristratified Site (identified by a specific code including all the cards of the sites that it is made of). The sheet related to the References has been then linked to the *UT* and *SITO* sheets, in order to make the extraction of a list of references possible at any time.

A final observation concerns the cataloguing of the historical settlement-infrastructure systems, consisting of roads, sheep-tracks, aqueducts, centuriation marks, but also by the system of coastal towers, divisions of land into farms, etc. These systems, mostly only in part known, have represented a problematic factor, which is difficult to include in the architecture described so far, made of *UT* and *SITO* (e.g. the emblematic case of the *Via Traiana*, whose route is well known thanks to a series of sources of different nature and of which has been identified and investigated only in some sections).

Such evidence cannot be treated as a Site, nor can be the

whole Via Traiana, which rather crossed many sites along its route. The methodological reflection conducted in the context of the project has led to a different solution, which provides the cataloguing of the single, partial evidence as *UT* (e.g. a stretch of road, aqueduct, centuriation marks), related not to *SITO* but, precisely, to settlement-infrastructure systems, equipped with a specific code and called *SISTEMA DI APPARTENENZA* (i.e. the system to which the *UT* belongs).

In the end, regarding the definition of the horizontal level of the logical structural system (i.e. the one characterized by co-evolutionary relationships), the sheet of *CTS* has been made (see *supra*). In response to a contextual vision of the historical landscape, this sheet collects all the useful data for the definition of a complex system, but in any case consistent system, including all those components (objects or traces) and those (environmental, temporal and spatial) data useful for the definition of a dataset or a set of complex data based on 'functional' type reports.

C. A.

The Processing and Cartographic Representation

As regards the more properly mapping aspect, the Map is characterized by a high level of precision in the location and perimeter of the assets, thanks to the possibility of using as a cartographic base the Regional Technical Map on a 1:5.000 scale and the original and high definition Orthophotomap. Such accuracy is essential for the purposes of protection and planning for which the Map has been promoted. The representation of the assets, in accordance with graphic possibilities offered by the G.I.S. platforms (point, line, polygon), has been made by tracing out the perimeter of polygons in the case of areas or buildings, the points in the case of reports of assets of which it is not possible to define the exact location and extension and the lines in the case of elements, mostly identified on the basis of aerophotographic tracks, of which the linear trend is known, but the largeness is difficult to define.

The guidelines useful for establishing the basic procedures of the graphic digitalization of the several UT and Sites, as well as a series of best practices for all the working groups involved have been defined. First of all, in order to provide InnovaPuglia, responsible for the final management of the Information System of the Region of Apulia, with the format of the final files, the shapefile format (.shp), standard established by OGC (Open Geospatial Consortium) has been chosen. The cartographic basis on which the several groups have digitized the graphic elements are the new Regional Technical Vector Map on a 1:5.000 scale, with WGS84-UTM 33 system, and the Orthophotomap 2006. The cadastral maps, useful for tracing out the perimeter of the Constraints carried out by the working group of the Regional Department for Cultural Heritage have been added to the ones mentioned before.

Analyzing more specifically the data to be graphically represented, it has been noted that actually, since most of the data has been acquired from references, the positioning of *UT* and *SITO* on a cartographic base so detailed and different from the one used by the authors has been often difficult. Actually, the cartographic basis used in publications are often very simplified and schematic, useful as a reference for the topographical positioning of Sites, and those few which are positioned, are mostly positioned on cartographic tables of IGM on a 1:25.000 scale. Therefore, a georeferenced mosaic of the 1:25.000 IGM tables related to regional territory has also been used as a simple 'functional reference level' for a quicker identification of the location areas of the sites.

For the definition of the modalities of graphic representations of UT and Sites, the parameters established show that they are expected to have the following forms:

- **Polygon**, well defined, positioned and georeferenced unit. It is the case of all data from on field activities (see field surveys and excavations) and from detections by Remote Sensing, and in particular aerial photographs at low level.
- **Point**, data whose location and position are uncertain, and have never been georeferenced (see data from references and historic archives).
- **Line**, all those elements of which it is possible to determine position, georeference and linear, but not dimensional development, or largeness (see data from photinterpretation and Remote Sensing, e.g. stretch of road, rests of walls, roads, paleo-riverbeds, etc.)

In this way a series of information levels that reflect the several types of data collected especially regarding the in-depth analysis of data, for each single UT and Site and the possibility to identify, with more or less accuracy, the position and area of the indicated entity have been produced. The G.I.S. exactly for its function of protection and prevention, as targeted to the implementation of the 'Map of risk', has allowed producing four levels reporting on the one hand the survival of the Cultural asset, on the other hand the possible risk to which it is subjected to. The first level corresponds to areas of archaeological interest in accordance with letter m of the article No. 142 of the Legislative Decree D. Lgs. 142/2004, the second one to archaeological risk areas (e.g. areas of fictile fragments with perimeter traced out as polygon), the third one to areas requiring urgent in-depth analysis (assets localized with punctual geometry), the fourth to areas with insufficient documentation that need to be subjected to in-depth analysis in case of interventions.

Conclusions

The Critical Points and the Limits of the Project

The project outlined so far has been complex and articulated not only for the content but also for the skills that have been involved in the work and have contributed to the implementation of the Project itself. From the start some critical points due to the dimension and complexity of the project have emerged. These critical points concern the presence within working groups of different disciplinary components, research traditions of each single University, operating with different approaches and on different territories, the need to identify assets of many types, each with specific traditions of studies, the necessity to meet adequately the needs of the Regional Department of Land Management.

However, the same elements of complexity have been an extraordinary wealth for the project, which can aspire to become a "best practice" for example in the field of the prior assessment of archaeological interest, representing a model of methodological, disciplinary, institutional integration (Malnati 2005, Volpe 2008, 455-458).

Regarding the content, it is necessary to highlight that the Map does not represent nowadays the complete description of the Cultural Heritage of Apulia, but the representation of the state of knowledge of this heritage. Therefore, while for some territories, as the Tavoliere Plain, the long tradition of systematic researches on the field, carried out also by the University of Foggia, revealed the extraordinary wealth in terms of Cultural Heritage, other territories now seem insufficient as regards the quantity of counted assets and the quality of descriptive data associated with them.

Moreover, as regards the reliability of data, 43% of the identified and described sites are represented not by polygons but by points and are mostly difficult to locate because the informations about assets are not enough.

Moreover, at the moment the network of historic roads is not included in the Map, except for the sheep-tracks, whose study the Secretariat of the Regional Territorial Landscape Plan (PPTR) has been entrusted with.

Future Development of the Project

One of the most innovative elements introduced by the Code of Cultural Heritage and Landscape seems to be the protection of the modern city, even if on the basis of criteria and regulation modalities perhaps different from the ancient city. Until now, the urban planning instruments have been entrusted with the protection and enhancement of the historical centres, except for single monuments, villas, historical gardens or monumental complexes reported by the Superintendence Offices, since the Landscape Planning as the *PUTT* (*ex lege* No. 431/85), is referred to open suburban spaces, therefore it excludes the cities.

However, the Code, referring to the definition of the landscape given by the European Landscape Convention (Florence, 20-X-2000) expects the Landscape Planning to deal with the whole regional territory, including historical city and contemporary urbanization. While for contemporary urbanization it is clear that, «except for some persisting assets of historical, artistic, cultural interest incorporated in the urbanization itself, the tasks of the regulatory and planning intervention concern mainly the Landscape Planning in relation to other planning tools» (Magnaghi 2008, 2), for the historic city the Map of Cultural Heritage may provide new regulatory instruments in order to integrate the policies and urban planning defining the prerequisites or the 'structural invariants', and giving these the force of a superordinate constraint.

Moreover, among the innovative instruments created there is the one that will allow each user who uses the Map on line to report personally known assets absent in the census. This procedure will enrich more the framework with the knowledge of the territory and will be an effective procedure for the involvement and participation of citizens in the project as provided for by regional, national and local documentation (*QSN 2007-2013*, 105, *DSR 2007-2013*, 118-119, *Linee guida Area Vasta*, 14-15).

It will also be desirable to define a procedure allowing any organization responsible for research, protection and enhancement of Cultural Heritage of Apulia (Superintendence, Universities, Region Authorities, Local Authorities etc.) to update the system through the transfer of possessed data to the Regional Observatory for the Quality of Landscapes and Cultural Heritage that will ensure standardization in phase of implementation. This procedure would be useful, in particular for the Municipalities whose Cultural Heritage is at the moment less known. In this way it will be possible, moreover, to ensure the continuation of a 'direct line' created among the Regional Authorities, Universities and Regional Central Department for Cultural and Environmental Heritage and therefore the constant updating of the Map, without which it risks a rapid obsolescence.

In the end, Universities may provide systematic projects dedicated to the verification on the field of the sites located as uncertain points, so they can further specify the degree of reliability of many Cultural Heritage assets listed in this stage.

A further and strategic line of development may be to build an archive of the material and immaterial Cultural Heritage of Apulia, integrating the Cultural Heritage assets counted in the Map with the other kind of local cultural resources. Therefore, it will be possible to carry out qualitative-quantitative type territorial analysis able to effectively support the policy makers and all the participants in local development (Touristic promotion agency, Local action groups, cultural districts, Local Tourist Systems, etc.) in planning investments, strategies and policies of development and enhancement of the territory, in particular in the tourism sector (Carta 2003, 2004, 2005)

A.D.Z.

Acknowledgements

Many thanks to Patricia Anna Clinca for the English translation.

References

BUORA, M. AND SANTORO, S. (EDS.) 2004. *Progetto Dürres. Strumenti della salvaguardia del patrimonio culturale: carta del rischio archeologico e catalogazione informatizzata. Esempi italiani e applicabilità in Albania (Villa Manin di Passariano – Udine – Parma, 27-29 marzo 2003). Alte tecnologie applicate all'archeologia di Dürres (Dürres, 22 giugno 2004)*, Atti del secondo e del terzo incontro scientifico, Trieste, Editreg Srl.

CAMBI, F., TERRENATO, N. 1994. *Introduzione all'archeologia dei paesaggi*, Roma, Nuova Italia Scientifica.

CARTA, M. 2003. *Pianificare nel dominio culturale. Strutture e strategie per l'armatura culturale in Sicilia*, Palermo, Dct.

CARTA, M. 2004. Strutture territoriali e strategie culturali per lo sviluppo locale, *Economia della cultura*, 1, 39-56.

CARTA, M. 2005. Indicatori e strategie per la formazione di distretti culturali, *Economia della cultura*, 2, 195-207.

DSR 2007-2013, Documento Strategico Regionale 2007-2013 della Regione Puglia; approvato con Delibera di Giunta Regionale n. 1139 del 1 agosto 2006 (Bollettino Ufficiale della Regione Puglia n. 102 del 9 agosto 2006).

FORTE, M. 2002. *I Sistemi Informativi Geografici in archeologia*, Roma, MondoGIS Srl.

Gelichi, S. 2001 (ed.). *Dalla carta del rischio archeologico di Cesena alla tutela preventiva urbana in Europa*, Firenze, Edizioni All'Insegna del Giglio.

GOFFREDO, R., VOLPE, G. 2007. *Fotografia aerea, archeologia globale e paesaggi antichi della Daunia. Esempi dalle valli dell'Ofanto e del Carapelle*, in G. Ceraudo and F. Piccarreta (eds.), *Archeologia aerea, Studi di aerotopografia archeologica II*, Roma, 219-246.

LINEE GUIDA AREA VASTA, REGIONE PUGlia, Assessorato alla Programmazione, Nucleo di Valutazione e Verifica degli Investimenti Pubblici, *Linee guida per la pianificazione strategica territoriale di Area Vasta,* Atto NVVIP n. 151 del 15.5.2007.

LUCCHESI, F., CARTA, M., DI ZANNI, A. 2008. *La storia, lo spazio, il paesaggio. Una ipotesi di rappresentazione multi temporale del patrimonio culturale stratificato per la tutela e la valorizzazione dei beni culturali pugliesi.* In P. Persi (ed.), *Territori contesi: campi del sapere, identità locali, istituzioni, progettualità*, Atti del IV Convegno Internazionale sui Beni Culturali Territoriali (Pollenza, 11-13 luglio 2008), Pollenza, 101-110.

MAGNAGHI, A. 2000. *Lo sviluppo locale*, Torino, Bollati Boringhieri.

Magnaghi, A. 2008. *Nota sulla proposta di scheda per i centri storici per la Carta dei Beni Culturali*, Documento interno al progetto del PPTR, 10 gennaio.

MALNATI, L. 2005. *La verifica preventiva dell'interesse archeologico*, Aedon. Rivista di arti e diritto on line, 3.

MANACORDA, D. 2004. *Prima lezione di archeologia,* Roma-Bari, Laterza

MANACORDA, D. 2007. *Il sito archeologico fra ricerca e valorizzazione*, Roma, Carocci.

MANNONI, T. 1997. Archeologia globale e archeologia postmedievale, *Archeologia postmedievale*, I, 21-25.

SETTIS, S. 2002. *Italia S.p.A. L'assalto al patrimonio culturale,* Torino, Einaudi.

VOLPE, G. 2007. *L'archeologia "globale" per ascoltare la "storia totale" del paesaggio*, *Sudest*, 20, 20-32.

VOLPE, G. 2008. *Per una 'archeologia globale dei paesaggi' della Daunia. Tra archeologia, metodologia e politica dei beni culturali.* In G. Volpe, M. J. Strazzulla, D. Leone (eds.), *Storia e archeologia della Daunia. In ricordo di Marina Mazzei*, Atti delle giornate di studio (Foggia 19-21 maggio 2005), Bari, Edipuglia, 447-462.

VOLPE, G., DI ZANNI, A., LAURENZA, S. 2008. *La Carta dei Beni Culturali della Regione Puglia: dalla lettura del paesaggio alla progettazione dell'Infrastruttura Informatica*, in Volpe G., De Felice G., Sibilano M. G. (a cura di), *Digitalizzare la pesantezza. L'informatica e il metodo della stratigrafia,* Atti del Workshop, Foggia, (6-7 giugno 2008), Bari, Edipuglia, 75-90

VOLPE, G., GOFFREDO, R., DI ZANNI, A. 2007. *Herdonia e l'archeologia dei paesaggi della Valle del Carapelle. Per un museo archeologico diffuso.* In L. Longo, V. Vecchione (eds.), *Sistemi locali e sviluppo. Lineamenti per un piano strategico*, Foggia, Claudio Grenzi Editore, 109-124.

Remote sensing and GIS application in the management and conservation of heritage properties at Agra

D. Dayalan

Archaeological Survey of India

Keywords Taj Mahal, Agra, GIS, Land Use Pattern, GIS layer

Remote sensing and GIS are complementary technologies. The remote sensing systems are used to observe the earth's surface through the sensors mounted on the satellite and aircraft and to collect information of an object, area or phenomenon. The geographic information systems (GIS) can be used to interpret remote sensing data more accurately by integrating them with data derived from various other sources and can analyse large volumes of geographically referenced data that would be overwhelming to process manually. The satellite pictures can also help to update efficiently certain data layers in the GIS.

Notwithstanding the application of remote sensing for various purposes in sporadic manner is reported elsewhere, the field of remote sensing took shape only during the 1960s as an outgrowth of aerial photography. William Pecora, Director of the US Geological Survey (USGS), took the lead in pushing for a spacecraft that could cover the entire world with false colour images. The first of them is Landsat 1 by the National Aeronautics and Space Administration (NASA) which reached the orbit in 1972. The French government launched the SPOT program, the Satellite Pour l'Observation de la Terre in 1978 and in 1982 established the firm of SPOT image to market its photos. Lockheed was the first company which obtained the license from the USA authorities for building commercial remote sensing satellite system and launched its IKONOS satellite in 1994. The other system followed this are: Orbview of Orbital Sciences, Quickbird for the firm of DigitalGlobe, etc. The photos taken by Quickbird which flew in 2001 show the detail as small as two feet i.e. 0.62 metre in size.

The new development in the technology like, microwave remote sensing, Global Positioning System (GPS), etc are also being utilized for various studies. As the satellites images are obtained in digital form it can be processed directly in the computer. Moreover the satellite data are more reliable, repetitive, non-invasive, rapid and cost-effective information. All information is exactly localized and gathered in one tool. The direct extraction of topographic and thematic maps for terrain use is also possible with this technology.

The geographic information system (GIS) is a combination of hardware, software and procedures for storing, managing, processing and reproducing spatial data with data coupled to them. The data can be obtained from ground surveys, GPS measurements and from remote sensing data.

Notwithstanding the remote sensing and GIS technologies are used in the fields such as town and country planning, watershed management, agriculture, forestry, conservation, mineral resource development and prevention and mitigation of natural disasters, the necessary for the utilization of remote sensing, GIS and related geo-information technologies for the conservation and maintenance of the cultural and natural properties has been felt only in the recent years. In fact the GIS and remote Sensing are the excellent tools to support the monitoring process of cultural and natural heritage sites

The activities wherein the remote sensing technology can be utilized for the management and conservation of cultural and natural properties are:

1. Creation of data-base and digital maps of monuments and their environments, ancient settlements, etc.
2. Evaluation and management of the monuments and sites
3. Study of chronological changes in the nearby area of the monument/site.
4. Measuring and monitoring of encroachment or any other changes in the ecosystem in the surrounding of the properties in order to take early corrective actions.
5. Preparing base maps for drawing master plan for integrated development in and around the heritage properties
6. Survey and monitoring of natural properties
7. Development of Natural Disaster Management System and monitoring of river pollution.
8. Survey and monitoring of natural properties
9. Development of Natural Disaster Management System and monitoring of river pollution

Study Area

Agra (270 11' N; 780 1'E) in the Uttar Pradesh State, India is the prime destination of tourism in India, located approximately 200 Kilometres south of New Delhi. It is not only an important urban centre in the medieval period but also blessed with three world heritage monuments including the world famous Taj Mahal (1631-1648). This miracle in marble constructed by the Mughal Emperior Shah Jahan (1628-1658) for his favorite wife Anjuman Banu Begam popularly known as Mumtaz Mahal is the culmination of the sepulchral architecture of Shah Jahan and indeed of the Mughals. Notwithstanding the contemporary sources refer to the mausoleum as "Illumined Tomb" (*rauza-i-munavvara*), it derived its name from the title of Mumtaz Mahal, distorted by popular etymology to Taj Mahal.

The Taj Mahal was inscribed on the World Heritage List in 1983. Its inscription as a cultural property of universal significance was justified in the following words: *"The Taj Mahal is the greatest architectural achievement of the whole range of Indo-Islamic architecture. Apart from its tremendous romantic appeal, it is a masterpiece of architectural style, in conception, treatment and execution, all alike".* Spread to an area of 22.44 hectare, Taj Mahal complex is planned in the form of a rectangle, aligned north-south and measuring 579.12m by 304.8m.. The central area is divided in to square garden leaving an oblong space at either end of the rectangular. In the south this is taken up by the gateway whereas on the north consists of a raised platform containing the main marble mausoleum and the auxiliary buildings like Mosque and Mehman Khana, all overlooking the river below. Taj Mahal has to be visualized along with the river bank, river Yamuna and adjoining monuments, gardens and landscape as they are the integral part of this monument.

The remote sensing and GIS techniques are applied for creating data-base, maps and various other purposes. These techniques are also used for the management of the buffer zone, preparation of the integrated scheme in and around the monument, conservation activities, etc. The satellites images are also useful to visualize the overall urban settlement pattern during the medieval period.

Fatehpur Sikri, another world heritage monument near Agra is the planned city of the Mughal Emperior Akbar (1556-1605) constructed on the bank of a large natural lake. The ancient city wall, which in its 9 kms run is pierced by nine gates, holds a large area with undisturbed archaeological remains in the form of palaces, mosques, *sarais, hammams*, water works and many other secular, religious and defense edifices. The department is the process of applying the remote sensing and GIS techniques for mapping the various structures, monitoring the monumental and the periphery areas and for the conservation activities.

Data Used

1. High resolution satellite images such as the Taj Mahal and its surrounding areas including Agra Fort, another World Heritage Monument and ancient gardens and edifices and Taj Mahal overlooking the southwest bank of the River Yamuna collected by Space Imaging's IKONOS satellite.
2. Aerial photography
3. The maps prepared by means of Topographical survey of the whole area by using the Total station. All the final survey drawing have been realized in AutoCAD or compatible industry standard digital vector files, i.e ESRI shape.
4. One inch Topography map by Survey of India.
5. Architectural Maps prepared by the Archaeological Survey of India

System Requirements for GIS

Hardware

RAM-256 MB (512 MB recommended)
Preferably P (4) CPU 3 GHz
Hard drive free space required 4 GB

Software

Arc GIS 9.0 or higher version
i_view32.exe (freeware exe already present in Images folder)

An Integrated digital database for Taj Mahal on designated GIS platform is created after site survey for proper monitoring of this World Heritage Monument.

The dataset is having three levels of details.

Location:

Overall area of interest has been divided into the following manner.

1. Fore Court
2. Char Bagh (NE, NW, SE, SW and Central)
3. Naubat Khana
4. Museum
5. Khan I Alam Nursery
6. Mehman Khana
7. Mausoleum
8. Mosque
9. Backyard

Land use Classification:
The entire site has been classified in the following 23 land use classes:

1. Ancient Water System
2. Backyard
3. Drainage System
4. Fence
5. Flower Beds
6. Fountain System
7. Lawns

8. Platforms
9. Steps
10. Walls
11. Cement/Modern Floor
12. Bench Mark
13. Built Structure
14. Facilities
15. Furniture and Fixtures
16. Foot Path
17. Garden
18. Pillar Post
19. Shrubs
20. Drainage
21. Water Features
22. Stone Lining
23. Inlay Works

The above classification is represented at two different levels as follows:

1. Land use classification at Macro level:

At this level the layers represent details at a Macro level. For instance, in the Footpath layer the complete footpath from edge to edge is represented as a single polygon without differentiating between the internal sub-divisions of different design patterns within the footpath.

2. Land use classification at Micro Level:

This level of information provides further detailed differentiation of the Macro representation by clearly identifying various internal patterns within the macro classification.

Schema

Following is the Schema of the GIS dataset for Taj ahal. It shows the layer classifications field structures along with the database depicted with the corresponding colour of the boxes.

Operational Guidelines of the GIS Programme

1. File structure
 \Tajmahal\Images*.jpg (Contains photographs in jpg format)
 \Tajmahal\Images\ I_view32.exe (Irfan viewer application)
 \Tajmahal\TajMahal_24022006.mdb (ArcGIS Personal Geo-database)
 \Tajmahal\TajMahal_24022006.mxd (ArcGIS project file)
 User Manual Ver 1.0
2. Open **<drivename>:\Tajmahal** folder and right click on TajMahal_24022006.mdb fil and then click Properties. Click to clear the
 Read only check box (if present), and then click Ok.
3. Following are the steps for setting I_view32.exe as a default image viewer.
 1) Start ***Windows Explorer*** and open **<drivename>:\Tajmahal\Images** folder and right click on any jpg file (in right hand side window in the explorer) and click on **'Open with'** and then click on **'Choose Program'** and then click on **'Other'** button and now select the application **I_view32.exe** present in the **<drivename>:\Tajmahal\Images** folder as shown in the picture below.
4. Open **<drivename>:\Tajmahal** folder and double click on the Tajmahal_2402006.mxd file to open your map project (Or open *ArcMap* and direct it to open an existing .mxd file which is located in **<drivename>:\Tajmahal** folder in the local machine) User Manual Ver 1.0
5. On File menu, click Map Properties, to set the values for 'Hyperlink base'
 to **<drivename>:\Tajmahal\Images** for e.g. if the data has been copied to
 d: drive hen write 'D:\Tajmahal\Images' (It defines the location of the folder
 in which the Taj Mahal snapshots are lying) and press **"OK"** button as
 shown in the picture below

The remote sensing and the GIS are the effective tools for the management of the heritage properties particularly for mapping, monitoring and analyzing. GIS is the integrated data-base management system that is increasing its use in many form of management. It gives the possibility of storing, analyzing and displaying large amount of data at the same time. Moreover this rapid growing technology is more user-friendly due to its advanced software and also provides more geographical information

Benefit of Remote sensing and GIS

1. The comparison of the Satellite images of recent years and the Aerial photographs taken few decades before provide excellent idea on the periodical land-cover changes in the surrounding area of the monument.At present Taj Mahal is surrounded by shops, green belt, emporium, offices, hotels etc. on the east; restaurant, shopping complex, parks, emporium etc. on the west; thickly populated Taj Ganj area, shopping complex, etc. on the south and Yamuna river, Mehtab Bagh, green belt area on the north. The aerial photograph taken few decades before shows that the greenery on the north and east side of Taj Mahal is not so dense in comparison to the situation today. But the building activities on the south side is increased

alarmingly in the recent years than what was the position earlier as depicted in the old aerial photography. Government of India in consultation with UNESCO proposed to formulate the Taj Mahal, the riverbank, river Yamuna and adjoining monuments and landscape as heritage zone to safeguard the landscape and the environment from any unregulated and inappropriate development. Archaeological Survey of India has done detail survey of the modern buildings within prohibited and regulated areas of Taj Mahal and prepared a data base by using the GIS tool. Each and every minute detail of the modern structures and other features in the surrounding area is stored in the GIS for analyzing and monitoring the activities in the area. The construction, mining, vehicle movement and pollution in this area are regulated by various govt. rules, gazette notifications and Supreme court directions from time to time. Remote sensing and GIS are very much useful in implementing the government and court orders.

2. The inlay of floral arabesques fashioned from precious/semi-precious stones in white marble and the inlaid black marble inscriptions that frame the central vaulted portal or *iwan,* are some of the prime attraction in Taj Mahal. The inlay work is seen on the cenotaphs, the borders of the marble screen, the dado panels, the spandrels of the arches, bottom of the dome and other places. In order to achieve the desired tonal variations the inlay workers used as many as forty-eight pieces of precious and semi-precious stones to compose a single flower.

   There are nearly 45 precious/ semi-precious stones are used for inlay work. They are: 1 .*Ajuba/ Ujuba* (It is mottled with green and yellow concretions in a brown ground-mass); 2. Alabaster; 3 Agate; 4. Black marble; 5.Bloodstone; 6.Carnelian of Arabia Felix; 7.Chrysolite; 8.Conch-shell;9.Coral;

   10.Cornelian;11..Crystal;12.Diamond; 13. Emerald; 14.Garnet; 15.*Ghar;* 16.*Ghori;*17.*Godar;* 18.Golden stone; 19.Green Marble; 20.*Gulabi;* 21.*Jade;* 22.Jasper; 23.*Khara;* 24.Lapis Lazuli; 25.Load stone; 26.Magnet Stone; 27.Melachite; 28.Mother-of-pearl; 29.*Nakhud;* 30.Onyx; 31.Onyx Marble; 32.*Pai-Zahar;*33.*Pankhni;* 34.Pearls; 35.Pink Marble; 36.*Pitunia;* 37.Redstone; 38.*Reg;* 39.Ruby; 40.Sapphire; 41.Slate (Calcareous shale); 42.*Sumaq* ; 43.Turquoise; 44.Topaz; 45.Yellow Marble.

   The data base comprising the details of location, number of pieces, name of precious/semi-precious stones, size and colour of stone and other details has been created with help of GIS programme for the purpose of conservation, documentation and checking of the inlay works.

3. Few years before there were big news that the minarets in the Taj Mahal is tilting and a portion of the monument is shrinking. The first scientific study regarding the foundation stability and verticality of the minarets was carried by the Survey of India in 1941-42 and fixed 104 bench marks in Taj Mahal complex for periodical cross checking. It was observed in the first scientific survey that all the minarets inclining outwardly perhaps they were designed originally like that and the detail of minaret wise inclination is as follows: South-west-21.6 cm; South-East-11.4 cm; North-East- 4.8 cm and North-West- 3.5 cm It was decided to check the bench marks ones in 3 or 4 years to ascertain the verticality of minarets and stability of structure. All the 104 bench marks fixed in the Taj Mahal complex are surveyed and documented. On this basis, a separate data layer has been created in the GIS programme which provides all details like location, size, year wise reading of the levels, present status and other details of the bench marks in nutshell. Now with the help of GIS it is scientifically proved with relevant data that all the minarets are stable as there is no change in the inclination of the minarets.

4. The details of the flora within the Taj Mahal complex including the Khan-i- Alam, the ancient water channel system associated with Taj Mahal has been documented and following information are stored in the GIS for the purpose of monitoring and analyzing.

   The details of the trees and shrubs available in the GIS are: 1.Object, 2.Shape, 3.Survey ID, 4.Photo ID, 5.Tree No, 6.Location, 7.Type, 8.Botanical name, 9. Canopy dia., 10.Trunk dia., 11.Height, 12.Tree age, 13.Planted date, 14.Image path, 15.Remark.

Digital documentation of Buddhist sites in Tamil Nadu

D. Dayalan

Archaeological Survey of India

Buddhism came to South India during the period of Emperor Ashoka (c. 273-236 BC). The rock edicts of him found in the bordering regions of South India invariably refer to the Buddhist dharma. However, the prevalence of Buddhism in the Tamil country till 3rd-4th century AD is not attested by any material evidences. The reason may be that consistent with pan-Indian usage, perishable materials like wood and brick had been used for the construction of secular as well as religious edifices of the early times. However, the discovery of large number of Buddhist vestiges datable to 5th-6th century AD and onwards at various places in Tamil Nadu and Pondicherry indicate the wide prevalence of Buddhism in this region. It appears that Buddhism was a flourishing faith during this period and has been patronized by the local people and the merchants and perhaps the rulers as well.

On the basis of the clue from the inscriptions, details given in the local and foreign texts and place names, a large number of sites yielding Buddhist relics are spotted all over the state. The relics include stone sculptures and bronzes, excavated remains, inscriptions and paintings The author of the present paper documented all these Buddhist vestiges and also prepared a map of Buddhist sites in Tamil Nadu with inputs such as detail of the location, nature, description, date and status of the relics, detail photographs, published references, etc. (Please see the Annexure for the List of Buddhist Sites in Tamil Nadu and Pondicherry). The mapping of the Buddhist sites not only revealed the distribution of the Buddhist sites in Tamil Nadu and Pondicherry but also the focal centres of Buddhism. Some of the epicentres of Buddhism are:

Kavirippumpattinam

Kavirippumpattinam (11^0 12' 50" N; 79^0 52' 50" E) in district Tanjavur, the celebrated capital and port city of the illustrious Cholas of the Cankam age (c. 200 BC to c. 200 AD) is said to have been situated on the confluence of the river Kaviri with the Bay of Bengal. This city was known by several names in ancient period, viz., Pukar (*Cilappatikaram* canto.1, line. 16 & 22. canto. 6, line 32 canto. 7, line 52, 56, 64, 163, 171 & 179 canto. 8, line 4; *Manimekalai* canto. 5, line, 109), Pumpukar (*Cilappatikiiram,* canto. 1, line 10), Kakanti (R. Nagaswamy, 1980, pp. 123-124), Campapati (*Manimekalai* Patikam, line 8), Kavirippumpattinam (*ARE,* 1991 nos. 261-273), etc. Ptolemy (c.1st century AD) refers to this place as "Kaberis Emporion" (G.E. Gerini, 1992, p.408) whereas Periplus of the Erthrian sea as "Camara (W.H. Schoff, 1974, pp. 46 and 242).

Tamil literature of the first three centuries AD gives a vivid account of this celebrate port city of the Cholas, its harbour, sailors, merchants, merchandise, etc. A poem from the 1st-2nd century AD states that big ships entered the port without slacking their sails and poured out on the beach precious merchandise from different overseas countries and also other ports of India *(Pattinappalai* lines 185-192). Flags were fluttering on the mast of the ships heaving in the port of Pukar like huge elephants chafing (*Pattinappalai,* lines 203-210). It also speaks of the tall light house on the coast summoning the ships to the harbour by the night *(Cilappatikaram,* canto. 7 line 3), yavanar-*irukkai* (colony of foreign traders) *(Cilappatikaram,* canto. 5 line 10), etc.

This metropolis was not only famous for its seafaring activities but also well known in times of yore as a glorious centre of Buddhism. Many texts indicate the existence and perhaps the prosperous condition of the Buddhist establishments at Kavirippumpattinam. The literature of Cankam period, the Augustan age which produced a bumper crop of Tamil, poetical work, do not make any direct reference to either Buddhist or Jain religions, although some references have been interpreted, often in a far-fetched manner, as an evidence of their presence. (T.N. Vasudeva Rao, 1979, pp. 68-72) The *Pattinappalai,* one of the Cankam works, refer to a *tavappalli* at Pukar. *(Pattinappalai* line 51-53) If the *tavappalli* is considered as Buddhist monastery-then it would have been the earliest reference in Tamil literature of Buddhist settlement in Tamil country in general and Kavirippumpattinam in particular.

There are literary references to a Buddhist monastery called Intira vikaram in this city. *(Cilappatikaram,* canto. 10 line 14, canto 27 line 92; *Manimekalai* canto. 26 line 55) According to the tradition, the construction of these viharas is ascribed to Makentra (Mahendra/ Mahindra), son or brother of Ashoka and also God Indra. (Smith Vincent, 1924, p.185) This *vikaram* as referred to in *Cilappatikaram* of c.5-6th century AD was not constructed by hand or machinery, but was a mind-born institution. (*Cilappatikaram,* canto. 27 line 92) It further mentions that Indra caused these sacred seven *viharas* to be built near the Mahapoti tree, sacred to Buddhist in the city of Pukar. *(Cilappatikaram,* canto 10. line. 11-14)

Associating any place or religious edifices with the divine heroes or great kings in order to boost its importance and antiquity is a common tradition in India. Following this tradition the *vikarams* at Pukar are also ascribed to divine authorship i.e. to Indra and to the great kings like Ashoka or Makentra. There is no probability at all of Ashoka or Makentra having built a *stupa* or monastery in this area. This narration is tainted by legendary accounts and cannot be relied upon for the purpose of history.

The mentioning of seven *vikarams* at Kavirippumpattinam is quite interesting. Number seven seems to be a very auspicious number to Buddhist religion. Buddha is said to have made seven steps immediately after birth. The Manuci Buddhas are also seven in number. The *Manimekalai,* composed with the intension of exaltation of Buddhism, is crucial for the study of the various Buddhist establishments at Kavirippumpattinam. It is, in fact significant that Manimekalai, the heroine of the poet-philosopher Cattanar, started as a Buddhist novitiate in Pukar, and is made to visit all the key centres of Buddhism in Tamilakam, Ilam (Ceylon) and South-East Asia. She adopted the robes of a *pikkuni* under the tutelage of Aravana atikal, the head of the Cankam of Pukar. (*Maniimekalai,* canto. 15 lines. 55-59)

The existence of seven *Intira vikarams* at Pukar is conspicuously mentioned in the epics. The reference to the *Intira vihara* of Pukar and Mapoti occurs in the *Cilappatikaram,* when Kovalan narrates a dream to the brahmana named Matalan. *(Cilappatikaram,* canto. 10 line 11-14) After leaving Pukar for Maturai, Kovalan and Kannaki are said to have crossed the outer gate of this city and then passed by the seven *vikarams* known as *Intira vikaram. (Cilappatikaram,* canto. 27 line 92) From this statement in *Cilappatikaram,* it is clear that this Buddhist establishment was located in the outskirts of the urban area. The epic further states that Macattuvan after hearing of his son Kovalan's unjust execution by the Pandya king disgusted of worldly life and joined the Buddhist Cankam establishments in the Intira vikaram of Pukar. *(Cilappatikaram,* canto. 27 line 90-95) This *vikaram* is also frequently mentioned in the Manimekalai. *(Manimekalai,* canto. 26 line 55; canto. 28 line 70) The *Intira vikaram* seems to be very popular at that time and was considered to be prestigious to compare it with other *vikarams.* While describing the Cera capital Vanci, Manimekalai mentioned a *vikaram,* which is said to be as beautiful as the Intira *vikaram* of Kavirippumpattinam. *(Manimekalai,* canto. 28 lines 69-71) Interestingly, Cattanar speaks of a small pavilion made of crystal in a park called Upavana. A replica of the Buddha's foot print was worshipped there. *(Manimekalai,* canto.3 lines 59-66; canto.5 lines 95-105; canto. 6 lines 11-12) The same author asserts that Killivalavan, a Cola king converted the prison into a Buddhist monastery at the request of the nun Manimekalai. *(Manimekalai,* canto 19. line 157-162)

In addition, there existed in the same city, the Cakkaravalakkottam, a Buddhist temple near the burial ground. *(Manimekalai,* canto. 6 lines 202-204; *Cilappatikaram,* canto. 9 line 20) The shrine was also perhaps known as Cutukattukkottam ("temple in the cremation ground") on the outskirts of Pukar. It was probably associated with the Kapalikas, a group of ardent Saivas, but later seems to have been converted into a Buddhist one. The existence of meeting ground of the Cutalai nonpikal (Kapalikas) in the Cutukattuk kottam is referred to in the *Manimekalai* (canto 6 line 86-89). The Cakkaravalakkottam seems to have been a symbolic representation of the universe as per Buddhist notions. *Manimekalai's* accounts on Cakkaravalam says that it had the depiction in clay; the Meru hill, with its seven allied hills, four island continents, two thousand islets and the celestials residents in them. *(Manimekalai,* canto. 6 lines 190-202) It is interesting to note that this description strikingly tallies with the plan of the remains unearthed from a site known as Pantuvasnuvara in Ceylon. The excavator of this site gives the following interpretation. "The plan reminds one of the manner in which the universe was conceived by the people of India and Ceylon in ancient times. The square structure corresponds to the mountain Meru, the ridges to the circles of mountains which are said to encompass it, and the depressions to the oceans in between these circles of mountains. Encircling all it was believed, there was Cakkravala, beyond which the light of the sun and moon does not travel. The central features being buried, the universe must have been here represented as seen from a point very much above the summit of Meru. There is thus reason to conclude that the circular site at Pantuvasnuvara is a representation in miniature of the universe..........the Cakkaravala..." (Gananath.Obeyesekere, 1984) The Cakkaravalakkottam at Kavirippumpattinam seems to be a model of a Vajrayana mantalam. In the same locality, there was a small Buddhist temple called Kuccarakkutikai which is stated to have enshrined Goddess Campapati. *(Manimekalai,* canto. 18 lines. 144-145, 152) The name Kuccarakutikai perhaps denotes a small temple of Gurjara (Gujarat) style. The temple was also known as Mutiyal Kottam (temple of elderly goddess). *(Manimekalai,* canto. 19 line 39) Campapati was considered the tutelary deity of the Buddhist at Kavirippumpattinam. *(Manimekalai,* canto. 25 line 160-215) In one of the pillars of the Campapati temple existed the pillar-deity known as Kantirpavai, a divine agent which has the spiritual power of explaining the past and future.

Barring the legendary portions and the miraculous elements, the details supplied by the twin epics about the Buddhist establishments at Pukar are reliable and are useful for the reconstruction of the history of Buddhism of that site.

Epigraphical and Archaeological Evidence of Buddhism in Kavirippumpattinam

Coming to the epigraphical and archaeological evidences, the allegedly earliest epigraphical references to Kavirippumpattinam comes from Barhut, the famous Buddhist centre near Satna in Madhya Pradesh, which flourished during the second and first centuries B.C. The inscription engraved on a railing of the stupa here says that it was the gift of a nun Soma from Kakamti. *(kakamtiya somaya bikkhuniya danam).* (Alexander Cunningham, 1962, p. 139) Kakamti is often identified by scholars with Kakanti, the other name of Kavirippumpattinam, which is said to have mentioned in the *Manimekalai. (Manimekalai, canto* 22 line37; R. Nagaswamy, 1980, pp. 123-124; Aruvar, 1975, p.3) There are few points yet to be settled before accepting this identification unanimously. Firstly there is no archaeological evidence available till date at Kavirippumpattinam to prove the existence of a Buddhist establishment of such an early time. Secondly the name Soma is not generally met with as a proper name in the Tamil land. Thirdly there is a less possibility of contact between the coastal town on the extreme south and the remote Buddhist centre in Central India, in the pre-Christian era. It is interesting to note in this context that the votive inscriptions found at Sanchi, a famous Buddhist centre in Madhya Pradesh mention that locality as Kakanaya, Kakanava, Kakanadapota, etc., (Sir John Marshall, 1982, pp. 297 -396). The word Kakandi referred to in the Barhut inscription may be taken as a corrupt form of Kakanaya or Kakanadapota. Of course, this assumption is based on mainly the nearness of Barhut with Sanchi than Kavirippumpattinam.

Buddhist Monastery

Excavations were carried out by the Archaeological Survey of India in many places at Kavirippumpattinam from 1962 onwards. Interestingly the excavations in a locality known as Pallavaneccuvaram at Kavirippumpattinam yielded an excellent evidence about the existence of Buddhist *vihara* and temple. (IAR 1964-65, pp.24-25; IAR 1972-93, pp. 32-33) The excavations at Kavirippumpattinam revealed that the Buddhist monastic traditions of Tamilakam were to some extent common with those noticed in lower Deccan and Andhra Pradesh. The *viharas* noticed in Andhra Pradesh usually consisted of elongated wings and each of them contained several cells with a common verandah for each wing. The wings led into a courtyard where the shrine unit was located. It is not clear in the excavation at Kavirippumpattinam whether there was any typical *chaitya* temple either with the replica of *stupa* or a statue of the Buddha inside as was common in most of the monasteries found in Andhra Pradesh in the Hinayana and Mahayana stages respectively.

One full wing of the monastery, consisting of nine oblong rooms, each measuring 3.10m x 2.40m was laid bare at Kavirippumpattinam. On the south of this row, at a distance of 2.25m a number of offsets were found in the wall indicating the presence of a common- verandah to the monastery. There was a provision of drainages of waste water which moved through a drain under the monastery unit towards north. The cells are laid in the south-east to north-west direction. The extent courses stand up to a meter or more and the size of the brick was 42 x 24 x 10cm average. A small bronze figurine of a seated Buddha in *dhyana* pose and a broken terracotta figurine of a goddess were found in one of the cells. The Buddha is shown on his head with the *usnisa,* the cranial protuberance symbolizing Buddhahood, in the form of a flame. The Lalitavistara, a Buddhist text states when the Buddha is in *Samadhi,* an ornamental ray called *jnana* "knowledge" proceeding from the opening in the *usnisa,* moves above his head. The shawl covered only the left shoulder leaving the right bare. Stylistically this figure is datable to the 5th century AD or little latter.

Buddhapada

There was a Buddhapada found away from the vihara to its south. It was tilted upside down at the time of discovery, showing thereby evidence of disturbance of life and desecration of the site temporarily. The pottery associated with this layer is dateable to around 3rd-4th century AD. The Buddhapada was of Palnad lime stone and it has a small receptacle cut between the long toes of the feet and intended for the relic offerings of bone fragments and gold flowers, which were customarily deposited in it, related to the acharyas or the Buddha as the case may be, which it venerated. The buddhapada of similar nature of course without receptacle socket noticed at Nagarjunakonda is dateable to 3rd century AD by its inscription. The buddhapada at Kaveripumpattinam is also seemingly of the Theravada school. The buddhapada pair shows the auspicious symbols such as the *purnaghata, svastika* and *srivatsa*, etc. The four important events of the Buddha's life ending with *chaitya* worship representing *mahaparinirvana* are also depicted on the slab It is datable to 3rd-4th century AD. The available evidence suggests that the monastery belonged to the fourth-fifth century AD. It went into disuse in 5th century AD, perhaps due to the effects of intrusion of the sea.

It is interesting to mention that in the concluding lines of Abhidhammavatara, Buddhadatta the author of the work mentions, *"In the lovely Kaveripattana, crowded with hordes of man and women, not belonging to impure castes, endowed richly with all the requisites of a town .... beautified with many gardens, and in a beautiful and pleasant monastery adorned with a mansion as high as the peak of Kailasa, built by Kanhadasa ....while I was living in an old house there, by me, who am shining with qualities, such as writing beautifully on good topics ....this was composed and propounded extensively."* (*Abhidhammavatara, Slokas* 1409-13, *Buddhadatta's Manuals*, parts. I and II) After closing the benedictory lines some later writer has added the colophon: This work named Abhidhammavatara was composed by Buddhadatta Achariya, inhabitant of Urugapuram

The second stage of buddhist activity at Kaveripumpattinam began in the form of a temple or *caitya.* The *caitya* or temple excavated on the south of the *vihara* was itself a best example of early brick architecture in Tamilakam. Notwithstanding mainly the foundation and ground floor pattern of the structure is preserved at present, it reveals about the techniques of the construction in brick, adopted at that time The temple, square on plan, built in *panca ratha* (five offsets) style stands on a lofty plinth of brick, moulded in *jagati, padma,* etc. The rectangular or 'L' shaped as well as the small square voids in the foundation, together with the large central squarish void, would seem to suggest that the structures stood in more than one storey height and perhaps the shrine was erected by a astylar corbelling principle with staircases leading to the upper floor from all the three sides except the east, towards which it was obviously facing. The sanctum was perhaps on the top of the central large square void, after spanned the void by corbelling method from four sides and provided with floor of the sanctum on its top. Except the traces of granite members, perhaps used as pillars the entire structure was of brick and stucco. The walls were originally decorated with the moulded bricks and stucco ornamentation. Traces of a few layers of paintings have also been noticed on the stucco pieces recovered in the excavation. There is no clear evidence about the height and number of storeys of the structure. But it could not be less than two or three storeys, as it was not uncommon in brahmanical temples of the age. The temple might have been surmounted by a vimana either on the form of kutakara or curvilinear. The available evidence suggests that the temple seems to be later than the *vihara* and can be dated to the sixth-seventh century AD. (IAR, 1978, pp. 32-33) The nature of some of the numerals engraved on the stone pillars, the style of the stucco heads of *vyalas* and *dvarapalas* etc. found in the site also indicate that they are of 6- 7th century AD.

Melaiyur, located very close to the site, where the Buddhist remains excavated at Kavirippumpattinam had yielded beautiful bronzes including a gilted Maitreya as a treasure trove in 1927. This place was a part and parcel of ancient Kavirippumpattinam. (R.Nagaswamy, 1992) Even today a locality in Kavirippumpattinam is known by the same name. This place is often wrongly located by scholars in the Nakappattinam region. (T.N. Ramachandran, 1954, p.52; C. Sivaramamurti, 1963, p.50) Dressed in royal attire and visualized as a crowned and bejeweled Boticattva, the gilted Maitreya figure has a socket beneath its hallow pedestal that suggest it would have fixed on either to the pedestal or *prabha* of a bigger image. This image is dated to 8th- 9th century AD. Perhaps the main image, by the side of which the beautiful Maitreya fixed originally is the magnificent gilded Buddha seated in front of a richly ornamented throne back, now in the collection of the Museum of Fine Arts in Boston. The Boston Museum acquired the screen on its own in the year 1967 and it was only some three years later that the Buddha image appeared in the market. When it became evident that the two belonged together, the Museum acquired the Bronze and placed it against its original backdrop. (*Marg,* vol. XXXIX No.4, pp. 54-74) Originally it was flanked by two Boticattva probably Avalokitesvara and Maitreya.

The Decline of Buddhism in Kavirippumpattinam

Buddhism seems almost wiped out of Kavirippumpattinam somewhere in the medieval period. Hardly have we got any traces of Buddhism after 8th-9th century. It is but natural; the reason propounded by the scholars for the vanishing of Buddhism in Tamilakam is also applicable to Kavirippumpattinam. In the subsequent period the maritime activities of southern Tamil land were shifted to Nakapattinam and it served as a major port city under the Imperial Cholas. The decline of commercial activity at Kavirippumpattinam perhaps forced the trading communities to abandon or lessen their activities in this city. This in turn perhaps brought to a crisis for Buddhism as it was mainly patronized by the mercantile community. After Pukar, Nakappattinam being a prominent maritime centre under the Imperial Cholas took the responsibility of fostering the Buddhism and trade.

References

CILAPPATIKARAM (U. Ve. Swaminataiyar, (ed), Kapir Accukkutam, Chennai, 1950)

MANIMEKALAI (Na. Mu. Venkatacami nattar, (ed), Caiva Cittanta Nurpatippu Ka!akam, Chennai, 1992)

PATTINAPPALAI (U.Ve. Caminataiyar (ed) *Pattuppattu Mulamum Naccinarkkiniyar Uraiyum),* Kesari Patippakam, Chennai, 1950

ABHIDHAMMAVATARA, SLOKAS. 1409-13, *Buddhadatta's Manuals*, Parts. I and II, Pali Text Society Publication.

ANNUL REPORT OF INDIAN EPIGRAPHY, Archaeological Survey of India, 1991

INDIAN ARCHAEOLOGY 1964-65-A REVIEW (IAR), A. Ghosh (ed) Archaeological Survey of India, Delhi, 1969; *Indian Archaeology 1965-66-A Review (IAR),* A. Ghosh (ed), Archaeological Survey of India, Delhi, 1973; *Indian Archaeology 1972-73- A Review (IAR),* M.N. Deshpande (ed) Archaeological Survey of India, Delhi, 1978;

MARG-A MAGAZINE OF THE ARTS, Vol. XXXIX No.4,

Mulk Raj Anand (General Editor), Marg Publications, Bombay

INDIAN ANTIQUARY: A Journal of Oriental Research, XXII (1878), 1984 (reprint).

EPIGRAPHIA INDICA, Archaeological Survey of India, Vol. XXII (1933-34), 1984

R. NAGASWAMY, 1980. *Art and Culture of Tamil Nadu.* Sundeep Prakashan, Delhi

R. NAGASWAMY. "Buddhist Icons of Tamilakam" (paper presented in the International Seminar on Tamil Buddhism held at Chennai during 25-31 May 1992).

G.E. GERINI, *Researches on Ptolemy's Geography of Eastern Asia,* Asian Educational Services, New Delhi, 1992 (Reprint)

W.H. SCHOFF, 1974 (2nd edition). *The Periplus of the Erythrean Sea (Travel and Trade in the Indian Ocean by a Merchant of the First Century),* Oriental Book Reprint Corporation, New Delhi,

T.N. VASUDEVA RAO, 1979. *Buddhism in the Tamil Country,* Annamalai University, Annamalai Nagar

SMITH VINCENT, 1924 (4th edition). *Early History of India,* Oxford,

GANANATH. OBEYESEKERE, 1984. .*Cult of the goddess Pathini,* Motilal Banarasi Das, Delhi,

ALEXANDER CUNNINGHAM, 1962. *The Stupa of Bharhut,* Indological Book House, Varanasi,

R. NAGASWAMY, 1980. *Art and Culture of Tamil Nadu,* Sundeep Prakashan, Delhi,

ARUVAR, *PUMPUKAR* (TAMIL), 1975. Tamil Nadu State Archaeology Department Publications, Chennai,

SIR JOHN MARSHALL, 1982. *The Mouments of Sanchi,* Swati Publications, Delhi

T.N. RAMACHANDRAN, 1954. *The Nagapattinam and other Buddhist Bronzes in the Madras Museum,* Bulletin of the Madras Government Museum, New Series-General Section, Vol. 7 No 1. Chennai

C. SIVARAMAMURTI, 1963. *South Indian Bronzes,* Lalit Kala Academy, Delhi

G.V. SAROJA,. "History of Buddhism in Tamil Nadu", paper presented in the International Seminar on Tamil Buddhism held at Chennai during 25-31, May 1992

C.MINAKSHI, 1977. *Administration and Social Life under the Pallavas,* University of Madras, Madras

Y. SUBBARAYALU, JULY 1993. "Nagappattinam Cudamani Viharattaic carntta oru puttar patimam" *Avanam, itai,* 3

HUIMIN BHIKKHU ET.AL. "Buddhist Eminent Monks' Tracks and the Sea Silk Road in Tang Dynasty", Vietnam in the East Asian Buddhist Traditions and Electronic Cultural Atlas Initiative (ECAI), Vietnam Buddhist University, Ho Chi Minh city August 21 2007.

N.P.UNNI (edited and translated), 1974. *Mattavilasaprahasana,* College Book House, Trivandrum

THOMAS WATTERS, 1988. *On Yuan Chwang's Travels in India 629-645 A.D.,* Asian Educational Services, New Delhi, Vol. II

K.R.SRINIVASAN, 1964. *Cave Temples of Pallavas,* Archaeological Survey of India, New Delhi

T.A.GOPINATH RAO, 1915 (Reprint 1985). "Buddha Vestiges in Kanchipura," *Indian Antiquary: A Journal of Oriental Research,*

Conservation and environmental issues of Taj Mahal

D. Dayalan

Archaeological Survey of India

Keywords Taj Mahal, Fatehpur Sikri, Mehtab Bagh, Lakhauri Brick, Marble, Sand Stone, Inlay Work, Minarets

India has rich and kaleidoscopic variety of cultural heritages right from Pre-historic period spread all over the nation. The diversity of cultures and heritages in the country is not only a matter of great pride but also an irreplaceable source of spiritual and intellectual richness for all human kind. The protection and enhancement of cultural and heritage diversity in India should be actively promoted as an essential aspect of human development. Few thousands of monuments and archaeological sites in India are protected by the Central and State Governments and other agencies. However there are considerable numbers of unprotected cultural properties need the care of the public.

The paper deals in detail about the conservation and management issues of Taj Mahal, the most important tourist destination in India as a case study for better understanding the conservation and management aspects of cultural properties. In Agra there are three World Heritage Monuments. i.e. Taj Mahal, Agra Fort and Fatehpur Sikri in addition to quite a good number of other monuments. Out of these three, the first two are located on the right bank of river Yamuna in the city of Agra within a distance of about 800 m. in between, whereas the last one is situated 40 kms from Agra.

The Taj Mahal (1631-1648) was inscribed on the World Heritage List in 1983. Its inscription as a cultural property of universal significance was justified in the following words, *"The Taj Mahal is the greatest architectural achievement of the whole range of Indo-Islamic architecture. Apart from its tremendous romantic appeal, it is a masterpiece of architectural style, in conception, treatment and execution, all alike".*

Taj Mahal is surrounding by shops, green belt, emporium, offices, hotels, etc on the east and west; thickly populated Taj Ganj area and shopping complexes on the south and Yamuna river, Mehtab Bagh and green belt area on the north. Taj Mahal has to be visualized along with the riverbank, river Yamuna and adjoining monuments and landscape as they are the integral part of this monument complex. It is essential that the buffer zone between Agra Fort, Taj Mahal and Mehtab Bagh should be formulated to safeguard the landscape and the environment from any unregulated and inappropriate development. Hence the modern construction, mining, vehicle movement and pollution in the buffer zone area are regulated by various govt. rules, gazette notifications and Supreme Court directions from time to time.

The main mausoleum is placed on the bank of river Yamuna. All along the river there are series of well foundations with filling of rubble-masonry in between. The main mausoleum is raised over the marble platform which in turn resting on a low height red sand stone platform known as *chameli* platform because of the floor design on it. The wall of the structure of the main mausoleum is constructed of *lakhauri* brick with lime mortar. The mausoleum has double dome. There are four minarets on four corner of the marble platform. The inner walls, the lower dome and the outer surface of the mausoleum and the minarets are totally clad with white marble. The intrados of the upper dome was plastered up to the height of 18.59 mt. and above that the intrados was veneered with sand stone slabs. The sand stone was used for veneering the inner face of the minarets and also for the steps. Covering the entire surface of the lower part of the drum there is a regular and continuous series of eight relieving arches which adds to the strength of the structure.

Taj Mahal has a long recorded history of its conservation. The earliest record of its repair available in a letter dated 1652 AD from Prince Aurangzeb to his imperial father Shah Jahan wherein he pointed out some defect in the dome and vaults of the Mausoleum. Since then the conservation work of the minor and major nature were undertaken to the various building of the Taj Mahal complex from time to time as per the necessity and requirement. Now it is in good state of preservation. However due to its age, minor defect such as cracking of stones due to corrosion of iron dowels and clamp, weathering effect on the stones, decay of the mortar of the roof, missing of inlay due to human vandalism and dead of the adhesive and other minor defects are bound to occur. The environmental effect on the building, wears and tears and some degradation due to the movement of large number of visitor are some of the major issues.

The conservation is defined as an action taken to prevent any further decay. It includes all acts that prolong the life of our cultural and natural heritage. The minimum affective action is always the best. If possible the action should be reversible and not prejudice possible future intervention. Before taking up conservation to any ancient monument, it is necessary to study and understand various aspects such as: its location, environment, material used, method / style of construction, cause of deterioration and status of deterioration. Some of the ethics of conservation followed are:

- All the methods and material used during treatment must be fully documented.
- The historical evidences must not be destroyed, falsified, or removed.
- Any intervention must be the minimum necessary and reversible. The existing material has to be retained maximum possible.
- Any intervention should be in conformity with the aesthetic, historical and physical integrity of cultural property.
- No conservation work to be undertaken by the Conserver/Restorer who is insufficiently trained or inexperienced, unless they obtain competent advice.

Keeping these principles in view, a through study of Taj Mahal has been carried out and prepared an inventory on its conservation related aspects. As stated elsewhere that Taj Mahal complex are built of brick and stone masonry with lime mortar and covered with marble / sand stone as a protective cladding to prevent deterioration of the masonry. The study of properties of the building materials are as follow:

Lakhauri Brick

The *lakhauri* bricks are perfectly baked and fed into hot fat mixed with some materials which were absorbed into the porous body of the bricks. This thin, almost imperceptible over coat made the brick adequately water proof. The bricks are examined and the test report is below:-

(a) Colour	orange to red
(b) Average water absorption	10.13%
(c) Average compressive strength	159.00 kg/m3
(d) Crushing load	33.00 MT
(e) Density	2077 Kg/m3
(f) Size:	(a) 18 x 11 x 2 ½ cm
	(b) 19 x 11 x 3 cm

Mortar

The mortar was prepared by mixing 1:1:2 (1 lime: 1 coarse sand: 2 *surkhi*), The lime used was hydraulic (slaked lime free from lumps & other possible impurities). Some ingredients like *gur* (jaggery/sugarcane juice), *urd* (kind of pulse), *tentile*, *batasha* (sugar bubbles), *baesan* (pulse), *rumi-mastagi*, *belgri*-water, jute, gum and *kankar* were also added. The cementing material, which had thus been prepared, was very strong and enduring with bounding material of best quality. The size of the brick used in the structures was reduced so that mortar could occupy a greater cubic volume of masonry. The thickness of mortar joints is observed to be 1 to 2 times the thickness of the bricks used in the construction.

Inlay Stones

The main decoration in the Taj Mahal complex consists of intricate inlay, relief carving and incised painting. The inlay of floral arabesques fashioned from precious / semi-precious stones in white marble and the inlaid black marble inscriptions that frame the central vaulted portal or *iwan*, commanding both in form and in content are some of the prime attraction in Taj Mahal. The treatment of the marble surface like soft muslin embroidered richly with *pietra-dura* work and multitude surface details are the testimony to the fine quality of the monument. The precious / semi precious stones used for inlay works are: Alabaster *(Sang-i-Rukham);* Agate (*Sang-i-gawa, sang-i-gheri, sang-i-panghan)*; Amethyst; Black marble (*Sang-i-Musa, sang-i- moosa)*; Bloodstone *(Sang-i-Gad; Sang-i-Dogasra);* Carnelian of Arabia Felix *(Yamani);* Chalcedony; Chrysolite/ Cat's eye stone (Lahsunnia); Conch-shell *(Sankh)*; Coral *(Moonga);* Cornelian *(Sang-i-Aqiq);* Crystal *(Billoor);* Diamond *(Hira);* Emerald *(Zamurrad);* Garnet *(Tamra, Sang-i-Tandra); Ghar; Ghori; Godar;* Golden stone *(Tilai);* Green Marble (Pazhar); *Gulabi; Jade (pitunia);* Jasper*(Sang-iBadal;Sang-i-Ratak;Sang-i-Galahra;Sang-i-Udha); Khara;* Lapis Lazuli *(Sang-i- Lajward);* Load stone *(Maqnatis);* Magnet Stone (Nakhod, Maknatis); Melachite *(Dana-i-Farang)*,; Mother-of-pearl *(Sip)*,; *Nakhud;* Onyx *(Sulaimani, Sang-i-Lahsania);* Onyx Marble (Cloud stone, *Badal, Sang-i-abri*); *Pai-Zahar; Pankhni;* Pearls *(Moti);* .Pink Marble; *Pitunia;* Redstone *(Surkh); Reg;* Ruby *(Yakut); Sang-i-Ajuba/ Ujuba,* a kind of Turquoise (It is mottled with green and yellow concretions in a brown ground-mass); Sapphire *(Nilam);* Slate (Calcareous shale); *Sumaq*,; Topaz *(Pukhraj);* Turquoise *(Firoza);* Turquoise *(Sang-i-Abri);* Turquoise *(Sang-i-Chitla)*,; Yellow Marble *(Khattu, sang-i-zard,* etc.

Marble

The marble used in the Taj Mahal is of varied typed although quarried from the same area, Makrana, with differences in colour, shade and grains. At least three to four difference in quality has been observed. The analysis of few pieces of marble indicated that they are not a pure variety of marble but, a dolomite one containing carbonates of calcium and magnesium as well as silica, iron, arsenic, tellurium etc. The veneering stone of the main mausoleum particularly the area of marble that was used for inlaying are of superior quality than the others. The spot or streaks of different natural colour are found on the slabs of marble mainly because of the presence of mineral impurities. The iron dowels or clamps were used for fixing the marble slabs in position. The different types of effects observed on the marble surface of the monument are:

1) Discoloration of the surface – either uniformly or in spots.
2) Breaking of marble slab edges.
3) Formation of cracks in marble surface.
4) Bulging of marble slabs.
5) Pitting or erosion of marble surface in some areas.
6) Loss of translucent shining.

Different types of studies were made to asserting the root causes of the above problem. They are:

1) Petrologic study to identify the elements present in the marble.
2) Study of crack formation pattern, porosity, soluble salts present in the marble, absorbing capacity and analysis of dust accumulation at various places on the surface of Taj Mahal, Mineralogical study, study of SPM, etc.
   Cracks of various types and magnitude are seen on the marble surface and the possible reasons for the cracks are as follow:

1. Inherent defects in the stone developed due to the residual stress at the time of quarrying, trimming etc
2. These vertical cracks in the marble could be due to the down load pressure exerted by the slabs placed above.
3. The mineral impurities in the marble have got preferentially leached by rain water and other factors leading to the formation of cracks.
4. Settlement of the foundation
5. Chemical weathering
6. Insulation weathering
7. Due to rusting of iron dowels Loss of translucent shining surface, formation of cavities and pits on the surface, creaking and flaking of the surface, etc. are happen due to various factors including environmental pollution.

Sand Stone

Sand stone in general consist of quartz, feldspar with mica. Sand stone (sedimentary rock) is being used in Taj Mahal of different tint & colour though reddish with or without spots are used for construction whereas buff colour are also used alternatively with red one for the floorings. The sand stone came from Tantpura, Paharpur, Karuli– Hindaun, Sar-Mathura and also perhaps Fatehpur Sikri. The sand stone deadly attacked by atmospheric pollution due to alteration of feldspar. Due to the atmospheric pollution the possible effect on sand stone are the weight loss, water absorption and loss of compressive strength. Sand stone became little dark due to the formation of Gypsum skin due to the effect of Sulphuric acid (H_2SO_4). It plays disaster role in the reduction of compressive strength and flaking. Attack of algae on sand stone surface creates black patches.

The structural stability, the nature of foundation and other constructional aspects of Taj Mahal have been studied in detail by various reputed organizations/ agencies in addition to the Archaeological Survey of India. Each and every part of the mausoleum has been measured meticulously and systematic drawings are also prepared. The height of the drum from top of the ceiling to the springing of the dome is measure to 11.45m and the thickness is 3.96m -1.07m. The dome above is 22.86m up to the base of the gilded finial. The finial of the dome is 9.89 metre high. The drum is circular on plan with internal and external diameters of 18.4 m and 26.52m respectively. The dome together with the drum is built up of brick in lime faced externally with alternate headers and stretchers of white marble and internally with plaster except for the top 10.40m where it is lined with red sandstones. The weights of dome and drum are calculated and the figures arrived approximately are 12,000 and 6,000 tons respectively, which would produce a compressive stress of 4 to 6 tons per square foot at base of dome and base of drum respectively.

It is necessary to know the type, dimensions and the depth of the foundations as well as to establish the boundaries of the subsoil strata beneath the structure for the analysis of the performance of the foundation of Taj Mahal. The Geo-technical investigation was conducted with the intention of obtaining the details like subsoil strata beneath the structure including the shear strength, compressibility and elastic properties of the soil strata; impounding of water on the behaviour of subsoil; probable type, dimension and the depth of the foundation; performance of the foundation, etc. The study of foundation and stability of structure revealed that Taj Mahal is raised on soiled foundation made of stone rubble and mortar. The approximate thickness of the foundation is 12m to 13m. Series of ring wells foundation are provided all along the river side to prevent erosion by the river water. The gaps in between the wells were filled with rubble-masonry. The soil mass below the masonry raft of 34.60m thickness consists of a sandy layer of 9.44m thick and a clayey stratum of 65.60m. thickness resting on rock. The soil below the monument is in the submerged state due to the high water table level. The settlement analysis based on consolidation test data indicates that perhaps 99.4% of the anticipated ultimate settlement may have taken place already and the future settlement, if at all, is likely to be insignificant. The stress analysis indicated that the entire foundation system is in the state of vertical compression. There are however zone of small tensile stresses in regions around the contact areas.

The four free standing minarets at the corners of the platform which has mausoleum in its centre look like ladies-in-waiting to the beautiful princess is not only a innovative design hitherto unknown to Mughal architects but also open a new avenue in the architectural vocabulary/ In order to monitor the verticality of the minarets and other edifices, a number of bench marks are fixed on the surface of different structure of Taj Mahal. Bench marks 1 to 88 are situated on the ground floor of Taj Mahal and 89 to 104 are located on the first floor. Geodetic and Research branch of Survey of India has carried out periodical surveys especially triangulation, traverse and precision leveling to check the verticality of its four minarets and other structures since 1940-41. From the above triangulation, traverse and leveling data of the years 1940-41, 1952-53, 1957-58, 1959-60, 1964-65, 1971-72, 1976-77, 1986-87, 1994-95 and 2004-05, .it can be concluded that all the four minarets, the dome and the other structures of Taj Mahal remained quite stable in plan as well as in elevation from 1940-41 to 2004-05. The changes found are within the standard errors of geodetic observations.

Taj Mahal and other monuments at Agra are not only the prime destination of tourism but also the source of economy of the nation in general and Agra in particular. These monuments are directly or indirectly responsible for the creation of many tourism based jobs/establishments like Hotel & Restaurants, Shopping complexes, Handicraft Industries, Tourist operators, Transport operators, Guides, Photographers, hawkers, etc. Taj Mahal is visited by 6000 to 8000 people every day. Further it is more on Saturday and Sunday due to holiday as well as closure of Taj on Friday. Tourist inflow is very high in winter months i.e. 10,000 to 15,000 visitors per day. Due to heavy tourist rush long queues can be seen at the ticket booking windows and at the security checking points. The Archeological Survey of India has proposed to establish a Visitors' Facilitation Centre at the Fatehabad Courtyard (popularly known as Gaushala) located near the Eastern Gate of the monument in consultation with the international and Indian experts after conserving the historic buildings of the Gaushala complex without any alteration. Basic tourist amenities like drinking water, lockers, toilets, departmental publication sales counter, touch screen information kiosks and First Aid Centre etc. will be provided in the proposed Visitors' Facilitation Centre.

Use of re-projected photos in the conservation of the Livonian order's castle in Cesis (Latvia)

Arturs Lapiņš

Architectural Investigation Group Ltd, Riga, Latvia

Abstract

Conservation of the medieval Livonian order's castle in Cesis, Latvia has been one of the largest conservation projects recently carried out in the territory of Latvia. Stone walls of the castle, originally designed to be protected by roofs, for almost 300 years have been exposed to the weather in the form of ruins and required consolidation. Planning of the conservation works required preparation of the construction documents. Until recently systematic metric survey of the castle was not carried out. Within the available time schedule, specifics of the recording of ruins and budget, the method of re-projection of photos was chosen, supplied with hand drawn sketches for specific details. According to the specifications, the conservation works of Cesis castle were tendered out and successfully completed in 2008. The experience gained in the preparation of the survey material and the conservation documents can be used in similar objects.
The realization of the conservation project of Cesis castle was co-funded by the European Foundation for the Regional Development.

Keywords Cesis, Latvia, medieval order's castle, architectural recording, metric survey, close range photogrammetry, re-projected photos, conservation of ruins, specification of conservation works,

Introduction

The ruins of the medieval castle in Cesis (Wenden) is the best preserved castle ruin complex within the territory of Latvia. The rich history of the castle and spectacular outlook place it among the most outstanding and interesting tourist attractions of the Baltics.

Construction of the Cesis Castle begun by the beginning of the 13th century. It has been reconstructed several times at the end of 14th to the beginning of the 15th century, and by the end of the 15th to the beginning of the 16th century. During the last reconstruction the castle acquired the layout of the so called Convent - house, where several wings are grouped around the central courtyard. The castle has been in ruins since the beginning of the 18th century (circa 1703), when it remained unrepaired after damage due to warfare and fire. Up until the mid 1800s the castle ruins and the surrounding area was used for utilitarian purposes. Since then the castle has become a beloved place for tourists, attracting thousands of visitors with the romantic view of the ruins. By the beginning of the 21st century, masonry walls, originally not intended for direct atmospheric impact, had gradually eroded and required consolidation.

Since 1970ies due to extensive archaeological excavations, the goal of conservation has been stabilisation of the uncovered constructions. Recent short - term conservation program of 2005 - 2006 included reconstruction of the three essential features of a medieval castle - the Bridge, the Entrance and the Yard (Lapins, 2005). In 2008 a more extensive conservation of the upstanding structures was planned. All of works dealt with conservation of medieval structures, therefore required preparation of conservation construction documents, specifications, commencement of works and authors' supervision. Ruined walls, due to their irregular character, have always presented a challenge to surveyors and architects for identification and reconstruction. A further challenge was how to present design information to the actual builder and to the conservation craftsmen.

Surveying of the Ruins

In general, the architectural recording of the historic buildings includes plans, sections and elevations, with drawings of architectural details. Conservation recording also includes cracks, levels, and damages of the surfaces. In case of ruins, conservation drawings also comprise considerations of the restorers regarding the technical state and the required actions for separate areas or even separate stones or bricks. It is necessary to be able to reveal both - broad view of the object and detailed look at the details. Furthermore, the conservation craftsmen has to be able to recognize the element, to accomplish the required actions. Over the course of preparation of the conservation documents, several methods of documentation have been tried out:

- architectural survey, based on direct measurements, has been used since the 1950s. Ongoing archaeological research and the changing landscape of the castle, has shown this method to be rather inefficient, mainly due to lack of permanent reference points. Therefore it has been difficult to overlay surveyed drawings made in one year with the drawings made, e.g. one year earlier. Nevertheless measured drawings, done by architects and archaeologists, comprise valuable observations for conservation and also present a kind of "evaluative" documentation, which will be discussed further on.

- delineated drawings, prepared by professional surveyors, on the basis of ortho-photos, proved to be of rather limited use. Conservation drawings require exact identification of building materials and correct interpretation of the remains of construction. Yet without special training the mission seemed to be impossible.
- ortho-photos, prepared by professional surveyors, on the basis of which theoretically an architect could do the delineation. Yet, although preparation of ortho-photos relies on a technical process, the process also implies photography, i.e. art of capturing the light and shadow patterns. Lighting of the ruins as an open-air object depends on the sun, so to achieve appropriate results, the surveyor has to be acquainted with basic concepts of architectural photography. Besides a certain degree of interest and a sound understanding of the subject is needed to depict the information required in the final material.

Fig.1. Traditional *versus* digital metric survey. Top - digital survey of the Brick structure.
Bottom - measured drawing of the arch. L.Stukmanis, 1980.

- Recent activities also included survey of the castle, using modern laser scanning technology. The acquired drawings, delineated on the basis of the point cloud, are claimed to be more accurate in global scale. Nevertheless extremely irregular ruinous constructions seem to be too complex for the chosen method or the accuracy is too low. Architectural details of the castle, openings, loose stones - all of them are Being extremely accurate, but rather "conditional", these drawings are of very limited use for the purposes of the preparation of conservation design documents, which can be judged from the Figure 1, where "traditional" drawing of the Bridge can be compared with the results of point cloud interpretation.

None of the listed documentation approaches could serve as a basis of self - explaining construction documents as well as effective communication with the craftsmen. Instead of preparing the 'total survey of the castle' or conditional plans, the task oriented documentation was necessary. Therefore documentation was included in the process of preparation of the design and commenced by the authors themselves. In case of Cesis castle, the method of re-projected photos with delineated additions and specifications turned out to be the most appropriate in terms of speed, handiness and within the available budget.

Re-projection of the photos

Until recent times, manipulation of the photos was a challenge. For re-projection of, e.g. aero-photos, huge mechanical devices were constructed. Computer assisted image manipulation has changed the methods. One of the earliest and most successful approaches were realized by professor Helmut Dersch of the University of Applied Sciences at Furtwangen, Germany, releasing a set of command line programs called Panorama tools (Dersch, 1998). Originally aimed at the transformations of perspective and the creation of virtual reality applications, the programs (or helpers, as they were originally called by Dersch) allow wide range of controlled image transformation (Figure 2).

Besides the program allows merging of several images into a single file, which later eases handling of both vector and raster files. Panorama tools have been already successfully used for the correction of the optical distortions of low-end digital camera photos in the remote sensing study on Easter Island (Bell, 2005). In case of Cesis castle, perspective corrections of the photos have be carried out, by defining vertical and horizontal lines within an image. On the opposite to stretching and correction of photographs in, e.g. Gimp or Photoshop, transformations by Panorama tools are based on the EXIF data of the image and the calculations of the inner geometry of the camera. Using graphical user interfaces like Ptpicker (Dersch, 1998) or Hugin (d'Angelo, 2009) it is possible to define the points directly on photographs. By pointing several horizontal and vertical lines the overall accuracy of the process can be improved. Least, but not last, is the option, that both Hugin and Panorama tools are issued under the GNU licences (Panorama tools - under the GNU Lesser Public Licence).

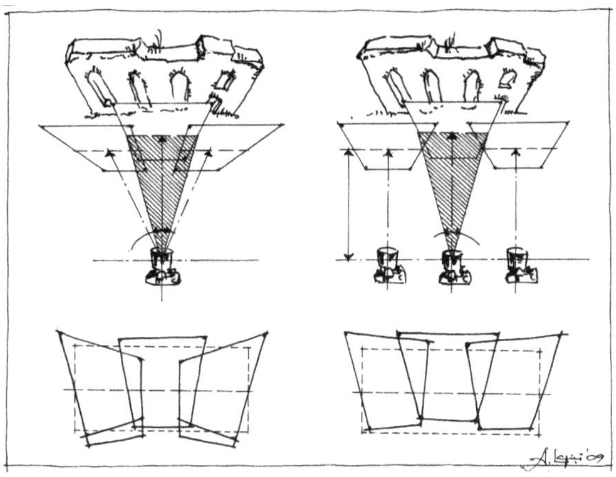

Fig. 2. Methods for acquiring the photos for re-projection and transformations by Panorama tools. From left to right - central point perspective and parallel projection.

Several commercial applications available on the market also allow similar transformations of the perspective of a single image. A German company Kubit (www.kubit.de) offers PhoToPlan - a commercial plug-in for AutoCAD, which operates from within AutoCAD. Cad Overlay ir a similar program Similarly, a plugin Phidias provides the same functionality for Microstation. Yet, compared with Panorama Tools, the applications do not seem to offer camera calibration, do not allow merging of several overlapping images and are commercial.

Building the System

The general workflow of the process of preparation of the conservation documents is given in Figure 3. The photos of, e.g. inner yard, measuring 20x30m, were taken by the conservation architect from the top of the wall, which is ca. 12m high.

The upstanding walls were photographed from the ground, at a distance approximately equalling the width of the wall, where possible. In case such a distance was not possible, several photos from the same viewpoint were taken, rotating the camera around optical axis of the lens to avoid parallax. Standard digital camera Canon 400D was used. The resulting image was re-projected in Hugin, either straightening single image or joining several photos in a single re-projected image. Vertical and horizontal lines were defined, referring to archaeological grid (small sticks, put into the ground) or, in case of upstanding constructions, vertical features (e.g. sides of window aisles) and levelled horizontal marks, glued on to the walls. All the actions were undertaken by conservation architects, sometimes returning back to the object several times to achieve the best possible results.

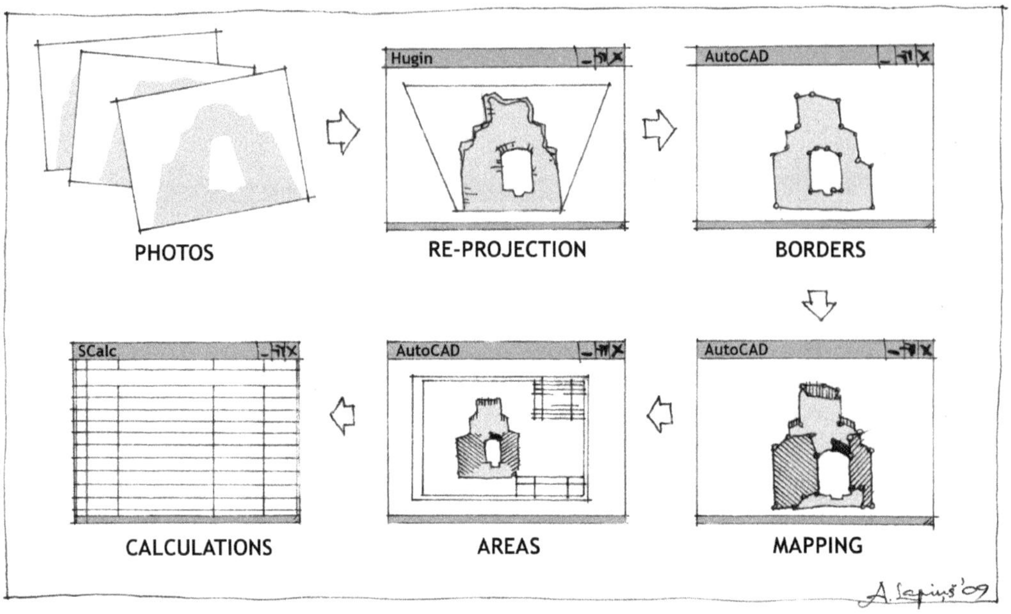

Fig. 3. Workflow for the preparation of the construction documents. The digital photos are re-projected in Hugin, inserted in an AutoCAD drawing, scaled, delineated, mapped and calculated. The resulted amounts are used to prepare the bill of quantities.

Software chosen for the preparation of the conservation drawings was AutoCAD.Although other vector applications have become available, AutoCAD can still be considered the standard vector drawing program in construction, and therefore also in conservation design field. Possibility to import the raster image into the drawing and to delineate over it was another crucial option. So the construction drawings for conservation works of the Castle were prepared on the basis of re-projected photographs and organized in the layers, as shown in Figure 4.

AutoCAD also historically provides relatively simple, but powerful scripting environment - Autolisp. Besides the possibility to manipulate drawing entities, Autolisp also allows creation of external data files. These tools were used to produce the construction drawings and accomplish the calculations for specifications. For quantity surveys custom scripts in Autolisp language were elaborated. These scripts had been previously used in mapping of stone decay in Riga Cemetery of the Brethren (Sidraba *et al.* - 2008). Mapping of the decay and the required conservation measures were defined by the conservation architect in collaboration with the building archaeologist. Each area was delineated with a continuous line (AutoCAD polyline), put on a separate layer. The names of the layers where standardised according to the following pattern: <three letter code> + <_(underline)> + <description of the conservation action> + <_(underline)> + <number, indicating whether the decay has to be calculated as area, or as linear distance (for cracks and loose joints)>. Therefore a typical layer name would read as follows: <AB1_Erosion of the core of the wall_2>. Using the Autolisp script, the total area of the decay and the amount of the conservation actions was calculated.

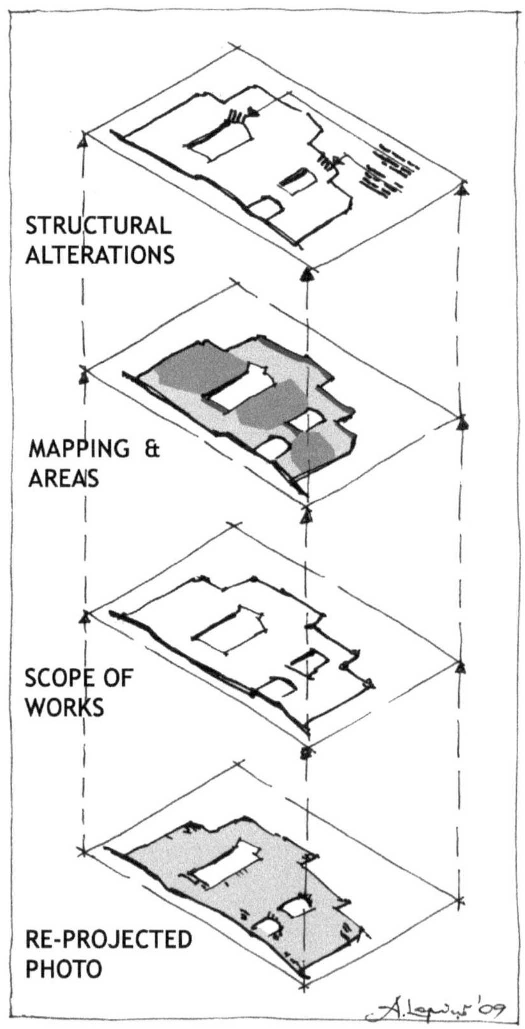

Fig. 4. Organization of the data in an AutoCAD file. Photo is.

Code	Type of decay	Amount	% of total
AA2	Erosion of the core of the wall	10.27m2	3.97%
AB2	Erosion of the shell of the wall	43.68m2	16.87%
CA2	Exposed horizontal plane	4.37m2	1.69%
Total area of decay:		58.32m2	
Total area of the wall:			258.95m2

Table 1. Example bill of quantities.

These figures were enough, to evaluate the overall state of the construction, calculate the necessary quantities and bill of required materials and to tender out the works.

Accuracy

The most important question - how much we can rely on re-projected images? How substantial in close range imaging is the distortion of the non-metric camera lens or errors from ill located points on the image? Looking at the image of the Yard, where re-projected photo is put atop the topographical drawing, one can observe two types of inaccuracies - in inconsistency between the photograph and the drawing and vice versa (Figure 5).

The first type shows the places, where the surveyors of the topography have simply misunderstood what they have measured. The other type shows deviation of the photo from the measured image, increasing as the Yard slopes down towards the upper side of the image. With the size of the Yard ca. 20x30m, the maximum inaccuracy in terms of exact location of some points lies between 10..12cm or 3..5%. This accuracy, ensuring, that what is drawn, is drawn correctly, is absolutely relevant in a conservation project, where overall impression is more important, than minute accuracy in details. During the realization of the project, all the restitutions were done according to the traces in the original stone work, rather than specific dimensions. These traces could be observed *in situ* by both - the architect and the restorer, and they were to be strictly followed. It can be stated, that in case of conservation of ruins, the correct interpretation of the clues is more important than absolute mathematical accuracy of the documentation.

Fig. 5. Yard of the castle. From top to bottom - photo of the Yard from the top of the wall of the East wing, photo, re-projected in Hugin and photo placed below the topographical survey.

Advantages and limitations

Modern laser scanner technology claim its accuracy being ca. 2mm. It is definitely true, judging if any single point of the so called point cloud is taken into considerations. Yet the point cloud is of very limited use in the conservation design process. So the extremely accurate data have to be projected and interpreted by an operator. Here we approach the "weakest link" in the process. So far no drawings of the ruins, delineated on the basis of point cloud, could have been used for conservation purposes. The 3D model - the by-product of point cloud scanning, whatever effective in presentation of the accomplished scanning works, cannot be directly interpreted by conservation specialists. Both the architect and the restorer deal with single planes and to communicate the information with them, traditional 2D drawing still is the most effective way of transferring the conservation design through the hands of the craftsmen to the restored object. Whatever detailed the drawings might be, in complex elements like ruinous walls, photo images, taken by the conservation architect or his team, most effectively help the actual restorer to identify the details described in the project.

As all single image methods, re-projected photographies can be used on relatively flat planes. Even small deviations in plane seem to give uncontrollable errors in terms of accuracy. Yet, even not so accurate, the photo of the wall serves as excellent reference not only for commencing the works, but also for documenting the changes. The experience gained in the preparation of the conservation documents and supervision of the conservation works can be used in similar objects.

References

D'ANGELO P. (2009). Retrieved from http://hugin.sourceforge.net/. Last accessed 13.06.2009.

DERSCH, H. (1998). Retrieved from http://www.all-in-one.ee/~dersch/. Last accessed 11.06.2009.

SIDRABA *ET AL.* (2008). Sidraba I., Purvins E., Lapins A., Strupule V., Kalinka M., Krage L. - Documentation System for Continuity and Traceability of State Condition of Cultural Heritage Objects. In *Preprints of the 8th Triennal Meeting for conservators of the Baltic States.* Tallinn, 2008. - pp. 60 - 65.

BELL M. (2005). Archaeology on Easter Island: Remote sensing for low-impact archaeological study of Rapa Nui's (Easter Island) monumental stone architecture. Retrieved from *www.spacegrant.hawaii.edu/reports/15_FA05-SP06/**Bell**_Matthew_FA.pdf.* Last accessed 15.06.2009.

LAPINS A. (2005). *Renovation of the castle ruins in Cesis. Program 2005 - 2006. Technical design.* Riga, Architectural Investigation Group, 2005.

Damage documentation on Shore Temple, Mamallapuram, India

Anupadma.R and M.S.Mathews

E-mail : anuthil@gmail.com, mathews@iitm.ac.in
Building Technology and Construction Management Division
Department of Civil Engineering
Indian Institute of Technology-Madras, Chennai – 600036, India

Abstract

Stone monuments occupy a significant role in the world's cultural heritage. These monuments and their artistic intricacies are the exhibits, which shed light on the ancient cultural heritage of the country. The severe damages and loss of these monuments due to environmental conditions, climatic factors etc. are on the increase. Especially in ancient countries like India, care of monument is very much required for the preservation of the built heritage. In order to have a clear understanding of the damages to the monuments, detailed assessments are required. Detailed assessments are further helpful in the experimental studies and in arriving at the appropriate preservative measures. One of the world heritage monuments in India, the Shore temple located at Mamallapuram, India has been taken as a case study for such an assessment. The monument is situated on the sea shore of Bay of Bengal. The damages due to wind erosion and salt decay in this monument are alarming. Detailed assessment is carried out on the worst affected wall facing the sea. The damages are categorized, mapped and quantified using digital mapping software called metigo MAP 2.2 and the results are discussed.

Keywords damage mapping, Shore temple, Alveolar, Pitting, weathering forms, Mamallapuram, Roughening, Back weathering, Rounding, soiling.

Introduction

Enduring nature of stone has made ancient civilisations to commemorate world's cultural heritage and historical events in the form of monuments. Monuments are not just historical records; they also embed a great form of art that visually communicates the intended message over many generations amidst historical linguistic changes. Though stone monuments are considered durable, they are susceptible to increasing damages caused by climatic conditions, pollution, saline environment etc. Hence a need arises for conservation, which requires thorough understanding of material properties and cause of these damages. A detailed assessment of these damages is further helpful in experimental studies that can determine the cause of these damages. A case study on Shore Temple at Mamallapuram, which is one among the world heritage monuments in India, is taken for study here

Fig. 1. View of Shore temple, Mamallapuram.

The Shore temple, Mamallapuram (Figure 1) is one of the first stone temples in the South Indian Dravidian style. It was built half in the sea and half on land, in the 7th – 8th century AD and was used for ceremonies in which water played a major role (Henri Stierlin 1994, 58). The circumstances in which the monument was built made it necessary to deviate in certain respects from the standard ground plan and vertical section.

The principle shrine has its entrance on the eastern side. In the circuit wall of the temple an opening was left at a point that could not be reached from the shore, through which mariners could catch a direct glimpse of the cella. This cella contains Shiva lingam and it is dedicated to lord Shiva. To the west of the main shrine is the small shrine with a statue of Vishnu in the cella, dedicated to Vishnu (Henri Stierlin 1994, 136). This monument also acted as a landmark for mariners. However the destructive actions of sea by wind erosion and shifting sand dunes are ruining the temple.

Fig. 2. Showing Various Major stone types and pointing work in southern facade of Big shrine, Shore temple Mamallapuram.

Fig. 3. Efflorescence.

Fig. 4. Soiling by droppings.

Fig. 5. Back weathering.

The ancient builders were relatively inexperienced in erecting the temple, but the basic principles of Hindu architecture have been well preserved. The best variety of locally available granite has been used to construct the temple, which is capable of withstanding wind and environmental forces for long years. The foundation and base were built with bedrock and a layer of stronger variety of granite stone has been laid as a damp proof coarse (Figure 2). The superstructure is built with relatively weaker granite variety.

Susceptibility of stone to the marine environment depends on the micro structural properties and the material properties. Keen mapping on various weathering forms and its intensity will give a clear picture on various types of damages, which in turn will help in further investigations. Here, the worst effected side (Southern side) of the main shrine is taken to study the damages due to marine environment.

Weathering forms

The following weathering forms are found in the Shore temple, Mamallapuram. Classifications are made as per Fitzner's classification of weathering forms (Fitzner et al 2001, 11)

Efflorescence
Loose salt deposits of salt aggregate found on the stone surface (Figure 3)

Soiling by droppings
Soiling due to deposits of droppings from birds, bat, etc. (Figure 4)

Back weathering
Uniform loss of stone material parallel to the original stone surface (the fresh stone surface of the building block) is called "Back weathering" (Figure 5)

Alveolar
Large depressions or pockets formed on the stone surface. The form is comparable to honeycombing. (Figure 6)

Fig. 6. Alveolar.

Fig. 7.Pitting

Fig. 8. Roughening.

Pitting
Small pits found on the stone surface due to corrosion of wind (Figure 7)

Roughening
Alteration of gloss due to corrosion or loss of smallest stone particles on smoothed stone surface. (Figure 8)

Rounding
Rounding of edges and features. (Figure 9) concave or convex soft forms of relief works

Fig. 9. Rounding.

Damage Mapping

Photomapping involves the process of drawing a map from a photographic base. In all likelihood, computer map making, including analogue processing, will have replaced manual map making for well over 90 percent of all maps Photomapping is the process of making a map with cartographic enhancements that have been drawn from a photomosaic. Photomosaic is nothing but the assemblage of individual photographs, which are rectified for tilt and brought to a common scale *(Estes et al 1977, 441).*

Fig. 10. Working stills for Photo montage.

Photo Montage

About 250 photographs have been taken at various levels and parts of the southern side of the big shrine and montage work (Figure 10) has been done to avoid skew or any other abnormality in the photographs. Montage is nothing but mounting a number of photographs of a single object brought to a common scale which have been rectified to tilts, skew or 3d projections.

Software Used

Metigo MAP is the ideal software solution for conservators, restorers and urban preservationists to perform digital mapping in the rectified image/photomap. In the context of restoring and conserving work the documentation forms a substantial focus. metigo MAP offers the possibility to create equalized pictures directly at the building site or in the studio.

MetigoMAP was developed in close cooperation with conservators and restorers. metigoMAP 2.2 is a software solution for digital rectification and subsequent mapping in the rectified photomap. metigoMAP is offered since 2001. Since 2004 an English version has been in use. It is a combination of image processing and CAD-functionality which the program. It involves digital rectification of the image, damage mapping and quantification. The screen shot of the mapping done with the help of metigoMAP is given below (Figure 11).

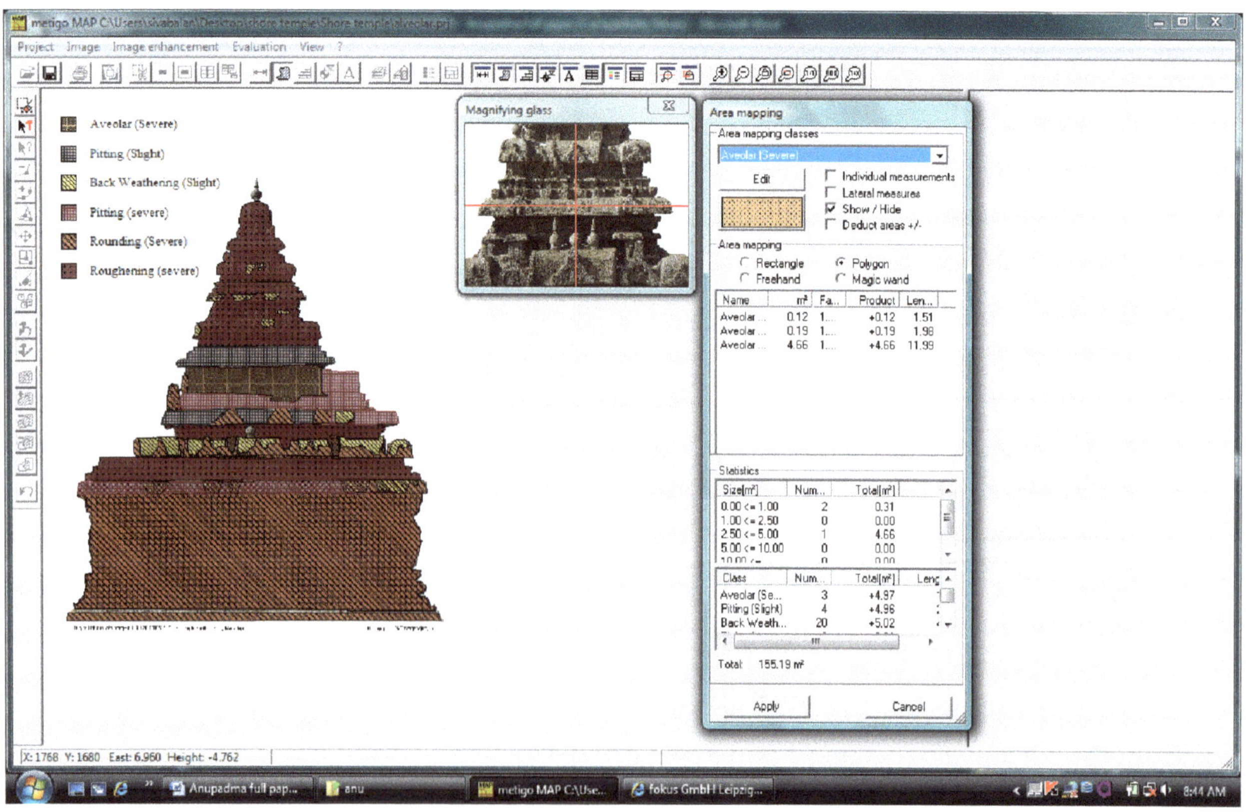

Fig. 11. Screen shoot of mapping with metigoMAP.

Mapping of the southern Façade

By mapping all weathering forms are grouped and registered systematically. Weathering forms are grouped into two. Group 1 (Figure 12) consists of weathering forms which leads to loss of stone material. Group 2 (Figure 13) consists of discoloration and deposits. Alveolar, pitting, rounding, roughening and back weathering (scaling) are the weathering forms grouped into group 1 and efflorescence and soiling due to dropping of birds bat etc. are grouped into group 2. The combinations and intensities of the weathering forms are mapped with different representations and the area of each damage type is quantified with the help of the software.

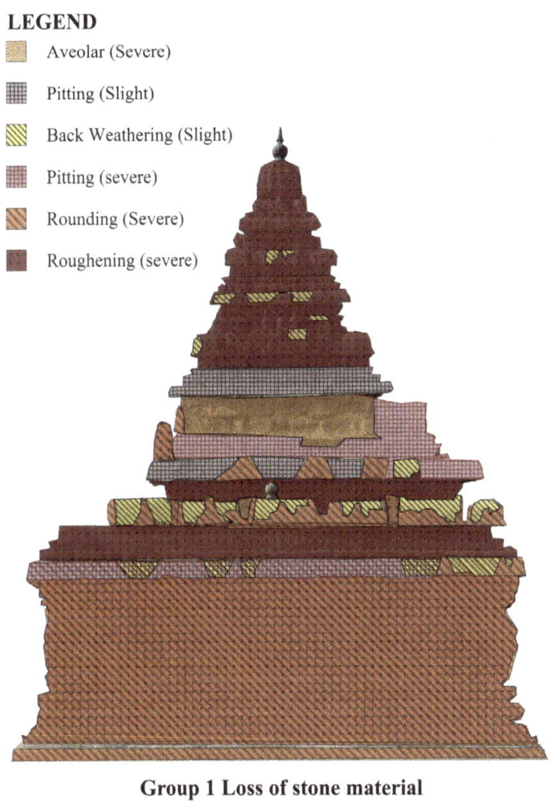

Fig. 12. Group 1 Loss of stone material.

The image is first rectified to the real scale and the damages/ weathering forms are mapped in order to have precise quantification. Damage mapping with metigoMap 2.2 involves the following steps, 1. Load the image to be mapped for damages, 2. Rectify the image to the true scale, 3. Map each type of weathering forms in separate layer, 4. metigoMap will quantify the damages (in cm or m) as we map the area of weathering forms, 5. Label the weathering forms and mark them with legend, 6. Quantified areas of weathering forms are taken to excel sheet and tabulation, bar charts etc. are plotted.

Quantification and Discussion on Weathering Forms

Quantification (Figure 14 and Figure 15) has been carried out and the results are given below. The quantification of the weathering forms using the metigoMAP software reveals that roughening and rounding which are the predominant weathering forms in this case. As the monument is situated in the marine environment the susceptibility of stone towards salt decay is also found prominent.

Fig. 13. Group 2 Discoloration and deposits.

Efflorescence is seen even after regular cleaning and maintenance works are carried out at intervals by ASI, Chennai Circle. Next level significant damages are pitting and alveolar. They are found at the tower level. Much less quantity of scales and soiling due to dropping by birds and bat are found.

S No	Weathering form	Area (m^2)	
		Slight	Severe
1	Efflorescence	28,26	6,52
2	Soiling	0	1,26
3	Scaling	5,02	0
4	Pitting	4,96	9,31
5	Alveolar	0	4,97
6	Rounding	0	60,13
7	Roughening	0	70,79
8	round n roughen combination	19,01	51,78

Fig. 14. Table showing the damage Quantification.

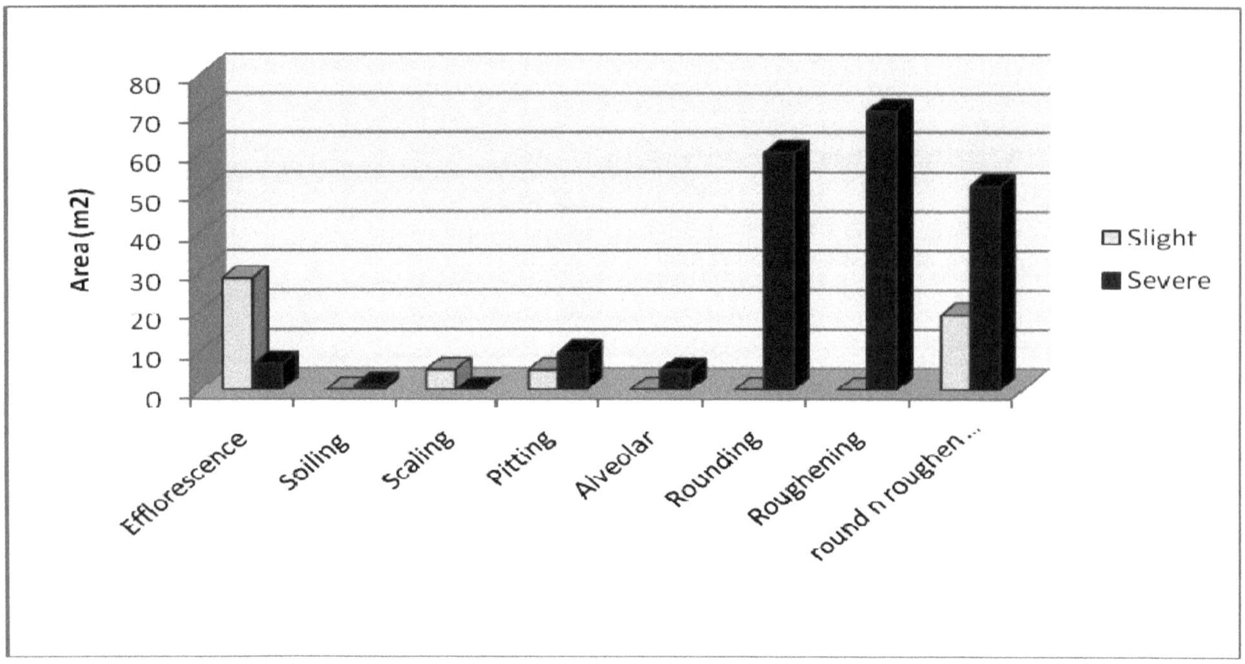

Fig. 15. Evaluation of damage mapping.

Conclusions

The most effected weathering forms are rounding, roughening and efflorescence, which are mainly due to the marine environment and driving breeze from the sea towards the shore. The wind action along with sand particles erodes the temple wall and roughens the surface and rounds the features of the sculpture. Almost all the sculptures and features on intricacies have been worn out due to wind action. This evidences the severity of the wind action on the monument.

Efflorescence is being regularly cleaned, but still it is found on the surface. This evidences that the pitting could have paved way for salt spray ingress. Pitting and alveolar is again due to wind action.

The above analysis and conclusions give a clear picture stating that the damages due wind and salt actions is severe and is on increase. Serious interventions and steps need to be taken in order to save our built heritage.

Acknowledgement

The author expresses their sincere thanks to Mr.N.Thyagarajan, Photographer for carrying out digital photo montage. We would also like to thank ASI, Chennai Circle for their co-operation during the photo session at site.

References

FITZNER. B. AND HEINRICHS. K. 2001. Damage diagnosis at stone monuments – weathering forms, damage categories and damage indices. *Proceeding of the international conference stone weathering and atmospheric pollution network (Swapnet 2001)*, 11-56

HENRI STIERLIN (ED.) 1994. *India, Architecture of the world,* Benedikt Taschen Verlag GmbH, Germany, 7, 58-9, 136-40

ESTES. J. E. , JENSEN. J. R. AND TINNEY. L. R. (1977), "The Use of Historical Photography for Mapping Archaeological Sites", *Journal of Field Archaeology,* 4, 441-47

Rethinking culture and heritage: cultural and natural heritage in NSW and southern India

Nick McClean

Phd Candidate
Dept of Archaeology and Natural History
Australian National University

Abstract

With theoretical developments in the humanities over the last 40 years slowly filtering through to heritage conservation practice and the increasing recognition over the same period of Indigenous rights through international and national laws, there is an increased requirement for established heritage conservation agencies and advocates to engage with indigenous and local communities in the protection and management of lands. One of the challenges facing conservationists working in this field is to engage with this new dynamic in ways that bring the recognition of contemporary culture back into the equation as a basis for collaboration and co-management. By looking firstly at the terms culture and heritage, re-thinking what they can mean in light of recent debates, and considering the implications of this rethinking with reference to two case study areas, north-west NSW and Southern India, this paper will explore some new ways of approaching heritage management that focus on the acknowledgement and protection of contemporary cultural processes as a part of best practice. Following from this, the second half of this paper will consider ways that these insights can be transferred into the field through practical collaborations with Indigenous communities, exploring emerging techniques of mapping cultural landscapes and geo-biographies as a tool for heritage assessment and protection.

Over the last 20 years, heritage conservation has undergone change, and new ways of considering what is meant by 'heritage' and 'culture' (and therefore 'nature') have emerged that have significant implications for how heritage protection agencies go about their work. In particular, increasing recognition of the contemporary relationships Indigenous people maintain with their lands has led to a need for heritage practitioners to rethink how they work with these communities, and seek new ways of doing so that value these ongoing relationships as an important part of heritage protection (see Kothari 2008; English 2002, Baker et.al. 2001). While these developments have been taken up most actively in the field of nature conservation, the challenges that this recognition makes cut across the traditional boundaries that have constructed heritage as 'cultural' or 'natural', and compel us to rethink these basic terms of conservation discourse and therefore how protection is conceived and implemented.

A good starting point for discussing some of these new developments is to consider the following definition of heritage NSW Department of Environment and Climate Change (DECC) website, one which attempts to incorporate some of these new approaches:

'Heritage is the environment, objects and places - both tangible and intangible - that we inherit from the past and pass on to future generations to use, learn from and be inspired by... Culture frames our understanding of the past and influences the decisions we make about what is worth keeping as our heritage. Our concept of 'what is worth keeping' changes over time - so heritage too is a living thing' (DECC 2009).

The first point to make is that there is no distinction made here between cultural heritage and natural heritage, and in the initial part of this paper I am interested to keep it on this general level so that any conclusions will apply regardless of whether we are trying to preserve ruins or protect an ecosystem. Following from this accepting the above notion of heritage leads us to the observation that heritage is influenced by and therefore reflective of contemporary cultures. Heritage is inherently cultural in a very contemporary sense, and it is possible to find out something significant about a present day culture by looking at what it considers to be its heritage and how it interacts with it. English goes further, when he states that 'heritage can derive much of its meaning from active interaction between people and a place. This interaction can be linked to past events or practices, but can also be a facet of contemporary life' (2002, 3). Heritage, therefore, is not only reflective of contemporary cultural values, but contemporary cultural practice can in fact be heritage in these terms.

While this emerging view of heritage as being suffused with contemporary cultural content is slowly becoming more widely accepted in mainstream heritage protection bodies, it nevertheless runs counter to some common trends that

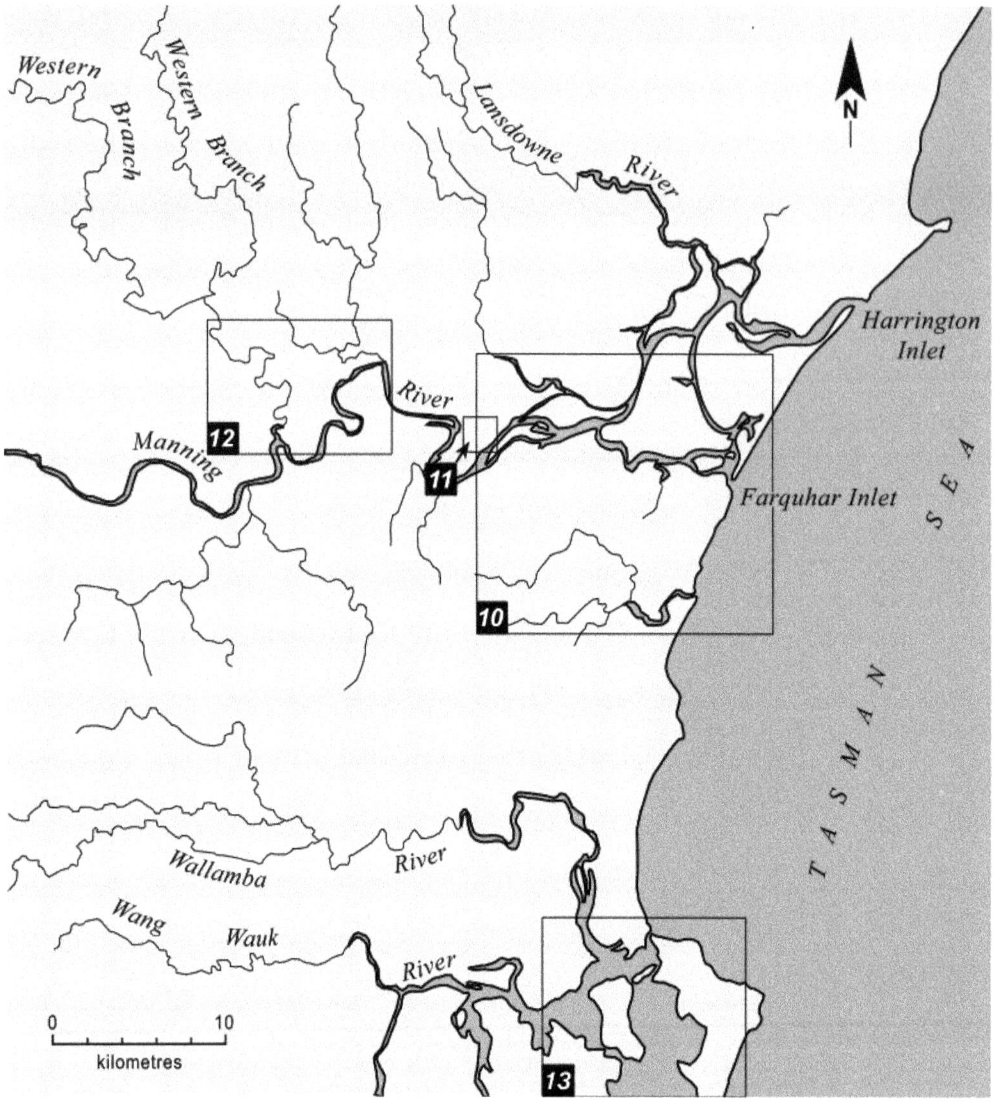

Fig. 1 Map of heritage landscapes in Forster region, NSW (Byrne and Nugent 2004, 72).

persist in heritage management practice on the ground. Byrne (2007) has made several observations about this, demonstrating how the approaches of archaeologists, anthropologists and nature conservationists working in Australia continue to combine in such a way as to result in a devaluation of contemporary cultural significance of landscapes as a legitimate part of heritage conservation practice. So the first implication of rethinking our terms is to state that any thorough heritage conservation process should take seriously the notion that contemporary culture plays a role in heritage protection and management.

Beyond this general level though, there must be some way of understanding what culture is that helps us turn this principle into an on the ground practice, and by understanding culture as a process, it is possible to incorporate contemporary culture into heritage protection processes. What is meant by this is that culture is not solely defined either by sets of values or beliefs, by cultural material such as objects, buildings or sites, or by particular actions, such as particular religious or subsistence practices. Culture in this view is a complex and active process that links all three in an attempt to make sense of the world and interact with it. This view of culture has become prominent in anthropology over the last 30 or so years, where traditional attempts to define cultures by identifying particular practices, objects or beliefs as iconic and therefore representative or definitive of a culture have been criticized, in preference of a view of culture as 'the active construction of meaning' (Street 1993, 23). In the theoretical sense, this active process of meaning making is the process through which cultural values, objects and practices are 'produced', or made apparent in the world, and therefore the outcomes of the process (a ceremony, a song, a social structure, a particular way of growing and harvesting food, a national park etc) will necessarily be reflective of the culture that produces it. To follow Thornton, rather than ask what culture is, we should ask 'what culture does' (1988, 26). This is important in the heritage protection context, because it allows us to move beyond the notion of cultural heritage protection as being constituted by the protection of sites and artifacts, the now well recognized 'stones and bones' approach, and begin to recognise the legitimacy of intangible

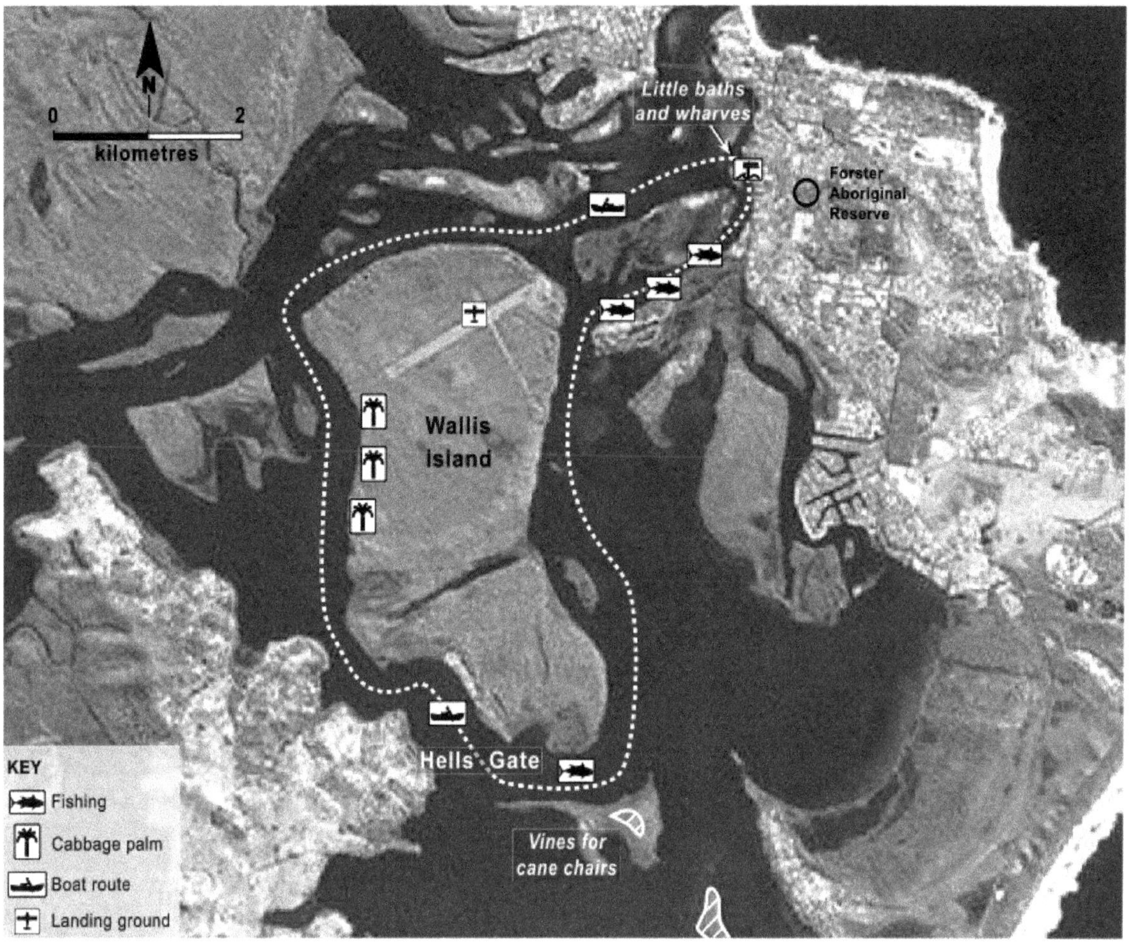

Figure 2. Map of Wallis Lake heritage landscape based on geo-biographical approach (Byrne and Nugent 2004, 118).

aspects of heritage, and of equal importance, of the ongoing meanings and values attached to places in the present day. This is particularly relevant in terms of working with Indigenous people, and English again touches on this when he notes that for Aboriginal people on the mid North Coast of NSW, 'the capacity to find and utilize wild foods has been described as integral to their sense of identity, morale and their cohesion as family groups... It is through these activities that people express and activate their associations with place. Such activities may not be encompassed by the protection of a midden or the salvage excavation of a stone artifact scatter' (2002, 4). That he mentions the protection of a midden is salient, for a midden is the material result of prolonged food gathering in a place over many generations. In this case the practice of wild food gathering is one that continues to make manifest Indigenous peoples values and social customs, and allows them to maintain their active relationship with their country and the material aspects of their culture. It maintains the cultural process. The conclusion here is that if we seek to maintain cultural processes, rather than simply to preserve the material remains of past cultural processes, then the active relationship that people have with their heritage, which we now know can actually be a part of that heritage, can be maintained.

Indigenous cultural heritage in NSW

While Aboriginal people have s been involved in heritage management in NSW since the 1970s and the first Aboriginal Sites Survey (see Kijas 2005), more recent collaborations have been documented by Rose, James and Watson (1998), English (2002), English and Gay (2005) and Byrne and Nugent (2004) that explore the contemporary cultural significance of country to NSW Aboriginal people and the dynamic nature of that culture. In particular, English's work is the first to deal in depth with the cultural issues detailed in the first part of this paper and consider their implications for Indigenous heritage management in a local context. While these early works and the slow development of jointly managed parks since 1996 represent an important shift in thinking for heritage managers in NSW, English (2002, 2; also Adams and English 2002) have pointed out that these conversations are still in their infancy in NSW, and the progression of these new ideas into practical conservation arrangements remains a challenge yet to be fully realized. A good example to illustrate how this shift might work out in practice is to consider the situation relating to Indigenous heritage conservation on the floodplain country of the Upper Darling River in NW NSW.

For many in white Australian society, Aboriginal people in south-east of the country have long been regarded as having 'lost' their culture due to dislocation from their lands over many generations. In the upper Darling region, many Murris took up droving jobs in the 1860's ('Murri' is a term Aboriginal people from NW NSW and much of Queensland use to refer to themselves), and then after the slow demise of the cattle industry through the first half of the 20th C., these families were moved onto government reserves and then into towns, where most Murris permanently reside today (Goodall 1996). Under these circumstances, the attitude of mainstream Australian society until very recently has been that Indigenous people experienced such a rapid change in lifestyle and culture that they have literally ceased to be 'Aboriginal' in any way that relates to traditional culture. Indeed, this is even supported by Australian native title law, which has consistently judged that many Aboriginal people in SE Australia have experienced a change in lifestyle to the extent that they are no longer considered to have rights over land that they are the recognized traditional owners of (Brennan 2003).

Archival and oral history research done by and with Murris over the last thirty years has shown, however, that Murris working in the cattle industry in this region between the 1860s and the 1950s were invariably choosing to work on stations that were on their traditional clan or tribal country, and that in the course of boundary riding, droving and the many other duties that required people to be out in the bush for days at a time, they would have the opportunity to visit places of significance, maintain connections to their country and pass on knowledge to younger generations (Goodall 2001, p.102; Goodall 1996; McClean 2007). This practice continued well after the last recorded tribal initiation ceremony in the region in 1910, and continues today, with many Aboriginal people continuing to visit areas with which they have a traditional association for the specific purpose of looking after their country, enjoying the right to hunt and gather food and for their own sense of spiritual wellbeing (McClean 2007, 2008).

Such a situation leads us to question whether Aboriginal culture really has disappeared in line with popular perceptions, and Australian law. True, most Murris in the region no longer undertake ceremonies or live permanently on their traditional homelands, but the process of looking after country through regular visits, spending time learning and utilizing knowledge of bush foods and recording and investigating the history of their people and homelands are practices which make manifest the continuing values that Aboriginal people maintain, in line with what they see as traditional cultural values. The lifestyle has changed, but the cultural values have not. The link between past cultural practices and contemporary ones is the continuing practice of visiting country, a central aspect of maintaining Aboriginal culture in the eyes of many Aboriginal people themselves, and one that allows other aspects of culture to be continually reproduced.

In this case, if we were to undertake to try and protect a site, for example an old mission site or perhaps a traditional campsite, it would mean legally and physically protecting the site from destruction, as well as supporting the processes that allow local indigenous communities to continue visiting that site and their country (egg supporting 'culture camps', and alternate forms of education that get Aboriginal kids out of the classroom and into the bush, ensuring legal rights to access land), to assist in documenting knowledge and stories, and to engage in discussions over shared responsibility for management. Importantly, the shift here not only involves seeing 'cultural processes' as a relevant part of heritage, but understanding that knowledge about country, bush foods, environmental patterns, experiential knowledge of biodiversity, as well spiritual beliefs, stories and memories about country, and the life patterns that enhance and maintain this knowledge are the constituent parts of these processes, and therefore valuing these as a part of the heritage protection spectrum. It is in this sense that Byrne and Nugent note that 'it is landscapes themselves that should be considered heritage, rather than discrete and dispersed 'sites' within them' (2004, 73).

Local communities and natural heritage in Southern India

In terms of working through the implications of these shifts in thinking for nature conservation, looking at the situation regarding turtle conservation in Southern India gives a good case study of two different approaches to conserving natural heritage. In this case there has been a widespread increase in turtle conservation initiatives, stemming from community based conservation initiatives in Chennai in the mid 1980s, which have slowly spread along the coastline in the form of village hatcheries and community conserved areas (see Shanker and Choudhury 2006). The most relevant aspect of this trend is the way traditional fisherfolk have been involved in conservation initiatives, and there are two places which make a worthwhile contrast in this regard.

To begin firstly in Kerala, where fisherpeople in the village of Kolavipalam have been protecting local turtle populations and revegetating mangrove areas since 1992 (for full accounts of this initiative, see Shanker and Kutty 2005, and Sharma 2002). At that time a group of local youths recognized the turtles on their beach as being the endangered Olive Ridley Turtle, and endeavored under the steam of their own resources and motivation, to begin patrolling beaches at nesting time, setting up a village hatchery to ensure adequate regeneration of local populations each year. As time went on, those involved in the initiative began to learn of the importance to turtles, and to their own wellbeing, of healthy mangrove systems. As a result, they also began regenerating mangroves, and lobbying for the cessation of sand mining in their estuary system. Realising that this local group was undertaking work of importance, the local forestry department began financially supporting the initiative and helped build facilities for their work. While this predominantly Hindu village is in a country that has a long tradition of turtle reverence, significantly, this new initiative has come at a time when villagers livelihoods are under threat from sand mining and economic change, and conserving turtles has become for them a way to invest their energies that also has direct flow on effects for their ability to maintain a fishing-based lifestyle. This is reflected in the willingness of other fisherpeople in Kolavipalum and surrounding villages to support the initiative (and even in some cases bringing back eggs and information about new nesting sites), with one village turtle conservationist observing that 'we began by saying that we are protecting the turtles, now they are the ones that are protecting us' (anonymous in Shankar and Kutty 2005, 225).

Despite this now well documented example of the potential confluence of the interests of traditional fisherfolk and turtle conservationists, in Orissa, on the eastern coastline of India, a debate over the impacts of trawler fishing in a region that has one of the world's largest nesting sites for Olive Ridleys has polarized the community and in fact increased opposition to turtle conservation from traditional fisherpeople. Shanker and Kutty cite a number of factors that have lead to a lack of success in working towards conservation after turtle numbers were declining rapidly with the advent of mechanized trawling in the region. Of particular interest here, however, has been the tendency for conservation activists to portray fisher people generally as being responsible for turtle deaths, without making the distinction between traditional and mechanized forms of fishing, and the emphasis among conservationists on enforcement of laws through patrols and providing legal support to those opposing fishing. While these measures are understandable in one light, the impact they have had on local perceptions of conservation is that 'traditional fishworkers associations... have begun to perceive conservation as anti-people, despite the fact that the Orissa Marine Fishing Regulation Act is designed to protect traditional fishing rights rather than turtles' and these associations have even begun publicly protesting and opposing new marine sanctuaries (Shanker and Kutty 2005, 230; Aleya 2005).

By illustrating these two scenarios I am interested to show that communities with very similar interests can react very differently depending on their situation, and depending on how conservationists interact with them in the course of going about getting conservation underway. In the case of Kolavipalam, local conservation bodies were able to realize that the local fisherfolk were the best possible asset for ensuring the ongoing protection of the local turtles, and so were able to support a situation that allowed for both conservation and local fishing activities to occur. In fact, the mutual dependence on healthy mangroves is what linked humans and turtles, and so realizing the interdependence of 'natural' and 'cultural' worlds here provides an example of how rethinking these relationships can actually lead to better environmental outcomes. In the case of conserving the Orissan turtles, the underlying assumption that humans and turtles cannot co-exist, that turtles must be conserved in a reserve that excludes human interaction, and the lack of attention to detail in acknowledging cultural and technical differences between different groups within the fishing community has in fact turned potential allies into active opposition. The importance of this for the long term maintenance of conservation measures cannot be underestimated, not only for the fact that in many parts of the world areas that are of high conservation value also have communities of people already living in them but also, as Rodriguez et. al. (2007) have discussed, if conservation is to be considered a long term strategy it must be ensured that there are people in situ who will support and willingly maintain the practice of conservation.

So heritage under these terms becomes less abstracted from the human world, and is something that humans actively choose to value and protect as heritage. In practical terms this realisation would ideally lead to a shift away from the classic conservationist approach that aims to protect nature or ancient sites from humans 'in its own best interests', to an approach that integrates protection with the maintenance of community ties to these areas. The dynamics of this in a globalised world are increasingly complicated, however, and conflicts over heritage conservation are no longer about 'western' notions of conservation versus the rest, as Kartik Shanker, an Indian ecologist and conservationist, has recently reflected on in the context of wildlife conservation in India:

'The western notion [of conservation and biodiversity] is not confined to the west. Parts of western culture have spread over the last century to other parts of the world. Traditional western notions have spread to India because we were colonized for 200 years and are an English speaking country. I would say that anyone who lives in urban India today already possesses a western notion, and that is why I would say it is more of an urbanisation of conservation than a globalisation of conservation [occurring in India]... The people who have the strongest voices in turtle conservation tend to be born and brought up in Delhi or Bombay or Chennai. In that sense I think there are deep divisions in India about who speaks for conservation and whose notion of wildlife and whose notion of biodiversity it is. These are issues that some of us have been trying to raise, that exclusive and probably very elite notions of wildlife and biodiversity are not the only ones out there.... I think the frustrations that many wildlife conservationists feel is due to prolonged attempts to impose their idea and their ideals on communities that are coming to it from a completely different angle, ironically often with a much much deeper connection to nature than we ever had.' (McClean, forthcoming).

Methods and tools for collaborative heritage management

In light of the acceptance of the role culture plays in heritage conservation, and the increase in both need and obligation for heritage protection agencies to engage with Indigenous and local communities, the next question is - how to develop collaborations with these communities to protect their heritage? English (2002), English and Gay (2005), Byrne and Nugent (2004) and Harrison (2004) have developed detailed methodologies for documenting and assessing contemporary heritage landscapes in NSW, providing a good reference point for any heritage practitioner, I want to outline two aspects that make up the core of a methodology for working with communities to document their knowledge and understandings of culture and heritage.

Cultural landscapes as an interpretive tool

One of the most useful concepts to come out of the theoretical debates in cultural geography in the 1980's is the development of the idea of cultural landscapes. Brown gives a good description of this approach:

'A cultural landscape approach integrates natural and cultural heritage conservation by examining them at a landscape level. This concept emphasises the landscape scale of history and the connectivity between people, places and heritage items. It recognises that the current landscape is the product of long-term and complex interrelationships between people and the environment' (2008, 2).

That this approach emphasises the interconnections between people and place, and acknowledges that change over time is a given factor of any heritage situation not only allows us to bring contemporary cultural practice back into the equation, but also allows us to move beyond the traditional 'sites based' approach that has dominated cultural heritage management to date (Brown 2008; Byrne and Nugent 2004) and focus on the long term dynamics that influence peoples' ability to maintain relationships with and knowledge about an area. For Horton cultural landscape is a conceptual tool capable of dealing with cultural evolution and change in the present day. 'The beauty of cultural landscape methodology is that it allows for a continual accretion of meaning' (2004, 180). Working at a landscape level allows us to take into considerations interconnections between sites and the intangible aspects of heritage, such as knowledge, stories and memories that may not be restricted to one place alone. Goodall (2007) has described in her timely discussion of Indigenous knowledge of water on the Darling River floodplains, that Aboriginal people have built up a unique body of experiential knowledge in post-contact times due to the particular constraints on land use and access in that region, highlighting that one of the necessary parts of an approach to collaborative heritage management based on a process orientation is to understand the dynamic interplay of a range of often parallel processes in determining what sort of relationship people have had or can have with their heritage. As Stratford (1999, 5) notes, 'a consequence of viewing culture as a dynamic 'way of life' and dynamic 'ways of human life' is that cultural landscape has been conceptualised as a process', and so taking a cultural landscapes approach provides us with an ideal template for understanding the many layers of cultural meaning as well as the relevant historical processes that may exist within a landscape, providing us with an interpretive framework for understanding the interwoven nature, the processes of interaction, between these different aspects..

Mapping Geo-biographies

Increasing recognition of the socially constructed nature of maps, and the acknowledgement of their power as a tool in planning anything from conservation reserves to warfare , has been an increasingly common theme in the social sciences over the last 20 years (see eg Woods 1993, Harley 2001, Pinder 2003, Byrne 2007). Of particular interest here is the growth since the late 1980s of 'counter mapping', in which local communities have sought to intervene in political processes surrounding natural resource management by mapping their own understandings of landscapes, allowing for their specific cultural, economic and social relations to land to be represented in a way that is readily useable in a management context (see Peluso 1995). Byrne points out that 'the relevance of counter-mapping to the Indigenous people of Southeast Asia and Australia derives from their shared need to be able to depict the topographic dimensions of their culture in terms that are intelligible to those empowered others with whom they must negotiate their continued existence'(2007, 261).

Likewise, oral history has emerged since the 1970s as a form of scholarship which has the potential to 'democtratise' history, opening up new fields of inquiry and including in the historical record 'the experiences and perspectives of groups of people who might otherwise have been "hidden from history"' (Perks and Thompson 1998, ix). Oral history has been identified as a useful and appropriate technique for working with Indigenous communities, where the embedding of social and knowledge networks in a primarily oral tradition means that 'hangin out' and 'yarnin' are essential parts of learning about community and country (Barker 2006). Passerini has advised against allowing the political potential of oral history to result in an 'alternative ghetto' for the voices of the oppressed (1979, 84), and in light of this, more recent scholarship has viewed oral history as a means of understanding the past from the point of view of the present. Drawing on parallel developments in environmental history, this approach sees different views of historical events as representative of different circumstances, motivations and aims rather than attempting to establish the truth and authenticity of any one version (Lucas 2004, Goodall and Lucas 1997, Cronon 1992). Such an approach is a useful method of inquiry in exploring these emerging views of heritage, where heritage is the result of a complex interaction between past and present, and can equally be about how the past is seen and carried on in the present day through the interactions of people and place, as it is about preservation of pristine ecosystems or ancient sites.

Applying these tools to heritage assessment and documentation, Byrne and Nugent (2004) have begun developing the technique of mapping 'geo-biographies', where satellite maps of a particular area are used as the basis for provoking discussion and then spatially representing the memories, experiences and knowledge of interviewees. The result is a series of individual maps which can be collated, cross checked and integrated with traditional archival and cadastral resources to identify 'heritage landscapes' that can be integrated into planning processes for heritage management. Despite their value in these broader processes, preserving the integrity of individual 'geo-biographies' and acknowledging their inherently personal nature is an important aspect for his process. 'Cultural landscapes do not have inherent meaning, their meaning is ascribed by the people who interact with them' (Byrne and Nugent 2004, 138), and so the 'geo-biographies' which represent individual views of particular cultural landscapes are made up of the interaction of lives and landscapes. Returning to the notion of cultural processes, thinking in these terms any cultural process is simply made up of a series of moments in the lives of people living in a community, and so in application, the view of heritage that I have been trying to demonstrate says that heritage is not only about things, it is also about the cultural processes that keep people attached to their heritage, and thus heritage is equally as much about peoples' lives as it is about objects or ecosystems.

Conclusions

In this paper I have tried to provide a basis for collaborative heritage management that goes beyond political and legal necessity, analyzing the underlying concepts of heritage and culture in such a way as to 'humanise' heritage, revealing it as a fundamentally social and cultural process. In light of this, heritage practitioners can begin to engage with the notion that present day culture is a fundamental part of heritage management, both on the level of present day cultural values and practices being a part of what heritage can be, and also in the sense that supporting a society and culture that values heritage conservation is an active way of engaging with culture that helps maintain the long term sustainability of conservation measures. Employing interpretive frameworks that allow for the flexibility inherent in cultural processes, and undertaking projects that value particular cultural viewpoints and individual experiences of these cultural realms are ways of transferring these insights into practice. As a final observation, these insights should of course lead us to an understanding of the importance of community participation in these processes and in light of Perks and Thompson's comment that the aim of many oral history endeavors has been 'the empowerment of individuals or social groups through the process of remembering and interpreting the past' (1998, ix), heritage practitioners should not only be aware of the facilitation role that needs to be played in such projects, but also that many of the underlying claims to recognition of culture go beyond a desire for simply understanding and documentation. In many cases embedded within such cultural perspectives are political claims to land ownership and responsibility for custodianship. Heritage managers must wherever possible consider these claims to be an honest desire for custodianship and caring roles and seek to identify common ground to work from as a starting point for shared responsibility for management.

References

ADAMS, M. AND ENGLISH, A. 2005. 'Biodiversity is a whitefella word': changing relationships between Aboriginal people and the NSW National Parks and Wildlife Service, in Taylor, L., Ward G., Henderson R. and Wallis L. A. (eds), *The Power of Knowledge, The Resonance of Tradition*, Canberra, Aboriginal Studies Press.

ALEYA, K. 2005. Perspectives of the Traditional Fishworkers on Sea Turtle Conservation. *Indian Ocean Turtle Newsletter* 1, 7.

BAKER, R., DAVIES, J. AND YOUNG, E. 2001. *Working on Country: Contemporary Indigenous Management of Australia's Lands and Coastal Regions*. Oxford, Oxford University Press.

BRENNAN, S. 2003. Native Title in the High Court of Australia a decade after Mabo. *Public Law Review* 14, 209-218.

BROWN, S. 2008. *Cultural landscapes and park management: a literature snapshot*. Hurstville, National Parks and Wildlife Service.

BYRNE, D. 2007. Counter-mapping: New South Wales & Southeast Asia. *Transforming Cultures eJournal* 3(1), 256-264.

BYRNE, D. AND NUGENT, M. 2004. *Mapping Attachment: a spatial approach to Aboriginal post-contact heritage*, Hurstville, National parks and Wildlife Service.

CAMPBELL, B. 2005. Changing protection policies and ethnographies of environmental engagement. *Conservation and Society* 3(2), 280-322.

CRONON, W. 1992. A Place for Stories: Nature, History and Narrative. *Journal of American History* 78(4), 1347-1376.

DECC 2009. Webpage www.environment.nsw.gov.au/cultureandheritage.htm accessed 25/5/09

ENGLISH, A., 2002. *The Sea and Rock Gives Us a Feed: Mapping and managing Gumbaingirr wild resource use places.* Hurstville, National Parks and Wildlife Service.

ENGLISH, A. AND GAY, L. 2005. *Living Land Living Culture: Aboriginal heritage & salinity*, Hurstville, National Parks and Wildlife Service.

GOODALL, H. 2007. Indigenous Knowledge, History and Water in a Changing Australia. Paper presented at *The History of Waters Conference.* Kolkata, Jadavpur University.

GOODALL, H. 2001. Renewing country: Aboriginal people and their lands in rangeland environments. *The Rangeland Journal* 23(1), 99-115.

GOODALL, H. 1996. *Invasion to Embassy: land in Aboriginal politics in New South Wales, 1770-1972*, St Leonards, Allen and Unwin.

GOODALL, H. AND LUCAS, D. 1997. 'Country' Stories: Oral history and sustainability research, in *Proceedings of 1997 Conference of the Australian Association for Social Research.* Wagga Wagga, Charles Sturt University.

HARLEY, J.B. 2001. *The New Nature of Maps: Essays in the History of Cartography*. Baltimore, Johns Hopkins University.

HARRISON, R, 2004. *Shared Landscapes: archaeologies of attachment and the pastoral industry in New South Wales,* Hurstville, National Parks and Wildlife Service.

HORTON, T. 2004. Reading the cultural landscape at Dyea, Alaska, in Harmon, D., Kilgore, B. M., and Vietzke, G. E. (eds), *Protecting our diverse heritage: the role of parks, protected areas and cultural sites.* (Proceedings of the 2003 George Wright Society/National Park Service Joint Conference.), 177–181. Hancock, The George Wright Society.

KIJAS, J. 2005. *Revival, Renewal and Return: Ray Kelly and the NSW Sites of Significance Survey*, Hurstville, National Parks and Wildlife Service.

KOTHARI, A. 2008. Protected Areas and People: The future of the past. *Parks* 17(2), 23-34.

LUCAS, D. 2004. *Shifting Currents: A history of rivers, control and change.* Unpublished Phd thesis, Sydney, University of Technology.

MCCLEAN, N. 2008. *Rivers, Lakes and Plains: a journey to the Narran Lakes* (radio documentary). Ultimo, Australian Broadcasting Commission.

MCCLEAN, N. 2007. *Rivers, Lakes and Plains: Stories from Yuwalaraay country – the Narran Lakes.* Unpublished honours thesis, University of Technology Sydney.

PASSERINI, L. 1979. Work Ideology and Consensus Under Italian Fascism. *History Workshop* 8, 84-92.

PELUSO, N. L. 1995. Whose Woods are These? Counter-mapping forest territories in Kalimantan, Indonesia. *Antipode* 27 (4), 383-406.

PERKS, R. AND THOMPSON, A. 1998. *The Oral History Reader.* London, Routeledge.

PINDER, D. 2003. Mapping Worlds: Cartography and the Politics of Representation, in Pinder, D., Blunt, A., Gruffudd, P., May, J., Ogborn, M. (eds) *Cultural Geography in Practice,* 171-190. London, Arnold.

RODRÍGUEZ, J. P., TABER, A., B., DASZAK, P., SUKUMAR, R., VALLADARES-PADUA, C., PADUA, S., AGUIRRE, L. F., MEDELLÍN, R. A., ACOSTA, M., AGUIRRE, A. A., BONACIC, C., BORDINO, P., BRUSCHINI, J., BUCHORI, D., GONZÁLEZ, S., MATHEW, T., MÉNDEZ, M., MUGICA, L., PACHECO, L. F., DOBSON, A. P. AND PEARL, M. 2007. Globalisation of Conservation: A view from the South. *Science* 10, 755-756.

ROSE, D., JAMES, D., WATSON, C. 2003. *Indigenous kinship with the natural world in New South Wales*, Hurstville, National Parks and Wildlife Service.

SHANKER, K. AND CHOUDHURY, B.C. (EDS) 2006. *Marine turtles of the Indian subcontinent.* Hyderabad, Universities Press.

SHANKER, K AND KUTTY, R. 2005. Sailing the Flagship Fantastic: Different Approaches to Sea Turtle Conservation in India. in *Mast* 3(2) and 4(1), 213–240.

SHARMA, S. 2002. *The Turtle People (Aamakar)* (Television documentary) Mumbai, Chrysalis Films.

STRATFORD, E. 1999. Australian cultural geographies, in Stratford E. (ed), *AustralianCultural Geographies*, 1–10. South Melbourne, Oxford University Press.

STREET, B.V. 1993. Culture is a Verb: Anthropological aspects of language and cultural process, in Graddol, D., Thompson, L. and Byram, M., *Language and Culture*, Clevedon, British Association of Applied Linguistics and Multilingual Matters

THORNTON, R. 1988. Culture: A Contemporary Definition. In Boonzaier, E. and Sharp, J. (eds), *South African Keywords: The Uses and Abuses of Political Concepts*, 18-29. Cape Town, David Philip.

WOOD, D. 1993. *The Power of Maps (Mappings).* New York, Guildford Publications.

Culture of excavation and sculptural geometry of the ground

Carlo Pozzi

Università degli studi 'G. D'Annunzio' Chieti-Pescara
Facoltà di Architettura, Pescara, Italy

Abstract

Sassi of Matera are able to reflect ways of building of the Mediterranean: a layer town, understandable in section more than in plan, leader of the *Murgia* family, analogous to subterranean towns and differently layered. The town is constructed on an intense relationship between built and hollowed and studded with rocky settlements where each time the drawing is renewed of the rocky front of the caves, through the geometric construction of the access panel. Culture of excavation and sculptural geometry of the ground are celebrated by the church/cathedral of St. Mary of the Valley with its mighty structure inside an enormous excavation showing itself through a thin wall juxtaposed to the tufa stone, traced by the succession of different geometries of portals showing the stylistic trends of the Christian architecture. These materials, so far the subject of specialist analysis of historians and restorers, are becoming a treasure for planners: today architects decide how to relate with the site, with a layer of land messed up by an intense and confused way of building; they search methods of transforming the ground with art through the teaching of artistic disciplines such as painting, photography, land-art.

Keywords ground, geometry, excavation, underground, new ground

Introduction

In various steps I studied the culture of excavation and sculptural geometry of the ground, trying to describe a *fil rouge* linking the historical, modern and contemporary experiences. Results of this research have been shown in international conferences:

- with the title *Building means cooperating with land* in Rome at the conference *The architectural project*, May 1998;

- with the title *Culture of excavation and sculptural geometry of the ground* in the international seminar in Pescara *The Mediterranean Medina*, June 2004, also introduced in the conference at Harn Museum in Gainesville, University of Florida – USA, November 2004.

What is excavated and what is built

The starting point of the research is the study of *Sassi of Matera*, extraordinary architectural settlement in southern Italy, whose first character is the natural site, that appeared available for human modification from the very beginning. The deep cut has been excavated since time immemorial in the platform that marks the bare geographic area, mainly in Puglia called *Murgia*, formed by homogeneous material, tufa, and easily subject to water erosion hence the form of ravines, like big valleys excavated in time by rivers whose coasts, sometimes overhanging, are drilled through the caves that have always offered shelter to shepherds and animals, hermits and whole communities of monks.

Fig. 1. Sassi di Matera, Basilicata, Italy.

The two coasts of the *Gravina*, facing each other out the level of *Sassi of Matera*, sum up in a meaningful point the story of human settlement, from the cave to the house to the town, in a unique relationship between nature and artificiality, the latter always generated by a condition of necessity that commands today sacred respect. The main quality of *Sassi* remains being a city, where the pit represents the first and more direct way to adhere to the site, lodging in its womb, looking for safety, protection, certainty in the oldest and most settled rules of living that have been handed down on us; and them through another pit and juxtaposition, the 'upper' streets and houses, in layers almost one above the other, in a complicated structure of 'vertical' city, upright as it is understandable in section more than in plan, or at least impossible to analyze if you don't consider both.

In this meaning of street in layers, the most interesting references go to other realities of excavated and built that are capable of reaching an urban configuration: this statement may represent a discrimination in a grading acceptation that, being scientific, cannot avoid a judgement. Therefore it is necessary to repeat that the most plausible analogies are possible among *Sassi* (as leader of the *Murgia* family including also the towns of Ginosa, Gravina, Massafra and Mottola) and other subterranean urban settlements of poor dimension such as Guadix in Andalusia or differently layered such as Göreme in Cappadocia. As regards the rocky settlements such as some necropolis (Mira, Petra) the reference could be better addressed on those rocky convent complexes, spread along the ravines of *Murgia*, rich of coenobies, houses, stables, hermitages, chapels, that face similar themes in the renewed design of the rocky front of the caves, first of all through the construction of the panel where the door is cut for access and to view the steep territory, whose sequence finally forms already a new foundation through the sculptural geometry of the site.

Speaking of culture of excavation you can't avoid thinking of the old Roman hydraulic works (*Piscina Mirabilis*), the catacombs, the subterranean towns, the cave of the Sibyl, the quarries, the mines: various forms of excavation, more or less complicated, more or less geometrical, according to the hardness of the rock was being carved, bringing inside these caves, with this heavy and arduous excavation, starting from a small cave that is enlarged in depth, themes of cultured architecture, which are immediately made almost rough by the difficulty of the material, by the resistance of the rock to modification. The underground basilica structure of *Piscina Mirabilis* is the geometry of a typology already known for its civil use, here used to define a big tank for the ships of the Roman fleet leaving from Baia: the same structure, moving with Christianity to define liturgical spaces, frames one really peculiar for its being underground: the church-cathedral of *Santa Maria della Vaglia* in Matera (commonly called *La Vaglia*); with its mighty structure inside a huge excavation, in a big tufa rock where it's impossible to reproduce the slim columns of noble architecture, therefore enormous pillars remain aiming to keep the covering rock, made graceful, as far as possible, by small corner columns. Outside, the church shows itself with a thin wall juxtaposed to the tufa rock, delineated by the succession of different geometries of portals forming a list of the stylistic trends of Christian eastern and western architecture: the facade adheres to the ground beneath which lies the excavation that generated the internal space. Another significant point of contact between outside and inside, excavated and built, is the rose window obtained in a circular stone monolith, roughly embroidered with round holes; eye of the architectural giant, it slept for years blinded by materials of a dump, only through recent restoration it has come back to cut the underground space with light.

When geometry acquires self-sufficiency underground, shaped spaces are proposed such as the extraordinary cave of the Sybil, with its elongated shape evocating the penetration into the ground to assume omens. More often geometry remains strong outside, an element of setting human work into the mountain, in whose womb the sacral character is closed, as in the tombs of Mira in Turkey, with the possibility for the wayfarer to perceive this extraordinary situation immediately through such an evident alteration. The same thing happens, with softer, almost baroque shapes, in the better known necropolis of Petra in Jordan, still capable of arousing, when going through the gorge that leads to the valley over which the rocky settlements looks. The sculptural shaping of the ground offers an anthology of variations on the theme in the many churches cut in the rock of the sacred complex of Lalibela (1100 BC) the 'new Jerusalem' in Ethiopia; from the monolithic excavation that shapes architecture separating with the cut the rocky stone on which it arises, to totally underground solutions, to a series of tunnels, trenches, excavated squares, churchyards with facades carved in massive stone, articulation of common space relating single artefacts.

In this rich Mediterranean circle the rocky presences in Abruzzo have a specific role, all linked to Pietro da Morrone's hermitage who, elected Pope like Celestino V, did the 'great refusal' of the pontifical throne, condemned by Dante in *Divina Commedia* as 'cowardice', rehabilitated by Ignazio Silone in *L'avventura di un povero cristiano* as the only possible solution to deny a papacy too compromised by economical and political power.

The Abbey of Santo Spirito a Majella is a fundamental point of a possible Mediterranean circle of the architectural relationship between built and excavated: it is a complex settlement, with the church and the convent at the lower level, a small hermitage on the top, connected by balconies and a long steep flight of stairs cut in the rock. A net of routes linked to the robbery paths coming from the top of the mountain.

On a not far away ridge, close to the stream which excavated it ages ago, there is the hermitage of San Bartolomeo in Legio: here the setting in the rock is perfected by the mimetic opposition of a wall made up with river stones. As in the case of Santo Spirito, the camouflage is broken by the imperfect juxtaposition between the rigid geometry of the front and the soft curve of the impending stone.

Sant'Onofrio is a 'nest of eagles' on Morrone and towers from a side the geometric convent of Badia, from the other the Roman temple of Ercole Curino, whose terraced construction on the side overlooking Sulmona valley, reminds the ascending settlement type of the temple of Fortuna Primigenia of Palestrina, that so much influenced Palladio, creator of Rotonda. The hermitage is built by a complex of volumes stuck to the rock, structured on rising routes; but its heart is the grotto, place of sacred concentration by definition; still today it is the place for ancient rites, such as rubbing against stone or picking up a stone and throwing it in the cliff, hoping it brings along all that you want to forget.

Fig. 2. The Abbey of Santo Spirito a Majella, Abruzzo, Italy.

In Sant'Angelo in Vetuli the relationship between architecture and earth's crust, the excavation, shows itself through a clashing juxtaposition between a channeled column, probably from Roman archaeological remains, and a naturally shapeless rock.

In San Michele hermitage in Pescocostanzo the relationship is in a decorative key: the white stone of Majella forms a balustrade whose refined articulation clashes with the compact shape of a grotto untouched by human hand.

The decorative element has a main role in the grotto of Sant'Angelo in Palombaro: a fragment of this hermitical settlement remains, with harmonious arches reminding the architectural culture of medieval Abruzzo, present in many convents such as San Liberatore a Majella and San Clemente a Casauria, recently damaged by the earthquake of 6th April 2009.

In Sant'Onofrio in Serramonacesca, the rock does not shelter the church but overhangs it; there still is this extraordinary relationship between a small set excavated object and the mountainous 'great mother'; inside it there is a series of tunnels creating mysterious hidden passages, therefore it is important to start the study from the sections.

Contemporary architecture finds refuge underground *(goes into the grotto)*

The underground planning recently shows a large number of experiences, often motivated by needs of environmental and bio-climatic compatibility, suggesting a tendency towards zero degree of shape, a sort of silence within the tiresome perceptive complexity shown in recent metropolitan processes.

The tradition of excavated architecture reminds the perspective invention of the Prisons by Piranesi, structured in intricate tangles of underground spaces: they are always large rooms where you can see any sort of devices and machines, wheels, wires, pulleys, levers, catapults; stairs and footways are the main characters of the scene, they cross the space in all directions multiplying, causing sensations of sublime distress.

Fig. 3. Prisons by Piranesi, tav. VII, second state.

The Soviet film director Sergei Eizenštejn offers an important reading of it, focusing the fact of advancing by jumps, with rips in proportions and in space, according to linked rings, with the intentional breakage of the figuration unit, through the overlapping of different perspective focuses, increasing in the second version of the same drawings, with the addition of new over-layered materials, stairs and footways, reaching a typically screen effect (as in the film October by the same author), the montage through juxtaposition.

Also complicated are the spaces under the thick city of Naples, that have seduced the numerous architects called to plan the hypothesis of new use for the exhibition *Sottonapoli, ideas for the underground city* (1988). The project by Aldo Rossi for the area of Monte Echia works on the idea of urban X-ray, discovering series of grottoes, recesses, quarries, trying to define a structured architectural relationship between the submerged and the emerged city, scratching a long and articulate sign, a sort of urban skeleton, in the compact pattern of the historical city. Today, twenty years after these unrealized projects, we see the paradox that, under the chaotic crust of the Neapolitan city, the last tidy remains are still the underground paths and the artistic spaces of the underground stations, completed or about to be completed.

Likewise in Paris, when they could not or did not want to resist the temptation to demolish the pavilions of Halles by Baltard, and afterwards they were unable to design again the urban space over the surface, didn't they take refuge in the underground eggshell colored reinforced concrete by Paul Chemetov, between fitness centers and malls, between RER and underground stations?

As a matter of fact when in 1946 Le Corbusier is called by Edward Trouin to design a basilica that should renew the myth of the landing of Maria Maddalena on the Provence coasts, for the first time in his huge and anthological career he uses the idea of the womb of the earth, looking for sacred, cutting articulate spaces, inclined connection routes and above all carrying on a special research on lighting techniques of underground spaces, through light pits: Sainte Baume will never be realized, but it remains a fundamental stage as concerns the involvement of underground spaces in modern architecture, their employment as main elements of the sacred.

The project for the Cemetery of Urbino (1974) by Arnaldo Pomodoro feels the hand of the artist who generated it; the earth's crust is cut with a violent and painful scratch, in search of the deep and tough sense of death, applying the teaching – and the distance – of land art to a relationship estranged by the pre-existence of the monumental Italian – style cemetery. We observe the application to the ground and to architecture of characters and signs already settled in the authors sculptural experience. The routes of the necropolis, the chapels, the tombs are organized in the jagged trenches: nothing appears above the profile of the hill.

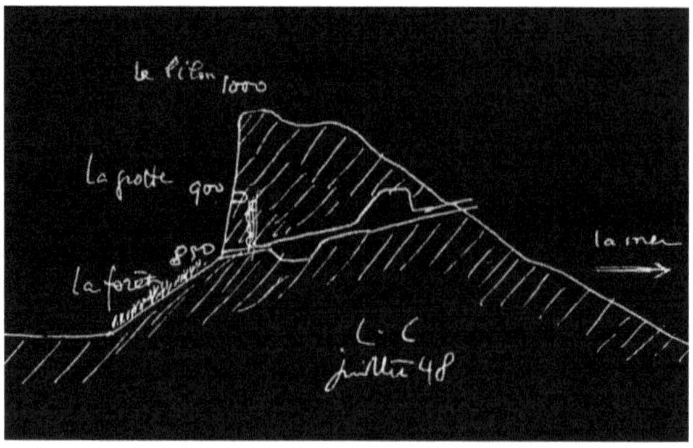

Fig. 4. La Sainte Baume, project by Le Corbusier.

The project for some residences of the Olivetti employees in Ivrea, realized by Gabetti and Isola in the 60s, is characterized by the geometric element in the shape of an incomplete circle with a side against ground, like an English 'crescent' going down into the earth's womb and whose curve opens up through glasses, in relationship with the landscape almost introjecting it, completely opening its concave side towards a tiny hill taking on the role of natural, whole, fire element of the intervention, point of view for the modern almost underground inhabitants.

In Stockholm the net of new underground stations, completed in 1976 with a project of P. Reimers and M. Grant, tries to juxtapose the clear technological perfection of the infrastructure and the artificial design of the sign system, floors, advertising panels highlighted by flashy colors to the foundation character of the rock not so much as a primitive element, but as the result of the excavation work of the building site: a grotto that with its ancestral and "natural" lack of shape contrasts with the modern technological geometry of the new interventions.

The Dune house realized by W. Morgan in 1974 at Atlantic Beach (Florida) foreshadows the dune house built by Future Systems twenty years later near Druidstone Haven (Wales): the latter delicately slips in the sloping ground above the coast *falaises*. It suggests a temporary high-tech refuge more than a lasting house, a sort of space cabin buried in the ground and completely provided with glasses only on the side looking the sea, with a sort of lenticolar façade, whereas its ancestor looked by means of two large portholes. The roof is covered with the same ground, softly reproducing the concavity of the ground with or typical land-art intervention.

Then there is a whole world of conceptual and visionary projects proposing large underground settlements, underground science-fiction cities, with the pragmatic speculative suggestions by P. Cook (multifunctional underground container in Montecarlo, 1966) and the ecological-utopian proposals by P. Soleri in the American Arizona desert (Cosanti since 1956, Arcosanti since 1970), all these proposals push history to mix with nature, aiming at a geographic process of territory transformation.

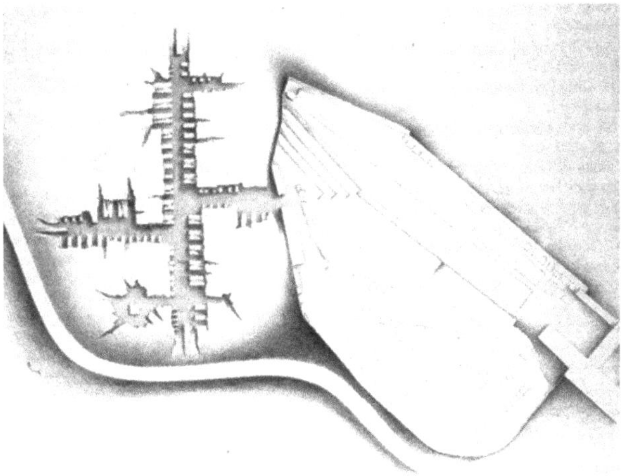

Fig. 5. New cemetery in Urbino, Marche, Italy, project by Arnaldo Pomodoro.

The research on underground architecture marks all Emilio Ambasz's metaphysical dream: the elements of his research vocabulary (the carvings in the topography of grounds, the walls, the underground spaces, the neat mirrored prisms) are assembled in the construction of Lucile Halsell Conservatory, in San Antonio, Texas, in a real mythic and initiatory process, remindful of ancient centre-American cultures that would like to reconcile nature and architecture, through an architectural concept based on excavation and void, more than on control of what is built on the ground.

The Catalan architect Carlos Ferrater explores the theme of underground space and of surface rippling, realizing a Fitness Center in Barcelona, annexed to a hotel he realized a few years before: he is not looking for a shapeless element such as a cave, but he works on well distinguished geometries. Some white reinforced concrete segments converge towards the lowered star-shaped court, which provides the underground spaces with light and air: they are splinters that cut the ground but also support it and form a new assembly. Finally only the top of segments and small skylights will emerge from the grass covering the underground building.

Enric Miralles in the Cemetery of Igualada (1991) reconstructs the entire front of the excavation using materials utilized for street edges with an *arte povera* approach, gabions of iron and stones: on the artificial rock of the necropolis he inserts the soft geometry of the portals of the subterranean chapels.

On the other shore of the Mediterranean Sea, in the competition for the Great Egyptian Museum (2003) two meaningful projects suggest alternative considerations on hypogeal building: the one by Aires Mateus uses an intense geometry of the plan in order to define pieces of construction alternating with courts (with a system which reminds the project by Le Corbusier for the hospital in Venice) until they melt away in the sand of Giza; the one by MM 26 group shapes the ground with "Etrurian" cuts on which emerging buildings are placed. The enormous hypogeal area wavers between reference to stone caves and the big archaeological excavation.

Fig. 6. Fitness Center in Barcelona, project by Carlos Ferrater.

Contemporary architecture builds a new ground
(builds the grotto)

In contemporary city the tendency is to make the architectural project react with the protection of a mantle of earth, an absolute artifice that wants to be read as a natural event, almost a sudden rippling of earth's crust due to a seismic or eruptive phenomenon that welcomes architectures with no composition protagonist, neither classical nor modern. Layers of ground appear on covers, façades are covered by grids on which trailers grow or even PVC simulations of vegetations. So nature mixes with architecture, covers it, crosses it, makes it hybrid: research moves towards forms of deep contaminations deriving from contemporary art.

Jean Nouvel, a much celebrated architect for his technical and technological skills, who would never give up a will of shape in favour of ambiguous mimesis, in the project of Museum for Human Evolution in Burgos, in Spain, introduces the theme of the "new" urban hill, which is at the same time cave welcoming museum spaces, conference and exhibition center, hotels, commercial centers. The cave, orientated towards the cathedral, marks the entrance: a huge star shaped eye lightens an internal water basin related to a near existing watercourse.

The museum front is formed by a wall made of ochre earth in metal square grids: the new settlement is declared externally only through these elements, the remainder is a homage to the site; in its landscape you can find the near archaeological excavations of Atapuerca. The bareness of a reconstructed nature contrasts with the whole urban center around it, wishing to take on the role of natural pre-existence, archaeological remains.
The mirror of this planning procedure is formed by the Museum and Study Center of the caves in Altamira, in Spain, realized by Juan Navarro Baldeweg; here, because of the necessity to reduce or almost forbid visits to the grotto walls, full of famous prehistoric graffiti, they built a cast of the same size of the original, hanging within the volume of the museum, softy plunging into the landscape, following its course and level curves. If in Nouvel's project inside the artificial hill you discover the world of technique and modernity, with all the necessary comforts, here a contrary operation happens: within a sober contemporary building, inside the ground of the original grotto, you get to the repetition of an underground space.

Most of the times the hybridation between architecture and nature through underground operations wishes to break the too strong separation between the surrounding landscape and the new building: Kengo Kuma, with the project for the Museum of Kitakami canal, just wants to recompose the distance between nature and constructional work. Here nature has already the sense of artifice, being formed by a canal, but the continuous flowing of water is considered a natural fact by definition; the navigability makes the canal an infrastructural work, taking on multiple aspects in balance between nature and device, able to influence the project, giving a sense in such directions to the new juxtaposing architecture. The new settlement appears as a green hill, suddenly emerged, where the pedestrian and cycling path coasting the canal – one of the oldest in Japan – is introduced without solution of continuity. The mysteries of the project that only section can reveal are concealed under the mantle of the ground, a soft uplift of the bank of the watercourse.

Zentrum Paul Klee in Bern, finished according to Renzo Piano's project in 2005, is defined starting from the idea that the ground, opposite Losanna-Berna motorway, arches through three gentle waves to shelter the greatest collection of Klee's pictures in the world. The shape of the cover, structured through the bending of large steel profiles, is inspired by the light sways of surrounding hills, but although the camouflage does not cover all of the grassy ground, the building appears a metamorphic geological transformation.
The theme of clods of a ground split by the sun, of a crevice (Burri takes his "poor art" canvases to urban scale wrapping with concrete the remains of Gibellina destroyed by the Belice earthquake in 1968) is reinterpreted by the big stirring up of the soil, which is the City of Culture by Peter Eisenman which is about to be completed in Santiago de Compostela, "sacred" city by definition as it is the finishing post of the pilgrims who cross Europe on foot. The operation of building a new ground and placing new volumes under it has a double aim: it tries to reproduce a continuity of landscape on the surface, and the emotion of an underground architecture internally.

The necessity to work on underground spaces protecting at the same time the archaeological remains – already excavated or not yet uncovered – is the backbone of the competition project for Iliria square in the center of Durazzo (2008) introduced by a group of architects of the faculty of Architecture of Pescara.

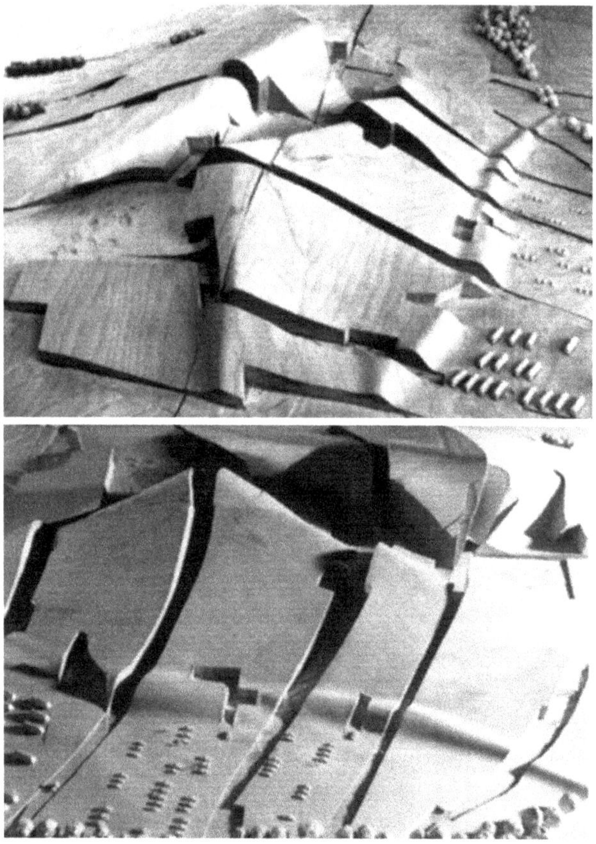

Fig. 7. City of Culture in Santiago de Compostela, project by Peter Eisenman.

Today Iliria square shows itself as the result of an accumulation of heterogeneous materials, whose stratification in time has to be interpreted as wealth within a contemporary project capable of building functional and formal hierarchies through the declared sequel of choices. That is to say it is necessary to give a new aware role to the surrounding architectures and that today are foreigners one to the other: the town hall, the mosque, the prefecture, the palace of culture, the recent many-floors building. It is necessary to evaluate the certain archaeological patrimony and the possible recoveries, giving continuity to a poetic route among the ancient remains, which must not feel under pressure for the presence of vehicular traffic or of impending old and new buildings. The first required operation is a

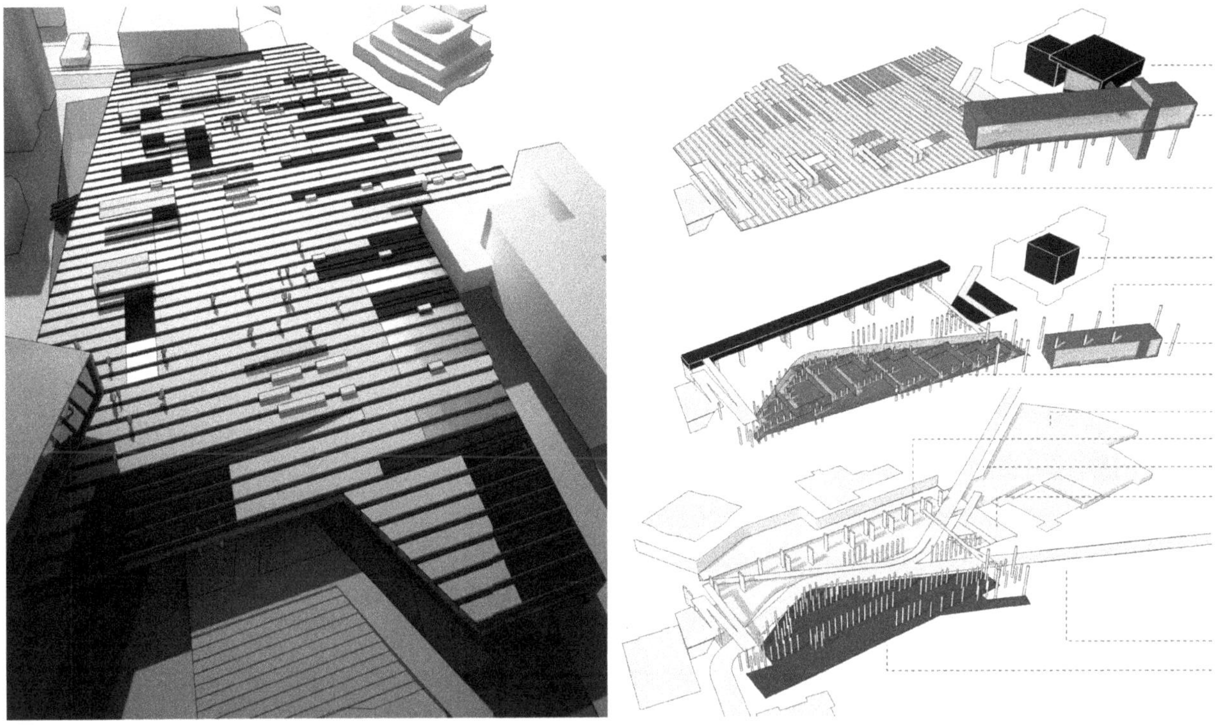

Fig. 8. Iliria square in the center of Durres, project by C. Pozzi - E. Vadini's team.

different and innovative management of vehicular traffic, separating it where it is possible from pedestrian paths, such as the most recent and important *boulevard* leading towards the harbour, recently paved and well lighted. The square, in order to be the site where to stay, needs first of all to be separated from vehicular traffic, to be characterized by the presence of architectures defining it, but to be able to be interesting enough to be crossed, suggesting many reasons to belong it, even only as an absent-minded crossing to reach a different area or to go shopping. We have thought that in order to obtain these characteristics, the only chance was to work in overlapping strata, defining a series of levels:

- the inferior level of streets, parking, separated by archaeological excavations;
- the intermediate level of commercial spaces, capable of evoking the character of *suq* which in the historic documents gives the name to urban areas next to the intervention area, and new exhibition spaces;
- the superior level, the space of the real square, calm, sunny and shadowy.

The constructional work granting the simultaneous presence of these levels is formed by a slab realized with prefabricated pre-cast concrete elements, resting on the foot of the hill facing the square, at the highest level of the terraces leading to the mosque and the prefecture; from the other side of the free space, the slab rests on *pilotis*, maintaining the same level of covering and forming a new front towards the least relevant side from the architectural point of view. From the slab hangs the underlying light volume which will lodge the commercial spaces, the contemporary *suq*, characterized by the many entrances and lighting systems. So the slab has the double role of a new ground of the square, at this point completely pedestrian, and of covering of an 'underground' world formed by streets, parking, the archaeological park, with openings and lighting systems on the most relevant parts of the excavations and of the commercial centre.

'Living' in overlapping strata is a typically Mediterranean way, if you think of *Sassi of Matera* and their description in *Cristo si è fermato ad Eboli* by Carlo Levi.

A forgotten heritage: Impact assessment studies at prehistoric sites in Tamil Nadu

Shanti Pappu,[1] Kumar Akhilesh,[1] Sudha Ravindranath,[2] and Uday Raj[3]

[1]Sharma Centre for Heritage Education
[2]Regional Remote Sensing Service Centre (RRSSC, Bengaluru)
[3]Indian Space Research Organisation (ISRO)
Email: pappu.shanti@gmail.com

Abstract

The destruction of India's prehistoric archaeological heritage is proceeding on a scale unforeseen in the past. Owing to rapid development, inadequate legislations, lack of public awareness and apathy; such sites will not remain intact over the next few decades. As part of our ongoing research project on prehistoric landscapes in Tamil Nadu, we built in a component of impact assessment studies, using a combination of satellite remote sensing, GIS and field studies. This paper presents an overview of this project, with a focus on impacts affecting prehistoric sites, and highlights the need for urgent conservation of India's 'non-monumental' heritage

Key words Satellite remote sensing, GIS, prehistory, impact assessment, South India

Invisible Elements in the Natural Landscape

As you read this paper, India's earliest prehistoric sites are being irreversibly destroyed by the unrelenting pace of human impacts on ancient landscapes. India has a wealth of prehistoric archaeological sites dating back to 1.2 mya, which tell a complex story of hominins adapting to varied ecological contexts and changing environments over the Pleistocene. In this respect, India forms a crucial area for investigating questions of hominin dispersals across Africa and Asia from the Lower or Middle Pleistocene onwards (Dennell 2009; Kennedy 2003; Korisettar 2002; Paddayya 1984; Paddayya *et al.*, 2002; Petraglia 1998; Petraglia and Allchin 2007; R.S.Pappu 2001; Mishra 1995, 2006-2007; Misra 1989; Sankalia 1974). Today, our prehistoric heritage is under urgent threat of being completely wiped off the face of India's archaeological record. This humble face of the past has been traditionally overshadowed by an overwhelming focus on the conservation of 'glamorous' elements of India's architectural and art heritage, a phenomenon one may term the 'Taj Syndrome' (Pappu 2006). This has led to a general apathy in conserving sites lacking larger structural/sculptural remains or evidence of art. In particular, Palaeolithic sites with their stone artefacts, fossils, features and sediments, form fragile and practically invisible elements of the natural landscape, ignored by the community and government authorities alike. In addition to impacts caused by infrastructure development, irrigation, agriculture, etc., destruction is also caused by random collection of artefacts by scientists, students and tourists. Existing legislations are outdated and do not specifically address issues related to archaeological impact assessment (Biswas 1999; Pappu 2007b, The Ancient Monuments and Archaeological Sites and Remains Act 1958 (24 of 1958); The Antiquities and Art Treasures Act, 1972 (52 of 1972). There is often an absence of coordination and understanding within and between government departments and non-governmental organizations/institutes and universities. Above all, there is widespread ignorance on the part of the community and of many government officials as to what constitutes a prehistoric site.

On a global scale, prehistoric sites (Palaeolithic/Mesolithic) form a small fraction of the 679 cultural World Heritage Sites listed by UNESCO as deserving special protection (http://whc.unesco.org/en/list/). These include sites renowned for either finds of hominin fossils with/without artefacts, or those marked by spectacular rock art. The same logic applies to the ICOMOS list of World Heritage Hominid Sites (http://www.international.icomos.org /centre_documentation); e.g. Zhoukoudian, the Awash and Omo valleys, etc. Bhimbetka is the only prehistoric site within the 22 World Heritage Sites listed in India (No.925, declared in 2003 as per criteria: (iii)(v); http://www.w3c.org/TR/1999/REC-html401-19991224/ loose.dtd; http: //whc.unesco.org/archive/advisory_body_ evaluation/925.pdf). In India, very few prehistoric sites feature in the list of monuments/sites declared protected by the Archaeological Survey of India. In Tamil Nadu, of the 403 protected sites, not a single one is Palaeolithic. The situation is slightly better in Madhya Pradesh where at least nine site complexes are listed, primarily owing to their wealth of rock art (http://asi.nic.in). This is not to deny occasional examples of large scale documentation and salvage of prehistoric sites conducted in areas threatened by major irrigation works, e.g. at Nagarjunakonda (Subrahmanyam 1975, xx-xxiii), and in the Narmada valley (Ota 2000). Efforts of individual archaeological officers have helped mitigate intense destruction occurring on smaller scales. The time has thus come for immediate documentation of sites and quantification of the cultural and environmental impacts to which they are being subjected (http://whc.unesco.org; http://www.icomos.org/icahm/).

Research and Heritage Management Planning: An Integrated Approach

In the course of our research project investigating prehistoric archaeology and Pleistocene palaeoenvironments in parts of Tamil Nadu, we built in a component of impact assessment and conservation, including a framework suggesting long-term strategies for managing these sites. The region studied is well known for the richness and diversity of its prehistoric heritage (Pappu 2001, 2007a). The current project drew on the foundation of excavations at the Palaeolithic site of Attirampakkam (Blackwell *et al.*, 2007; Gunnell *et al.,* 2006; Pappu and Kumar 2006, 2008; Pappu *et al.,* 2003a,b, 2004; Sreedhar *et al.*, 2008); and was then expanded to investigate regional patterns in hominin landuse in relation to changing Quaternary environments. The first phase of the project considers an area of 8000 km^2 in northern Tamil Nadu (Figure 1), following which studies are now being initiated in the southern parts of the state. There have been relatively few attempts to use satellite remote sensing in archaeology in India (Balaji *et al.,* 1996; Chakrabarti *et al.,* 2007; Nageswara Rao *et al.,*1997; Ramasamy *et al.,* 1992, 2006; Ray and Ravindranath 2007; Tripathy 2005). Our project is the first attempt in Indian prehistory, to use a combination of satellite images, GIS, field and laboratory studies; integrating research aims with those of impact assessment, management planning and predictive modelling (Allen *et al.*; 1990; Mehrer and Wescott 2005). We provide here an overview of this work, with a focus on impact assessment studies on prehistoric sites.

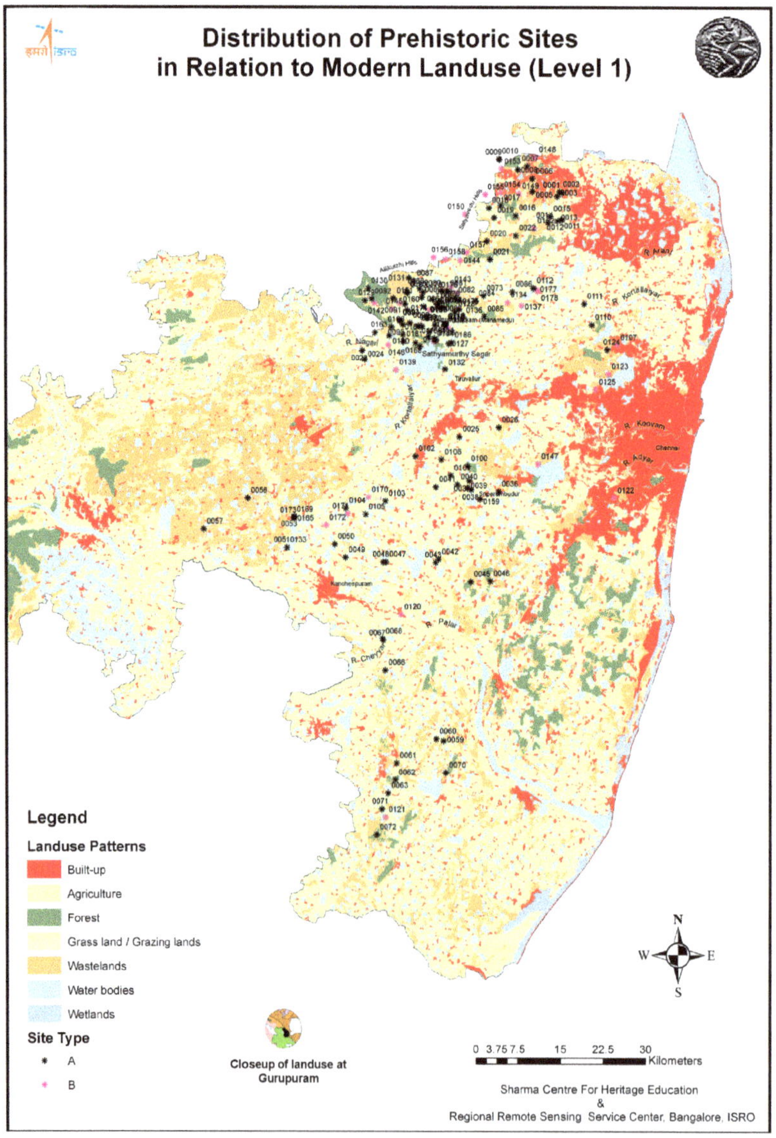

Fig.1. Location of Prehistoric Sites in the Study Region in Relation to Landuse Patterns (derived from satellite data: 50 m buffer) and inset: closeup of landuse at the site of Gurupuram. (Site type A: discovered by the team/previous scholars and re-visted/restudied; Site type B:site description based on existing literature (work of previous scholars).

Satellite Remote Sensing, GIS and Fiels Studies

As regards satellite remote sensing, a three tiered approach was adopted for creating a database of the study region. The regional context of each site was studied through medium resolution satellite data of IRS 1D LISS 3 (Nov 2004) of 23 m. To understand the local context of sites, IRS 1D PAN + LISS 3 (Nov 2004) merged satellite data of 5.8 m resolution was used. High resolution IKONOS 1 m satellite data (April 2005) was used to get an in-depth understanding of the site of Attirampakkam and its surroundings. Following geo-referencing of the images and use of image fusion techniques, enhanced images could be generated. This was accompanied by digitisation of relevant information from reference maps related to topography, hydrology, etc. The satellite images were interpreted and updated and thematic maps were prepared on various themes like landuse/landcover, geomorphology, geology, elevation, transport network, modern settlement patterns, surface water bodies, canal networks; drainage, etc. Archaeological sites and important sections studied for palaeoenvironmental purposes were examined in relation to these parameters. A codification scheme was evolved for various thematic maps. A database was created for all environmental, landuse, and archaeological parameters. Village cadastral maps for key sites were also geo-referenced and tiled together. GIS based spatial analysis techniques like overlay, proximity and query were used for generating various combination thematic maps. This relational database was then organised with systematic structuring of the tables and records with one to many relation. The multiple tables were linked through unique identification numbers and concatenation (as required). The entire RDBMS was linked to the geospatial database for prognostic analysis. In addition to this, predictive locational modelling was also attempted to infer the location of archaeological sites. This was used to address specific archaeological research questions and for Heritage Management studies. Sites studied included those discovered by our team, those discovered by previous scholars and re-examined by us. The work of previous scholars was also referred (Pappu 2001).

Field studies included surveys, test-pits, studies of Quaternary geomorphology; analysis of artefact typology and technology, sedimentological and palynological studies, geochronology, etc. Data from excavations at Attirampakkam was included to provide stratigraphic and chrono-cultural controls. At each site/section, GPS readings were taken with the help of Garmin Etrex handheld GPS (error around 10 m), followed by a DGPS survey which recorded accurate locations of key sites in the study area. Although salvage operations are the last strategy in any management program, we were forced to adopt them at sites which were under immediate threat by quarrying, e.g. Senrayanpalayam, Sirunai quarry, Singadivakkam, etc. Sampling strategies ranged from random collections to grid-based sampling and synchronized walking across transects.

Impact Assessment Studies

In addition to generating a database on sites, we quantified impacts affecting site integrity as a first stage for heritage management planning. Parameters considered in this study (Figure 2) include agriculture, construction and infrastructure development, water bodies (natural and man-made), forested areas, wastelands and quarried areas. Other impacts include persistent disturbance by random collection of artefacts, and trampling by students and visitors. The magnitude and severity of these impacts was assessed by expert judgment and were assigned weights on a scale of 0 (no impact) to 1 (maximum impact). Impacts were analysed at several levels; *viz*-: a) based on field studies (2005-2009); b) based on satellite data of the actual site area sampled and studied; c) based on satellite images within a buffer radius of 50 m, 500 m and 1 km around sites, for purposes of standarised comparisons. Broad 'Prehistoric Zones' were demarcated, within which landuse patterns were analysed. These comprise areas with clusters of prehistoric sites (areas with low to high density scatters in clusters; those with low density scatters across several kilometers, and those with single or a few artefacts only). Impact indices were generated using a generic 50 m buffer from satellite data and from field studies; the index was calculated as a product of weight and area affected for each type of landuse impact. A long term perspective on impacts could be obtained by plotting sites on the 1917 topographic sheets (published under Colonel Sir S.G. Burrard, K.C.S.I., R.E., F.R.S, Surveyor General of India; with a scale of 1 inch is to 1 mile or 1:63,360), and studying observations in the writings of Foote (1870, 1916a, b), and other sources (Crole 1879).

The Prehistoric Archaeological Record

The study area chosen covered an area of 8000 km^2, in the districts of Thiruvallur, Kanchipuram and Chennai; i.e. the basins of the rivers Arani, Kortallaiyar, Cooum, Adyar, Palar and Cheyyar (12° 15' 00" to 13° 30' 00" N; 79° 30' 00" to 80° 22' 00" E). The geology of this region has been discussed elsewhere (see Geological Survey of India 2006; Gunnell *et al.,* 2006; Kumaraguru and Trivikrama Rao 1994; Muralidharan *et al.,* 1993; Pappu 2001; Pappu *et al.,* 2003 b; Sreedhar *et al.,* 2008; Subramanian and Selvan 2001). Palaeolithic sites occur within Quaternary formations consisting of a range of ferricretes, lateritic/ferricreted sediments and gravels, reworked ferruginous and lateritic cobble to boulder beds, within silty-sands and clays, and on the surface of Cretaceous formations and Cenozoic laterites. A total of 197 sites/sections within 154 site complexes, were studied for both archaeological purposes and for generating palaeoenvironmental data. A complex is a broader term for a locality which may comprise one or more sections or sites (defined as low to high density clusters of artefacts within limited areas; low to medium density scatters of artefacts across several kilometers, or spots with individual or a few artefacts) (Pappu 2001). Of these, 165 are prehistoric sites within 138 site complexes. Lower Palaeolithic Acheulian assemblages were noted at 95 sites within 82 complexes. Although concentrated in the basins of the rivers Arani and Kortallaiyar, occasional scatters of Acheulian artefacts are found further south, up to the north bank of the river Palar. Such sites occur within lateritic gravels and sediments, on the surface of Cretaceous deposits and Cenozoic laterites, and within argillaceous sediments at Attirampakkam (Pappu *et al.* 2003b, 2004; Gunnell *et al.,* 2006). Middle Palaeolithic sites number 133 within 119 complexes; they overly Acheulian horizons occurring within a range of lateritic/ferricreted gravels; on the surface of Cenozoic laterites or Cretaceous formations; and are also noted within non-lateritic/ferruginous fluvial gravels, silts and sands, and on the surface of charnokites. They have a wider distribution extending further south of the river Palar. Raw materials now include occasional use of gneiss, quartz, chert and chalcedony, in addition to quartzites. Late Palaeolithic sites comprising assemblages with a distinct blade element and with/without microliths (regional variants of the Indian Upper Palaeolithic), occur at 43 sites, in 34 complexes. There are only 7 sites with a truly microlithic assemblage. Although fewer in number, they have a greater range of distribution as compared to earlier phases, and extend far south of the river Palar. They occur on the surface of and within lateritic gravels and fluvial deposits, and on the surface of various other formations. There is an increase in the diversity of raw material used, including silicified wood, in addition to types noted earlier. There are very few truly microlithic assemblages. In addition to surveys, excavations at Attirampakkam (locally known as Manamedu; 13°13 50 N and 79°53 20 E) yielded a stratified cultural sequence ranging from the Lower to the late Middle Palaeolithic. This aided in establishing controls by which assemblages across the region could be correlated (Pappu and Kumar 2008).

Assessing Impacts on Prehistoric Sites

Results of impact assessment studies at prehistoric sites are presented below (Pappu, 2008; Figure 2).

Agriculture

Most sites are affected to some degree by agricultural practices (Figure 2). In 1917, only 34 sites fell within zones classified as comprising agricultural land. Recent satellite images indicates that 121 sites are affected by some form of agriculture, of which crop lands predominate, followed by plantations (Figure 2).

Landuse as per Satellite Images		**Actual Site Area**		**50 m buffer radius**		**1 km buffer radius**		**Prehistoric Zones**	
Discr_L1	**Discr_L4**	**N**	**Area (km²)**	**N**	**Area (km²)**	**N**	**Area (km²)**	**N**	**Area (km²)**
Agriculture	Coconut	2	233.65	0	0.00	6	578.86	7	1716.91
	Aquaculture	0	0.00	1	1.88	2	24.30	1	8.23
	Permanent Fallow	1	221.69	1	7.35	9	1946.52	3	565.54
	Current Fallow	24	1744.31	34	187.83	423	44784.91	129	39726.52
	Kharif	18	720.84	14	54.80	267	27851.54	69	27581.17
	Kharif + Rabi (double cropped)	20	2862.01	18	100.52	211	32545.75	70	25816.98
	Rabi	28	3241.99	28	127.94	479	54769.99	80	52460.77
	Plantations	28	1133.33	23	133.05	426	63159.17	123	49742.62
Agriculture Total		*121*	10157.82	*119*	613.37	*1823*	225661.05	*482*	197618.74
Built-up	Mixed Built-up land	0	0.00	1	6.33	3	1176.74	2	112.52
	Open Spaces/Vacant Land	1	63.66	2	4.38	12	3311.07	8	2682.70
	Residential	1	14.04	5	34.53	19	7187.78	5	2887.49
	Villages (Rural)	13	589.28	12	54.81	328	29855.20	133	14054.40
	Airport	0	0.00	0	0.00	3	65.92	1	156.08
	Industrial	0	0.00	0	0.00	20	671.55	21	586.83

Landuse as per Satellite Images		Actual Site Area		50 m buffer radius		1 km buffer radius		Prehistoric Zones	
Discr_L1	**Discr_L4**	**N**	**Area (km^2)**	**N**	**Area (km^2)**	**N**	**Area (km^2)**	**N**	**Area (km^2)**
Built-up Total		*15*	*666.98*	*20*	100.05	*385*	42268.25	*170*	20480.03
Forest	Forest (Indeterminate)	1	516.62	1	7.81	1	1282.58	3	2450.04
	Forest Plantations	5	160.63	3	3.59	26	6364.35	8	31561.59
	Open Forest	1	21.20	1	7.08	8	742.15	3	1182.84
	Scrub Forest	13	2024.18	8	32.41	63	15403.82	15	45986.33
	Dense/Closed Forest	0	0.00	0	0.00	2	131.21	1	1914.47
Forest Total		*20*	2722.62	*13*	50.90	*100*	23924.12	*30*	83095.27
Wastelands	Barren Rocky/Stony Wastelands	0	0.00	1	7.81	3	1258.78	3	4371.14
	Land with gullies/ravines	3	28.66	7	22.25	53	2035.67	8	2046.22
	Land with scrub	32	6719.38	33	226.11	211	52162.59	86	107708.64
	Land without scrub	16	2157.37	18	119.31	91	18159.32	51	23401.59
	Salt Affected Land	15	2179.13	12	81.99	85	21130.11	37	19313.80
	Mining process	0	0.00	0	0.00	3	255.96	1	115.91
	Mining/Industrial waste	0	0.00	0	0.00	5	608.97	1	19.07
Wastelands Total		*66*	11084.55	*71*	457.47	*451*	95611.41	*187*	156976.38
Water bodies	River bed Vegetation	1	22.62	0	0.00	53	1847.35	5	455.01
	River bed Cultivation	0	0.00	0	0.00	4	153.60	2	104.90
	Reservoir with Vegetation	0	0.00	0	0.00	20	1104.77	5	399.02
	Tank bed vegetation	22	524.93	2	11.06	440	18974.17	250	12965.49
	Tank bed cultivation	0	0.00	0	0.00	1	33.98		0.00
	Water channel area	1	11.97	0	0.00	24	1373.98	1	23.36
	Water spread area	13	1303.51	1	0.22	230	11371.64	97	10011.87
	Sandy area	4	151.23	2	8.36	152	9509.37	54	1715.81
Water bodies Total		*41*	2014.26	*5*	19.64	*924*	44368.86	*414*	25675.45
N=counts of areas within sites									

Fig. 2. Landuse patterns at Prehistoric Sites as derived from Satellite Data. The counts indicate the number of areas within sites affected by different landuse patterns. Each site may have one or more types of landuse within the same broad category (1 km buffer data updated from Pappu *et al.* 2008, which lists average area impacted).

In contrast to these observations, field studies point to intense agriculture at only 10 sites. This discrepancy arises from factors related to time and scale. Field studies were of a very high resolution and assessed impacts directly affecting the actual areas of artefact distribution. This comprised lateritic sediments/gravels, boulder conglomerates, etc.; where agriculture is difficult; cultivation occurring generally on the surrounding fluvial fine grained deposits. Recent cultivation, post-dating satellite image data, is noted at five sites.

Infrastructure Development: (Construction/Built Up Areas)

In 1917 only 28 sites were within built up areas, chiefly representing villages. As compared with data from satellite images, field studies point to a higher degree of destruction (n=110 areas within sites from field data) (Figure 2). This is a result of two factors: a) the time gap between recent field work and satellite data; b) higher scales of resolution of field studies which enabled observation of small scale but destructive activities such as graves, small wells/canal cuttings, plots demarcated for houses, field boundaries, etc. Satellite data indicates that most sites are disturbed by built up areas associated with villages (Figure 2). Occasional artefacts are found within larger cities such as Chennai. Intensive development in recent years is seen largely in Kanchipuram and Thiruvallur districts (http://www.kanchi.tn.nic.in/industries.html; http://tiruvallur.nic.in/); with the former being one of the largest industrial areas of Tamil Nadu, with production bases of international industrial groups such as Ford, Hyundai, etc. Roads (national and state highways, major and other district roads); are a major impact at all sites, and were studied within a 500 m buffer radius around each site. Ancillary activities such as quarrying for lateritic deposits which contain artefacts, stone crushing (including artefacts), and the impact of vehicular traffic, also affect sites.

Quarries

Rapid and often illegal quarrying is commonly seen in areas with charnokites, Cenozoic laterites, Quaternary lateritic sediments, boulder conglomerates, and fluvial silts, sands and gravels; within and on whose surface, prehistoric sites occur (Figure 2, Figure 3). Satellite data indicates 11 major quarried areas, destroying 60.11 km^2 (50 m buffer radius) within prehistoric sites; more recent field studies point to 54 sites being quarried. Quarries are often filled with sediments from neighbouring areas, leveled, and used for cultivation. Quarrying of sediments for the construction of the Telugu Ganga Project canal (Krishna Water Supply Project), or embankments for the Sathyamurthy Sagar reservoir resulted in destruction of numerous sites in Thiruvallur district. The Red Hills, Erumaivettipalayam and Sholavaram areas near Chennai have been totally destroyed by quarrying of alluvial deposits, laterites and lateritic sediments.

Wastelands

A high percentage of sites occur on wastelands, where animal grazing, cutting of firewood and burials, affect sites to some extent. The high site occurrence in these contexts (Figure 2) may be attributed to four reasons: a) wastelands occur generally in areas with Cenozoic laterites; lateritic sediments, gravels or other formations, which constitute artefact horizons; b) most raw material sources are located here; c) they lack intensive agriculture or dense vegetation resulting in high site preservation and visibility; d) they are often dissected by seasonal streams and gullies which erode Pleistocene sediments and expose artefacts. From 1917 onwards (n=96), most wastelands at sites, have been encroached on (Figure 2).

Fig. 3. Recent destruction of an Acheulian site via quarrying and leveling.

Forested areas

In the 1917 toposheets, 39 sites were under forest cover, some of which were Reserved Forests. Both satellite and field studies note few sites in these zones, owing to a bias in survey arising from low visibility and difficulty of access. Sites are affected by root action, burrowing, erosion, as well as illegal deforestation and cultivation. Afforestation schemes affect spatial distribution of artefacts via digging pits for planting, for drainage ditches and firebreaks, and construction of low stone circular barriers around saplings. There is some overlap with wastelands owing to illegal deforestation, and most of the 14 major Reserved Forest zones in this region marked in the 1917 maps have been encroached. Field surveys noted forests at 20 sites, while nine had some form of plantations (mango, eucalyptus, casuarinas).

Water Bodies

As per the 1917 topographic maps, 23 areas within sites were affected by major water bodies-dams, tanks, river/streams. Major reservoirs (Sathyamurthy Sagar, Red Hills,

Sholavaram, Chembarambakkam and Madurantakam), have submerged areas where sites must have existed. The maximum impact is that of Sathyamurthy Sagar which occurs in the immediate vicinity of the famous sites of Pundi and Attirampakkam; with a capacity of 2573 Mcft. Severe damage to sites was caused by the construction of the Telugu Ganga Krishna water project (http: //en.wikipedia.org/wiki/Telugu_Ganga_project), bringing water over 434 km from the Srisailam reservoir in Andhra Pradesh; to reservoirs at Red Hills, Sholavaram and Chembarambakkam. Impacts resulting from groundwater fluctuations, water release, sedimentation, etc.; were noted but require further studies.

Scales of Management: From Single Sites to Prehistoric Zones

IKONOS data (1 m resolution) aided in studying landscapes around the excavated site of Attirampakkam (Figure 4). In addition to micro-level mapping of landuse; overlaying of village cadastral maps helped plan management strategies at the level of the site (Figure 4). On a broader scale, estimates of impacts within a 1 m buffer radius around sites also highlighted potential threats (Figure 2). For heritage management purposes, consideration of landuse patterns within broader 'Prehistoric Zones' as defined earlier, was important to plan regional level strategies. These zones were graded into areas having lower or higher intensity impacts, based on generation of impact indices. It is hoped that these zones may eventually be developed into 'Prehistoric Parks'.

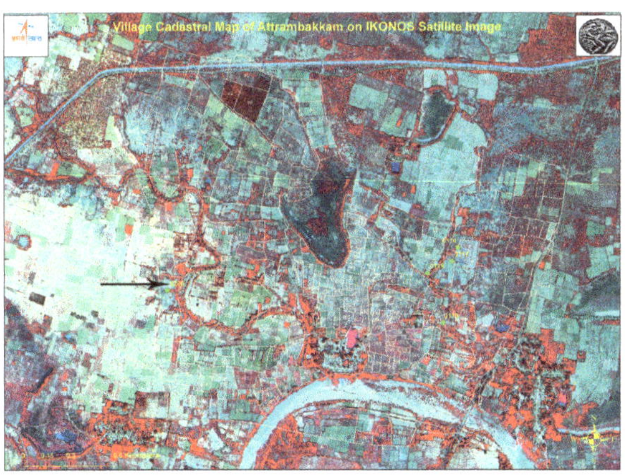

Fig. 4. IKONOS image (1 m resolution) showing area around the site of Attirampakkam (Manamedu; Attrambakkam); overlaid with the cadastral (revenue) maps.

Grading Sites on the Basis of Impacts

Following calculation of impact indices, sites were graded into two categories above and below a median value; *viz.* those with low-medium impacts and those with high impacts (Figure 5). Discrepancies between field and satellite data arising from variability in time and scales of observation were factored in. Studies of palaeoenvironmental impacts were conducted by Pappu (1999); and there is considerable future scope for examining effects of climatic change (Case Studies on Climate Change and World Heritage, http://whc.unesco.org.).

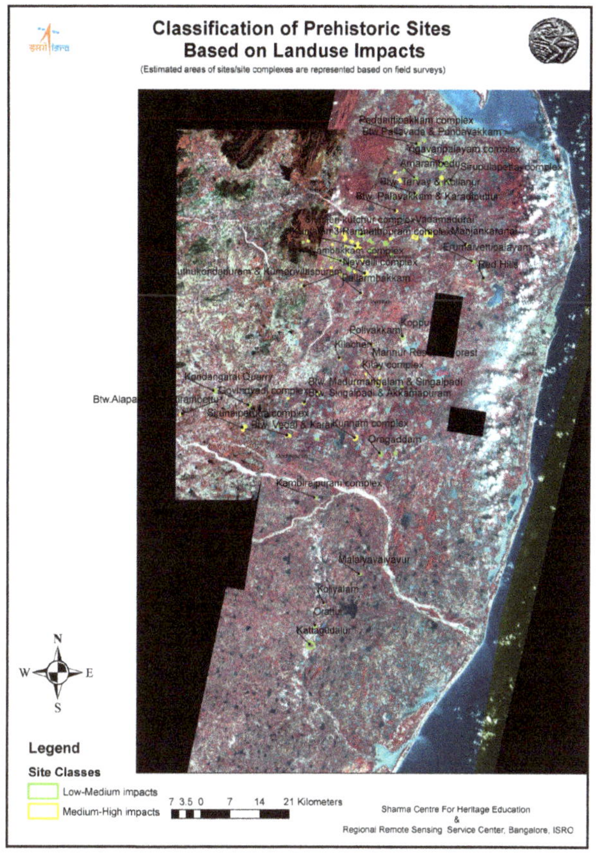

Fig. 5. Map showing gradation of prehistoric sites as per the degree of impact suffered.

The Scientific and Educational Significance of Prehistoric Sites

For research purposes, all sites were considered important irrespective of their level of destruction. However, for management purposes, the scientific potential of sites required to be assessed in order to grade sites, and focus on those which required immediate investigation. There was also a need to assess the role sites could play in community educational programs (http://www.tsa.gov.bc.ca/archaeology/docs/impact_assessment_guidelines/assessment_and_review_process_part3. htm).

Scientific Significance

For heritage management purposes parameters considered for assessing the scientific significance of a site included :-1) significance in terms of understanding long-term hominin behavioural changes. The cultural history and stratigraphy of the site was considered; i.e. whether the site has evidence of stratified single or multicultural phases, and its significance in terms of understanding changes in the use of the landscape in relation to long environmental records; 2) significance of surface sites or shallow buried sites (upto 10-20 cm), in terms of informing on single or multiple cultural phases, and their potential in understanding behaviour; 3) The nature of artefact assemblages in terms of assemblage structure, density per unit area, unique/rare typo-technological features; spatial patterning; and other features; all of which aid in understanding the site and its position in the regional perspective; 4) importance of the site/section for yielding information on palaeoenvironments and for chronology. Sites were then graded into categories to facilitate implementing various strategies for conservation and management. Following consideration of impacts suffered and scientific significance, remedial measures were suggested for each site/section studied (Figure 6).

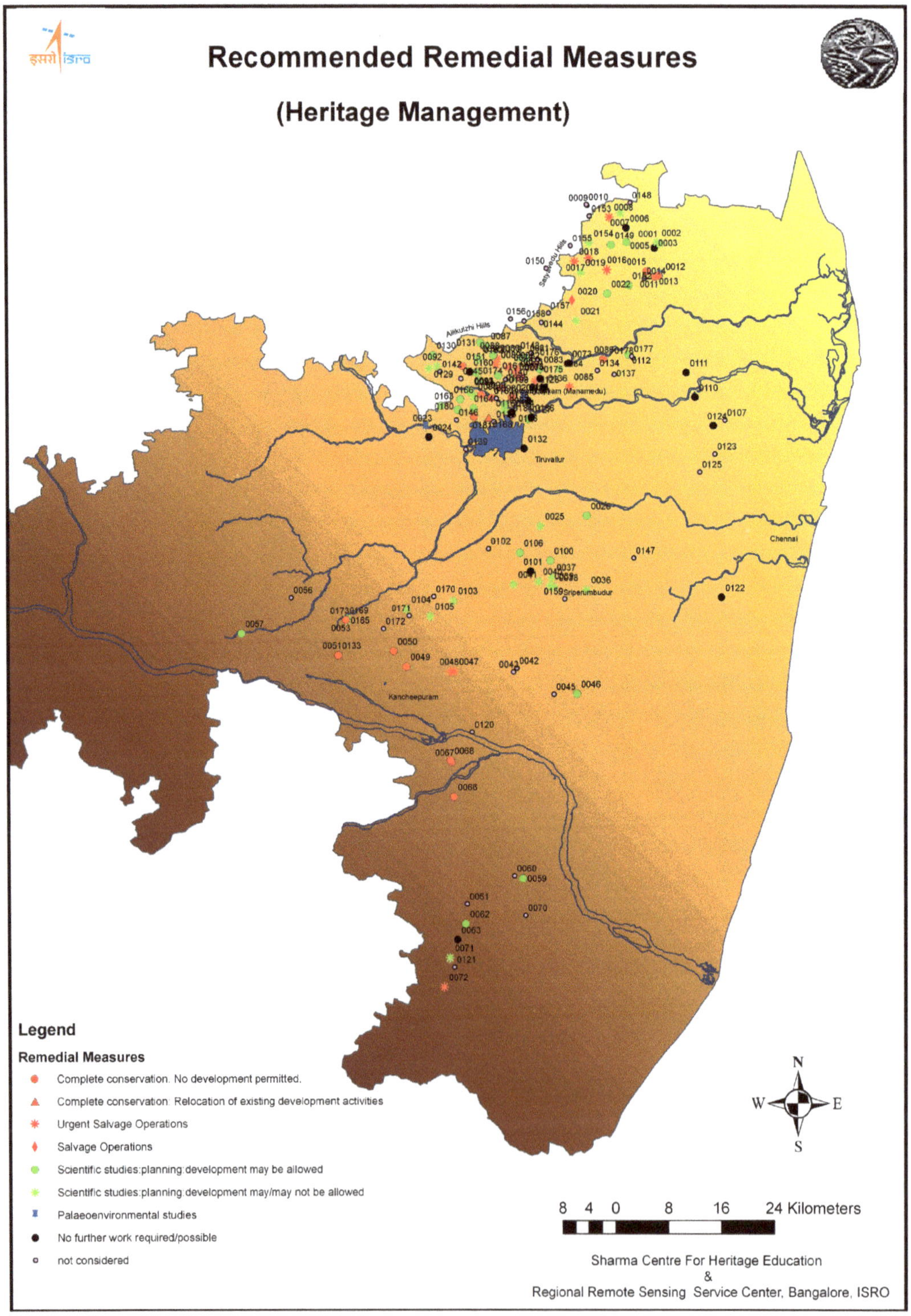

Fig. 6. Remedial measures recommended for individual prehistoric sites.

Significance for Community Education

Irrespective of the severity of impact or scientific significance, a site may be of importance from the point of view of developing public awareness on prehistory and human evolution. This is essential for the success of any long term management strategy (e.g. ICCROM Awareness Projects). At Attirampakkam, villagers were involved in the process of excavation, and were introduced to prehistory and human evolution as related to their own local heritage, and helped disseminate this information to visitors. Villagers, school and college groups and government officials visited the excavations and were introduced to the subject through lectures and workshops. Similar programs in the Sharma Children's Museum in Chennai generated an interest in prehistory (Pappu and Kumar 2005; Kumar and Pappu 2008; http://www.sharmaheritage.com). Despite the immense value of site museums for education, we believe that, at present, they should not be established at prehistoric sites in India. Without continuous vigilance, site museums and tourism will lead to uncontrolled trampling on fragile deposits, collection of artefacts as 'souvenirs', building of ancilliary shops, roads, etc.; which will destroy the site. Alternately, it is more effective to establish small village/town museums, site replicas/virtual museums in central locations, organise displays within schools or government offices. These should be multidisciplinary, including both archaeology and other topics of local interest. Awareness and generation of interest should go hand in hand with implementing remedial measures.

Conserving India's Prehistoric Past: An Uphill Task?

The fragile and non-renewable nature of prehistoric sites (see ICOMOS Charter of 1990), calls for urgent, yet carefully structured methods in impact assessment, conservation and management. Adapting international recommendations (http://www.icomos.org/icahm/) to suit local conditions, we suggest an Integrated Management Plan for Prehistoric Archaeology. Such a plan encompasses archaeological and environmental dimensions of a site, within a broader social and economic milieu. It suggests coordination of several government and non-government agencies and the local community (see similar strategies in South African Heritage Resources Agency; SAHRA, http://www.sahra.org.za/nat_heritage.htm). It recommends legislative reforms, and suggests that their implementation adopts a decentralized balanced approach, strict where necessary, and yet compassionate towards local needs. The Indian social, environmental and archaeological landscapes form a complex integrated whole, and any attempt to ensure that prehistoric heritage is conserved for future generations, should adopt a practical policy where research, conservation and development may proceed in mutual harmony. A change in attitude towards the prehistoric past is required by generating awareness both in local communities and at the level of senior government officials responsible for decision making.

It is time to wake up before India's prehistoric past is reduced to dry words in research papers and forgotten stones and bones on dusty museum shelves.

Acknowledgements

This article is based on an ISRO-GBP project for which we thank Dr. C.B.S. Dutt, Dy-Project Director ISRO-GBP; Professor R. Ramesh, PRL, Ahmedabad, and Dr. P.G. Diwakar, Director, RRSSC-Bengaluru. We thank the Archaeological Survey of India and the Chennai Circle, and the Department of Archaeology, State Government of Tamil Nadu, for granting us licenses. Research at Attirampakkam was funded by the Homi Bhabha Fellowships Council, the Leakey Foundation, and the Earthwatch Institute, USA. Community education programs and work on children's museums were funded by the Earthwatch Institute, and the Ministry of Culture, Government of India. We thank Mr. Saleem Sheikh for his help in fieldwork. S. Pappu and K. Akhilesh thank the Sharma Centre for Heritage Education, our field staff, and the villagers of Manamedu, Krishnaveram and Pundi.

References

ALLEN, K.M.S.; GREEN, S.W. ; AND ZUBROW, E.B.W. (EDS.) 1990. *Interpreting Space: GIS and Archaeology*. London, Taylor and Francis, Ltd.

BALAJI, K., SURESH, L.S., RAGHAVASWAMY, V. AND GAUTA, N.C. 1996. The Use of Remote Sensing in Monitoring Changes in and around Archaeological Monuments-A Case Study from Hyderabad, Andhra Pradesh. *Man and Environment* XXI(2), 63-70.

BISWAS, S.S. 1999. *Protecting the Cultural Heritage (National Legislations and International Conventions)*. New Delhi, Aryan Books International.

BLACKWELL, B.A.B., MONTOYA, A., BLICKSTEIN, J.I.B. , SKINNER, A. R. , PAPPU, S., GUNNELL, Y., TAIEB, M. , KUMAR, A., LUNDBURG, J. 2007. ESR analyses for teeth from the open-air site at Attirampakkam, India: clues to complex uptake and paleoenvironmental change. Radiation Measurements, 42 (6-7), 1243-1249.

CHAKRABARTI, C., BHATTACHARYYA, B. , CHAKRAVARTY, P. , BANERJEE, S.N. , GANGOPADHYAYA, K. AND SENGUPTA, G. 2007. Preliminary Observations on the Growth of Early Historic Settlement of Chandraketugarh, West Bengal-A Geomorphological Approach. *Man and Environment*, XXXII(2), 47-60.

CROLE, C.S.1879. *The Chingleput, Late Madras District. A Manual Compiled under the Orders of the Madras Government.* Madras, Lawrence Asylum Press.

DONNELLY, R. 2009. *The Palaeolithic Settlement of Asia.* Cambridge, Cambridge University Press.

FOOTE, R.B. 1870. Notes on the Geology of Neighbourhood of Madras, *Records of the Geological Survey of India* III (1), 11-17.

FOOTE, R.B. 1916a. *The Foote Collection of Indian Prehistoric and Protohistoric Antiquities. Catalouge Raisonne.* Chennai, The Commissioner of Museums. Government Museum, Chennai.

FOOTE, R.B. 1916b. *The Foote Collection of Prehistoric and Protohistoric Antiquities. Notes on Their Ages and Distribution.* Chennai, The Commissioner of Museums. Government Museum, Chennai.

GEOLOGICAL SURVEY OF INDIA. 2006. *Geology and Mineral Resources of Tamil Nadu and Pondicherry.* Government of India Press.

GUNNELL, Y., RAJASHEKAR, C., PAPPU, S., TAIEB, M. AND KUMAR, A. 2006. On the Depositional Environment of Lower Palaeolithic Horizons at the Prehistoric Site of Attirampakkam, Tamil Nadu. Current Science 91,114-118.

KENNEDY, K.A.R.2003. *God-Apes and Fossil Men: Palaeoanthropology in South Asia.* Ann Arbor, University of Michigan Press.

KORISETTAR, R. 2002. The Archaeology of the South Asian Lower Palaeolithic:History and Current Status, in S.Settar and R.Korisettar (eds.), *Indian Archaeology in Retrospect*, Vol.1., 1-65. New Delhi, Manohar, Indian Council of Historical Research.

KUMAR, A. AND PAPPU, S. 2008. Perspectives on Archaeology from an Indian Village: Examples from Attirampakkam'; paper presented in 'Perspectives on the Past: Archaeology Beyond Academics' (Shanti Pappu and M.Bezerra de Almeida organisers), at the World Archaeology Congress, WAC-6, Dublin, Ireland: 29th June-4th July, 2008.

KUMARAGURU, P., TRIVIKRAMA RAO A. 1994. A Reappraisal of the Geology and Tectonics of the Palar Basin Sediments, Tamilnadu. In *Ninth International Gondwana Symposium, Hyderabad, Jan. 1994*, 2, 821-831. Rotterdam, Geological Survey of India and Balkema.

MEHRER, M.W. AND WESCOTT, K.L. 2005. *GIS and Archaeological Site Location Modeling* London, Taylor and Francis.

MISHRA, S. 1995. Chronology of the Indian Stone Age: the Impact of Recent Absolute and Relative Dating Attempts. *Man and Environment* 20(2), 11-16.

MISHRA, S. 2006-2007. The Indian Lower Palaeolithic. *Bulletin of the Deccan College Post-Graduate and Research Institute*, 66-67:47-94.

MISRA, V.N. 1989. Stone Age India:An Ecological Perspective. *Man and Environment* XIV (1):17-64.

MURALIDHARAN, P.K., PRABHAKAR, A., KUMARGURU, P. 1993. Geomorphology and Evolution of the Palar Basin, in *Workshop on Evolution of East Coast of India*, April 18-20, *Abstract of Papers.* Tanjore, Tamil University.

NAGESWARA RAO, K., UDAYABHASKARA RAO, Ch. and Srinivasa Rao, K.1997. Significance of Remote Sensing for Geomorphic Mapping in Archaeological Investigations. *Man and Environment* XXII(2), 67-72.

OTA, S.B. 2000. Cultural Heritage Management Vis-A-Vis Dams: The Narmada Issue, India, in *World Commission on Dams and Cultural Heritage Management Final Report August 2000 Working paper submitted to the WCD* (A.Brandt and F.Hassan Eds.), pp. 59-61. (pdf). http://www.dams.org.

PADDAYYA, K. 1984. India: Old Stone Age, in O. Bar-Yosef (ed.), *Neue forschungen zur Altsteinzeit. Sonderdruck* , 345-403. Munich, Verlag C.H.

PADDAYYA, K., BLACKWELL, B.A.B, JHALDIYAL, R., PETRAGLIA, M.D., FEVERIER, S., CHADERTON, D.A, BLICKSTEIN, J.I.B, AND SKINNER A.R. 2002. Recent findings on the Acheulian of the Hunsgi and Baichbal Valleys, Karnataka, with special reference to the Isampur excavation and its dating. *Current Science* 83(5), 641-7.

PAPPU, R.S. 2001. *Acheulian Culture in Peninsular India.* New Delhi, D.K. Printworld Ltd.

PAPPU, S. 1999. A Study of Natural Site Formation Processes in the Kortallayar Basin, Tamil Nadu, South India. *Geoarchaeology, an International Journal* 14(2), 127-150.

PAPPU, S. 2001. *A Re-examination of the Palaeolithic Archaeological Record of Northern Tamil Nadu, South India.* Oxford, British Archaeological Reports (BAR) International Series 1003.

PAPPU, S. 2006. Prehistory in Tamil Nadu:The Needs for Links and Communication, in M.Kannan and C. Mena (eds.), *Negotiations with the Past: Classical Tamil in Contemporary Tamil*, 1-24. Puducherry, Institut Français de Pondichéry and Berkeley, University of California.

PAPPU, S. 2007A. Changing Trends in the Study of a Palaeolithic Site in India: A Century of Research at Attirampakkam, in M.D.Petraglia and B.Allchin (eds.), *The Evolution and History of Human Populations in South Asia, Inter-disciplinary Studies in Archaeology, Biological Anthropology, Linguistics and Genetics Series*, 121-135. The Netherlands, Springer.

PAPPU, S. 2007B. *A Study of Laws Governing India's Cultural Heritage*, unpublished Project Report, II Year, LL.B.Pune, Shri Shivaji Maratha Societies Law College.

PAPPU, S. AND KUMAR, A. 2005. Prehistoric Archaeology and Community Education: earning from excavations at Attirampakkam, Tamil Nadu, *Humankind* 1, 131-145.

PAPPU, S. AND KUMAR, A. 2006. Preliminary Observations on the Acheulian Assemblages from Attirampakkam, Tamil Nadu, in Naama Goren-Inbar and Gonen Sharon (eds.) *Axe Age. Acheulian Toolmaking from Quarry to Discard,* 155-180. London, Equinox Publishing Ltd.

PAPPU, S. AND KUMAR, A. 2008. The Acheulian of the South-East Coast of India: Attirampakkam and Beyond, in H. De Lumley (ed.), *Les Cultures A Bifaces Du Pleistocene Inferieur Et Moyen Dans Le Monde. Emergence Du Sens De L'Harmonie, in press.*

PAPPU, S. ,GUNNELL, Y., TAIEB, M., BRUGAL, J-P., ANUPAMA, K., SUKUMAR, R., AND KUMAR, A. 2003A. Excavations at the Palaeolithic Site of Attirampakkam, South India. Antiquity, 77 (297), September 2003.

PAPPU, S., GUNNELL,Y., TAIEB,M., BRUGAL,J-P., TOUCHARD, Y. 2003b. Ongoing Excavations at the Palaeolithic Site of Attirampakkam, South India: Preliminary Findings, *Current Anthropology* 44(4), 591-597.

PAPPU, S., GUNNELL,Y., TAIEB,M.,AND KUMAR A. 2004. Preliminary Report on Excavations at the Palaeolithic Site of Attirampakkam, Tamil Nadu (1999-2004), *Man and Environment* XXIX (2),1-17.

PAPPU, S., KUMAR, A., RAVINDRANATH, S., UDAY RAJ 2008. A Preliminary Note on Impact Assessment Studies at Prehistoric Sites in Tamil Nadu, paper presented at *Conference on Emerging Trends in the Study of Indian History and Archaeology*, organised by the Department of History, School of Social Sciences and International Studies Pondicherry University, 14-15th March 2008, *in press.*

PETRAGLIA, M.D. 1998. The Lower Palaeolithic of India and its bearing on the Asian record, in M.D. Petraglia and R. Korisettar (eds.), *Early Human Behaviour in Global Context. The Rise and Diversity of the Lower Palaeolithic Record,* 343-390. London, Routledge.

PETRAGLIA, M.D. AND ALLCHIN, B. 2007. Human evolution and culture change in the Indian subcontinent, in M.D.Petraglia and B.Allchin (eds.), *The Evolution and History of Human Populations in South Asia, Inter-disciplinary Studies in Archaeology, Biological Anthropology, Linguistics and Genetics Series,* 1-20. The Netherlands, Springer.

RAMASAMY, S.M., VENKATASUBRAMANIAM, V. , RIAZ ABDULLAH, S., AND BALAJI, S.1992. The Phenomenon of River Migration in Northern Tamil Nadu-Evidences from Satellite Data, Archaeology and Tamil Literature, *Man and Environment* XVII (1), 13-25.

RAMASAMY, S.M; SARAVANAVEL, J., YADAVA, M.G., RAMESH, R. 2006. Radiocarbon Dating of some Palaeochannels in Tamil Nadu and their Significance, *Current Science* 91(12), 1609-1613.

RAY, H.P. AND RAVINDRANATH, S. 2007. The Archaeology of Ritual Spaces: Satellite Images and Early Chalukyan Temples, *Man and Environment* XXXII (1), 89-101.

SANKALIA, H.D. 1974. *Prehistory and Protohistory of India and Pakistan.* Pune, Deccan College.

SREEDHAR, A. K., BALAKRISHNAN, S., PAPPU, S. AND KUMAR A. 2008. Clay mineralogical studies of sediments and strontium isotope analyses on calcretes at the prehistoric site of Attirampakkam, Tamil Nadu, Current Science 94 (6), 2008, 779-786.

SUBRAHMANYAM, R. 1975. *Nagarjunakonda (1954-1960).* New Delhi, Archaeological Survey of India.

SUBRAMANIAN, K.S. AND T.A. SELVAN. 2001. *Geology of Tamil Nadu and Pondicherry.* Bangalore, Geological Society of India.

THE ANCIENT MONUMENTS AND ARCHAEOLOGICAL SITES AND REMAINS ACT 1958 (24 of 1958) along with The Ancient Monuments and Archaeological Sites and Remains Rules 1959. The Ancient Monuments Preservation Act 1904 and The Madras Ancient and Historical Monuments and Archaeological Sites and Remains Act 1866. Bare Act with short notes 2005. Delhi, Universal Law Publishing Co. Pvt.Ltd.

THE ANTIQUITIES AND ART TREASURES ACT, 1972 (52 of 1972) along with The Antiquities and Art Treasures Rules 1973 Bare Act with short notes 2005. Delhi, Universal Law Publishing Co. Pvt. Ltd.

TRIPATHY, A. (ED.). 2005. *Remote Sensing and Archaeology.* New Delhi, Sandeep Prakashan

Heritage management of Mattancherry Palace, Kochi, Kerala (India) - a case study

K.K. Ramamurthy[1] and K.P. Mohandas[2]

[1]Superintentding Archaeologist (Rtd), Archaeological Survey of India, Thiruvananthapuram- Kerala
[2]Assistant Archaeologist, Archaeological survey of India, Chennai Circle, Fort St.George, Chennai

Abtract

Mattancherry palace picturesquely located in the coastal town of Kochi was once the seat of Cochin maharajas. This was a famous port town known since ancient times where traders had come here from far to establish colonies. It had accepted many cultural transformations and lived harmoniously. The rich cultural legacy represented by the monumental heritage forms an important tourist destination. In light of this it becomes necessary to protect and preserve the authenticity along with its original environs. Mattancherry palace an important nationally protected monument for it contains the best of murals depicting scenes from Hindu mythology was put to adaptive use by establishing a museum displaying objects of Cochin Maharajas like palanquins, coins, dress, swords etc. The legislatives both Central and State have been formed considering the management of this heritage.

Keywords Mattancherry palace, Cochin Rajas, Synagogue, heritage, monuments.

Mattancherry Palace, popularly known as Dutch Palace is a protected monument of National Importance, by the Archaeological Survey of India, under the provision of Ancient Monuments and Archaeological sites and Remains Act 1958 and Rules 1959.

Mattancherry (Lat 9° 57'N; Long 76° 15'E), is in the District of Ernakulam, in Kerala State. The place is about 10km from Ernakulam and the nearest Airport is at Nedumbacherry which is about 40kms.

The origin of the Cochin Royal family is mentioned in the literary texts like *Keralokpatti, Keralamahathmyem* or *Perumpatappu Granthavari*, trace the origin of Kerala ,but mostly legendary, coupled with local tradition and with little historical facts. During *c.*1400 AD, the Perumpadappu family who once had their capital at Mahodayapura, shifted to Cochin and thenceforth known as Cochin Royal family. The family had the residence at Kalvetti palace and in AD 1555 shifted their occupation to Mattencherry. By AD 1755 the family changed their seat to Tripunithura and gained prominence in the eighteenth century.

The history of Cochin Rajas can be traced back from the beginning of sixteenth century:

Unni Raman Koil I (c 1503-10 AD): Staying at palace Kalavatti. He was the ruler at Cochin when Portuguese arrived and attacked by Zamorin. Died in 1510.

Unni Raman Koil II (1503-37): Foundation for the Portuguese fort at Cochin was laid in 1503 and arrival of Francis De Almeda, the first Viceroy in 1505. Presented gold crown to the king.

Vira Keralavarman (1537-65): Visit of St.Xavier to Cochin in 1542. During this period the Mattancherry palace was built. In 1549 the Portuguese soldiers looted the temple and totally destroyed the structure. Later, the Portuguese constructed a new palace and gifted to the king in 1555.

Kesava Ramavarma (1565-1601): The Jewish community shifted from Cranganore (modern Kodungallur in Thrissur District), to Cochin, during his reign and constructed the Synagogue in 1567. Donated land to Konkan Brahmins to build the Tirumaladevan temple. World conference of Christians was conducted in 1599, at Udayamperur, by the Arch Bishop of Goa. The conference was known as the Synod of Diamper.

Vira Keralavarman (1605-15): Arrival of first Merchant ship from England. The king annexed the land Mathilakam near Kodungallur from Zamorin of Calicut.

He was followed by **Ravivarma (1615-24), Vira Keralavarma (1624-37), Godavarma (1637-45)** and **Vira Rayiravarma (1645-46).**

Vira Keralavarma (1646-50): Witnessed a lot of political dispute during his period. He was succeeded by the adopted prince **Ramavarma**, and followed by **Rani Gangadhara Lakshmi (1656-58).**

Ramavarma (1658-62): Adopted prince of Rani Lakshmi. During his period the Dutch sieged the Mattancherry in 1662, under the Commander Goen and attacked the Cochin fort built by the Portuguese. During the period of **Godavarma (1662-63),** witnessed the end of Portuguese power in Cochin history and the emergence of Dutch power.

Vira Keralavarma (1663-87): With his accession began the Dutch history of Kerala and particularly in Cochin and the king bore the insignia of the Dutch East India Company. He was followed by **Ramavarma (1687-93)**, **Ravivarma (1693-97)**, **Ramavarma (1697-1701)**, and **Ramavarma (1701-21)**, considered to be the most important ruler. He was followed by **Ravivarma (1721-31)**, **Ramavarma (1731-46)** and **Vira Keralavarma (1746-49)**. During his period, the battle at Kulachal was fought between the Travancore Kings and the Dutch, in which the Dutch Commander Dilhoy was defeated and later became the head of Travancore army.He was followed by **Ramavarma (1749-60)** and **Vira Keralavarma (1760-75)**. During his reign Cochin was invaded by Hyder Ali of Mysore and Cochin became a vassal State of Mysore. The history of modern Cochin starts with the accession of **Ramavarma,** popularly known as Saktan Thampuran in 1790. He was the architect of the modern Cochin State and accepted the suzerainty of East India Company. The 19th century witnessed the end of Dutch power and the beginning of British domination in Cochin history. After many rulers **Ramavarma (1895-1914),** ascended the throne. He was one of the most striking rulers in the long line of rulers. He was popularly known as Rajarishi. The last ruler was **Ramavarma Parikshat Thampuran (1948-64)** and was responsible for the integration of Travancore and Cochin. A great Sanskrit scholar wrote many works and an authority on *Nyayasastra.*

Mattancherry palace is one of the oldest structures of the Portuguese model in the oriental style. Around 1665 the building had a major repair and modification by the Dutch. The palace is known for its Indo-European architecture. It is a two tired quadrangular structure, housing Bhagavathi, Krishna and Siva temples at the lower level and the upper storey has Coronation hall, Bed chamber, Dinning hall Assembly hall, Ladies chamber etc. The palace is known for its mural paintings, depicting Hindu religious themes executed in effective and elaborative style. The *palliyarai* or Royal Bed Chamber contains the Ramayana painting, narrated in synoptic way from the birth of Rama and brothers and their return to Ayodhya after the victory over Ravana. Likewise the stair case room has the individual Hindu deities representing the Saiva-Vaishnava-Sakta tradition. The underground Ladies chamber has also mural paintings depicting Uma Mahesvara, Krishna with Gopis, Govardhanadhari and Siva-Mohini.

The palace is being utilized as a Museum and housing portraits of Cochin Maharajas from 1864 onwards, Howdah, Royal Palanquins, royal dresses, iron swords etc. The details of the portraits of the Maharajas are:

Sl.No.	Name of the King	Period	Artist
1.	Sri Ramavarma	1864-88	K.Sekhara Warriyar, 15th June 1886.
2.	Sri Keralavarma	1888-95	-
3.	Sri Ramavarma	1895-1914	P.J. Cherian
4.	Sri Ramavarma	1914-32	-
5.	Sri Ramavarma	1934-41	Y.J.Cheriyan,1935
6.	Sri Keralavarma	1941-43	Vaidyanathan
7.	Sri Ravivarma	1943-46	Vaidyanathan
8.	Sri Keralavarma	1946-48	Vaidyanathan
9.	Sri Ramavarma	1948-64	Cherian Artworks,1960

The other important monuments/structures are enumerated hereunder:

Adjacent to Mattancherry palace, lies the Jews Synagogue built in 1567 by Portuguese and rebuilt by the Dutch in 1664. In 1805, Travancore Maharaja presented a golden crown to the Synagogue. The canto titles paved in the Synagogue was presented by Ezekiel Rahbai in 1763.

The St. Francis Church at Fort Cochin is also a protected monument of National importance by the Archaeological Survey of India built in sixteenth century, originally a wooden structure, constructed by five friars, who accompanied Albuauerque to Cochin in 1503. In 1506 the Portuguese Viceroy Almedia commenced a new city and the present edifice was completed in 1561 and dedicated to St. Anthony. From 1516 to 1663, the church was officially called conventional church of the order of St. Francis of Assisi, by the Portuguese. Later it came under the control of Dutch and carried out repairs in 1779. It has become Government protestant church under British from 1819-46. Major restoration works were carried out in 1886-87 and the edifice came to be known as St.Francis Church.

The uphill task is to preserve the monument and environs as Heritage Zone and this task is not only to this particular monument but to all Heritage monuments/structures/ palaces, both protected and non-protected, widely spread all over India. As referred to earlier, Mattancherry Palaxe is a protected monument of National Importance under the provision of Ancient

Monuments and Archaeological sites and Remains Act 1958 and Rule 1959, by the Archaeological Survey of India. Under 32 of AM & ASR Rules 1959 the Government of India, Archaeological Survey Of India issued Notification in 1991 to declare areas up to 100m from the protected limits and further beyond it up to 200m near or adjoining protected monuments to be prohibited and regulated areas respectively for purposes of both mining operation and construction. By this act there has been to a large extant a restriction on a unplanned constructional activities in monumental area.
Similarly, the Kerala State Department of Archaeology Act 1968 also is helpful in checking the construction in the

monumental area. The Kerala Government has taken steps in this direction. Based on Kerala building rules 1984, the Local Administration and Social Welfare Department of the Kerala State, appointed Art Commission to identify the areas of architectural and monuments to be protected in 1990. Taking all the salient features of the centre and different States acts, the Kerala government has taken steps to issue Heritage Zone Act, which will definitely serve the purpose. The Mattancherry area, temple area at Tripunithura and East fort area at Thiruvananthapuram have been identified as Heritage zone in the first phase and will be extended to all areas of Kerala in different phases. The Act awaits the legal sanction by the Government by way of issuing Government notification.

The above legislations have gone a long way in regulating the fast urban growth of the city but, the modern construction should blend with the heritage structure so as to maintain the original character of Mattancherry. A general awareness among the public has to be created to understand the importance of their cultural legacy to get their co-operation in keeping the original authenticity of Mattancherry's heritage.

Pattanam excavations and explorations 2007 & 2008: an overview

J. Cherian

Director, Pattanam Explorations and Excavations
V. Selvakumar & K.P .Shajan, Co-Directors

Research context

The sources for the study of Indian Ocean exchanges on the Malabar Coast have so far been the textual references in classical writings, early travelogues, a few coin-hoards and legends on cultural transmission. Archaeological evidence has been elusive; numerous field studies and two earlier archaeological excavations failed to unravel the mystery surrounding the ancient port of Muziris or the phenomenal trade between the Malabar Coast and the Roman Empire (Anujan Achen 1946; K.V. Raman 1976; Chedambath 1997; PJ Cherian 1998; Gurukkal and Whittaker 2001).

The archaeological significance of the Pattanam area was identified through surface surveys and salvage operations (Shajan 1998) which led a team of researchers to put forward the hypothesis that Pattanam could be the ancient port of Muziris (Shajan et al. 2004). Trial excavations by Center for Heritage Studies (Selvakumar et al. 2005), and surface surveys (Abraham 2005) provided supportive evidence for the latent potential of the site.

The inter-disciplinary archaeological research undertaken by KCHR at Pattanam had the collaborative support of major research institutions inside and outside the country and researchers drawn from the disciplines of history, archaeology, geology, paleo-botany, archaeo-zoology, physical anthropology, geo-physics, chemistry, marine archaeology, chemical oceanography, metallurgy, and epigraphy and conservation sciences

Location

Pattanam, located 7 km south of Kodungallur (N. Lat. 10° 09.434'; E Long 76° 12.587') is a hamlet in Vadakkekara revenue division in Paravur taluk of Ernakulam District in Kerala (Fig. 1). It is located in the delta of the river Periyar, about 25 km north of Ernakulam/Kochi. The river Periyar flows about 6 km to the north of the site; the Paravur Todu, a distributory of the Periyar, flows about one km to the south and the Arabian Sea is about 4 km to the west. About one km from the western boundary of the site is the Tattapally Kaayal, a backwater body that runs parallel to the Arabian Sea separating the Vaipin Island from the mainland.

Objectives

The main objectives of the research at Pattanam are to:

- understand the cultural and chronological sequence of the site
- study the spatio-temporal organization of the settlement
- reconstruct the paleo-environment
- unravel cross-cultural movements
- map the maritime heritage of the West Coast
- evaluate the role of Pattanam in the Indian Ocean trade
- locate Muziris, the ancient trade emporium on the West Coast
- establish a site museum at Pattanam

The KCHR research routines during the two seasons were:

1. Surface surveys in the Kodungallur - North Paravur region.
2. Underwater exploration in the water bodies surrounding Kodungallur
3. Survey of oral and written evidences and
4. Archaeological Excavations at Pattanam.

1. Surface Surveys in the Kodungallur- North Paravur Region

The surface survey in the Periyar delta was carried out by thorough plot-by-plot surface search for archaeological remains. The location of the remains of important artifacts and significant archaeological features were determined with precision, using Geographical Positioning System (GPS). The survey did not identify any noteworthy Early Historical settlement remains. A few urn burials were reported around Kodungallur, but their chronological context was not certain since no Iron Age pottery had been found. The major finds in the survey were medieval archeological material and remains of temples and architectural vestiges including a possible medieval fortification wall at Mathilakam. Students from various universities in Kerala participated in the survey.

2. Underwater Exploration in the Water Bodies Surrounding Kodungallur

The Naval Hydrographic Team of the Southern Naval Command of the Indian Navy carried out underwater explorations during both seasons. The objective was to identify archeological remains like ancient shipwrecks, debris of structures etc. from rivers, backwaters and off-shore areas. The naval hydrographers organized the bottom profiling using the Side Scan Sonar method. The diving teams physically explored the spots identified by the Side Scan Sonar. This needs to be followed up by underwater photography and excavation. Though the explorations did not yield conclusive evidence, it helped in understanding the sedimentation pattern which evolved over centuries through accretion. Besides the Indian Navy, the Centre for Earth Science Studies, Trivandrum and the Department of Underwater Archaeology, Tamil University, Thanjavur, collaborated in the exploratory studies.

3. Survey of oral and written evidence

Researchers scrutinized the Settlement registers in the taluk and village offices to elicit historical information related to settlement patterns, ownership and place names. Specific data sheets were used to gather topographical and anthropological information. The surveys brought out interesting oral and anthropological evidences on the antiquity of Pattanam. The earliest available written evidence with a reference to Pattanam is a Portuguese document of 1603-04, by Bishop Francisco Ros, SJ (Translated by Jacob Kollaparambil, Historical Sources on Knanites, Kottayam, 1986)

4. Archaeological Excavations at Pattanam

This section presents an overview of the research methodology, chronology, findings, and significance of the Pattanam excavations.

Excavation Methodology

The excavation method adopted here may be termed as locus (context) based. The precise context of the objects and features are of prime importance for this methodology. Their association/relationship with nearby objects and features are to be meticulously documented using specific data/locus formats. The Harris Matrix system is then employed to study and explain the relationship between various loci.

Tentative Chronology and Related artefacts

Pattanam is a multi-cultural site that seems to have been occupied from the beginning of the first millennium BC. The ca 3 m thick cultural deposit can be tentatively divided into four cultural periods, based on the stratigraphic, artefactual and chronometric analyses. They are the Iron Age-Early Historic transition [ca.10th C BC to ca.2nd C BC]; Early Historic [ca.1st C BC to ca.4th C AD]; Early Medieval [ca.5th C AD to ca.10th C AD] and Modern [ca.17th C AD to the present].

Period I: The Iron Age-Early Historic Transition [ca.10th C BC to ca.2nd C BC]

The earliest cultural period can be termed as Iron Age-Early Historic and is represented by the deposit above the aeolian sandy layers. This period represents the terminal Iron Age and the beginning of the Early Historic, prior to the Roman/West Asian contacts.

<u>Related Artefacts</u>
It is marked by typical Iron Age/megalithic ceramics, iron objects and fragmentary human bones. A Russet Coated Pottery (RCP) sherd with wavy designs and a red ware sherd with two Brahmi characters were significant finds of the 2008 season. The nature of findings suggests sparse occupation without external contact at this point of time.

Period II: The Early Historic [ca.1st C BC to ca.4th C AD]

This period seems to be the peak activity phase of the site when foreign and indigenous trade flourished concomitant to which "urban" and maritime features evolved at the site

Related Artefacts

A plethora of imported pottery was found. Roman Amphora, Terra Sigillatta, West Asian pottery and Turquoise Glazed Pottery (TGP) are the most important identified categories. Huge quantities of local pottery including rouletted ware suggest that the Pattanam trading community had been engaging with emergent regional trade networks. The burned bricks, triple-grooved tiles, brick structures and postholes are comparable to the Early Historic architectural features found elsewhere (Begley,V. et al. 1996). Other interesting finds include semi-precious stone beads, cameo blanks, gold ornaments, Roman glass ware, glass beads and early Chera copper and lead coins.

Period III: Early Medieval [ca.5th C AD to ca.10th C AD]

Early Islamic and West Asian contacts seem to have been active during this period.

Related Artefacts

The Early Medieval period is represented mainly by the blue glazed early Islamic pottery and enormous quantity of glass beads. Some of the architectural features of the Early Historic Period seem to have been re-used in this period.

Period IV: Modern [ca. 17th C AD to the present].

The very scanty evidence for Late Medieval occupation at the site probably indicates desertion. There is lack of evidence for the 11th to 16th Century period.

Related Artefacts

The blue-on-white Chinese ceramics and oral sources suggest that the site witnessed re-occupation possibly after the arrival of Portuguese traders.

AMS C14 Dating

Five samples analysed provide an error weighted AMS mean age of 500±80 B.C. (Ravi Prasad and Koushik Dutta 2007). The charcoal samples from the Iron Age layer places the beginning of the Iron Age in the first half of the first millennium B.C., and the wooden samples from the wharf context mark the maritime features of the site to the latter half of the first millennium B.C.

Major Findings

Roman Amphora

One of the most exciting categories of finds from Pattanam is fragments of amphorae jars that were made in South Italy and on the eastern coast of the Red Sea. Amphora jars were used for storing and transporting wine, olive oil, fish and fish sauce across the Roman world. During the two seasons a total of 401 (168+233) shreds were excavated from Pattanam. The Roman amphora shreds together with other types of Mediterranean provide the clinching archaeological evidence for the Indo-Roman trade (Tomber 2007).

Terra Sigillata

Terra Sigillata, yet another imported Roman ceramic, was earlier referred to as Arretine ware after Arrezzo, a place in Italy. This high quality, stamped tableware has thin fabric, red slip, smooth waxy surface and fine gloss. This pottery was earlier discovered in India at Arikamedu and was helpful in determining the chronology of Arikamedu (Wheeler et al. 1946). A total of 8 (2+6) sherds of terra sigillata were found in the 2007 & '08 excavations. Besides Pattanam and Arikamedu, Alagankulam on the east coast is another site in India, which has yielded terra sigillata sherds.

Turquoise Glazed Pottery (TGP)

These ceramics belonging to the Early Historic/Medieval layers are of light buff colour in the core with a heavy turquoise blue or green glaze on one or both sides. They are attributed broadly to North Africa (Egypt) and West Asian regions. The precise provenance of Pattanam TGP needs to be analysed by petrographic studies. A total of 161 (55+106) shreds of TGP were recovered from the 2007 &'08 excavations

Apart from the glazed pottery discussed above, 224 (43 + 181) fragments of possible Yemenite and Mesopotamian ceramics, several of them with black bitumen coated interior, were found. These ceramics highlight the contacts between the West Asia and the Malabar Coast which might have begun even before the Roman contacts.

Glass Beads

Large numbers of glass and semi-precious stone beads and bead manufacturing wastes were found at Pattanam. The glass beads also known as Indo-Pacific beads are the most common type found in South and South East Asia. The glass beads are classified according to their chemical composition into three [(a) potassium-silica glass, b) lead-barium glass and c) soda-lime glass] and each has distinct provenance.

The glass beads from Pattanam vary from ca. 0.1 mm to 30 mm in length. Transparent and translucent beads in white, black, blue, brown, green and yellow colours and cylindrical, octagonal, tubular, rectangular, circular and square shapes are found. Glass waster was also found at the site in 2008. Around thousand & two thousand glass beads were recovered in 2007 and '08 excavations respectively.

Semi Precious Stone Beads

Pattanam yielded semi precious stone beads made of cryptocrystalline materials like quartz, crystal, agate, chalcedony, carnelian, onyx, garnet and amethyst (Ravi Mohanty and Tilok). Finished beads, bead blanks (rough-out of the beads before drilling and polishing) and waste products like flakes and chips were found in large quantity. This suggests manufactory of stone beads at Pattanam. 40 semiprecious stone beads along with substantial quantity of debitage, and 25 semiprecious stone beads with scanty debitage were found in 2007 & '08.

Cameo blanks

Cameos as a jewellery item was very popular in ancient Rome and it seems raw cameo blanks were exported from Pattanam. Cameos were made through a process where multiple copies of a master design could be produced very quickly by pressing a master die onto an agate *cameo blank.* Cameos were also carved directly out of a material with integral layers or banding, such as (banded) agate/carnelian or layered glass, where different layers have different colours. A total of 7 (1+6) cameo blanks were found at Pattanam.

Roman Glass Bowls

Around a dozen of dark blue or green bowl fragments were

found at Pattanam from the two seasons. Authentication is awaited. The tableware bowls made of Roman glass, so far found in India, were identified to be of the first century BC/AD and later.

Terracotta Lamps

Two Roman terracotta lamps or their local variation were found in the 2008 excavations at Pattanam. Authentication is awaited. Roman terracotta lamps though found less frequently, are culturally significant and are tentatively dated to c.100 BC- AD 100.

Iron Objects

Iron objects like nails, knives, hooks and slag were found in large numbers. Of these, nails outnumber others. A total of 1338 (575 + 763) iron objects were found in 2007 &'08. The samples from Pattanam reveal a higher carbon steel micro-structure where as Roman samples were of pure wrought iron with no carbon. (Sharada Srinivasan, 2007) Comparative analysis is necessary on the indigenous ferrous technology of the times.

Chera Coins

Heavily corroded copper objects including early Chera coins, were found for the first time in Kerala from a stratigraphic context. A total of 20 (9+11) heavily corroded copper coins were recovered. Most of them are identified as Early Chera coins. These coins are either circular or square with elephant on obverse and bow and arrow on the reverse. These coins are important indicators for the study of the "state and polity" of the Early Historic Period.

Gold Ornaments

The gold artefacts so far reported from Kerala were mainly Roman aurei (coins) and a few ornaments in grave goods assemblages of Iron Age burials.

In 2007 part of a gold ornament, weighing 1.16 gm and of 21 karat was found in the Early Historic layer at Pattanam. In 2008 a gold string weighing 0.0345 gm and 22 karat and a thread or *jari* 0.030 gm of 0.2 karat were found (Sharada Srinivasan, 2008). The *jari* could have been part of the value addition to woven clothing which was a major export item as mentioned in the Vienna Papyrus document of the 2nd century AD. Authentication is awaited.

Inscriptions and Graffiti

The ancient inhabitants of Pattanam could have been literate as indicated by the evidence. A total of 8 (5+3) sherds of coarse ware with post firing marks of Tamil-Brahmi letters were found in 2007 &'08 excavations.(Varier, 2007) Similarly large numbers of sherds with post-firing graffiti were also found. These finds are being studied by experts. Compared to the sites of Tamil Nadu the number of sherds with inscriptions and graffiti are less. Heavy rains and soil conditions in Kerala seem to have hastened the decomposition of graffiti / inscriptions on pottery.

Indian Pottery

Rouletted ware, a fine Indian pottery is being reported for the first time on the West Coast (Gogte, 2007). Rouletted ware has been found extensively on the east coast, especially at Arikamedu and Alagankulam. These ceramics indicate the active contact of the Malabar Coast had with the east coast. A total of 2380 (1125+1255) rouletted ware were excavated from Pattanam. The local pot sherds from Pattanam amounted to nearly one lakh seventy thousand in 2007 and over three lakhs in 2008 (Selvakumar, 2008).

Russet Coated Painted (RCP) Ware

This pottery is characteristic to the lower Deccan and parts of South India. It is also termed as Andhra ware. They are usually associated with Early Historic period (c 300 BC- AD 300), along with rouletted ware. This has also been reported along with Iron Age burials in Coimbatore- Palakkad-Madurai region. A single sherd of RCP found from Pattanam Excavations, 2008 is significant in relating Pattanam with the South Indian human geography of the times.

Chinese Ceramics

Pattanam produced several sherds of blue and white Chinese ceramics belonging to the 17th / 18th centuries. (N. Karashima, 2007) It is made of thick Chinese porcelain with white core and has a feldspar glaze of blue or green tone. Paintings are sometimes noticeable on one or both sides and common designs include plants, flowers, fruits, horizontal lines, scrolls etc. The types represented at Pattanam are the bowl, dish and shallow cup varieties with ring-base, probably used as tableware.

A total of 60 (29+31) sherds from 2007 & 08, support the assumption that there was a discernible desertion or abandonment of the site after or around the10th century AD and re-occupation probably during the colonial times. It might be recalled here that the Pallipuram fort, the first ever European fort in India is less than four km west of the site.

Marine Features

The most striking structural feature found at Pattanam is the wharf complex resembling a port, with a canoe, bollards and an adjacent commercial /ware house area. The wharf (007) was found in the northeastern boundary of the site close to a waterlogged deposit in Trench PT07. The wharf was a huge structure (6 +m in length and 7.30 m in width) made of a mixture of laterite granules, lime, pottery and clay with an outer 19 course brick lining. The water logged area adjacent to the brick-lining had a dugout canoe made of single log of *anjili* wood (PT 07 –IV, 011a). (KM Bhatt.2007) The intact portion of the highly decayed canoe measured ca.6 m in length and the AMS dates puts the ^{14}C cal. age range from 1300 BC to 100 BC, [i.e., 700±600 cal. bc with 95.4% probability](Ravi Prasad and Koushik Dutta 2007). Of the nine bollards meant to secure boats the AMS dates for the teak sample bollard (PT07-IV, 011a B7/H,) puts the ^{14}C Cal. age range from 1100 BC to 1300 AD, [i.e.100±1200 cal. ad, with 95.4% probability] (Ravi Prasad & Koushik Dutta 2007). The layer below the canoe yielded amphora, rouletted ware, TGP and other ceramics and their provenience are under study.

Architecture

Pattanam excavations revealed interesting Early Historic 'urban' architectural features. The brick structure comparable to a warehouse, exposed near the wharf in Trench PT07 III included a platform (006) with postholes (67 in number) and brick walls belonging to at least three different phases. The postholes do not show any particular pattern. It appears that they were dug in different points of time and indicate repeated use for a long duration. A ring well, made of three terracotta rings was also found close to the structure.

The architectural features and related artefacts like brick walls, clay platforms, and multiple layers of floor and storage jar found in trenches PT07 II and PT07 I suggest residential context.

The 2008 excavations also revealed similar architectural features including ring well, foundations, brick walls, soakage pits and burnt clay floors belonging to Early Historical Period. A complex architectural feature exposed in trench PT08 XI had seven brick-structures, which included a burnt brick wharf (?), platform (locus 005=006), a tilting brick wall probably part of the brick lining of the wharf (locus 039), a semi-circular structure etc. In Trench PT08 IX, adjacent to the 2007 warehouse like structure, a brick structure with three enclosures, perhaps of residential or commercial use was found. A disturbed brick structure (locus 018) and two sets of soakage pits (loci 031 and 033) were found in Trench PT08 VIII.

Grooved Roof-tiles

Burnt clay roof-tiles of Early Historic period were found at Pattanam in large numbers. The tiles have three grooves/channels for draining the rainwater and two holes for fixing them on the roof. Similar tiles have been reported from many Early Historic sites across India.

Human bones

Highly brittle and decayed bone fragments were found in the earliest sand layers in a horizontal stretch of over a 160 cm. This is an unusual find in Kerala context where such evidences are lost either due to the mortuary practices or the poor preserving condition of the moist and acidic soil. The bone remains were too fragile and fragmented making the DNA analysis almost impossible in Indian laboratories. Taphonomic analysis has revealed that the bone fragments have burn marks indicating exposure to fire in 'flesh on' condition and suggestive of secondary burial practices. (Walimbe, 2007)

Botanical Remains

A rich harvest of botanical remains from the water-logged area at a depth of three meter is another remarkable find of the Pattanam Excavations. A 25 cm to 30 cm thick layer of clay prevented oxidation in the lower layers and that seems to have ensured the preservation of the palaeo-botanical remains from decay.

The 2007 yielded black pepper, bread fruit seed and grape pips, teak, coconut fronds, coconut shells, prop root base of arecanut, bamboo pieces and indeterminate wooden tissues. In 2008 the main items were black pepper, cardamom, rice, wheat, green gram (*Cheru payar*), lentil (*Cheera*), mango seeds, *nellikka* seeds, Indian jujube (*Thudali*), *anjili*, Borassus palm (*Pana*), Brinjal (*Vazhuthananga*), ladies finger (*Vendakka*), Gourds (*Vellarikka*) arecanut and coconut shells. (Kajale, 2007 & 08). In 2008 season, four out of five trenches yielded frankincense (*Kunthirikkam).* A rope made from an unidentified plant fibre was also obtained. The wood identified in both seasons include, teak (*Tectona grandis L.f.*), aanjili (*Artocarpus hirsutus Lamk.*), punna (*Calophyllum inophyllum L.*) , kadukka (*Terminalia chebula Retz.)* and Karimaruthu / Thenmavu (*Terminalia crenulata Heyne ex Roth*) (KM Bhatt, 2007 & 08).

Indirect botanical evidences in the form of leaf impressions on potsherds and grain impressions (rice husk) on potsherds, tiles and bricks were also found.

Significance

The Malabar Coast has been archaeologically barren despite the sporadic efforts of many researchers to get tangible evidence for the much-referred Indian Ocean exchanges. But now Pattanam has become the first ever site on the Kerala Coast to yield a wide variety of archaeological evidence for the Mediterranean, North African and West Asian contacts since the Early Historic Period. The Roman presence is dominant and in all probability the site could have been part of an Indo-Roman settlement/commercial centre with international trade networks.

It is also the first site, which yielded striking features of an inland port, throwing some light on the ancient maritime history and technology. The "urban" features are crucial in understanding the socio-economic formations of the early historic Kerala. The coins suggest the presence of 'polity' and probably 'state'.

In the history of the Kerala archaeology, Pattanam is the first habitation site of the Iron Age / Early Historic Period. Though innumerable sites with amazing range of burial monuments with mortuary remains have been unearthed in Kerala, habitation-contexts have been elusive so far. The site seems to offer a fascinating transition story of an Iron Age–Early Historic habitation context.

How Kerala could become the melting pot of different cultures and religions may be etched from the findings of the Pattanam excavations. If Kerala has a more composite and cosmopolitan culture than any other part of India, probably she owes that to her early international exposures. The trade contacts gradually led to affable cultural interfaces, thus various belief systems ran their roots deep in this soil.

The enormous amount of artefacts recovered from the excavations, unique features of the site and the significance of the location in relation to the palaeo-coastline and water bodies, suggest that Pattanam could have been part of the ancient port of Muziris. Its location and the architectural features indicate that Pattanam was a planned port town. More research is necessary to unravel the mystery of this early historic settlement and its cultural past.

References

[*The reference to experts in the text denotes their analytical reports submitted to KCHR*]

ABRAHAM, S. 2005. The Malabar Regional Archaeological Survey: Introduction, Goals, and Prospects. *Journal of the Centre of Heritage Studies* 2: 1-5.

ACHAN, A.P. 1946. *Annual Report of the Archaeological Department, Cochin State for the year 1121 M.E. (1945-46)* Ernakulam Cochin.

BEGLEY, V.1988. Rouletted Wares at Arikamedu: a New Approach. *American Journal of Archaeology* 92: 427-40.

BEGLEY, V.ET AL. 1996. *The Ancient Port of Arikamedu. New Excavations and Researches 1989-92.* Pondicherry: Ecole francaise d'Extreme Orient.

CASSON, L. 1986. P. Vindob. 'G 40822 and the Shipping of Goods from India'. *Bulletin of the American Society of Papyrologists* 23: 73-9.

CASSON, L. 1989. *The Periplus Maris Erythraei. Text with Introduction, Translation and Commentary,* Princeton, Princeton University Press.

CHEDAMBATH, R.1997. 'Investigations into the Megalithic and Early Historic periods of the Periyar and Ponnani River Basins of Kerala'. Unpublished Ph.D. Dissertation. Pune: University of Pune)

CHERIAN, P.J. 2007. *Puravasthu patanam orupichaveppinte ormakuruppu, Suvarnakeralam Smaranika,* University of Kerala, Trivandrum 170-74.

CHERIAN, P.J., SHAJAN, K.P. AND SELVAKUMAR, V., 2007A. 'In Pursuit of Missing Links - Pattanam and the Maritime History of Malabar Coast: A Report of 2007 Excavations', *EASAA Revanna Conference Proceedings.*

CHERIAN, P.J., SELVAKUMAR, V AND SHAJAN, K.P., 2007B. The Muziris Heritage Project: Excavations at Pattanam- 2007, *Journal of Indian Ocean Archaeology,* 2007, no:4, 1–10.

GURUKKAL, R. AND C.R. WHITTAKAR. 2001. 'In search of Muziris', *Journal of Roman Archeology* 14, 335-50.

MCCRINDLE, J.W.1927 (2000REPRINTED). *Ancient India as Described by Ptolemy.* New Delhi: Mushiram Manoharlal Publishers.

RAMAN, K.V. 1976. Archaeological Excavations in Kerala, *Souvenir of the 37th Annual Meeting of the Indian Historical Congress, Calicut University, India,* Dec. 29-31, 6-10. Calicut: Calicut University.

SELVAKUMAR, V., P.K. GOPI AND K.P. SHAJAN. 2005. Trial Excavation at Pattanam: a Preliminary Report. *Journal of the Centre for Heritage Studies* 2: 57-66.

SHAJAN, K.P. 1998. *Studies on Late Quaternary Sediments and Sea Level changes of the central Kerala Coast, India,* Unpublished PhD. Dissertation., Cochin Univ. of Science and Technology, Cochin.

SHAJAN, K.P.TOMBER, R, V. SELVAKUMAR AND P.J.CHERIAN. 2004. Locating the Ancient Port of Muziris: Fresh Findings from Pattanam. *Journal of Roman Archaeology* 17: 312-320.

TOMBER, R.S., 2000. Indo-Roman Trade: the Ceramic Evidence from Egypt, *Antiquity* 74, 624-31.

TOMBER, R. S. 2004. Rome and South Arabia: New Artefactual Evidence from the Red Sea, *Proceedings of the Seminar for Arabian Studies* 34, 351-60.

WHEELER, R.E.M., A. GHOSH AND KRISHNA DEVA 1946. Arikamedu: an Indo-Roman Trading-station on the East Coast of India, *Ancient India* 2: 17-125.

Resolution invariant content-based retrieval of artistic images

C. Menaka,[1] S. Maheswari,[1] B. Sathya Bama[2] and S. Raju[2]

[1]Student, [2]Faculty, Department of ECE, Thiagarajar College of Engineering, Madurai-15.

Abstract

Content-Based Image Retrieval (CBIR) of cultural heritage images is an emerging field of research bridging society, culture and information technology. Querying by example databases of paintings, sculptures, photographs, and documents of historical value from different civilizations, would facilitate both educational and research and enable the exploration of unknown inter and intra cultural associations. It would be preferable that the system was "text query free" for two reasons: first, the creation of the descriptor is more difficult because the database must be indexed; second, the text query could become fundamental for the retrieval. The above situation exists in artistic image database which contains photos of monuments, hand-manufactured, paintings, sculptures which are present as a collection of cultural heritage of a country. The creation of a sophisticated descriptor requires many time for its execution: this is not a problem during the descriptor database creation, but it is a problem if a fast comparison among the query image and every image inside of the database is desired. In this paper we propose a resolution invariant descriptor based on color and a data fusion solution to compare the descriptors among the images which is easy to implement.

Key words Content-Based Image Retrieval, cultural heritage image, Resolution Invariant descriptor, data fusion and wavelet.

Introduction

To manage and utilize large image databases, content based image retrieval (CBIR) tools have emerged recently for diverse applications like automated inspection, database management and web based searches. CBIR can be extended to cultural and heritage applications also. The interpretation of objects, scenes and other data in the archaeological images is useful for visualizing scenes from viewpoints impossible in the real world due to size or accessibility issues, and providing virtual tourism and virtual museum exhibits. Most of the existing systems such as ADL, BDLP, CANDID [1] use color and texture features, shape features, and still less use layout features. The retrieval on color and texture usually yields images with similar colors and textures unless the database contains many images with a dominant texture and color. However from the literature survey we performed, resolutions of the images are not taken in to consideration. Our work deals with the resolutions of the cultural data to retrieve images more efficiently from a database. Archeological images taken using different sensors may vary in spectral and radiometric resolution drastically which may end up with ambiguous results after applying CBIR algorithm. Moreover the query image and the relevant image in the database may have different spatial resolution and those images might have been taken at different years which account in to temporal resolution. So to eliminate this we have used techniques considering these resolution constraints for obtaining perfect retrieval system.

Proposed CBIR System

We have proposed the architecture for CBIR system in which we have eliminated the dependency on resolution for retrieving remotely sensed data. The query image may have different sensor characteristics with respect to the images in the database.

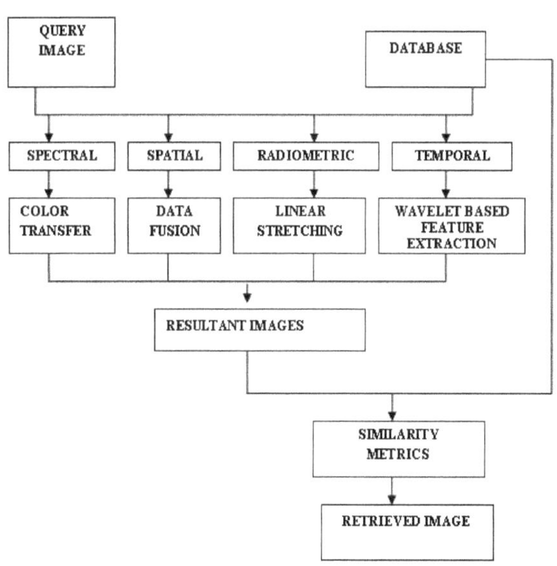

Fig. 1.Work Flow.

This issue degrades the performance of image retrieval systems. The work flow of our CBIR System is shown in figure1. In order to convert the panchromatic images to multispectral images, the color of the reference image is transferred to the target image without altering the image contents. For this purpose, we have basically considered color as main low level feature of the images. For extracting the relevant image having different spatial resolution to that of the query image; we have used data fusion techniques. The dependency of the radiometric resolution is removed by means of linear stretching method. At last dependency of temporal resolution is eroded by k means clustering technique feature extraction, comparing query image with all the images in the database and the relevant is retrieved.

Algorithm to Overcome Resolution Constraints

a. Spectral resolution:

Color can be added to panchromatic images in order to increase the visual appeal of images and to spectrally match with the images in the database. To overcome the spectral resolution constraints the color content from the source image from the database is transferred to the query image. Color from the source image is transferred to the query image by converting both images to lαβ color space from RGB space. After normalization the luminance comparison the corresponding α, β color space values are transferred to the closest match. After color processing, we must transfer the result back to RGB to display it. So irrespective of the color of the query image the relevant image can be retrieved by overcoming spectral resolution using **color transfer algorithm** [4].

b. Spatial resolution:

The relevant image in the database may have different spatial resolution to that of the query image which may fail to match the correct image with the corresponding query called data fusion. The low resolution image can be fused with high resolution image to obtain the high spatial resolution. The fusion algorithm we followed is called **wavelet fusion technique** [5] [6]. Generally panchromatic have high spatial resolution and multispectral have high spectral resolution. Fusion technique involves the process of histogram matching of panchromatic image with R, G and B of multispectral image, replacement of detailed coefficients by the real low-resolution multispectral image bands (B, G, R), respectively and inverse wavelet transform consequently. The resultant image has the high resolution and high spectral resolution which offers correct retrieval independent of spatial facts.

c. Radiometric resolution:

The sensor data can have any radiometric resolution say 6 bit, 7 bit. So the retrieval process may fail to give back the relevant images corresponding to the query image. The dependency of radiometric resolution is eliminated as follows. All the images in the database are converted to 8 bit images. The input image is also converted to 8 bit image by means of **linear stretching** [7] before the retrieval process is carried out. This avoids retrieval of irrelevant images and makes the search algorithm simple.
Formula for linear stretching is given as

$$DN' = \frac{DN - MIN}{MAX - MIN}\ 255 \quad (1)$$

Where:

DN' = digital number assigned to pixel in output image
MIN = minimum value of input image, to be assigned to a value of 0 in the output image.
DN = original digital number of pixel in input image
MAX= maximum value of input image, to be assigned a value of 255 in the output image

d. Temporal resolution:

The system should also have the capacity to account for seasonal variability. Interpretation of changes in the images due to temporal resolution is done by **wavelet based feature extraction technique** [8]. This technique involves the process of discrete cosine transform using Haar wavelet and applying maxmin composition rule for the resultant coefficients (A,H) by considering 8-nearest neighbors. It yields 16 histograms, 2 for each direction i = 1, 2... 8. Construct the cumulative histogram for each histogram obtained in above step and normalize it. From the sample points on the cumulative frequency curve.

* Mean of the sample points

*Mean deviation of the sample points.

Repeat the above last two steps for the pair of images (A, V), (A, D), (A, abs (V-H-D)) for feature extraction. The above process generates 192 features (1 (image) x 4 (combinations) x 2 (features) x 16(histograms)) which constitute the feature vector f. minimum mean square distance is found between the features and similar textures are identified.

Results and discussion

To justify the better performance of our proposed work compared to the previous work, we have shown the results of our experiment undertaken on 400 images in the database in figures 2, 3,4,5,6 & 7. The simulations were performed using the image processing Toolbox of Matlab7.

Fig. 2.Query Image

Fig. 4. Figure 4. Spatial Resolution.

Fig. 5. Radiometric Resolution.

Fig. 3. Spectral Resolution.

Fig. 6. Temporal Resolution.

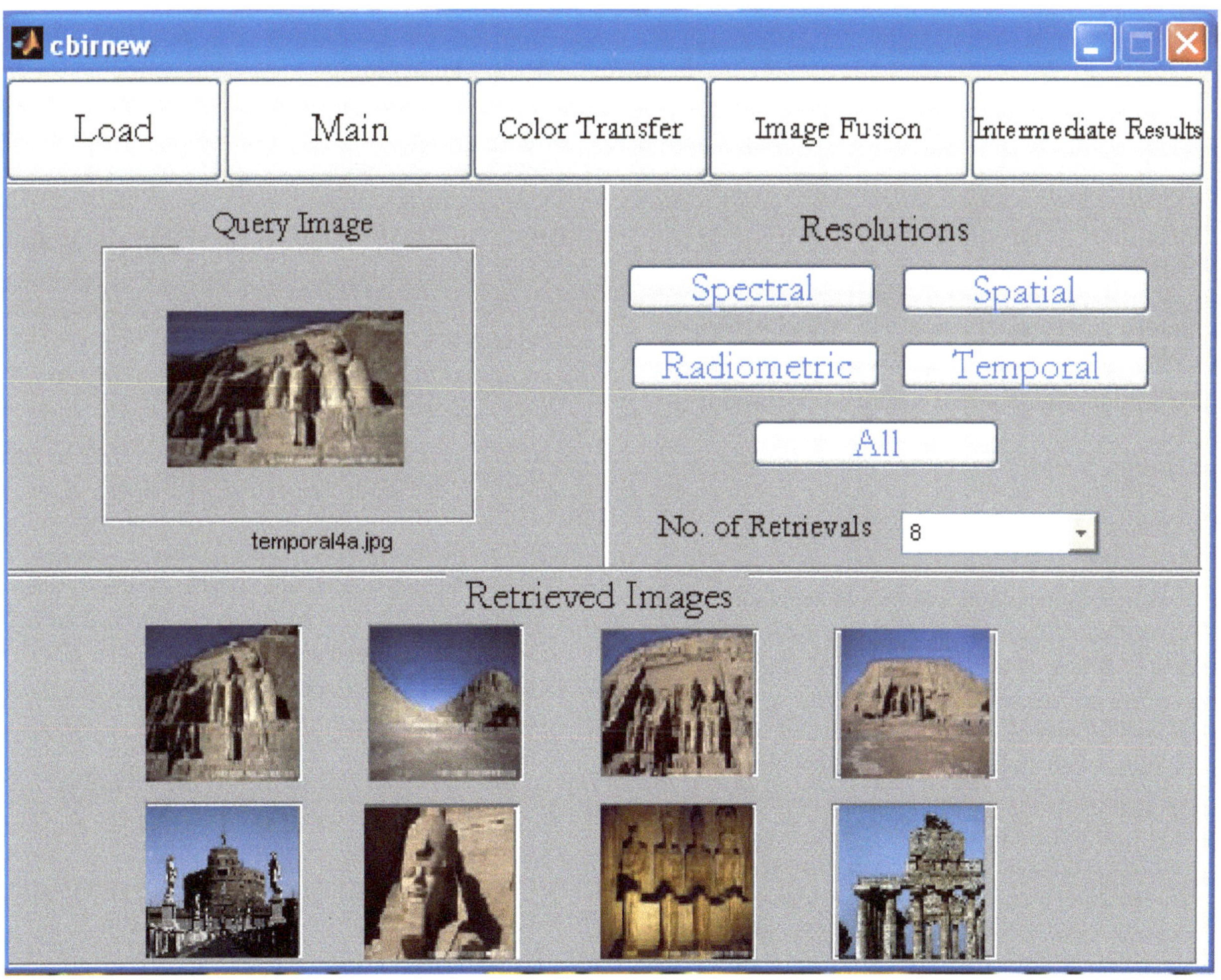

Fig. 7. Proposed CBIR System

Performance Evaluation of CBIR System

Evaluation of retrieval performance is a crucial problem in content-based image retrieval (CBIR) [2] [3]. Many different methods for measuring the performance of a system have been created and used by researchers.

$$\text{Precision}=\frac{No.\ of\ relevant\ images\ retrieved}{Total\ no.of\ images\ retrieved} \quad (2)$$

$$\text{Recall}=\frac{No.\ of\ relevant\ images\ retrieved}{Total\ no.\ of\ relevant\ images\ in\ the\ database}$$

(4)

$$R.E=\begin{cases}\dfrac{No.\ of\ relevant\ images\ retrieved}{Total\ no.\ of\ images\ retrieved} & No.\ retrieved <= No.\ relevant\\[2ex] \dfrac{No.\ of\ relevant\ images\ retrieved}{Total\ no.\ of\ relevant\ images} & otherwise\end{cases}$$

(R.E= Retrieving Efficiency)

No. of Retrievals	4	8	20	40
Precision	1	0.8750	0.75	0.7
Recall	0.133	0.233	0.5	0.933
Error Rate	0	0.125	0.25	0.3
Retrieval Efficiency	100	87.5	75	70
Retrieval Time	5 sec	9 sec	14 sec	20 sec

Fig.8. Performance Evaluation of Proposed CBIR System.

Conclusion

CBIR at present is still very much a research topic. The extent to which CBIR technology is currently in routine use is clearly still very limited. There are several organizations and systems that store or provide access to archaeological data. In particular, CBIR technology has so far had little impact on the most general applications of archaeological data. The project work has proposed an efficient method for retrieval of images relevant to the input query image from a large database by overcoming the four resolution constraints. This system also provides an efficient retrieval. The results are encouraging for the wide range of images that have been used which makes the proposed work to be possibly used for studying the information contained in the raw imagery captured by the sensors without the use of expensive software and complex algorithms. Our future work in this project is to develop powerful query processor and also to include the rotation invariant image retrieval for achieving effective CBIR system.

Acknowledgement

We thank everyone who contributed images to this article. We also thank our college administration for supporting us and the help available in the department. We would like to express our heartfelt gratitude to our faculties and friends for their continuous and consistent encouragement.

References

REMCO C. VELTKAMP, MIRELA TANASE, *Content-Based Image Retrieval Systems: A Survey*, UU-CS-2000-34, October 2000.

JORMA LAAKSONEN, MARKUS KOSKELA,AND ERKKI OJA, *Content-Based Image Retrieval using Self-Organizing Maps* VISual'99, Amsterdam, The Netherlands, June 1999.

ANDREW BERMAN, LINDA SHAPIRO. "A flexible image database system for content-based retrieval". *Computer Vision and Image Understanding*, 75(1/2):175–195, 1999.

ERIK REINHARD, MICHAEL ASHIKHMIN, BRUCE GOOCH, AND PETER SHIRLEY, "Color Transfer between Images", *IEEE Computer Graphics and Applications*, September/October 2001, 34-40.

DR. YUN ZHANG , "Problems in the fusion of commercial high-resolution satellite images as well as Landsat 7 images and Initial solutions" , *ISPRS, Vol. 34, Part 4, "GeoSpatial Theory, Processing and Applications"*, Ottawa, July 8 12, 2002.

JH PARK, KO KIM AND YK YANG, "Image Fusion using Multiresolution Analysis", IGARSS, vol. 2, pp. 864-866, 2001.

THOMAS M.LILLESAND, RALPH W.KIEFER, JONATHAN W.CHIPMAN, "Remote Sensing and Image Interpretation".

R.HARALICK, K.SHANMUGAM, AND I.DINSTEIN. "Textural features for image classification". IEEE Transactions on systems, Man, and Cybernetics, SMC-3(6):610-621, November 1973.

Socio-economic impact analysis as a strategic management tool for cultural heritage sites

Jaime Kaminski, Jim McLoughlin and Babak Sodagar

Brighton Business School, University of Brighton, UK

Abstract

Most impact studies undertaken to-date have usually focused on a single impact dimension (such as an economic study using a local GDP measure), at one moment in time. This research takes a broader view of impact through the use of a holistic impact model. This attempts to capture the complex, multi-dimensional nature of impact, and perspectives on impact, the multiple influences on impact, and offers a guide to which impacts should be examined, given the specific circumstances of a cultural heritage site. This does not set out to be an exhaustive account of all the influences that affect impact at heritage sites but a thematic overview of such influences. Furthermore, this model highlights the role of impact assessment as a strategic management tool.

Introduction

Understanding the meaning, significance, and contribution to society of our tangible heritage is fraught with complexity. Learning from the past is central to our understanding of the present. It contributes to a sense of cultural and community identity, and in a wider sense, to our intercultural understanding and appreciation of other cultures and civilisations.

Conserving and preserving our tangible heritage and historic environment, provides a vital conduit between the present, the past and future generations. Tangible heritage, according to Lowenthal (1985), provides a concrete way to connect with and touch the past and, in a sense helps to validate memories. At the same time heritage sites are cultural assets which need to be managed, but their multi-dimensional impacts pose numerous conceptual and practical challenges. There are multiple perspectives which, in turn, will have different implications for management decision making. For example, heritage impact can be viewed individually and collectively (from local to global community interests depending on the site); or as having intrinsic and extrinsic value; as an industry and driver of economies; intergenerational responsibility for preservation; social, educational and environmental outcomes; and holistically thinking, sustainability issues which need to bind and balance all of the above.

Therefore impacts are diverse, multi-layered, multi-dimensional, dynamic and complex and always contingent on purpose, use, and who is examining the issues. But to-date most impact studies undertaken have usually focused on a single impact dimension at one moment in time. This research takes a broader view of impact through the use of a holistic analytical model (see Figure 1). It attempts to capture the complex, multi-dimensional nature of impact, the multiple influences on impact, and offers a guide to which impacts should be examined, given the specific circumstances of a cultural heritage site. This is not an exhaustive account of all the influences that affect impact at heritage sites but a thematic overview of such influences.

By adopting a strategic approach to impact it is important to recognise that it is a dynamic concept, influenced by numerous elements, both internal and external to each heritage site. The holistic impact model, in Figure 1, comprises the impact drivers, the organisational strategies, the impacts and a feedback loop. Such a feedback loop improves understanding of the strategic implications of the workings of organisations and their impacts, which can lead to long-term improvements.
Socio-economic impact embraces many possible impact dimensions (e.g. economic, individual, social, environmental, etc). Within each dimension there are a number of possible methodologies which can be employed to identify and 'measure' impact, each method having advantages and disadvantages.

The holistic impact model consists of five elements: the cultural heritage site (CHS) impact context, the site mission and objectives, the site stakeholders, and the site organisational context/management and decision making context, which all influence and contribute to the potential socio-economic impacts of a heritage site. The different components of the holistic impact model and its relationship to impact evaluation are examined below.

The heritage site (CHS) impact context

The impact context is interpreted broadly as the specific macro-contextual influences and micro-contextual (such as organisational) influences on a cultural heritage site. Macro contextual influences can include:

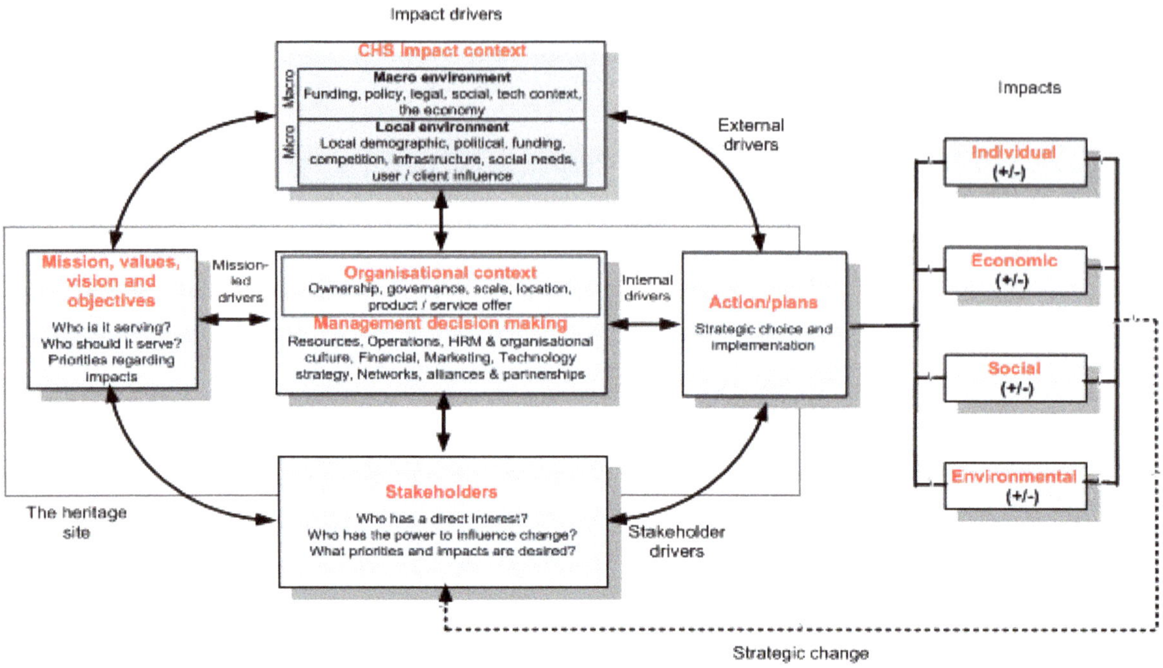

Fig. 1. A dynamic holistic impact model for cultural heritage sites (CHS).

- Macro-economic environment
- Policy context
- Legal framework
- Cultural context and values
- Technological context.

The micro-contextual influences exist in the local environment of the site (of course, some elements within the impact context (such as policy, funding and legal frameworks) can straddle the border between macro and micro influences.) These influences include elements such as economic, political, funding, demographic, legal, competition, infrastructure, etc. For heritage managers the impact context creates opportunities and threats for their organisations and can impose constraints on decision making. Most of these factors are beyond the direct control of cultural heritage managers, but nevertheless affect heritage site strategies and final impacts and outcomes. Furthermore, many of the factors are inter-related and so for example, local economy could affect heritage site funding or the policy context could affect the legal framework.

The macro environmental context

Each site operates in a macro-national context (and wider European and global context). A number of influences from this context affect heritage sites, these include:

- *Macro-economy*: The macro-economy (regional, national and international) affects, for instance, tax revenues, disposable income, and policy funding priorities. The macro-economy has a major influence on the heritage sector.
- *Policy context*: The macro-policy context is another important determinant for potential outcomes and impacts at heritage sites. Policy is fundamental to understanding impact; it influences heritage sites at multiple levels. It determines what gains funding and what does not, what is conserved and what it is not, it influences local authority policy, and it can also affect national legal structures which influence the heritage sector, etc. (Mignosa and Rizzo 2004, Rizzo and Mignosa 2006).
- *Cultural context and values*: The 'cultural context' and values of a society in supporting heritage, will in turn affect practical policy and funding priorities. For example, the cultural context helps define heritage. As Ashworth and Howard (1999: 11) note "Heritage is whatever people want to conserve, preserve, protect, or collect". As such, definitions of what should be preserved can differ between countries (i.e. market squares in Germany, country houses in the UK, or heritage coastlines in France). Furthermore, the definition of what constitutes 'heritage' is not static but dynamic. In the developed world the definition of heritage has broadened considerably in the later half of the twentieth century. Early definitions of cultural heritage that encompassed the monumental remnants of cultures were gradually extended to include new elements from non-artistic sectors of activity such as industrial heritage, or from specific contexts such as the underwater heritage. Today, the notion of heritage is much more open, and is used to reflect living culture rather than just our past. UNESCO now includes in its' definition of cultural heritage historic cities, cultural landscapes, natural sacred

sites, underwater cultural heritage, museums, movable cultural heritage, handicrafts, documentary and digital heritage, cinematographic heritage, oral traditions, languages, festive events, rites and beliefs, music and song, the performing arts, traditional medicine, literature, traditional sports and games, and culinary traditions. It is important to acknowledge that as time passes the definition of heritage will continue to change according to different political aspirations, and the increasing input of communities and groups outside of the traditional field of 'experts'.

- *Technological context*: It is important to consider technological developments and how these might affect the visitor experience. New ICT hardware, software and their associated standards are being developed continually. The applicability of these technologies and standards to the heritage sector is dependent on economic and social factors such as cost and user acceptance (e.g. in the last fifteen years websites have become an integral part of heritage marketing and presentation, this has only become possible through the global advances in ICT and the acceptance and penetration of the PC and Web use in households across Europe).

In a dynamic model such as this some allowance must be given to the ability of a heritage site to influence the impact context. However, much of the macro context is beyond the sphere of influence of cultural heritage sites. Such sites are unable to exert any influence on the regional, national and international economies. Similarly, the development of technology such as ICT takes place outside of the cultural heritage sphere (usually in the commercial or military sectors) and gradually migrates to the heritage sphere.

The micro environmental context

The micro environmental context includes the local demographic profile, economy, policy and political context. For example, numerous local authorities and governments have developed strategies, with accompanying funding, targeting heritage as a key element in regeneration programmes.

In heritage sites with a strong orientation towards tourism, a principal element of a site's economic impact will depend on the total visitor experience which itself is dependent on numerous off-site factors (e.g. coordinated local tourism strategy, the presence of other visitor attractions, quality of facilities such as transport, restaurants, hotels, etc). It is rare for a heritage site to be immune to these factors.

Competition or complementarity: The degree of competition or complementarity with other attractions can also influence impact. For example, a heritage site within a historic urban centre (such as Rome, Venice or Paris) could face competition from numerous alternative heritage attractions; however, the nucleation of heritage sites within a town or city can act as a stimulus to attract visitors. In such cases the visitors would be more likely to be interested in heritage tourism. Such situations have been given the label 'co-opetition'. Of course, the competition is not limited to other heritage sites, any attraction which could divert tourist footfall away from heritage represents potential competition, but the creation of a diverse tourist product offering is likely to be beneficial for attracting a more diverse range of visitors.

In the dynamic model there is the potential for sites at the micro-contextual level to have some influence over the 'impact context'. Cultural heritage sites have a greater potential to influence and have an impact on the micro context compared to the macro context. Some heritage sites can make a (sometimes significant) contribution to the local economy through increased visitor numbers, capital expenditures, or brand value.

Mission, vision and objectives

The *mission* can be thought of as a heritage site's overriding purpose. It outlines the broad general directions that an organisation should and will follow. Questions that need to be asked include; what is the site there for? Whom is it serving? Who should it serve? Why is it being funded?

All heritage sites have a sense of their mission, either explicitly or implicitly, which partly reflect the macro and micro impact context (the culture, the national system, and corporate governance and legal system) and also the power and interest of the stakeholders. As a process, not least to guide an impact evaluation, it is useful to know who decides the mission and how it is decided (see Figure 2).

The *objectives* of an organisation represent a more specific commitment, often over a specified time period, consistent with the mission (this may be quantified, but this can be inappropriate in some circumstances). Objectives take the generalities of the mission and turn them into more specific commitments: usually this process will cover what is to be done and its timing. Different kinds of objectives are possible: some will be quantified, some not. Of course, there may be conflict between objectives, particularly between the long-term and short-term interests of the organisation. Typically, objectives should be *challenging but achievable.*

Objectives may be seen as more of a heritage management tool, being statements of specific outcomes to be achieved which may or may not be measurable. For example, in the UK there is a national benchmarking process for museums which offers comparisons against certain prescribed criteria. There are also performance indicators being employed by various museums. Despite the growing targets/objectives culture in the public sector of the UK there is a need to be sceptical of

Fig. 2. The influences on the formulation of mission and objectives of a site.

their role and aware of the potentially distorting effects in delivering a service.

Most heritage sites express their *values* in two key ways: Through their services (what they do): such as addressing educational and social needs and their organisational practices (how they do it): who owns, controls and benefits from the value created by the heritage site, employment practices (who is employed, participation, job design, etc), and the relationships between different stakeholder classes.

One role of a socio-economic impact analysis would be to evaluate the extent to which the mission, values and objectives are being achieved. Are the intended outcomes being delivered? Should the mission and objectives be revised? Which impacts should be evaluated?

Stakeholders

The holistic model places stakeholders as a separate dimension of impact because they are a key consideration for all the other components of the model because they either directly or indirectly influence final impacts.[1] There are numerous definitions of stakeholders; "Stakeholders are those individuals or groups who depend on the organisation to fulfil their own goals and on whom, in turn, the organisation depends" (Johnson et al. 2006). Broadly speaking stakeholders encompasses all those who have an interest in the site and its running. Stakeholders are critical for impact evaluation, because they have:

- A key role in forming a site's mission and objectives
- Strong influence over strategy and management decision making
- A high importance in establishing which impacts are priorities and which are selected for measurement, and
- Influence over the cost and complexity of the impact measures chosen.

[1] Of course, stakeholders are not separate entities they are integral to all the components of the framework from management decision making, mission and objectives, to impact context.

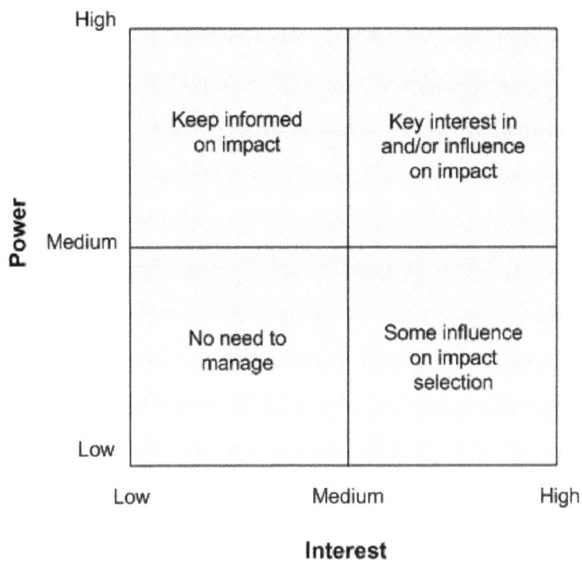

Fig. 3. A power/interest matrix applicable to cultural heritage stakeholders

Stakeholders can be divided into a number of groups:

- General public, press
- Heritage interest groups, heritage support networks, the UNESCO World Heritage organization.
- Local authority, trade associations, wider client groups
- Funding organisations, narrower client groups, trustees, the board of directors, staff, and family of staff.

Of course each of these groups has differing levels of power and interest in the heritage site. For example, a museum curator, or the board of directors would have both a high interest in the impacts of the site and a high power to influence change. Many heritage sites are publicly financed and are required to be accountable to a range of stakeholders. When difficult strategic or investment decision are being considered it can be useful to map stakeholders in terms of their power and interest in the outcomes of a decision to aid the political process of the decision making (see Figure 3)

The heritage site's organisational context

Organisational context of the heritage site is central to understanding impact. The impact of any site is heavily dependent on its location, quality, significance and the scale of the heritage site itself. As sites vary in their local, regional, national and global significance then so will their relative impacts. Some factors to consider include:

Ownership: The ownership of heritage sites is a principal determinant of the impact that a site will have. Ownership influences funding sources, governance structures, objectives, etc. However, ownership of heritage sites is not static. For example, because cultural heritage sites can have high maintenance costs – especially in countries with strictly enforced legislation regarding the upkeep of such sites – there is a tendency to see the movement of ownership from private to public hands. (Although, there are limits to the size of the public purse and without sustainability it is questionable as to how long the transfer of assets from private to public hands can take place).

- *Corporate governance*: Heritage sites can have a wide range of governance structures ranging from private and public, to not-for-profit and charities. Each of these will influence the impacts and outcomes of a heritage site. While it would be simplistic to assume that all sites under private ownership have a greater profit motivation than sites in public ownership there is a trend towards this scenario that cannot be ignored.
- *Location*: Location is paramount for the impact of a cultural heritage site. The location determines factors such as accessibility to transport networks, proximity to population centres links with other potential attractions. Surprisingly, location can be a dynamic entity. Although cultural heritage sites are fixed entities within the landscape or urban fabric the significance of the surrounding locality can change over time. A rundown part of an urban centre can become a popular tourist zone increasing the potential of the heritage sites within that area (such as Barcelona's Gothic quarter, the Barri Gòtic, preserved through neglect and now one of the principal tourist magnets in the city). Alternatively, the creation of new transport links such as low cost airline routes, or motorway and train-links can radically change the accessibility of a heritage site.
- *Quality of the cultural offer:* This exists at two levels. The significance of the site to society, and the quality of the 'visitor offer'.

  - *Significance*: The significance and importance of a site is a difficult entity to define. Sites have significance at multiple levels such as local, aesthetic, regional, and national. Of course, as with so many elements of the dynamic impact context the significance of cultural heritage sites is not a static element, it can change over time. It can change because of changes in the political system, technology, etc. Even at a single point in time a site may hold alternative significance to different elements of the population – this can determine who visits a particular site. For example, the post-World War II cultural heritage from Eastern Europe is seen by many from the old Communist Bloc countries as symbols of Communist oppression and is being rapidly destroyed and erased from the cultural landscape. To many academics especially, this rapidly vanishing heritage is an important historical resource that requires preservation.

- *Quality of the visitor offer*: The quality of the visitor offer at a heritage site or experience can be determined by a number of factors such as the level of preservation, which lies outside the scope of the heritage site, however, site maintenance, level of restoration and visitor facilities tend to fall within the potential control of a site, finance depending, as can the actual or perceived authenticity of the site. Contemporary Western society is a consumer society. In this society the public have been exposed to progressively more sophisticated products, services and marketing, and as a direct result they have become much more sophisticated consumers. It is these same consumers who visit heritage sites and they will judge those sites accordingly. Facilities and services at heritage sites need to be at a standard commensurate with contemporary consumer ideals otherwise sites risk alienating many of their visitors.

- *Scale*: Scale can act as a guide to the potential impact of a cultural heritage site (although, no more than a guide). Larger sites have the potential to induce a greater impact than smaller sites, because of their ability to support a greater throughput of visitors, sustain larger potential capital costs, higher staff requirements and other running costs. Of course, concentrations of smaller heritage sites can have a similar effect.

These factors have a strong influence on the site – and feed into the management decision-making context. Furthermore, it is argued that contextual factors are immensely important determinants of the socio-economic impact of heritage sites. Placing a heritage site in context will guide what impacts that should be evaluated. For instance, there would be little point in doing a full, and often costly, economic impact analysis of a small museum based in a large city that was designed to serve the local community and foster local cultural identity. In such a context impact assessment may be aimed at issues of community integration and social inclusion, etc.

1.1. Heritage management decision making

The quality of the heritage management decision making will have a strong influence on final impacts. Many impact studies either ignore this or treat it as a black box. The principal components of the management decision making element include: operations, financial, marketing, human resource management and organisational culture, and technology strategy.

- *Operations*: Operations management considers how resources such as capital, people, information and materials are converted into outputs such as services, and information (products). Cultural heritage site operations management is made more complex by the lack of funds.
- *Financial*: The cultural heritage sector is perennially short of funds. Increasing competition for central funds, increasing operational costs, and an increasing number of heritage sites makes this situation unlikely to change. The management of financial resources within heritage sites is crucial.
- *Human resources*: The effective management, training and motivation of personnel within heritage sites are other factors that can influence impact. This affects all levels within a site from management to voluntary staff. The drive and determination of heritage site managers is crucial to the long-term success of sites. With funding being such an issue in the sector, financial incentives for high-end managers are poor and considerable reliance is placed on the dedication, and devotion of managers in the sector. Correspondingly, heritage sites often have highly qualified staff with postgraduate qualifications who are paid correspondingly less than in other sectors. Furthermore, many heritage sites increasingly use voluntary labour to support many functions in the day-to-day running of sites. This also requires careful management. Human resource management issues also exist beyond the level of the site. For example, the decline in skilled crafts persons capable of maintaining and preserving heritage sites could have long-term implications for the sustainability of some sites. There is a widespread lack of traditional craft skills in certain parts of Europe. The increasing use of modern building techniques and materials has reduced the need for traditional craft people such as stone masons, thatchers, plasterers, etc. And it is not only ancient buildings that are affected by lack of relevant skills. The Big Pit mining museum in the Blaenavon World Heritage site in South Wales now has to consider how the underground mine workings that closed in the mid-1980s will be maintained in the future. This is because collapse of the coal mining industry in the UK has resulted in a lack of properly trained mining engineers in the UK.
- Marketing: The marketing strategy will influence the number and type of visitors which determines key final economic impacts. With often limited marketing budgets, a key marketing strategy will be the extent that sites can leverage advertising and promotion by combining with other interested partners in joint marketing campaigns. The increasing number of undergraduate and post-graduate courses, and academic books devoted to heritage marketing suggests that there is a growing awareness of the critical nature of marketing in the cultural heritage sector. Furthermore, the growing number of heritage special interest groups in the marketing sector further reinforces this trend.

This raises the need for regular impact reports based on efficient accumulation of relevant data to feed back into strategic decision making and influencing future investment decisions. For example, few sites have a *detailed* breakdown of their visitor profile. Not only is this useful impact data but it can enable more targeted marketing strategies to take place – further enhancing desired impact. This presents socio-economic impact as a dynamic rather than static notion.

Strategic choice and implementation

The quality of leadership and strategic thinking not only defines the mission and objectives but sets the visions, makes the key directional choices and innovations, and implements and manages strategic change. The key questions to any heritage site needs to answer are:

- Where is the site positioned now?
- Where does the site want to be positioned? and
- How can that be achieved?

This last question is where management decision-making comes to the forefront of site strategy.

1.2. Socio-economic impacts and outcomes

The socio-economic framework developed (see McLoughlin, Sodagar, and Kaminski 2006) is summarised in Figure 4 below. The impact drivers and strategies generate a site's impact and reveal the dynamic nature and complexity of impact measurement. Choices need to be made regarding which impacts should be prioritised for measurement, taking into particular account the mission and stakeholder analysis. The impact framework can help prioritisation by identifying the principal impact categories. For example if economic impact is not a priority then there is no need to devote resources to measuring it.

Once the priority impact categories (e.g. economic, social and environment) are identified there remains the key issue of determining the appropriate measurement methodology. Numerous factors will influence this:

- The fundamental purpose of the measures.
- Impact for whom? On whom?
- Specialist research method (normally one-off impact projects) versus an ongoing management Impact information.
- Cost/complexity.
- Capabilities and resources of the management team.

The feedback loop and strategic decision making

With reference to the Holistic Impact model, Figure 1, it is essential to utilise the impact results not only for external reporting but also to inform management decision making and future strategic directions. This emphasises that impact analysis is part of the strategist's toolkit and needs to be fully integrated in to management reviews, business and strategic planning processes. This vital usage of the impact data will encourage continuous improvement and positive strategic change.

Conclusions

This chapter has presented a broad practical impact framework that can guide heritage professionals through the thought process and enable them to draw from a 'menu' of possible measurement methodologies that can be selected and applied depending on their specific organizational needs.

This is significant, because while it would be convenient, and even perhaps desirable, to identify a single generic measurement methodology that could capture, the range of socio-economic impacts of cultural heritage sites and that could be easily transferable to different contexts, such a goal is unlikely. This is because of the heterogeneous nature of the cultural heritage sites. For example, variations in scale and significance (local, national or global), geographical location, local competing activities, policy and funding contexts, mission, governance and stakeholder influences as well as approaches to strategic management and marketing.

There are a number of uses for the model presented in this chapter:

- Such a holistic approach will provide a useful basis from which heritage managers can conceptualise socio-economic impact.
- If managers can begin to see how various elements come together to influence impact they can increase their understanding of heritage impact and this could form the platform from which site managers can influence positive outcomes.
- By looking at sites using the same criteria the models allow managers to compare sites.
- It can help to enhance internal performance.
- It can support better reporting to stakeholders and funding bodies.
- It is fundamentally a strategic management tool to help drive continuous improvement and positive strategic change.

The holistic impact model presented in this chapter as a conceptual management tool that is applicable to heritage sites.

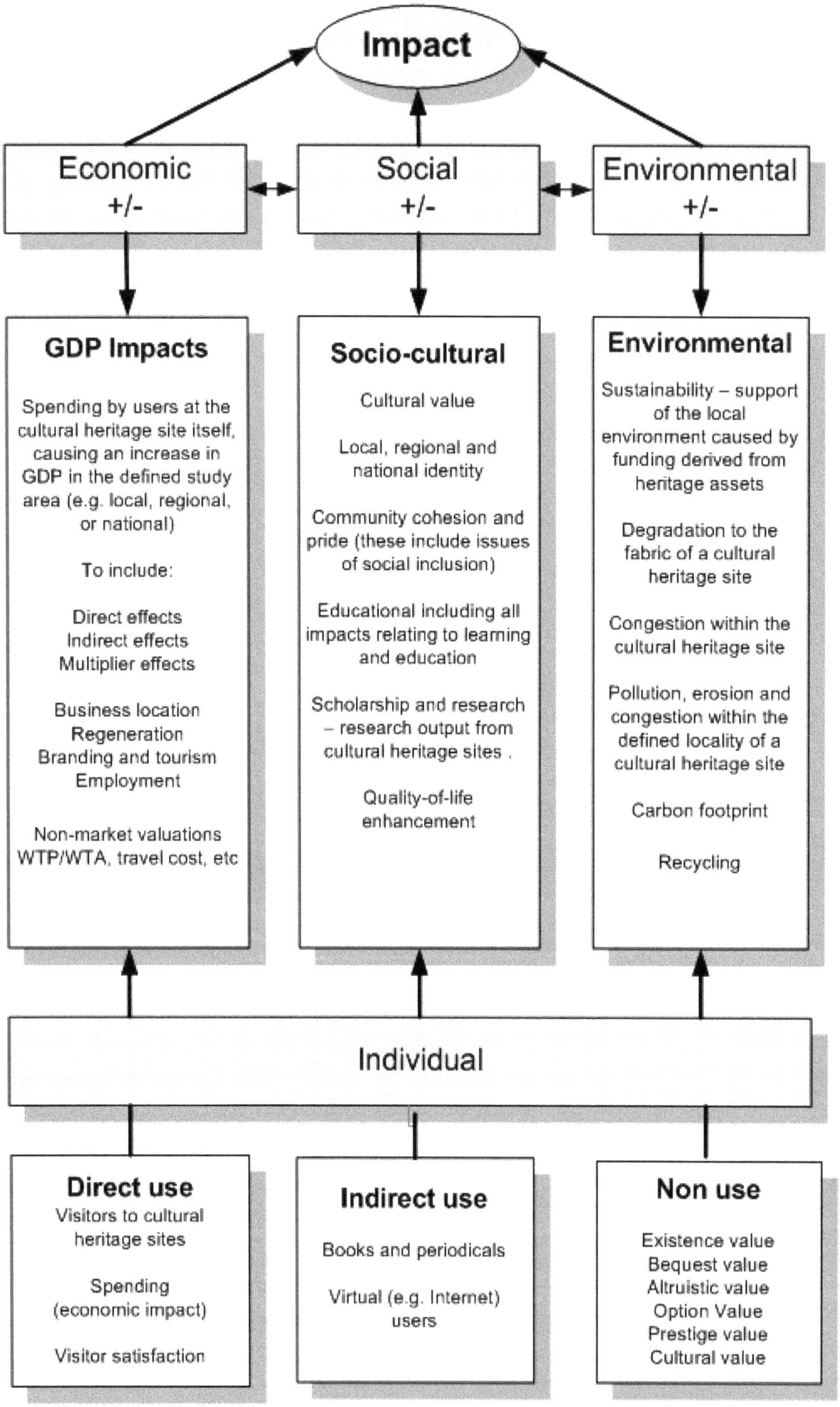

Fig. 4. A socio-economic benefit and impact model.

References

ASHWORTH, G. AND HOWARD, P. 1999 *European heritage planning and management*. Exeter: Intellect.

JOHNSON, G., SCHOLES, K., AND WHITTINGTON, R. 2006 *Exploring corporate strategy*, London: FT Prentice hall.

LOWENTHAL, D. (1985) *The past is a foreign country*, New York: Cambridge University Press.

MCLOUGHLIN, J., SODAGAR, B. AND KAMINSKI, J. 2006 Dynamic socio-economic impact: a holistic analytical framework for cultural heritage sites, in McLoughlin, J., Kaminski, J. and Sodagar, B. (eds.) *Heritage Impact 2005: Proceedings of the first international symposium on the socio-economic impact of cultural heritage*. Budapest: Archaeolingua, 43-57.

MIGNOSA, A. AND RIZZO, R. 2004 Heritage and information, in Cain, K., Chrysanthou, Y., Niccolucci, F., PLETINCKX AND SILBERMAN, N. (EDS.) *INTERDISCIPLINARITY OR THE BEST OF BOTH WORLDS – THE GRAND CHALLENGE FOR CULTURAL HERITAGE INFORMATICS IN THE 21ST CENTURY*, SELECTED PAPERS FROM VAST 2004, Budapest: Archaeolingua.

RIZZO, R. AND MIGNOSA, A. 2006 POLICY DECISIONS AND CULTURAL HERITAGE IMPACT, IN MCLOUGHLIN, J., KAMINSKI, J. AND SODAGAR, B. (EDS.) *HERITAGE IMPACT 2005: proceedings of the first international symposium on the socio-economic impact of cultural heritage*, Budapest: Archaeolingua, 58-68.

Geospatial database of Ahichchatra using geoinformatics techniques

Amit Tare,[1] Onkar Dikshit,[1] G. K. Rai,[2] R. S. Fonia,[3] V. N. Prabhakar,[3] Bhuvan Vikram[3]

[1]Department of Civil Engineering, IIT Kanpur
[2]Department of Ancient History, Culture and Archaeology, University of Allahabad
[3]Archaeological Survey of India

Abstract

This work involves the analysis of the geospatial database created for Ahichchatra, an ancient site of national importance near Bareilly (Uttar Pradesh) to draw inferences that are significant from archaeological point of view. The work can be divided into five components (a) Processing of preliminary data (b) Field work (c) Post-processing of field data (d) Preparation of geospatial database (e) Analysis and interpretation of database.
Preliminary survey for Ahichchatra was planned using the images from Google Earth and SRTM data. A detailed survey using the state-of-art instruments like motorized total station and geodetic quality Global Positioning System (GPS) receivers was conducted. Total station and GPS data were processed along with satellite images from LISS-III, LISS-IV sensors using digital image processing system and converted to a common coordinate system.
On the basis of interpretation of this database and associated archaeological literature, the following inferences were drawn about the past environment of the study site:

1. There should be a water body close to the site, associated with least elevation.
2. High elevation areas having small footprints are associated with temples.
3. Fortification can be associated with higher elevation and higher slope values.

Keywords geospatial, database, archaeology, interpretation, survey, data processing.

Introduction

Geoinformatics is defined as the technology for collection, processing, and dissemination of geoinformation for managing and monitoring of objects and phenomena. The two basic components involved are data collection and data management. Conventional data collection is generally carried out using land surveying and analogue photogrammetric techniques, which are time-consuming and have limitations in terms of accessibility, speed, storage and dissemination. However, since a few years tremendous advancement in data collection techniques have taken place with the advent of electronic surveying, remote sensing, Global Positioning System (GPS), airborne and terrestrial LiDAR, digital photogrammetry and sub-surface sensing Ground Penetrating Radar (GPR).

Archaeological data in a digital form is very useful. The management of data acquired by the aforementioned technologies can be carried out using GIS, which has the potential to integrate geospatial and attribute data for any required application. GIS can be used as a tool for outlining and understanding the ways in which people have thought about and utilized their surroundings.

Objective

The objective of the study is to develop a geospatial database of Ahichchatra region, an ancient archeological site near Bareilly, and use this database to draw inferences about the past environment of the site. The present work is perhaps the first systematic attempt in India to use Geoinformatics technology to investigate an archeological site of national importance.

Overview of study site

The study site is located at Ramnagar in Aonla Tehsil situated in Bareilly District of Uttar Pradesh. It is situated around 55km away from the Bareilly city in north-west direction. The nearest river at the site is Ramganga. Ahichchatra, located 11km away from Aonla, is one of the most extensive and early historical sites in India. The site is located between latitude 28°21'52"N to 28°22'42"N and longitude 79°07'38"E to 79°08'54"E. Ahichchatra can be approached by rail or by road. The nearest railway station is Aonla. When approached by road, the nearest bus station is at Ramnagar.

Data resources used for investigation

Data resources used for the purpose are topographic sheets, satellite imageries (from PAN and LISS-4 sensors), Shuttle radar topographic mission (SRTM) data C band, Google Earth snap shots. Imageries of the region were obtained from NDC, NRSA. The list of satellite data set is given in Figure 1.

The site lies in the topographic sheet $53\frac{P}{3}$, at a scale of 1: 50,000. The SRTM data is available free of cost and gives coarser topographical information on a near-global scale at 90m spatial grid.

Sensor	Resolution (m)	Swath (km)	Dates	Satellite data
LISS -4	5.8	23.9	1-04-2005	IRS P6 Sensor L4MX Orbit 7560 Segment 2 Strip 2 Scene 045
PAN	5.8	70	29-10-2006	IRS1D Sensor PANC PATH 098 Row 051

Fig. 1. Satellite images used for the study.

Methodology

The methodology for the work can be divided into the five components (a) Processing of preliminary data (b) Field work or survey work (c) Post-processing of field data (d) Preparation of geospatial database (e) Analysis and interpretation of geospatial database.

As the extent of the site was known from the archaeological records, the site location was searched with the help of Google Earth. Snap shots of the same were taken which helped in planning the preliminary survey of the site. Also, the SRTM data for the same location was searched and analyzed. The field work involved surveying using the state-of-the-art technologies and instruments like auto level, total station and GPS. The survey was carried over a period of 12 days. Survey work was initiated by setting up the base station and a large number of control (or reference) stations. These control stations were used as starting points for the total station survey. The reduced levels were transferred from available bench mark at the site to these control points using fly leveling. The coordinates in total station were recorded in Universal Transverse Mercator (UTM) zone 44 map projection with WGS-84 as the reference ellipsoid. The laboratory work comprised processing all the data gathered by total station and GPS during field surveying. Satellite images, Google Earth's snap shots were also processed using an image processing system to convert all data to a common coordinate system (UTM Zone 44 with WGS-84 as the reference ellipsoid).

Preliminary interpretation

Preliminary interpretations were carried out by using images from Google Earth and SRTM data. In the initial stages, site related data was not available. In order to plan the preliminary survey, a snap shot was taken from the Google Earth covering the entire site. It was georeferenced in UTM WGS-84 using the coordinates available in Google Earth. SRTM data was also used due to the fact that it is available free of cost which can be used for analysis on macro level. Non-availability of finer geospatial data also required field survey. Using the total station point data, a digital elevation model (DEM) was created. This DEM was used for interpreting certain results which are important from archaeological point of view.

Field work or survey work

The various surveys carried out on the study site involved (a) Fly leveling (b) GPS survey and (c) Total station survey. The fly leveling at Ahichchatra was carried out using Nikon auto level AC2S auto level. The survey was carried out for transferring the MSL available at site to different control (or reference) points. The elevation of the bench mark at the Ahichchatra was marked as 174.015m above the mean sea level (MSL). The specifications of the instrument used for this purpose are given in Figure 2.

S. No.	Specification	Description
1	Weight	1.25kg
2	Telescope power	24X
3	Image	Erect
4	Minimum focusing Distance	0.75 m
5	Accuracy 1km double run	±0.4mm

Fig. 2. Specifications of AC2S (Nikon instruction manual, 1996).

The GPS survey was started by setting up a base station at the site. Several control points were collected each for 30 minutes at each station. These reference points were processed against the base coordinates to eliminate various errors. The GPS survey at Ahichchatra was done using GPS instrument SR 530. The specifications of the same and various setting on which the survey was carried out are listed in Figure 3.

The observations were recorded in Static, Stop and Go and On the Fly mode. The configuration settings of SR 530 for different methods are given in Figure 4. The total station survey was carried out using the Trimble 5600 DR200+ equipment which has the capability to work in direct reflex mode also. The specification of this instrument is given in Figure 5.

The total station survey was conducted by assuming the UTM grid as an idealization of a square grid so that the error propagated can be assumed to be negligible. The basic approach involves collecting GPS coordinates at certain reference points from where area to be covered has maximum visibility and converting it in a UTM WGS-84 grid. These coordinates form the basis for the total station survey. The survey was started by feeding the northing, easting and elevation information for the reference point and then

S. no	Particulars	Specification	Description
1	Satellite Reception	Dual frequency	L1 and L2 channels
2	Receiver channels	L1 channels	Carrier phase, P1 code, C/A code
		L2 channels	Carrier phase, P2 code
3	Base line precision (Differential Phase)	Static	5mm +1ppm
		Rapid static	5 mm + 1ppm
		Stop & Go	10 mm + 1ppm
		Kinematic	10 mm +1ppm
4	Base line precision (Differential code)	Static	30 cm
		Kinematic	30 cm
5	Weight	*SR 530*	1.25 kg
		AT 502	0.4 kg
6	Power Consumption	*SR 530*	7 Watt Maximum
7	Supply Voltage	*SR 530*	12 Volts
8	Temperature	Operating	-20°C to +55°C
		Storing	-40°C to +77°C
9	Humidity	Non condensing	Upto 95 %

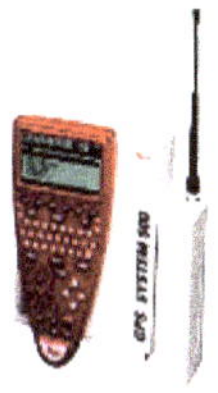

Fig. 3. SR 530 specifications (Leica Geosystems, 2001).

S. No.	Particulars	Static		Stop and go		On the fly	
1	Mode	Standard Advanced (Customizable)		Standard Advanced (Customizable)		Standard Advanced (Customizable)	
2	Vertical offset	Pole	0	Pole	0	Pole	0
		Tripod	0.3600	Tripod	0.3600	Tripod	0.3600
3	Update rate (in sec)	1		1		1	
4	Coordinate System	WGS-84		WGS-84		WGS 84	
5	Quality Type	GDOP		GDOP		GDOP	
6	Log static observation	Yes		yes		yes	
7	Log moving observation	No		yes		yes	
8	Observation rate (in sec)	10		1		1	

Fig. 4. Configuration setting for SR 530 (Leica Geosystems, 2001).

sighting another reference point whose coordinates are already known. A User Defined Sequence (UDS) was created in the total station for collecting the topographical information. UDS is a program which collects the user defined information a number of times until the program is ordered to halt.

Post processing of field data

All data generated from the survey was processed in the laboratory and was used for the generation of DEM. The laboratory work for the same comprised processing and manipulating all the data. It involved usage of different

S. No.	Particular	Property
1	Light source	• Pulsed laser diode 870 nm
2	Beam divergence	• Horizontal. 0.4 mrad (4 cm/100 m) (0.13 ft/328 ft) • Vertical. 0.8 mrad (8 cm/100 m) (0.26 ft/328 ft)
3	Atmospheric correction	• –60 to 195 ppm continuously
4	Leveling	• Circular level in tribrach 2 mm • Electronic 2-axis level in the LC-display with a resolution of 6" • Clamps and slow motions
5	Centering	• Trimble 3-pin • Optical plummet • Magnification 2.4x
6	Telescope	• Magnification 26x (30x Optional) • Aperture 40 mm (1.57 in) • Field of view at 100 m is 2.6 m • Shortest focusing distance 1.7 m to infinity • Illuminated crosshair • Operating temperature –20 °C to +50 °C
7	Power Supply	• Internal battery. Rechargeable NiMH battery 12 V, 1.8 Ah, Operating time approximately 3 hrs • External rechargeable NiMH batteries 12 V, 3.8–11.4 Ah, Operating time approximately 11 hrs
8	Weight	• Instrument (not including controller) 6.6 kg • Tribrach 0.7 kg • Internal battery 0.4 kg

Fig. 5. Trimble 5600 DR 200+ specifications (Trimble user guide, 2001).

softwares like Ski Pro, Geodimeter, ERDAS IMAGINE 8.7 and ArcGIS 9.1. Programming languages MATLAB and Visual Basic 6.0 were also used.

(i) GPS data

SR 530 GPS data was downloaded and processed using Ski Pro software. The base station was established by collecting GPS data for 12 hour. This data was assumed as the base station for further GPS data processing. The GPS coordinates of base station in WGS-84 after processing were latitude 28° 22' 23.04269" N, longitude 79° 08' 14.04871" E, and elevation 114.4284m (geodetic height). Reference points were corrected using the given base station coordinates in the Ski Pro software.

(ii) Total station data

Total station survey was carried out using the GPS coordinates. Approximately 7000 points were collected. This point data was downloaded using the Geodimeter software available with the Trimble 5600DR 200+ series total station. These downloaded data files are in raw format which were converted to ASCII point file format. This ASCII file can be imported in ERDAS IMAGINE 8.7 to form a DEM. To form a Triangular Irregular Network (TIN), the ASCII point file is converted to comma separated value (*.csv) format. This *.csv file is converted to Point feature class. Point feature class is then converted to TIN. This TIN can be converted to any contour map or a DEM can also be generated.

Preparation of geospatial database

For the preparation of geospatial database, major steps carried out were generation of DEM (using survey data, contour lines and benchmarks), coordinate system transformation, archaeological grid prediction, georeferencing of topographic sheets, forming their mosaics, registration of satellite imageries, forming Google Earth mosaic, digitization of different features (boundary, mounds, contours, etc.).

(i) Generation of DEM

Initial DEM was created using contour information available with the topographic sheet having a 10m contour interval. Thus, finer interpretation was difficult. This information can however, be used for analysis on macro level. This aspect also necessitated field survey work as most of the data available till now was coarser in nature. DEM was created using the total station point data. This DEM can be used for interpreting certain results which are important from archaeological point of view.

(ii) Coordinate system transformation

Archaeological Survey of India (ASI) uses traditional surveying instruments like levels and theodolite for laying out the grid points while excavating. But this system is a local system and does not hold any relationship with any global system like WGS-84 used with GPS. A programme was written in MATLAB to interconvert between ASI grid and global grid in UTM map projection in WGS-84.

(iii) Topographical sheets

Georeferencing was carried out to attach geographical information to an image. This information is in the form of Degrees Minutes Seconds (DMS) in the curvilinear coordinates and in meters or km if in a projected coordinate system. For present work, we have georeferenced all the topographic sheets in Lambert Conformal Conical (LCC) projection system using Everest as the reference ellipsoid. For this purpose, coordinates were calculated by using coordinate calculator of ERDAS IMAGINE. Each sheet was then reprojected from LCC Everest to the UTM zone 44 WGS-84 system.

(iv) Satellite images

PAN and LISS-4 sensor images were registered using the reprojected topographic sheets in UTM zone 44 WGS-84 system. Around 40-60 well distributed time invariant GCPs were taken to georeference the images.

(v) Mosaic

16 Google Earth snap shots of the entire site were taken in parts keeping some overlap. These individual parts were first georeferenced and then mosaiced. The resulting mosaic yields high resolution georeferenced imagery of site.

(vi) Digitization

Digitization of the prominent features was done using ArcGIS 9.1. Features like water bodies, road and rail network, contour lines and bench marks, etc. were marked using topographic sheets. Features like mounds outside the test site, boundary of the test site, road connectivity, bastions were marked from the Google Earth mosaic.

Analysis and interpretation of geospatial database

Google Earth imagery and SRTM data was processed to make certain interpretations discussed later. In Google Earth mosaic (Figure 6), created by ERDAS Imagine software, bastions are clearly identified using red lines. Out of 32 bastions mentioned in the archaeological literature, 18 can be visually interpreted. SRTM gives topographic information at a spatial grid of 90m. But the raw data is difficult to interpret. After colour coding of DEM, some features can be interpreted very easily as seen in Figure 7. Flood plain of the river is depicted by blue colour and the elevated area in red shades. In Figure 7, the site location is marked with a circle where the whole site and its neighboring region can be distinguished by different colour shades.

In Figure 8, the enlarged view of the site is shown in three different colours. The region bounded within the triangle like portion highlights low lying area of the site. From the DEM, it can be concluded that the irregularly bound region was likely to be prone to flood than the remaining part of the city. Whereas the region marked within a circle shows elevated land. Further, the Google Earth imagery in Figure 9 shows a structure bounded within red lines which the existing archaeological literature confuses with an ancient road or a wall. As this structure partitions the adjoining low lying and elevated area, the chances of it being a fortification for protection against floods is higher than a road network.

The survey work carried out at Ahichchatra yielded several outputs. For example, fly leveling provided MSL of all control points. The GPS survey yielded coordinates of all control points in UTM system. These coordinates were later on used by the total station survey of the site.

Figure 10 provides location of these control points. Some of the major features such as road ways, boundary, and temples were also plotted with the help of GPS can also be seen in the Figure 11.

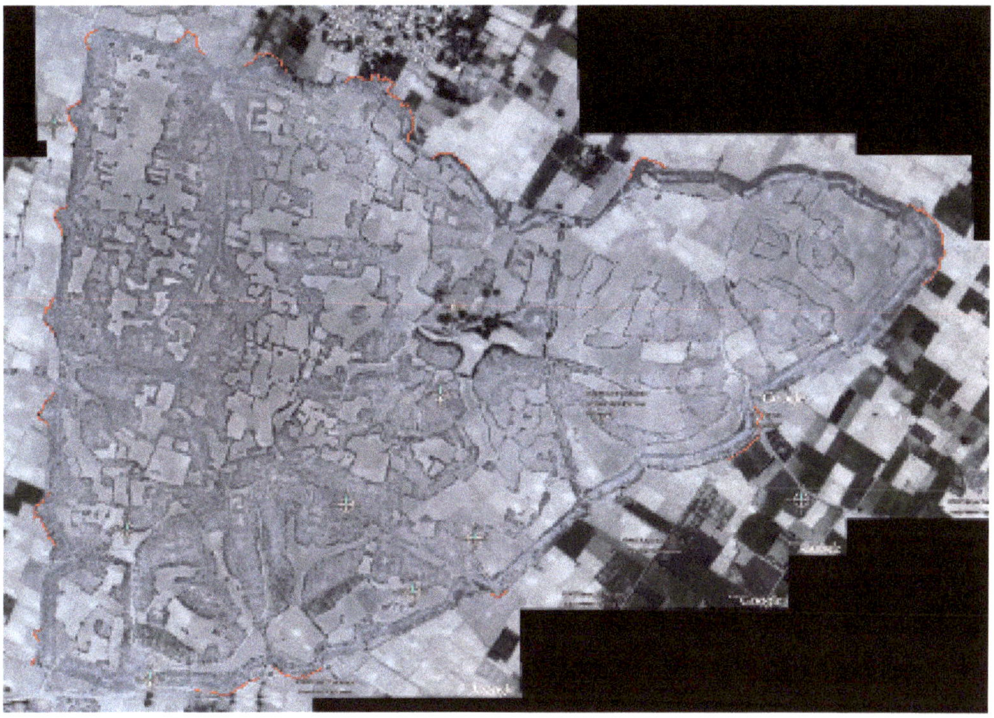

Fig. 6. Google Earth imagery

Fig. 7. Colour coded SRTM data.

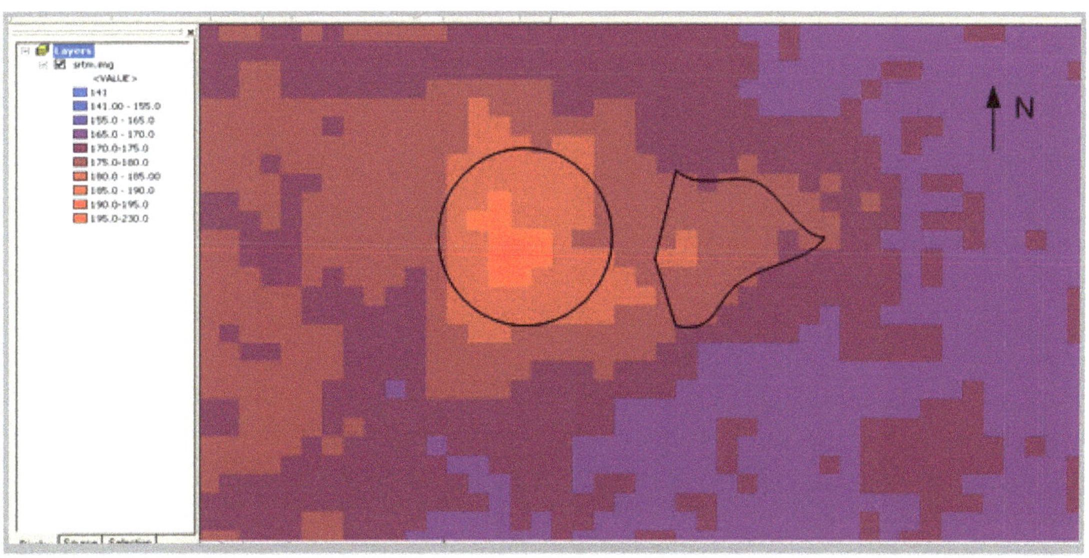

Fig. 8. Classified DEM.

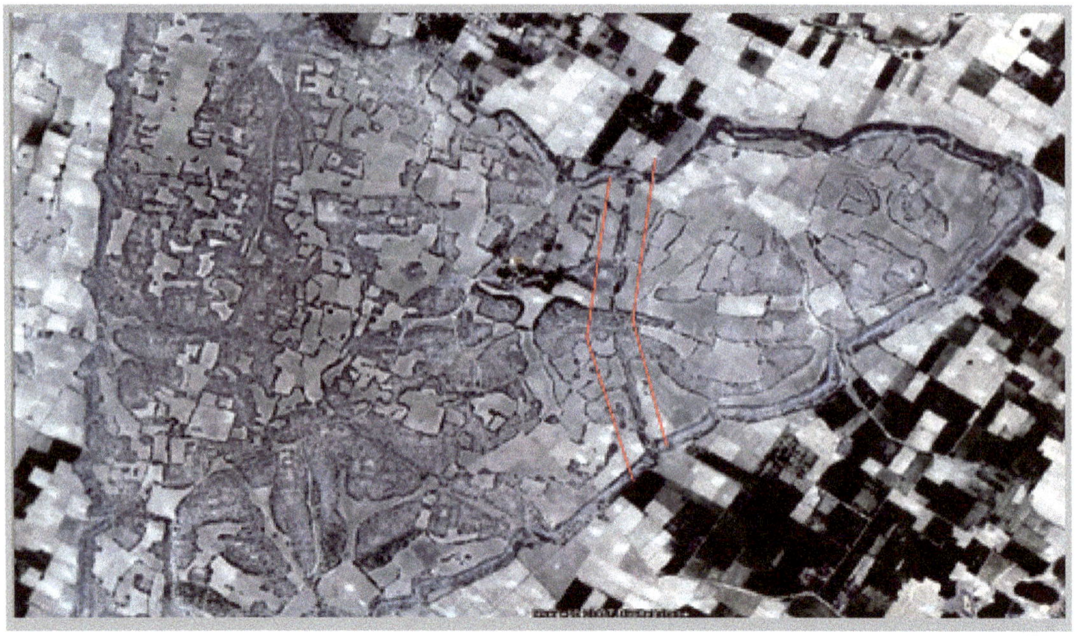

Fig. 9. Google Earth imagery marked with the confusing structure.

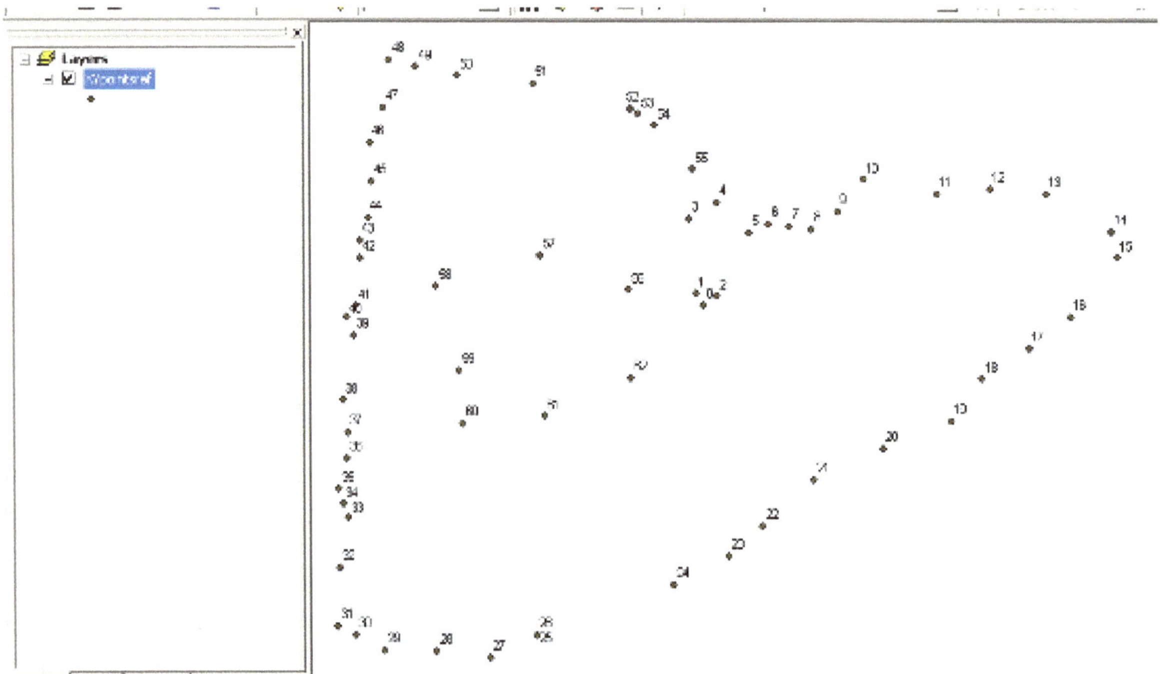

Fig. 10. Plot of reference points.

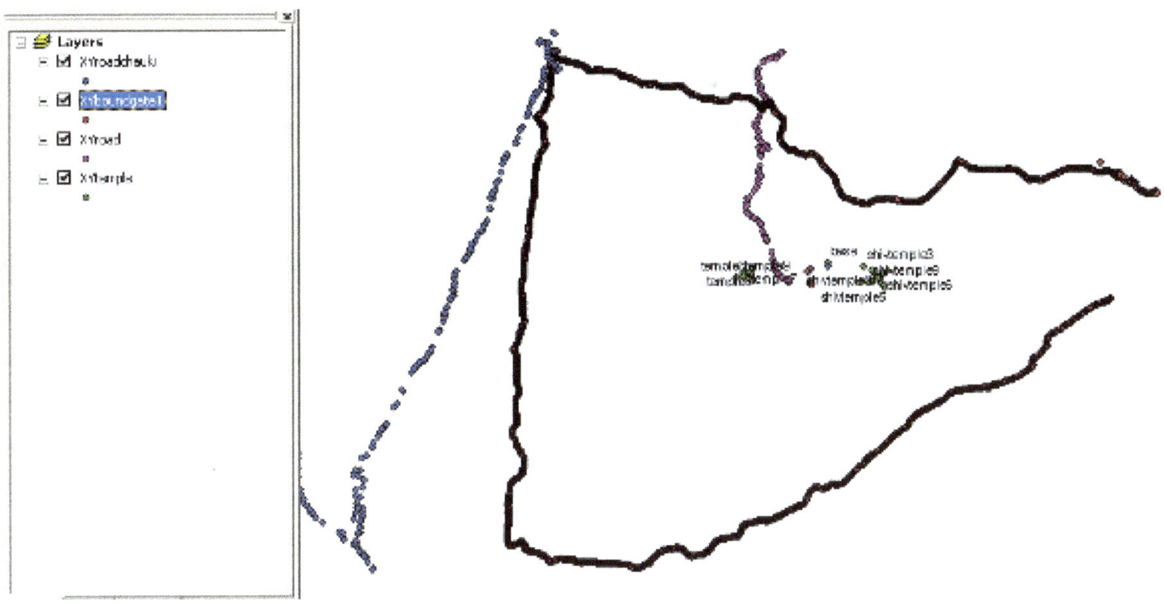

Fig. 11. Features plotted through GPS.

Total station survey resulted in the formation of a DEM and contour map for the area. Figure 12 shows the Triangular Irregular Network (TIN) model of the study site prepared in GIS. In TIN, the fortification can be seen very clearly. It appears to be a raised feature above the ground with a considerable elevation difference from the outside area. Figure 13 shows a slope map created from the DEM indicating that most of the fortification is associated with the high slope value with all parts of fortification. One can also identify red spots inside the fortification area. These red spots correspond to the temples which have the highest elevation difference from the neighboring area. In early times, temples used to be situated at the highest elevation and most of the city used to be visible from there.

The contour map of the site is given in Figure 14. Closely placed contour lines near the boundary denote the fortification. The classified DEM in the Figure 15 highlights various features.

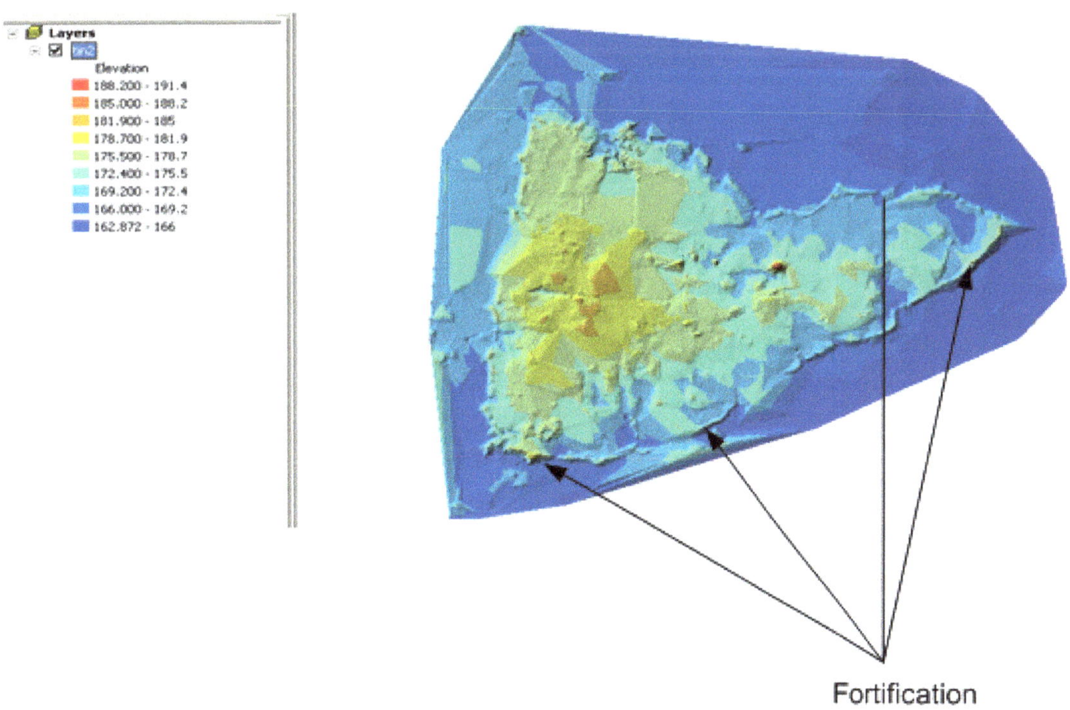

Fig. 12. Triangular Irregular Network (TIN) model of the site.

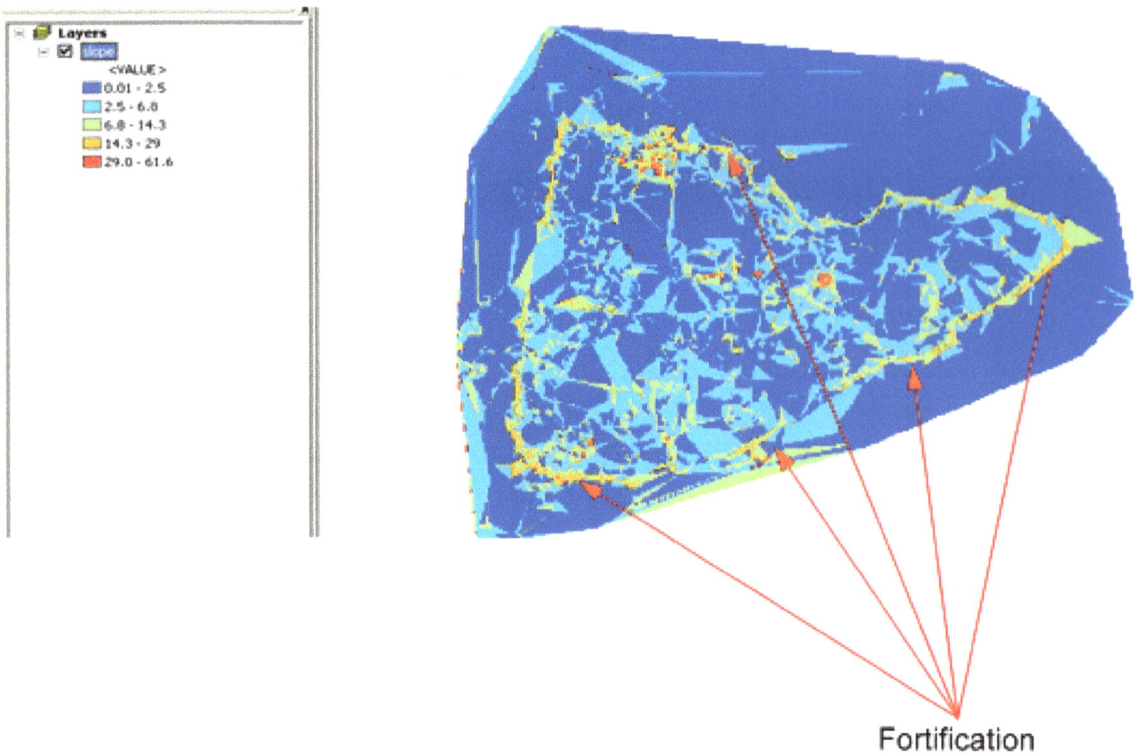

Fig. 13. Slope map of the study site.

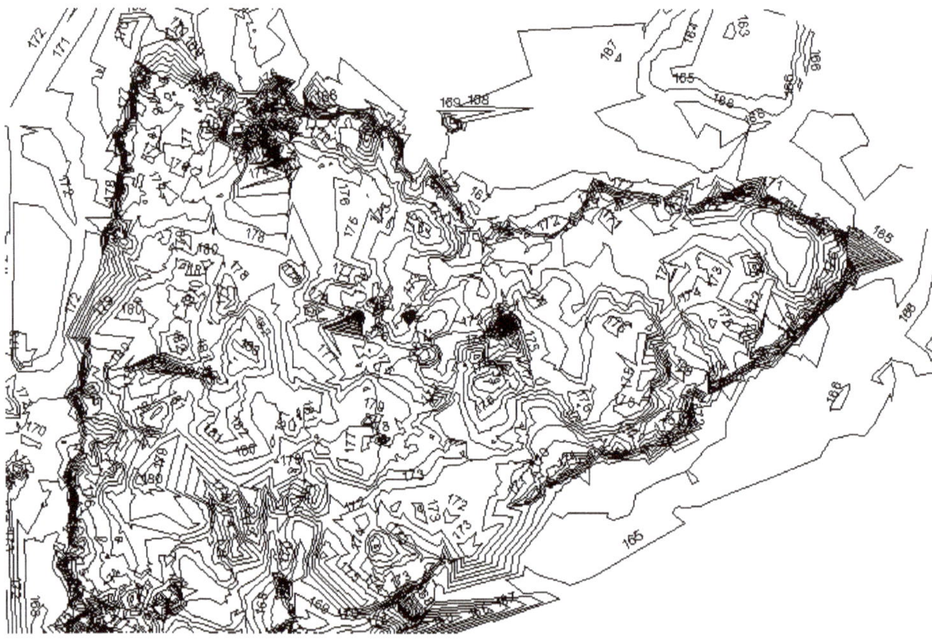

Fig. 14. Contour map at 1m contour interval.

The classified DEM in the Figure 15 shows location of probable water body in dark blue. The green color denotes the fortification and the yellow portion denotes the higher level areas mainly for habitation. The red area denoting the higher elevation points may be temples or structures resembling the colonies. The lower part of the triangle in green colour shows lower elevation areas which were inhabited for only short durations. According to archaeological reports, this area contains depositional records of only two periods. Normally deposition of each period would be found accumulated one above the other. This is due to the fact that when a structure fails, the next generation just levels the debris and builds their dwelling above it. Therefore, red portion in the upper part marked by a big circle with higher elevation can be interpreted to represent more periods of archaeological depositions. Archaeological records also confirm that this part contains deposition of all the five periods.

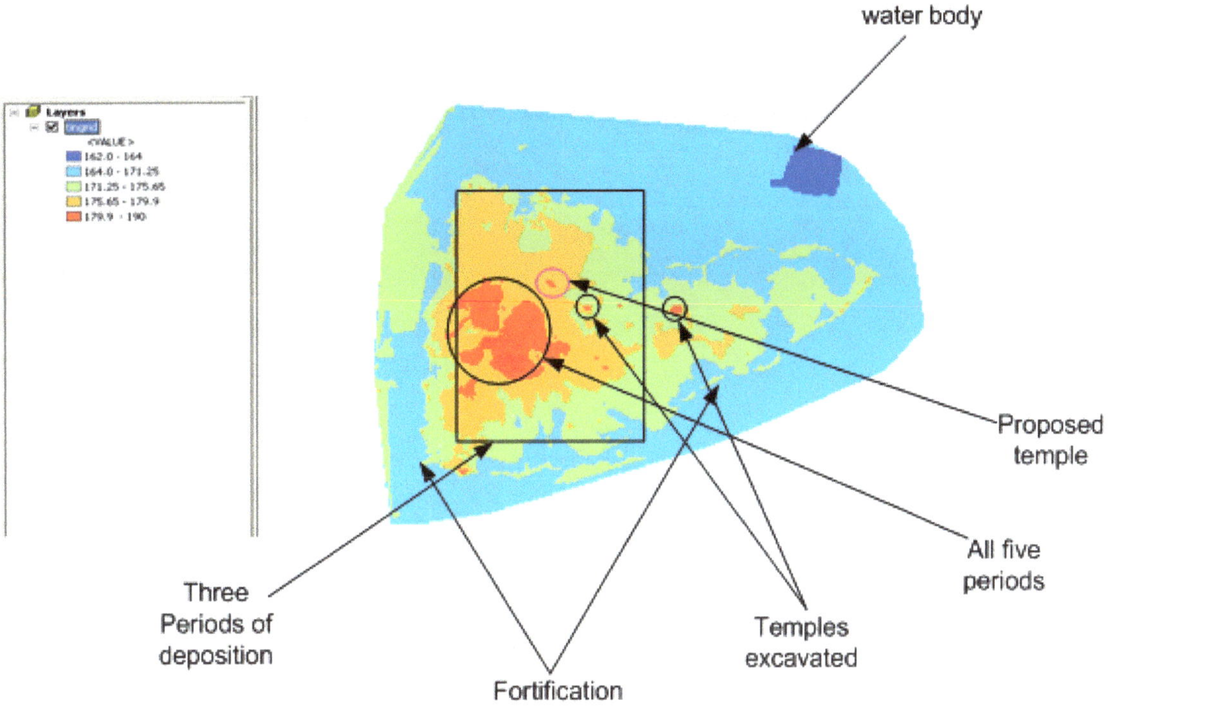

Fig. 15. Colour coded DEM.

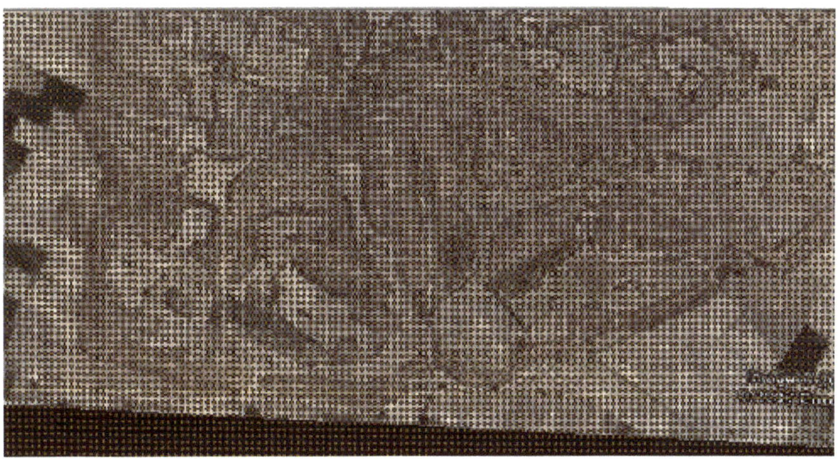

Fig. 16. Grid at Ahichchatra overlaid on Google Earth imagery.

Human beings always located settlements near a water body. From the prepared maps of the site, it is clear that the nearest river to the study site is situated about 11km away from Ramnagar which is quite far off from the site for sustenance of civilization. Hence, a water body was expected to be situated somewhere near the test site for satisfying normal water requirements in the past. As low lying areas tend to get flooded easily, the regular shaped blue feature shown in Figure 15 has greater chance of being a water body. The archaeological literature also confirms the presence of water body in that particular area (Cunningham Archaeology report, 1862-63 p. 262).

The plan of the predicted grid in UTM projection system is shown in Figure 16. The inputs to the program were coordinates of two initial points in UTM projection system and the rest of the points were calculated by the trigonometric formulation developed in MATLAB. This program will convert the grid formed by ASI while surveying into GPS coordinates. This will become handy when the artifact records found at a particular location will be entered into the database.

Nine topographic sheets were georeferenced and images were registered using these topographic sheets. In PAN (Figure 17) and LISS-4 (Figure 18) images, the site is easily distinguishable from the rest because of its shape. PAN shows fewer details as compared to LISS-4 even though their resolutions are same.

Fig. 17. PAN imagery.

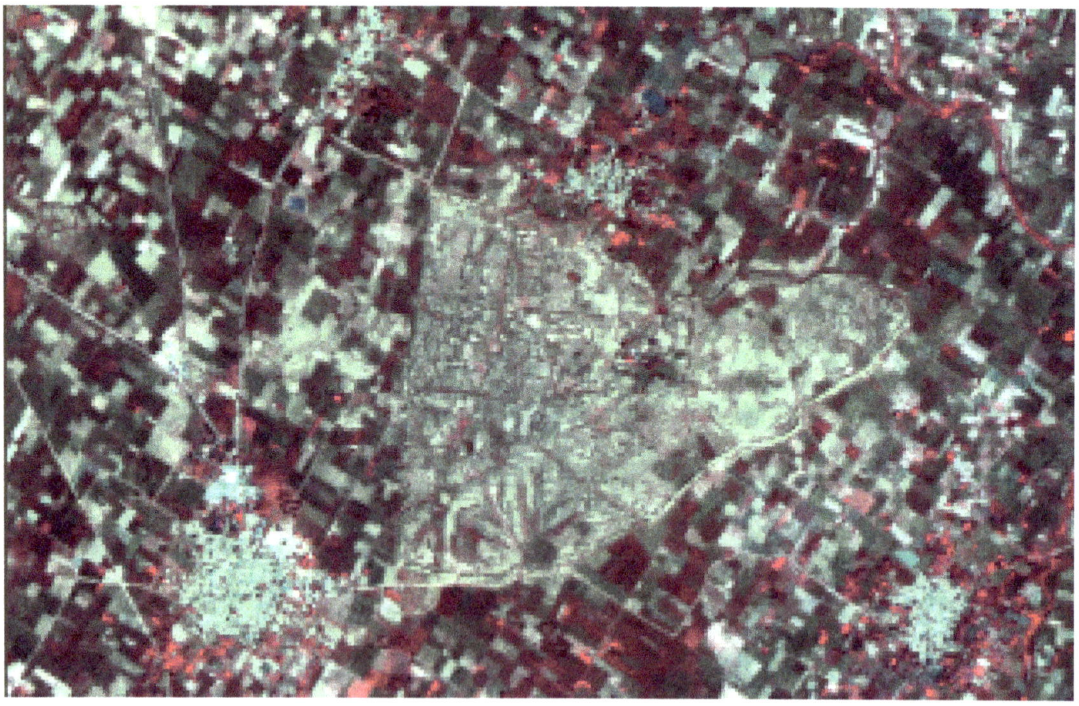

Fig. 18. LISS-4 imagery.

This is due to the fact that multispectral LISS-4 can be used to prepare colored false colour composite (FCC). Figure 19 shows a perspective view of Google earth image draped over the DEM of site. It clearly shows the high raised structures still intact and the plain areas being used as farms by local people. Draping can also be used to prepare a flythrough over entire site.

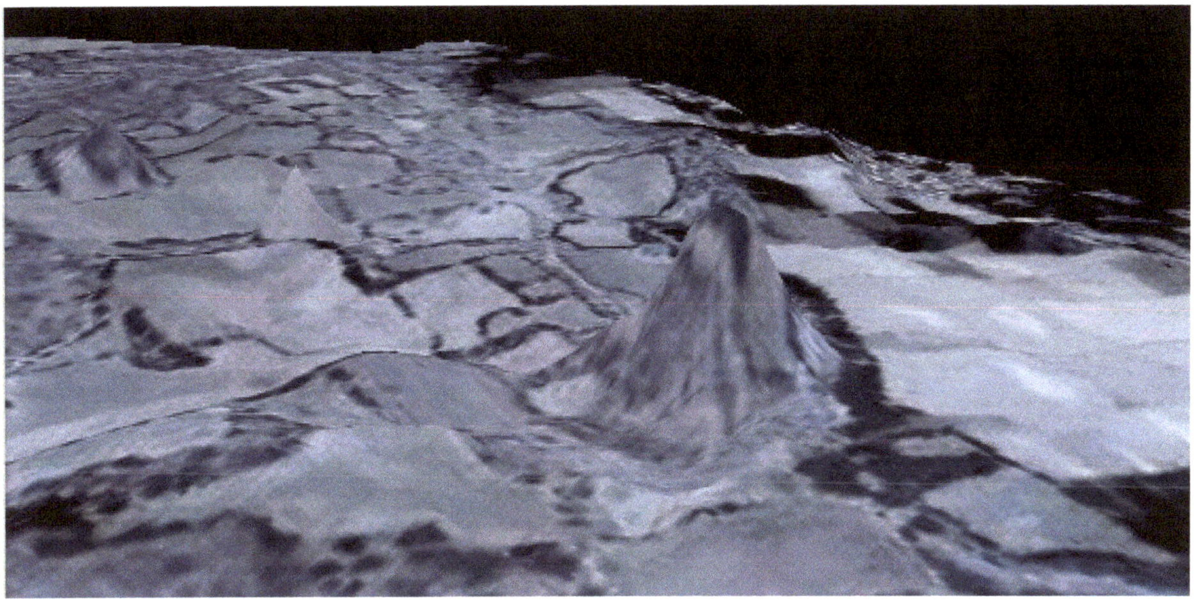

Fig. 19. Google Earth image draped on the DEM.

Conclusions

The present work has highlighted development of a geospatial database of Ahichchatra, an ancient archeological site of national importance near Bareilly (U.P.), using Geoinformatics technology. This database was used to draw inferences about the past environment of the study site. As the spatial referencing scheme used by the archaeologist is quite different, a conversion function was developed which can establish a linkage between an archaeological grid scheme and the projected coordinate system. Some of the inferences drawn from the preliminary study of the geospatial database of the study site are as follows:

1. Location of a water body close to the site and associated with low elevation was identified.
2. High elevation areas with small footprints were identified to be associated with locations of temples.

References

ARCHAEOLOGICAL SURVEY OF INDIA. 1963, Indian Archaeology: A Review.

ARCHAEOLOGICAL SURVEY OF INDIA. 1964, Indian Archaeology: A Review.

CUNNINGHAM. A. 1862, Archaeology report, *Archaeological Survey of India*, Vol.1.

HISTORICAL BACKGROUND ,GOVERNMENT OF INDIA MINISTRY OF COMMUNICATION & INFORMATION TECHNOLOGY, VIEWED 31ST SEPTEMBER 2006, <http://bareilly.nic.in /hist.htm>

JAIN MANDIR TEMPLE RELIGION GUIDE TEERTH, VIEWED 2ND OCTOBER 2006, http://www.jainjagat.com/viewtemple.php/Alpha/97

LEICA GEOSYSTEMS. 2001, *GPS equipment user manual version 3.0*

LEICA GEOSYSTEMS. 2001, *User manual / Getting started with SKI PRO version 2.0*

NIKON CORPORATION. 1995, Instruction manual automatic levels AC-2/AC-2S

POLITICAL HISTORY. GOVERNMENT OF UTTAR PRADESH OFFICIAL WEB SITE, 12TH SEPTEMBER 06, http://www.upgov.nic.in/upinfo/poli_hist.html

TRIMBLE. 2001, *User guide general Geodimeter CU and 600 CU*

Geoarchaeology of central west coast of India

Busnur R. Manjunatha

Department of Marine Geology, Mangalore University, Mangalagangothri-574 199, India,
E-mail omsrbmanju@yahoo.com, Fax 0824 2287 289

Abstract

Here, an attempt has been made to link up pre-historic human settlements along with fluctuations in climate and sea-level along the central west coast of India during the late–Quaternary. Ancient human settlements seems to be very few which are mainly reported from banks of rivers and laterite plateaus ~50m above the current sea-level. This was probably to avoid inundation of land because the sea-level fluctuated ~140m between glacial and interglacial periods. The sea-level curve constructed for the west coast of India is generally in good agreement with the mean global sea-level indicating that the coast is relatively stable. Archaeological artifacts are rather simple and particularly made out quartz pebbles. Sharp hunting tools which are dominant in most of the archaeological sites of the Indian sub-continent are very rare along the coast, suggesting that prehistoric-early humans were dependent on vegetarian/aquatic source of food. The sparse distribution of prehistoric human settlements along the study area could have been due to high rainfall, dense Western Ghats (mountains) forest, lack of wood cutting tools, sea-level fluctuations and perhaps some epidemic like malaria. Furthermore, it is urged for detailed sub-surface explorations to trace links with prehistoric period.

Keywords Stone Age human settlement, climate, rainfall, sea-level, Quaternary period, west coast of India, Mangalore

Introduction

The west coast of India is a narrow strip of land exists between the long chain of Western Ghat Mountains towards the east and Arabian Sea towards the west. The coast is neither submergent nor emergent, but has features of both types due to long-term sea-level fluctuations, lithological variations and local crust movements (Wagle, 1987). However, the submergent features are dominant (Wagle, 1987), because of the sea-level rise during the Holocene. In contrast to the east coast of India, the west coast is shaped by the dominance of marine processes leading to formation of Ria-type of estuaries, estuarine marshes, lagoons, tidal flats and mud flats, spits and bars, backwaters, etc. (Ahmed, 1972; Bruckner, 1989). Ria-type of estuaries are common along the Konkan coast (19° 30' N - 16° N), belong to macrotidal category where seawater intrudes up to 30-40 km inland and river mouths are widened to form broad creeks. Estuaries of the Karnataka coast (15° 30" N - 12° 45" N) are shallow, narrow river mouths and meso to micro tidal range where sea water intrusion rarely seen beyond 15-20 km inland (Shankar and Manjunatha, 1997).

Area of investigation

The central west coast of India and the adjacent Western Ghats, particularly the Karnataka coast (~12° 45' N - 15 ° N latitude; Figure 1) is the zone of receiving intense monsoon rainfall particularly during the summer season. The climate prevailing along the study area is the Monsoon and trade wind coastal climate belongs to Köppen's Af and Am type. The Karnataka coast has about 300 km long coastline covering an area of ~16,000 sq km (Figure 1). The annual rainfall varies 200-400cm along the coast with pockets of 400-800cm along the Western Ghats. Generally the rainfall increases from coast to the Western Ghats. About 85 % of annual rainfall is received during southwest monsoon. Due to high rainfall, there are more than ten small rivers originated in the Western Ghats and drain into the Arabian Sea. The area of investigation is important sites for fisheries and navigation.

Geology of the study area

The coastal land of Karnataka is an extension of the Peninsular Indian shield comprising of principally of Archaean rocks such as gneisses, granites, amphibolites, etc. The gneiss is served as a basement with enclaves of older/supracrustal rocks and emplaced by granitic batholiths. These are confined to the central and northern sectors of the coastal tract (Manjunatha and Harry, 1994). The Archaean rocks have been intruded by dolerite dykes during the Proterozoic eon and lateritised during the Tertiary period. Infilling of boulders, pebbles and fine-grained sediments has been noticed in paleo-channels of some of the present rivers such as Kalinadi and Netravati-Gurpur. Deposition of recent to sub-recent alluvium in river banks and beaches is also significant along the coast. Some of unconsolidated sediments and plateau surfaces comprise archaeological artifacts that are discussed in this paper.

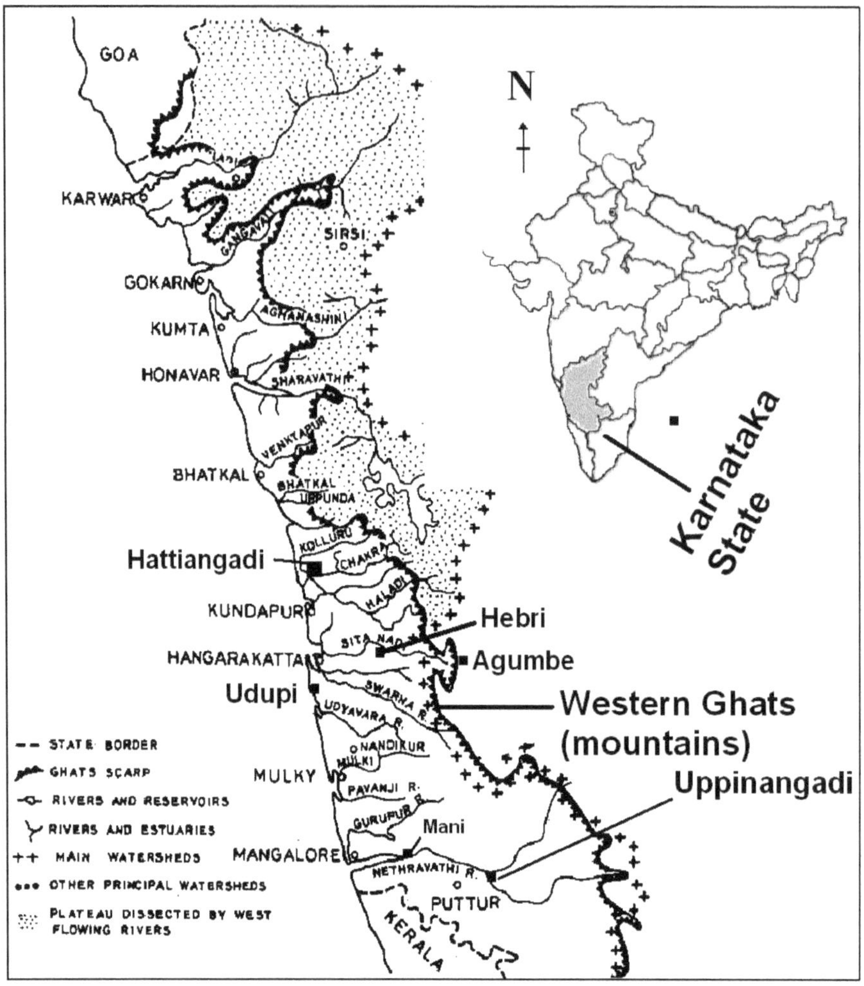

Fig. 1. Map of Karnataka coast showing important rivers and important locations which are discusses in the Text with an inset map of india showing the Karnataka State.

Prehistoric human settlement

In contrast to the upland of Peninsular India, information on prehistoric and early human settlement is very sparse along the Karnataka coast. The paleolithic artifacts have been mainly noticed near the banks of relatively large rivers such as Sitanadi near Udupi, Netravati-Gurpur at Uppinangadi and Mani, near Mangalore and Gokarn near Karwar (Joshi *et al.*, 1983; Sundara, 1991). The archaeological artifacts are quite simple and made out of quartz pebbles that are generally comparable to those reported along the Konkan and the Kerala coastal regions such as Manori, Mandovi, Chevayur (near Calicut; Marathe, 1995). This also indicates that the prehistoric-early humans along the Karnataka coast were dependent mainly on aquatic source of food such as fishes and molluscans, and perhaps vegetarian type. These artifacts are in contrast to the sharp hunting tools which are generally found in most of the Stone Age archaeological sites in the upland of the India Peninsula. This suggests that there was no link of ancient human beings of the coast and the main land because of the existence of long chain of Western Ghats (mountains) with dense forest. The available prehistoric information along the Karnataka Coast has been compiled in the Table 1. It is clear from this Table that the human settlements were not widely distributed during the major part of the Quaternary period. These probably suggest that climate was unfavorable for human settlement. Generally, high precipitation rate with increase in rate of sea-level rise and dense forest are characteristic features of the interglacial period, particularly during the early Holocene. Malaria is known from a long time (2700 BC). It is the world's largest killer of children and pregnant women (http://www.malaria.com/). Malaria and mosquitoes related diseases are quite intense in the high rainfall regions of the tropics.

Paleoclimatic information

The west coast of India, in particular, the Karnataka Coast and the adjoining Western Ghats is the region of second of highest rainfall in India. The rainfall over the past 135 years (http://www.tropmet.res.in) varies from 160cm/yr to 340cm/yr with an annual average of 240cm ± 30cm/yr. For rainfall information beyond the 18th century, can be known

Geologic period (Time ~kyr. BP)	Archaeological stage	Archaeological information
Late Holocene (2.35-1.80)	Early historic period	Began of Raman trade; Pottery, vessels and ash, earthy materials, brick shrine with Shivalinga noticed at Hattiangadi.
Late Holocene (~1.9-3.2)	Iron Age or Megalithic	Burial sites have been reported around Hebri, Udupi and Uppinangadi.
Late Holocene (2.8-4.2)	Neolithic or Neolithic in Chalcolithic	Polished stone axe reported from Sita Valley, Agumbe
Mid Holocene – Early Holocene (3.0-10.0)	Mesolithic	Tiny/pigmy size stone tools called microliths from laterite plateaus of Mani
Early Holocene to Pleistocene (10.0-200.0)	Early Palaeolithic	Bifacial quartz pebble tool from Uppinangadi,

(Archaeological information from Sundara, 1991; Marathe, 1995).

Table 1. Summary of time spans of prehistoric and early human settlement along the Karnataka coast.

from historical records as well as various types of proxies preserved in geological materials, such as tree-rings, speleothems, etc. Rainfall reconstructed from a stalagmite of the cave near Ulvi village, Karwar, Western Ghats shows that the year AD 1666 has been identified as the most intense monsoon year, whereas AD ~1900 was the draughtiest year over the past three centuries (Yadava *et al.*, 2004). Information before this comes from a pollen study of a sediment core off Karwar which indicates that the climate was arid from 3500 yr B.P. to 2200 yr B.P. (Caratini *et al.*, 1994). This phase can be correlated with the regional aridity that has been documented from the Indian sub-continental as well as neighbouring oceanic sediments (Caratini *et al.*, 1994; Sukumaran et al., 1993). The climate prevailed during the early Holocene to mid-to-late Holocene appears to have been very humid as indicated by an increase of mangrove pollen in the marine sediments off the west coast of India (van Campo, 1986).

Studies on sea-level fluctuations are not many along the west coast of India (Hashimi *et al.*, 1995; Kale and Rajguru, 1985; Manjunatha, 1994; and references therein). These studies indicate that the sea-level, particularly south of Mumbai (formerly Bombay) has been controlled predominantly by eustatic/global changes in the sea level. The late-Quaternary sea-level curve constructed for the west coast of India, in deed, is in fairly good agreement with that of the global sea-level. Generally, the sea-level along the west coast of India was lower by about 138m during the last Glacial maximum (LGM) than the current sea-level (Kale and Rajguru, 1985). It rose at a high amplitude during the early Holocene to mid Holocene than that during the late Holocene (18 mm/yr and 0.3-0.8 mm/yr respectively; Kale and Rajguru, 1985), while attained higher level than the current level during late Holocene.

Acknowledgements

Prof. Dr. Claudia Liuzza, Secretariat Coordinator, ICOMOS International Scientific Committee on Interpretation, and Presentation (ICIP), University of California, Merced, School of Social Sciences, Humanities and Arts, 5200 North Lake Road, Merced, CA 95343 for encouragement to prepare this paper and participation in the conference.

References

AHMED, E. 1972. *Coastal Geomorphology of India*. Orient Longman Press, Bombay.

BRUCKNER, H., 1989. Late Quaternary shorelines in India., in D.B. Scott, P.A. Pirazzoli and C.A. Honing (eds.), *Late Quaternary Sea-Level Correlation and Applications,* 169-194. Kluwer Academic Publishers, Dordrecht/Boston/London.

CARATINI, C., BENTALEB, I., FONTUGNE, M., MORZADEC-KERFOURN, M.T., PASCAL, J.P. AND TISSOT, C. 1994. A less humid climate since ca. 3500 yr BP from marine cores off Karwar, western India. *Palaeogeography, Palaeoclimatology, Palaeoecology* 109, 371-84.

JOSHI, R.V., RAJGURU, S.N. AND PAPPU, R.S. 1983. Quaternary studies in the Karwar-Gokarn region, Karnataka State, in *Professor Kelkar Volume*, 97-106, Pune. Indian Society for Earth Sciences.

KALE V.S. AND RAJAGURU, S.N. (1985) Neogene and Quaternary transgressional and regressional history of the west coast of India. *Buletin Deccan College Research Institute, Pune* 44, 153–165.

HASHIMI N H, NIGAM R, NAIR R R AND RAJAGOPALAN G. 1995 Holocene sea level fluctuations on western India continental margin: An update. *Journal of Geological Society of India* 46, 157–162.

MANJUNATHA, B.R., 1994. Coastal geomorphology and late Quaternary sea-level changes in Karnataka, in B.M. Ravindra and N. Ranganathan (eds.), *Geo-Karnataka: Mysore Geological Department Centenary Volume,* 327-336. Karnataka Assistant Geologists Association, Department of Mines and Geology, Bangalore, India.

MANJUNATHA, B.R. AND HARRY, N.A. 1994. Geology of Western Karnataka, in B.M. Ravindra and N. Ranganathan (eds.), *Geo-Karnataka: Mysore Geological Department Centenary Volume,* 109-116. Karnataka Assistant Geologists Association, Department of Mines and Geology, Bangalore, India.

MARATHE, A. 1995. Prehistory and Quaternary sea level changes along the west coast of India: a summary, in S.Wadia, R. Korisettar and V.S. Kale (eds.), *Quaternary Environments and Geoarchaeology of India: Essays in Honour of Professor S.N. Rajguru,* 405-413. Memoir 32, Geological Society of India Bangalore.

SHANKAR, R. AND MANJUNATHA, B.R. 1997. Onshore transport of shelf sediments into the Netravati-Gurpur estuary, west coast of India: geochemical evidence and implications. *Journal of Coastal Research* 13, 331-340.

SUKUMAR, R., RAMESH, R., PANT, R.K. AND RAJAGOPALAN, R. 1993: A $\delta\,^{13}C$ record of late Quaternary climate change from tropical peats in southern India. *Nature* 363, 703-706.

SUNDARA, A. 1991. Prehistoric and protohistoric cultures of Dakshina Kannada. In: Ramachandran, C.N. (ed.), *Mangalore University Decennial Volume 40-63.* Mangalore University, India.

VAN CAMPO, E. 1986. Monsoon fluctuations in two 20,000-Yr B.P. Oxygen-isotope/pollen records off southwest India. *Quaternary Research* 26, 376-388.

WAGLE, B.G., 1987. Geomorphology and evolution of the coastal and offshore areas of Maharashtra and Goa, India. Ph.D. thesis, University of Bombay.

YADAVA, M.G. RAMESH, R. AND PANT, G.B. 2004. Past monsoon rainfall variations in peninsular India recorded in a 331-year-old speleothem. *The Holocene* 14, 517-524.

Session 5

INTEGRATED REMOTE SENSING TECHNOLOGIES FOR THE INTERPRETATION OF LANDSCAPE ECOSYSTEMS: CONFRONTING SCALES OF DETAIL

Detection and excavation of remains of underground burial by geophysics and remote sensing methods

Peng Jia,[1] Guoding Song,[2] and Yueping Nie[1]

[1]Joint Laboratory of Remote Sensing Archaeology, Institute of Remote Sensing Applications, Chinese Academy of Sciences, Beijing, 100101, China
[2]Graduate University of Chinese Academy of Sciences, Beijing, 100049, China
E-mail: jiapengff@hotmail.com

Abstract

Shenmingpu Site, locating in Henan Province in China, will be a part of the flooded areas during the magnificent South-North Water Transfer Project. According to previous investigations, there might be some remains of underground burial, so it is urgent to conduct a rescue excavation. Due to a large size of the site area, it will cost high expense, substantial labor and time if we totally rely on traditional methods to detect them.
Both geophysics and remote sensing are powerful tools for detecting surface and underground archaeological information, especially when they are combined. The former can detect the underground structural compositions based on the differences of density and resistivity. The latter is useful when we suppose that the remains are buried not very deep and make the soil above them different from the surroundings in various ways, which makes discrepant features appear on the satellite imagery.
A collection of CBERS-02 images were used to find the anomaly in the site area. For the anomalous area on the image, geophysical prospecting was done on the corresponding ground to confirm the anomaly. After an in-situ excavation on the chosen anomalous ground, a group of graves including brick-chamber graves and shaft graves were unearthed.

Keywords Geophysics, Remote Sensing, Archaeology, Underground Burial, Grave

1 Introduction

Shenmingpu Site sits in the north of Shenmingpu Village, which, according to the administrative division, belongs to Xichuan County, Nanyang City, Henan Province in China. Topographically speaking, it is located on the mesa along the south bank of Danjiang River. The center coordinates of the site, recorded in the form of latitude and longitude, is 33°00 47 (N), 111°17 47 (E) and the elevation ranges from 158m to 163m.
In 1976, for the first time, Shenmingpu Site was found in the Cultural Relics Investigation organized by Cultural Relic Management Committee in Xichuan County. Then, in 1984, it was rechecked in the Cultural Relic General Survey of Henan Province. In 1985, it was formally placed in the list of Province-level Cultural Relic Protection Units. Through the results of several investigations, it is estimated that the approximate area of the site is 20,000 square meters and the remains are buried within a two-meter depth, but on the ground there are no obvious clue showing the existence of the remains. However, some remnants of ceramics and bricks are found around the site, according to which, it is concluded that the age of the site may date back to Shang Dynasty, Zhou Dynasty and Han Dynasty.
The South-North Water Transfer Project is a strategic project in order to relieve the scarcity of water resource in northern China and optimize the distribution of water resource throughout the whole country. As adjacent to Danjiang River, which is regarded as Danjiang Reservoir Area where the migration is in progress, Shenmingpu Site will be a part of the flooded area during the magnificent project, so it is urgent to conduct a rescue excavation. Due to a large size of the site area, for the purpose of low cost and high efficiency during the detection stage, on a basis of traditional methods we introduce geophysics and remote sensing methods, both of which are powerful and effective tools for detecting surface and underground archaeological information, especially when they are combined.

2 Remote Sensing Detection

2.1 Remote Sensing in Archaeology

Remote sensing is an effective tool to record the energy reflected and emitted from the ground in a broad region. One side, the ability of recording the reflected energy shows that we could know the actual ground situation from the energy recorded because different types of objects can reflect, absorb and transmit the incident radiation in their own proportions. It is useful when we suppose that the remains are buried not very deep and make the soil above them different from the surroundings in various ways (porosity, weathering degree, water content, surface temperature, etc.), which makes discrepant features appear on the satellite imagery directly, or sometimes through a series of processing. Another side, the feature of covering a broad region at one time leads it to be widely used to look for the remains of ancient magnificent projects, such as the Great Wall, the Beijing-Hangzhou Grand Canal, etc.. Based on these two features, remote sensing has been introduced to the field of archaeology and is being progressively applied to more and more archaeological projects.

2.2 Image Selection

According to the principles of remote sensing in archaeology, ancient remains buried underground can make the soil above them discrepant from the surrounding due to moisture content, porosity, fertility, etc., which consequently leads to the differences of growth and distribution of the wood and shrub, or the differences of height, density and color of the crop and grass. All these discrepancies will make each type of objects appear on the image with its own characteristic different from others, which is recognized as the vegetation features of underground remains and phenomenon.

As for the obvious differences of the features on the image, herbaceous plant is better than woody plant because it looks fine-grained on the image during the vegetation growing time and this phenomenon will appear repeatedly each year. Cereal crop is quiet useful because if there are rammed earth, brick, rubble, ancient road underground, the crops will be mature later than the surrounding during the harvest time.

Based on the analysis above, the most ideal covering of the ground is wheat because it not only belongs to herbaceous plant but also belongs to cereal crop. It is fortunate that it is winter wheat that is cultivated on the ground of Shenmingpu Site. Winter wheat is seeded in the autumn and harvested in the next summer. In addition, the wheat contains the spring wheat, which is seeded in the spring and harvested in the same autumn.

Through detailed investigations, the winter wheat on site is seeded in the early October, begins growing in about November or December, flourishes in about March or April, gets mature and is harvested in the late May or the early June. According to the detailed growth cycle and the existing remote sensing images, we choose two scenes both from the CBERS-02 taken respectively on 12 June 2005 and 11 June 2006.

CBERS-02 is manufactured jointly in China and Brazil, which belongs to the first generation of transmission remote sensing satellite of earth resource. The multispectral camera taken with CBERS-02 can take images in five bands, that is, B1 (0.45~0.52μm), B2 (0.52~0.59μm), B3 (0.63~0.69μm), B4 (0.77~0.89μm), B5 (0.51~0.73μm), all with 19.5m spatial resolution.

2.3 Study Area Recognition

FIGURE 1 was taken on 12 June 2005. The area in the small red square on it contains the site area and the cyan line passing through the red square is Danjiang River. The red square area is magnified as FIGURE 2 and through color stretching the anomaly is found out. It is possible that the wheat in the anomalous area got matured later than the surrounding, so it still existed when the surrounding have been harvested.

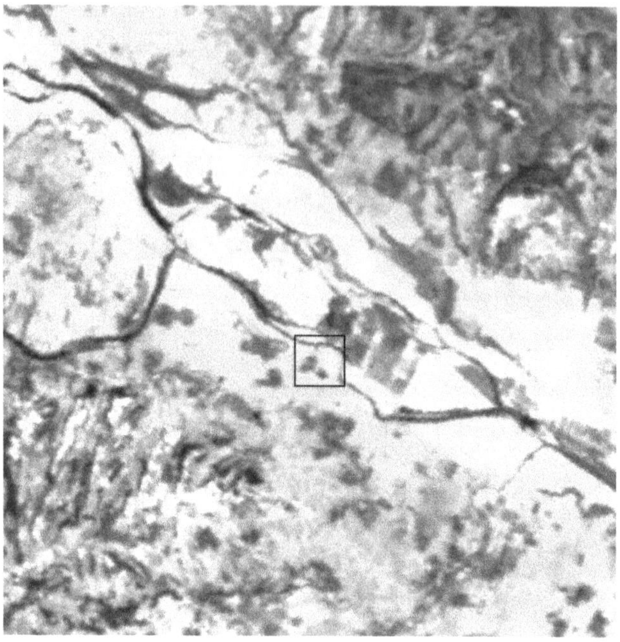

Fig. 1

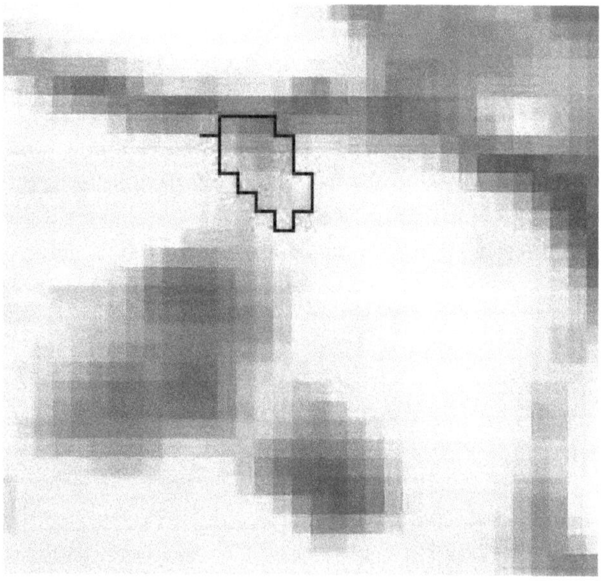

Fig. 2

FIGURE 3 was taken on 11 June 2006 and the red square area on it was magnified as FIGURE 4. As the same to FIGURE 2, we also find out similar anomaly in the same location on FIGURE 4.

Therefore, we choose the area marked by red polygon on both FIGURE 2 and FIGURE 4 as the study area, where geophysical prospecting will be conducted.

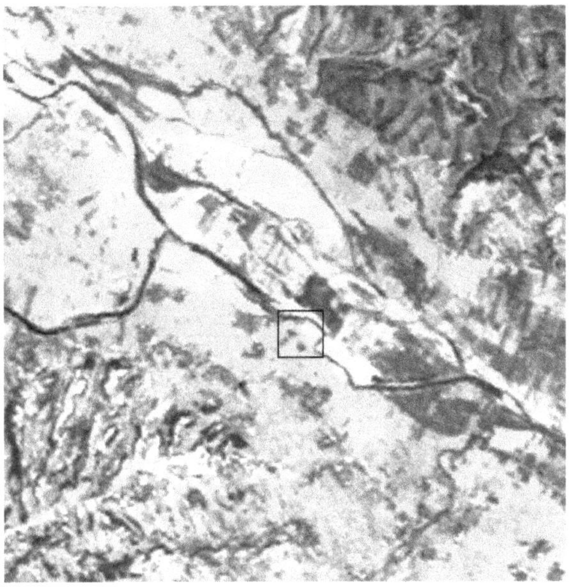

Fig. 3.

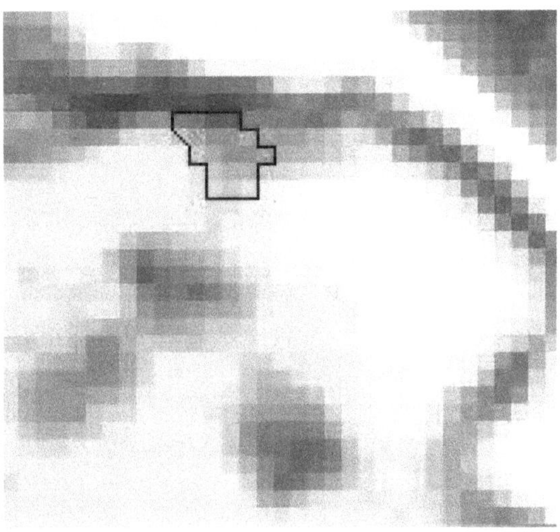

Fig. 4

3 Geophysical Detection

3.1 Geophysics in Archaeology

Geophysical prospecting is a sort of theories and methods which applies physical principles to prospect underground resources and research geological structure. Because non-destructive, it is usually used as an advanced archaeological technology to directly detect underground sites or archaeological remains. It can detect the underground structural compositions based on the differences of physical properties of objects, such as density, resistivity, etc.

3.2 Method

According to landform and geophysical condition, the natural electric field method and high-density resistivity method are chosen. First of all, we use the natural electric field method to conduct a general investigation in the whole study area. Then, we carry out the high-density resistivity measurement to confirm on the anomalous areas in the course of general investigation. According to the result of geophysical field test, electric parameters of the lithology are shown in FIGURE 5.

Lithology	Range of resistivity values (Ω·m)	Range of natural field values (mV)
Silty clay	16.5~22.5	0.6~0.8
Grave	42~66	1.6~1.9
Pebble	90~150	1.0~1.2

Fig. 5.

3.2.1 Natural Electric Field Method

This method is a kind of alternating current exploration method. Based on the differences of the resistivity of underground rocks and minerals, on the ground we measure the changes of several electric field components in different frequency produced by the geo-electromagnetic field, which is made use of as the source of working field, to study the electrical changes of underground geo-electric sections, in order to solve the geological problems.

Because of portable equipment, easy working methods, high efficiency, considerable information obtained and easily explainable data, this method is appropriate to general investigations. There are eight power levels, one of which can be selected to work each time, for the equipment we use. According to the analysis of the strata in the sampling holes and the estimation for buried depth, we choose the power level which can explore underground about 0.8~1.0m deep.

3.2.2 High-density Resistivity Method

As for the principle, the same as the natural electric field method, this method is also a kind of electric exploration method on a basis of the conductivity of rocks and minerals. Under the effect of the artificial electric field, the distribution of electric current passing through the underground reflects the distribution of geological objects.

In this method dense measurement points are distributed on the ground. A two-dimensional exploration process (i.e. vertical and horizontal) can be completed once and abundant geological information can be obtained from it. Practice has proven that because of high observation precision and reliable data obtained, it is a better method in engineering geology exploration, hydrogeology exploration and underground burial exploration, for example, underground grave exploration.

3.3. Survey

According to the site condition, we evenly set up eight measurement lines covering the whole study area to conduct the general investigation (FIGURE 6).

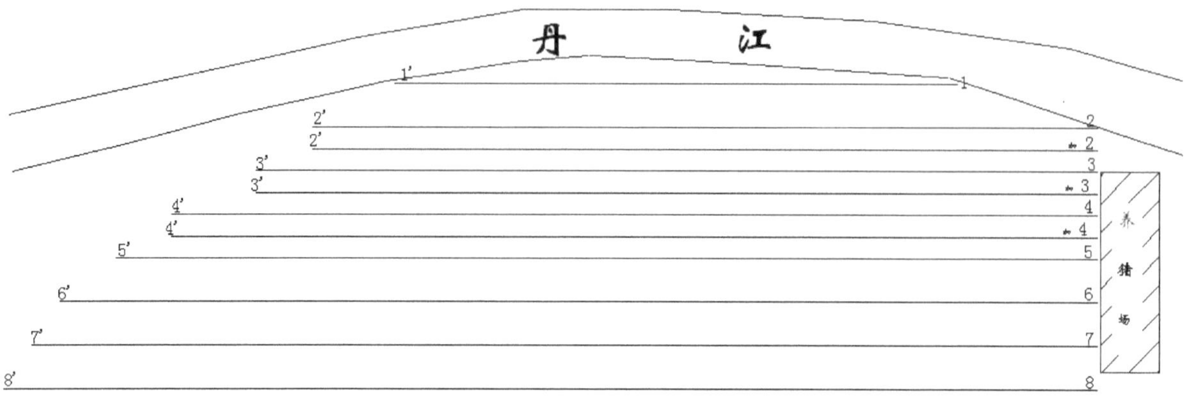

Fig. 6.

They are orderly named 1-1', 2-2', ... , 8-8'. After analyzing the investigation result, we add three lines respectively next to the three measurement lines (2-2', 3-3', 4-4') showing strong anomaly to meet the need of detailed investigation. According to each locality, three added lines are respectively named Add 2-2', Add 3-3' and Add 4-4'. So there are eleven measurement lines in all evenly set up in the whole study area. For the purpose of comparing with the natural electric field method, we conduct the high-density resistivity exploration on three added lines. After balancing the degree of the anomaly and project funding, we first decide to excavate the area on seven measurement lines (i.e. from 2-2' to 5-5'). The data obtained through two methods will be analyzed in the next section.

4 Geophysical Data Analysis

4.1 Result of Natural Electric Field Method

The profile 2-2' is 112m long and the background value (normal value) is 1.5mV; the maximum is 1.7mV and the minimum is 1.4mV. It is inferred that there are graves (ΔV=1.7mV) where the distance value equals 88m in the horizontal axis (FIGURE 7).

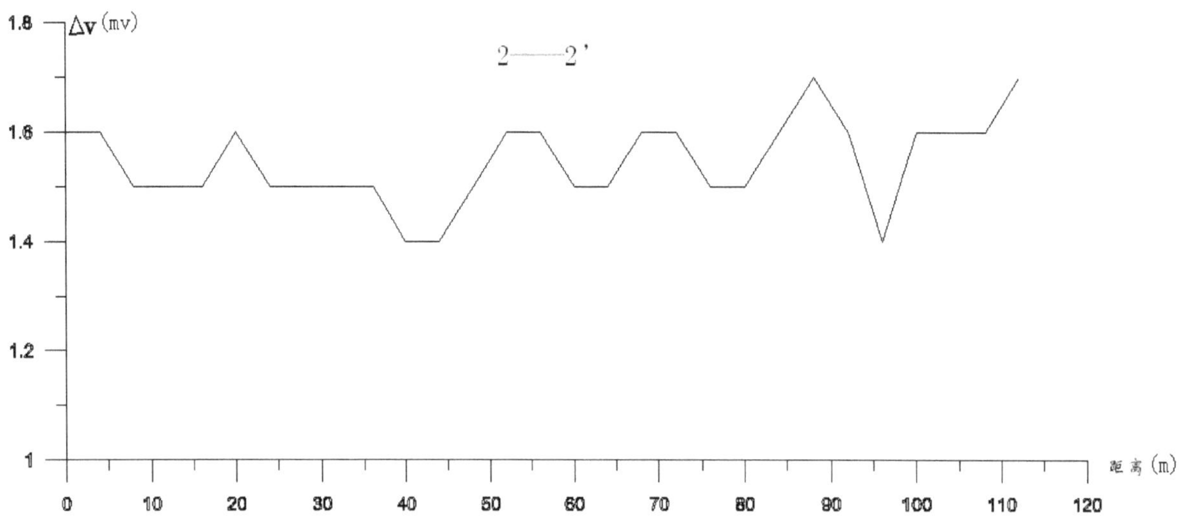

Fig. 7.

The profile Add2-2' is 112m long and the background value is 1.6mV; the maximum is 1.8mV and the minimum is 1.5mV. It is inferred that there are graves (ΔV=1.8mV) where the values respectively equal 24~28m, 36m and 48m in the horizontal axis (FIGURE 8).

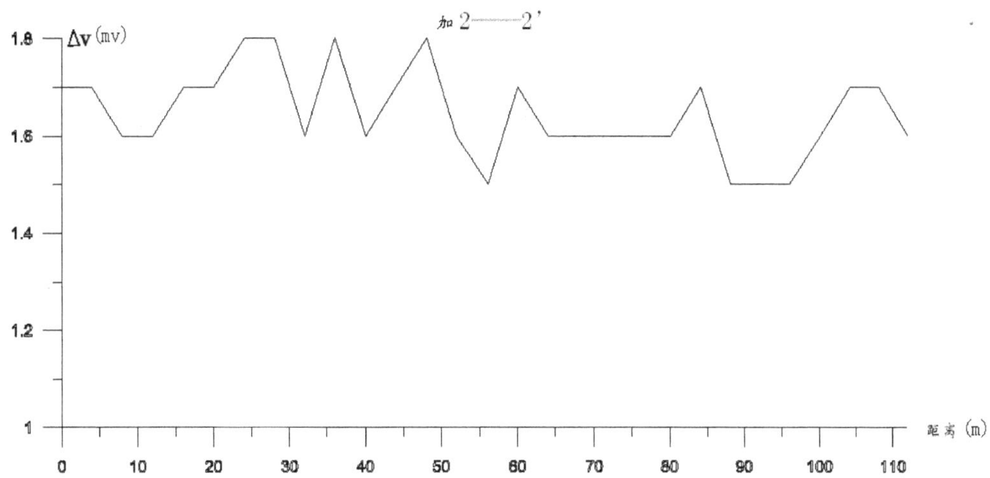

Fig. 8.

The profile 3-3' is 120m long and the background value is 1.2mV; the maximum is 1.9mV and the minimum is 1.0mV. It is inferred that there are graves (ΔV=1.6~1.9mV) where the values respectively equal 32m, 48m and 64~68m in the horizontal axis (FIGURE 9).

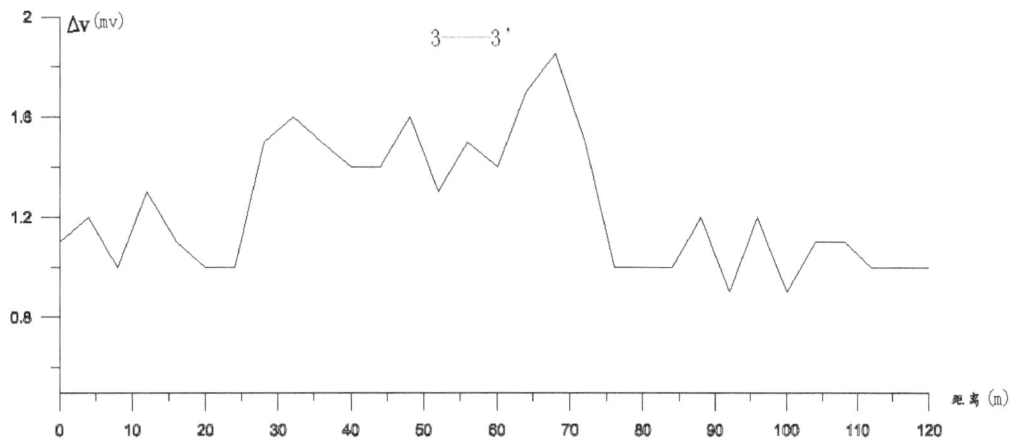

Fig. 9.

The profile Add3-3' is 120m long and the background value is 1.6mV; the maximum is 2.0mV and the minimum is 1.4mV. It is inferred that there are graves (ΔV=1.8~2.0mV) where the values respectively equal 40m and 92m in the horizontal axis (FIGURE 10).

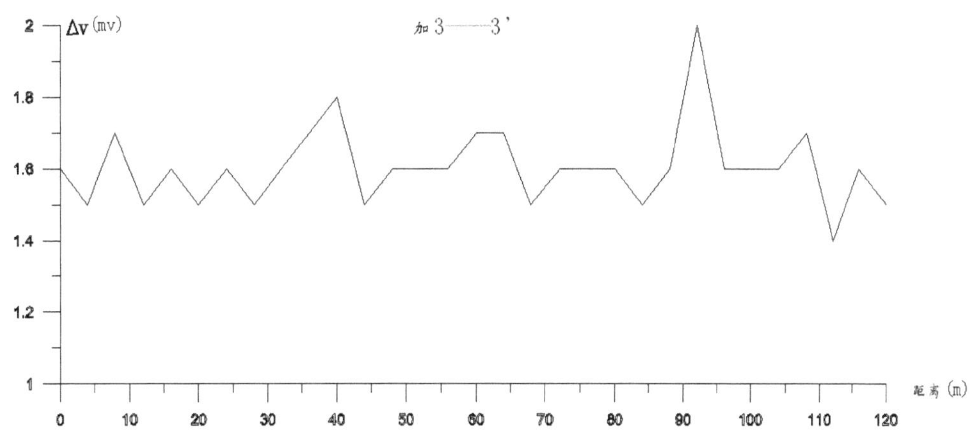

Fi

g. 10.

The profile 4-4' is 130m long and the background value is 1.2mV; the max is 1.6mV and the min is 0.6mV. It is inferred that there are graves (ΔV=1.5~1.6mV) where the values respectively equal 20m, 44m and 68m in the horizontal axis (FIGURE 11).

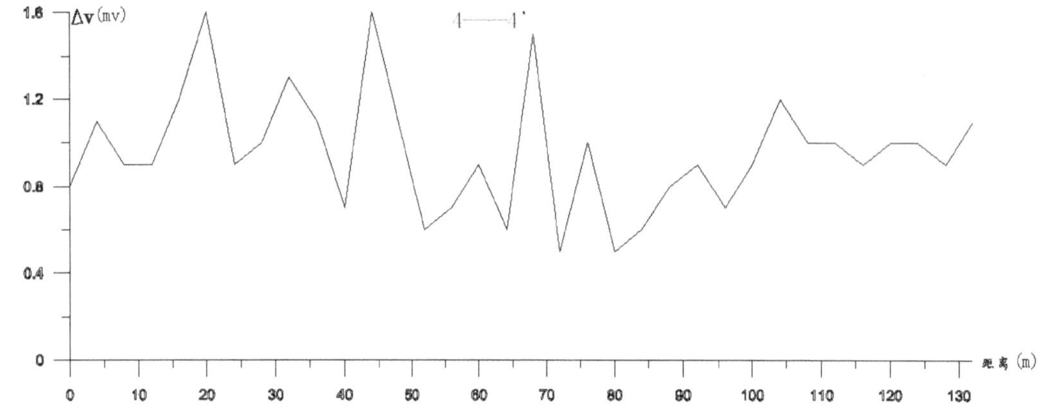

Fig. 11.

The profile Add4-4' is 130m long and the background value is 1.5mV; the maximum is 1.8mV and the minimum is 1.5mV. It is inferred that there are graves (ΔV=1.8mV) where the values respectively equal 48m, 72~76m and 92m in the horizontal axis (FIGURE 12).

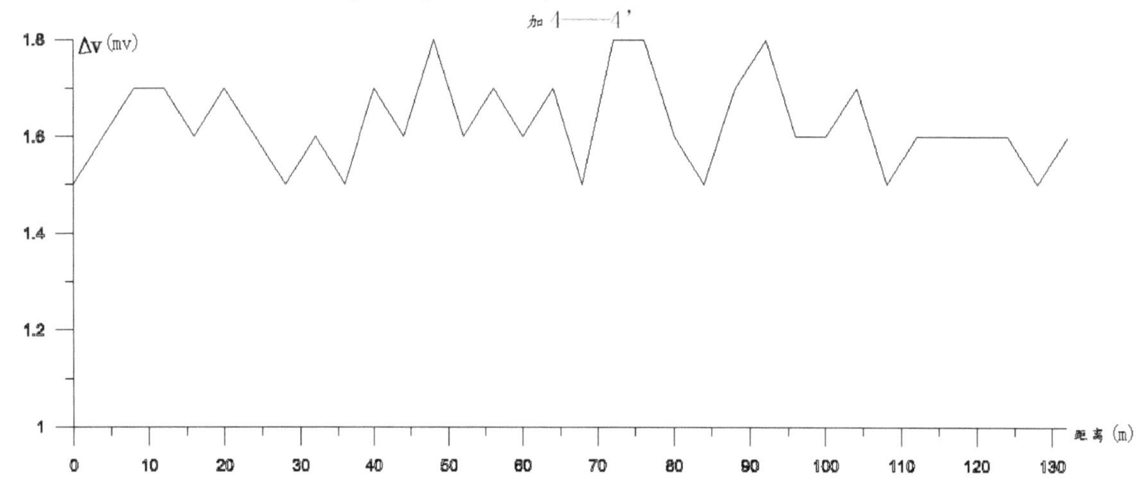

Fig. 12.

The profile 5-5' is 140m long and the background value is 1.2mV; the maximum is 1.4mV and the minimum is 0.9mV. It is inferred that there are graves (ΔV=1.3~1.4mV) where the values respectively equal 52m, 60m and 76m in the horizontal axis (FIGURE 13).

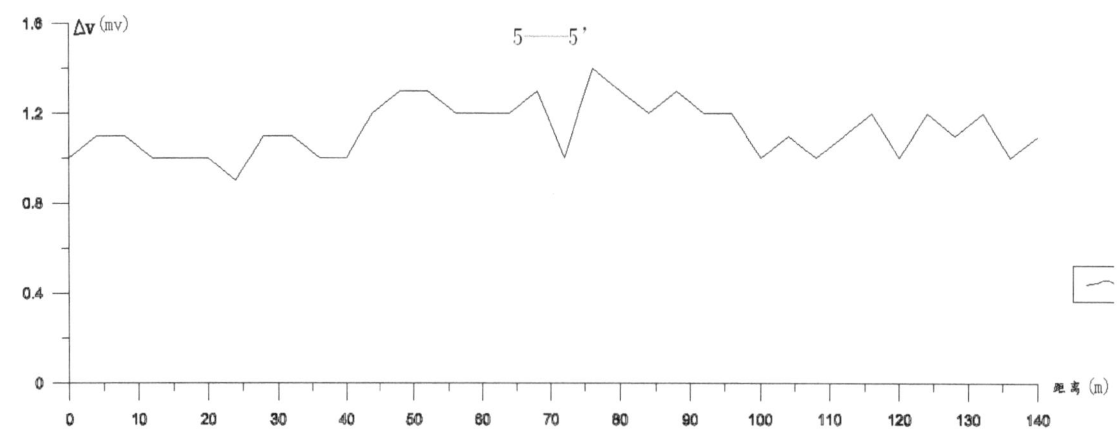

Fig. 13.

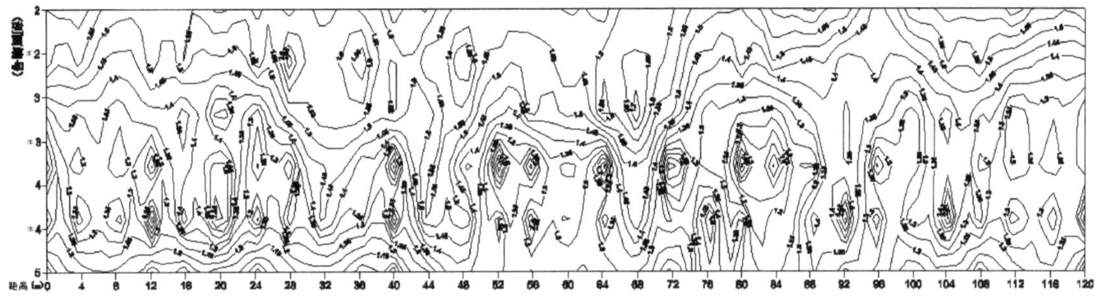

Fig. 14.

4.2 Isogram

All the data obtained in the area to be excavated are plotted in the form of the isogram (FIGURE 14). It is inferred that there are graves where the isograms with high values (1.5~1.7mV) are closed, for example, 1.65mV circle and 1.7mV circle in the upper of FIGURE 15, not where the isograms are only dense, for example, numerous circles with the values between 1.2~1.3mV in the lower of FIGURE 15. FIGURE 15 is the partly magnified image of FIGURE 14.

Therefore, through analyzing the isogram (FIGURE 14), we can infer that there are graves 1 where the values respectively equal 26~30m, 34~36m and 46~52m in the horizontal axis for the profile between 2-2' and Add2-2'; 2 where the value equals 66~72m in the horizontal axis for the profile between Add2-2' and Add3-3'; 3 where the value equals 20~22m in the horizontal axis for the profile between 4-4' and Add4-4'; 4 where the values respectively equal 46~48m and 74~78m in the horizontal axis for the profile between Add4-4' and 5-5'.

4.3 Result of High-density Resistivity Method

The profile Add2-2' is 100m long and it is found that the resistivity value is high (ρ=30~53Ω·m) within the depth of 1.2m. It is inferred that there are graves where the distance values respectively equal 16m, 23m, 30m, 55m, 62m, 82m and 87m in the horizontal axis (FIGURE 16). Under the depth of 1.5m, it scatter silty clay, silt, silty-fine sand and pebble.

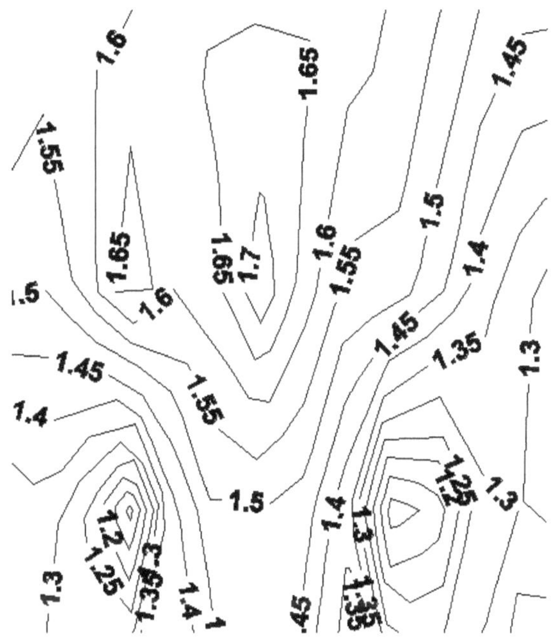

Fig. 15.

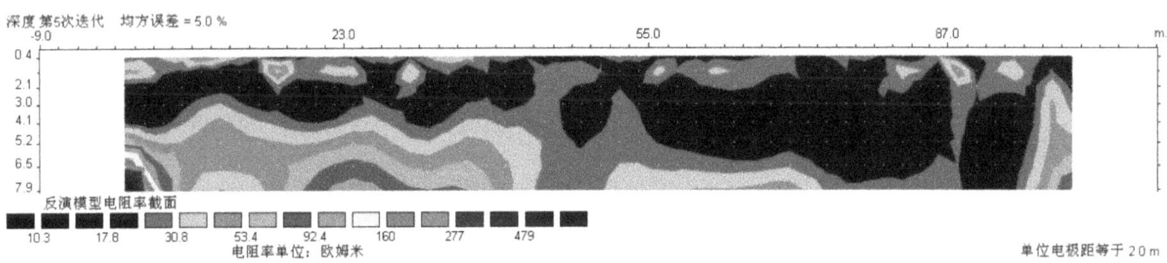

Fig. 16.

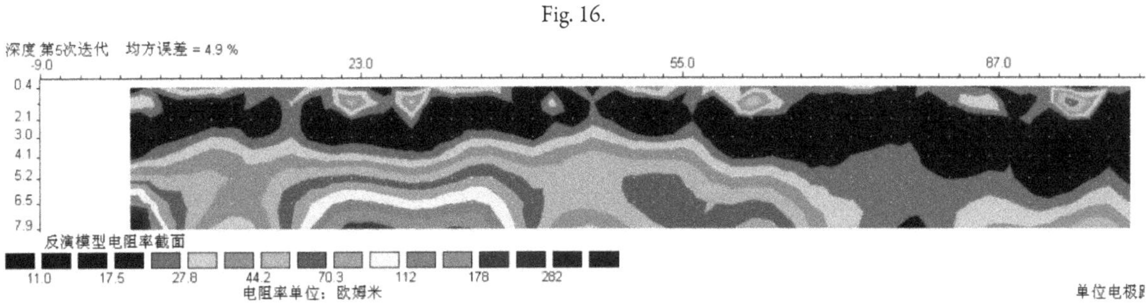

Fig. 17.

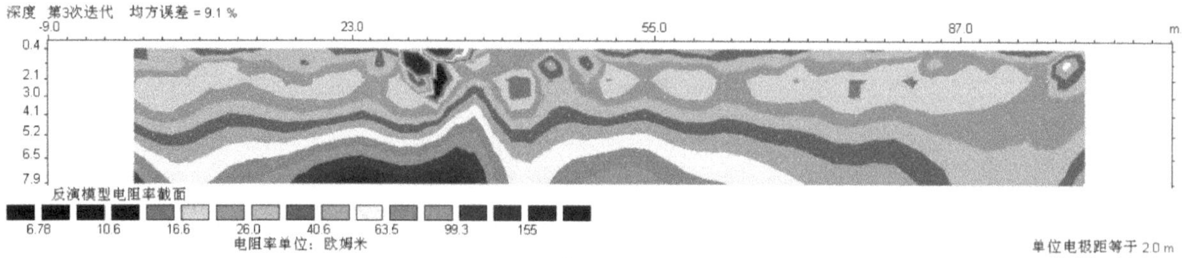

Fig. 18

Fig. 19

The profile Add3-3' is 100m long and it is found that the resistivity value is high (ρ=24~70Ω·m) within the depth of 1.2m. It is inferred that there are graves where the values respectively equal 22m, 28m, 40m, 62m, 85m and 95m in the horizontal axis (FIGURE 17). Under the depth of 1.5m, it scatter silty clay, silt, silty-fine sand and pebble.

The profile Add4-4' is 100m long and it is found that the resistivity value is high (ρ=26~155Ω·m) within the depth of 1.2m. It is inferred that there are graves where the values respectively equal 26m, 30m, 48m and 97m in the horizontal axis (FIGURE 18).

Under the depth of 1.5m, it scatter silty clay, silt, silty-fine sand and pebble.

4.4 Comparison

In this subsection we call the natural electric field method "Method 1" and call the high-density resistivity method "Method 2" for conciseness. For the profile Add2-2', the anomaly (22~28m) detected through Method 1 is basically consistent with the anomaly (23m, 30m) detected through Method 2.

For the profile Add3-3', the anomaly (40m, 92m) detected through Method 1 is basically consistent with the anomaly (40m, 95m) detected through Method 2.

For the profile Add4-4', the anomaly (48m, 92m) detected through Method 1 is basically consistent with the anomaly (48m, 97m) detected through Method 2.

5 In-situ Excavation

5.1 Strata Analysis

The stratum of the sites can be divided into four layers from the top to the bottom: the first (top) layer is the modern cultivated soil layer; the second layer is the modern disturbed soil layers; the third layer is the cultural layer of Han Dynasty;

the fourth layer is the cultural deposit of Eastern Zhou Period; and under the fourth layer is the immature soil layer.

5.2 Excavation Site

According to the terrain we plot nine square areas to be excavated on seven measurement lines and each square area is 10×10 square meters. After excavating, the aerial photo of the excavation site was taken by the low-altitude unmanned aerial photography system (FIGURE 19).

5.3 Unearthed Remains

Up to now, there have been 26 brick-chamber graves unearthed. It is inferred that they can entirely date back to Eastern Han Period or later except one dating back to about Tang Dynasty. In addition, there are 39 shaft graves unearthed, among which ten graves probably date back to Qing Dynasty and the others date back to a period of time spanning from the late Spring and Autumn Period to the middle Western Han Period. Because the ground was leveled up many times and the bricks were ever taken away for other purposes by local residents, most brick-chamber tombs have been severely wrecked. Comparing with brick-chamber graves, shaft graves remain better.

6 Summary

□ From the excavation result we can see that the site area is not large. It is difficult to find the anomaly of such a small site on the image, especially on the satellite imagery with 19.5m resolution, so on the premise that the fund of the project is enough, we will try to buy some imageries with higher resolution such as SPOT, ALOS, etc..

□ The difference of electric potential energy in electric field between silty clay and grave is the premise of the natural electric field prospecting.

□ The difference of resistivity between silty clay and grave is the premise of the high-density resistivity prospecting.

□Advanced technical methods, which are advocated in modern archeology field, played an important role in the detection stage of the whole project. The remains of underground burial, though shallow, are buried underground and as for the proportion, they just cover a rather small portion of the whole site area, so the research in this paper belongs to the detection and extraction of weak underground information based primarily on geophysics and remote sensing methods.

Acknowledgement

The research was supported by the Key Technologies R&D Program Project (2006BAK30B01) "Spatial Informational techniques Application in the Preservation of Sites, A Case Study in the Grand Canal of China," which was funded by the Ministry of Science and Technology of China.

References

JAMES WISEMAN AND FAROUK EL-BAZ, 2007. Remote Sensing in Archaeology. Boston, Boston University.

Reconstruction possibilities of the ancient Roman villas – based on air photographs, geodesical and geophysical surveys

Firnigl Anett

Landscape architect, PhD student, Hungary
Corvinus University of Budapest, Department of Garden Art H-1118, Budapest, 29-43. Villányi Str.
anett.firnigl@uni-corvinus.hu

Abstract

The scientific objects of the landscape archaeological researches are mapping and placing of the archaeological objects to the natural environment, based on archaeological excavations and remote sensing methods (aerial photography, geodesical and geophysical surveys).
Roman villas were farm units with landed property, the biggest group in former Pannonia was on the Hungarian Balaton Highland. The two main unshattering remote sensing methods in the examination of the archaeological objects and its environments are the geodesical surveys, and examination of the air photographs. These methods are inseparable, the achievements of which complement each other, as a result a closer image could be drawn about the Roman villas. Geophysical measuring could also help archaeological researches: the various procedures based on the physical quality-difference of the earth and inside hiding objects. The Roman villa of Gyulafirátót was suited for this examination.
After all the theoretical model of the Roman villas could be created from several separated data: with the superimposition of the measurement-results and delineations we could demonstrate the reconstruction mass of the buildings, added to the environmental data of the villas

Keywords environmental archaeology, unshattering methods, geodesical surveys, geophysical surveys, resistance, direct current

Introduction – Environmental archaeology and its methods

Environmental archaeology is a dynamically developing multidisciplinary science of nowadays, the researches of which are based on comparing historical and archaeological data, and also historical maps and the terrain. The connections of so-called earth covered monuments (for example buildings, roads) could be explored for example with the help of the modern relief, from which conclusions to the presumable ancient Roman historical landscape could be drawn (see Aston 1997, Bender and Wolff 1994, Firnigl 2007, Tóth 2006, Hahn and Prammer 1992, Küster 1995, Schreiber 1985, Spivey and Squire 2005, Sümeg *et al.* 2003).

A gradual approach change can be observed in the Central European archaeology in the past decades: not only the archaeological objects, but the environment of these has come to the front (opening from the object level to the scale of landscape). The Roman cities and rural settlements as objects also take a vital part in these researches, because these places often remain in the later ages (for example many Roman villas were inhabited in the Age of Great Migrations in the 5th century AD). It is really important, because the environment adjusted to the prevailing way of life: the landscape wears the marks of the interventions of the different Ages. The human intervention basically modifies the Landscape. The Romans' was the first culture in Hungary, which consciously intervened in the environment: they built cities, rural settlements, villas, roads, and aqueducts, dikes as well. The traces of the former objects are often visible, and because the landscape preserves the tracks of interventions, it could be considered as archaeological source: the landcsape is a complex occurence, it has both natural and human elements (these often could not be separated). These make possible the examination of the different historical layers, or the archaeological or natural protection, and monument heritage of these places.

Environmental archaeological methods apply the different sciences during the examination of the historical landscape: by means of these methods the find place could be examined on object-level and also in wider scale. In the course of this research after the excavation and the collecting and analysis of the historical data, and maps, remote sensing (unshattering instrument researches, for example air photography, geophysical measurements, LIDAR surface radaring) and mapping of the quarries and its environments are needed alike. Nevertheless, other multidisciplinary researches (for example archaeobotany, pollenanalysis, archaeozoology) help the hypothetical reconstruction of flora, fauna and hydrography, and also sift out the tracks of human activities, and sundering the environmental layers of the different Historical Ages from each other (Grynaeus 2003, Jerem 2003).

This study discusses the remote sensing methods of landscape archaeology, presenting it on the example of the Roman villa of Gyulafirátót.

Remote sensing methods in the examination of Roman villas

The three main unshattering remote sensing methods

(Scollar 1990) in the examination of the archaeological objects and its environments are the geodesical surveys, the air photographs and the geophysical surveys. These methods are inseparable, the achievements of which complement each other, as a result a closer image could be drawn about the Roman villas.

Archaeological geodesy

The geodesical survey is the basic condition of a later excavation, or of the continuation of a started excavation. The task of the geodesy in archaeology is the documentation of the find place or archaeological occurence (for example building, road, ditch, hole, find), moreover the spatiality and results of the archaeological research, i.e. the determination of the position comparing to each other and to the environment of the finds and the observed occurences. The geodesical survey means both assigning the coordinates of the measuring sections and surveying the extension of the archaeological object, and the leveling of the area too.

Luigi Ferdinando Marsigli, officer of the engineer corps, made the first, archaeologically significant map about Hungary, at the turn of the 17-18th century: he indicated on this map many Prehistorical earth mounds and Roman fortresses. Thereafter Sámuel Mikoviny surveyed the remains of the Roman aqueduct (above the surface – *supra terram substructio*) in Brigetio (today Komárom-Őszőny). Also in the 18th century worked István Schönwisner. The activites of Flóris Rómer, Ákos Szalay and Kornél Seitl are the most important ones from the later centuries (see Derdák 2002).

Air photographs in archaeology

The „disturbings" under the surface could not be perceived sufficiently on the spot, because the surface could be seen only in small angle. That is why the air photographs mean such a great help in archaeology: the determination of the extension of a find place is very important, because, for instance the complete excavation of a village from the Middle Age should be very labour- and money-intensive, and also time-consuming without a help of an air photograph. Some objects become identifiable on the air photographs, which could be examine with geophysical surveys.

The occurences which render the observation of former buildings from the air possible are diverse and variable: the results could have an effect on the nature of the objects (e.g. trench, wall), the geological and soil qualities (the thickness, type and quality of the strata), other circumstances (climate zone, season, part of the day) and the vegetation.

C. F. Close second lieutenant wanted to make the first archaeological air photographs about the ruin field near Agra at the end of the 19th century, to compile the archeological ma pof this area with the help of the photographs. But Close's plan came to grief, it could be realized only after the Second World War.

P. H. Sharpe second lieutenant made the first archaeological air photograph about the Stonehenge at the beginning of the 20th century, originally with a military purpose. Mamny more photographs were made in Italy about the Emperor Forums in Rome and the harbour of Ostia Antiqua in the first years of the 20th century. Even more possibilities were in the aerial photography with the improvement of the aerotechnics: many photos were made, which often depicted also archaeological objects as well. The air photographs were used in Central Europe at the polish castle researches, and at the researhes of the Prehistoric monuments in Germany.

The aerial photography was forced back in the Second World War, but individual researchers, like D. N. Riley made flights. More significant wide examinations began after the war, for example the tracks of Roman town planning (*centuriatio*) were mapped in Tunisia and in Syria (see Erdélyi 2002).

Nowadays three institutes make archaeological air photographs in Hungary today: the Eötvös Loránd University (Zoltán Czajlik), Hungarian Academy of Sciences (Zsuzsa Miklós), and University of Pécs (Zsolt Visy; he is occupied in the Roman objects of Sopianae-Pécs, and the *limes*).

Geophysical surveys

The human life always has effects on the environment: it could be changed according to the Ages, cultures and technologies, in different forms and measures (see Aitken 1982, Clark 1990, Kis and Meskó 1980, Milsom 1989, Puszta 2002). The environmental changes, demanded from the former circumstances of life, devolve on people with the disappearance of the different using modes. Few written records has remained in the most case before the Middle Age (the Roman *auctor* sources generally did not dwell on the accurate depiction of the objects like provincial small villas), even there are only some visible tracks on the present surface because due to the weather conditions (mainly wind, rainfall, and after these erosion) and the activities of the people form the later ages (for example settling on, ploughing). In such cases the geophysical surveys, which are founded on the measuring of the specific resistance of the soil, could give help.

So the activities and landscape transformation of the people from the Historical Ages could be examined especially with the help of field-works. The geophysical measuring procedures make the faster and cheaper recognition of the large-sized areas possible, so they considerably facilitate and establish the later excavations. The traditional excavation of large areas is carried out with trial trenches because of the cutbacks, on the points that are characteristic and encourage with promising results. The geophysical instruments have helped the archaeological excavations for decades, but the general using in Hungary ensued only in the past decade.

On the other hand the whole area could be mapped relatively easily with geophysical surveys. The various procedures are based on the physical quality-difference between the earth and objects hiding inside (resistance, magnetism). There is no need to dolly the field, because the instruments can collect data under the surface. Thus the part, which could be more interesting, can be seen before the excavation. It is particularly important in the examination of the former cities and villages, and as also the Roman villas are large-sized, the use of the geophysical methods is absolutely proposed in the examination of villas as well.

The underground parts of the objects in demand for appear in physical changes compared to the environment: the changes of the physical of the materials hidden by soil could be shown by the measurings on the surface. Four different methods could be sundered in geophysical surveys, which function on the principles of the different mechanisms: gravitational, magnetic, seismic, and electrical (it has two sub-groups: used by us is direct current, and the high frequency surface radar) research methods.

The different geophysical methods are able to show objects with diverse attributes (e.g. the magnetic methods are reliable by the examination of trenches, while the electrical methods are reliable by the identification of the remained walls or their foundations of buildings). The most general and to the most used instruments measure the specific electrical resistance in the different points of soil. We used such an instrument by the examination of the Roman villa of Gyulafirátót.

The method used by us is based upon the advance ability of the electrical waves. We used Geoplot RM 15 (UK) instrument. The direct current geophysical measuring method was introduced by Schlumberger in France, in 1912-1913. In the course of the direct current geoelectrical research, electric power is lead into the soil and we measure the voltage on the points of the surface. The resistance could be evenly measured on the midpoint of the electrodes with the help of the sections.

A regular square (it has usually 20x20 or 10x10metres dimension) is needed to be laid down by the using of RM 15 instrument: we need to systematically (normally in a 0,5x0,5 or 1x1metres grid) measure inside this square, in the direction of towards or zig-zag. We could to tighten only a 10x10metres grid in Gyulafirátót because of the on-site conditions (hill, military trench) and the weed shrubs: we advanced with 0,5 meter distance, in zig-zag, with Schlumberger method.

The application of geophysical surveys at the Roman villas in Pannonia: the Roman villas

The Roman villas were cultivation and stock-raising adapted farm units with landed property, the biggest Pannonian group of villas was on the Balaton Highland (B. Thomas 1964, Bakay *et al.* 1966, Éri *et al.* 1969, Gabler 1994, Gabler 2003a, Gabler 2003b, Hajnóczy 1974, Hajnóczy 1987, Mócsy and Fitz 1990, Fitz 2003, Palágyi 2003, Póczy 1976). On the basis of the ideal villa model two bigger parts could be separated: the so-called *pars urbana* (the living part) and the *pars rustica* (the farming part). But this detachment often did not succeed, for example in the case of Gyulafirátót. Still the dimensions of the estate (named *fundus*) could be concluded from the type and size of the buildings.

The farm buildings were often light-structured, constructed of wood, therefore these did not have visible marks on the surface. The fencing walls usually played a role only in bordering, because these were so slim, that they could not be used for protection of the villa. The fencing walls bordered not only the whole villa, but these walls could separate smaller yards in the fenced area as well, according to the different functions (for example flower garden, vegetable garden, yard of the animals).

Geophysical surveys were carried out at several villas in Hungary, for example in Nemesvámos-Balácapuszta, Szebény, Keszthely-Fenékpuszta, Biatorbágy and Nemesbőd, accomplished with Kristóf Fatsar's leadership (Corvinus University of Budapest, Department of Garden Art). Short recapitulations of the most important results can be read in the following.

The first period of the villa of Nemesvámos-Balácapuszta could be dated in the 1st century AD, but the next period in the 2nd and 3rd century was more significant: the villa obtained its final form, that is known today, at this time. The villa had a fencing wall: geophysical surveys were done at three smaller sections in this impounded area, which has momentous results in Garden History. In the section 1 – which was reconstructed as an unbuilt yard before – was a north-south oriented wall. The results of the survey in the section 2 reveals presumably an orchard, which was deployed in a diagonal quincunx (see Fatsar 2005).

This Inner-Pannonian fortress of Keszthely-Fenékpuszta was established in the Late Roman Age, in the first half of the 4th century: it was used for the protection of the province and an important ferry on the Lake Balaton. Several teams made geophysical surveys in Fenékpuszta, which were drifted to the not excavated areas, and also to the clarification of the known areas, which were excavated in the 19th century. The surveys were successful: more villa-like buildings could be traced in the fortress.

The villa of Szebény supposedly flourished in the 3-4th century: the traces of the villa were stumbled in 2006, at the construction of the track of the M6 freeway (Friderika Horváth from the Hungarian Academy of Sciences carried out the excavation). In the course of the geophysical survey at the inferential district of the villa, cleared up, that there were several big buildings under the surface, south-eastward from the known building.

The application of the remote sensing methods at the Roman villa of Gyulafirátót

The Roman villa of Gyulafirátót on the Balaton Highland is on a humid place westwards from the village of today. The villa is surrounded by wells from the east and south, and by the Hill Bakony from the north. A military territory was established on the field of the villa and towards the north after the second World War, in that time was the trench network established, which cuts through two bigger buildings of the Roman villa (B. Thomas 1964, Éri *et al.* 1969, Palágyi 2003).

Archaeological excavations at the villa were last at the beginning of the 20th century, under Gyula Rhé's leadership: the quantity and the ground-plans of the buildings and kilns became known at that time. The Roman villa of Gyulafirátót was built presumably at the turn of the 1st and 2nd century or at the beginning of 2nd century and it was continuously inhabited to the 4th century. According to its buildings, it had a medium-sized estate. The area has three terraces. The size and direction of the extension of the estate is not known, but the humid area towards the southeast, south and southwest and also the rocky soil of the Hill Bakony on the north side sets a limit to the estate. So this field is suitable rather to handicrafts (see also the kilns) than agricultural production. The two parts of the villa-schemes, the pars rustica and pars urbana could not be sharply separated here. The main building No. I. could be considered as the pars urbana, which has a small inner courtyard (*atrium*) and heated rooms. The other four, smaller buildings (presumably a bath, storehouses and a workshop to lead-work) and the kilns belonged to the pars rustica. So this construction – a main living house and more farming houses – shows the picture of a typical veteran settlement.

The buildings of Gyulafirátót

The villa of Gyulafirátót had five known buildings and four kilns, but no fencing walls, which was common at the villas, were excavated: no walls can be seen on the air photographs too, only a part of the buildings (Figure 1.).

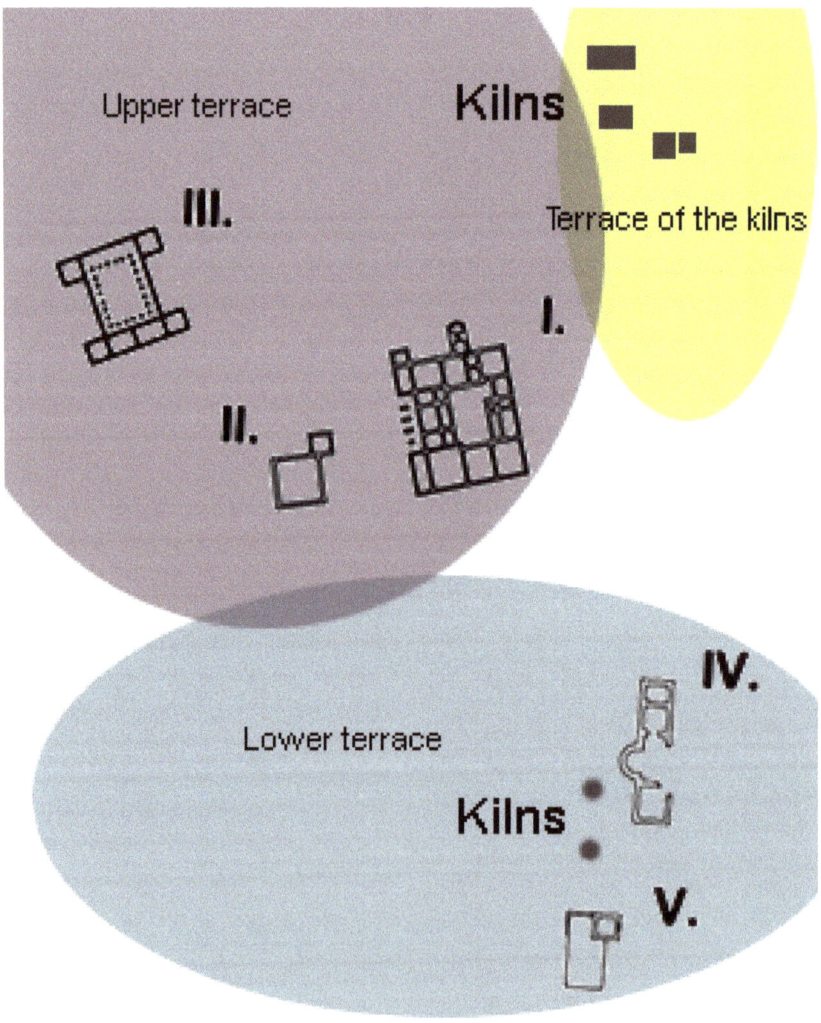

Fig. 1. The buildings of the Roman villa in Gyulafirátót.

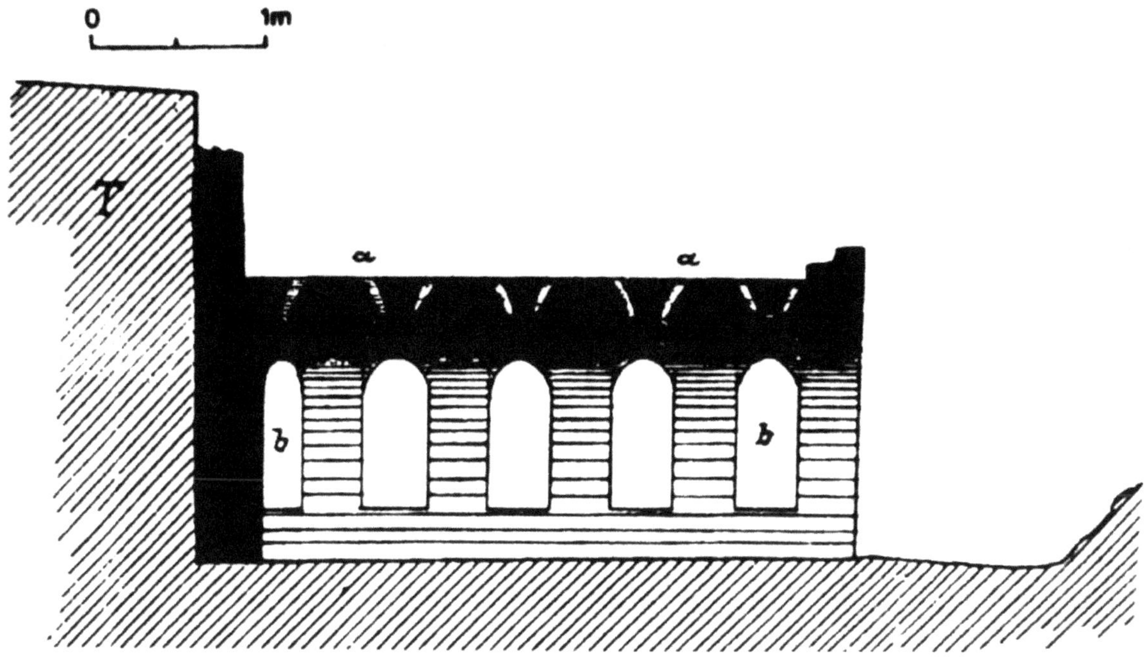

Fig. 2. The Roman kilns of the villa in Gyulafirátót after Gyula Rhé.

The building No. I. had 24 rooms, the north-eastern wing of which with bathrooms was heated. This house had two periods: it was a typical Italian house in the first period, but because of the colder and windy Pannonian climate, a rebuilding was needed in the second period.

The smaller building No. II. had only two rooms, but it had presumably also a storey. This house was probably a workshop to lead-work, because lead objects, lead waste and bungled pieces were excavated here. The excavated pledge-souvenirs of the goddesses Isis-Fortuna (agriculture) and Silvana (forestry) could refer to the cultivation forms of the countryside.

A paved road connected building No. II. with building No. III., which was a farming house by reason of excavated working tools. This house was reinforced with four corner towers because of the barbaric invasions at the end of the 4th century, which caused the fall of Pannonia province. This building (and the marks of the former excavations) can still be seen on the spot nowadays.

Building No. IV. had four rooms and an apsis and it was probably an independent bath, but only a small canal to the southern fountain was excavated, without a water-conduit.

The building No. V. had 2 rooms and it was built of wood, functionating probably as the drying house of the kilns.

The Roman villa had four kilns: two smaller round kilns westwards from the buildings No. IV. and No. V., and two bigger square kilns (Figure 2.) northwards from the building No. I., beside a well. There were several material quarries here. The two bigger ones could be brick burning kilns by reason of its shape: the vaulted furnace chambers were built of bricks. Gyulafirátót is the only known Pannonian villa which had kilns from bricks.

As mentioned above, buildings ruined in barbaric invasions were also inhabited in the Age of Great Migrations, like in many other Roman villas of Pannonia.

The character of the terrain, the position of buildings

The buildings No. I., II. and III. are situated on a common terrace, on a plateau rising over the springs. The buildings No. IV. and V. are under this plateau, at the northeastern corner of the area sorrounded by the plateau and the springs. The terraces of the kilns are on two levels: the smaller ones are westwards from the buildings No. IV. and V., the bigger ones are on a probably artificial terrace between the two levels of the buildings.

The remote sensing examinations in Gyulafirátót

The area of the Roman villa is recorded partly as archaeological field and partly as an agricultural area (at the area of the building No. IV. and No. V.) since the excavations in the 20th century. There are only a few air photographs from the period of the military drill ground (which was developed sometime in the Soviet Period

between the middle of 1940's and the Change of Regime in 1989-1990), and they are partially secret. Complete air photographs could be made from the 3rd Millennium, but the line of the trench could be seen on a photograph from 1985.

The tracks of the building No. III. were perceptible on the spot, and it was also observable, that the modern, in some places two metres deep military trench (Figure 3.) cuts through some buildings, because building rubbles were found in there (with the help of the air photographs, these rubbles could be identified later as the remains of the buildings No. I. and No. II.).

We made geodesical surveys in the autumn of 2006 and in the spring of 2007, when we measured the still visible tracks of the archaeological objects and other characteristic elements, for example a part of building No. III. and the modern, zig-zag shaped trench in 2006, and other military elements, shooting-stands in 2007. This survey and the results (i.e. the separation of the layers of the Historical Ages) constituted the basis of the following researches: the analysis of air photographs and geophysical surveying. Hereby it is easier to identify the environmental connections of the buildings, for example the relationship with the springs, or the detachment of the surface levels.

On the air photographs, the tracks of the former excavations from 1903-1904 at building No. III. are visible, and also the buildings No. I. and II. are identifiable, in spite of the fact that the trench cuts these buildings. The buildings No. IV. and V. are unidentifiable nowadays, because an agricultural area, deep-ploughing is here. The object of geophysical survey was building No. III. that is observable on the spot and air photographs (Figure 4.). This testing survey helped us to recognize the reliability of the chosen process, which always depends on the local conditions (for example soil humidity): reference signs (with well-known meanings) could be given with this survey, which help to explain other data, survey results (Figure 5.). After that we measured the presumable area of No. IV. and V. that are undefiniable on spot. This survey was carried out in the autumn of 2008, using Geoplot RM 15 instrument.

We pointed out nine survey areas: from which three were on the upper terrace at the building No. III., four were on the lowest terrace at the building No. IV. and V., one was at the upper kilns, and one was northwards, at a presumable Roman road. As I mentioned above, we could tighten only 10x10metres grids because of the on site conditions and the thick weeds (e.g. *Rosa sp.*). The weather was favourable, the soil was properly humid.

The walls, or the former archaeological trial trenches nearby the walls of building No. III. were well visible on the measuring. But the area of the buildings No. IV. and V. was really ploughed, so this surveying did not succeed; however, the darker stains of the measuring are probably the concentrated rubbles of the buildings.

Neither the survey at the bigger kilns was successful, probably because of the erosion and the strongly rocky subsoil. The presumable ridged Roman road appeared rather as a protrusion on the rock.

Fig. 3. The modern military trench

Conclusions

The measuring results reveal that the above-mentioned methods are adaptable in Hugary too, although in the respect of the geophysical surveys the circumstances of relief and vegetation are not so advantageous, like in the home of this method, in the United Kingdom.

The results of the geophysical surveys based on electric current are affected by not only the vegetation, but from the soil too, so the hot and dry summers and also the wet periods are not proper for the measurements: the cycles in November-December and February-March are the best, when the weather is mildly wet, and the vegetation is immature.

There are more conditions even in one measuring area (for example the villa of Gyulafirátót could be devided into a rocky soiled, bushy area with worse results, and beyond the modern trench to a soft soiled agricultural area), however this technology is well worth applying in the general excavations.

Acknowledgements

I offer my thanks for all those who took part in the geodesical and geophysical surveys in Gyulafirátót, and helped to extend the knowledgment about the villa with their work, and thus contributed to this publication coming into being.

So I owe thanks to Béla Bálint and Dóra Firnigl, as well as the students of the Study Program for the Conservation of Historic Gardens at Corvinus University of Budapest: Bálint Balatoni, Krisztina Dancsuly, Boglárka Dücső, Bernadett Erdélyi, and Mária Klagyivik for their efforts.

Fig. 4. During the geophysical survey at the building No. III.

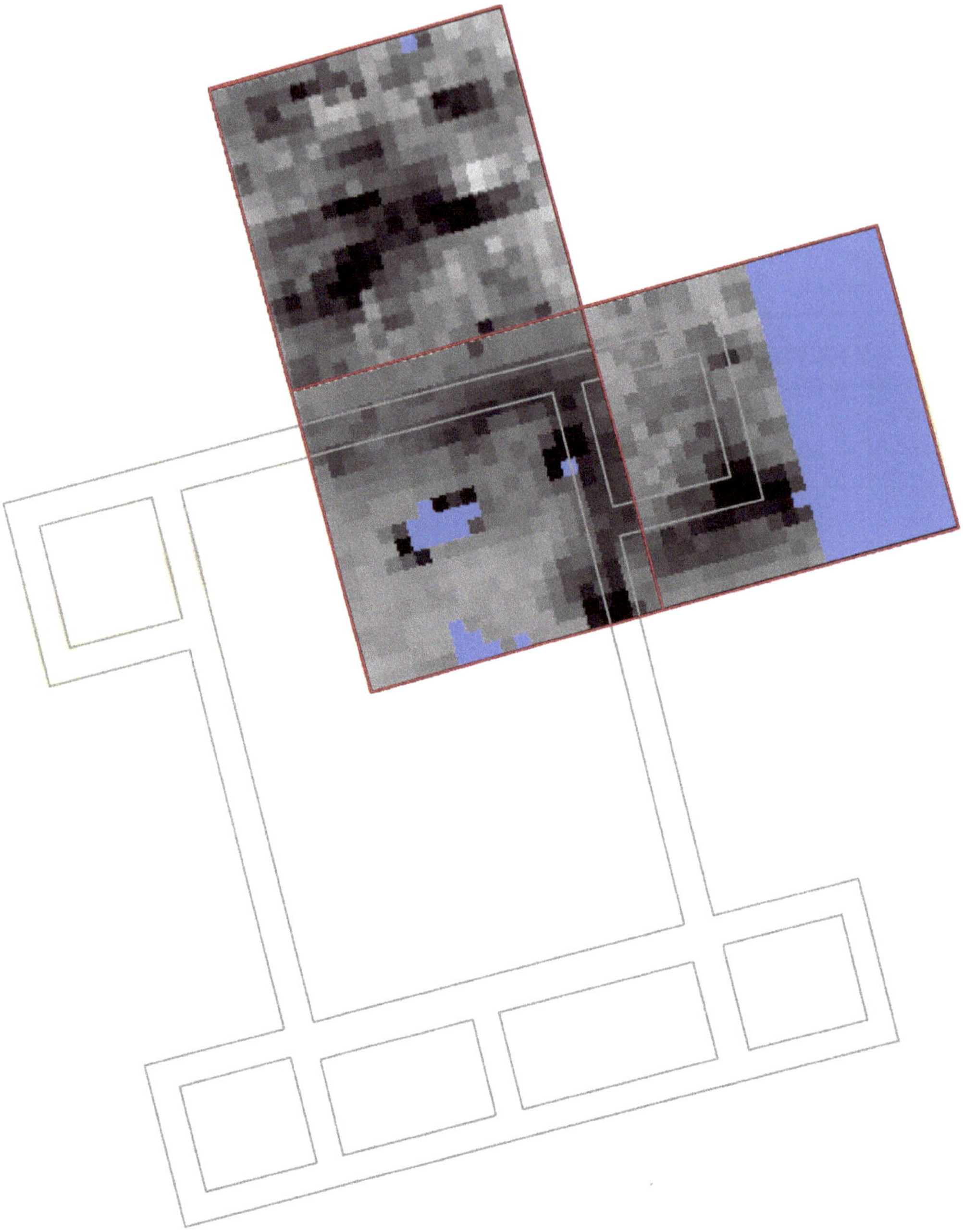

Fig. 5. The results of the geophysical survey at the building No. III.

References

AITKEN, M. J. 1982. *Fizika és régészet* [Physics and archaeology]. 223-239. Budapest, Akadémiai Kiadó

ASTON, M. 1997. *Interpreting the landscape – Landscape archaeology and local history*. London, Routledge

B. THOMAS, E. 1964. *Römische Villen in Pannonien – Beiträge zur pannonischen Siedlungsgeschichte*. 13-127. Budapest, Akadémiai Kiadó

BAKAY, K. *ET AL.* 1966. *Magyarország Régészeti Topográfiája 1, Veszprém megye régészeti topográfiája – A keszthelyi és tapolcai járás [Archaeological Topography of Hungary 1. Archaeological Topography of Veszprém County – Townships Keszthely and Tapolca]*. Budapest, Akadémiai Kiadó

BENDER, H. AND WOLFF, H. 1994. *Ländliche Besiedlung und Landwirtschaft in den Rhein-Donau-Provinzen des Römischen,* Passauer Universitätsschriften zur Archäologie 2. Espelkamp

CLARK, A. 1990. *Seeing Beneath the Soil. Prospecting methods in archaelogy*, London

DERDÁK, F. 2002. Geodéziai alapismeretek [Geodesical fundamentals], In: Ilon, G. (ed.), *Panniculus Ser. B. No. 6.*, 87-105. Szombathely

ERDÉLYI, B. 2002. Régészeti célú légifényképezés [Aerial photography for archaeological purpose]. In: Ilon, G. (ed.), *Panniculus Ser. B. No. 6.*, 47-60., Szombathely

ÉRI, I. *ET AL.* 1969. *Magyarország Régészeti Topográfiája 2, Veszprém megye régészeti topográfiája – Veszprémi járás [Archaeological Topography of Hungary 2: Archaeological Topography of Veszprém County – Township Veszprém].* Budapest, Akadémiai Kiadó

FATSAR, K. 2005. Geofizikai mérésekkel a balácai villagazdaság kertjeinek nyomában. [With geophysical surveys on track of the gardens of the villa of Nemesvámos-Balácapuszta]. In: K. Palágyi, S. (ed.): *Balácai Közlemények* IX., 99-109.

FIRNIGL, A. 2007. Traces of the Roman Road Networks in the Historical Landscape of Balaton Highland. In: *4D Tájépítészeti és Kertművészeti folyóirat* 7. 43-51.

FITZ, J. 2003. A római kor történeti vázlata [The historical survey of Roman Age in Hungary]. In: Visy, Zs. (ed.): *Magyar régészet az ezredfordulón.* 204-208. Budapest, Teleki László Alapítvány

GABLER, D. 1994. A Balatontól északra lévő terület római kori településtörténetének néhány kérdése [Some questions of the history of Roman settlements of the Northern part of Lake Balaton]. In: *Veszprém Megyei Múzeumok Közleményei* 19-20. 149-155.

GABLER, D. 2003A. Vidéki települések Pannoniában – Villák [Rural settlements in Pannonia – Villas]. In: Visy, Zs. (ed.): *Magyar régészet az ezredfordulón.* 235-238. Budapest, Teleki László Alapítvány

GABLER, D. 2003B. Falusi jellegű települések Pannoniában [Country-like settlements in Pannonia]. In: Visy, Zs. (ed.): *Magyar régészet az ezredfordulón.* 241-243. Budapest, Teleki László Alapítvány

GRYNAEUS, A. 2003. Dendrokronológia és erdősültség [Dendrochronology and forests]. In: Visy, Zs. (ed.): *Magyar régészet az ezredfordulón.* 44-51. Budapest, Teleki László Alapítvány

HAHN, M. AND PRAMMER, J. (eds.) *Bauern in Bayern – Von den Anfängen bis zur Römerzeit*, Gäubodenmusem Straubing, 4. Juni – 1. November 1992.

HAJNÓCZY, GY. 1974. Pannónia fürdőépítészete [The bath-building of Pannonia], In: *Építés- és Építészettudomány* VI. 63-100.

HAJNÓCZY, GY. 1987. *Pannónia római romjai [Roman Ruins of Pannonia].* Budapest, Műszaki Könyvkiadó.

JEREM, E. 2003. Környezetrégészet – Egy új tudományág születése [Environmental archaeology – The birth of a new discipline]. In: Visy, Zs. (ed.): *Magyar régészet az ezredfordulón.* 38-44. Budapest, Teleki László Alapítvány.

KIS, K. AND MESKÓ, A. 1980. *Alkalmazott geofizika. Gravitációs és mágneses kutatás [Applied geophysics. Gravitational and magnetic researches].* ELTE Egyetemi jegyzet, Budapest.

KÜSTER, H. 1995. *Das südliche Mitteleuropa als Provinz des römischen Weltreiches*, Geschichte der Landschaft in Mitteleuropa – Von der Eiszeit bis zur Gegenwart, München, 152-162.

MILSOM, J. 1989. Field geophysics. John Wiley & Sons.

MÓCSY, A. AND FITZ, J. (EDS.) 1990. *Pannónia régészeti kézikönyve [Archaeological Reference Book of Pannonia].* 215-233. Budapest, Akadémiai Kiadó.

PALÁGYI, S. 2003. Római kori villák a Balatontól északra [Roman villas northward from the Lake Balaton]. In: Visy, Zs. (ed.): *Magyar régészet az ezredfordulón.* 238-241. Budapest, Teleki László Alapítvány.

PÓCZY, K. 1976. *Pannoniai városok [Pannonian countries].* 11., 30-36. Budapest, Corvina Kiadó.

PUSZTA, S. 2002. Geofizikáról régésztechnikusoknak [For archaeologist- technicians about geophysics]. In: Ilon, G. (eds): *Panniculus Ser. B. No. 6.*, Szombathely, 61-83.

SCHREIBER, H. 1985. *Auf Römerstraßen durch Europa*, München.

SCOLLAR, I. 1990. *Archaeological prospecting and remote sensing.* Topics in Remote Sensing 2. Cambridge

SPIVEY, N. AND SQUIRE, M. 2005. *Az antik világ panorámája [Panorama of the Classic World].* 44-47., 158-161. Budapest, Officina '96 Kiadó.

SÜMEGI, P. *ET AL.* 2003. Magyarország rövid környezettörténete [The short environmental history of Hungary]. In: Visy, Zs. (ed.): *Magyar régészet az ezredfordulón.* 51-56. Budapest, Teleki László Alapítvány

TÓTH, E. 2006. *Itineraria Pannonica – Római utak a Dunántúlon [Itineraria Pannonica – Roman Roads in Transdanubia].* 5-66. Budapest, Magyar Nemzeti Múzeum.

Application of integrated digital technologies in the study of settlement archaeology of Kausambi Region

S. S. Rai

University of Allahabad
India

Abstract:

Kausambi is one of the largest fortified city sites of early historic India. Due to its rich archaeological wealth and strategic location, it can be termed as a key site in the middle Ganga Valley for understanding the origin and development of urban centers in the middle Ganga Valley. Settlement archaeology constitutes a major key approach for understanding the man environment relationship. An integrated application of satellite imagery, GPS based survey and data visualization in GIS can make the most effective way in the analysis of settlement archaeology of particular region. An attempt is being made in this paper first to present spatial data on site location, their relative sizes and chronologies within and outside the fortified township and then to superimpose the same on toposheets and satellite imagery towards GIS based analysis of the change in the settlement system of this region.

Introduction

When the Buddha expressed his desire to die in the town of Kusinara, his disciple Ananda said: "Let not the Exalted One die in this wattle and daub town in the midst of the jungle, in this branch township. For, Lord, there are other great cities such as Campa, Rajagaha, Savatthi, Saketa, Kausambi and Varanasi." Kausambi is thus described among the six great cities in the time of the Buddha, in *Mahaparinibban Sutta* of *Digha Nikaya*. Situated on the northern bank of the Yamuna, it was a rich commercial city at that time. Like Taxila, Sravasti, Varanasi, Rajagriha and Vaisali, Kausambi was a wealthy city in which millionaire merchants, lesser merchants and middle men resided (Rhys Davids 1909). Kausambi is now represented by modern Kosam, Garhwa and other neighbouring villages. Alexander Cunningham was the first archaeologist to have identified Kausambi with the above villages (Cunningham 1871).The exploration and further excavations by Prof. G. R. Sharma finally proved beyond doubt this (Sharma 1960, 1969).

The ruins of Kausambi are situated on the left bank of river Yamuna. It is about 48km, south-west from Allahabad in Manjhanpur subdivision of district Kausambi. This vast area is divisible into several mounds of different dimension and elevation (Sharma 1998). The chain of mounds has a peripheral circuit of about 6km. The habitations of Kausambi are encircled by a rampart with towers and bastions which has an average height of 9m from the surrounding of the field level. This fortified city forms an irregular oblong on the plain (Figure 1). The fortified area has gates on three sides – east, west & north. Unfortunately the location of the southern gate cannot be determined because of the erosion caused by Yamuna. Most probably the people of this area had no need of this gate because Yamuna in herself was providing defense from this side. The city was encircled by a moat on three sides. The moat, though filled up at several places, is still visible on the eastern side. The defense at Kausambi betrays an advanced knowledge of fortification (Rai *et al.* 1988). The entire city thus encircled by the rampart wall and moat is littered with a huge mass of brick bats indicating the density of structure in the city. With the passage of time the habitation level inside the city went on rising and an approximate thickness of 16m of habitation deposits furnishes the vertical record of city's life and antiquity.

There are so many archaeological sites in the hinterland of Kausambi which developed during the course of time. In the fortified area of Kausambi, several chronological stratifications can be observed. These small settlements played a prominent role in the development of Kausambi as an important city site in the 6th century BC. A GIS-based study of these settlements, the main objective of this paper, has the potential to provide a meaningful explanation for the settlement hierarchy; site location and resource distribution pattern; ecological niche present in this area; probable routes of transportation and detailed overview of the landscape.

Geographical Setting

The entire Gangetic plain was formed by the uplifting of Himalayan chain and the changing of river courses. Due to this process the vast alluvial plain was formed which is pedologically and topographically homogenous. As a result local variations are nevertheless present and have lead to zones of markedly different agricultural potential, in turn reflected in pattern of settlements. The entire area is formed by the streams of Ganga and Yamuna as well as it has a few minor streams as Sasur Khaderi, Kilnahi etc. with a range of water bodies and thus it has a variety of localized environmental zones. The river Ganga in this belt has a maximum width of 3km and the Yamuna by 2km. The Ganga and Yamuna form the apex of triangle which is known as the *Sangam*, a sacred place of Hindus. In this belt along the lower reaches of river Sasur Khaderi has high banks marking

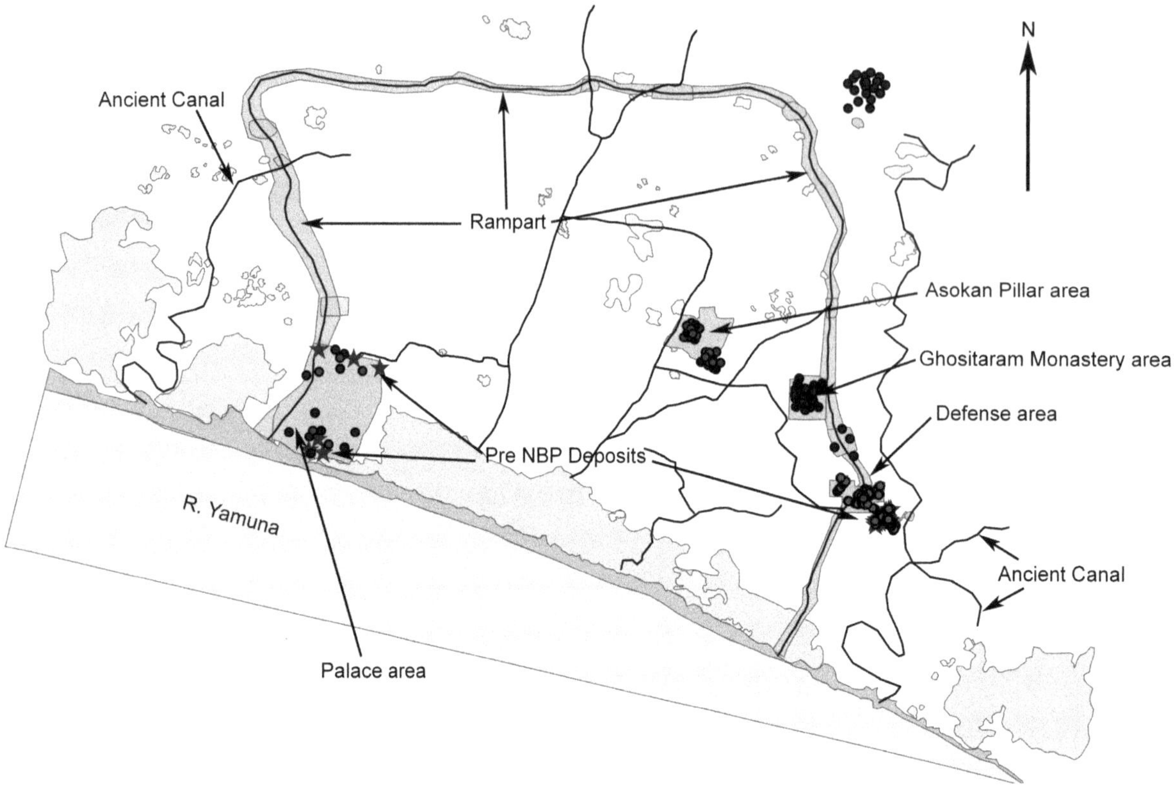

Fig. 1. Map from GIS database showing excavated area and defense of Kausambi

the limits of the recent movement of the channel. The entire plane is amazingly homogeneous and has hardly the contour elevation of 50cm. High banks have been favored to the sitting of habitation for a variety of regions. This area is well protected from floods and the soil is very fertile. The high banks also give access to the major transport through its courses. All these factors are important for the study of settlement pattern of this region.

Excavation at Kausambi

The Archaeological Survey of India excavated a large portion near the Asokan Pillar area and laid bear walls and rooms of a living complex. The ASI could publish no detailed excavation report of the excavation. University of Allahabad excavated at Kausambi from 1949 to 1965 under the direction of G.R Sharma. As a result four major areas have been unearthed:

1) Near the Asokan Pillar which laid bare a part of the residential area of the city,
2) The Ghoshitaram Monastery,
3) The Defense area, eastern gateway and towers at the northwestern corner, and
4) The stone fortress/Palace area.

Excavation near the Asokan Pillar Area

The excavations in the area near the Asokan pillar, revealed for the first time the plan of an ancient Indian township, especially a residential area, distinct from monastic establishment or a residence of royalty. A remarkable feature of the excavations was the light it has thrown on the drainage system prevalent in the Ganga valley. The various drainage devices in vogue include ring wells, soakage jars, terracotta pipes, open brick drains, covered brick drains and brick tanks. The lay-out of the excavated area indicates the existence of a number of cross-walls in the lane. The cross-walls were provided with single-leaf doors and open drains. Considerations of security from trespassers and thieves must have prompted the erection of these cross-walls. The houses were aligned with roads and lanes and generally maintained correct orientation with the cardinal points. The excavations yielded a number of seals, including a unique sealing of king Kanishka, a very large number of coins, four coin-hoards containing coins and *cowries*; terracotta figurines of which the Saka-Parthian ones constitute a unique group; sculptures; and iron objects used in agriculture and industrial work. According to the available evidence, this area of the city, in which the common citizens lived, had a continuous settlement from the middle of the fourth century BC to the early part of the 4th, century AD.

Excavation of Ghositaram Monastery Area

The Ghoshitaram monastery, frequently mentioned in early Buddhist literature, was built by one of the leading bankers of Kausambi, Ghoshita by name (Rhys Davids 1881). During the excavations a number of inscriptions have been found.

They include the seals of the monastery, bearing the legend of the great monastery of Ghoshitaram of Kaushambi. In addition there is the famous *Ayagapatta* inscription inscribed below the footprints of the Buddha, the border being decorated by stylized swastika

The excavation has reveled that the monastery was a combined *Chaitya* cum *Vihara*. The main *stupa* measures about 24.69×24.69m. The central area, measuring 7.32×7.32m, enclosed two cross-walls which formed four triangular brick platforms to the four sides. These triangles were packed with layers of mud alternating with layers of bricks forming what looks like regular floors. The entire packing, from bottom to top, contained 'Northern Black Polished Ware' (NBPW). There is clear evidence of two structural phases of the *stupas* antedating the enlargement, presumably by Asoka, suggesting thereby that the original *stupa* was built in the 5th century BC just after the Great Demise of the Buddha. A number of C-14 dates obtained from the monastic area clearly indicate its existence from *c.* 5th century BC to 6th century AD.

Excavation of the Defense area

The city of Kausambi when founded was not provided with any defense. The excavation of the defenses has brought to light two pre-defense sub-periods. The beginning of the massive defenses, however, marks a major change in the life of the city. The core of Rampart-I was built of hard, compact and thoroughly rammed clay evidently excavated from the natural soil. It is significant that the first rampart is completely devoid of pottery and other material indicating human association (Sharma 1960, 27). In the fifth century BC we see the instant building activity which brought into being a number of major structures in this area and a complete change in the plan of the defenses and the top of the rampart was raised by about 2m. The next major stage in the evolution of the defense system was the building of the tower and guard rooms at the top of the rampart and flank walls ranging from top to the toe (Sharma 1960, 34). Its external face of rampart was 2.74m thick at the base and its existing height is 12.31m. The bricks measure 49.5×33×6.98 cm. The revetment was covered with plaster, 5-8 cm thick, composed of mud and lime. The moat surrounding the township ran parallel to the rampart. Besides, there was an underground tunnel built on the principle of corbelled arch. The next interesting structure recovered from this area is a sacrificial alter.

It is brick built on plain and looks like a bird *Syena* or hawk. It was located outside the rampart in an enclosed area. It was highly disturbed and only the top representing the head, the left wing and a part of tail now survive. This structure has been identified as *syena-chiti* of the *Agni-chayana* where a human sacrifice *Purusamedha* was performed. It was probably built by Mitra kings on the occasion of his victory over the Indo-Greek rulers in the 2nd century BC. These defenses were finally destroyed by the *Huna* king Toraman.

Excavation in the Palace area

In the south-western corner of the ancient walled city of Kausambi, an area on the Yamuna is clearly marked out from the rest of the site by its well-defined contours. The entire area was littered with chips of stone, fragments of plaster and shreds of the NBPW and associated wares. On the Yamuna two small prominent mounds covering an overall area of almost 75×45 m were included within this complex. The excavations have laid bare a massive stone fortress on the Yamuna measuring 320×150 m. On plan, it is barrel shaped. The northern and southern sides are parallel and the eastern and western curvilinear. The corresponding tower on the south-western side and also a considerable part of the palace complex has been washed away by the river. The earliest stage is represented by undressed stone boundary wall which has been exposed beneath the dressed wall of the second stage on the northern and the western sides. The Palace was destroyed by Menander sometime in the 2nd century BC, when the stone boundary-walls and towers were razed to the ground. The debris layers, which covered the ruins of wall, yielded shards of NBPW along with a sealing bearing an inscription in the *Brahmi* script of 2nd century BC. The layer has also yielded double-tanged arrow heads ascribed to the Indo-Greek invasion during the third stage. The Palace was rebuilt immediately after the destruction. But there is a noticeable change in the alignment and in the method of construction. The boundary-wall was no longer of stone alone, the core being made of stone and the bricks being used for the facing.

The entire settlements of Kausambi are divisible into four cultural periods. Period I (1165 BC to 600 BC) is a pre-NBP deposit and can be further subdivides in to pre-PGW (pre-Painted Grey Ware) and Painted Grey Ware phases. Period II (600 BC to 400 BC) is characterized by Black & Red ware and early-NBP ware. The most distinctive pottery tradition of the third period (400 BC to 100 BC) is the NBPW, which comes in different varieties such as black, grey, brown, orange, silver and golden in color and the period IV (100 BC to 590 AD) roughly covers the *Saka, Kushan* and *Gupta* periods.

Data Sources

In this region explorations of archaeological sites have been carried out by University of Allahabad under 'village-to-village survey scheme' and the details of findings have published in the concerning years of Indian Archaeology – A Review (IAR). In addition, exploration by George Erdosy, brought many new archaeological sites in this area (Erdosy 1988). With an objective to know the condition and location of the sites, this area has been surveyed by this author in three stages. As a result, ground reality of the sites has been found and some new archaeological sites have explored (Figure 2).

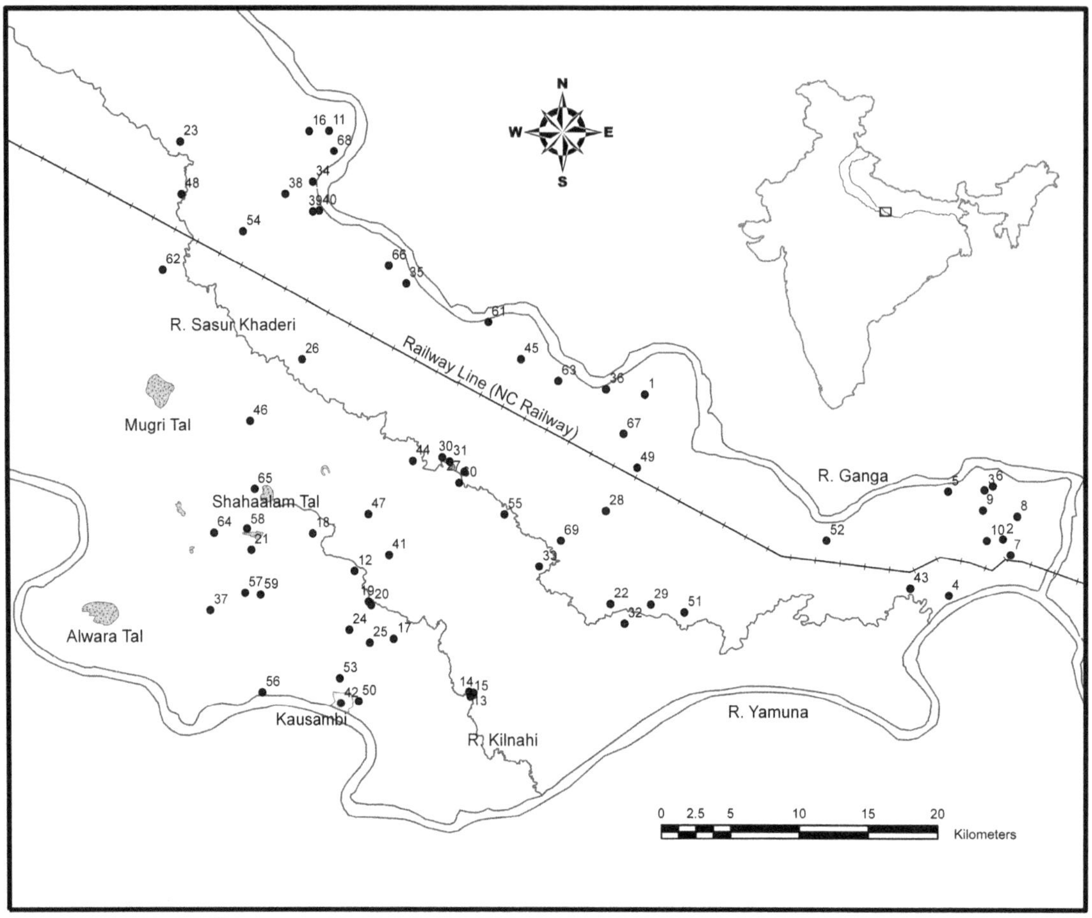

Fig. 2. Map of sites present in study area (after Erdosy 1980).

Hand held GPS has been used to get coordinates of the sites. In the year 2003, a GIS database 'ArchaeoGIS' for Kausambi has been developed by Prof G. K. Rai, Dr. R. D. Gupta and C. Bala Bhaskar in order to prompt the research and development in the archaeology of Kausambi. This database carry information about spatial distribution of artifacts, lay out plan of excavated area, geohydrology, etc., inside the fortified area of Kausambi. The 'ArchaeoGIS' has been found suitable for management of large volume of archaeological data, efficiently at Kausambi site. This data base has been used for the study of settlements inside the fortified area.

The Allahabad and Kausambi district topographic map of scale 1:5,00,000 were obtained from Survey of India and has been used as base map and topographical map '63G7' at the scale of 1:50,000 has been used for locating sites and features. The rest of data as aerial photograph of Kausambi, LISS-III satellite imagery of the area etc. were obtained from the Centre of Advanced Study, Department of Ancient History, Culture and Archaeology, University of Allahabad. Google Earth imagery has been used to see the details of sites on the landscape.

The GIS software ILWIS 3.3 Academic and ArcGIS 9.1 were used for developing and analyzing the data base. The image processing software ERDAS has been used for processing of LISS-III satellite imagery.

Use of GIS

The extraordinary pace of development in computer technology has transformed the potential of field survey and spatial archaeology by borrowing again from techniques employed in the study of geography. Settlement patterns are distinct special expression of cultural adaptation and archaeologists share with geographers a concern to evaluate the degree to which certain factors condition settlement distributions. Every phenomena studied by archaeologists took place inside a spatial temporal dimension. Therefore, this science is largely involved in the study and understanding of the evidence that can be synthesized in distribution schemes. As a result of this, archaeological research has always demanded a careful and meticulous approach to distributions on space. And it is for this reason archaeologist had always searched for better methods and instruments to achieve this goal in the most comprehensive and objective manner. With the continued growth of the survey database, the contemporary era has faced new challenges in managing the enormous data flow from the cumulative regional studies. Geographic Information Systems has become a basic part of the surveyor's analytical repertoire for database management (Balkansky 2008). GIS provides an important new dimension to field archaeology that is very widely applicable to the complex problems of location and distribution. As with other forms of computer-based data handling, an initial investment of time and energy in recording data in an appropriate form is repaid by the flexibility that it allows at the analytical stage. GIS can be used as a tool to manage and assemble spatial layers including site location, settlement hierarchy and the available resources within surrounding area. GIS allows the physiographic surface analysis which can lead to the development of the precise catchment boundaries for the study of the relation among each sites in the local geographic setting. Sites of a specific period could also be added to test ideas about their relationship to this phenomenon. GIS allows archaeologists, for the first time, to deal with large amount of spatial data and its analysis in a reasonable amount of time. Actually GIS can be used as a management tool that provides means for data collection and presentation as well as an analytical tool. GIS has been operational in archaeological research on a regional scale as a tool to integrate observations from field studies of depositional environments, site specific interpretations and associated cultural materials (Mandry and Crumley 1990). The storage and analytical capability of spatially referenced information obtained from various sources makes GIS well suited for using in this type of multi-disciplinary context (Bauer *et al.* 2004).

Methodology

In this work exploratory visualization of the data has been carried out. Exploratory visualization is the process that involves a discipline expert creating maps and other graphics while dealing with relatively unknown geographic data. These maps are generally for a single purpose, and function as an expedient in the expert's attempt to solve a (geo) problem. While dealing with the data, the expert should be able to rely on cartographic expertise, which allows him to look at the data from different perspectives to create a fresh view on the data. As such, the maps and graphics, which should be available in an interactive viewing environment, stimulate the visual thinking of the expert and should support (spatial) decision making.

The maps obtained for Kausambi were in paper format and hence they were scanned at 300dpi. The topographic map which was used as base map is first georeferenced based on the given latitudes and longitude values. The other maps were georeferenced using the registered topographic map by image to image registration. The features on these maps were digitized in ArcInfo in suitable shapefiles. The discovered archaeological sites in the study area were compiled into the GIS database by creating point shape files for each cultural period. The district boundary, alluvium deposits, *tals*, palaeochannels etc. were digitized as polygon layers and the rivers were digitized as polyline layer. The layers thus created were analyzed using ArcGIS 9.1 and the results were tabulated (Figure 4). The maps describing the distribution of sites across different strata of alluvium and distribution with respect to a water source and deposits were also developed in this work.

All these works have been completed before venturing to the field. Standard procedure in many archaeological survey involve collection of locational data using GPS equipments, that's why in the field, hand held GPS receiver was used in the mapping of sites within a landscape and feature within a site. These maps have been used to interface directly with topographic maps and satellite imagery. Satellite imagery contributed to a stronger appreciation of the essential integration of finds, sites and their natural setting, and the potential significance of these relationships. To view the sites in higher resolution Google Earth satellite imagery were also used. These mapped data were then incorporated into GIS database to understand site location, resource patterning and to obtain multiple view of objects and feature within a landscape. I also took the opinion of the local inhabitants regarding the sites and settlements. And thus a database containing the details of 69 early historic sites, was developed and mapped using GIS (Figure 2). The distribution of early historic sites was studied with respect to geology and geomorphology also.

Result and Discussion

Archaeological sites of the first period are mostly located on the bank of river Ganga, although some sites are present at the banks of Sasurkhaderi and Kilnahi as well. The sites along the river Ganga can be divided into 2 levels. Kara & Sanjaiti represent the upper echelon and rest of the sites make lower echelon of the settlements. No site of lower echelon has been excavated in this region. But their function of agriculture & craft activities is clear because of use of mud bricks in construction, pottery, presence of microlithic blade & flakes and iron slag on sites like Bahadurpur, Bidaon, Diha, Gangsari, Mahgaon & Ark Mahabiram (IAR 1994-95, 1997-98; Erdosy 1988). The high soil fertility permitting small growing areas can be traced as agents of agglomeration of settlements. Access to recent alluvial deposit and to routes of transportation, are crucial factor for making settlements along the permanent banks of rivers.

Bahadurpur is an important site which is close to a route from Sanjaiti to Kausambi in present days. From Kausambi & Bahadurpur core and flakes of chalcedony as well as chert from Kausambi has been reported (Sharma 1960, 69, Erdosy 1988) and the pottery of Ist phase recovered from all the three sites are identical. This typology suggests that the banks of Ganga & Yamuna were well connected to utilize the sources in and around the region. It is worth mentioning that in the fortified area of Kausambi, heavy construction was started between the Ghoshitarama monastery and Yamuna river as well as in fort/palace area (Figure 1). The construction of defense had begun in this phase (Sharma 1969). The construction along high bank of river Yamuna indicates some kind of trade of raw material via the river course.

In the second phase four tier hierarchies of settlements were originated. In this area at the bottom of hierarchy, settlements lie between 0.42 to 2.0 ha. We can assume agricultural function from their appearance. In this period no sites are located away from water sources. Uninscribed cast copper coin from Draupati Ghat site; square uninscribed cast coins and punch marked coins from Kausambi (Sharma 1960) indicate motivation towards trade as well as pottery and presence of a few surface finds indicate some craft activities.

Kara & Jai Chand Ka Qila are two such sites which acquire area more than 20ha. Jasper and Chert, reported from earlier stratum of these sites, imply their connection to the hilly areas of Vindhya & Kaimur. Iron slag has also been reported which indicate the development of metallurgical technology. The dominant settlement in this hierarchy is Kausambi where the deposits of Ghositaram monastery and Asokan pillar give a much reliable date of 600 BC to 300 BC for this period. In this phase construction can be seen in the whole fortified area. This is the time when Kausambi developed as a capital city of Vatsa Mahajanpada and as an important trading centre. The specialization of craft & trade could be suggested to be characteristic of this phase.

In phase three, the emergence of a network of towns and crystallization of settlements, already emerged in phase II, can be seen and we find that some sites acquired larger areas as Adampur, Kara and Sanjaiti. At the low level of hierarchy the most amazing development is the establishment of 12 villages/small settlements away from the rivers (Baresa, Bidaon, Chak Alai Niyaz, Awna Alampur, Garhi, Kamalpur, Karari, Kothari Purab, Manhai, Pura Kotia, Purab Sarira and Sarsawan) which indicates colonization of the forested upland area. One thing is very important that all the newly originated settlements are located within the buffer zone of 25km of Kausambi which reflect the need of fulfillment of requirements for a large urban centre. The settlement hierarchy is clearly visible in this phase as well as regulatory role of higher order settlement can be seen very clearly. The rapid & uneven growth of urbanization in the capital city, growing at the expense of smaller centre, the steady expansion of rural population and the limited extension of central into rural areas are all dramatical, brought out by the evidences.

In this phase, the impression given in all respects is of a period of peak prosperity which is the culmination of the centuries of development. This phase witnessed the period of massive construction within the fortified area of Kausambi. In this phase most of the population seem to have withdrawn behind the rampart and the strengthening of the defense is also evident (Sharma 1980). Two sites earlier existing adjacent to Ghoshitarama monastery immediately outside were abandoned during this period (Figure 3) as well as 3 sites have developed on northern and western city gates, which was due to the need of defense.

Fig. 3. Location of two sites Abandoned in pd. III.

In phase IV the five tier hierarchy are visible. The important sites developed along the bank of the river Ganga are Kara, Jaichand Ka Qila, Adampur, Faridpur Parsakhi & Farhanpur kalesar Mau. Out of them Farhanpur Kalesar Mau & Faridpur Parsakhi disappeared in phase III and to the north-west of these sites, three sites Ambai Bujurg, Aureni and Tiwaripur emerged. Aurani is a large settlement. Farhanpur Kalesar Mau was highly affected by cut and erosion of Ganga; that's why these subsidiary settlements originated in phase IV. For the same reason, Kamalpur, Koh Inam and Nidhaura came into existence. One reason for the development of these new settlements could be an increase of population.

Along the bank of Sasur Khaderi river, sites 6, 24, 27, 28, 33, 50, 54, 64 and 66 (Figure 2 & 4) are important which maintained its position in period 2 and period 3. Diha is another site which is located on right bank of Sasur Khaderi and emerged in period IV. No change could be detected in patterns associated with towns, as the area in this period did not increase. Two of the minor settlements along Sasur Khaderi (Chak Khuda Ganj and Sanai) disappeared which reflect the decline of importance of Sasur Khaderi stream. Here a continuous & gradual decline of intermediate sites can be observed. In this period some sites were found in clusters. All these data related to that sites suggest that they were not providing services but were established due to pressure of population, as well as a political uncertainty reflected by destruction level at Kausambi can be seen (Sharma 1980).

The highest concentration of archaeological sites is visible under buffer zone of 20km from the main site Kausambi, between the river Kilnhi & river Yamuna. The evidence recovered from Bahadurpur establish the relation of this area with the hill of Banda and Kaimur and the evidences from Bidaon (IAR 1997-98) cross certify it. The people of this region were utilizing the resources from the begining. Paschim Sarira initiated the development of Purab Sarira.

In Allahabad city, 45km east of Kausambi some archaeological sites have came into light which are Draupadi Ghat, Mirapur, Bharadwaj Ashram, Chilla, Govindpur, Salori, Medical College and T.B. Sapru Hospital (IAR 1960-61). The evidence from Bhardwaj Ashram (IAR 1960-61, 1978-79, 1982-83) and Draupadi Ghat indicate that human habitation here started during the period of first construction at Kausambi.

During the first structural period of Kausambi there was developing a chain of human habitation in the hinterland of Kausambi. The people of this region have reach to the resources of Vidhyan Mountain and were utilizing it for tool making. The fertile area of *Doab* prompted the increase of population and thus an agglomeration of settlement originated. Chert, chalcedony & iron slag reported from the fortified area of Kausambi prove that a certain kind of craft specialization was present at that time. A site Manhai has been developed adjacent to Kausambi in period III which was a satellite town of Kausambi containing a community of craftsmen. Buddhist literatures of this period also cross certify these sets of evidence. The Buddhist texts authored during 3rd century BC to 1st century AD are very important for throwing light on the commercial activities taking place in Kausambi in the period of Buddha. The *Suttanipata* mentions that Kausambi was the most important *entrepot* of goods and passengers coming to Kosala and Magadha from the south and west. The *Vinaya Pitak* mentions that it was well connected with Kosala, Magadha and Campa via waterways and right up to the mouth of the Ganges and thence either across the Bay of Bengal or along the coast to Burma (Rhys Davids 1909). Roads coming from the southwest and northwest also converged on Kausambi for import and export of goods from those quarters. Thus Kausambi in the time of the Buddha was the greatest river port for import and export of goods for the whole of North and Mid-India and had commercial relations even with Burma (Rhys Davids 1909).

SITES

Sr. No.	Site	Hight	Area
1	Adampur	6m	2.31ha
2	Allahabad (Bhardwaj ashram)		
3	Allahabad (Chilla)		
4	Allahabad (Dariabad)		
5	Allahabad (Draupadi Ghat)	2.5m	2 ha
6	Allahabad (Govindpur)		
7	Allahabad (Mirapur)		
8	Allahabad (Salori)	2m	4 ha
9	Allahabad (T. B. Sapru Hospital)		
10	Allahabad Medical College		
11	Ambai Bujurga	3m	1.5 ha
12	Asadha	3m	2.16 ha
13	Atarsueya1	3.5m	0.95 ha
14	Atarsueya2	5m	2.82 ha
15	Atarsueya3	2m	2.12 ha
16	Aureni	2m	4.15 ha
17	Awna Alampur	7m	3.79 ha
18	Bahadurpur	2m	1.62 ha
19	Bahman Kanti 1	5m	1.56 ha
20	Bahman Kanti 2	6m	1.8 ha
21	Baresa	3m	2.16 ha
22	Bariaawan	2m	3.14 ha
23	Bherawan	6m	1.60 ha
24	Bidaon	8m	5.15 ha
25	Chak Alai Niyaz	1m	1.3 ha
26	Chak KhudaGanj	5m	1.92 ha
27	Chapaohan	5m	1.95 ha
28	Chirwa	4m	2.01 ha
29	Daryapur	1m	1.3 ha
30	Daryapur Majjhiyawan 1	6m	3.14 ha

31	Daryapur Majjhiyawan 2		1.60 ha
32	Diha	7m	3.78 ha
33	Faisalabad	2m	1.26 ha
34	Farahampur Kaleshar Mau	1m	0.42 ha
35	Faridpur Parsakhi	3m	1.19 ha
36	Gangsari	5m	2 ha
37	Garhi	4m	1.55 ha
38	Kamlapur		1.65 ha
39	Kara	10m	25 ha
40	Kara (Jai Chand Ka Qila)		20 ha
41	Karari	4m	1.23 ha
42	Kausambi		
43	Kesari Masari	3m	1.54 ha
44	Khajuapur	5m	1.95 ha
45	Koh Inam	6m	3.8 ha
46	Konder	8m	3.46 ha
47	Kotari Purab	5m	2.31 ha
48	Madhia mai	5m	1.21 ha
49	Mahgaon	6m	4.15 ha
50	Maihai	3m	3.14 ha
51	Makhupur	3m	1.02 ha
52	Maria Dih	7m	1.56 ha
53	Mustafabad	7m	0.28 ha
54	Nidaura	6m	2.55 ha
55	Pabaiya	7.5m	3.8 ha
56	Pabhosa	2m	2.4 ha
57	Paschchim Sarira	3m	2.5 ha
58	Pura Kotia	3m	3.46 ha
59	Purab Sarira	6m	1.62 ha
60	Sanai	3m	6.75 ha
61	Sanjaiti	8m	9.17 ha
62	Sankha	1m	1.8 ha
63	Sapaha	4m	1.65 ha
64	Sarsawan	5m	1.68 ha
65	Shahaalamabad	6m	2.3 ha
66	Shahzadpur	5m	1.95 ha
67	Sukhda	6m	1.8 ha
68	Tiwaripur	3m	2.16 ha
69	Udhatu	7m	3.46 ha

Fig. 4. List of sites studied in this work

References

BALKANSKY, A.K. (2008). Settlement Pattern Analysis. In Deborah M Pearsall (Ed.), *Encyclopedia of Archaeology* (pp. 1978-1980). Elsevier.

BAUER, A., KATHLEEN, N., PARK, L. AND MATNEY, T. (2004) Archaeological Site Distribution by Geomorphic Setting in the Southern Lower Cuyahoga River, Northeast Ohio: Initial Observations from a GIS database. *Geoarchaeology*, **19**, 711-729.

CUNNINGHAM ALEXANDER (1871). *The Ancient Geography of India*. London: Trubner and Co.

ERDOSY, G. (1988). *Urbanisation in Early Historic India*. Oxford: British Archaeological Reports.

INDIAN ARCHAEOLOGY - A REVIEW 1960-61 (pp. 33)

INDIAN ARCHAEOLOGY - A REVIEW 1978-79 (pp. 56)

INDIAN ARCHAEOLOGY - A REVIEW 1982-83 (pp. 90)

INDIAN ARCHAEOLOGY - A REVIEW 1994-95 (pp. 71-72)

INDIAN ARCHAEOLOGY - A REVIEW 1997-98 (pp. 170-73)

MADRY, S.L.H. AND CRUMLEY, C.L. (1990). An application of remote sensing and GIS in a regional archaeological settlement analysis: The Arroux River Valley, Burgundy, France. In K. M. S. Allen, W. G. Stanton and B. W. Z. Ezra (Ed.), *Interpreting Space: GIS and Archaeology* (pp. 364-381). London:Taylor & Francis.

RAI, G.K., MISHRA, B.B., SRIVASTAVA, R.K. AND DUGGAL, S.K. (1988). Geotechnical Investigations at Kaushambi. *Indian Geotechnical Conference (IGC-88)* Allahabad.

RHYS DAVIDS, T.W. AND OLDENBERG, H. (1881). *Introduction to Vinaya Texts Part I, Sacred Books of the East XIII*. Oxford.

RHYS DAVIDS, T.W. (1909). *Buddhist India*. London.

SHARMA, G.R. (1960). *The Excavations at Kaushambi (1957-59)*. Allahabad: Department of Ancient History, Culture ans Archaeology, University of Allahabad.

SHARMA, G.R. (1969). *The Excavations at Kausambi (1949-1950)*. : Memoirs of the Archaeological Survey of India, No. 74.

SHARMA, G.R. (1980). *History to Prehistory: Archaeology of the Ganga Valley and the Vindhyas*. Allahabad: Department of Ancient History, Culture and Archaeology, University of Allahabad.

SHARMA, G.R. (1998). *Kushan Studies*. Allahabad: Department of Ancient History, Culture and Archaeology, University of Allahabad.

Investigating megalithic astronomy: the role of remote sensing

Srikumar M. Menon,[1] Mayank N. Vahia[2]

[1] Lecturer, Faculty of Architecture, Manipal Institute of Technology, Manipal
[2] Professor, Department of Astronomy and Astrophysics, Tata Institute of Fundamental Research, Mumbai and Centre for Excellence in Basic Science, Mumbai

Abstract

Megaliths in the Indian subcontinent are believed to be the constructions, sepulchral and otherwise, of iron-using cultures that date between 1500 BC to 200 AD. They are an enigmatic class of monuments, varying in scale, form and typology, but exhibiting similarities over large geographical regions. They vary from the simple marked burial to large alignments that comprise of a few thousand stones spread over a few acres. We have been investigating these monuments for about two years in order to understand the knowledge systems possessed by their authors, especially in mathematics, geometry and space-time concepts. We have studied monuments that fall into the major typologies that exist, out of the nearly 3000 known megalithic sites that exist in south India. We present the preliminary results from some of these surveys and make the case for studying as many of the remaining sites as early as possible. In today's scenario of increased pressure from development, many megalithic sites are vanishing rapidly. There is a clear need to document these sites as quickly as possible. Since many of these sites are in remote areas, physical surveys of these are taxing and time-consuming. Also, several of the alignment sites, like Hanamsagar are spread out over a very large area, which makes it time-consuming to survey. Hence a case is made for studying these sites initially from high-resolution remote sensing imagery.

Another point that emerges is the importance of remote sensing in the landscape level study of the planning of these monuments, site selection criteria etc. We also make a case that studying the remote sensing imagery before undertaking the ground survey also helps plan the survey better, incorporating even other features near the site that otherwise would have been missed.

Keywords Megaliths, Iron Age, Archaeoastronomy, Ancient knowledge systems, Ancient astronomy.

Introduction – the megaliths of south India

Megaliths have captured the imagination of the lay public and the archaeologist and antiquarian alike for ages because of their 'curious and bold appearance on the surface' of the earth (Sundara 1975). Megalithism seems to be a 'world phenomenon' – their occurrence from the shores of England to that of Japan testifies to this (Moorti 2008). Almost throughout the Old World, these ancient stone structures have been noticed and various theories have been propounded to account for these. However, Moorti (2008) cautions that "it may be safer to outline the distribution of megalithic monuments but forming opinions on their spread is nothing less than walking on a hotbed of controversies."

On the Indian subcontinent, the area between the Vidarbha region of Maharashtra and the tip of the southern peninsula constitutes the major zone of the burial style denoted by the various types of megaliths (Chakrabarti 1999). Till recently, megaliths were supposed – quite wrongly – to have formed an independent cultural entity in this region. However, it is now clear, especially after the excavations of sites like Watgal in Karnataka and Bhawar in Vidarbha, that it is nothing more than a burial style which emerged in the context of the Neolithic-Chalcolithic of its distribution area and formed part of its cultural milieu for a long time. Generally, it is associated with iron but there is also a possibility of its beginning in the pre-iron stage.

It is well-known that in India, the Deccan or Peninsular India contains a very large number of megalithic structures – since the first reports of "Pandu Coolies" in Kerala by Babington (1823). In the decades that followed, hundreds of megalithic sites have been discovered in southern India. Several of them have been excavated, studied and classified; among them the most celebrated study being that of Brahmagiri. Brahmagiri has been intensively explored by M. H. Krishna in 1940 (Krishna 1942, Ghosh 1989). Later on R. E. M. Wheeler excavated the site on behalf of the Archaeological Survey of India in 1947 (Wheeler 1947, Ghosh 1989). Since those days, studies of Indian megaliths have come along way and a lot of these sites have been systematically studied, though megaliths continue to be one of the most perplexing problems in South Asian archaeology, as mentioned earlier (Ehrich 1992).

The etymology of the word "megalith" comes from the words "*megathos*" – referring to the scale and "*lithoi*" – referring to the material. Thus, megalith literally means "built of large stones". This terminology arose because the earliest monuments belonging to this category to be noticed were the ones with the most impressive surface markers. As we shall see, not all the monuments that are today categorized as "megaliths" are built of large stones! The usage of the term "megalithic" can, however, be justified because of its antiquity and continued popular use. The term denotes in the present context a socio-religious expression of burying the deceased in a grave (which may or may not have a lithic appendage) accompanied by certain specific cultural traits of the period under reference (Moorti 2008, Moorti 1994).

A large majority of these monuments are funerary in nature. However, the funerary aspect of this tradition is not entirely a new feature of the Iron Age. The antiquity of burial practice in India dates back to the Mesolithic period and marked burials begin in the Neolithic (Agrawal (ref)). Though evidence for an antecedent stage of "megalithism" is found in the pre-Iron Age context, this tradition became very popular in the Iron Age and continued to survive into the Early Historic and even later periods.

Though a few megalithic sites have been reported from North India, (e.g. Burzahom, Gurfkral in Kashmir; Gagrigol in Kumaon area), by far the large majority of megaliths are found in the southern part of the country. The various megalithic types encountered are stone circles, dolmens, dolmenoid cists and cist burials, pit burials etc., apart from menhirs, stone alignments and avenues, as well as rock-cut chambers and the unique topikals and kudaikals of Kerala. In fact, in spite of this apparently bewildering variety of typologies one finds distributed all over the subcontinent, they can be classified into a few broad categories.

For a detailed classification of Indian megaliths, the reader is referred to Moorti (2008) and Moorti (1994). Basically, he classifies all megaliths into **sepulchral** – which contain the mortal remains of one or more human beings and **non-sepulchral**, which has no human remains associated with them. In the former category belong **pit burials** (unlined excavations where the remains are buried with a variety of surface markers like cairns or boulder circles etc.) and **chamber burials** (pits lined with stone slabs, often with a surmounting cap-stone, again with a variety of surface markers). In the non-sepulchral category fall megaliths like dolmens, menhirs, stone alignments and avenues. However, it may be pointed out that sometimes variations may be seen – such as menhirs being non-sepulchral in Karnataka and markers for pit burials in Kerala (Sundara 1975).

In India, the megalithic period is supposed to coincide with the Iron Age – roughly 1500 BC to around 200 AD, with some amount of overlap with the early historic period. This seems to be based primarily on the occurrence of iron tools and implements among the burial goods found along with human remains in a large number of megaliths. This is by no means a sound means of dating all types of megaliths – especially those non-sepulchral in nature, such as stone alignments and avenues. In fact, Sundara (1975) even wonders whether the above mentioned typologies belong to the south Indian megalithic complex at all.

Our project – the knowledge systems of the megalith builders

Though the technology possessed by these megalith builders – such as smithery, carpentry, bead-making etc. and the socio-economic systems extant in those times have been studied earlier (Moorti 2008, Moorti 1984), not much is known about the knowledge systems of these ancient engineers. That they possessed refined engineering skills (and hence some knowledge mechanics, visualization and, by implication – of geometry and mathematics) there can be no doubt – one has to only see the elegant conception of a *kudai-kal* (figure 1) or a laterite rock-cut chamber burial (figure 2) to grasp this.

Fig. 1. A *kudai-kal* of Kerala.

However, to formally understand the knowledge systems extant in megalithic society, one has to study these monuments from this point of view. We started this project in 2007 – with the objective of examining megalithic monuments to understand the knowledge systems their builders possessed, especially in astronomy.

Fig. 2. A rock-cut chamber burial with high degree of workmanship.

We wanted to investigate if the megaliths followed any orientation patterns, as also whether indeed some of them

Site visited	Typologies of megaliths on site
1. Brahmagiri, Karnataka	Chamber burials, pit burials
2. Marayoor, Kerala	Dolmens, rock art
3. Burzahoma, Kashmir	Neolithic/megalithic pit burials, menhirs
4. Bandipur, Karnataka	Chamber burials (previously unreported site)
5. Hire Benkal, Karnataka	Chamber burials, dolmens
6. Onake Kindi, Karnataka	Rock art
7. Nilaskal, Karnataka	Avenue (?)
8. Baise, Karnataka	Avenue (?)
9. Hergal, Karnataka	Avenue (?)
10. Mumbaru, Karnataka	Avenue (?)
11. Thrissur, Kerala	Rock-cut chamber burials, urn burials
12. Aihole, Karnataka	Chamber burials, dolmens
13. Bachinagudda, Karnataka	Chamber burials, dolmens
14. Hanamsagar, Karnataka	Avenue
15. Kyaddigeri, Karnataka	Chamber burials (destroyed)
16. Chik Benkal, Karnataka	Chamber burials (dilapidated)
17. Vibhutihalli, Karnataka	Avenue
18. Bheemarayanagudi, Karnataka	Avenue (disturbed)
19. Gudde Maradi, Karnataka	Avenue (?) (Destroyed)
20. Santhekkatte, Karnataka	Menhir(?)

were designed to keep track of celestial events on the local horizon.

Our plan was to study all classes of megalithic monuments and examine their orientations. From literature we short-listed sites typical of each category of megaliths, preferably in different geographical regions. Reconnaissance visits were to be made to each of these sites to assess the condition in which they survived and to make preliminary measurements, especially orientations. Promising sites were to be short-listed and detailed measurements made, including horizon surveys. From the above, we would try to conclude if any orientational preferences existed and indeed, if any class of monuments were intended to keep track of celestial events like solstices and equinoxes.

Sites visited so far

We have mainly studied the following classes of megalithic monuments, mainly in Karnataka and Kerala so far.

The orientations of the various types of megaliths were measured. Some of the sites were found to be heavily disturbed (Chik Benkal, parts of Brahmagiri) or completely destroyed (Kyaddigeri), making visits to these worthless from the point of collecting data.One site at Bandipur, which turned up during our studies, was subsequently confirmed as previously unreported. The sites at Nilaskal, Baise, Hergal and Mumbaru were suspected to be avenues after reconnaissance, in spite of claims to the contrary in literature (hence the '?' above), so Nilaskal and Baise were visited twice more and surveys made of the menhirs.

Preliminary results

As mentioned earlier, orientations were measured at all the sites, wherever possible. In the case of sites with very large number of monuments (Hire Benkal, Hanamsagar, Vibhutihalli), only a few random monuments were measured.

In the case of the sepulchral megaliths, it was found that they were largely oriented towards east and west within the annual ranges of sunrise/sunset. This means that the tombs faced sunrise or sunset on some day of the year. Maybe they were oriented to face sunrise or sunset at the time of construction. However, it must be mentioned that the megaliths on Meguti Hill at Aihole faced every direction on the compass – indicating that maybe they were just intended to face downhill – a case of topography rather than celestial events dictating orientation. Stylistically, the Aihole megaliths seem to be older than the Hire Benkal type.

Among the non-sepulchral type, the avenue seems like a potential candidate for astronomical use. Two kinds of avenues were encountered during our surveys. The first consisted of grids (according to Allchin (1956), Paddayya (1995), Rao (2005) etc.) of rough, undressed stones of varying dimensions in large numbers (Hanamsagar, Vibhutihalli etc.). The second was made of quarried stone slabs or natural elongated stone slabs of lenticular cross-section, erected oriented north-south and in formation (Nilaskal, Baise, Hergal and Mumbaru). The latter were recorded as menhirs, 'erected haphazardly, unlike those of Vibhutihalli or north Karnataka' (Sundara 2004, Sundara 1975). They are also recorded as being about 20 in number, though our studies have thrown up the remains of more than 100 stones. We have recorded evidence for many pairs of stones forming sight-lines to the sunrise and sunset points on the local horizon during both the solstices, at Baise as well as Nilaskal (figure 3, 4). However, in the case of Hanamsagar and Vibhutihalli, we do not have enough data to put forward any case.

Fig. 3. Showing a sight-line defined by two menhirs at Nilaskal.

Fig. 4 The winter solstice sunset along the same sight-line at Nilaskal.

The case for remote sensing

It is in the case of monuments like those at Hanamsagar and Vibhutihalli, especially the former, that remote sensing can play a major role. Hanamsagar was first reported by Mahadevan (Allchin 1956) and subsequently described in detail by Paddayya (1995). It is reported as a diagonal grid of rough, undressed stones of varying size, roughly spaced 15 to 40 feet from each other. Paddayya (1995) describes the number of stones as 2500 while Sundara (1975) and Allchin (1956) reports 1000 menhirs. Our own reconnaissance visit to the site showed lot of disturbance at the southern end of the avenue, resulting in inability to fix the limits of the avenue. One needs to have a good fix on the extents of the monument to hazard astronomy as the purpose of the avenue; conjectures such as Rao (2005) based on an approximation of the form of the avenue derived from Paddayya (1995) are found to be grossly erroneous upon a visit to the site.

Also, Sundara (1975) and Paddayya (1995) report the rows to be aligned to the cardinal directions. Our random checks of the orientations in the avenue show that, though the north-south orientation of the menhirs is accurate to within one degree of arc, the east west is off by about seven degrees. An aerial view taken from the hill on the west seems to suggest a curvature in the alignment of the rocks (figure 5).

Fig. 5. A view of a portion of the avenue at Hanamsagar from a hill on the west.

Hence, the very shape of the avenue, which has been studied for long, is under question. A ground survey is arduous given the remoteness of the location as well as the fact that farmers have set up thorn fences between different plots within the avenue. A high-resolution satellite image of the region would settle the matter beyond dispute trivially. Also the monument, which has been disturbed in recent times, can be studied from satellite images of even ten years ago – to see it in undisturbed condition.

Another point that emerges is at the landscape level study of the planning of these monuments, site selection criteria etc. Many a time, it is noticed that the clues to the understanding of the structure in totality is much larger than the immediate surroundings of the monument, embedded in the topography around on a much larger scale. This is another case we make to study remote sensing imagery of the region around each of these sites on a scale of a few kilometers to few tens of kilometers. We do not recommend this as the only study to be carried out on megalithic sites, but as an important and early component in the study, to be followed by intensive surveys on the ground of promising sites. We also make a case that studying the remote sensing imagery before undertaking the ground survey also helps plan the survey better, incorporating even other features near the site that otherwise would have been missed.

Conclusion

The study of megalithic monuments to understand the astronomical knowledge of their builders, as evident from their form and orientation, is a challenging task – both because of the nature of the task itself, and because of the fact that most of these are unprotected – thus heavily disturbed by development activities. Hence there is an urgent need to study and document as many sites as possible immediately. Remote sensing helps us to zero in on potentially promising sites, plan the ground-based study of the sites better as well as understand the monument in the larger scale of its setting.

Acknowledgements

Srikumar M. Menon wishes to gratefully acknowledge the Prof. D. S. Achyutha Rao Memorial Trust and Manipal University for a grant to pursue research on megaliths. Mayank N. Vahia wishes to thank the Sir Jamsetji Tata Trust for a research grant to investigate Ancient Indian Astronomy. We wish to thank Dr U. S. Moorti for valuable advice and guidance. We also wish to thank Dr Sudhakara G. and Mr Kailash Rao M. for stimulating discussions and valuable help in site surveys.

References

AGRAWAL, R. C., The *Megalithic Culture of India: its Genesis, Spread and Continuity*, Keynote address.

ALLCHIN, F. R., (1956) The Stone Alignments of Southern Hyderabad, *Man*, Vol. 56, 133-136.

BABINGTON, J., (1823) Descriptions of the Pandoo Coolies in Malabar, *Transactions of the Literary Society of Bombay*, Vol. III, 324-330.

CHAKRABARTI, D. K., (1999) *India: an Archaeological History (Palaeolithic Beginnings to Early Historic Foundations,* Oxford University Press, New Delhi.

EHRICH, R. W. (ED.), (1992) *Chronologies in Old World Archaeology Vol. I,* The University of Chicago Press, Chicago and London.

GHOSH, A. (ED), (1989) *An encyclopaedia of Indian Archaeology, II,* Munshiram Manoharlal Pub. Pvt. Ltd. New Delhi.

KRISHNA, M. H., (1942) *ARMAD*, 100-109.

MOORTI, U. S., (1994) *Megalithic Culture of South India: Socio-Economic Perspectives*, Ganga Kaveri Publishing House, Varanasi.

MOORTI, U. S., (2008) Megaliths, in Pearsall, D. M. (ed.), *Encyclopedia of Archaeology*, Academic Press, New York.

PADDAYYA, K., (1995) The Stone Alignment at Hanamsagar, District Gulbarga, Karnataka, In Srinivasan, L. K. and Nagaraju, S., Eds., *Sri Nagabhinandanam,* Dr. M. S. Nagaraja Rao felicitation committee, Bangalore, 23.

RAO, N. K., (2005) *Bull. Astr. Soc. India*, **33**, 499-511.

SUNDARA, A., (1975) *The early chamber tombs of South India: a study of the Iron Age megalithic monuments of N. Karnataka*, University Publishers, Delhi.

SUNDARA, A., (2004) Menhirs in Mid-western Karnataka: Further Notices, In Sundara A., Ed., Proceedings of R. B. Foote *Memorial National Seminar (1995) on Indian Prehistory and Protohistory (Recent Studies),* Directorate of Archaeology and Museums, Govt. of Karnataka.

WHEELER, R. E. M., (1948) Brahmagiri and Chandravalli 1947: Megalithic and Other Cultures in the Chitaldrug District, Mysore State, *Ancient India*, IV, 81-308.

Chronological aspects of functioning of the set of early mediaeval north Indian epigraphic sources

Alexander A. Stolyarov

Institute of Oriental Studies of the Russian Academy of Sciences, Moscow, Russia

Abstract

The temporal dynamics of early mediaeval North Indian epigraphic sources has been traced on the base of two lists (corpuses) of North Indian inscriptions published in 1899-1906 (F. Kielhorn) and 1928-36 (D.R. Bhandarkar). Two tables and graphs were built and compared. The third table and graph was built on the base of the newly done (1995-2009) database of early mediaeval North Indian copper-plate grants. It is considered to be the newest master sample of the chronological dynamics of early mediaeval North Indian copper-plate grants.

Keywords Early mediaeval North India, copper-plate grants, database, chronological dynamics, representativeness

At the moment the set of Indian inscriptions consists of various texts that differ very much both by their exterior and interior forms and also by their content. Edicts, documents and literary works can appear in inscriptions as well. They can be incised on stones, rocks, walls, monuments, statues, metal plates, etc.

This situation creates difficulties for the scholar dealing with certain samples of inscriptions and the first of these difficulties is to proving the representativeness of his samples. Otherwise the value of the epigraphic material as a mass source might be seriously doubted.

The most reliable and available criteria for proving the representativeness of epigraphic sources are temporal and spatial characteristics. Of these two the first one is the most important because every epigraphist in the final analysis proves the spatial representativeness of his sample more or less "individually"; it is subject to the pattern by which the inscriptions are distributed within the region studied. By contrast for proving the temporal representativeness of the epigraphic set some master sample is needed. In this master sample the chronological dynamics of epigraphic sources that had been found over the territory of the space studied (North India) for all periods of their functioning will be fixed. Once the scholar has such a sample, his task of defining the temporal representativeness of the epigraphic set under research would be to compare its temporal dynamics with the corresponding part of the sample (either in a graph or in a table).

One of the requirements for having such a sample is to create a corpus or at least a list of epigraphic sources. The earliest list of North Indian inscriptions of the 4th – 19th centuries was published by F. Kielhorn (Kielhorn, 1899-1906). It consisted of 716 inscriptions. The next and most recent list of the North Indian epigraphic sources of the 3rd – 19th centuries was published in 1928-36 by D.R. Bhandarkar (Bhandarkar, 1928-36). It consisted of 2114 inscriptions. In his own words, D.R. Bhandarkar had "endeavoured to make this List as complete as" he could.

Now in view of the lack of efforts to make the new list of the North Indian mediaeval epigraphic sources it would seem that the availability of two alternative lists of one and the same subject area based on mutual modes of description, would make the task of developing the master sample described above much easier. It is possible to consider each of two lists as an ideal spontaneous sampling of hypothetic array of all inscriptions that were produced in mediaeval times. Hence there arises the possibility of obtaining the master sample of temporal distribution of North Indian inscriptions by comparing the chronological dynamics of both lists.

Before dealing with the temporal distribution of inscriptions forming each of the lists it is necessary to define the most appropriate units of the subject of research and also of time periods. The matter is that there are several inscriptions in the list that were dated in two or more Indian eras or systems of dating. Each of these datings was separately numbered in the list though the inscription contained information of the single action. By contrast some other inscriptions contained information of a series of alternative single actions (for instance of several grants to some family of Brahmans and so on) which were not always given a separate number in the list. That's why it would be logical to introduce the category of "simple entry" or "simple inscription". A simple inscription would be characterized as a "connected text written on a nonrecurring basis by one and the same person on this or that material that form a single document (several pages of text, text on several copper-plates, on stone, on different sides of a pillar, etc.)" (Deopik, 1977, 32).

It is appropriate to consider a century as a unit of a chronological periodization. On the one hand it is a sufficiently representative characteristic. On the other it is the least possible one because some of the inscriptions of the Bhandarkar List are dated roughly – according to the time of this or that ruler. Sometimes there is no date at all but only the name of the ruler. It is clear that in these cases the inscriptions can only be dated within the century.

Regarding the special role that copperplate grants play among Indian epigraphic sources while it is possible to use them for fixing the spatial representativeness of epigraphic sample it also makes sense to study along with the chronological dynamics of all inscriptions also the same dynamics of the copperplate grants.

The result of the foregoing statements is the two graphs (See) that reflect the chronological dynamics of North Indian epigraphic sources according to the Kielhorn List (Figure 1) and that of Bhandarkar (Figure 2).

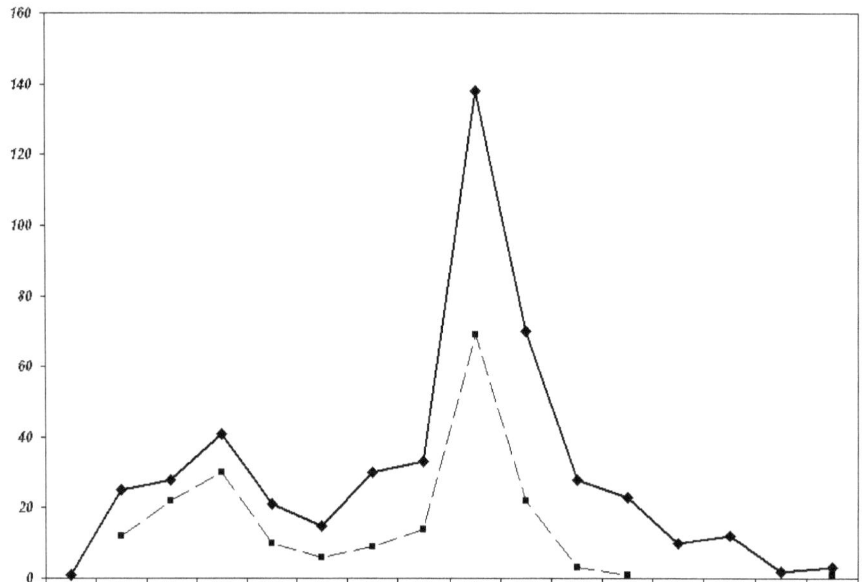

	IV	V	VI	VII	VIII	IX	X	XI	XII	XIII	XIV	XV	XVI	XVII	XVIII	XIX
KLINI	1	25	28	41	21	15	30	33	138	70	28	23	10	12	2	3
CP KLINI		12	22	30	10	6	9	14	69	22	3	1				1

Fig. 1. Chronological distribution of inscriptions of Kielhorn List.

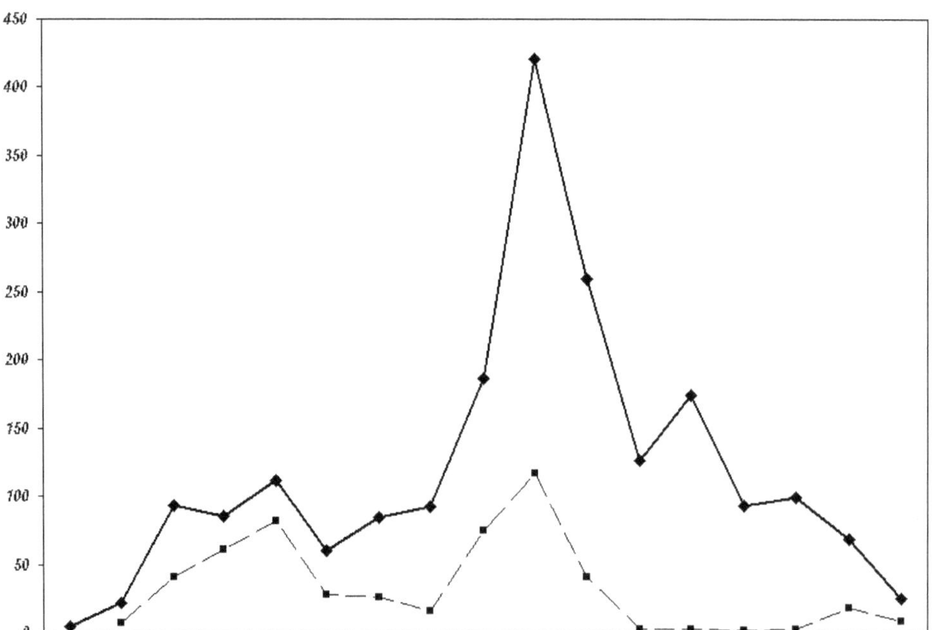

	III	IV	V	VI	VII	VIII	IX	X	XI	XII	XIII	XIV	XV	XVI	XVII	XVIII	XIX
BLINI	4	22	93	85	111	60	84	92	186	420	259	126	174	93	99	69	25
CP BLINI		7	41	61	82	28	26	16	75	117	41	3	3	2	3	18	9

Fig. 2. Chronological distribution of inscriptions of Bhandarkar List.

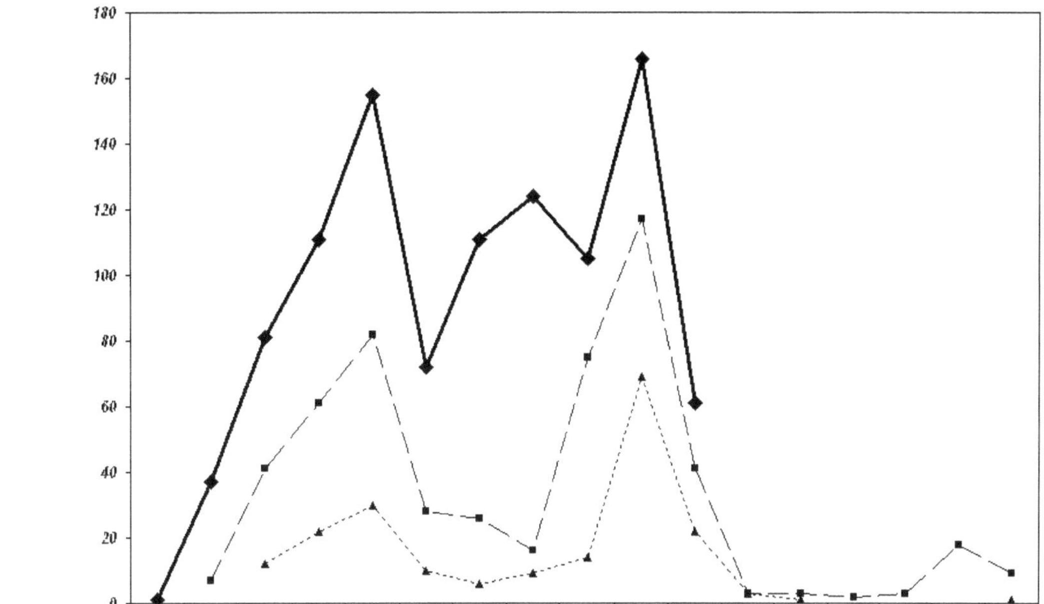

	III	IV	V	VI	VII	VIII	IX	X	XI	XII	XIII	XIV	XV	XVI	XVII	XVIII	XIX
CP DI	1	37	81	111	155	72	111	124	105	166	61						
CP BLINI		7	41	61	82	28	26	16	75	117	41	3	3	2	3	18	9
CP KLINI			12	22	30	10	6	9	14	69	22	3	1				1

Fig. 3. Chronological distribution of CP grants of Diplomatica Indica Database, Kielhorn and Bhandarkar Lists.

The solid line in both graphs is used to designate the chronological dynamics of all inscriptions. The dashed line is used to designate the chronological dynamics of the copperplate grants. The total number of inscriptions included in the Graphs and Tables is 475 for Kielhorn List (or 66% of the total number of the entries of the List) and 2004 for Bhandarkar List (or 95% of the total number of the entries of the List). Both percentages tend upwards owing to the above-mentioned duplication in different eras of one and the same date of one and the same simple inscription. The total number of copper-plates is 199 for Kielhorn List (or 42% of the total number of inscriptions included in the Graphs and Tables) and 532 for Bhandarkar List (or 27% of the total number of inscriptions included in the Graphs and Tables).

While comparing the graphs the high level of their isomorphism is striking. From this we can argue that both graphs in principle can be considered as master samples of the chronological dynamics of the epigraphic sources under research. However the second graph should be given the greater preference because it is based on larger evidence.

The broken line that reflects the chronological dynamics of the CP grants should be given special attention. On the whole it is smoother than the base line. Furthermore, beginning with the 12th century the clear tendency to the sharp isolation of the epigraphic tradition proper from that the CP grants can be seen. This circumstance seems to be connected with a change in the social function of the epigraphic sources and also with the possible emergence of different modes of fixing events such as land grants in particular.

At the moment, the new database of North Indian CP grants has been compiled contains some 1020 documents. Its name is 'Diplomatica Indica' (DIDB). It attempts to be as complete as possible. It is very interesting to compare the chronological dynamics of the DIDB with that of the Kielhorn's and Bhandarkar's lists. This is shown in the Figure 3.

The solid line in this graph is used to designate the chronological dynamics of DIDB. The dashed line is used to designate the chronological dynamics of CP grants of the Bhandarkar's List. The dotted line is used to designate the chronological dynamics of CP grants of the Kielhorn's List. It could be seen in the Table 3 that the number of CPs published for more than seven decades passed since the BLINI publication, increased disproportionately by centuries. Nevertheless the level of the isomorphism of all three sets shown in the Graph 3 is very remarkable.

The foresaid permits to consider the DIDB set as the newest master sample of the chronological dynamics of one of the main epigraphic sources of India – early mediaeval North Indian copper-plate grants. Of course this master sample can be applied not only to solving auxiliary problems. At some extent it reflects the dynamics of historical development of the North Indian region in whole. However it is only one of the tendencies of this development. That's why it would not be wise to connect it directly with this or that tendency of political and social economic development of the region in different periods of the epoch under research.

References

KIELHORN, F., *A List of Inscriptions of Northern India from about A.D. 400,* in: "Epigraphia Indica". – Calcutta, 1899. – Vol. 5. – App. – P. 1-121; 1906. – Vol. 8. – App. 1: Suppl. to the List of Inscriptions of Northern India. – P. 1-19.

BHANDARKAR, D.R., *A List of Inscriptions of Northern India written in Brahmi and its Derivative Scripts from about A.C. 200,* in: "Epigraphia Indica". – Calcutta, 1928. – Vol. 19. – App. – P. 1-42; 1930. – Vol. 20. – App. – P. 43-266; 1932. – Vol. 21. – App. – P. 265-296; Index to App. – P. 297-310; 1934. – Vol. 22. – App. (Index to App.). – P. 311-62; 1936. – Vol. 23. – App. – P. 363-406

DEOPIK, D.V., 1977, *Problems of Methods of Study of Epigraphic Set Concerning the Tasks of Social-Economic Analysis (the example of Burmese Epigraphy),* in: "Herald of Moscow University". –Ser. "Oriental Studies". – No 2. (In Russian: Деопик Д.В. Проблемы методики исследования эпиграфического комплекса применительно к задачам социально-экономического анализа (на материале бирманской эпиграфики) // *Вестник Московского университета. – М., 1977. – Серия Востоковедение. – № 2*).

The archaeology of ritual spaces and the temples at Aihole, Karnataka

Himanshu Prabha Ray

Centre for Historical Studies,
Jawaharlal Nehru University
New Delhi

Abstract

The spectacular temple complexes of Aihole, Badami and Pattadakal located along the 25 kilometre long fertile valley of the Malaprabha river, in north Karnataka have long been admired for their distinctive architecture and sculptural exuberance. The hills and rocky outcrops of reddish-golden sandstone provided the stone used for the temples and the large number of inscriptions linked several of these to the early Chalukya rulers. This paper shifts the focus of analysis to space as a conceptual category to locate the temples in their geo-physical habitat. The attempt is to articulate reasons for the clustering of the more than one hundred and twenty shrines in a four square kilometer area at Aihole at one end of the Malaprabha valley. Since little of the archaeological context of these temples survives, the use of satellite images provide a constructive technique of investigation.

Keywords Malaprabha valley, Chalukya, Badami, Pattadakal, megaliths

High-resolution satellite images are an invaluable source for situating religious architecture within its geo-physical habitat and it is the results of this collaborative venture that are presented here. In keeping with the objectives of the conference that visualises space, place, landscape or "context" as a conceptual framework for archaeological investigation, this paper locates the Early Chalukyan temples at Aihole in the Malaprabha valley both within the physical and cultural landscape and also highlights the inter-linkages between the different shrines at the site. Time, in this paper, is refracted through several chronological periods including the time of construction of the temples at Aihole that ranges from the seventh to the twelfth century, as the temples continued to be added to and modified, but also the *longue durée* of cultural adaptation within a prehistoric landscape culminating in the discovery of the temples at Aihole in the nineteenth century and their identification as archaeological monuments, which distanced them from their social moorings.

At both ends of the 25 kilometre long fertile valley of the Malaprabha river, (a tributary of the Krishna) in Bagalkot district, the hills close in and Badami lies some distance outside the southern entrance to the enclosed terrain (Figure 1).

Here, between the cliffs there is a huge tank and temples are scattered around this water-body. Aihole (Aryapura/ Ahivolal) is located at the opposite entrance to the valley and in the middle of the valley lies Pattadakal (ancient Kisuvolal), where a complex of temples was built on the banks of the Malaprabha, which at this point runs auspiciously from south to north. Mahakuta lies five km east of Badami as the crow flies in a cleft in the hills and continues to be a place of pilgrimage, with most of the shrines built around a sacred tank within a walled enclosure. Other significant places include the site of Naganatha temple, in the forested area, rock shelters at B.N. Jalihalla and Siddhanakolla in the hills near Aihole, Sandur in Bellary district, Banavasi, the ancient Kadamba capital, Kudaveli Sangam, Satyavolu and Alampur, 150 km or so to the east. Elsewhere I have discussed these temples within the broader cultural landscape of the Malaprabha valley and analysed diverse uses of ritual space by highlighting the cult of Lajjā Gaurī – often termed a fertility goddess, which shared sacred space with the Hindu, Buddhist and Jain temples, as also with memorial shrines (Ray, 2007: 89-101). In this paper, I focus only on the site of Aihole, where more than one hundred temples were built over a period of nearly five hundred years and include not only the early cave temples at Ravalphadi, but also shrines affiliated to a variety of religions, such as Buddhism, Hinduism and Jainism. I start this paper with the discovery of Aihole and then move on to a discussion of the contexts of the Aihole temples, while in the final section, the scope of the paper expands to cover the archaeology of the Malaprabha valley.

The Discovery

The architectural ruins of Aihole and Pattadakal were first noticed and photographed in the middle of the nineteenth century by Colonel Briggs. Meadows Taylor and James Fergusson published these photographs, but at this time all that could be said about the temples was that these were pre-tenth century AD (Taylor and Fergusson, 1866). The connection with the dynasty of the Chalukyas had yet to be established. In 1874, James Burgess visited Badami and reports the story narrated to him by the villagers that the Jina image in one of the caves was that of the raja, who had built the tank at the site (Burgess, 1874: 15). He made an estampage of the inscription in cave III and dated the record to AD 578. Even though the caves and the temples at Badami

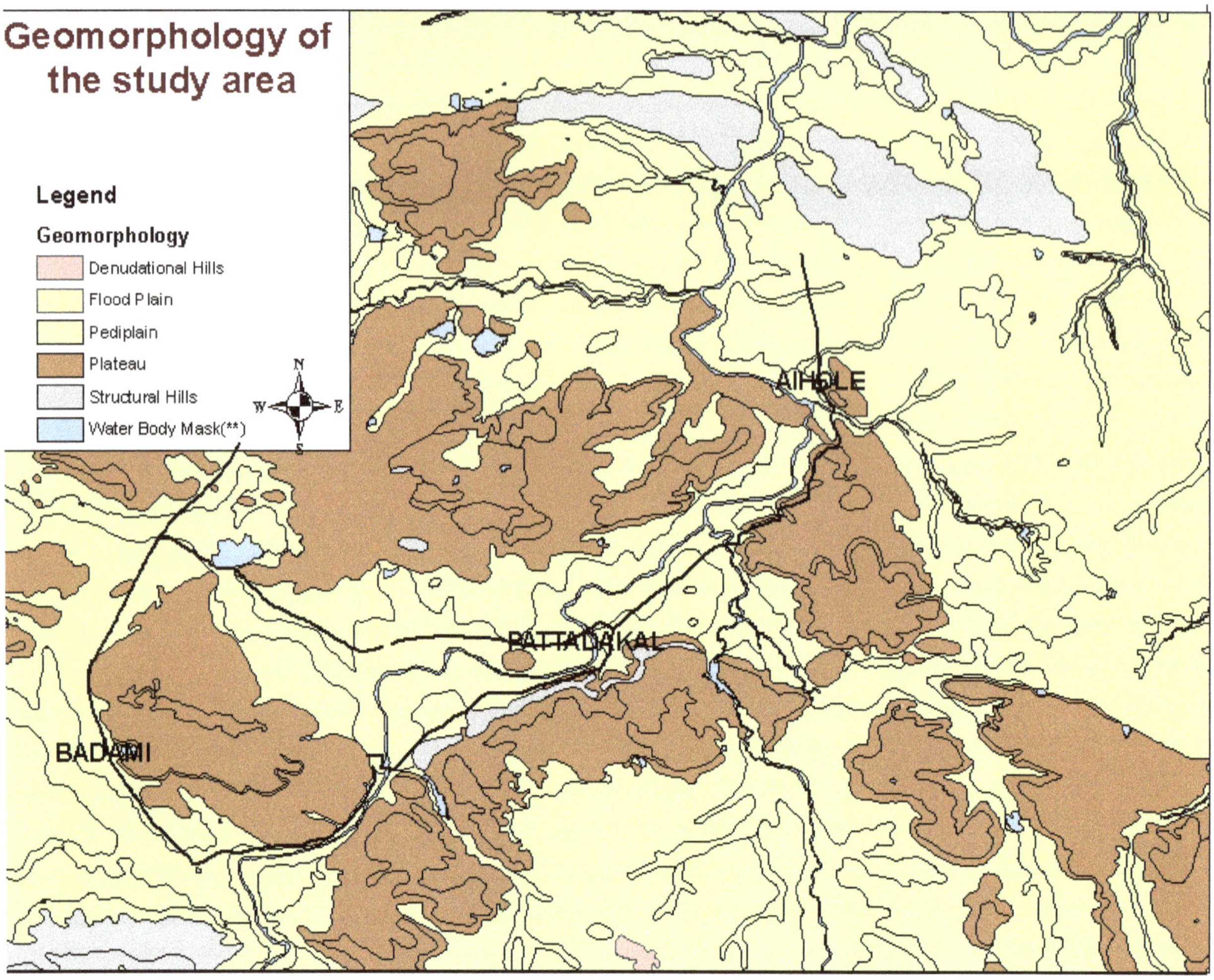

Fig.1. Geomorphology of the study area.

were no longer in worship, Burgess records the annual pilgrimage at Banashankari, a short distance to the south of Badami. A prominent structure at the spot was a tank about 400 feet square and surrounded on all sides except the west by a colonnade or covered walk (Burgess, 1874: 27). At Pattadakal, opposite the village of Katapur, Burgess mentions brisk iron-smelting activity carried out by villagers as well as the continuation of worship in the largest temple at the site.

Burgess also described several megalithic monuments termed 'dolmens' in the region, especially south east of the village of Aihole on a rocky hill, which we now know to be Meguti hill. To the west of Aihole on rocky rising ground on the banks of the Malaprabha he refers to more such structures "among a group of very old temples in the same place is one raised on four unhewn pillars – but this may have been the shelter of some bygone devotee" (Burgess, 1874a: 308). This structure is in the Galaganatha group of temples, which date stylistically to the second quarter of tenth century AD (Meister and Dhaky, 1986: 147). At the time of their discovery, however, the temples had long been abandoned.

The Ancient Monuments Preservation Act was passed in 1904 under Lord Curzon, "to provide for the preservation of ancient monuments, for the exercise of control over traffic in antiquities and over excavation in certain places and for the protection and acquisition in certain cases of ancient monuments and of objects of archaeological, historical or artistic interest" and this formed the basis of interest in the monuments by the Archaeological Survey of India.

Henry Cousens visited Aihole and several other Chalukyan temples in Dharwar district and focussed his discussion largely on the Dravidian style of architecture of these temples following Fergusson's racial-religious categorisation of architecture (Cousens, 1926: 19). Nevertheless Cousens identified numerous features that provided a unitary style to the temples, such as the use of massive blocks of locally available sandstone, "an exuberance of lace-like carving," profusely decorated doorways and the adoption of domical ceilings. He suggested that the size of the temples was governed by the size of the pillar, which was made of the most convenient length of block available in the quarries (Cousens, 1926: 23). The temples were erected with no cementing material and no clamps. Another significant observation related to the sculpted panels and the fact that while the earlier temples carried narratives from the Epics, this practice was given up in the later tenth-eleventh century structures, where the walls were decorated entirely with images of deities (Cousens, 1926: 26).

"Within its (the old village of Aihole's) crumbling walls, which enclose a space little more than five hundred yards across, it has more than thirty ancient stone temples, all jostling one another for room. They are embedded amongst closely-packed dwellings, and are as a rule, desecrated and ruined, many having been converted into dwellings and cowsheds.... Around the village, without the walls, are many more dilapidated shrines, embosomed in prickly-jungle or standing uncared for in outlying fields" (Cousens, 1926: 28).

It is not evident when the temples were abandoned and why or when the present village was settled. We do know that by the fifteenth century the core of the settlement had shifted to Bidar, further north under the Bahmanis and by the seventeenth century the Mughal Emperors had annexed the region.

Ten of the temples at Pattadakal were protected in 1914 and one in 1918 and the others in 1930. Of the monuments at Aihole, 65 were notified in 1914 and one in 1918. At the time they were first declared protected, except for two at Pattadakal the others were in indifferent condition and from 1913-14 to 1946-47 the Archaeological Survey of India spent a sum of Rupees 6,383/- on repair and conservation at Pattadakal and a sum of Rupees 18,109/- for the same purpose at Aihole. These sums were obviously inadequate for the efficient maintenance of all the protected monuments over a period of thirty-three years and another constant problem was the lack of infrastructure at the site (Archaeological Survey of India File no. 13/3/47). An additional problem was that the temples were situated in close proximity to houses and the village of Aihole had almost no infrastructure – a problem that continues well into the present.

More problematic for archaeological purposes was the obliteration of the context of the temples during the process of discovery and conservation. This has meant that there is very little information on the non-religious habitation structures at the site or the history of te site after the twelfth century. Archaeological work conducted by S.R. Rao at Aihole brought many temples to light and revealed the relative chronology of some of the temples. His excavations also revealed earlier structures, such as a brick shrine underlying the stone plinth of the Ambigergudi (Rao, 1972, 1973, *Indian Archaeology – A Review*, 1972-73: 90). Thus a major issue relates to unearthing the context of the temples and it is here that satellite images provide useful details.

The Temples at Aihole: Defining the Context

Aihole located at a bend in the river Malaprabha has perhaps the largest number of temples but there is no evidence of any royal inscription nor of any *Mahatmya* linking the site to the *Puranas*. Unlike Alampur and Pattadakal, these temples do not form a cohesive complex, but are instead scattered in a four-square kilometre area (Figure 2). There is enormous diversity in the plans of the temples with the rock-cut caves coexisting with the ornately sculpted Durga temple. Aihole has a marked Jaina presence and the inscriptional record indicates patronage provided by merchants and guilds, the best known being the Ayyavole guild whose scope of operations included both South and Southeast Asia.

The temples at Aihole are situated along a north-south axis on the banks of the Malaprabha with Meguti hill forming one end of the axis, past the caves and temples at Ravalaphadi to Tarabasappa at the northern end. Among the three or four small shrines in front of Ravalaphadi cave, the oldest is the west facing structure to the south (Mesiter and Dhaky, 1986: 22). In the centre of the ancient town of Aihole stand four temples: Gaudargudi or Lad Khan to the north-west; Suryanarayana to the north; an ancient tank to the south; and Cakragudi close to the southern edge of the tank (Mesiter and Dhaky, 1986: 42). The presence of water-bodies close to Mallikarjuna temple at the foot of Meguti hill, Tarabassappa and Huccimalliguddi is significant and is in keeping with the close association between religious architecture, both Buddhist and Hindu, and the community as evident from the archaeological record of shared spaces at a range of other sites (Ray, 2004: 350-375).

The beginnings of temple construction may be traced to Meguti hill and its extension, in which the Ravalaphadi caves are carved (Figure 3). Another early locale was in the centre of the present village at Lad Khan where wells were dug for the purpose of providing a source of water. The natural source, viz. the Malaprabha river does not appear to have been a favoured early core area as only one small temple in the Ramalinga group near the Malaprabha river, viz. the Puspabhadra temple to the south of the village of Aihole dedicated to the goddess Sri in the form of a lotus and oriented to the west to face the river is dated to the first phase of construction (Mesiter and Dhaky, 1986: 64).

A characteristic feature of early religious architecture at Aihole is its diversity in terms of religious affiliation and this is evident both from the earliest constructions on Meguti hill. The Jaina cave at Meguti hill at Aihole faces southwest and comprises of a *garbhagrha*, two side chambers, a common hall and a rectangular verandah. Another two-storey Jaina shrine is located on top of the hill and bears an inscription in Sanskrit on stone on the outside east wall. It mentions the kings Jayasimha I, Ranaraga, Pulakesi I, Kirttivarma and

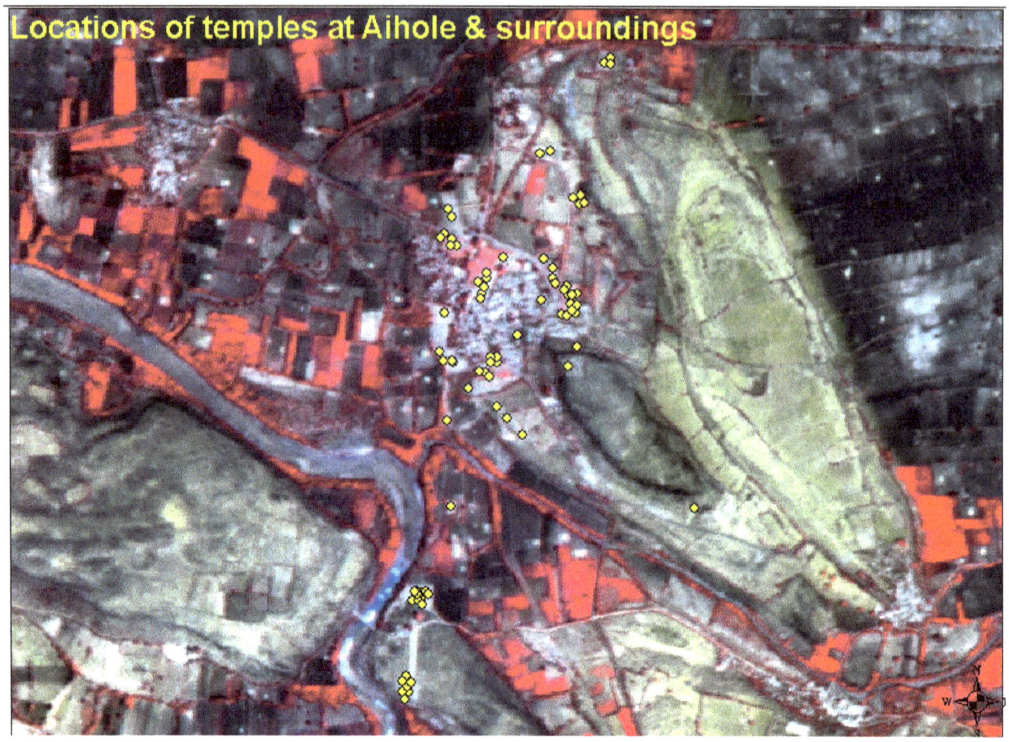

Fig. 2. Locations of temples at Aihole.

Fig. 3. Ravalaphadi cave at Aihole.

Mangalisa (whose horses were picketed on the shores of the oceans of the east and the west – *pūrva-paścima-samudratatositāśvah*) and Pulakesi II. The objective is to record the erection of a stone temple of Jinendra by a certain Ravikirtti, the composer of the eulogy who has attained the fame of Kalidasa and Bhairavi by his poetry, during the reign of Pulakesi II in Saka year 507 (AD 585-6) equal to the year of the *Mahabharata* 3731 (Fleet 1876: 67-76). The upper structure of the temple bears two inscriptions in Kannada, which read Sri Rajakomara and Sri Bai (Rajasekhara, 1985: 119). The temple walls are plain and there is an absence of decoration on the pillars and door-frames as well.

On the northern slope of Meguti hill is a Buddhist vihara, which is part rock-cut and partly structural. It has an upper storey and depicted on the ceiling of the latter is a seated figure of Buddha, while scenes from the life of the Buddha are sculpted on the door-jambs of the cave. A short seventh century inscription on one of the pillars of the ground-floor mentions mahendra of the Pinthavadi (Pindapati) school and a disciple of Sthavira Ananda (Mesiter and Dhaky, 1986: 22). The inscription from Gaudargudi temple dedicated to Durga Bhagavati has been referred to earlier, thus making it one of the temples affiliated to the goddess. The affiliation of other temples at Aihole is varied. The Durga temple is among the largest Chalukya buildings and the most finished of all the Chalukya temples. The temple has a donative inscription of the time of Vikramadutya II (AD 733-745), which records a gift to a sun temple founded by a kumara, possibly a revenue officer. On the north face an engraved word "Jinālayan" in seventh-eighth century characters denotes a personal name, perhaps of an architect and does not denote a Jain sanctuary as once supposed (Meister and Dhaky, 1986: 49-52).

Several phases of temple construction have been identified at Aihole starting from the late sixth century, which continued until the twelfth century with several renovations and additions. In the Durga temple, for example, a step-well and a shrine were added in the later Kalyani Chalukya period (AD 973-1198), while Konti-guddi group received additional structures in the

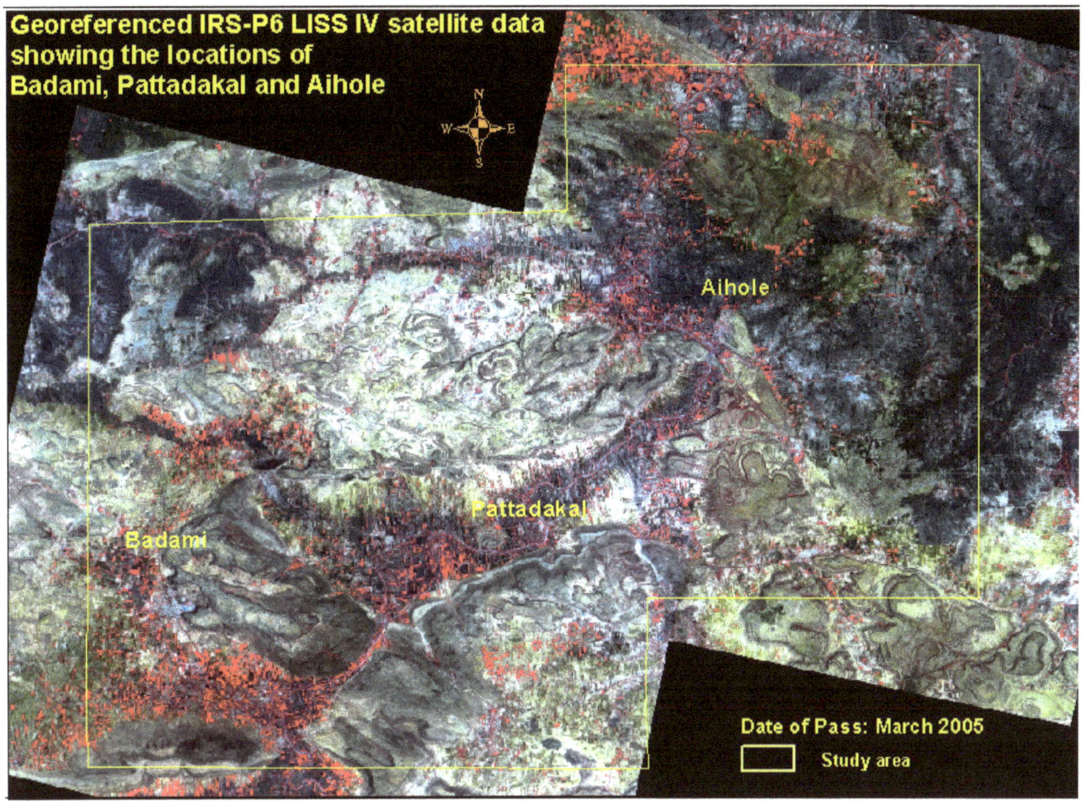

Fig. 4. The Malaprabha valley from Aihole to Badami.

Rashtrakuta period (AD 757-973). No stone temples were, however, created at Aihole after the fall of the Kalyani Chalukyas at the end of the twelfth century. Thus there is a concentration of early shrines in the Malaprabha valley. How is this concentration to be understood?

The Malaprabha Valley: The Macro Region

The Malaprabha is characterised by broad floodplains covered with black cotton soil (Figure 4). Sandstone outcrops and rolling sandstone hills overlying Precambrian metamorphic rocks and covered with thin soil cover provided raw materials for stone architecture and limited dry farming. Comprehensive archaeological investigation carried out along the 160 kilometre stretch of the Malaprabha valley since the 1950s resulted in the discovery of a large number of Palaeolithic sites, as well as an ancient lake in the Badami group of rocks (Korisettar and Petraglia, 1993: 44-46). Lower Palaeolithic tools, such as handaxes of quartzite were collected near Rachi Gudi temple at Aihole (*Indian Archaeology – A Review 1975-76*: 21). Extensive outcrops of laterite have been observed in an area of about 15 to 20 sq km around Badami town, while finds of Lower Palaeolithic tools are associated with alluvial formations.

These rich Acheulian sites around Badami represent open-air occupation-cum-factory sites. Rock paintings and Mesolithic occupation is reported from the Siddhalaphadi rock shelters near the town, which show evidence of prolonged use as indicated by finds of microliths, pottery, graffiti, and inscriptions on wall in white over red background dated to fifth-sixth century AD (*Indian Archaeology – A Review1955-56*: 5; *1980-81*: 27-29).

There is evidence for division of space between human activity, such as places for production of tools and what may be termed ritual space, such as the use of sandstone ridges and shelters for prehistoric rock paintings. The hill at Hiregudda, six km west of Aihole is rich in rock paintings and hundreds of petroglyphs are to be found along a dolerite cliff. Amidst the petroglyphs at this site are also round, polished grooves that emit musical ringing tones when struck with granite stones, as demonstrated to the archaeologists by a local informant, who referred to the inscribed boulders as 'musical stones' (Boivin, 204).

Megalithic monuments constructed of large slabs of sandstone often with sepulchral associations are conspicuous features of the landscape of the Malaprabha valley (Figure 5). About 224 localities with megalithic remains are known in north Karnataka (Sundara, 1975: 7) and within this a prominent group comprises of the megaliths at Aihole, Badami, Pattadakal and Guledgudda, about 25 kilometres north-east of Badami. At Guledgudda, Taylor had reported a cairn and three cromlechs near a tank to the south of the village when he visited the site in the nineteenth century (Taylor 1941). These were no longer present in the 1970s, but Sundara did report six cairns (Sundara, 1975:30).

Fig. 5. Megaliths on Meguti hill.

Another early site is that of Aihole. About forty port-holed dolmens are found on a sandstone hillock extending from the Jaina Meguti temple to the rock-cut Jaina cave at Aihole and this belt continues towards another village of Kyaddigeri about 1.50 kilometres south-east of Aihole. At Kyaddigeri about thirty-five passage chamber tombs are scattered in an area of about ten to fifteen acres. Another sixty passage chambers and cairn stone circles are located in the revenue limits of the village Chilapur, only a few metres from Kyaddigeri (Sundara, 1975: 31). Although based on few radiocarbon dates, this group of megaliths gives a date range of 800- 600 b.c. This is an uncalibrated C-14 date from wood charcoal samples within the Aihole group (Sundara 1975: 215). The period between the second-first centuries BC and fourth century AD was a period of creativity and experimentation as far as religious architecture was concerned. It was at this time that the concept of the superstructure evolved and can be traced through several formative stages. Recent archaeological research has shown that the Malaprabha valley formed a part of the larger settlement network in the Deccan in the early centuries AD. The site of Cotta Chandor in Chandor district about 300 kilometres from Aihole towards the west coast was subjected to excavation for two filed seasons in 2002-2004. The complete plan of a brick temple complex datable from third to eleventh century AD was unearthed and five phases of structural activity were identified. In the earliest Phase I the temple was cruciform in plan and this marks it out as perhaps one of the earliest structures to adopt this plan of construction. Though three phases of construction were identified, these were marked by continuity of religious beliefs and in the last phase the sculpture of Nandi was added to the temple complex (Kennet and Rao, 2001: 97-107). Other sites with religious architecture are also known from the neighbouring districts. Vadgaon-Madhavpur is a forty-hectare site in the suburb of Belgaum, which was discovered and excavated in 1945 by the Kannada Research Institute, Dharwad, though the report has yet to be published and again from 1972 to 1977. A hexagonal stone pillar from the site carries an inscription dated to first century BC and records the performance of Vajapeya and other sacrifices by a Brahmana of the Kasyapa gotra. Based on an analysis of ceramics from the site, three phases of occupation have been identified dating from third century BC to first century AD. The site has yielded a rich hoard of coins and pottery, but relevant for this paper are the find of two brick structures, one apsidal and the other consisting of a spacious hall, which have been identified as religious in nature (Ghosh, 1989: 456). The University of Mysore in collaboration with the State Department of Archaeology excavated Banavasi near Sirsi in district North Kanara in the 1970s and unearthed an apsidal brick temple of the Satavahana period (*Indian Archaeology – a Review* 1970-1: 29). A number of mounds spread across one square kilometre area along the river Varada were explored and these yielded ceramics, such as Rouletted Ware and Russet-coated Painted Ware, in addition to providing information on brick structures. A square inscribed pillar bore the names of Kakusthavarman, Santivarman and Mrgesavarman of the Kadamba dynasty and their victory over the Pallavas. Thus it is evident that temple-building in the region dates to the early centuries of the Christian Era, but it is around the middle of the first millennium AD that the temples define the cultural identity of a region such as the Malaprabha valley. This identity included the demarcation of a territorial boundary within the enclosed terrain of the valley, adoption of a new mode of worship in the temple, the use of the Kannada language on coins and inscriptions and assumption of a sculptural programme based on the Epics and the Puranas. Scholars have studied the architecture and iconography of these temples since the nineteenth century. The attempt in this paper has been to locate the religious architecture at Aihole within its cultural and geo-physical environment. It is evident that fertile agrarian stretches were avoided for temple construction. Instead the shrines were built in upland areas such as the Meguti hill or along its base. Expansion over five hundred years resulted in clustering of the temples within the Aihole area, rather than spatial expansion into the valley, indicating perhaps increased value attributed to the land. It is only when studied in the wider context of the landscape that religious architecture provides indications of the changing needs of the community. Religion here is accepted as the structuring principle for the lives of past societies and is seen as influencing the constructed environment within which communities lived and functioned (Insoll, 2004: 22). This aspect of archaeology has been generally neglected, but the lacuna needs to be redressed. This paper is a modest attempt in this direction. It has not been possible to place the methodology adopted here within a historiographic context, though this is an exercise that has been undertaken elsewhere (Ray, 2009).

Acknowledgement

This paper is the result of collaborative work with Sudha Ravindranath of the Regional Remote Sensing Service Centre Bangalore. Two adjacent scenes of satellite images IRS P6 LISS IV of March 2005 covering 23 kilometres x 23 kilometres of each scene have been used for preparing the layout of the derived maps. Multi-spectral data of 5.8 metres spatial resolution provides the base map of the region, while panchromatic data of 2.5 metres spatial resolution allows location of the temples at Aihole. Additional information on the temples has been geo-referenced and a series of maps generated that indicate the spatial context of the variables used. The author gratefully acknowledges a grant from the Jawaharlal Nehru University under the UPOE (University with Potential of Excellence) Scheme that enabled fieldwork and purchase of satellite images.

Chronological List of Temple complexes at Aihole

Name of temple	Date	Nature of plan	Data from Inscription – if available
Jaina cave	During regency of Mangalesa		
Ravalaphadi	During regency of Mangalesa in reign of Pulakesi II	Contemporary structures: column in front of Ravalahadi; Banantigudi shrine Mahakuta; Hire Mahakutesvara near Mahakuta; column near Mahakutesvara of 601/2	
Meguti	AD 634	Garbhagrha & upper shrine, pradaksinapatha, ardhamandapa & pillared porch	Long Sanskrit inscription on east wall mentions that it was dedicated to Jinedra and founded in 634/5 by Ravikirtti in the time of Pulakesi II
Buddhist vihara	Palaeo-graphically of Pulakesi II		Inscription on pillar "Mahendra" and some other names
Gaudargudi	Pulakesi II	Has a rectangular pillared hall within which a sanctum and pradaksinapatha are located	May have been dedicated to Durga as inscription on entrance lintel refers to donation by citizens of Aryapura to Durga Bhagavati
Sarangi matha (no.4 in Kuntigudi group)	Stylistically in 7th century		Graffiti in garbhagrha "Sri Surama-nama"
Cikkiguddi	Stylistically in 7th cent.	Garbhagrha, pradaksinapatha, pillared hall, porch	
Lad Khan	In reign of Vinayaditya (Meister & Dhaky 1986: 6)	Garbhagrha 2.58 x 2.46 metres with an open porch, upper chamber	Inscription of Rashtrakuta period refers to the temple as ghatikālaya (assembly hall) for the town elders.
Surya-narayana, north of Lad Khan			
Tank south of Surya-narayan			
Cakragudi, south of tank			
Durga temple	In reign of Vinay-aditya (Meister & Dhaky 1986: 6)	Apsidal garbhagrha, pradaksinapatha, pillared hall, porch and outer apsidal pradaksinapatha. Gateway on southern side.	"Jinālayan" on the north kapota face denotes a personal name, perhaps an architect. Does not refer to Jaina structure

Puspabhadra (in Ramalinga temple group)	In reign of Vinay-aditya (Meister & Dhaky 1986: 6)	Flat roofed shrine with four-pillared porch	
Sulesvara pratoli, Sulebhavi (five and a half miles east of Aihole to the southeast of the village of Sulebhavi)	AD 705 Local legend tank and pratoli built by a prostitute (Sule in Kannada)		
Rama lingesvara temple, Hungund (12 miles east and north of Aihole past Amingad, Bijapur district)			Inscription on temple of 11th century
Ramalingesvara group, Kelur (3 miles south of Aihole)		Five small east-facing temples with a huge tank to north. Possibly low budget village sponsorship during the reign of Kirttivarma II (AD 750)	
Huccappayya-matha		Square garbhagrha with huge linga & rectangular hall with pillars.	
Huccappayyagud i		Garbhagrha, pillared hall and porch	On right pilaster in porch "svasti Jambūdvīpāntarekaści[d] vāstu prāsāda tadgatah Narasobba samo vidvān na bhūto na bhavisyati" (IA 9: 74)
Meena basadi			On rock to the right of cave inscription mentions Narasoba, disciple of Binjadi, well versed in sastras and in the preparation of vimanas and sculptures (Rajasekhara, 1985: 28)
Huccimalli II		Garbhagrha, antarala & porch	
Surya temple		Garbhagrha with standing Surya image (later Calukyas), pillared hall & porch	
Ambigergudi	Late 10th century		
Nadargudi	10th century		
Racigudi	10th century		
Jaina temples	10th century		
Galaganatha group	Late 9th century		
Kuntiguddi I	Stylistically third decade of 8th century	Pillared hall & garbhagrha	
Kuntiguddi II	Rashtrakuta period	Square hall with garbhagrha (probably added later as in Lad Khan, Rajasekhara, 1985: 146). Sikhara typical Rashtrakuta vimana.	
Tārappa temple		Garbhagrha, sabhamandapa and porch. Nandi image. Well.	
Jyotirlinga group		Nine temples & pillared hall	

		to northeast of Durga temple	
Jyotirlinga V		Square Garbhagrha & porch; nandimandapa	
Jyotirlinga VI		Plinth of garbhagrha and steps; nandimandapa formed by four pillars	
Jyotirlinga VIII		Single-celled garbhagrha	
Hallibasappa temple	Last quarter of 9th century		
Aihole 52 & 53 Galaganatha group	Second quarter of 10th century		

References

ARCHAEOLOGICAL SURVEY OF INDIA FILE NO. 13/3/47: Question in the Legislative Assembly regarding ancient temples at Aihole and Pattadkal. Starred question no. 531 by Shri DP Karmakar.

BOIVIN, NICOLE, 2004, Rock art and rock music: Petroglyphs of the south Indian Neolithic, *Antiquity* vol.78, no. 299, March.

BURGESS, JAMES. 1874. *Report of the First Season's Operations in the Belgam and Kaladgi districts*, New Delhi, Archaeological Survey of India.

BURGESS, JAMES. 1874A. The Dolmens at Konur and Aihole, *Indian Antiquary* 3, 308.

COUSENS, H. 1926. *Chalukyan Architecture of the Kanarese Districts*, Archaeological Survey of India, New Imperial Series XLII.

FLEET, J. F. 1876. Sanskrit and Old Canarese Inscriptions, *Indian Antiquary*, V, no. XIII, 67-76.

GHOSH A. EDITED. 1989. *Encyclopaedia of Indian Archaeology*, volume II, New Delhi, Munshiram Manoharlal.

INSOLL, TIMOTHY. 2004. *Archaeology, Ritual, Religion*. London and New York, Routledge.

KENNET D. AND RAO, J.V.P. 201. The Early Historic Brick temples at Chandor, *South Asian Studies*, XVII, 97-107.

KORISETTAR, RAVI AND PETRAGLIA, MICHAEL. 1993. Explorations in the Malaprabha valley, Karnataka, *Man and Environment*, XVIII, no. 1, 43-8.

MEISTER MICHAEL W. & DHAKY M.A. EDITED. 1986. *Encyclopaedia of Indian Temple Architecture*, South India: Upper Dravidadesa, Delhi, Oxford University Press & American Institute of Indian Studies.

RAJASEKHARA, S. 1985. *Early Chalukya Art at Aihole*, New Delhi, Vikas Publishing House.

RAO, S.R. 1972. A Note on the Chronology of Early Chalukyan Temples, Lalit Kala 15, 9-18.

RAO, S.R. 1973. Recent Discoveries in Aihole and Pattadakal, *Srikanthika*, Mysore, 36-9.

RAY, HIMANSHU PRABHA. 2009. The Shrine in Early Hinduism: The Changing Sacred Landscape, *The Oxford Journal of Hindu Studies*, 2, 76-96.

RAY, HIMANSHU PRABHA. 2007. The Archaeology of Ritual Spaces: Satellite Images and Early Chalukyan Temples, *Man and Environment*, XXXII (1), 89-101.

RAY, HIMANSHU PRABHA. 2004. The Archaeology of Sacred Space: Introduction, Himanshu Prabha Ray & Carla Sinopoli edited, *Archaeology as History in Early South Asia*, New Delhi, Indian Council for Historical Research & Aryan Books International, 350-75.

SUNDARA, A. 1975. *Early Chamber Tombs of South India*, Delhi, University Publishers.

TAYLOR MEADOWS AND FERGUSSON, JAMES. 1866. *Architecture in Dharwar and Mysore*, Madras, Asian Educational Services, 1989 [reprint].

TAYLOR, P. MEADOWS. 1941. Megalithic Tombs and other Ancient Remains in the Deccan, republished by the Department of Archaeology, Hyderabad.

Applications of archaeological remote sensing in Cambodia: an overview of Angkor and beyond

Damian Evans

Postdoctoral Fellow, Department of Archaeology, University of Sydney, Australia

Abstract

Early Cambodian settlement sites provide excellent opportunities for aerial archaeology. Settlements were highly and consistently structured and can be identified from the air from their spatial patterning. Furthermore, remains of occupation and infrastructure can generally be found on the surface of the landscape, or are near enough to the surface to be identified by surface characteristics. After two decades of intensive and systematic remote sensing research at Cambodian archaeological sites, the contribution of aerial archaeology has been substantial. Not only has an enormous amount of new information been generated, but applications of remote sensing have provided the foundation for major multidisciplinary projects and have addressed important historical issues and debates. This paper summarises applications of remote sensing in archaeology at the site of Angkor undertaken by members of the Greater Angkor Project, and provides preliminary results for a range of ongoing studies at other early settlement sites in Cambodia including Banteay Chmar, Koh Ker, Preah Khan of Kompong Svay, Beng Melea and Sambor Prei Kuk.

Keywords Angkor, Cambodia, landscape archaeology, hydraulic systems, urbanism, remote sensing

Introduction

For at least a hundred and fifty years, the great temple sites of Southeast Asia have been studied in considerable detail, first by colonial archaeological agencies and more recently by a variety of modern archaeological projects. Most of this work has been devoted to the urgent task of preserving and restoring the monuments, and at the World Heritage site of Angkor, for example, the majority of the numerous projects underway are still directly related to the conservation of monumental architecture. This longstanding focus on temple architecture in Southeast Asia, though obviously important, has led to a situation where the understanding of these great temple sites as urban environments – as a place where people actually lived – is superficial at best. A colonial emphasis on formal systems of belief and the material context of 'ritual' behaviour has persisted until the present day, and has led to an empirical void outside of the largest of the Hindu and Buddhist structures. This is particularly true of the great monument complexes of Cambodia, where decades of civil strife have largely prevented the systematic application of modern, large-scale archaeological survey techniques, developed in the 1950s and 1960s, that have been used to great effect in other parts of the world (see e.g. Billman 1999, 144-5).

Over the last twenty years, with the end of decades of conflict and the re-opening of Cambodia to archaeological teams, an enormous amount of new information on Angkor and the Khmer Empire has come to light. A series of archaeological projects, using techniques as diverse as traditional excavations and cutting-edge remote sensing technologies, have transformed our understanding of Angkor, in Cambodia, as an urban space. In broadening the traditional focus of archaeological research beyond the temples, archaeologists have mapped thousands of square kilometres of the urban landscape; acquired palynological data to enable the reconstruction of the medieval environment; conducted excavations which shed light on the everyday life of the people who built the massive sandstone monuments; and traced enormous water management systems surrounding the temple complexes. A range of different platforms and sensors, including declassified Corona spy satellite imagery from the 1960s, high resolution imagery from Ikonos, Quickbird and OrbView-3 satellites, radar instruments such as AIRSAR and TerraSAR-X, conventional aerial photography, and the recently launched GeoEye-1 satellite, have been critically important components of this research. In addition, many traditional data sources such as historic maps, survey data and inscriptional evidence have been digitised and geographically referenced against the base derived from remote sensing.

This paper will provide an overview of that work and examine some of the implications for longstanding debates about Angkor, including the nature of the hydraulic system and the role it played in the demise of the settlement complex at some time in the middle of the second millennium AD. Research at Angkor is ongoing, and is now being complemented by a new study of diverse sites across Cambodia that have previously been understood simply as temple complexes rather than urban environments (Figure 1). Surveys at the sites of Banteay Chmar, Koh Ker and other locations over the previous three years have resulted in a range of interesting results and discoveries, while major projects are now underway at sites such as Preah Khan of Kompong Svay, and planned for Beng Melea and Sambor

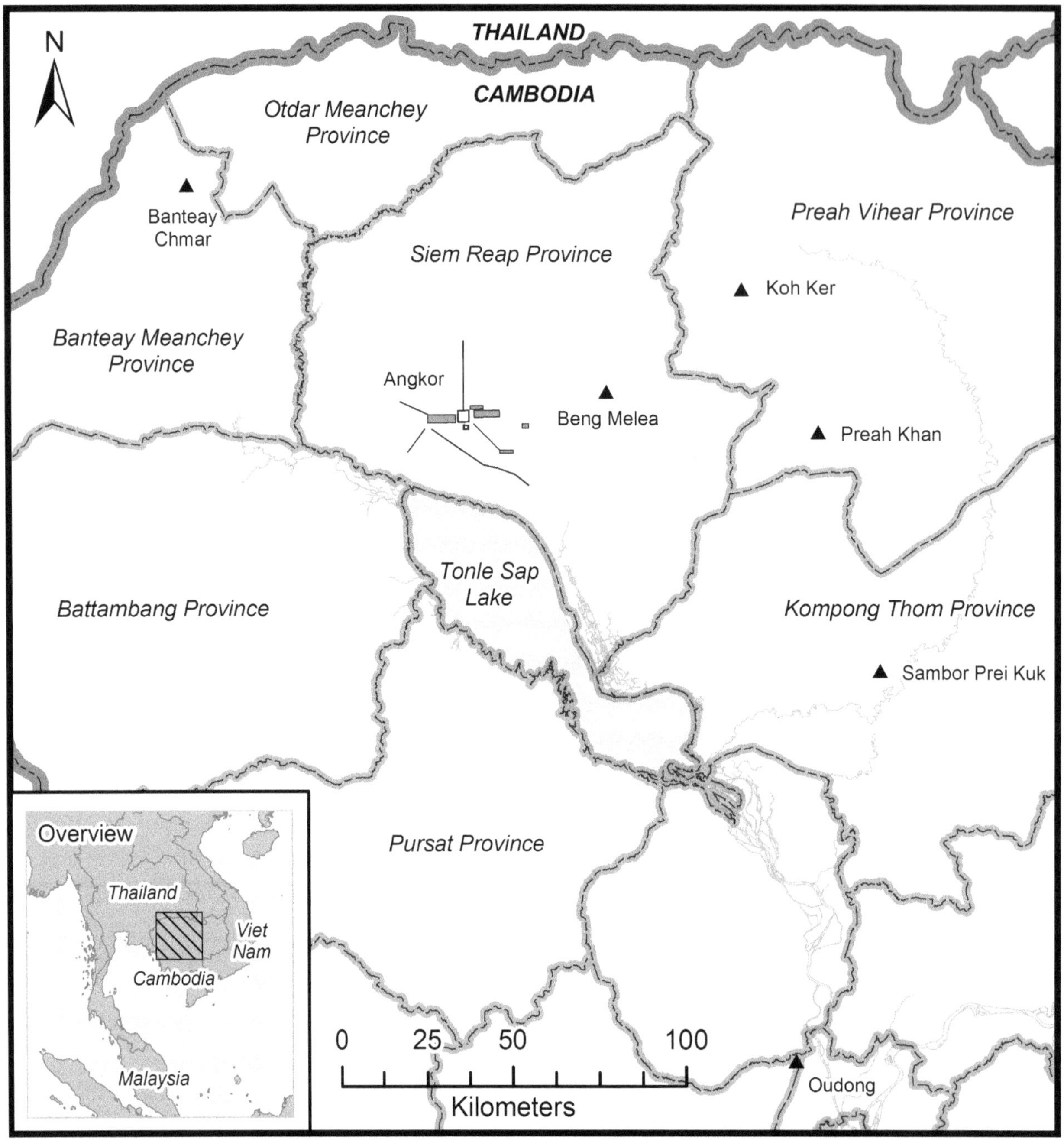

Fig. 1. Early Khmer temple complexes and cities mentioned in this paper.

Prei Kuk. Looking beyond Angkor, this paper will outline research in progress as well as future potential for archaeological remote sensing in this region, in particular using LIDAR and high-resolution spaceborne SAR instruments, and provide some preliminary observations about the nature of early urbanism in Cambodia.

Background

Theorists have long recognised the potential global significance of Khmer archaeological sites, and theroblem of the collapse of Angkor has engaged scholars since at least the 1970s (Groslier 1979). Theories of Angkor's decline, however, have been difficult to evaluate given the dearth of evidence about the settlement context of the monuments. In recent years, research undertaken by members of the Greater Angkor Project (GAP), a multidisciplinary and multinational (French, Australian and Cambodian) research program, has largely remedied this situation at Angkor itself. Using a variety of technologies including radar remote sensing and geographic information systems (GIS), archaeological maps have been produced which, for the first time, provide a holistic picture of the settlement (Evans 2007; Pottier 1999). The maps reveal the full extent to which religious institutions were connected with the great water management networks, and the role that this infrastructure played in the development of residential space, of agriculture, of networks of transportation and communication; the new maps also indicate the extent of the impact of this built environment on the local ecology (Evans 2007; Fletcher *et al* 2004). As a result of this work Angkor has been confirmed as a site of global historical significance, perhaps the largest urban complex of the pre-industrial era (Evans *et al* 2007). Yet Angkor is not the only great 'engineered landscape' of the medieval Khmer world: across Cambodia, antecedent and contemporary settlements remain largely unstudied by major archaeological projects, in spite of compelling evidence that similar systems may have functioned at some of those sites as well.

Because Angkor is a vast palimpsest of different settlements with a poorly-defined boundary, the archaeological record there is extremely complex and difficult to interpret; many of the secondary centers, on the other hand, are well-defined, single-period settlements where historical and cultural processes can be identified with greater clarity. Further disentangling the palimpsest at Angkor remains a core objective, but by leveraging the work completed at Angkor into a truly regional-scale, comparative study, further research will uncover crucial new data and provide the first real opportunity to address these important questions, which cannot be resolved from a study of Angkor alone. This is the focus of an important new research program funded by the Australian Research Council for 2008-2010, and based at the University of Sydney.

The project is concentrating most resources on the study of the three largest temple complexes in Cambodia (Jacques & Lafond 2007) : Sambor Prei Kuk (Isanapura), the great pre-Angkorian capital in Kompong Thom province which flourished in the middle of the first millennium AD; Angkor, the capital of the classical Khmer empire and the site of continuous, temple-centred urban development up until the middle of the second millennium AD; and Koh Ker, a temple complex often considered to be a 'temporary' capital of the Angkorian state, in an interlude from Angkor, in the 10th century AD. In addition to the three major temple complexes, a number of secondary sites are also being mapped and studied in detail: Beng Melea, a satellite temple complex to the east of Angkor in which the principal monument dates from the reign of Suryavarman II (AD 1113-1150); Banteay Chmar, a complex in the far north-west of Cambodia in which the principal monument dates from the reign of Jayavarman VII (AD 1181-*c*.1219); and Preah Khan of Kompong Svay, a site in which the major elements date from the reigns of both Suryavarman II and Jayavarman VII. These particular sites have been chosen in order to provide a spatially, ecologically and temporally diverse set of data that will lay the appropriate foundation for a comparative analysis. Several of them have markedly different historical trajectories: Beng Melea, for example, shows evidence of several centuries of continuous occupation prior to the reign of Suryavarman II and the construction of its 'ceremonial centre', while the site of Koh Ker is thought to have had a very brief lifespan as the seat of Khmer power, lasting no more than two decades.

There are substantial differences in the amount of attention each of these sites has received from archaeologists, and in the adequacy of their current archaeological maps: the pre-Angkorian capital of Sambor Prei Kuk, for example, shows strong evidence of a water management system that rivals the scope and scale of Angkor's (relative to the size of the settlement and the number of temples) and its map has developed somewhat in recent years due to research by Waseda University, although that remains unpublished; at the Angkor-era capital of Koh Ker, however, no such system has been revealed, and the latest published maps of the site (Jacques & Lafond 2007, 108; Ministère de la culture et des beaux-arts & École française d'Extrême-Orient 2007) are essentially identical to the map published seventy years ago by Parmentier (1939, 16). Continuing research should by no means be limited to the capitals: analysing the spatial organisation of the secondary centres will provide outcomes that are equally significant to analyses of the great cities. Pottier's recent (2004) aerial investigation of Banteay Chmar is a case in point: although it has not produced a map that is comparable in detail to that of Angkor, the study confirms various early hypotheses of George Groslier's regarding the existence of an urban context to the monuments and of a water management system at that site. The differences in archaeological knowledge between the sites are often starkly apparent: at Angkor, for example, mapping work by GAP has provided archaeological maps of 1:10,000 scale over several thousand square kilometres (Evans 2007), while maps of the highly significant site of Preah Khan of Kompong Svay show little more than the massive sandstone temple at its centre, along with its outer wall and reservoir (Jacques & Lafond 2007, 180; Ministère de la culture et des beaux-arts & École française d'Extrême-Orient 2007). The mapping component of the project described here addresses these disparities in knowledge, and will result in a new archaeological map for every site that, for the first time, will permit rigorous comparative studies of Khmer settlements.

Methodology

The mapping process is now underway at each site except Angkor, for which an adequate archaeological map is available from GAP. At all other sites, the final study area for mapping will, ideally, be defined by a geographical characteristic that is appropriate to an inquiry on water management systems: the boundaries of the catchment(s) which feed the water resources available to a particular ceremonial centre. At each site, the settlement is narrowly defined in the first instance by the spatial distribution of the known major monuments. A preliminary analysis of large-scale remote sensing datasets such as Landsat and ASTER, as well as point-based maps of temple locations is first undertaken to determine the spatial distributions of secondary sites and the broader limits of habitation in a given area. Hydrological datasets and elevation data are then analysed in a GIS in order to model the boundaries of the catchment for the river systems in the area, using essentially the same methodology used to define water catchment areas (and thus an appropriate study area) for Angkor (Evans 2007; Kummu 2003). A systematic 1km grid square analysis of all of the available remote sensing data is then being undertaken for each site, following the methodology established by Evans (2007), using in particular stereo pairs of aerial photographs. Digitisation of archaeological features takes place over the top of a precise geographic base in a GIS, at a nominal scale of approximately 1:10,000. The features mapped include occupation mounds, temple sites (both moats and mounds), village ponds (banks and water), agricultural field systems, roads, canals (embankments and areas of flow), and other features of interest, with comments and uncertainty about interpretations captured within the database.

All interpretation is conducted 'manually' through subjective visual interpretation by an experienced archaeologist/analyst. A number of experiments have been conducted in relation to automated feature recognition routines, but all have proven problematic to some extent. Conventional image classification routines are generally unsuited to the identification of relatively small, discrete artefacts in the archaeological landscape. Temples, for instance, are found in a range of different natural environments with different land use and land cover regimes (both past and present), and the classic spatial structure of their mound, moat and reservoir can be identified by any number of spatial or spectral characteristics across different datasets (differential elevation between the moat and mound, differences in crop growth, differences in soil moisture content, canopy height, vegetation density, patterning of modern residence over an ancient site, and so on). Such variability defies straightforward classification. Attempts at linear feature recognition have proven more effective at identifying modern roads and field systems than fragmentary, indistinct Angkorian canal or road features, and provide no effective solution to the perennial problem of differentiating the two. Although there are basic rules to do with the spatial patterning of temple sites – the existence of a rectangular pond to the east is one example – a significant number of exceptions to this rule further complicates attempts at object-oriented feature recognition and the development of logical, hierarchical structures to describe surface archaeology. Creation of effective models for automatically identifying features is therefore incredibly difficult and time-consuming; this, in addition to the intensive quality control process that would be required, and the fact that models developed for one site may not be generally applicable to other sites, is considered reason enough to abandon attempts to automate the mapping process. Instead, data products which highlight particular aspects of archaeological significance – variation in crop growth, for instance – are generated in order to assist the interpretation by a skilled analyst. These vary depending on the availability of datasets for a given site, and the particular characteristics which make a given sensor useful for archaeological purposes in a given environment.

In order to facilitate a temporal GIS analysis, known inventories of inscriptions and monuments are digitised and carefully related to the features being mapped (Lustig, Evans & Richards 2007). Particular attention is paid to the capture of surface stratigraphic data that might define temporal and/or functional relationships between features, noting for example that rice field systems often have exactly the same orientation as the temples to which they belonged (Pottier 2000b) and looking specifically for outlets from reservoirs and temple moats into canals, and clusters of occupation mounds around temple sites. Once captured in a GIS, these spatial, temporal and functional relationships can be used to model the development and the interrelatedness of the various components of the settlement structure (the agricultural, the residential and the politico-religious) over time (Evans 2002).

The fieldwork component of the project consists almost entirely of verifying the mapped features on the ground, an obviously important part of any archaeological application of remote sensing. A particular emphasis is placed on longitudinal studies – multiple annual seasons of field verification, carried out over a period of some years – in order to maximize the potential recognition of features in landscapes that alter significantly between seasons (e.g. the deciduous forests of Koh Ker) or that are subject to rapid change (e.g. the newly urbanized areas of the Angkor plain). However, several of these sites are in remote and forested areas replete with landmines, presenting substantial difficulties for complete and systematic ground verification, and remote sensing therefore continues to play a critically important role in their exploration and mapping.

Angkor

In the 12th and 13th centuries AD, Angkor sprawled over an area of over 1000km2 and may have carried a population approaching 750,000 (Fletcher *et al* 2003, 116-7). A vast water management system connected the monumental religious complex in the centre of the Angkor plain to the Kulen Hills in the north and to the great lake, the Tonle Sap, in the south. For reasons that remain unclear, the settlement then entered a period of decline. By the later 16th century AD, the hydraulic system fell into disuse, most of the religious monuments were left to the jungle, and the population diminished to insignificance. Conventional histories tend to locate the settlement's demise in the political or religious developments of the time. Many of these explanations, however, are built on very tenuous empirical foundations (see Fletcher 2000-2001, 4; Vickery 1977), and not one of them is able to explain the magnitude and the finality of Angkor's fall, nor the collapse of the settlements which preceded it (such as Sambor Prei Kuk) or were contemporaneous with it (such as Koh Ker). Throughout the 1990s, Pottier's (1999) aerial survey of Angkor documented a system of house mounds, local shrines, pre-modern field systems and household ponds stretching between the great monumental structures and the water management system. It was revealed that the entire southern half of Angkor consisted of an array of residential features and religious establishments that revolved around the management of small-scale seasonal water resources. Pottier's maps were digitised into a GIS dataset in 2001 at the University of Sydney, and in recent years, Groslier's (1979) model of environmental decline has been fleshed out and transformed into a concrete research agenda as part of the Greater Angkor Project (Fletcher 2000-2001; Fletcher & Pottier 2002).

The basic empirical requirements for GAP had already been met by Pottier (1999) for the southern half of Angkor. The acquisition of airborne imaging radar (AIRSAR) data over Angkor in 2000 by the Jet Propulsion Laboratory (JPL), under contract to NASA, was designed to provide a similar perspective on the northern half of Angkor. As a method of archaeological data collection, radar imaging has some key advantages over conventional remote sensing techniques. It is an excellent method of identifying subsurface features, either directly or through the detection of subtle surface manifestations. However, in spite of a series of radar imaging projects around the world in the 1990s, including a Space Shuttle mission over Angkor in 1994, AIRSAR data had never been rigorously or systematically applied to an archaeological problem, and evidence of its utility as a method of data collection remained anecdotal. It was therefore necessary to assess the strengths and weaknesses of the method, to gain an understanding of radar interaction with the archaeological landscape and to develop a methodology for image interpretation (Evans 2002). Using colour composite images derived from the raw radar data, archaeological features were then systematically catalogued across the north of Angkor. Throughout 2002 to 2007 the maps were further refined and expanded to include the entire watersheds of the three major rivers that were manipulated by the population of Angkor (Evans 2007). A diverse range of datasets were used including Quickbird data, ASTER, aerial photographs taken during the 1940s by Williams-Hunt (Moore 1985), a complete aerial coverage of Cambodia at 1:25,000 scale taken in 1992-3 by the FINNMAP company, and the 2.5m resolution SPOT 4 imagery that was used as a cartographic base for heads-up digitising. The resulting archaeological map information was stored into a GIS database in accordance with the typology developed by Pottier (1999), and the combination of all datasets was used to generate a comprehensive archaeological map of the Angkor area covering 3000km2 (Figure 2).

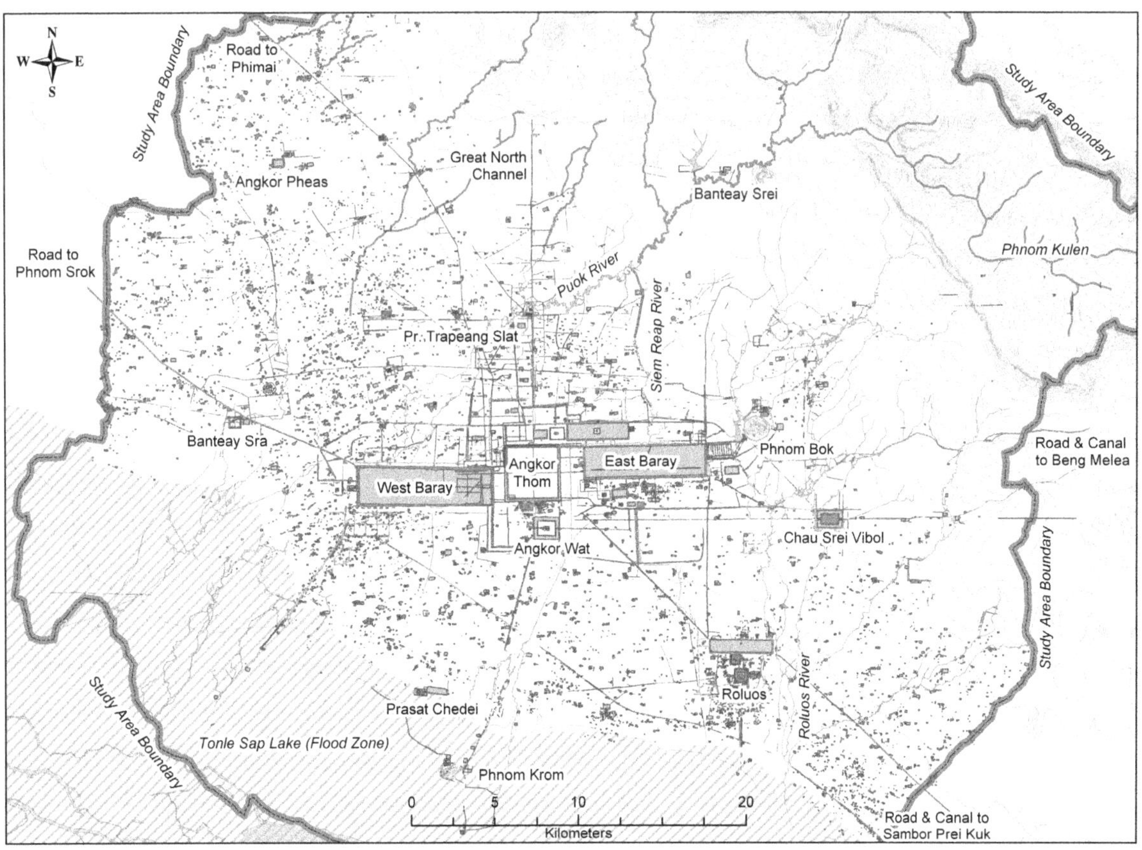

Fig. 2. An overview of the comprehensive map of Angkor generated from remote sensing and survey.

A series of spatial analyses were undertaken on the many thousands of ponds, local shrines and linear features relevant to the theoretical issues described above. Using ponds as an indicator of residential behaviour, for example, the extent, density and patterning of Angkorian population was analysed and compared to the spatial patterning of the water management system. This showed that the large-scale material infrastructure, which displayed a high degree of regularity and symmetry in the south, became increasingly disordered and chaotic as it stretched northward over time into areas which were at least as densely populated as the south of Angkor. In fact, the water management network in the north showed evidence for the kind of breaches, ad hoc adaptations and functional problems one might expect from a system that ultimately failed due to increasing sedimentation, complexity and erratic water flow, and our analysis suggested a high likelihood of disruption to essential water supplies in areas of high residential and agricultural activity.

This process, in addition to refining a number of technical and methodological innovations (Evans 2002; Fletcher, Evans & Tapley 2002), has resulted in a number of significant theoretical outcomes and substantially increased our understanding of the processes which led to the decline and eventual abandonment of Angkor (Evans *et al* 2007; Fletcher *et al* 2003; Fletcher *et al* 2004). In particular, the results facilitated a critical assessment of the arguments for and against the now-famous but still-controversial 'hydraulic city' hypothesis articulated by the French archaeologist Bernard-Philippe Groslier, who proposed that the success and failure of the Khmer capital was due to a reliance on Angkor's great reservoirs for rice surpluses (Groslier 1979). Not only was Angkor perhaps the largest settlement complex in the pre-industrial world, but it clearly possessed a water management system of the scale and sophistication necessary to play a significant role in the rise and fall of Angkor as a regional centre. Angkorian hydraulics were enormously complex, and the maps showed evidence of repeated failure and re-engineering within that system. Several key elements of Groslier's were therefore supported by this research: Angkor was a vast and populous urban centre, and it had a water management system that was capable of producing very significant agricultural surpluses, even as it grew increasingly unwieldy and prone to systemic failure over the course of Angkor's history. The various outcomes at Angkor in recent years clearly demonstrate the potential of these kinds of archaeological methods for assessing specific historical questions.

Beyond Angkor

Banteay Chmar

The temple complex of Banteay Chmar, located approximately 100km to the northwest of the centre of Angkor, is one of several great provincial centres established or redeveloped by Cambodia's most prolific temple builder, king Jayavarman VII. Until recently the temple complex was considered relatively remote and not easily accessible; little archaeological work has been done there, and notwithstanding the enormous size, the Bayon-style face towers and the galleries of bas-reliefs that feature in the principal temple, it is principally famous for the wholesale looting of its sandstone carvings in the late 1990s (Pottier 2004, 144-5).

Following on from many years of successful research on the urban context of the temples at Angkor, a pilot program of surface survey, using much the same methodology, was undertaken in the Banteay Chmar in 2006-2007 by Richards (2007). Although only partially verified on the ground, the results of that survey using aerial photos and Google Earth Quickbird data revealed numerous potential temple sites and abundant evidence for the type of distributed, low-density, pond-based settlement system that had been documented at Angkor. Further field verification work conducted by Richards in conjunction with the author in 2008 has confirmed the existence of an additional 27 temple sites beyond those documented in the most recent archaeological map , and the study area has since been expanded by the author throughout the course of further surveys in 2009 (Figure 3). Work is ongoing in this extended area, assisted by a grant of 50cm resolution GeoEye-1 data from the GeoEye Foundation, by oblique scenes acquired from low-level ultra light aircraft survey in 2009, and by the acquisition of declassified Corona spy satellite imagery from 1967 in order to assess the nature of the agricultural and water management systems prior to the extensive (and largely undocumented) Khmer-Rouge era modifications of the 1970s (see Figure 4).

A number of important outcomes have already been achieved as a result of this research. The new map of the Banteay Chmar region shows a distributed low-density settlement system similar to that discovered at Angkor, but not penetrated by an infrastructural network to the same degree. Like Angkor, a number of prehistoric (likely Iron Age – see Richards 2007) sites dot the landscape and indicate a sequence of occupation spanning at least two thousand years, up until the present day. The great reservoir has a system of intake canals and embankments as well as a distributor system which links the water storage system to nearby agricultural areas and – again, as with Angkor (Boulbet 1979; Evans 2007) – to quarry sites. Sandstone quarries have been located at the foot of Dangrek range nearby, and laterite quarries have been located 11km south of the main temple complex from aerial photos (Figure 3). The latter is potentially a sandstone quarry as well: although all Angkorian quarries conclusively identified so far in Cambodia show evidence of stone extraction using the channelling method, recent surveys at Koh Ker (see below) and this site reveal evidence (in the form of split stones, lever holes and wastage) of sandstone outcrops being quarried by leveraging natural fault lines, by splitting stones using levers or wedges into holes, or simply by removing naturally-eroded sandstone blocks with little or no attachment to the underlying bedrock. Elements of the infrastructural network connecting these quarried areas to the temple sites, and the nature of the water flow through and around the temple itself, can be clearly identified in the remote sensing data and have been successfully differentiated from Khmer Rouge developments by analysis of the Corona data (Figure 4).

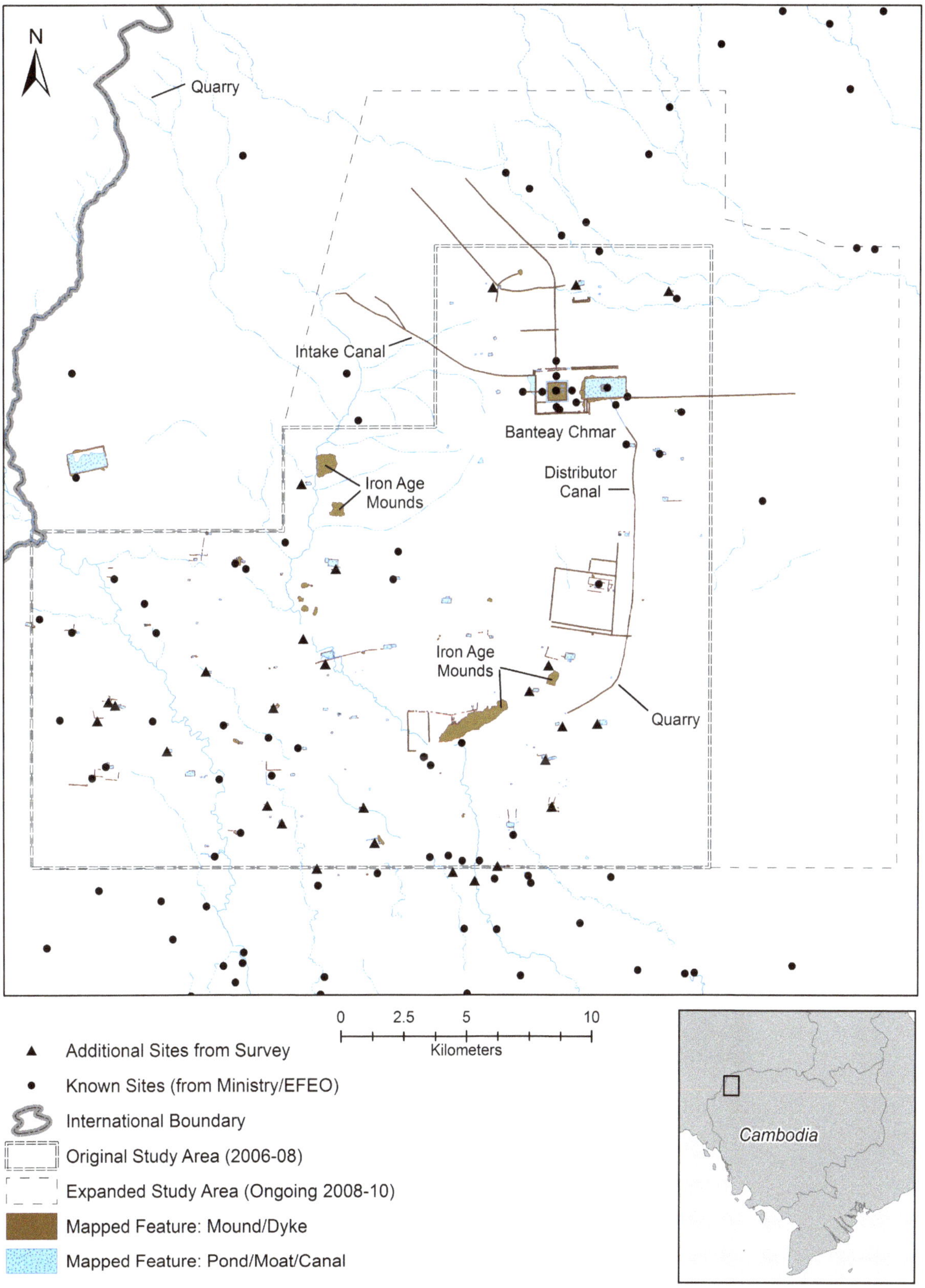

Fig 3. Preliminary map of the Banteay Chmar temple complex and surrounds.

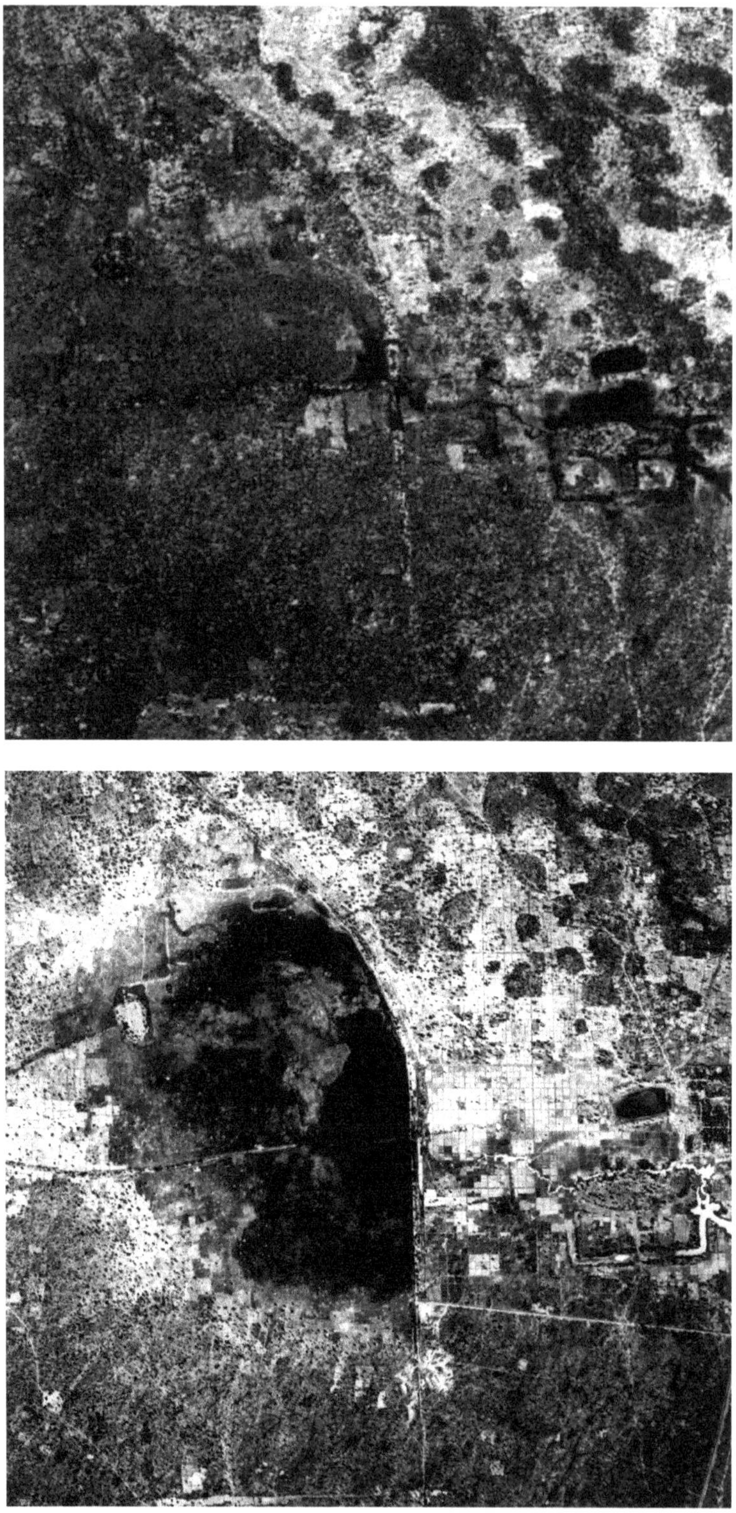

Fig. 4. Corona spy satellite image from 1967 (top) compared to the aerial photo coverage from 1992 (bottom), detailing an area immediately to the north of the Banteay Chmar temple. An Angkorian embankment runs from north to south across the images; note its conversion in the 1970s, during the Khmer Rouge era, into a dam, and the 1ha square-block field system developed to the east

Koh Ker

The site of Koh Ker, at least as presented in the classical accounts, presents something of an anomaly in the history and archaeology of Cambodia. Conventionally, Koh Ker is considered to have been the capital of the Khmer Empire in the 10th century AD in a brief hiatus from the Angkor area, where the seat of power otherwise remained between the 9th to 15th centuries AD. Considering that the inscriptions at Koh Ker are generally ascribed to the 10th century (with the exception of inscription K.682 dated to AD 1002), and that the 50-60 temples of the area present a substantial degree of uniformity in architectural and decorative style, it is commonly assumed that Koh Ker is fundamentally a single-period site developed during its tenure as capital (Chandler 2000; Jacques & Lafond 2007, 110; Parmentier 1939). Like Banteay Chmar, Koh Ker is generally considered to be an arid and inhospitable site by scholars, in that it has little access to year-round sources of water (Chandler 2000, 40; Jacques & Lafond 2007, 107; Parmentier 1939, 17); unlike Banteay Chmar, however, no system of water catchment or distribution has been documented beyond the central reservoir. In fact, no original contribution to the landscape archaeology of the site has been published at all since Parmentier's classic (1939) work (Figure 5), although existing data have been reappraised (Jacques & Lafond 2007, 103-33) and temple inventories have been modernized and updated by national teams from Japan (Nakagawa *et al* 2005), France (Ministère de la culture et des beaux-arts & École française d'Extrême-Orient 2007) and Hungary (Royal Angkor Foundation, unpublished) in partnership with Cambodian authorities. With the exception of various anomalies and questions raised by Jacques and Lafond, Parmentier's interpretation of Koh Ker as a single-period ceremonial city bounded by city walls essentially stands firm in the literature.

Even without a detailed study, however, a cursory examination of the available Landsat 7, ASTER and aerial photo collections at the University of Sydney revealed that Koh Ker was likely much more complex than originally thought, and that there were numerous extensions to the known 'city wall' fragments which were more consistent with water management structures or roads than with the formal boundary of 'a city'. Given the ostensibly single-period nature

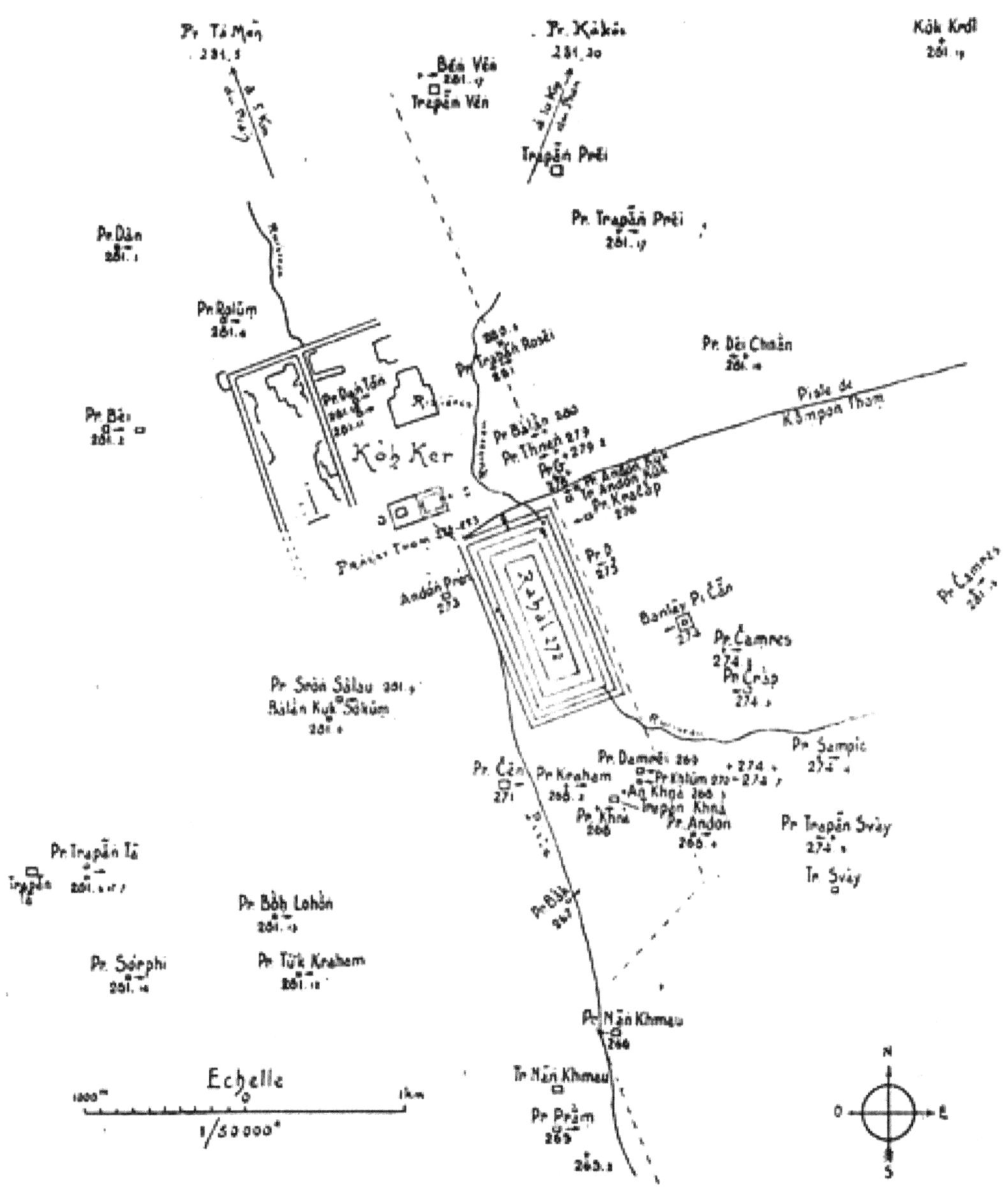

Fig. 5. Parmentier's (1939) map of Koh Ker, which has remained the standard map for 70 years.

of the site, Koh Ker was therefore seen as an ideal location for uncovering evidence about the nature of Khmer water management systems in the 10th century AD, given the difficulties in untangling the palimpsest of overlapping systems that had accumulated throughout various periods at Angkor itself. However, no immediate evidence could be seen for the existence of a low-density settlement system surrounding the dense ceremonial core of Koh Ker. With a view to addressing these and other issues, members of the Greater Angkor Project began preliminary field work in 2003 (surveys on the ground and by ultra light aircraft), which has been extended into a comprehensive survey program from 2007-2009 under the auspices of the project described here.

Data sources for the landscape survey include a 2009 acquisition of 50cm resolution TerraSAR-X spaceborne radar imagery and two scenes of 1m resolution Ikonos imagery acquired in 2001 and donated by the GeoEye Foundation, in addition to conventional aerial photography and oblique scenes acquired from multiple low-level ultra light aircraft surveys. Imagery analysis and feature mapping over 300km2 has been complemented by several months of ground verification surveys throughout 2007-2009; the proposed work at the site is nearly complete, a draft map has been produced (Figure 6), and some very preliminary conclusions can be presented here.

In terms of the built environment the major conclusion to be drawn from the mapping work is that the embankments identified by Parmentier (1939, 16-7) as the boundary walls of a city are highly unlikely to have fulfilled any such function; they are, in fact, much more consistent in morphology and placement with an intended function as water channelling and damming devices, and as key components of a broader Angkorian hydraulic system. As with Goloubew's imagining of a 9th century 'city wall' at Angkor (Pottier 2000a), Parmentier's (1939) analysis and mapping of the embankments at Koh Ker offered an idealised image of their layout, and he was obliged to explain away very substantial gaps in the evidence for an enclosing wall. In fact, contrary to Parmentier's map, it is clear that the embankments are generally not straight, and that there is no evidence that they continued at any stage to delineate a bounded area. On the contrary, the two north-south embankments, both of which show evidence of canalisation and/or the diversion of streams along their length, curve in accordance with local topography, thereby fulfilling an apparent requirement to funnel water towards the northern east-west 'wall', which would, in fact, given its height, morphology and cross-slope orientation, clearly have functioned as a dam. Indeed, were it not breached in three locations by modern streams, it would continue to serve that purpose.

Nor is this the only evidence for the damming of watercourses at Koh Ker. A huge, 4.5km long embankment stretches northwards from the embankments mapped by Parmentier (Figure 6). At the halfway point of this feature, it interrupts the natural eastwards flow of a number of important watercourses with a massive, enlarged section which clearly functioned as a dam. Immediately to the south of this dam, the eastern (down slope) flank of embankment is reinforced with a large, sloping masonry structure 190m in length; although incorrectly interpreted by local people as the remains of an Angkorian bridge, it is clearly the remains of a laterite spillway and bears many similarities to that discovered at the site of Bam Penh Reach at Angkor (Fletcher *et al* 2008). At the far northern end of the embankment, where it turns to the west and meets a river, are the remains of two dam constructions likely dating from different periods. One consists of a series of postholes carved into the sandstone bed of the river, which would have held support beams for a structure made of non-durable materials such as wood and perhaps reinforced with earth. The other presumed dam consists in fact of a westward extension to the embankment which at one stage would have crossed the river and blocked flow in this area; now breached, its remains can nonetheless be seen abutting the stream's edge on both sides, while the bed of the river at this point contains a substantial collection of the laterite rubble and debris commonly used to reinforce the upslope flank of Angkorian dykes at Koh Ker.

Just to the east of these latter dam remains is located a laterite-paved outlet canal which again bears a strong resemblance in width, height, and method of construction to the Krol Romeas outlet of the East Baray at Angkor (Fletcher *et al* 2008). What is particularly remarkable about this outlet is its length: on the east side, the wall of outlet can be seen to extend for 450m, while the west wall traces it in parallel for 250m, compared to the 100m long parallel walls of Krol Romeas. When the 1:50,000 scale contours of the area (Japan International Cooperation Agency 1999) are also considered, all of the key structural elements for an exceptionally large dam – earthen wall, masonry-reinforced spillway and outlet, and watershed – can be clearly identified at the site of Koh Ker. Planning is currently underway for a LIDAR acquisition in 2010, which will produce a high-resolution (0.5m or better) elevation model of the area and permit analysis of the hydrology of the area – including the holding capacity and available watershed of the numerous dams and reservoirs – in considerable detail.

Parmentier's (1939) inventory of temple sites at Koh Ker was very thorough, and the new map presents only one major additional temple site, immediately to the west of a large Angkorian reservoir known as Trapeang Russei located northeast of the main complex. The existence of some features previously identified at Koh Ker, such a Parmentier's putative north-south road to the East of the Rahal reservoir (Figure 5), and the island-temple or *mebon* in the middle of the Rahal denoted by a recent map (Ministère de la culture et des beaux-arts & École française d'Extrême-Orient 2007), is deemed to be highly questionable given the lack of evidence. A range of additional features, however, have been uncovered during the surveys and documented (Figure 6): several laterite quarries and potential sandstone quarries beyond the two (Moriai, Matsumura & Akama 2002, 25; Parmentier 1939, 77) already documented; a series of secondary canals and embankments for water control; evidence for change, adaptation and multi-period construction within this network of infrastructure; remnants of pre-modern rice field systems concentrating in natural drainage channels and basins; two circular sites surrounded by radial field systems potentially indicating prehistoric occupation; and areas where concentrated ceramic scatters suggest occupation at Koh Ker from the early Angkorian period until at least the 15th century AD. Aerial archaeology has, therefore, uncovered the first new information on landscape and settlement at Koh Ker in 70 years, and will now underpin a complete reappraisal of the nature, significance and longevity of this critically important site.

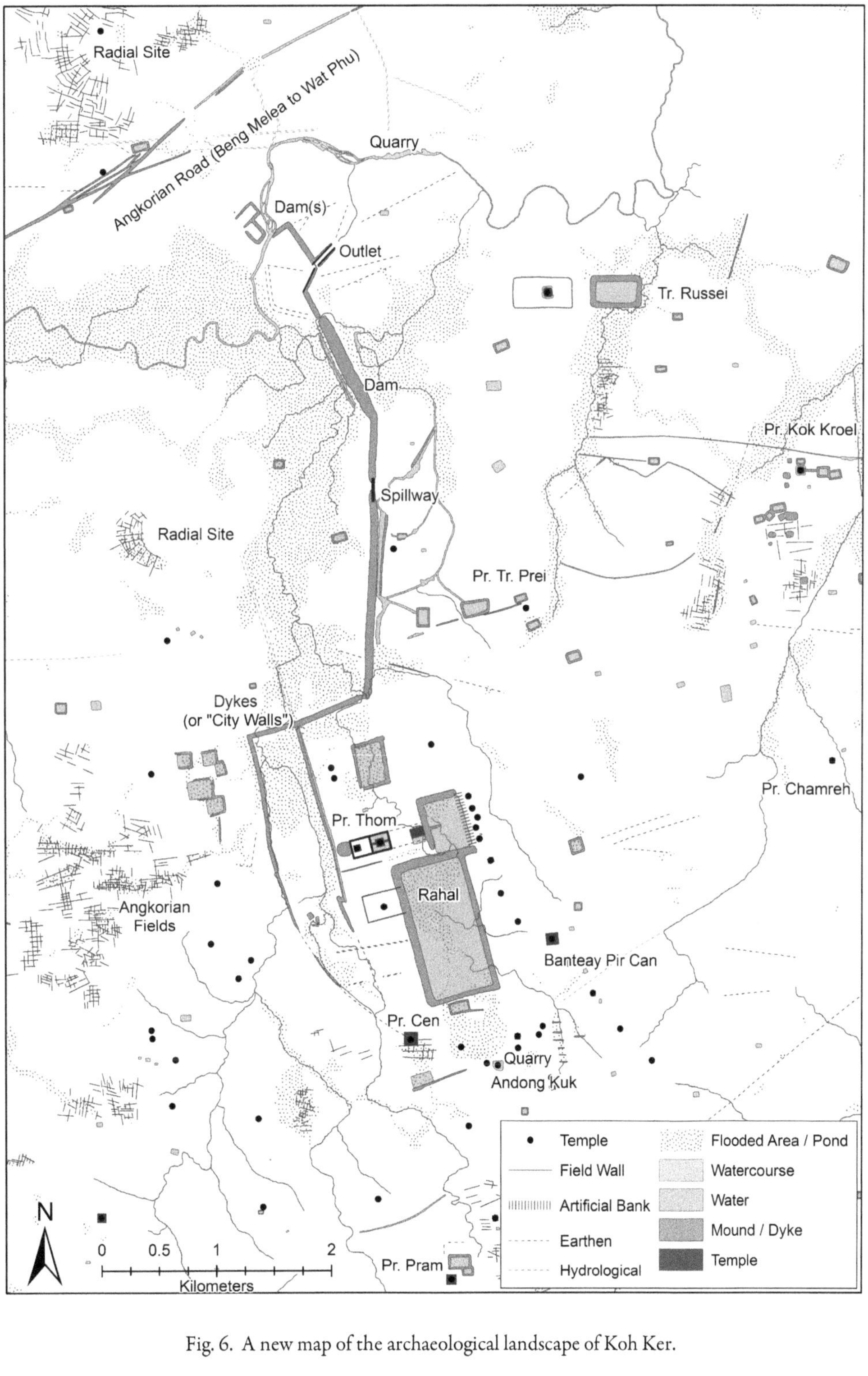

Fig. 6. A new map of the archaeological landscape of Koh Ker.

Further Potential

Preliminary fieldwork has also been undertaken in 2009 at the site of Preah Khan of Kompong Svay, in Preah Vihear Province, in close collaboration with Mitch Hendrickson and other team members of the Australian Research Council-funded Industries of Angkor Project (INDAP). Here, as with the other sites mentioned above, the GeoEye Foundation has played an important role in the provision of up-to-date, high-resolution imagery for analysis and use as a topographic base: in this case, OrbView-3 data provides the best coverage. Results of the 2009 field seasons are currently being collated, but a number of preliminary conclusions may be advanced based on the analysis of OrbView-3 data and air photos that has been completed at this point. Perhaps most significantly, Preah Khan provides little evidence for an extended, temple-based agrarian settlement system of the type that has been discovered at sites like Angkor, Banteay Chmar and Koh Ker. Ceramic evidence for occupation here is also sparse, and the various water tanks or *trapeang* show far more surface evidence for industrial usage (i.e. iron smelting) than for domestic residence. Additionally, although Preah Khan possesses a relatively sophisticated intramural water management system fed by a natural stream entering the enclosure to the northwest, there is little or no evidence, at this stage, for an extended system of water collection or distribution beyond the enclosure walls.

Beyond the scheduled fieldwork at Beng Melea and Sambor Prei Kuk in 2009-2010, various other sites across Cambodia present a number of opportunities and challenges for archaeological research. Of particular interest to various scholars at the University of Sydney are the post-Angkorian capitals (15th century AD and later) located in the region of what is now Phnom Penh. These poorly understood and sparsely documented sites are critically important not only for an understanding of post-Angkor but also the end of the classical Angkorian period in the 15th century AD. For example, diachronic analyses of urban form during the genesis and development of post-Angkorian capitals may speak to longstanding debates about whether a collapsing Angkor was 'abandoned' en masse for settlements further south, or whether the coastal cities of the post-Angkorian period rose gradually, in their own right, to rival Angkor's influence due to flourishing maritime trade in Southeast Asia's "Age of Commerce" (Reid 1988). No studies on urbanism at these sites have been undertaken, but even a cursory analysis of the available remote sensing datasets suggests that such studies might prove very fruitful. At the post-Angkorian capital of Oudong, for example, 30km north of Phnom Penh, Quickbird imagery served in Google Earth and Google Maps reveal a pattern of remnant ponds and landscape structures which, although they are clearly pre-modern, demonstrate significant differences in orientation, size and structure from those typically associated with the Angkorian and pre-Angkorian periods (Figure 7).

Fig. 7. Potential archaeological sites in the vicinity of the post-Angkorian capital of Oudong, north of Phnom Penh (image courtesy Google Earth).

A recent analysis by Hendrickson (2007) of medium-resolution ASTER data, although focused primarily on mapping Angkorian roadways, has identified a number of reservoir clusters and secondary road networks that are worthy of closer consideration.

The greatest two challenges for large-scale surface surveys of pre-modern settlements in Cambodia remain forest cover and the existence of anti-personnel landmines in vast swathes of the country. Mine clearance is ongoing, and most of the major temples have been demined in recent years, but the immediate areas around those temples and their extended hinterland at sites like Koh Ker, Beng Melea and Banteay Chmar remain very dangerous. Forest cover at sites such as Koh Ker and Beng Melea limits the potential of remote sensing for site discovery from the air, even as landmines limit the possibilities for ground-based survey methods. Paradoxically, the extensive deforestation now underway in these areas serves, on the one hand, to clarify the aerial view of features which cannot be safely accessed on the ground and opens the way for agricultural expansion and demining; on the other hand, as experience in the Angkor area has shown, forest clearance is very often a prelude to the kind of intensive urban development which rapidly obliterates the subtle traces of settlements that have been preserved on the surface for a millennium or more. Currently, several different platforms and sensors are being canvassed as a potential solution to these difficulties in detecting and mapping surface features. The potential for a LIDAR acquisition has already been mentioned in relation to the site of Koh Ker (above). However, the development of advanced, high-resolution radar instruments capable of producing precision elevation models also presents some interesting opportunities. The GeoSAR radar instrument, the next generation in development from the AirSAR instrument, is reportedly capable of leveraging its longer-wavelength P-band to produce bald earth models with high accuracy and resolution. A less expensive solution may be provided by the imminent launch of the TanDEM-X satellite, whose elevation products, although they will represent a first-bounce or 'tree-top' DEM, could potentially be processed to generate a bald earth model given the very high resolution of the output data and the likelihood of ground strikes within a reasonable quantity of the postings, particularly in winter acquisitions at sites like Koh Ker with deciduous forests.

After two decades of intensive remote sensing research at Cambodian archaeological sites, the contribution of aerial archaeology has been substantial. Not only has an enormous amount of new information been generated, but applications of remote sensing have provided the foundation for major multidisciplinary projects and have addressed important historical issues and debates. Classical techniques such as the subjective, visual interpretation of aerial photo prints in search of spatial patterning in elevation and vegetation remain the mainstay of such research; however, a range of new sensors and platforms have contributed substantially to studies of particular areas, and the ability to store, analyse and distribute spatial data by means of a GIS has greatly assisted archaeological research and diverse other fields such as heritage management (Fletcher *et al* 2007). Given the very limited timeframe during which the archaeological landscape remains exposed in Cambodia between the time of forest clearance and modern urbanization, very high resolution satellite imagery in particular offers an excellent opportunity to capture surface information before it is lost. Ideally, future applications will harness the repeat-pass capability of such platforms in order to provide real-time updates to heritage managers on rates of clearance, encroachment on protected areas, and potentially even identify areas were valuable cultural heritage is being destroyed in contravention of the law and contrary the long-term interests of the Cambodian people.

Reference

BILLMAN, B. R. 1999. Settlement Pattern Research in the Americas: Past, Present and Future. In B. R. Billman and G. M. Feinman (eds.), *Settlement Pattern Studies in the Aericas: Fifty years Since Viru*, 1-9. Washington, Smithsonian Institution Press.

BOULBET, J. 1979. *Le Phnom Kulen et sa Région: Carte et Commentaire*. Paris, École française d'Extrême-Orient.

CHANDLER, D. P. 2000. *A History of Cambodia*. Chiang Mai, Silkworm Books.

EVANS, D. 2002. *Pixels, Ponds and People: Urban Form at Angkor from Radar Imaging*. Unpublished BA(Hons) Thesis, University of Sydney.

EVANS, D. 2007. *Putting Angkor on the Map: A New Survey of a Khmer 'Hydraulic City' in Historical and Theoretical Context*. Unpublished PhD Thesis, University of Sydney.

EVANS, D., POTTIER, C., FLETCHER, R., HENSLEY, S., TAPLEY, I., MILNE, A. AND BARBETTI, M. 2007. A comprehensive archaeological map of the world's largest preindustrial settlement complex at Angkor, Cambodia. *Proceedings of the National Academy of Sciences of the United States of America* 104, 14277-82.

FLETCHER, R. 2000-2001. A.R. Davis Memorial Lecture. Seeing Angkor: New views of an old city. *Journal of the Oriental Society of Australia* 32-33, 1-25.

FLETCHER, R., BARBETTI, M., EVANS, D. H., THAN, H., SOKRITHY, I., CHAN, K., PENNY, D., POTTIER, C. AND SOMANEATH, T. 2003. Redefining Angkor: structure and environment in the largest, low density urban complex of the pre-industrial world. *Udaya, Journal of Khmer Studies* 4, 107-21.

FLETCHER, R., EVANS, D. H. AND TAPLEY, I. J. 2002. AIRSAR's contribution to understanding the Angkor World Heritage Site, Cambodia - Objectives and preliminary findings from an examination of

PACRIM2 datasets. *Proceedings of the 2002 AIRSAR Earth Science and Application Workshop, NASA/JPL*. WWW Resource. URL: http://airsar.jpl.nasa.gov/documents/workshop2002/papers/P1.pdf

FLETCHER, R., EVANS, D. H., TAPLEY, I. J. AND MILNE, A. K. 2004. Angkor: Extent, settlement pattern and ecology. Preliminary results of an AIRSAR survey in September 2000. *Bulletin of the Indo-Pacific Prehistory Association* 24, 130-8.

FLETCHER, R., JOHNSON, I., BRUCE, E. AND KHUON, K.-N. 2007. Living With Heritage: site monitoring and heritage values in Greater Angkor and the Angkor World Heritage Site, Cambodia. *World Archaeology* 39, 385-405.

FLETCHER, R., PENNY, D., EVANS, D., POTTIER, C., BARBETTI, M., KUMMU, M., LUSTIG, T. AND AUTHORITY FOR THE PROTECTION AND MANAGEMENT OF ANGKOR AND THE REGION OF SIEM REAP (APSARA) DEPARTMENT OF MONUMENTS AND ARCHAEOLOGY TEAM. 2008. The water management network of Angkor, Cambodia. *Antiquity* 82, 658-70.

FLETCHER, R. AND POTTIER, C. 2002. The Gossamer city: A new inquiry. *Museum International* 54, 23-7.

GROSLIER, B.-P. 1979. La Cité Hydraulique Angkorienne: Exploitation Ou Surexploitation Du Sol? *Bulletin de l'École française d'Extrême-Orient* 66, 161-202.

HENDRICKSON, M. 2007. *Arteries of Empire: An operational study of transport and communication in Angkorian Southeast Asia*. Unpublished PhD Thesis, University of Sydney.

JACQUES, C. AND LAFOND, P. 2007. *The Khmer Empire: Cities and Sanctuaries, Fifth to Thirteenth Century*. Bangkok, River Books.

JAPAN INTERNATIONAL COOPERATION AGENCY. 1999. *Cambodia Reconnaissance Survey Digital Data*. Japan, Japan International Cooperation Agency (JICA).

KUMMU, M. 2003. The Natural Environment and Historical Water Management of Angkor, Cambodia. *5th World Archaeological Congress*. Washington, D.C., USA. WWW Resource. URL: http://users.tkk.fi/~mkummu/publications/kummu_WAC_WashingtonDC_2003.pdf

LUSTIG, E., EVANS, D. H. AND RICHARDS, N. 2007. Words across Space and Time: An Analysis of Lexical Items in Khmer Inscriptions, Sixth–Fourteenth Centuries CE. *Journal of Southeast Asian Studies* 38, 1-26.

MINISTÈRE DE LA CULTURE ET DES BEAUX-ARTS AND ÉCOLE FRANÇAISE D'EXTRÊME-ORIENT. 2007. *Carte archéologique du Cambodge: Province de Preah Vihear*. Phnom Penh, Kim Long.

MOORE, E. H. 1985. The Williams-Hunt Collection of Aerial Photographs of the Archaeological Research Trust for Thailand. *Journal of The Siam Society* 73, 252.

MORIAI, T., MATSUMURA, Y. AND AKAMA, Y. 2002. Geological Study on Angkor Monuments: Part 3. *Memoirs of the Tohoku Institute of Technology. Series 1. Science and Engineering* 22, 23-39.

NAKAGAWA, T., SATO, K., SHIMODA, I., SO, S., CHAN, V., SENG, K., VA, S., SOKUAN, D., YAMASAKI, T. AND YOSHIKAWA, M. 2005. *Provisional Inventory of Monuments at Koh Ker*. Tokyo, Laboratory of Architectural History, Waseda University.

PARMENTIER, H. 1939. *L'art Khmer classique : monuments du quadrant nord-est* Paris, Les Éditions d'art et d'histoire.

POTTIER, C. 1999. *Carte Archéologique de la Région d'Angkor. Zone Sud*. Unpublished PhD Thesis, Université Paris III - Sorbonne Nouvelle.

POTTIER, C. 2000A. À la recherche de Goloupura. *Bulletin de l'École française d'Extrême-Orient* 87, 79-107.

POTTIER, C. 2000B. Some Evidence of an Inter-relationship between Hydraulic Features and Rice Field Patterns at Angkor during Ancient Times. *The Journal of Sophia Asian Studies* 18, 99-120.

POTTIER, C. 2004. À propos du temple de Banteay Chmar. *Aséanie* 13, 131-50.

REID, A. 1988. *Southeast Asia in the age of commerce, 1450-1680*. New Haven, Yale University Press.

RICHARDS, N. 2007. *Prehistoric and early historic settlement around Banteay Chmar, north-west Cambodia*. Unpublished BA(Hons) Thesis, University of Sydney.

VICKERY, M. 1977. *Cambodia after Angkor: the chronicular evidence for the fourteenth to sixteenth centuries*. Unpublished PhD Thesis, University of Michigan.

Archaeology and landscape features in magnetometer data

Immo Heske[1] and Martin Posselt[2]

[1]Seminar für Ur- und Frühgeschichte Georg-August-Universität Nikolausberger Weg 15, Göttingen, Germany
[2]Posselt & Zickgraf Prospektionen GbR Fürthweg 9 D-64367 Mühltal, Germany

Abstract

While magnetometry is well accepted for detecting archaeological features it also allows to map details of the natural environment of the site. Theese structures can be integrated as elements of the landscape. Additionally to research at the sites archaeologists are interested in the criteria people have decided to settle in a certain landscape. The subject is mainly pictured by the case study of a site in Lower Saxony (Hünenburg, Gevensleben-Watenstedt), consisting of a hillfort and an unfortified large scale settlement during Bronze Age. Magnetometer survey has enlarged the known extension of the unfortified settlement from a small spot of surface findings up to about 15 hectars as minimum, a multiple of the size of the hillfort. Furthermore it presents its features extraordinarily well preserved due to the existence of remainders of a medieval agricultural systems. Furthermore a system of silt up creeks was revealed by magnetometry. The natural system of drainage is supposed to have had a certain impact on the decision of man where to place settlements and how to use the landscape. In detecting ancient landscape features as well as the extension of the prehistoric settlement magnetometry changed the archaeologists' perspective from site to landscape while researching the large-scale site.

Keywords archaeological feedback, magnetometry, landscape archaeology, neolithic, bronze age, iron age.

Geophysics, especially magnetometry, right now has become famous for detecting archeological features and mapping entire archaeological sites (Gaffney 2008). It allows to map pits, fortifications, buildings, post holes, graves etc. without excavation i. e. destruction. Nevertheless, since magnetometry plots any type of magnetic anomalies including modern disturbances and natural e. g. geological structures, it also maps the natural environment of the site. Geophysics allow to detect features concerning the history of the landscape generally (Kvamme 2003). On the one hand human beings always have shaped the landscape they lived in and their decision to settle in a certain place depended on special landscape features also. On the other hand todays face of the natural environment in fact is highly artificially shaped or has changed due to natural processes. Recently within landscape archaeology researchers are not interested in the sites merely but also in why and how people have settled in a certain landscape (Schade 2000). The paper at hand aims to picture that recent geophysical surveys have detected archaeological features of sites and also features of the landscape the site is placed in: silt up ancient streams, areas of erosion and accumulation, former systems of cultivation of soils etc. Archaeological interpretation of the geophysical data and its integration into the results of other disciplines changes the archaeologist`s point of view as well as the research strategy (Meyer 2007). Also within the framework of landscape archaeology geophysics does not detect archaeological features merely but also elements of the natural setting inbetween settlements. While geophysics maps structures buried underneath invisible to modern men but which were part of the ancient landscape which prehistoric men found for placing his settlements, an assessment of the initial landscape can be gained and conclusions concerning the criteria for placing a settlement can be found (e. g. water supply, situation of functional areas (areas of agriculture, metallurgical areas, limitations of the settlements).

This subject is pictured by the case study of a site in Lower Saxony, Germany, called "*Hünenburg*" (Gevensleben-Watenstedt) which is well known as a hillfort fencing in an area of about 2,5 ha defined by a rampart as well as steep slopes. Right now the site is investigated in the course of a research project by the University of Göttingen financed by the Deutsche Forschungsgemeinschaft. The *Hünenburg* has been settled on from Neolithic until Early medieval periods (Heske 2000, 2003). It is situated on a spur of a high ridge ranging up to 100 m above its surroundings including the large low-lying areas of the *Großes Bruch* to the south. In its beginning research at the *Hünenburg* focused on the hillfort with its rampart and fenced in settlement areas. Its environs bear further sites comprising graves, deposits and areas of else surface findings. Especially the latter were found at the southern slope of the hillfort but were not seen as a remnant of a separate settlement outside the ramparts. They were interpreted as the product of erosion of the hillfort down its steep slopes.

In 2001 a magnetometer survey was initialized to prove weather the surface findings from the southern slope of the hillfort are produced by a comtemporary settlement outside the ramparts instead (Posselt 2001). An answer to this question would allow conclusions concerning the function and place in the hierarchy of the settlement system throughout the surrounding landscape. Finally the initial magnetometer surveys as well as further campaigns until 2008 were successful in detecting an extensive coherent settlement area to the south and west of the *Hünenburg*. Until now an area of about 18 hectares is surveyed (Figure 1).

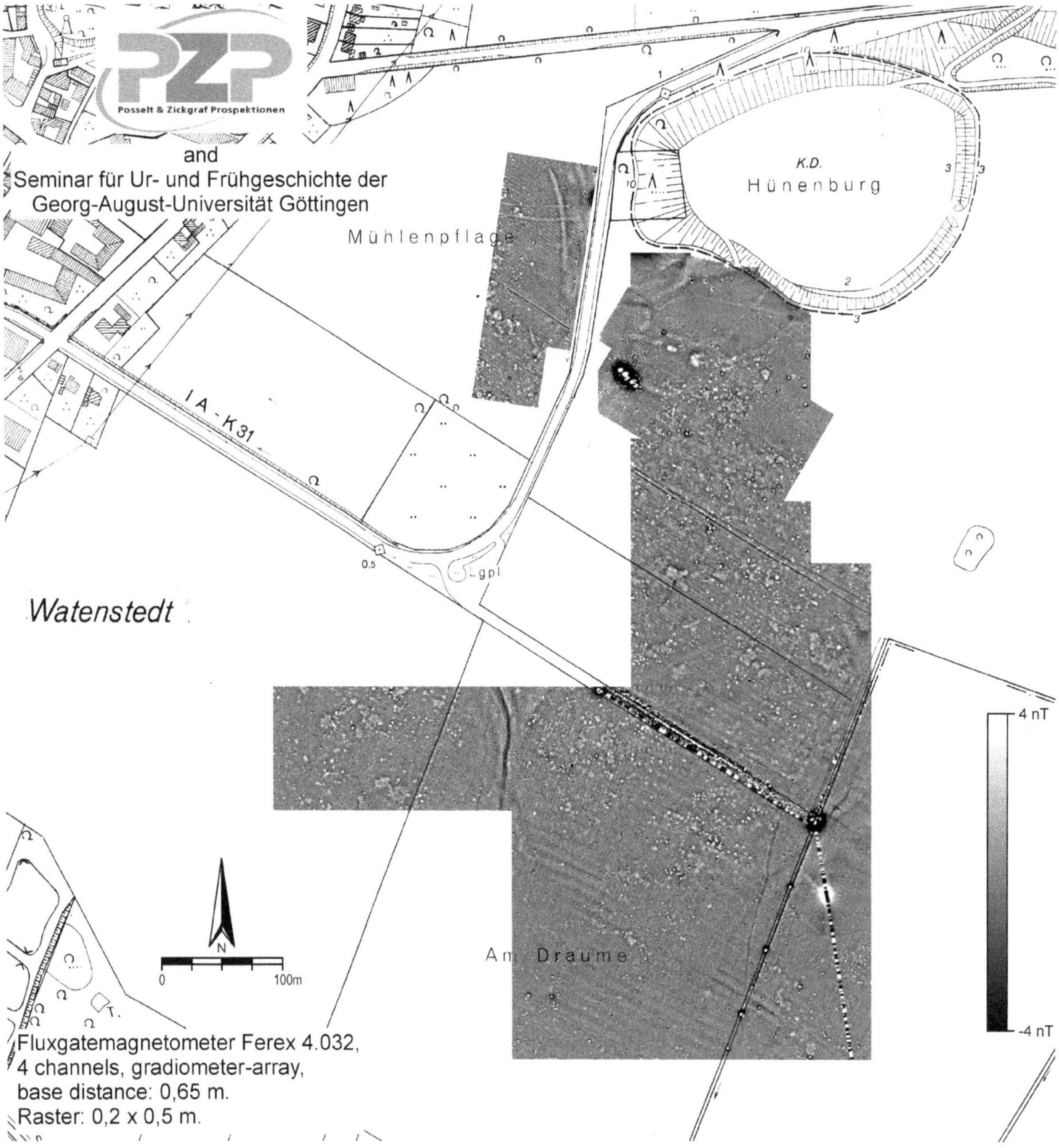

Fig. 1. Gevensleben-Watenstedt. Hillfort and settlement "Hünenburg". Total magnetogram from the settlement south of the hillfort. State: September 2008 (map courtesy of PZP/Seminar für Ur- und Frühgeschichte Göttingen).

The survey was done using a four channel Fluxgate-magnetometer (Förster Ferex). The archaeological features extend throughout almost the whole investigation area. According to magnetogramm the settlement seems to continue to the west and to the south. Yet it is not proven of what age the sections of the settlement are. The detailed small scale excavations (fig. 2) yet undertaken have gained some information about character and age of the features.

Beside numerous small features (waste pits etc.) the magnetogram (Figure 3) shows large anomalies, which usually would be interpreted as geological patterns. But excavations have proven, that such anomalies are produced by cultural layers below the topsoil of some decimetres in thickness covering further small features (Heske et al. 2009). Some of these cultural layers extend up to hectares in size. Excavations could identify the numerous small features as the usual waste pits, storage pits, post holes etc. Unusually also graves were found within the settlement area.

Fig. 2. Gevensleben-Watenstedt. Hillfort and settlement "Hünenburg". Airphoto summer 2008, from west (photo courtesy of Andreas Gruettemann).

Furthermore in the places of high magnetic amplitudes conglomerations of stones have been found, which then could be interpreted as hearths or ovens (Heske 2002, 2007). Several groups of these ovens can be identified throughout the ca. 18 ha magnetogram.

On the one hand the site presents its features extraordinarily well preserved. On the other hand the huge size of the settlement area has remained unrecognized for long time. The reason therefore could be, that a medieval agricultural systems (*Wölbäcker*), also detected by magnetometry as broad weak lineaments, have prevented a huge amount of the site from destruction by ploughing. In fact there seems to be a negative correlation of the visibility of the lines of the medieval agricultural system and the prehistoric features in the magnetogram. This phenomenon allows to conclude that areas of excellent visibility of *Wolbäcker*-structures may hide supposed archaeological features completely. For such areas magnetometry needs to be substituted by other fieldmethods, but in case archaeological features do exist they should be preserved in a quite good state. The excellent state of preservation of archaeological features is quite seldom within middle-Europe especially in the case of soils used for intensive agriculture.

Furthermore a system of stream beds, partly silt up or changed in course heavily due to natural processes or for meliorization reasons was revealed by magnetometry. Today the surroundings of the *Hünenburg* are coined by extensive acres and willows. The nearest body of water is the stream Soltau situated some 900 m southwest of the hillfort, while the area inbetween is an acre prepared for up to date industrial agriculture without any natural structures and obstacles. In one case further east of the site a ditch for water drainage runs dead straight aside a margin of an acre. But the results of magnetometer survey have changed our view of that part of the landscape. It was used for settlements and according activities throughout several periods of prehistoric ages according to magnetic data. A system of a huge settlement area towered above by an hillfort. Furthermore – as magnetically detected stream beds prove (Figure 3) – the settlements were placed in a natural system of water supply and drainage much more detailed than the recent landscape gives the impression.

The reconstruction of the initial situation of the landscape which prehistoric men found at the scene might allow further arguments concerning history and function of the settlement. Which elements of such a settlement system and the landscape are of the same age and what function they have has to be researched in much more detail yet. Nevertheless in the case of the *Hünenburg* magnetometry changed the perspective of the archaeologist when investigating why men decided where to place settlements and how to use the landscape concerning water supply, use of natural resources and areas useful to place dwellings.

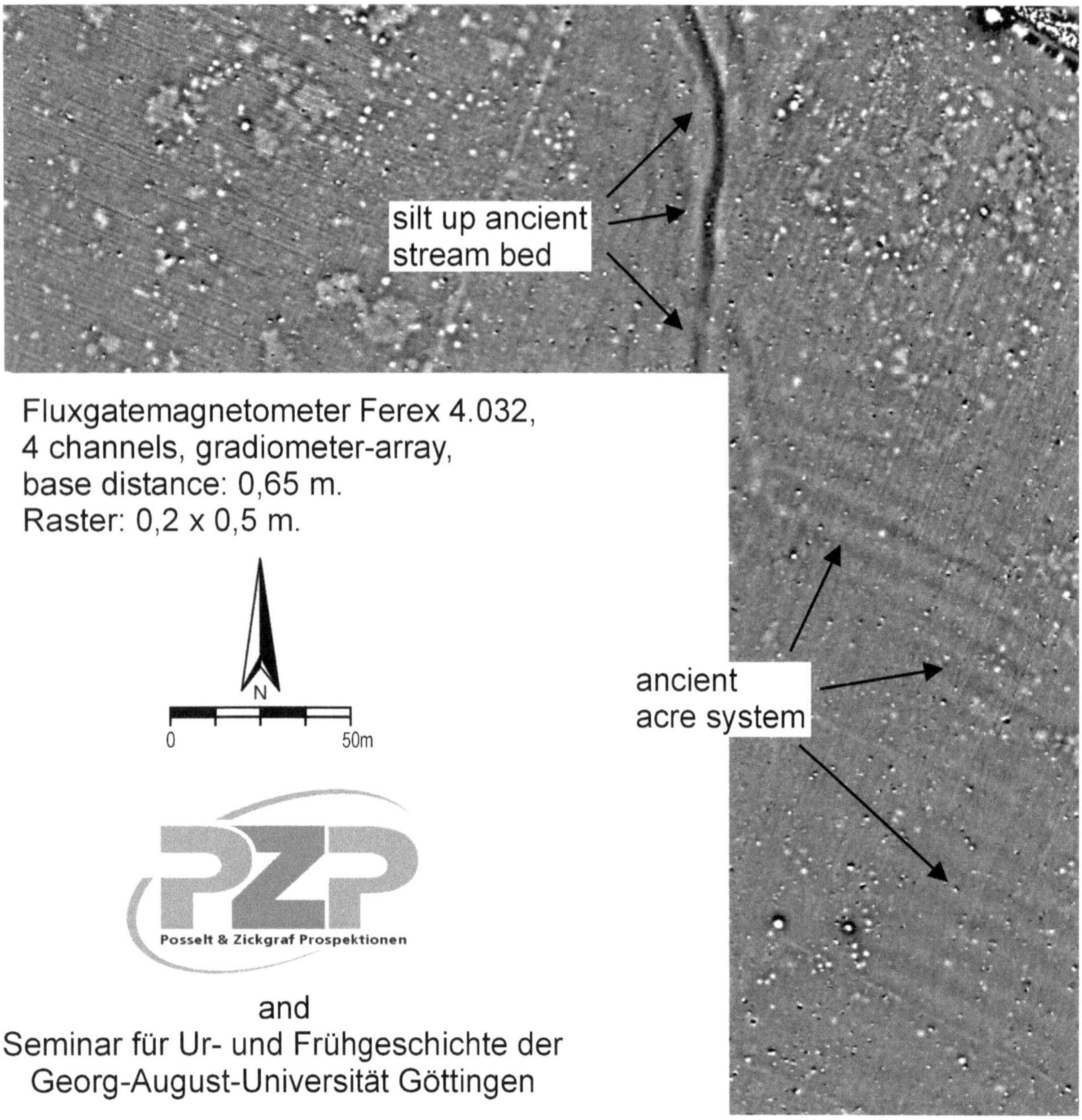

Fig. 3. Gevensleben-Watenstedt. Hillfort and settlement "Hünenburg". Detail of the magnetogram from the settlement south of the hillfort including archaeological features, landscape features and medieval acre systems. State: September 2008 (map courtesy of PZP/Seminar für Ur- und Frühgeschichte Göttingen).

References

GAFFNEY, C., 2008, Detecting trends in the prediction of the past: a review of geophysical techniques in archaeology. *Archaeometry,* 50, 313-336.

HESKE , I., 2000 Ein bewehter Platz an einem bewährten Ort. *Archäologie in Niedersachsen,* 2000, 17-19.

HESKE , I., 2002 „...und der leuchtende Schild auf der Schulter". *Archäologie in Niedersachsen,* 2002, 12-14.

HESKE, I., 2003, Die Hünenburg bei Watenstedt, Ldkr. Helmstedt. Vorbericht über die Prospektionsgrabungen der Jahre 1998 bis 2000. *Nachrichten aus Niedersachsens Urgeschichte,* 72, 15-27.

HESKE, I., 2007, Eine steinerne Sichelgussform aus der jungbronzezeitlichen Außensiedlung der Hünenburg bei Watenstedt, Ldkr. Helmstedt. *Nachrichten aus Niedersachsens Urgeschichte,* 76, 29-39.

HESKE, I., POSSELT, M., 2009, Bericht über die Ausgrabungen im Außenbereich der jungbronzezeitlichen Höhensiedlung „Hünenburg" bei Watenstedt, Kr. Helmstedt, in den Jahren 2005-2006 (prepared manuscript).

KVAMME, K.L., 2003, Geophysical surveys as landscape archaeology, *American Antiquity,* 63, 435-457.

MEYER, J.-W., 2007, Veränderungen der Grabungsstrategie in Tell Chuera (Syrien) aufgrund der Ergebnisse der geomagnetischen Prospektion. In: Posselt, M., Zickgraf,

B., Dobiat, C. (eds.), *Geophysik und Ausgrabung. Einsatz und Auswertung zerstörungsfreier Prospektion in der Archäologie. Internationale Archäologie. Naturwissenschaft und Technologie 6* (Rahden/Westf.) 223-236.

POSSELT, M. 2001, Spurensuche mit dem Magnetometer an der Hünenburg. In: G. Biegel/A. Klein (Hrsg.), *Die Hünenburg bei Watenstedt. Ausgrabungsergebnisse 1998-2001. Braunschweigisches Landesmuseum, Informationen und Berichte* 3-4, 2001, 26-29.

SCHADE, C., 2000, *Landschaftsarchäologie – eine inhaltliche Begriffsbestimmung. Studien zur Siedlungsarchäologie II Universitätsforschungen zur prähistorischen Archäologie 60* (Frankfurt/M.) 135-225.

Session 6

DIGITAL CULTURAL ATLASES (PROJECTS AND PROTOTYPES) AND CULTURAL ATLAS COMPONENTS (GAZETTEERS, TIME-PERIODS, BIOGRAPHY AND SOCIAL NETWORKS, THESAURI, TECHNICAL INFRASTRUCTURE, CONTENT SOURCES AND DISPLAY MODELS)

Digital Historical Atlas of South India

F. Borne, Y. Subbarayalu, G. Muthusankar, K. Iswarialakshmi and D. Balasubramanian

IFP (French Institute of Pondicherry), UMIFRE 21 CNRS-MAEE,
11, St.Louis Street, Pondicherry 605001, INDIA

Abstract

The Digital Historical Atlas of South India is a desktop/web-based application. It has been prepared using the latest Geographic Information Systems (GIS) technology. Its geographical coverage includes entire states of Tamil Nadu and Kerala and two pilot areas in the states of Andhra Pradesh and Karnataka. It covers vast time range from prehistoric times to AD 1600 organized into seven major periods and under four major themes. Altogether, there are at present 7000 sites of archaeological and historical importance put on the maps with all the related theme-based information. Facilities are provided to navigate through, and interact with, by querying, the entire database by time, space, and themes. Maps have been illustrated by three different versions, namely Scalable Vector Graphics (SVG) version, Google Earth version and Map Server version, the last one being in the experimental stage.

Keywords Digital atlas, South India atlas, Prehistoric, Historic, GIS, SVG, DOM, XML, Google Earth

Introduction

The Historical Atlas of South India (from prehistoric times to AD 1600) is a joint venture of the French Institute of Pondicherry in collaboration with four south Indian Universities, namely Tamil University of Thanjavur, Mahatma Gandhi University of Kottayam, University of Mangalore and University of Hyderabad. A large team of historians, archaeologists and GIS experts was involved in the preparation of the atlas during 2005-2008. The Ford Foundation supported the project with a liberal grant. The atlas is a computer application based on CD/DVD and has been prepared using the latest Geographic Information Systems technology. It could be accessed using popular browsers and it is made available through Internet at: http://www.ifpindia.org/hatlas/.

At present the entire area included in the present states of Tamil Nadu and Kerala has been put on maps. In addition to that, the mapping of two pilot areas that were selected in the states of Karnataka and Andhra Pradesh has also been completed. The atlas covers a vast time range from prehistoric times to AD 1600 and is organized into seven major periods, three encompassing the prehistoric times and four the historical times. The information is organized under four major themes, namely political, social, economical and cultural. Altogether, there are at present 7000 sites of archaeological and historical importance put on the maps with all the related theme-based information. Facilities are provided for the prospective users to navigate through, and interact with, by querying, the vast historical and archaeological database by time, space and themes. In addition to the conventional method of indicating the geographical information by self-explanatory symbols, text and visual material (about 10,000 image files) are also linked to the sites wherever necessary. For making the atlas, in addition to the material available in standard secondary works, like the archaeological reports, inscriptions, and literature, and old maps, field surveys were made by the research staff of the project to check the geo-coordinates and make photo documentation of several known sites. In the process, nearly 3000 new archaeological sites were freshly identified and documented. This is an additional benefit of the project, as many of the archaeological sites are fast disappearing due to expansion of agricultural activities, increasing urbanization, etc.

Infography of the atlas

The static DVD version of Historical atlas application uses Scalable Vector Graphics for maps, XML for data representation, XPath for data navigation, and DOM for updation of data dynamically and client-side language javascript.

Represention of maps

Maps have been illustrated through Scalable Vector Graphics (SVG), a language for describing two-dimensional graphics in XML. SVG is a vector technology for providing high-quality, interactive maps, which is a World Wide Web Consortium (W3C) recommendation. It provides many advantages compared to the use of a raster-based presentation, such as the quality of the graphical representation, maintenance, interactivity, and extensibility through other Web programming languages. The main advantage of SVG graphics is that the users can infinitely zoom without losing cartographic quality. SVG is a promising technology for delivering high-quality, fully-vector topographic maps, via the Internet or local system, both in terms of graphic quality and interactivity.

The file size of the SVG document is an important aspect in publishing the maps through internet. The file size is quite smaller and loads much faster in a Web browser. Further these svg documents can also be compressed (to svgz) to reduce the file size.

The number of interactive features that can be implemented in SVG, using the client-side language Javascript, is basically unlimited. Zooming, turning layers on and off, panning, measuring the distance between two points, providing coordinate readouts, and displaying attributes through a mouseover function represent basic functions that would allow interactive manipulation with SVG maps.

DOM

The Document Object Model is a platform- and language-neutral interface that will allow programs and scripts to dynamically access and update the content, structure and style of documents. It provides a structured, object-oriented representation of the individual elements and content in a page with methods for retrieving and setting the properties of those objects. It also provides methods for adding and removing such objects, allowing one to create dynamic content. The DOM also provides an interface for dealing with events, allowing us to capture and respond to user or browser actions.

Database

Database is required to store the information pertaining to site details, while interacting with the maps and responding to the requests. Extensible Markup Language (XML) is used for storing the information about sites, site co-ordinates, and other site details such as images and their description

XML representation of site data:

```
<?xml version="1.0" encoding="utf-8"?>
<siteData>
      <row>
           <sitecode>420</sitecode>
           <site>mayiladumparai</site>
           <hsite>0</hsite>
           <lat>12 25 12</lat>
           <long>78 18 48</long>
           <elev>461</elev>
           <photo>OBS</photo>
           <sitetext>DES</sitetext>
                 <assocperiods>Period
             II (Cultural, Economical);
             Period   III   (Cultural,
             Social,       Economical);
             Period   IV    (Cultural,
             Social,        Economical,
             Political)</assocperiods>
           <themes>Cultural,
      Economical</themes>
      </row>
</siteData>
```

Images XML:

```
<?xml version="1.0" encoding="utf-8"?>
<imageData state_code="tn"
site_name="mayiladumparai" period="1">
<record object="obs">
       <row>

      <imgName>tn_0420_1_sa_mayiladumparai_
      obs_01.jpg</imgName>
          <imgCaption>Mayiladumparai:
           Microlithic tool found in the
           excavation</imgCaption>
          <imgDesc>~~~</imgDesc>
       </row>
       <row>

      <imgName>tn_0420_1_sa_mayiladumparai_
      obs_02.jpg</imgName>
          <imgCaption>Mayiladumparai:
          Microlithic tools found in the
          excavation</imgCaption>
          <imgDesc>~~~</imgDesc>
       </row>
       <row>

      <imgName>tn_0420_1_sa_mayiladumparai_
      obs_03.jpg</imgName>
          <imgCaption>Mayiladumparai:
          Microlithic tools found in the
          excavation</imgCaption>
          <imgDesc>~~~</imgDesc>
       </row>
</record>
</imageData>
```

XPATH

The XPath language is based on a tree representation of the XML document. It is used to navigate through elements and attributes in an XML document and also selecting nodes by a variety of criteria. Both XML and XPath are recommended by W3C.

Explore the digital atlas

If you look at the Figure 1, on the top right-hand side, you can see the overview map. Its purpose is to help you locate where your current map view is relative to the full extent of the map. It is also possible to use the overview map to change the location in the map. From the top menu, user can switch to different period or knowledge. On the left-hand side, the user can view the legend details. Options are also provided to query the database and view all the attributes in the database. Numerical scale and scale bar is available below the title. Apart from pan tool, the user can change the location using the combo box.

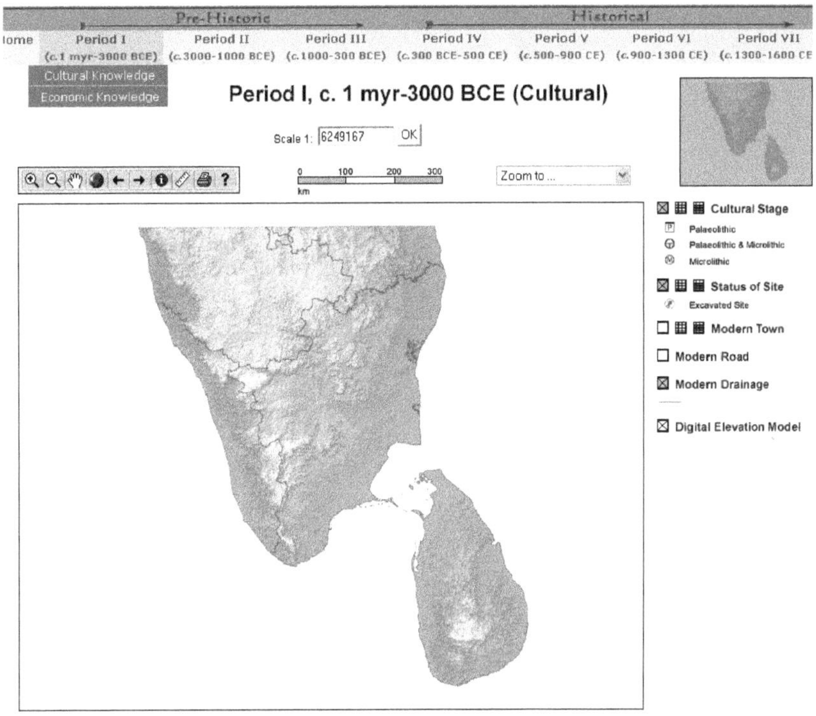

Fig. 1. View of Period I, cultural map.

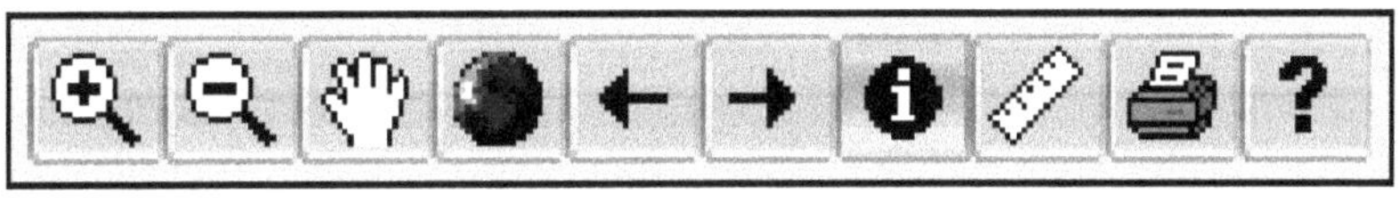

Fig. 2. Tool box in svg map.

The toolbox (Figure 2) provides options to interact with the map, namely zoom in, zoom out, pan, back to original view, previous view, back view, object information, measure tool, print and help.

When the users click any site, they can see the popup window as Figure 3. The site window has three tabs namely, Site Details, Images and Text. Site Details tab provide the details like modern and ancient site name, latitude and longitude, elevations and its associated periods and knowledge (Figure 3). Images tab shows the images associated to the site, if any. The user can also navigate to other images in the site using hyperlinks or previous/next buttons (Figure 4). Text tab shows the important information about the site (Figure 5).

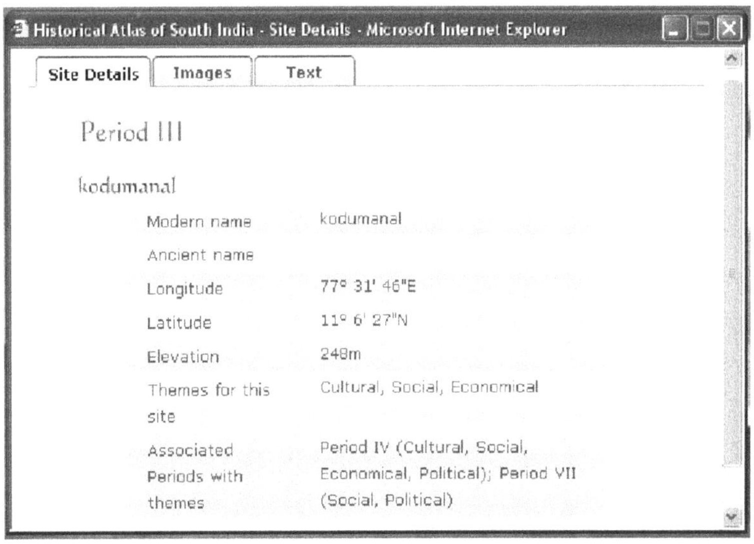

Fig. 3. Kodumanal site details.

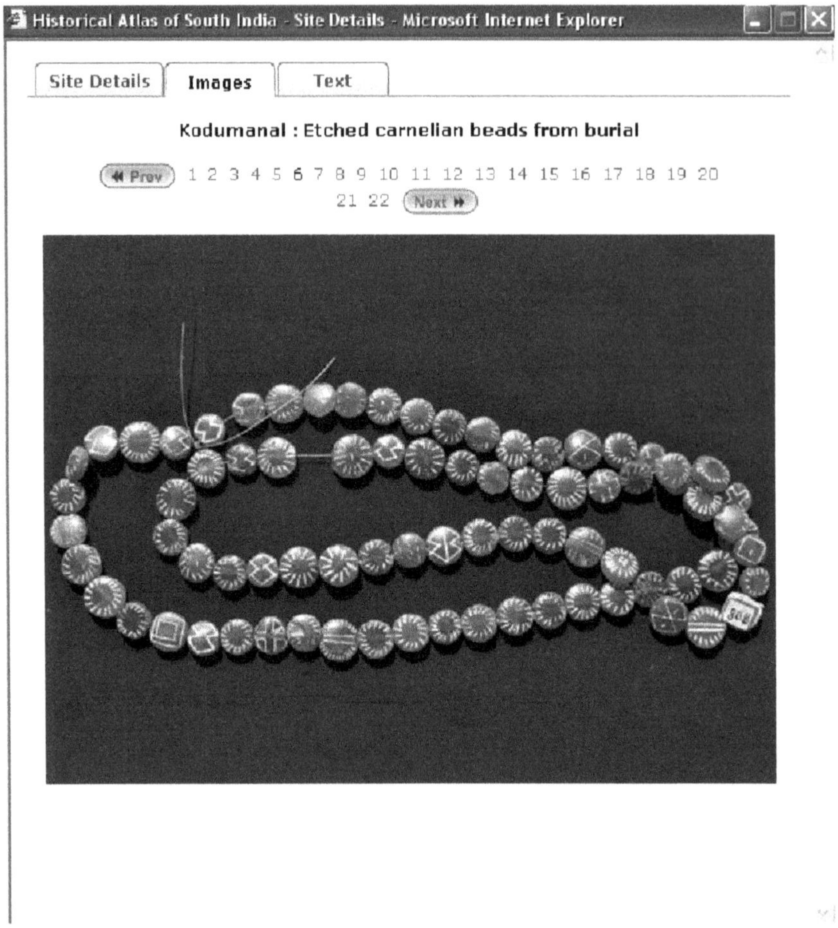

Fig. 4. Images of Kodumanal site in Period I, cultural knowledge.

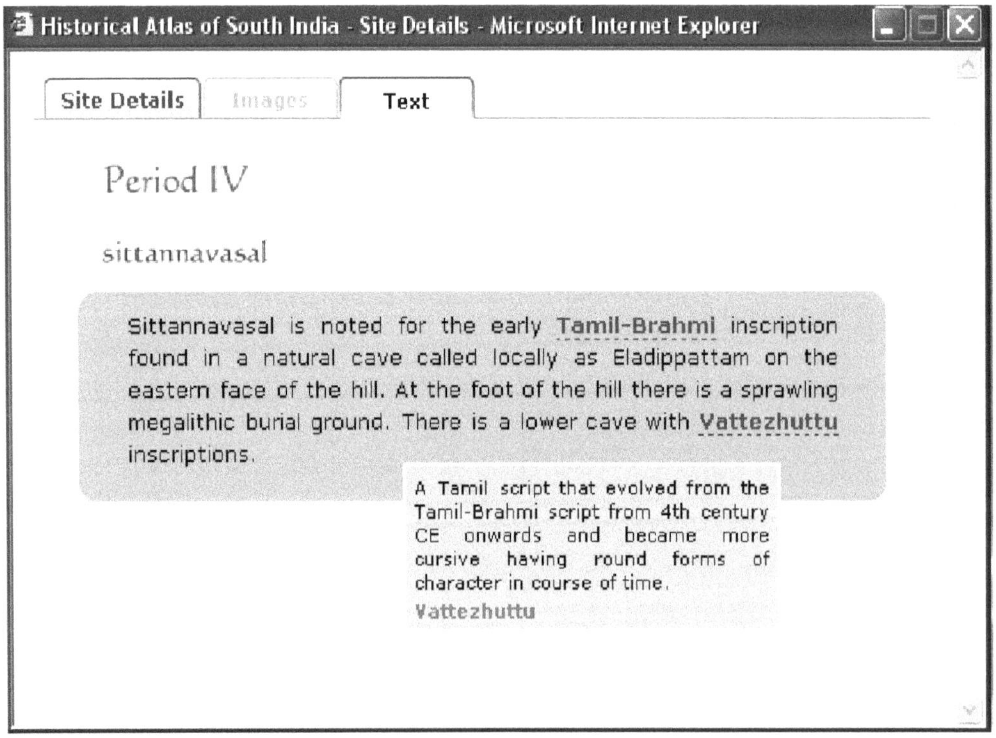

Fig. 5. Description about Sittannavasal site, Period IV, Social knowledge with glossary.

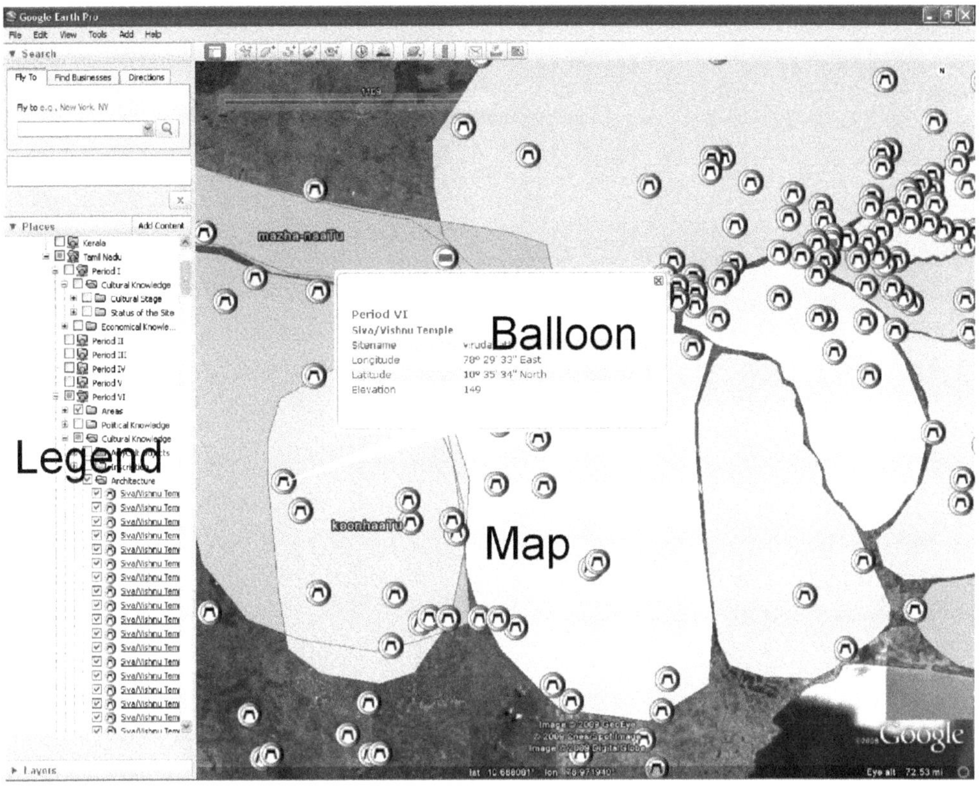

Fig. 6. Digital Atlas in Google Earth.

Google Earth version of the Historical Atlas of South India

Introduction

Google Earth is client software; it reads KML or KMZ files in the same way a web browser reads an HTML page. The data can be situated on the local system or can be accessed from a remote machine; similarly to dynamic web pages, you can generate dynamic KML data too. We chose the latter solution to implement the Historical Atlas.

Principle

On the user's machine, in Google Earth, the user opens a KMZ file downloaded from the French Institute website. Google Earth then calls a PHP program developed by the LIAG situated on the Apache web server of the Institute. This program reads the data stored in the database, which is also situated on the server of the Institute, and formats it into a proper KML code, containing the data (latitude, longitude, description, etc) and the way to display it (icons, size, balloon color, etc). Finally, the web server sends the data back to the user's machine, to display it in Google Earth.

Example of KML code for a site:

```
<Placemark>
      <name>Microlithic Tools</name>
<snippet>Anakkara</snippet>
<Snippet
maxLines="2">Anakkara</Snippet>
      <!-- description -->
 <description><![CDATA[<table
 width='400px'>
 <tr><td>Period I</td></tr>
 <tr><td>Legend</td><td>Microlithic
 Tools</td></tr>
 <tr><td>Modern Name</td><td>Anakkara
 </td></tr>
 <tr><td>Ancient
 Name</td><td></td></tr>
 <tr><td>Longitude</td><td>76° 2' 33"
 East</td></tr>
 <tr><td>Latitude</td><td>10° 49' 00"
 North</td></tr>
 <tr><td>Elevation</td><td>28
 m</td></tr>
 <tr height='6px'></tr>
 </table>]]></description>

 <!-- time information -->
 <TimeSpan>
      <begin>-5000</begin>
      <end>-3000</end>
 </TimeSpan>

 <!—style definition -->
 <styleUrl>#mapm_tl</styleUrl>
 <!— geometry information -->
 <Point>
```

```
	<altitudeMode>relativeToGround</
altitudeMode>
	<coordinates>76.04272,10.81694,0
</coordinates>
</Point></Placemark>
```

Features

It consists of points, standing for the objects discovered on the sites of study, and polygons standing for ancient administrative boundaries/political areas. By clicking an object, a balloon will appear, showing information about the site (name, longitude, latitude, etc) and, eventually, a link to display images or texts related to the object (Figure 6).

Software used

- Google Earth (client mapping software)
- PostgreSQL (database server)
- Apache (web server)
- PHP (server-side program interpreter)

Conclusion

This web application will give easy access to varied historical and archaeological information and it can be used as a powerful research tool, besides of course as a pedagogic tool. Being an application based on GIS, updating and modifying the contents of the atlas becomes relatively easier.

Bibliography

Website of the Digital Historical Atlas of South India : http://www.ifpindia.org/hatlas/

2007, BORNE F., SUBBARAYALU Y., G. MUTHU SANKAR, D. BALASUBRAMANIAN, *Historical Atlas of South India in Digital Form*, Map World Forum, 22 – 25 January 2007, Hyderabad, INDIA

Documentation and visualization of archaeological data

Abhijit Pillai,[1] Onkar Dikshit,[1] G. K. Rai,[2] R. S. Fonia,[3] V. N. Prabhakar,[3] Bhuvan Vikram[3]

[1]Department of Civil Engineering, IIT Kanpur
[2]Department of Ancient History, Culture and Archaeology, University of Allahabad
[3]Archaeological Survey of India

Abstract

Cultural Resource Management (CRM) has been widely used for solving numerous problems in archaeology. In India, however, developments in this area have been virtually non-existent. Conventional methods of recordkeeping followed, have limited scope for archaeological applications. This work is a pioneering effort for development of CRM system for documentation of Indian monuments and antiquities.
Work involves documentation of archaeological data into digital form that can be used for GIS based analysis. Separate CRM tools are built for works related to antiquities and monuments that are capable of performing data recording, modification, and manipulation in the form of non-spatial queries. It is possible to catalogue the outputs of these queries into reports, datasheets and tabulated records with spatial information for antiquities and monuments. One of the prominent features of RISHI is the conversion from ASI Grid-Square-Trench-Quadrant system to a global coordinate system.
A spatial analysis tool is also developed for antiquities using which various spatial queries can be performed on outputs of the non-spatial queries. These queries are supported by statistical reports. The outputs of the spatial queries are brought into a three-dimensional scale for visual analysis. Along with the antiquity data, stratigraphic layer information is also incorporated for better analysis.

Keywords Cultural Resource Management, archaeology, visualization, spatial analysis, RISHI, DHAROHAR, antiquities, monuments, grid conversion

Introduction

Archaeology, as a practice, has been well served in India. However, archaeologists in India still follow the age old conventional techniques of data recording, data handling, manual drawings and coarse interpretations. According to Gupta (2002, 5), a total systematic documentation has never been attempted in India. Archaeology is an erudition discipline where the knowledge grows up in necessarily referencing the precious documents already gathered. This makes archaeological data voluminous (Meyer *et al.*, 2007). The comprehensive data format is recorded manually and is in a hard copy form that makes data accessibility, modification and retrieval very difficult. The present data format is also unorganized and lacks structural flexibility.

Once out of the excavation pit, an archaeological material loses its spatial significance. It is necessary to recreate this spatial information in the form of some virtual model in order to have some useful derivations out of it. However, the present representations of manually prepared cross-sectional and planimetric drawings can hardly be put into any constructive use.

These limitations in spatial and non-spatial data representation retard the purpose of archaeological interpretations. Hence, it is necessary to switch over from the present hard copy form to the digital soft copy format that will facilitate the automatic processing, analysis and identification of archaeological data from the voluminous dataset.

Objectives and Scope

The objective of this work is to develop a Cultural Resource Management System for documentation of Indian monuments and antiquities that will help build a framework to support the future archaeological studies.
The scope of the work includes:

1. Documentation of archaeological data
2. Geospatial and non-spatial data connectivity
3. Visualization and distribution analysis of archaeological data on a map
4. 3D visualization of archaeological data

Initial usage of the CRM tools for monuments and antiquities is restricted to the archaeologists of the country, who hold access to data entry and data modification. After a complete database is build and web based connectivity is established, these tools will be of interest to several classes of people right from general public to the educational institutions and archaeological researchers. With all the available archaeological data brought onto a single screen, a wider vision would help researchers (or so called cultural resource managers) and statisticians in better analysis of the archaeological data.

Study areas

Ahichchhatra is an archaeological site located at Ramnagar in Aonla Tehsil situated in Bareilly District of Uttar Pradesh that stretches from $28^0 21'52"$ N to $28^0 22'42"$ N in latitude and $79^0 07'38"$ E to $79^0 08'55"$ E in longitude. Surrounded by rivers Chambal, Ramganga and Gangham and Himalayas on the north, Ahichchatra was strategically important in the past, especially during the Panchala rule around 6th century BC. First excavation in Ahichchatra took place between 1940 and 1944. Painted grey wares of the Aryan period were found then. In the later excavations a wide range of artifacts belonging to periods between 2nd millennium BC to 11th century AD were identified. With good amount of terracotta materials, nearly five thousand coins of pre-Guptan period, pottery shreds and past historical records, Ahichchatra is one of the most significant archaeological sites in India.

Alongside the antiquities, CRM tool is developed for monuments belonging to the Agra Circle. Many historically significant places like Mathura, Kanauj, Fatehpur, Bareilly, and Meerut are a part of the Agra Circle. Presently, the work on CRM for monuments is primarily concentrated towards data management and preservation. Future works will involve detailed work on individual monuments.

Issues related to Indian archaeology

Indian archaeology is going through the initial phase of transition from age old archaeology to modern archaeology. Following are some priorities that were considered before proceeding with the problem at hand.

i) Good user interface that can facilitate easy data entry, modification and retrieval,
ii) The CRM tools capable of performing complex queries on monuments and antiquities; and it must be possible to store the reports and datasheet of the queried monuments and antiquities into a soft copy form as well as a hard copy form,
iii) CRM tools should have some access control, authentication and in-built security system,
iv) Transformation of spatial information for antiquities from the local ASI GSTQ system to a global projected system.

Development platform

CRM tools are developed with VisualBasics.Net 2005 as the front end and MS Access as the back end for database entries and manipulations. Database management system (DBMS) can be easily changed from Access to any other system like SQL Server or MySQL with minor changes in the connectivity. Installation setups have been created for both the CRM tools.

Outputs of the queries within the CRM tools are brought into the ArcGIS environment (developed by ESRI) for spatial analysis. ArcObjects is used as a programming tool for the spatial queries. ArcMap is applicable as a 2D visualization tool for both querying and displaying, while ArcScene supports the 3D representation of the data.

Approach

Instead of creating separate CRM tools for antiquities of different types (say coins, graves, etc.), a generalized approach is followed where every antiquity (or monument) is considered to be a part of a single set. Indian archaeology, as the case is with many other countries, is in many ways different from the western archaeology. Specifications, chronology, timeline, methodology vary. Hence, a generalized approach is a better way to first analyze this difference and make change accordingly. This helps in establishing the basic platform for all the antiquities from where specific tasks like statistical analysis of coins can be worked out.

A generalized approach is therefore advantageous in a number of ways. It helps to:

(i) relate two or more antiquities of different types according to their provenance, or even between two or more sites,
(ii) reconstruct the past by associating between structure (like wall), antiquities and the place of find,
(iii) determine the range of the antiquities at a particular place or source,
(iv) determine the density of all the antiquities at different locations for source or site identification,
(v) statistical analysis of antiquity data,
(vi) understand the socio-economic-political system present in the past, etc.

On similar lines, work is initially generalized for all the monuments. Subsequently, the work can be focused on individual monuments and management of entities associated with it.

Data

The work of converting archaeological data for Ahichchatra into digital form was initiated in 2006 by Tare (2007). He obtained topographic sheets, satellite imageries (PAN and LISS-4 sensors), Shuttle Radar Topographic Mission (SRTM) data (C band), and Google Earth images for Ahichchatra and neighboring areas. Topographic sheets were procured from Archaeological Survey India (ASI) and satellite imageries from National Remote Sensing Agency (NRSA now known as National Remote Sensing Centre (NRSC)).

Survey data

A variety of field surveys – Fly leveling using automatic level, GPS, Total Station were carried out for Ahichchatra by Tare (2007). Fly leveling was used to setup a benchmark from where other levels like that of the highest peg within a grid (datum for depth measurements) were calculated. GPS survey was carried out to establish precise ground control points (GCPs). These GCPs were used as reference points for high precision total station survey. About 7000 survey points were collected within the site using which a Digital Elevation Model (DEM) was generated to represent the surface topography. In 2008, a detailed survey was again carried out in the western parts of the site.

Vector data

Important features like rivers, roads, surrounding villages, bastions, etc. are a part of Ahichchatra and its premises. Digitized data of these features are available from the previous works.

Planimetric drawings of excavated features (like ancient walls) from plane table survey carried out at Ahichchatra hold detailed information about the structures found within an excavated area. This is very important contextual information for the archaeologists that can be used to relate the antiquities to their provenance. These drawings were scanned and vectorised.

Raster data

Raster data is used as a background to associate the vector data and run spatial queries. Google Earth imagery is also superimposed on the Digital Elevation Model (DEM) generated using total station data that replicates the site topography in ArcScene.

Synthetic data

Sample records for both antiquities and monuments were obtained from ASI. These records are not complete as yet. Some very essential data like stratigraphic layer information are not available. It is expected to take some time to devise a method for collecting these data. In order to accommodate stratigraphic layer information in the present work, synthetic layers were generated using the surface information i.e. total station data (Example 1).

Data organization

The present dataset is a simple long list of features that does not classify the data into proper subsets. This adds to the complications in relating different data. The data is organized into a Relational DBMS with data for antiquities organized into nine tables (Figure 1) and for monuments into 13 tables (Figure 2).

Sr. No	Table	Feature(s)	Functionality
1.	tblSite	Accession number, site name, locus, quadrant, depth, period, date of find	Location and period of the artifact identified by accession number and site
2.	tblLayer	Layer	Layer information
3.	tblDump	Dump, dump sealed by	Dump information
4.	tblPit	Pit, pit sealed by	Pit information
5.	tblObjMat	Object name, object category, material name, material category, shape and weight	Non-spatial properties of the artifact
6.	tblPeg	Peg1 number, peg1 distance, peg2 number, peg2 distance	Planimetric location within the ASI grid
7.	tblThreeD	Length, width, height	Dimensions (regular shape)
8.	tblTwoD	Major axis, minor axis	Dimensions (irregular shape)
9.	tblTime	Last modified at, date of recording	Recordings

Fig. 1. Antiquities: Classes and features.

Sr. No	Table	Feature(s)	Functionality
1.	tblIdentification	UniqueID, Monument name, Common name, Circle office, Street, Region, District, State, Topography, Part of, Site-in-charge, site officer, nodal officer	Location & administration, identification by UniqueID, monument name and common name
2.	tblNature	Nature, Architectural Style, Period, Timeline, Typology, Builder or Patronage, Building material, Construction method, orientation, length, width, height	Archaeological significance
3.	tblUsage	Office, museum, store, guest house, institution, other	Current usage of the site
4.	tblProtection	Year of protection, area of protection, notification number, declaration date, whether demarcated,	Whether protected or not, if yes - details.
5.	tblDeprotection	Already deprotected or proposed for deprotection or decision pending	Whether deprotected, if yes - details
6.	tblVisitors	Average, peak season, during functions. Ticketed or not ticketed	Statistics on number of visitors and tickets
7.	tblNecessaryAmenities	Toilets, drinking water, handicap provisions, tourist information centre, dustbin, shoe cover, shoe racks, cloak rooms, benches, others	Important amenities necessary at the site
8.	tblSupplementaryAmenities	Signages, approach road, parking, flood lights, cafeteria or shops, sound and light, lawns or gardens, guide book, cultural notice board, publication counters, others	Additional amenities at the site
9.	tblOwnership	ASI, central govt., state govt., other govt. agencies, religious trust, private party, other	Responsible agency for site ownership
10.	tblCoordination	State govt., other govt. agencies, local authorities, religious trust, private party, other stockholders	Division of work
11.	tblTraditionalPractices	On site, in the vicinity, festivals, continued or discontinued, living or non-living	Any traditional practices associated with the site
12.	tblPublishedReferences	Names of published references	Additional information about a monument
13.	tblCourtcases	Case number, court name, dispute, result, status	Information about court cases for the site

Fig. 2. Monuments: Classes and features.

Database normalization

Normalization achieved for the current database is of the 3NF (third normal form) as every attribute in every table is directly linked through the primary key of the relation.

Thus, the access to an antiquity (or a monument) is directly through the accession number. Changes are made via accession number alone. This avoids unwanted modifications to the antiquities and monuments, those are not referred to.

Functionalities of the CRM tools

The CRM tools, RISHI and DHAROHAR, are capable of new data entry (and deletion), data modification, data retrieval and complex queries on antiquities and monuments respectively. Figure 3 and 4 show the main window forms for RISHI and DHAROHAR. It is possible to generate reports of the queries (Figure 5) and datasheets that hold information for individual monument (Figure 6) and antiquity.

A security system is built which has the following facilities

- administrative rights
- new user creation
- add and change user details
- change password
- identify the last user who modified the database

RISHI (CRM tool for antiquities) has a special functionality that converts the spatial information of every antiquity from a local system (ASI GSTQ system) to the global projected system (UTM-44) for GIS based applications. In case of monuments, spatial information is manually entered.

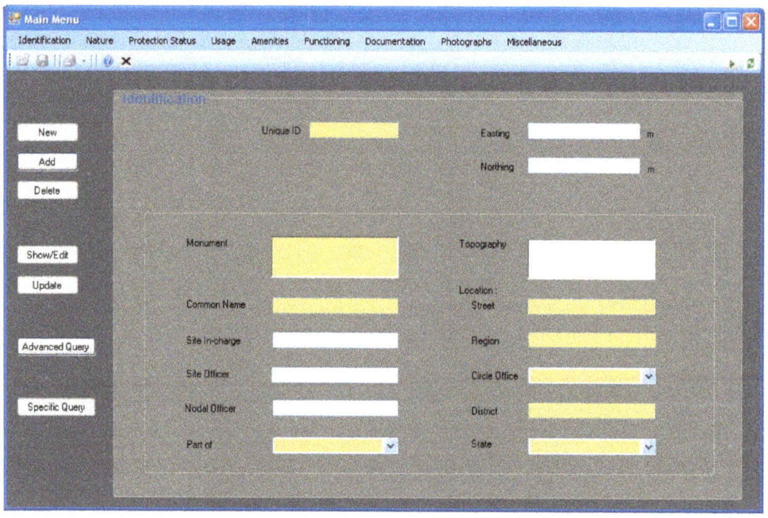

Fig. 3. Main window form for DHAROHAR

Fig. 4. Main window form for RISHI.

DHAROHAR
Monument Query Report

Circle Office : Agra | Cultural Period : Later Mughal | Guide Book : No

State : Uttar Pradesh | Time Frame : 17 AD

Region : Tajganj | Ticketing : Yes

UniqueID	Monument	Yr.of Prot.	State	Circle	District	Partof	Period	Time Line	Archi. Style	Typology	Construct. Method	Building Material	Guide Book
101	Drinking Fountain-Taj Garden	1920	Uttar Pradesh	Agra	Agra	Taj Group	Later Mughal	17 AD	Mughal	Fountain / wall			No
102	Gateway Khan-I-Alam Bagh	1920	Uttar Pradesh	Agra	Agra	Taj Group	Later Mughal	17 AD	Mughal	Gateway / wall			No
103	Fatehpur Masjid	1920	Uttar Pradesh	Agra	Agra	Individual	Later Mughal	17 AD	Mughal	Mosque			No
104	Kali Masjid & Enclosure Wall	1920	Uttar Pradesh	Agra	Agra	Individual	Later Mughal	17 AD	Mughal	Mosque of wall			No
105	Khan-I-Alam Bagh & New Tank	1920	Uttar Pradesh	Agra	Agra	Taj Group	Later Mughal	17 AD	Mughal	Bagh			No
106	Old Mughal Aqueduct	1920	Uttar Pradesh	Agra	Agra	Taj Group	Later Mughal	17 AD	Mughal	Water structure			No
107	Sahelion-ka-Gumbaz No.1	1920	Uttar Pradesh	Agra	Agra	Taj Group	Later Mughal	17 AD	Mughal	Bastion			No
108	Sahelion-ka-Gumbaz No.2	1920	Uttar Pradesh	Agra	Agra	Taj Group	Later Mughal	17 AD	Mughal	Bastion			No
109	Sahelion-ka-Gumbaz No.3	1920	Uttar Pradesh	Agra	Agra	Taj Group	Later Mughal	17 AD	Mughal	Bastion			No
110	Sahelion ka Gumbaz No.4	1920	Uttar Pradesh	Agra	Agra	Taj Group	Later Mughal	17 AD	Mughal	Bastion			No
111	Sirhi Darwaza - Taj Mahal	1920	Uttar Pradesh	Agra	Agra	Taj Group	Later Mughal	17 AD	Mughal	South gate of Taj			No
112	Taj and its Premises	1920	Uttar Pradesh	Agra	Agra	Taj Group	Later Mughal	17 AD	Mughal	Tomb			No
100	Dalan Round Taj Quadrangle	1920	Uttar Pradesh	Agra	Agra	Taj Group	Later Mughal	17 AD	Mughal	Dalan (cloistered)			No

Fig. 5. Report for Queries in the CRM tool, Text in blue: Queried fields, Tabulated text: Output of the query and additional information associated with each out, Blank: Fields for which data is not available or entered.

UniqueID	1254
Monument	Lucknow Building

Location	
District	Lucknow
State	Uttar Pradesh
Circle Office	Lucknow
Region	Lucknow
Nature and Typology	
Part of	Individual
Time Line	15-16 AD
Cultural Period	Mughal
Typology	Fort
Architectural Style	Mughal
Construction Details	
Construction Method	Mixed
Build. Material	Ashler Stone
Protection Status	
Ownership	
Year of Protection	1968
Area of Protection	350 acre
Visitors	
Ticketing	Yes
Guide Book	Yes
Basic Amenities	
Supplimentary Amenities	

Fig 6. Datasheet for individual monument

Grid conversion

Spatial data for antiquities on the ASI GSTQ system is complete, but latitude and longitude information (i.e. on a global coordinate system) is coarse. Precise latitude-longitude information is obtained by applying an affine transformation from ASI Grid system to the Universal Traverse Mercator (UTM) coordinate system by the sequence of operations in Figure 7.

Direct conversion from ASI GSTQ system to the UTM-44 system is not possible. Hence, a virtual grid system was designed by Srivastav (2008) to facilitate transformation from locus information to virtual grid system to UTM-44 coordinate system (Figure 8). 1km x 1km areas are marked such that coordinates at north-west corner of grid A is (1000m, 1000m). The site of Ahichchatra is covered under the four grids A, B, C and D. Origin is always considered to be the north-west corner, irrespective of whether it is for a grid, a square or a locus.

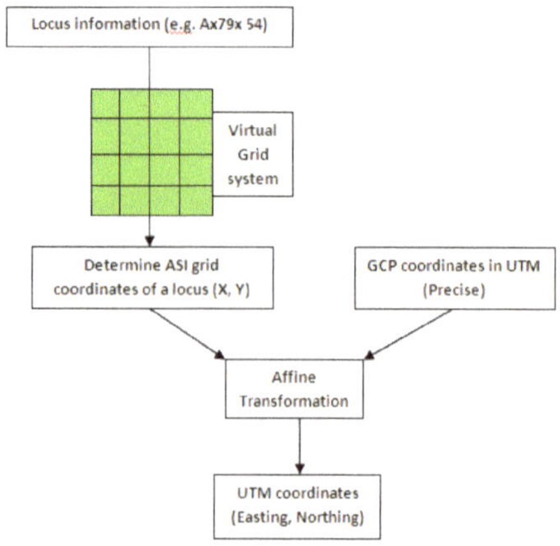

Fig. 7. Flow chart: Grid conversion.

Virtual Grid system

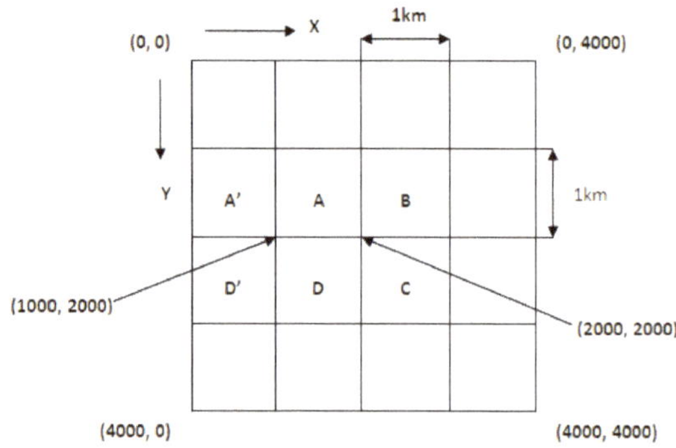

Fig. 8. Virtual Grid system

ASI GSTQ system

Figure 9 is a zoom from grid A to the trench no. 67 within the 79th square of the grid. Each grid is divided into 100 squares, each of 100m x 100m. Every square is further divided into 100 trenches of 10m x 10m. Every trench is again partitioned into four quadrants. The coordinates of locus are calculated as described in Example 2.

Locating antiquities

Knowing the coordinates of the locus, the X and Y coordinates of any antiquity can be measure (Figure 9). Depth is measured from a datum established with respect to the highest peg within a grid.

Affine transformation

A two-dimensional affine coordinate transformation was used in order to fit the regression line of Easting and Northing on X and Y.

$$Easting = a_0 + a_1 X + a_2 Y \quad (1)$$

$$Northing = b_0 + b_1 X + b_2 Y \quad (2)$$

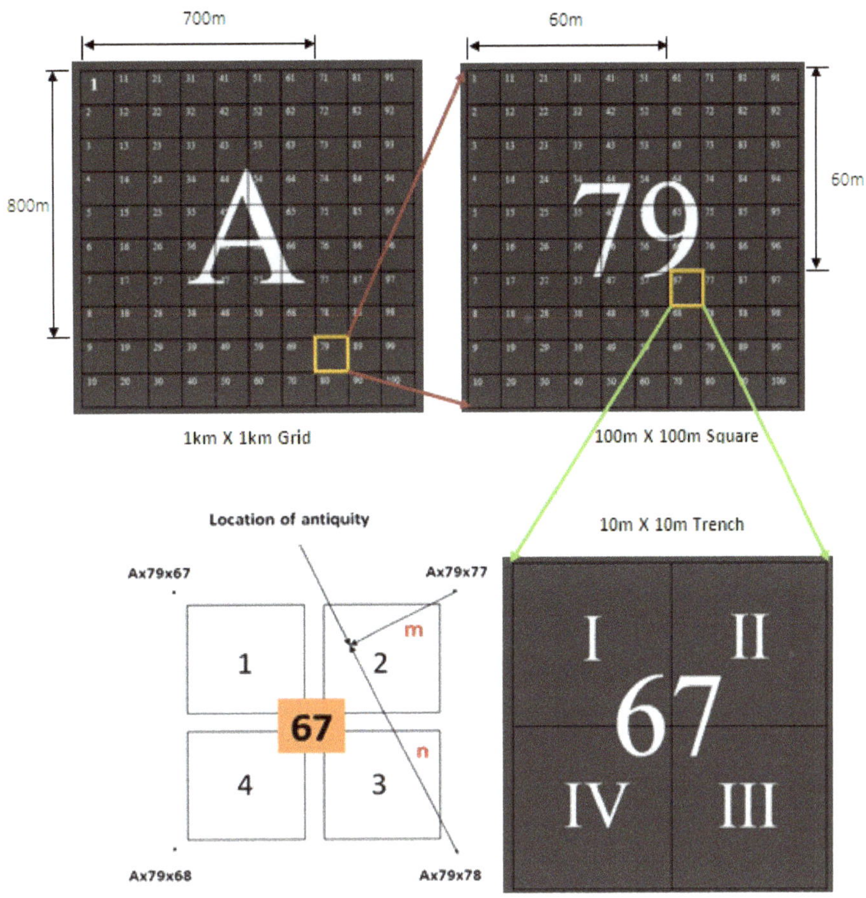

Fig. 9. Getting coordinates (*X*, *Y*) in the virtual system (a) 1000m X 1000m Grid (b) 100m X 100m Square (c) 10m X 10m Trench (d) Survey of coordinate computations in Virtual Grid system.

The minimum number of GCPs required to solve affine transformation are three i.e. (*E1*, *N1*), (*E2*, *N2*) and (*E3*, *N3*). CRM tool for antiquities (RISHI) provides the facility to enter maximum of ten GCPs (Figure 10). Once approximate coordinates are obtained, it is important to check the deviation of the results based on the GCP coordinates that can be considered accurate. Hence, Root Mean Square Error (RMSE) i.e. the resultant error of deviation in *X* and *Y* is calculated as

$$RMSE = \sqrt{\frac{(\sum Residual_{Easting})^2 + (\sum Residual_{Northing})^2}{n - d.o.f}} \quad (3)$$

Where, *d.o.f* is degree of freedom (three in this case)

$$\sum Residual_{Easting} = \sum (approx.Easting - accurate\ Easting)_{GCP}$$

$$\sum Residual_{Northing} = \sum (approx.Northing - accurate\ Northing)_{GCP}$$

For checking the consistency of the coordinates, the CRM tool provides a facility to accept or reject a GCP reading based on the

$\sum Residual_{Easting}$ and

$\sum Residual_{Northing}$.

Figure 10 shows entry for eight GCPs of eight different locus points. Easting and Northing coordinates are estimated based on affine transformation using GCPs. Comparing these coordinates to the actual coordinates, residual in both Easting and Northing are calculated. These are used to further calculate the total RMS Error given in the figure as *Sigma*. If the residual value for any GCP in either direction is very high (as in GCP 3), then that GCP is rejected (i.e. zero in the column *Acc/Rej* as for GCP 9 and 10); while others are retained. Based on the retained GCPs (minimum four GCPs) transformation is carried out in order to identify the coefficients a_0, a_1, a_2, b_0, b_1 and b_2 based on which Easting and Northing coordinates can be determined.

Non-spatial and geospatial connectivity

Once spatial information for antiquities (or monuments) is known, tabulated report of any query in the form of comma separated value (*.csv) file are also generated that hold spatial information about every queried antiquity (or monument) (Figure 11). This *.csv file can be imported into the ArcGIS

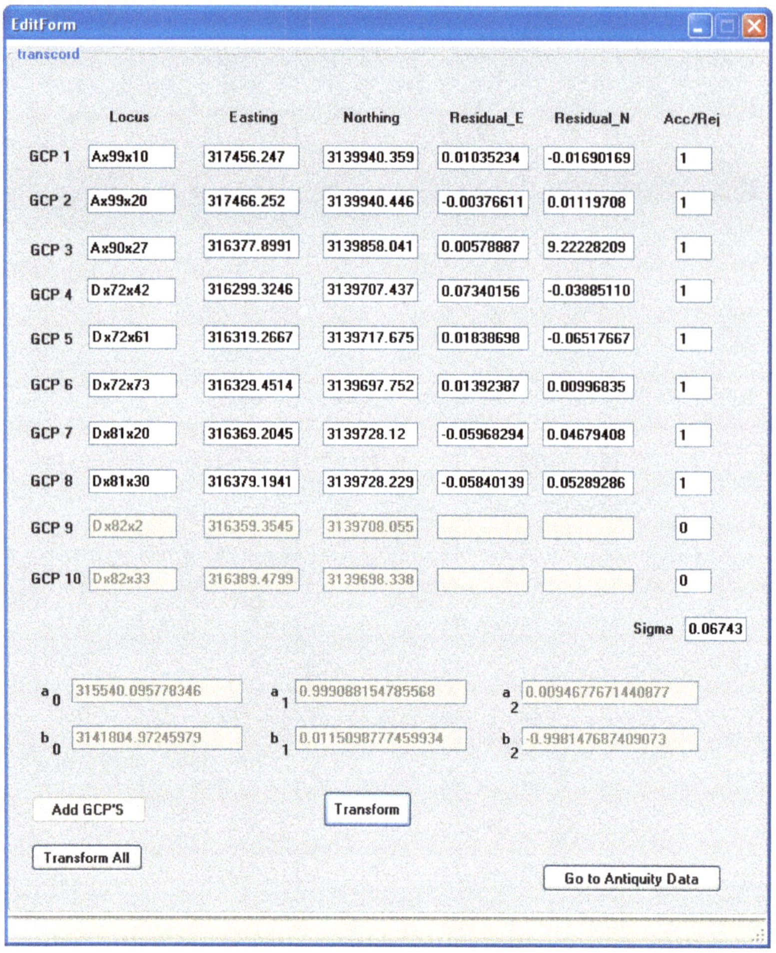

Fig. 10. Form for Grid conversion.

UniqueID	Easting	Northing	RL	Period	ObjectName	ObjectCategory	MaterialName	MaterialCategory
101	317396.8	3139580	171.43	Mughal	Ornaments	Beads	Stone	Crystal
102	317386.5	3139322	172.76	Mauryan	Weapons	Dagger	Bone	
103	317345.5	3139119	186.43	Mughal	Figurine	Animal	Stone	Quartz
104	317407	3139601	180.53	Mughal	Sculpture	Demi-Gods	Stone	Agate
105	317298.2	3139424	179.23	Mughal	Inscribed Objects	Seal	Wood	
106	317300.1	3139886	167.34	Gupta	Inscribed Objects	Inscription	Wood	
107	317376.1	3139432	188.24	Mauryan	Weapons	Dagger	Metal	Iron
108	317343.3	3139236	190.43	Mughal	Tools	Crucibles	Terracotta	Red
109	317396.5	3139009	175.61	Mughal	Tools	Spatula	Metal	Silver
110	317360.5	3139976	178.67	Gupta	Weapons	Razors	Bone	
111	317218.6	3139346	182.3	Mughal	Tools	Knife	Metal	Bronze
112	317267.3	3138765	176.43	Mughal	Ornaments	Bangles	Glass	
884	317245.7	3139877	177.75	Gupta	Coins	Gold		
883	317400.2	3138645	187.34	Mughal	Inscribed Objects	Stamped Object	Stone	Agate
882	317456.1	3138655	170.89	Mughal	Tools	Dies	Metal	Lead
881	317292	3139111	172.34	Mughal	Ornaments	Ear studs	Stone	Emerald
880	317493.7	3139655	176.34	Gupta	Sculpture	Ayudhpurush	Stone	Topaz
879	317177.7	3139580	190.23	Mauryan	Weapons	Chakra	Wood	
801	317287.5	3139777	184.34	Mughal	Coins	Copper		
802	317640.3	3139098	173.23	Mughal	Coins	Copper		
803	317512.1	3139984	166.44	Mauryan	Ornaments	Beads	Glass	
804	317446.5	3139128	160.12	Mughal	Weapons	Sword	Metal	Iron
805	317597	3139765	168.35	Mughal	Sculpture	Deities	Stone	
806	317375.7	3139575	169.34	Mauryan	Sculpture	Deities	Stone	

Fig. 11. Tabulated report (.csv file) for spatial attachment.

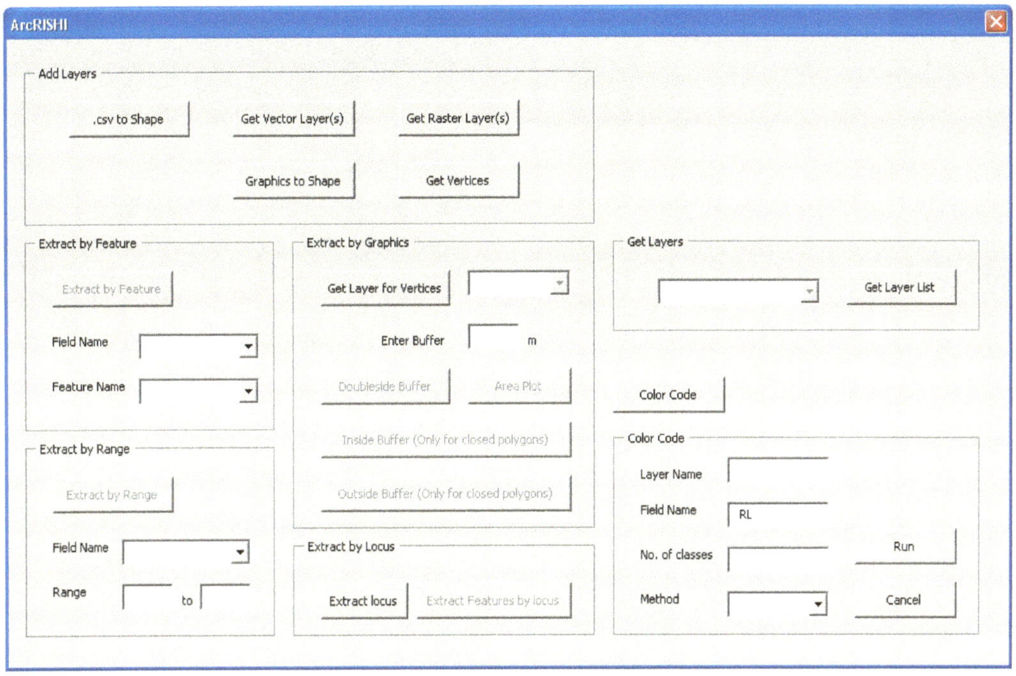

Fig. 12. Spatial analysis tool – ArcRISHI.

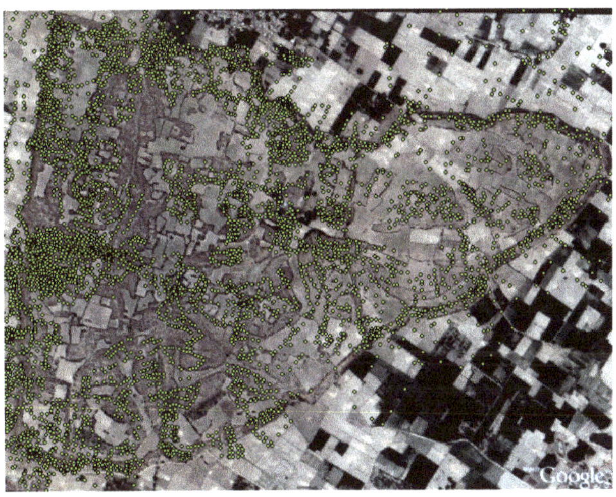

Fig. 13. Artifact distribution at Ahichchatra.

using the spatial analysis tool, ArcRISHI (Figure 12), created using ArcObjects. Thus, it is possible to obtain the spatial distribution of the queried antiquities (Figure 13). Similarly monuments can be brought onto a map.

Spatial analysis

Spatial queries can be performed on the distribution of antiquities using ArcRISHI. Four methods (illustrated below) are used for this purpose.

Extract by shape

Extract by shapes method involves user defined shapes with respect to which spatial queries can be performed to identify artifacts within a region or a buffer zone as explained below.

(a) Polyline plot

Many times it may be necessary to know 'what artifacts are found along a wall', 'what weapons are found just inside or outside the boundaries of the site', etc. Based on topographic details available similar queries can be built. For such queries, a polyline plot is designed. Following steps-wise procedure is followed in implementing the polyline plot.

Step 1: User draws a polyline along the area of interest say a part of a road passing through the site.

Step 2: Vertices (or nodes) of the polyline are identified and their coordinate information is attached. Using the vertices (or nodes), equations of each segment of the polyline are obtained.

Step 3: User defines the buffer zone (say 5m) i.e. the width in meters on either side of the polyline. Spatial query (using the tool ArcRISHI) is run in order to identify artifacts within the buffer zone (Figure 14).

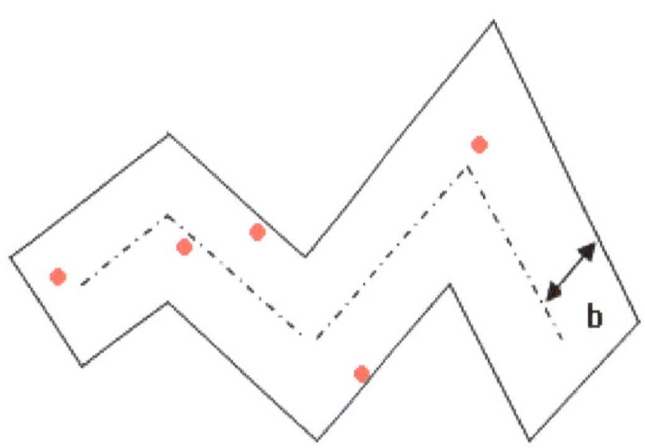

Fig. 14. Artifacts within the buffer zone of *b* meters.

(b) Boundary plot

Similar to a polyline plot, spatial query can be run for the boundary of a closed area. The methodology applied is same. In this case buffer can be of two types:

(i) Two-side buffer: In a two-side buffer, the buffer is set on the either side of the boundary line similar to the polyline plot (Figure 15).
(ii) Single-side buffer: In this case, buffer may be either inside the area or outside the area (Figure 15).

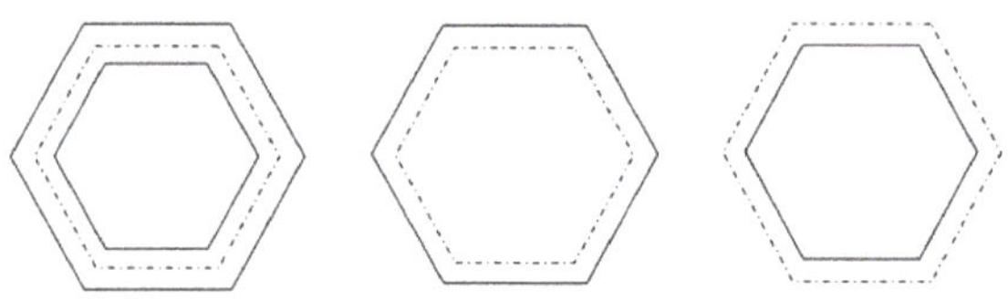

Fig. 15. Results of a spiral plot with a 4m buffer.

(c) Spiral plot

Unlike a wall, many times irregular features like the shore of a paleochannel or a river, an ox-bow lake, a pond, or foothills of a mound are also observed. In these situations, spiral plots are more practical than polylines. The difficulty with any spiral plot is that identification of vertices is not possible. Hence, another approach is to trace the spiral using a number of points on the plot. Figure 16 shows a spiral plot with 30 points on the plot. Considering these points as the vertices of a polyline (approximation of a spiral plot), artifacts within the buffer zones are identified.

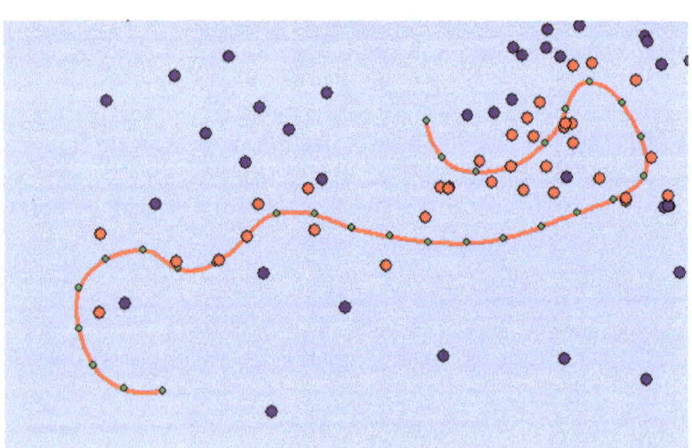

Fig. 16. Excavated wall structure.

(d) Polygon plot

A query may be to identify artifacts within a closed wall structure (Figure 17). A polygon plot serves the purpose of such queries. Following operations are followed for such a plot.

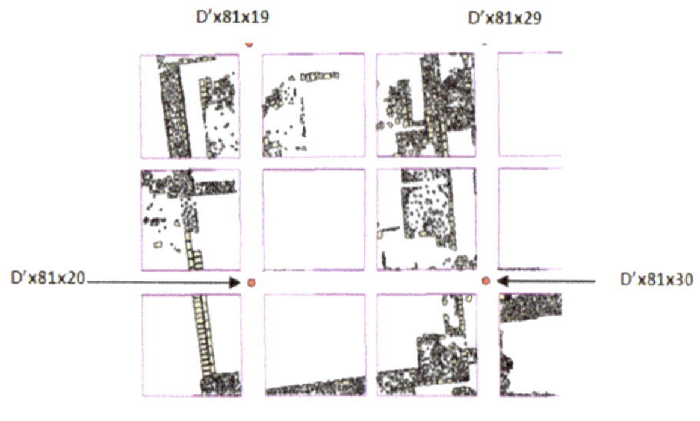

Fig. 17. User defined polygon plot with wall represented by a polygon.

Step 1: User plots a polygon along the boundary of the area of interest (say the outer wall of a closed brick structure) as in Figure 18. Vertices (or nodes) of the polygon are identified and their coordinate information is attached.

Step 2: Antiquities within a polygon are identified using the Plumbline approach (http://cgm.cs.mcgill.ca/~godfried/teaching/cg-projects/97/Octavian/compgeom.html). This method is used for its simplicity and less calculation involved for problem solving. It is also feasible because vertices and equation of the lines are already known from the previous steps.

On similar lines, other plots can be (a) box or rectangular plot for individual trenches or quadrants, (b) area or region plot for irregular shaped closed areas like a temple premises, a mound, an ancient pond, etc.

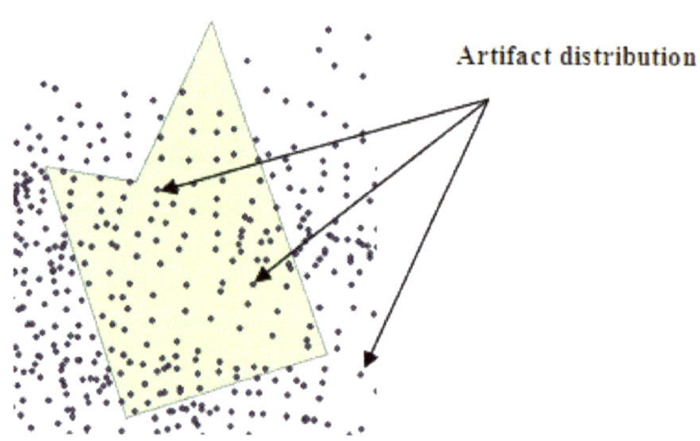

Fig. 18. Layer-wise and object-wise distribution of antiquities for the area plot.

Extract by feature

Extract by feature is of particular interest for the Indian archaeologists as it resembles their present methods of data interpretation and association. The spatial queries based on this are extract by layer, object name and category, material type and category, period, etc. and their combination. Such queries may either be applied directly over the entire artifact distribution or may follow after 'Extract by shape' queries.

For example, an area plot identifies 176 artifacts within a temple premise. 'Extract by feature' can be used to sort figurine (object name) of deities (object category) of Mauryan period (period) out of the 176 artifacts.

Extract by range

Sometimes archaeological predictions are made based on physical evidences like weights, precise shapes, etc. For example, artifacts like pottery are available in various sizes. Say at a particular period of time some pottery was used for religious activities while other for other activities in and around Ahichchatra. Pottery used for religious purpose may vary from that used for storing water. Thus variations may be observed in sizes, weights, shapes of the pottery. Considering weight as one of the criteria, pottery is clustered into a number of ranges say A to B grams, B to C grams, C to D grams, etc. Predictions are made based on the statistics like mean, mode, standard deviation, median values for each cluster. This can also be used for understanding the social and economic condition present at that particular period of time. Distribution statistics like number of pottery per unit area are also useful in detecting the source(s) of production. Thus, 'Extract by range' method is used to query on the features like artifacts weights, heights, lengths that hold numeric values. On similar lines, range can be specified for vertical depth at which an object is found. This range may be (a) from a fixed datum or (b) from the top surface.

Data classification

Range based queries can further be divided into sub-ranges. For example, 400-700 grams of weight can again be divided into 400-500, 500-600 and 600-700 grams. On a planimetric scale, color variations or symbolic variations can be used to distinguish these sub-ranges. Different object types, material types, etc. can also be shown by the color variations. The sub-ranges can be assigned to the artifacts based on the following classification methods –

(iii) Equal interval or equal step
(iv) Natural breaks
(v) Quantile

Extract by locus

ASI grid is very dense and involves a large number of trenches. 'Extraction by locus' is used to identify the individual trenches and artifacts within them. This is similar to a box plot but the size of a box plot is not restricted to a trench as in this case.

Reports and statistics associated with spatial queries

The statistical information attached to the artifact distribution is also vital in a number of ways. Statistics like mean, mode and standard deviation are computed along with the layer-wise distribution of individual artifact types and their proportions (Figure 19 and 20). Such statistics can be used to get an idea about a numerous archeologically significant factors like range of the objects, range of the materials found, periodic distribution of the artifacts, etc.

Antiquity	Layer 1	Layer 2a	Layer 2b	Layer 3a	Layer 3b	Layer 4	Layer 5		Total	%(Antiquities)
Stamps	0	0	0	0	0	0	0		0	0
Daggers	0	0	0	0	0	0	0		0	0
Pendants	0	0	0	0	0	0	0		0	0
Coins	0	0	0	0	3	0	0		3	3.409
Swords	0	0	0	0	0	0	0		0	0
Figurine	6	0	0	0	1	6	28		41	46.59
Sculptures	0	0	16	0	0	0	0		16	18.18
Pottery	0	0	0	19	0	0	0		19	21.59
Beads	0	0	0	0	0	0	0		0	0
Spatula	0	0	0	0	0	0	0		0	0
Knives	0	5	0	1	0	0	3		9	10.22
Total	6	5	16	20	4	6	31		88	
%(Layer)	6.81	5.68	18.18	22.72	4.54	6.81	35.22			100

Fig. 19. Range-wise grouping of potteries.

Weight (gm)	Layer 1	Layer 2a	Layer 2b	Layer 3a	Layer 3b	Layer 4	Layer 5	Total
<50	0	0	0	0	0	0	0	0
50-100	0	0	0	0	0	0	0	0
100-150	0	0	0	0	0	0	0	0
150-200	0	0	0	0	0	0	0	0
200-400	19	0	0	0	25	13	0	57
400-1000	0	0	47	23	17	45	22	154
>1000	0	0	0	0	10	8	0	18
Total	19	0	47	23	52	66	22	229
Mean (gm)	45	0	75.55	68.45	74.98	80.43	85	
Std. Dev (gm)	0	0	13.32	6.5	38.75	12.23	0	

Fig. 20. Layers in 3D

Visual Analysis

Planimetric distribution analysis does not always solve the problem of interpretations because it is difficult to have a perception of the depth and layer-wise distribution on a map. Although color variations are used, it is not possible to directly visualize the separation between artifacts and an artifact to its provenance. Hence, a 3D visualization system in ArcScene is developed which allows the visualization of artifacts in relation to the surface of the past.

Layers in 3D

Figure 21 shows the superimposition of top stratigraphic layer over the layer below it. Stratigraphic changes can be detected based on such 3D layer plots.

Fig. 21. Site formation process.

Site formation process

Layers in 3D can be applied for detecting change in the settlements, mounds and other topographical features. Figure 22 shows a similar study on the site of Ahichchatra where settlements are believed to have shifted towards the west because of regular floods in the east. This is visible from the figure where top layer is visible more towards the west. This result is based on the synthetic data generated for layers and therefore, is just a sample representing the actual work that can be carried out in the future.

3D box plot (or 3D area plot or 3D polygon plot)

Once a spatial query is run, it is possible to bring the outputs into ArcScene for 3D visualization. Figure 23 show a similar 3D plot for antiquities identified using a box plot. A fair idea about the depth-wise distribution of the artifacts is available from the plot. Layer information is very important in order to differentiate artifacts belonging to different layers. Hence, layers are created based on the topographic information of each layer (Figure 24). It is difficult to know which artifact belongs to which object type or object category. Therefore, the distribution is also separated based on the object type or object category (Figure 25). Similarly, other plots like 3D area plot and 3D polygon plot can also be generated.

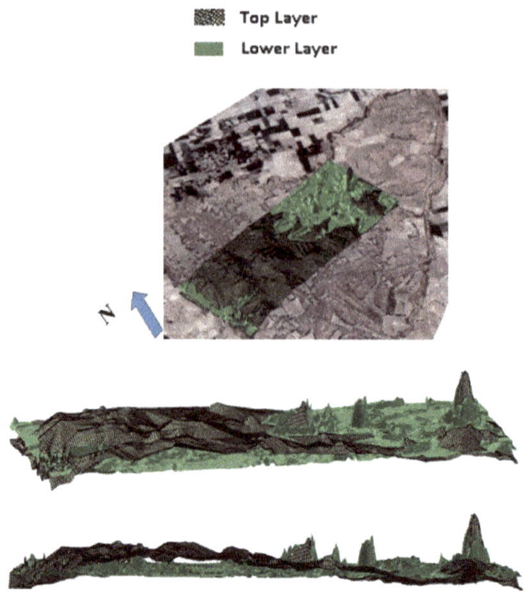

Fig. 22. Artifact distribution within a box plot - Isometric view and Cross-sectional view

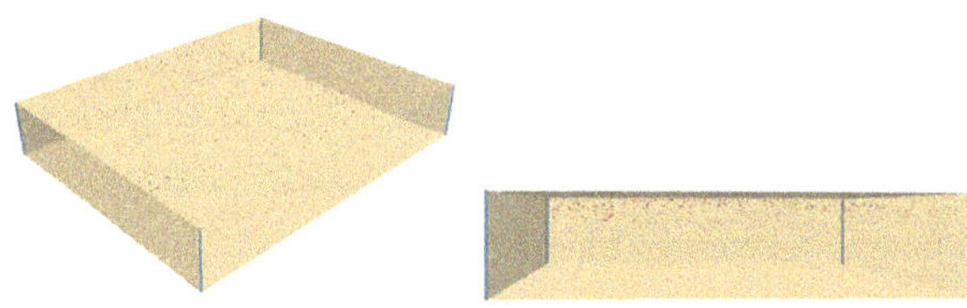

Fig. 23. 3D box plot - Layer-wise separation (isometric view and cross-sectional view).

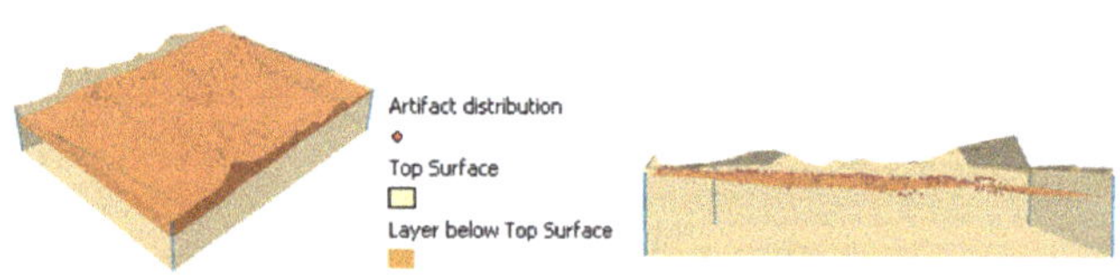

Fig. 24. 3D box plot - Separation based on object type - an isometric view.

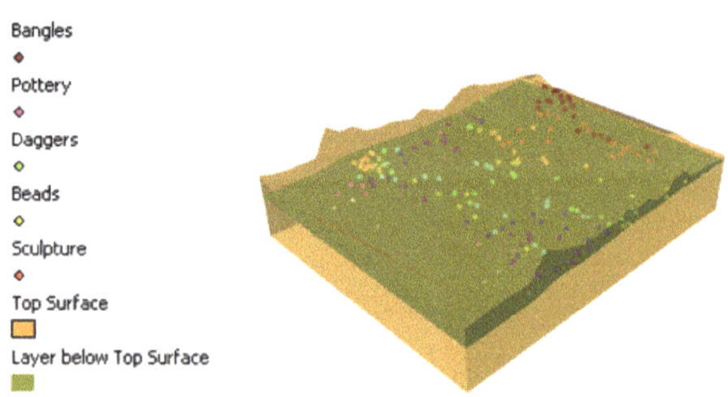

Fig. 25. 3D Spiral plot with Google Earth imagery draped over the DEM and three stratigraphic layers

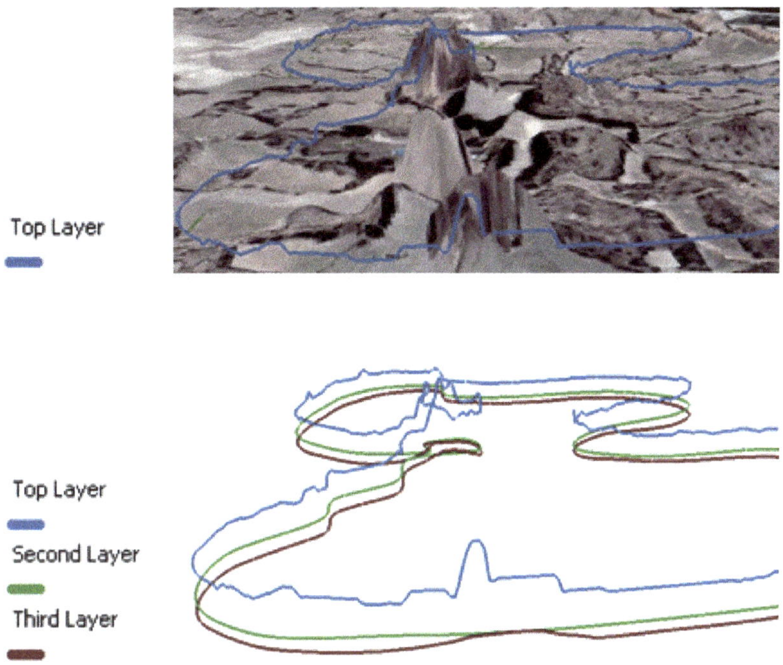

Fig. 26. 3D Spiral plot – antiquities distribution based on depth variation

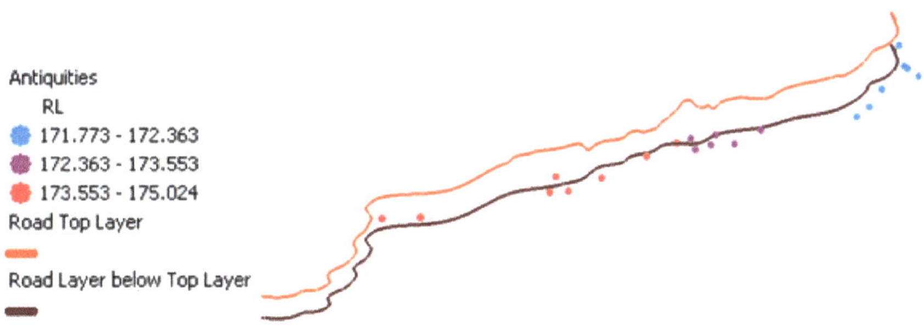

Fig. 27. Similarly, 3D plots can be generated for polyline plots and boundary plots.

3D Spiral plot

A 3D spiral plot represents the distribution of the layers as shown in Figure 26. Similar to the 3D box plot, antiquity distribution can be separated based on the depth at which the antiquities were found (Figure 27).

Example 1

Suppose we have seven total station points with RLs 175.8m, 177.54m, 180.76m, 169.24m, 172.54m, 181m and 173.4m. Some random value (floating point number) was deducted between fixed intervals from the actual total station value, say Layer1: 0.1m to 1.1m, Layer2a: 1.1m to 1.65m, Layer 2b: 1.65m to 2.3m and so on (Figure A and Figure B). This was done for other total station points. Considering this as the point cloud for each layer, interpolation was carried out to generate layer information (i.e. DEMs) of past surfaces.

	RL – (0.1m to 1.1m)	**RL – (1.1m to 1.65m)**	**RL – (1.65m to 2.3m)**
RL/Layer	**Layer 1**	**Layer 2a**	**Layer 2b**
175.80m	175.80-0.60 = 175.20m	175.80-1.60 = 174.20m	175.80-2.11 = 173.69m
177.54m	177.54-0.78 = 176.76m	177.54-1.26 = 176.28m	177.54-2.13 = 175.41m
180.76m	180.76-0.54 = 180.22m	180.76-1.54 = 179.22m	180.76-2.00 = 178.22m
177.24m	177.24-1.05 = 176.19m	177.24-1.55 = 175.69m	177.24-1.90 = 175.34m
174.54m	174.54-0.75 = 173.79m	174.54-1.34 = 173.20m	174.54-1.87 = 172.67m
181.00m	181.00-0.84 = 180.16m	181.00-1.20 = 179.80m	181.00-1.70 = 179.30m
173.40m	173.40-10.0 = 172.40m	173.40-1.45 = 171.95m	173.40-2.21 = 171.19m

Figure A Sample layer calculations from total station points.

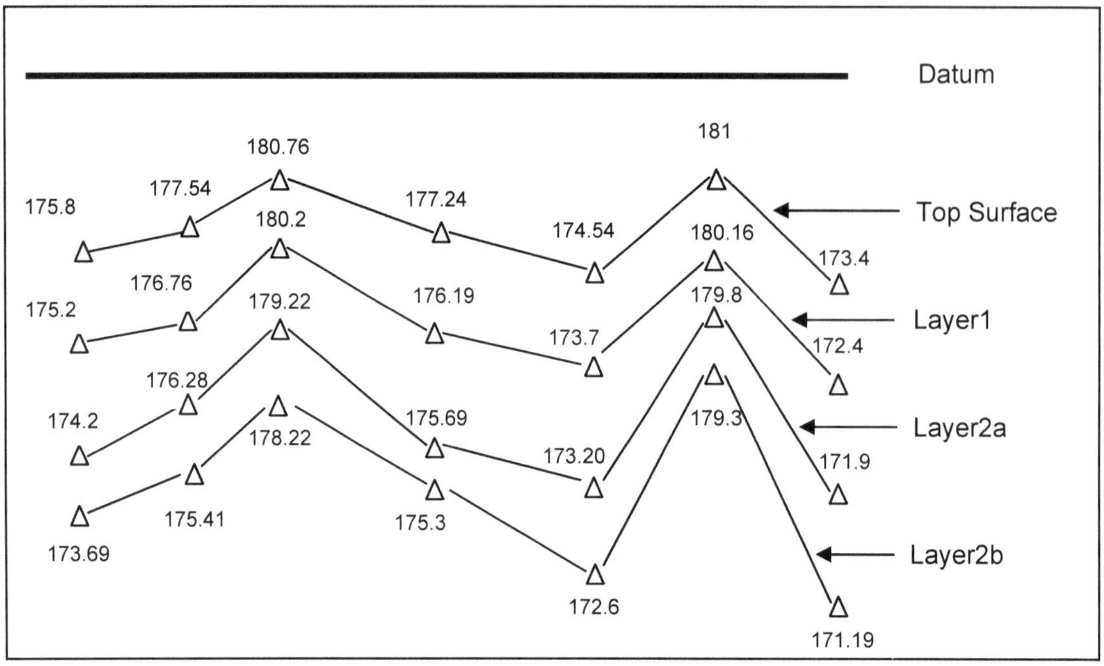

Figure B Generated layers.

Example 2

The examples given below are for the calculation of coordinates of locus Ax79x67 and locus Bx24x37

<u>**Locus: Ax79x67**</u>

Origin of A: (1000, 1000)

Origin of 79: (700, 800)

Origin of 67: (60, 60)

⇨ X=1000 + 700 + 60 = 1760
⇨ Y= 1000 + 800 + 60 = 1860

Coordinate (*X*, *Y*) of locus Ax79x67 is (1760, 1860)

<u>**Locus: Bx24x37**</u>

Origin of B: (2000, 1000)

Origin of 24: (200, 300)

Origin of 37: (30, 60)

⇨ X=2000 + 200 + 30 = 2230
⇨ Y= 1000 + 300 + 60 = 1360

Coordinate (*X*, *Y*) of locus Bx24x37 is (2230, 1360)

References

GUPTA, S. P. 2000. Some problems of heritage documentation and management in India, *The European Archaeologist*, No. 13, pp. 5-6.

MEYER, E., GRUSSENMEYER, P., PERRIN, J. P., DURAND, A. AND DRAP, P. 2007. A Virtual Reality Environment for Archaeological Data Management, Visualization and Documentation, Acceté pour publication dans les actes du "Computer Applications and Quantitative Methods is Archaeology" CAA2007. *35e conférence du CAA "Layers of perception. Advanced technological means to illuminate our past"*, 2 – 6 Avril 2007, Berlin, Allemagne, p. 6.

PILLAI, A. M. 2009. *Cultural Resource Management System for Indian Archaeology*, M. Tech. Thesis, Indian Institute of Technology Kanpur, Kanpur, India.

SRIVASTAVA, A. 2008. *Development of a Cultural Resource Management System*, B. Tech. Project Report, Department of Civil Engineering, Indian Institute of Technology Kanpur, Kanpur, India.

TARE, A. 2007. *A GIS based approach to study of Archaeological Site- A case study of Ahichchatra,* M. Tech Thesis, Indian Institute of Technology Kanpur, Kanpur, India.

Online References

http://cgm.cs.mcgill.ca/~godfried/teaching/cg-projects/97/Octavian/compgeom.html as accessed on 23rd February, 2009.

Mapping archaeological sites of Tamil Nadu

K. Rajan

Professor and Head, Department of History. Pondicherry University. Pondicherry

Abstract
More than hundred years of intensive and untiring field works undertaken by various scholars since the early part of 19th century in different eco-zones played a guiding role in understanding various cultural processes that had taken place in the long span of human life, say, from Palaeolithic period down to Historical times. These explorations were and are being made in different segment of the discipline such as archaeological, epigraphical and numismatic. Some of the discoveries are unique in nature helping to fill the crucial cultural gaps and some are as significant as they provided a vital input to understand inter and intra regional cultural linkages. Besides, there were several accidental discoveries like coin hoards, copper plates, inscriptions and archaeological vestiges. In some occasions, these findings were totally transformed our understanding of the fast and in a few cases changed our perception. The growth of science further accelerated and changed our approach towards the subject. The outcome of these findings are documented or published at different level, say, international, national and regional. The regional level publications dominate the scene. Some of the publications reflect the contemporary practices and theories and a few of them were provided altogether a new direction. A brief survey on historiography the research would help to understand the fathom of archaeological activities that was achieved in the past.

Keywords Tamil Nadu, Historiography, Archaeological Sites, Prehistoric, Proto-historic, Early Historic, Atlas

Introduction

More than hundred years of intensive and untiring field works undertaken by various scholars since the early part of 19th century in different eco-zones played a guiding role in understanding various cultural processes that had taken place in the long span of human life, say, from Palaeolithic period down to Historical times. These explorations were and are being made in different segment of the discipline such as archaeological, epigraphical and numismatic. Some of the discoveries were unique in nature helping to fill the crucial cultural gaps and some are so significant as they provided a vital input to understand inter and intra regional cultural linkages. However, majority of the explorations were sporadic in nature and a few of them were very systematic in their approach. Some of the explorations and excavations were designed to meet the expectation of the policy makers and a few of them were made to satisfy the contemporary society. Besides, there were several accidental discoveries like coin hoards, copper plates, inscriptions and archaeological vestiges. In some occasions, these findings were totally transformed our outstanding of the fast and in a few cases changed our perception. New findings and accumulation of data opened a new area of research. The outcome of these findings were documented or published at different level, say, international, national and regional. The regional level publications dominate the scene. Some of the publications reflect the contemporary practices and theories and a few of them provided altogether a new direction. Several thousands of archaeological sites had been discovered and documented through the years. Researchers and students find it difficult to measure the fathom of these findings. The non-availability of data in certain areas restricted their area of research. To fill this gap, a small attempt is made to compile all these archeological sites so far discovered in Tamil Nadu. A brief survey on historiography would help to understand the fathom of archaeological activities that achieved in the past.

Historiography

Archaeological investigations and their subsequent interpretations are subject to the paradigm of archaeological thought. These archaeological concepts determined the method of documentation of a site. The cursory look on study of history of archaeological research would help to situate the data in a given social context. The history of archaeological research in India has been extensively dealt with much focus on archaeological works of British India. In all these works, the historiography of Tamil Nadu is very limited that prevented integrating history of Tamil Nadu into the mainstream history of India. Irrespective of its long history and cultural richness, one rarely finds its right place in the arena of Indian history. This may be due to the availability of research material in vernacular language. This situation prevented the scholars giving a complete picture on the history of archaeology of Tamil Nadu. Till date, the situation remains unchanged. Tamil Nadu witnessed a continuous research in all front particularly in the field of archaeology and epigraphy. The numismatics studies are also received a fresh impetus in recent years. Unlike other states, Tamil Nadu experiences the involvement of field-based researchers that resulted with accumulation of enormous fresh data. Each area of studies namely archaeological, epigraphical and numismatic is progressed well in terms of research. These studies have been analysed to expose the richness of the data and also to estimate the value of these archaeological sites.

Prehistoric studies

The prehistoric research in Tamil Nadu starts with the discovery of a Palaeolithic tool at Pallavaram by Robert Bruce Foote on 30th May 1863 (Foote 1866). Subsequently, Foote completed an extensive survey between 1863 and 1884 and made exciting discoveries. But, these discoveries with far-reaching implications received neither the attention of scholars of then Asiatic Society nor the Archaeological Survey of India. However, his collections were published posthumously by the Government Museum, Madras and still are considered as a reference book. Since then, the prehistoric research is continuing with certain gaps. Till date, the old Palar river basin, popularly known as Kortallayar valley, alone received much attention of scholars and the remaining area did not receive any attention. Shanti Pappu who studied extensively on historiography of prehistoric research in Tamil Nadu highlighted several important phases of research. According to Shanti Pappu, the prehistoric research could be divided into four broad phases. The first phase (1863-1912) was focused mainly in establishing the human antiquity in the subcontinent. Robert Foote and William King played a prominent role in this phase. Foote took all initiative in understanding these cultures in proper archaeological context rather than merely documenting the site. The artefacts were analysed in the background of land formations, typology, and their transport over the landscape (Foote 1866, 14; 1916, 8-9). Though his classification system was influenced by that of John Evans, his innovative use of ethnographic data to understand the functionalities of the tools were the greatest contribution of Foote. However, the spectacular discovery of Indus sites and other protohistoric and Early Historic sites in India during the early part of the 20th century led to a general shift in prehistoric research.

The second phase of research (1930-1960) was mainly concentrated in placing the tools in proper chronological base. The concept of evolution influenced the prehistoric research in India particularly by the works of de Terra and Paterson. This phase was represented by the works of Cammiade and Burkitt (1930), V.D.Krishnaswami (1947, 11-57) and K.V.Soudara Rajan and culminated with the Yale-Cambridge Expedition. These works were essentially based on geological framework by building river terrace sequences caused due to the Pleistocene pluvial and inter-pluvial (Dennel and Rendell 1991, 91-92). The identification of boulder-conglomerate, primary pre-Pleistocene laterites, secondary Pleistocene laterites, river terraces and eustatic sea level changes were some of the important areas materialised in this phase. These formulations helped to place the tools in proper chronological frame and also created a pan-Indian broad cultural sequence. The classification of tools into three series based on patination, rolling and refinement in technology was one of the important contributions of this phase. The difference witnessed between the pebble tools found in Soan and core tools in Kortallayar basin led to coin a new term called Madras hand axe tradition.

The third phase (1960-1980) was marked with intensive excavation works carried out by the Archaeological Survey of India under the leadership of K.D.Banerjee in the sites like at Gudiyam, Attirampakkam, Vadamadurai, Poondi and Neyveli (*IAR* 1962-63 to 1966-67). The results of the series of excavations questioned the previous claims like the existence of pluvial terraces, pre-laterite industry and boulder-conglomerate. Banerjee put forward a new hypothesis like marine terraces in place of river terraces and he also saw that the early Acheulian at Vadamadurai found in the boulder-conglomerate was possibly reworked from elsewhere. The subsequent analysis made by the scholars like Vidula Jayswal (1978), A.Swamy (1976), P.Gupta and P.C.Dutta (Gupta and Dutta 1961) on the tools collected from explorations and excavations could not give any direction other than typology.

The prehistoric research of fifth and contemporary final phase (1990-2000) took a shape after a decade. The main contribution was of Shanti Pappu (Pappu 1996, 1-23). She took all initiative in reviving the prehistoric research in Tamil Nadu. Her intensive explorations between 1991-1995 supplemented by laboratory analysis of sediments, satellite image studies, lithic analysis, experimental manufacture of tools and ethnographical research opened a new vista in the prehistoric research of Kortallayar basin. She questioned the so called four-fold terrace sequences and the cultures associated with it. The site formation theory was put into test by studying the distributional pattern in the background of geomorphic processes. The carefully designed excavations undertaken by Shanti Pappu (Pappu 2004, 1-17) between the years 1999 and 2004 at Attirampakkam reveal high resolution data to understand the hominid behaviour and palaeo-environmental change over the Pleistocene and the continuous occupation of man from Lower Palaeolithic times down to Microlithic times.

All the above studies have been concentrated on a single geographical zone namely Kortallayar basin. Efforts to identify the prehistoric sites beyond Kortallayar basin received a less attention. However, the collection of Palaeolithic tools at Varatanapalli near Dharmapuri and Kilkolattur near Ariyalur by the author (Rajan 1977). Subsequently, M.S.Ramji (2007) documented several Middle Paleolithic and Microlithic sites in Ariyalur region. Research scholars like K.Kumar (2002, 134-142), Singaravelan and Jinu Koshy collected palaeolithic tools respectively in and around Vellore, Chingleput and Sivaganga. Recently, M.Prasanna (2008) noticed a site near Gangaikondan in Tirunelveli district. These recent investigations turned the attention of prehistoric research beyond Kortallayar basin. V.Selvakumar by taking the link from these recent discoveries surveyed the entire Vallam table tableland in Thanjavur region and made a trail dig in Tamil University campus in

April 2009. The excavation explicitly revealed two cultural phases namely Palaeolithic and Microlithic. Palaeolithic artifacts are unearthed in stratified context found below a sterile layer under the layer that produced Mesolithic artifacts signifying the chorological gap between these two cultures. It is important to note that for the first time such stratigraphic evidence is found in Thanjavur region. The Palaeolithic cultural evidence could go beyond 10,000 years. Attempts are on to get scientific dates.

Besides the study of Palaeolithic tools, there is hardly any research on other prehistoric cultures. There is barely any research work on Microlithic tools that recovered in *Teri sites* in and around Sawyarpuram by Aiyappan in 1950 (Aiyappan 1950) and by Zeuner in 1956 (Zeuner and Allchin 1956, 4-20). The excavations at Teriruveli and Mangudi clearly placed the Microlithic phase before Iron Age phase without any trace of Neolithic phase. The absence of Neolithic phase between the Microlithic and Iron Age phase in the southern part of Tamil Nadu throws much debate on the nature of cultural process that witnessed in southern Tamil Nadu. V.Selvakumar's work in Gundar basin south of Madurai also reflected the same phenomenon. However, the recent excavations at Mayiladumparai near Dharmapuri cleared the mist revealing four cultural phases in continuous order namely Microlithic, Neolithic, Iron Age and Early Historic. This is the first and the only site which had shown such cultural sequence. The cultural difference encountered in the south and in north Tamil Nadu is yet to validate with more data. The excavations at Modur (Selvaraj 2005), Guttur (Raman 1982-83, 72), Mayiladumparai (Rajan 2004, 74-89), Appukkal (*IAR* 1976-77, 47-48), Paiyampalli (Rao 1967-68, 26-30) and Mangudi (Shetty 2003) throw much light on various aspects of Neolithic culture.

Another important recent development in prehistoric research is the discovery of rock paintings and rock engravings. The identification of rock paintings at Mallapadi in Dharmapuri district by K.V.Raman in 1977 (*IAR* 1977-78, 50) opened a new vista in prehistoric art of Tamil Nadu. Since then more than fifty rock art sites have been discovered and most of them were associated with Iron Age (Rajan 1991, 51-69). However, the occurrence of red ochre painting super imposed with white ochre clearly demonstrated their cultural affinity. It seems, in northern Tamil Nadu, red ochre is associated with Neolithic phase and white pigment with Iron Age. This identification helped to place the paintings in chronological phase which in turn helped to see changes in the cultural theme. Likewise, rock engravings identified at the entrance of the rock shelter associated with Neolithic culture at Mayiladumparai (Rajan 2004, 74-89) kindled interest among the scholars. Recently, rock art sites were reported at many places in Palani hills. The study of Perumukkal engravings, one of the biggest and largest in Tamil Nadu, has been seen under this light (Rajan 1996, 3-10). The paintings and engravings opened a new area of prehistoric research in Tamil Nadu.

Protohistoric studies

The Iron Age phase (also called Megalithic phase or Black and Red ware phase) is considered as the proto-historic phase of Tamil Nadu. Tamil Nadu has not witnessed any exclusive chalcolithic phase, probably, due to non-availability of commercially and technically exploitable copper ore. Within Tamil Nadu, there are two distinct cultural phases, the first one is the transformation of the Microlithic phase into Iron Age phase without entering into the Neolithic phase. This situation is witnessed in the southern part of Tamil Nadu in the sites like at Mangudi and Teriruveli (Shetty 2003). The northern Tamil Nadu revealed the usual cultural sequence. The identification of stratigraphical position of various cultures in Tamil Nadu is the result of pioneering works of so many dedicated scholars in the field of proto historic studies. The protohistoric phase of Tamil Nadu is variously known as Iron Age culture or Black and Red ware culture and Megalithic culture. It is, because, the monuments essentially belong to the Iron Age and are largely sepulchral in character with dominant use of black and red ware. Though the term 'megalith' is a misnomer but still being used due to the structural impressiveness of the monument and the quality and quantity of the cultural material exposed in the graves. It is a system of burying the dead, which is followed in different cultural context.

Research on these monuments was initiated in the beginning of the nineteenth century, when Babington (1823, 324-330) unearthed an interesting group of burial monuments at Bangala Motta Paramba in the northern part of Kerala. Since then, the impressive visible features and contents of these monuments, and the issues related to origin, chronology, authorship, social structure, technology and other aspects had received the attention of scholars and antiquarians and as a result vast amount of literature has been published. The history of these investigations could be broadly placed two phases namely pre-independence phase and post-independence phase.

In pre-independence era, many megaliths were discovered and many of them were dug out by colonial administrators, officers, missionaries, treasure hunters and others. Important works include that of H.Congreve (1847, 77-146), William Fraser (1861, 122-159), J.W.Breeks (1873), Rev Phillips Maurice (1873, 223-228), M.J.Walhouse (1875, 17-34), H.G.Turner (1879, 207), B.R.Branfill (1880, 8-10; 1887, 47-100), H.O.D.Harding (1889-94, 13-20), Sandford (1901, 461-471), A.H.Longhurst (1913-14, 43-45), Alexander Rea (1915), and F.J.Richards (1924, 154-165). Among these works, Breeks's work in Nilgiris and Rea's works at Adichchanallur and Sandford's work in Coimbatore region brought to light on the nature of monuments spread out in different geographical zones of Tamil Nadu. Many of these attempts were largely antiquarian in nature. Several hundreds of burials were literally robed. As pointed out by Wheeler 'Megalithic monuments had been ransacked rather than excavated, sometimes with the help of dynamite. Urn fields had been gathered up like rice crops' (Wheeler 1948, 181). In the midst of these exploitations, some of the investigators went beyond antiquarian interest and raised various questions pertaining to the origin, authorship and chronology of these monuments. In the early stage, the diffusionist theory dominated in their writings. Some of the human skeletal remains studied also primarily used to give a racial affinity (Zuckerman 1930, 1-24). In spite of a large number of so-called excavations conducted during the pre-independence era, the megaliths could not be placed in proper chronological framework. However, the accumulated data provided a foundation for subsequent studies.

In post-Independence era, archaeological research got new impetus with the appointment of Mortimer Wheeler as Director General of Archaeological Survey of India in 1944. After getting some idea from the earlier works of M.H.Krishna at Brahmagiri, Wheeler ventured into the act of placing the Early Historic and Megalithic culture in definite chronological order through his excavations at Brahmagiri (Wheeler 1948). This excavation becomes a turning point in studies of all south Indian megaliths. Krishnaswami's detailed investigation in Chingleput district resulted in the discovery of over 200 Megalithic sites. Krishnaswami classified and defined the Megalithic types based on the exposed morphological features. In line of his suggestion, all the subsequent discoveries have been brought under uniformed nomenclature. K.R.Srinivasan's excellent publication on Tamil literature, particularly Sangam literature and megalith monuments give a new thrust in this study. Irrespective of these studies, the excavations at Sanur, Amirthamangalam, Kunnattur and Paiyampallai did not yield a desired result due to non-availability of habitation mounds. Most of the researchers involved in explorations failed to document the habitation mound but rather impressed with huge graves. The works of Gururaja Rao (1972) and B.Narasimhaiah (1980) could be cited for this. This resulted with the formulation of a theory that the megaliths were nomadic people (Leshnik 1974). However, the discovery of more than thousand archaeological sites through out Tamil Nadu in different ecological zones changed the view of scholars (Rajan 1997). Nearly half of the Megalithic sites are associated with habitation mounds. The excavation at Kodumanal, a habitation-cum-burial site, became a turning point in the studies of Megalithic society (Rajan 1994). The subsequent excavations at Mayiladumparai (Rajan 2004) and Thandikudi (Rajan 2005) further cleared the mist. The recent excavations at Adichchanallur and Mangudi further augmented the data.

As an offshoot of the development, the iron, steel, gemstone, pearl, chank industries have been extensively studied. Sharada Srinivasan and B.Sasisekaran's work on steel, the author's study on traditional gem stone industry in the background of Kodumanal excavations, N.Athiyaman's work on traditional pearl and chank diving, S.R.Walimbe's study on skeletal remains, Kajale and Dorian Fuller's study on palaeobotanical samples could be seen as an important development in the history of science. Now it is very clear that the Iron Age people led a very advanced culture by giving equal importance to both agriculture and industry. The accumulated data served as a strong base to understand the cultural process. It further suggests that the cultural transformation from Iron Age to Early Historic is smooth and strong in Tamil Nadu.

Early Historic Studies

The emergence of Early Historic period is not uniform through out India and one need not to expect such scenario due to varied environmental settings. However, it is not averse to have a comprehensive view of this situation and bringing them with in a specified time bracket. In general term, the availability of deciphered written document is considered as the beginning of history. In that sense, the decipherment of Brahmi script datable to 3rd century BC of Asokan times is considered as the beginning of historic period in India. This demarcation line between the Early Historic and proto-historic times also stands on thin evidence which could tilt the balance at any time on the occurrence of new evidence or new interpretation of old data or any breakthrough in the decipherment. As far as Tamil Nadu is concerned, the occurrence of Tamil-Brahmi script is considered as the beginning of Early Historic times. Therefore, the date of Tamil-Brahmi is very crucial to understand the beginning of Early Historic period in Tamil Nadu. The historical input received from these Tamil-Brahmi script based data (like inscriptions, coins and inscribed potsherds) is considered as the chronological anchor. The appearance of script along with so many other social parameters like the usage of coins, formation of trade routes, the size of the settlement, the formulation of various clan groups, the emergence of state and so on so forth is the product of previous Iron Age culture that flourished before the Early Historic period. The remarkable development taken place in different spheres of the society during Iron Age is culminated into the Early Historic period.

The study of Early Historic period in Tamil Nadu is still a nascent discipline, particularly in pre-independence era. There is hardly any excavation worthy of mentioning which unfolds all aspects of Early Historic period. Much of the energy has been drained in opening the Megalithic monuments and identifying the antiquity of man. However, the scenario has been changed by the excavation of port town Arikamedu (Wheeler 1946, 17-124). The University of Madras, Tamil Nadu State Department of Archaeology,

Archaeological Survey of India and Tamil University channelised their energy in locating Early Historic sites after Independence. In ture sense, the Sangam literature served as a base to identify those sites. Almost all the important capital cities, trade centres and port towns of Sangam Age have been identified through intensive explorations and excavations. Among them, the sites of Korkai, Adichchanallur, Kaveripattinam, Alagankulam, Mangudi, Perur, Kodumanal, Karur, Uraiyur, Vallam, Kanchipuram, Thandikudi and Modur (Tagadur) need to be remembered for the wealthy of information. Some of the sites like Kodumanal and Alagankulam were excavated for many years. The epigraphical, numismatic and archaeological evidences that encountered in these sites go well with literary evidences. The underwater archaeological explorations under taken in Kaveripattinam (Poompuhar) waters further validated the literary references. Iravatham Mahadevan's work on Tamil-Brahmi, R.Krishnamurthy's work on Sangam Age coins and S.Suresh's work on Roman coins added to the knowledge on the nature of the society of Early Historic period. The wealth of data that had accumulated in recent years placed Early Historic studies of Tamil Nadu in a better position when compare with the neighbouring states.

As discussed above, these archaeological studies accumulated enormous data to understand the various social forces that played a greater role in shaping the cultural mosaic of Tamil Nadu. These studies provide a platform to understand the region with certain clarity. Still, there are certain important cultural gaps but that need to be clarified through intensive explorations and excavations and subsequent rigorous analysis. Several thousands of archaeological sites have been discovered and documented through the years. Researchers and students find it difficult to measure the fathom of these findings. The non-availability of data in certain areas restricted their area of research. To fill this gap, a small attempt is made to map all archaeological sites so far discovered in Tamil Nadu.

The mapping of archaeological sites began in the year 1989 with generous financial grant of various organizations like University Grants Commission, Indian National science Academy, Indian Council of Historical Research, Archaeological Survey of India and Universities. The twenty years of sustained exploration works brought to light more than two thousand archaeological sites. The archaeological information on these sites was integrated and a digital atlas was prepared under the aegis of Ford Foundation in the year 2008. Altogether there are at present seven thousand sites of archaeological and historical importance put on the maps with all related theme-based information. Facilities are provided for the prospective users to navigate through, and interact with, by querying, the vast historical and archaeological database by time, space, and themes. In addition to the conventional method of indicating the geographical information by self-explanatory symbols, text and visual material (image files) are also linked to the sites wherever necessary.

For making the atlas, in addition to the material available in standard secondary works, like the archaeological reports, inscriptions, and literature, and old maps, field surveys were made by the research staff of the project to check the geo-coordinates and make photo documentation of several known sites. In the process, nearly three thousand new archaeological sites were freshly identified and documented. The research staffs, drawn mainly from history and archaeology disciplines, were also given training in GIS and computer mapping programme, and it is gratifying to note that some of them have become adepts in GIS techniques.

Reference

AIYYAPPAN, A. 1933. Rock-cut Cave Tombs of Feroke, South Malabar. *Quarterly Journal of Mythic Society* 23(3), 299-314.

BABINGTON, J. 1823. Descriptions of the Pandoo Coolies in Malabar. *Transactions of the Literary Society of Bambay* 3, 324-330.

BRANFILL, B. R. 1880. Rude Megalith Monuments in North Arcot. *Journal of the Royal Asiatic Society of Bengal* 49, 8-10.

BREEKS, J. W. 1873. *An Account of the Primitive Tribes and Monuments of the Nilgris.* Delhi, Cultural Publishing House.

CAMMIADE, L. A. AND BURKITT, M. C. 1930. Fresh Light on the Stone Age of Eastern India. *Antiquity* 5(15), 327-339.

CONGRAVE, H. 1847. The Antiquities of Neilgherry Hills. *Madras journal of Literature and Science* 14, 77-146.

DENNEL, R. W. AND RENDELL, H. M. 1991. De Terra and Paterson and the Soan Flake Industry: A New Perspective from the Soan Valley, Northern Pakistan. *Man and Environment* 16(2), 91-99.

FOOTE, R. B. 1866. On the Occurrence of Stone Implements in Lateritic Formations in Various Parts of the Madras and North Arcot Districts. *Madras Journal of Literature and Science,* 3rd Series, Part II 1-35.

FOOTE, R. B. 1916. *The Foote Collection of Prehistoric and Protohistoric Antiquities-Notes on Their Ages and Distribution.* Madras, Government Museum.

FRASER, WILLIAM. 1861. Antiquities in the Coimbatore District. *Madras Journal of Literature and Science* 22, 112-159.

GUPTA, PABITRA AND PRATAP CHANDRA DUTTA. 1961. Lower Palaeolithic Industry of Amirdhamangalam, Chingleput District, Madras. *Man in India* 41(1), 45-54.

GURURAJA RAO, B. K. 1972. *The Megalithic Culture in South India.* Mysore, Prasaranga, University of Mysore.

HARDING, H. O. D. 1889-94. Notes on some old graves in the Coimbatore District, Madras. *Journal of Literature and Science*, 13-20.

JAYSWAL, VIDULA. 1978. *Palaeohistory of India.* New Delhi, Agam Kala Prakashan.

KRISHNASWAMI, V. D. 1947. Stone Age in India. *Ancient India* 3, 11-57.

KUMAR, K. 2002. Kanchipuram Mavattat Tolliyal Kandupitippukkal (Archaeological discoveries in Kanchipuram District), *Avanam, Journal of Tamil Nadu Archaeological Society* 13, 134-142.

LESHNIK, L. S. 1974. *South Indian Megalithic Burials - Pandukal Complex.* Wiesbeden, Franz Steiner Verlag GMB H.

LONGHURST, A. H. 1913- 14. A Report on the excavations of certain cairns in the Coimbatore District. *Annual Report of the Archaeological Department, Southern Circle, Madras and Goorg*, 43-45.

MAURICE, REV. PHILLIPS. 1873. Tumuli in the Salem District. *Indian Antiquary* 2, 223-228.

NARASIMHAIAH, B. 1980. *Neolithic and Megalithic Cultures in Tamil Nadu.* New Delhi: Sundeep Prakashan.

PAPPU, SHANTI 1996. Reinvestigation of the Prehistoric Archaeological Record in the Kortallayar Basin, Tamil Nadu. *Man and Environment* 21(1), 1-23.

PAPPU, SHANTI, YANNI GUNNELL, MAURICE TAIEB AND KUMAR AKHILESH. 2004. Preliminary Report on Excavations at the Palaeolithic Site of Attirampakkam, Tamil Nadu (1999-2004). *Man and Environment* 29(2), 1-17.

PRASANNA, M. 2008. Archaeology of Tamiraparani River Valley. Unpublished M.Phil thesis, Pondicherry University.

RAJAN, K. 1991. Rock Paintings in Dharmapuri District, Tamil Nadu. *Man and Environment* 19(1), 37-52.

RAJAN, K. 1994. *Archaeology of Tamil Nadu.* Delhi, Book India Publishing Company.

RAJAN, K. 1996. Petroglyphs of Perumukkal. *Purakala* 7(1-2), 3-10.

RAJAN, K. 1997. *Archaeological Gazetteer of Tamil Nadu.* Thanjavur, Manoo Pathippakam.

RAJAN, K. 2004. Excavations at Mayiladumparai, Tamil Nadu. *Man and Environment* 29(2), 74-89.

RAJAN, K. ATHIYAMAN, N. AND YATHEES KUMAR, V. P. 2005. Further Excavations at Thandkudi, Tamil Nadu. *Man and Environment* 33 (2), 60-70.

RAMAN, K. V. 1977-78. Excavation at Mallapadi, District Dharmapuri. *Indian Archaeology Review* 50.

RAMAN, K. V. 1982-83. Excavation in Gutturmalai, Guttur, District Dharmapuri, *Indian Archaeology Review.* 71-72.

RAMJI, M. S. 2007. *Archaeology of Perambalur Region.* Unpublished M.Phil thesis, Tamil University.

RAO, S. R. 1967-68. Excavations at Paiyampalli, District North Arcot. *Indian Archaeology - A Review* 26-30.

REA, ALEXANDER, 1915 *Catalogue of the Prehistoric Antiquities from Adichchanallur and Perumbair.* Madras, Government Press.

RICHARDS, F. J. 1924. Note on Some Iron Age Graves in the North Arcot District, South India. *Journal of the Royal Anthropological Institute of Great Britain and Ireland.* 54, 154-165.

SANDFORD, F. R. 1901. Notes on the Recent opening of some Prehistoric burials in the Coimbatore District, Madras Presidency. *Journal od the anthropological Society, Bombay* 3, 461-471.

SELVAKUMAR, V. 1996. Archaeological Investigations in Upper Gundar Basin, Madurai District, Tamil Nadu. *Man and Environment* 21(1), 26-42.

SHETTY, ASHOK VARDHAN 2003. *Excavations at Mangudi*, Chennai, Department of Archaeology, Govt. of Tamil Nadu.

SRIDHAR, T. S. 2005. *Exacavtions of Archaeological Sites in Tamil Nadu – Modur (2004-2005),* Chennai: Dept. of Archaeology, Govt. of Tamil Nadu.

SWAMY, A. 1976. Archaeology of Chingleput District, Unpublished Ph.D thesis, University of Madras.

TURNER, H. G. 1879. Cromlechs near Anamalais. *Indian Antiquary* 8, 207.

WALHOUSE, M. J. 1875. Notes on the Megalithic monuments in the Coimbatore District. *Journal of the Royal Asiatic Society of Great Britain and Ireland* 7, 17-34.

WHEELER, R. E. M. GHOSH, A. AND KRISHNA DEVA. 1946. Arikamedu: An Indo-Roman Trading Station on the East Coast of India. *Ancient India* 2, 17-124.

WHEELER, R. E. M. 1948. Brahmagiri and Chandravalli 1947: Megalithic and Other Cultures in the Chitaldurg District, Mysore State. *Ancient India* 4, 200-202.

ZEUNER, F. E. AND ALLCHIN, B. 1956. The Microlithic Sites of Tinnevelly District, Madras State. *Ancient India* 12, 4-20.

ZUCKERMAN, S. 1930. The Adichanallur Skulls, *Bulletin of the Madras Government Museum. 2, 1-24.*

An 'Encyclopaedia of Cultural Heritage'? The Encyclopedia of Life as a model for a comprehensive, online archaeological database

Shawn A. Ross[1] and Adela Sobotkova[2]

[1]The University of New South Wales, Sydney, Australia
[2]The University of Michigan, Ann Arbor, USA

Abstract

In February 2008 the first demonstration pages of the Encyclopedia of Life (EOL) website went online. In less than six hours, it received 11.5 million hits, and has since become a valuable resource for educators, researchers, and the general public around the world. Looking at another metric, the EOL has attracted over $50 million in funding, and eventually expects to receive twice that amount. Has the time come for a similar project in archaeology, an 'Encyclopedia of Cultural Heritage'? The situation in archaeology resembles that in the life sciences several years ago. A number of valuable online archaeological databases exist, and attempts to improve interoperability between databases have been initiated (with limited success). Existing resources provide the raw material for an Encyclopedia of Cultural Heritage, which in turn could accelerate the development of interoperable databases and improve data management practices across the discipline. This paper provides an overview of the EOL and a discussion of existing archaeological databases. We consider the EOL as an example for archaeology, and explore the obstacles to and potential benefits of an Encyclopedia of Cultural Heritage.

Keywords Encyclopedia of Life, archaeological databases, web-based research tools.

Imagine an online resource that contained, say, one million web pages representing archaeological sites and landscapes around the world, with excavators and experts providing and authenticating the content. The precedent for such an undertaking has already been set in the discipline of biology by the Encyclopedia of Life (EOL). The EOL serve as a valuable example for archaeology. It has developed strategies for collecting, amalgamating, and certifying data. An Encyclopaedia of Cultural Heritage could begin with existing electron and print resources, would encourage interoperability between existing databases as well as improve data management in archaeology. Obstacles to such a project exist - the nature of archaeological information differs from biological data - but the potential benefits of an Encyclopaedia of Cultural Heritage to archaeologists, educators, cultural resource management professionals, and the public make its construction worthwhile.

The Encyclopedia of Life

In 2003 Dr. Edward Wilson, a Harvard University biologist, observed that a wealth of biological information had recently become widely available from a variety of disconnected databases, and argued that the time had come to amalgamate that information into a single source. In February 2008 the first demonstration pages of the EOL website went online (Wilson 2003, Marris 2008; 'Encyclopedia of Life' 2009).

The EOL is animated by a clear, powerful mission statement: its goal is to host one standardised, extensible, authoritative webpage for each species of life known to science ('Who We Are' 2009). Information in fully developed pages includes an introduction, taxonomy, images, and sections containing a description, ecology and distribution, evolution, relevance (essentially, human interaction with the species), and references and further information ('Exemplar Pages' 2009). Sections are further subdivided to assist retrieval of specific information. Eventually, the EOL hopes to build 1.8 million comprehensive species pages.

To accomplish such an ambitious goal, the EOL has taken a multipronged approach to building its store of information. The EOL aggregates and distributes data partly through the direct contribution of many experts, and partly by drawing upon information contained in the online databases of its partners. To accomplish this task, the EOL is divided into a constellation of five subprojects or 'groups': the Species Pages Group, the Biodiversity Informatics Group, the Scanning and Digitization Group, the Education and Outreach Group, and the Biodiversity Synthesis Group.

The Species Sites Group works with contributors to build and authenticate each species page. Inspired in part by Wikipedia and other collaborative endeavours, the Species Sites Group recruits contributors. It also works with 'curators' from the scientific community to authenticate information ('Institutional Partners' 2009; 'Encyclopedia of Life' 2009). Software has been developed to assist contributors and curators with the generation and authentication of species pages (LifeDesks 2009).

To assist the Species Sites Group collect data, an Informatics Group develops the software to mine existing data in order to populate species pages. Much of the data displayed in the

EOL is collected automatically from partner organisations. Proprietors of existing life sciences databases are invited to become EOL partners, and provided with a 'transfer schema' to facilitate integration ('EOL Taxon Resource Transfer Schema' 2009). Realising the centrality of these partners to the success of the EOL, the project offers a commitment that they will provide partners with visibility, increased traffic, data management tools and services, and other benefits ('Commitment to Content Partners' 2009). The EOL will even assist scientists or groups who have built off-line databases ready their work for integration into the EOL ('Help Build EOL' 2009). At the time of writing, the EOL had 30 content partners, including database like FishBase, the Tree of Life web project, and the World Register of Marine Species, to choose just three ('Institutional Partners' 2009; 'Content Partners' 2009).

A Digitization Group aims to digitise and widely distribute the huge body of relevant literature now available only in print. This group is lead by the Biodiversity Heritage Library, which aims to build a repository of digitised literature and open it to worldwide use under a Creative Commons license ('Institutional Partners' 2009; 'Biodiversity Heritage Library' 2009).

The final two groups promote the project. The Education and Outreach Group targets educators and the general public, seeking to increase awareness of the project and build a large user community. The Biodiversity Synthesis Group works with scientists to develop new ways that the EOL can be used for biodiversity research ('Institutional Partners' 2009; 'Biodiversity Synthesis Group' 2009).

It should be emphasised that the EOL relies heavily on its institutional and content partners, which contribute the bulk of information contained in its database. When the EOL launched in 2008, only two dozen of the pages had been fully developed and authenticated, but several tens of thousands had been partially filled by automated data collection from partners such as FishBase, and the 'skeleton' of over a million records had been produced in the same manner (Marris 2008, 1038). Existing databases would likewise swerve as the heart of any similar effort in archaeology.

In its first year and a half of existence, the EOL has been a success. When the EOL website débuted, it received 11.5 million hits in under six hours (Shorthouse 2008). Using another metric, the EOL has attracted over $50 million in funding - of a projected ten-year cost of over $110 million (Marris 2008, 1038; 'Encyclopedia of Life' 2009). Although Marris (2008) has noted the difficulty of maintaining funding over the long term, for now the EOL has attracted enviable resources, site traffic, and coverage in the popular and academic press.

Existing Online Archaeological Databases

Has the time come for a similar project in archaeology, an 'Encyclopaedia of Cultural Heritage'? The concept has been proven by the EOL. The raw materials for an Encyclopaedia of Cultural Heritage may not be quite as extensive as those available in the life sciences, but they are sufficient to begin considering its merits. Such a project would, furthermore, yield significant benefits to the discipline of archaeology beyond the obvious utility of a more or less universal database of sites and landscapes.

The situation in archaeology resembles that of biology several years ago; a number of valuable online archaeological databases exist. Some of these databases are limited to a particular country or region (FastiOnline). Others are site-specific (the Athenian Agora Excavations database). Still others are thematic (the Digital Archaeological Archive of Comparative Slavery). A few have wider ambitions of linking diverse information through single portals to facilitate analysis (the Digital Archive Network for Anthropology and World Heritage). Some distribute documents such as reports (the National Archaeological Database), while others (Arachne) provide direct access to an online database of artefacts ('FastiOnline' 2009; 'Resources' 2009; 'Digital Archaeological Archive of Comparative Slavery (DAACS)' 2009; 'The National Archaeological Database (NADB)' 2009; 'Arachne' 2009). Many online databases of archaeological information exist, providing ample raw material for an Encyclopaedia of Cultural Heritage. Figure 1 contains a list of selected databases; associated websites are included in the References.

However their content is delimited, online databases use some combination of two strategies for collecting information: (1) manual entry of data into a single database, by means of either the digitisation of existing paper records or the input of new data submitted via paper or electronic forms, or (2) development of a distributed, interoperable network of databases accessed through a shared portal. The latter strategy falls on a range from 'hard' to 'soft'. Hard interoperability requires all contributors to use the same database or database structure (as is the case with DANA-WH; see 'Operating Structure' 2009). Soft interoperability is based on the promulgation of metadata standards or transfer schema that amalgamate selected data from, or search across and link to, any database that adopts them (as is the case with Pleiades / Concordia; see 'Concordia Project' 2009; 'Concordia - Pleiades Project Wiki' 2009; Gillies 2009). As a comparison, the EOL combines manual entry and authentication of information (by creating of new records from diverse sources and the digitising existing information) with soft interoperability; any online database that can produce output adhering to its published transfer schema can have its contents incorporated into the EOL ('EOL Taxon Resource Transfer Schema' 2009).

Site	Description
Arachne	A database of artefacts (sculpture, coins, etc) and some texts held by a number of institutions in Europe developed by the German Archaeological Institute (DAI). Content created primarily through digitisation of institutional records, although interoperability with DAI excavation GISs has also been implemented. Interoperability with the Perseus Project is being developed.
Archaeological Data Service (ADS)	Digital repository for archaeological data in the UK; promotes best practices for archaeological data management.
Archaeological Records of Europe - Networked Access (ARENA)	Collects data digitally from partner institutions in six European countries, archives and distributes it electronically, and also provides a portal to further information on the websites of partner institutions.
ArchAtlas Atlas & Gazetteer	Digitised information drawn from published and unpublished gazetteers. Links to further information.
The Athenian Agora Excavation database	A database of publications, plans, drawings, and artefacts spanning decades of work in the Athenian Agora.
Concordia	An 'interoperability framework' designed to automate integration of diverse online datasets into Pleiades by promoting a 'shared vocabulary' or metadata standard.
Digital Archaeological Archive of Comparative Slavery (DAACS)	An archive of archaeological materials (sites, features, artefacts) relevant to the study of slavery in the New World (America and the Caribbean). Updated manually.
Digital Archive Network for Anthropology and World Heritage (DANA-WH)	A 'federation of distributed, interoperable databases' that together create a repository of data about human cultural and biological heritage. Includes 3D models of many objects. DANA-WH distributes table schema that may be implemented on any database system.
FastiOnline	A database of archaeological sites in selected European and North African countries. Updated manually.
The Jordan Archaeological Database & Information System (JADIS)	A database of Palaeolithic through modern archaeological sites used by researchers and the Jordan Department of Antiquities. Data is entered by projects through an online form.
Mediterranean Archaeology GIS (MAGIS)	A database containing information about archaeological surface survey projects in the Mediterranean (not actual survey data). Updated manually.
The National Archaeological Database (NADB)	A database of documents (reports, maps, and permits) administered by the United States National Park Service.
Pleiades	An enhanced and extended online version of the Classical Atlas Project that also provides a system for the collaborative use, creation, and dissemination of historical geographic information

Fig. 1. A Sample of Existing Online Archaeological Databases.

Two examples of existing online archaeological databases will illustrate differing strategies of data collection. FastiOnline, the electronic successor to the *Fasti Archaeologici* published until 1987 by the International Association for Classical Archaeology, records and distributes geographical, chronological, and other basic information about archaeological excavations from12 European and North African countries. FastiOnline collects information on forms, submitted on paper or via email, which are entered manually into its database - a time-consuming process, but one that produces consistent, authenticated, multilingual data. Visitors to the site can perform various text-based searches or use a map interface ('FastiOnline' 2009). FastiOnline publishes more detailed information about sites in its own peer-reviewed online journal, *FastiOnline Documents & Research*, which is linked from related FastiOnline entries (*FastiOnline Documents & Research* (*FOLD&R*), 2009).

Concordia takes a different approach. Concordia is an extension of Pleiades, an enhanced, digital version of the Classical Atlas Project. The Concordia project inherits from Pleiades a system for the collaborative use, creation, and dissemination of historical geographic information ('Pleiades Beta Portal' 2009; 'Pleiades Phase 1 - Pleiades Project Wiki' 2009). Concordia aims to extend Pleiades by building 'an interoperability framework for research materials related to the history of Greek and Roman Egypt, Libya and surrounding areas' ('Concordia Project' 2009). Through the use of a shared 'vocabulary' (essentially a metadata standard), Concordia plans to link many discrete digital resources from within Pleiades, turning the latter into a portal for knowledge about the Greek and Roman antiquity ('Concordia - Pleiades Project Wiki'2009; see also Gillies 2009). Links would be created automatically since data adhering to the standard could be manipulated by 'software agents such as web crawlers or spiders, and indexed by search engines' ('Concordia Project' 2009). Although at present Concordia / Pleiades is limited to a particular geographic area and emphasises certain kinds of information (e.g. sites, coins, papyri), any interoperability framework it developed has the potential for wider application.

Despite these existing online databases, the amount of archaeological data available electronically constitutes a vanishingly small percentage of published - let alone unpublished - data, and interoperability of a substantial subset of existing databases remains distant. The state of archaeological data management is one of the chief culprits in this situation, and what initially motivated our interest in the EOL.

Archaeological Data Management

In an ideal world, every archaeological project would have a powerful relational database capable of managing all information produced by the project. Data would be entered into this database directly in the field, eliminating redundant input. It would be structured enough to produce all data relevant to a particular question quickly and reliably, yet flexible enough to serve rather than dictate the methods and practices of the project. It would integrate seamlessly with a project GIS, providing underlying data for sophisticated geospatial analysis. Finally, it would incorporate metadata standards or a transfer schema that would allow it to feed information into online databases with as little human intervention as possible.

Yet at a time when data warehousing and data mining are transforming businesses and some academic disciplines, too few archaeological projects have deployed effective relational databases. Not having information stored systematically in an accessible electronic format hinders interpretation, slows publication, and impedes the development of online resources. This slow adoption and shallow penetration of database technology is handicap for the entire discipline.

Although statistics are hard to come by, the state of data management practices in archaeology is poor. The results of a review conducted by arts-humanities.net found that of projects listed in its ICTGuides showcase of digital arts and humanities projects, only 11 of 15 used a database, with the database management systems in use were divided as follows:

- Microsoft Access (6)
- AdLib (2)
- MySQL (2)
- Filemaker (1)

Of these database management systems, AdLib and Filemaker are rather limited, leaving only eight of 15 projects with reasonably robust systems (projects needing multi-user capability will eventually find Access limited as well, although migration to MS SQL Server is possible). Bear in mind that arts-humanities.net is a website promoting the use of information technology in the arts and humanities, and that projects listed its ICTGuides are more likely take a keener interest archaeological applications of IT than the average project ('Wikipage: Databases (Archaeology)' 2007; 'ITCGuides' 2009).

Our own experience on about a dozen archaeological projects over the past decade corresponds reasonably well with the arts-humanities.net review. None of these projects had a comprehensive relational database integrated with a GIS at the time of our association (and some of those projects were run by major universities and had relatively generous funding). Some had partial databases, including limited information about some activities (e.g., photography) or some classes of artefact (e.g., ceramics). Most relied on flat or lightweight 'databases' (typically Excel spreadsheets or Filemaker), or at best Microsoft Access. Some of the projects did have extensive GISs, and had developed geodatabases using ESRI ArcGIS. Where relational databases had been deployed, quality of design varied considerably, but tended to be poor.

Two factors combine to produce this situation. First, the quantity, complexity, and diversity of archaeological data requires the deployment of flexible yet robust databases that must, in most cases, be built from scratch or extensively customised to fit the needs of each project. Second, few projects have the resources to invest in time-consuming development and ongoing modification of complex databases, especially considering that what might be considered 'core' activities such as excavation, survey, and scientific analysis, always seem to be funded and manned before 'supporting' activities like data management.

Over the past several years, we have been involved with a number of workshops and discussions about how to improve data management in archaeology. These exchanges have returned to two possible solutions. The first centred around the idea of a flexible, modular, spatially-capable relational database, such as the Archaeological Recording Toolkit developed by LP Archaeology ('Archaeological Recording Toolkit (ARK)' 2009). The problem of encouraging widespread adoption, however, led us to conclude that distributing even a well-designed database was, at best, only a partial solution. The second proposed solution involved the promotion of metadata standards or design protocols, like the CIDOC CRM, a metadata standard for cultural resource management developed in Europe ('CIDOC Conceptual Reference Model (CRM)' 2006; see also Dally and Foertsch 2007; Adida and Birbeck 2008; Gillies 2009; 'Standards in Archaeology' 2009). A widely adopted metadata standard could improve design and allow interoperability, facilitating the creation of online databases, but two problems with this approach were immediately flagged: (1) achieving widespread adoption (again) and (2) providing tools that allow individual projects to apply those standards. Indeed, existing databases have run into problems incorporating the CIDOC CRM (Gillies 2009).

Our discussions ended inconclusively, but allowed us to survey the landscape of archaeological data management. As it stands, components that might revolutionise data management and dissemination in archaeology exist: flexible, modular databases (ARK), metadata standards (CIDOC CRM), and online databases (FastiOnline; Concordia / Pleiades). All aspects of this emerging universe are, however, limited. No standard, let alone particular database, has been widely adopted. 'Middleware' allowing archaeologists to adapt existing databases or apply metadata standards - without employing an expensive team of developers beyond the means of most projects - is lacking. Managers of databases can find fault with any given set of standards, and as each chooses her own, vocabularies proliferate. Our final impression was that while slow development is taking place, some further initiative to improve practices and promote convergence is desperately needed in the discipline.

The Encyclopedia of Life as a Model for an Encyclopaedia of Cultural Heritage

The EOL provides a valuable example of constructing and administering a universal information clearinghouse. We see four principal lessons that can be learned from the EOL:

1) Existing databases represent valuable resources that can be mined for information to populate, at least in part, a universal project like the EOL. These existing resources will, in turn, be better utilised if linked through and promoted by a high-profile site like the EOL.

2) A combination of wiki-style collaboration and peer review by page 'curators' can marshal distributed knowledge but maintain high standards. A peer review process also increases the likelihood that contributions will be taken seriously and applied towards the career advancement of authors and 'curators'.

3) A generous commitment to contributing institutions and individuals, including publicity and technical assistance, can be attractive enough to promote participation.

4) A large-scale project like the EOL can attract significant funding and capture people's imagination, resources and interest can in turn 'trickle down' to contributing partners. In particular, some of the tools developed to facilitate collaboration and database interoperability might improve data management across the discipline

Problems

There are, of course, differences between biology and archaeology, and adapting the model offered by the EOL will raise certain problems:

1) Classification of archaeological materials is more complex than species taxonomy. Although certainly controversies remain, any type of life can usually be placed in a tree ranging from kingdom to species. No such (relatively) objective hierarchy exists for archaeological sites.

2) Archaeological data is more complex than species data. Recording basic information about a site is similar to recording basic information about a species, and even contains some of the same data types (geographic data, images, text descriptions, conservation information, etc.), but there is no analogue in the EOL to the archaeological artefact. The fact that sites may yield tens or hundreds of thousands of artefacts increases the complexity of archaeological data by orders of magnitude.

3) Data from landscape archaeology does not easily fit into the site-based model. The results of surface survey, remote sensing, geological and palaeo-environmental investigation, and other types of distributed information are critical to modern archaeology, but will not fit into the 'site' data container or anything quite equivalent to a species in the EOL.

Adapting the EOL Model

We believe that most of the problems of adapting the model provided by the EOL can be mitigated. Although better solutions probably exist, we propose the following modifications to the EOL model in order to address the issues raised above:

1) Limit taxonomy of sites. In order to clearly distinguish between data and interpretation, definitive placement of sites in particular culture groups should be avoided. Instead, emphasis should be placed on presenting evidence for cultural change and exchange (e.g., features and artefacts that indicate internal evolution and external contact or influence). Such an approach would not rule out the possibility of including an interpretive section on each site page, we only argue that the overall structure of the database should not be built on interpretations that are likely to change or that stifle debate (cf. Eiteljorg 2005).

2) Organise information based upon the most objective criteria available: geographical location and (more problematically) chronology. Such parameters will produce more of a matrix of sites than the hierarchy found in the EOL. Surveys, etc., might be entered into the database by project and year (providing a snapshot of a changing cultural landscape), but viewable and searchable geographically.

3) Strengthen the collaborative nature of the project. An Encyclopaedia of Cultural Heritage that operated on the level of the site would, we think, be similar in size to the EOL (several million records), and could be approached using some of the same techniques that the EOL employs. Extending the ECH to include artefacts would require a far greater degree of collaboration, including aggressive promulgation of metadata standards or transfer schemata for use with existing databases, and provision of model databases to projects that do not yet have them. The 'wiki' nature of the project would have to be enhanced, ideally without forfeiting peer review.

Overall, the differences between the EOL and an ECH centre around two issues: (1) the greater scale and complexity, and (2) the more subjective nature of archaeological classification. There are ways to address both problems centring around (1) an organisation based upon geography and possibly chronology of sites; (2) an even more distributed and collaborative strategy (especially where artefact catalogues are concerned), and (3) a very clear distinction between data and interpretation, one which leaves room for the users of the system to rethink issues such as cultural change and affinity, regional grouping, and periodisation.

Conclusions

Considering the various efforts currently underway to amalgamate and distribute archaeological information, the slow but ongoing transition to digital recordkeeping, the example provided by the EOL, and the increasing tendency to turn to the internet as a resource, we are prepared to venture that something resembling the Encyclopaedia of Cultural Heritage will eventually exist.

Not only would an Encyclopaedia of Cultural Heritage inspire the public and open new avenues of research for archaeologists, but it could contribute to improving data management across the discipline. It would not replace existing online databases, but utilise and strengthen them. An Encyclopaedia of Cultural Heritage could lead by example, providing incentives for adherence to standards as well as developing tools and supplying resources that would foster improved data management and interoperability. Projects and partners participate because they want to - participation would have tangible benefits, and it would be rewarding to contribute to such an inspiring project.

We would like to conclude with a practical suggestion about initiating an effort to build an Encyclopaedia of Cultural Heritage. Establishing a limited prototype (perhaps including a few selected projects) seems a logical first step - with the help of proprietors of archaeological databases, interested project directors, database professionals, staff from the EOL and other interested parties. We invite members of the audience (and readers of this paper) to contact us and assist with such a beginning.

References

ADIDA, BEN, AND MARK BIRBECK, 'RDFA PRIMER', 2008, W3C, accessed 15 June 2009 <http://www.w3.org/TR/xhtml-rdfa-primer/>.

'ARACHNE (ARCHAEOLOGICAL DATABASE)', 2009, Wikipedia, accessed 15 June 2009 <http://en.wikipedia.org/wiki/Arachne_(Archaeological_Database)>.

'ARACHNE', 2009, DAI, accessed 15 June 2009 <http://arachne.uni-koeln.de/>.

'ARCHAEOLOGICAL DATA SERVICE (ADS)', 2009, ADS, accessed 15 June 2009 <http://ads.ahds.ac.uk/>.

'ARCHAEOLOGICAL RECORDING TOOLKIT (ARK)', 2009, LP Archaeology, accessed 15 June 2009 <http://ark.lparchaeology.com/>.

'ARCHAEOLOGICAL RECORDS OF EUROPE - Networked Access (ARENA)', 2009, ARENA, accessed 15 June 2009 <http://ads.ahds.ac.uk/arena/index.html>.

'ARCHATLAS', 2009, ArchAtlas, accessed 15 June 2009 <http://www.archatlas.org/Home.php>.

'BIODIVERSITY HERITAGE LIBRARY', 2009, Biodiversity Library, accessed 15 June 2009 <http://www.biodiversitylibrary.org/>.

'BIODIVERSITY SYNTHESIS GROUP', 2009, Biodiversity Synthesis Center (BioSynC), accessed 15 June 2009 <http://synthesis.eol.org/>.

'CIDOC CONCEPTUAL REFERENCE MODEL (CRM)', 2006, CIDOC CRM, accessed 15 June 2009 <http://cidoc.ics.forth.gr/>.

'COMMITMENT TO CONTENT PARTNERS', 2009, Encyclopedia of Life, accessed 15 June 2009 <http://www.eol.org/files/pdfs/docs/EOL_Commitment_to_Partners.pdf>.

'CONCORDIA - PLEIADES PROJECT WIKI', 2009, Pleiades Project, accessed 15 June 2009 <http://www.atlantides.org/trac/pleiades/wiki/Concordia>.

'CONCORDIA PROJECT', 2009, Concordia Project, accessed 15 June 2009 <http://concordia.atlantides.org/>.

'CONTENT PARTNERS', 2009, Encyclopedia of Life, accessed 15 June 2009 <http://www.eol.org/content/partners>.

DALLY, ORTWIN AND REINHARD FOERTSCH, 2007, 'The Arachne Object-database', Center for the Study of Architecture / Archaeology (CSA) Newsletter, accessed 15 June 2009 <http://csanet.org/newsletter/spring07/nls0703.html>.

'DIGITAL ARCHAEOLOGICAL ARCHIVE OF COMPARATIVE SLAVERY (DAACS)', 2009, DAACS, accessed 15 June 2009 <http://www.daacs.org/>.

'DIGITAL ARCHIVE NETWORK FOR ANTHROPOLOGY AND WORLD HERITAGE (DANA-WH)', 2009, DANA-WH, accessed 15 June 2009 <http://www.dana-wh.net/home/>.

EITELJORG, HARRISON, II, 2005, 'CSA Database Discussion', Center for the Study of Architecture / Archaeology (CSA), accessed 15 June 2009 < http://csanet.org/inftech/dbdisc.html>.

'ENCYCLOPEDIA OF LIFE', 2009, Wikipedia, accessed 15 June 2009 <http://en.wikipedia.org/wiki/Encyclopedia_of_Life >.

'EOL TAXON RESOURCE TRANSFER SCHEMA', 2009, Encyclopedia of Life, accessed 15 June 2009 <http://www.eol.org/files/pdfs/docs/EOL_Transfer_Schema_Documentation.pdf >.

'EXEMPLAR PAGES', 2009, Encyclopedia of Life, accessed 15 June 2009 <http://www.eol.org/content/exemplars>.

FASTIONLINE DOCUMENTS & RESEARCH (FOLD&R), 2009, FastiOnline, accessed 15 June 2009 <http://www.fastionline.org/folder.php?view=home>.

'FASTIONLINE', 2009, FASTIONLINE, ACCESSED 15 JUNE 2009 <http://www.fastionline.org/index.php>.

FOSS, PEDAR, ELIZABETH FENTRESS, STEPHEN SAVAGE, *ET AL.*, 2008, 'Workshop: Web-Based Research Tools for Mediterranean Archaeology' (abstract), 2008, Archaeological Institute of America Annual Meeting, accessed 15 June 2009 <http://www.archaeological.org/webinfo.php?page=10248&searchtype=abstract&ytable=2008&sessionid=5I&paperid=1501>.

GILLIES, SEAN, 2009, 'Concordia, Vocabularies, and CIDOC CRM', Concordia Project, accessed 15 June 2009 <http://concordia.atlantides.org/docs/concordia-crm.html>.

'HELP BUILD EOL', 2009, Encyclopedia of Life, accessed 15 June 2009 <http://www.eol.org/content/page/help_build_eol>.

'INSTITUTIONAL PARTNERS', 2009, Encyclopedia of Life, accessed 15 June 2009 <http://www.eol.org/content/page/institutional_partners>.

'ITCGUIDES', 2009, Arts-humanities.net: Digital Humanities and Arts, accessed 15 June 2009 < http://www.arts-humanities.net/node/ict_guides>.

'THE JORDAN ARCHAEOLOGICAL DATABASE & INFORMATION SYSTEM (JADIS)', 2009, JADIS, accessed 15 June 2009 <http://gaialab.asu.edu/Jordan/JADISGIS.php>.

'LIFEDESKS', 2009, LifeDesks, accessed 15 June 2009 <http://www.lifedesks.org/>.

MARRIS, EMMA, 2008, 'First chapter of book of life goes live', *Nature* 451, 1038.

'MEDITERRANEAN ARCHAEOLOGY GIS (MAGIS)', 2009, MAGIS, accessed 15 June 2009 <http://cgma.depauw.edu/MAGIS/>.

'THE NATIONAL ARCHAEOLOGICAL DATABASE (NADB)', 2009, National Parks Service Archaeology Program, accessed 15 June 2009 <http://www.nps.gov/history/archeology/TOOLS/Nadb.htm>.

'OPERATING STRUCTURE', 2009, DANA-WH, accessed 15 June 2009 <http://www.dana-wh.net/opstruct/>.

'PLEIADES BETA PORTAL', 2009, Pleiades, accessed 15 June 2009 <http://pleiades.stoa.org/>.

'PLEIADES PHASE 1 - PLEIADES PROJECT WIKI', 2009, Pleiades Project, accessed 15 June 2009 <http://www.atlantides.org/trac/pleiades/wiki/PleiadesPhase1>.

'RESOURCES' [DATABASE PORTAL], 2009, Athenian Agora Excavations, accessed 15 June 2009 <http://www.agathe.gr/resources.html>.
SHORTHOUSE, DAVID, 2008, 'We're too Popular!', Encyclopedia of Life Blog, 26 February, accessed 15 June 2009 <http://blog.eol.org/2008/02/>.
'STANDARDS IN ARCHAEOLOGY', 2009, ARENA, accessed 15 June 2009 <http://ads.ahds.ac.uk/arena/links/standards.html>.
'WHO WE ARE', 2009, Encyclopedia of Life, accessed 15 June 2009 <http://www.eol.org/content/page/who_we_are>
'WIKIPAGE: DATABASES (ARCHAEOLOGY)', 2007, Arts-humanities.net: Digital Humanities and Arts, accessed 15 June 2009 < http://www.arts-humanities.net/wiki/databases_archaeology>.
WILSON, EDWARD O., 2003, 'The Encyclopedia of life', *Trends in Ecology and Evolution*, 18.2, 77-80.

Computer tools for creation of electronic atlases of the historical monuments state

Stanislav B. Shchigorets, Albina A. Doos, "GREC" research team, St.-Petersburg, Russia, Valery M. Grishkin, Alexey P. Zhabko and Dmitry Yu.Vlasov

St.-Petersburg State University, Russia

Abstract

The proposed are intended for monitoring monument condition by means of a regular digital photography of monuments followed by a computer-aided analysis of the images. The output electronic atlas is produced as a result of processing digital image series, using a special computer system. This approach significantly reduces monitoring costs, facilitates and speeds up creation of electronic atlases of cultural heritage objects. The methods are based on revealing information attributes and comparing them with some master values. The system works as follows. The input data are series of digital color images of a deteriorated monument. First, the images are processed to ensure the adjustment of shooting conditions and image unification in terms of color. Then, special computer algorithms are applied to analyze color parameters of the images and detect deterioration areas due to biodegradation and mechanical damage. The methods are based on revealing information attributes and comparing them with some master values.

Keywords Cultural heritage, monitoring, biodeterioration, image processing, electronic atlases

To preserve the cultural heritage monuments it is especially important to have the full information about the current state of the object as well as an opportunity to display the changes of the monument in time. The obvious decision of this question - current monitoring of cultural heritage objects and preservation of results in the electronic form (detailed atlases of objects). The purposes of monitoring is the estimation of biological, physical and chemical influence on object, revealing of the most dangerous factors of the environment, type and sizes of the destruction and accumulation of the data, allowing to predict a situation and to develop the recommendation for the object preservation.
This work it is offered to spend monitoring by means of regular electronic photographing of the objects. The information on destruction areas can be received by the processing of object color images with the special software. The processing results are the basic data for the electronic atlas of object creation. Such approach allows to lower essentially expenses for carrying out of monitoring and to automate the getting data about the object. The electronic atlas of the object represents a full map of mechanical and biological destruction areas with their geometrical parameters. It also includes the information about types of destructors, their aggression and the degree of influence on a material of object. The basic idea of the computer system is the segmentation of the objects images on some color and brightness areas and their correlation with the types of destruction.

The initial sorting of destructors is carried out by experts. The data are accumulated in the database, which is the basis for the subsequent classification. First stage is the teaching of the system. At the second stage (recognition) the images of the segmented objects areas are shown to system for analyses. This procedure is based on the comparison with the etalon data about the types of the destructors.

The creation of the computer system is connected with the following algorithms:

- algorithm of the images processing;
- algorithm of the preliminary processing;
- algorithm of the allocation color and brightness areas;
- extraction of the informative features;
- recognition of the destruction areas;
-definition of destruction areas parameters and preservation of results.

The described system is realized in Matlab with the tools of Image Processing Toolbox and Statistics Toolbox as well as the means of graphic interfaces Guide creation. It is explained by the convenience and simplicity of realizations of images processing described algorithms. Software structure of the computer system is shown on figure1.

The system has already examined on the some historical objects of the Saint Petersburg and Chersoneses of Tauride (Crimea, Ukraine). It is necessary to note, that the images which were checked by the system, have been received in field conditions without application of any special illumination of objects. Preliminary results have been shown the legitimacy of the offered computer system.

The application of this system was very effective on the some monuments the Saint Petersburg. Many of the famous Saint Petersburg monuments are intensively decaying under the influence of the external factors such as atmospheric humidity and temperature changes, air pollution, salts,

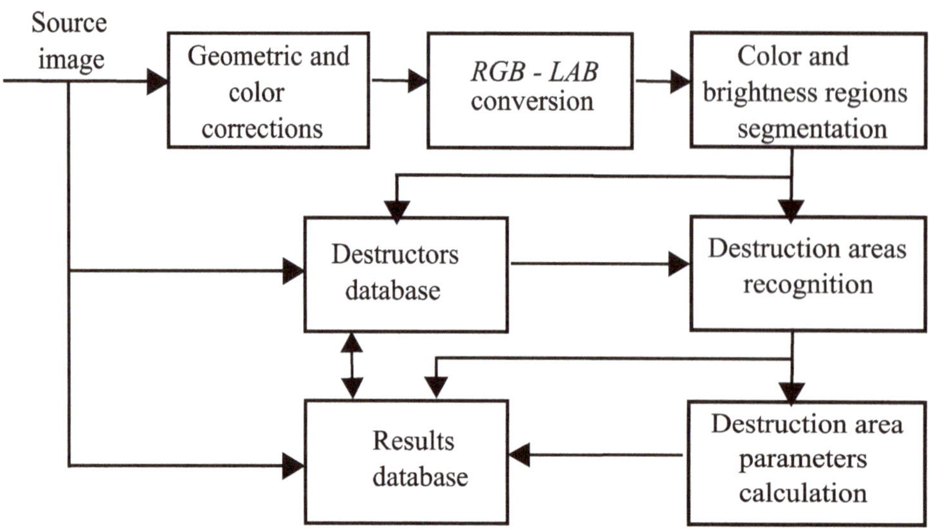

Fig. 1. Software structure of computer system.

aggressive microbial communities. The surface of many historical monuments is colonized by different biological objects: lichens, bacteria, fungi, moss and seed plants. They change the monument surface state by biofouling processes. The biofouling maps were made for some objects using buy the computer system CONSOLIDATOR-BIO.The creation of the biofouling maps includes the following stages:

- Detailed digital photofixing of the various biofouling forms on a monument surface;

- Exact identification of the basic biofouling types, analysis of their allocation on the monument surface;

- Creation of the biofouling map with use of computer technology CONSOLIDATOR-BIO.

The samples of the some Saint Petersburg monuments maps are shown on fig. 2 and fig. 3.

Figure 2 shows the changes of the memorial monument state (in the spring and autumn seasons). The monument surface is covered by biofilm with alga domination. After the half of year from the first observation the biofouling area have changed visibly. The maps give the possibilities to analyze these changes and to explain the biofouling dynamic. Another sample is from the Saint Petersburg suburb – Petershoff (figure 3). On the map of the granite head it is shown the location of the biodeterioration agents (lichens, mosses, alga, fungi) on the monument surface. This map can be used for the monitoring observation on this monument. The covered areas can be changed very rapidly in connection with the changes of the environment.

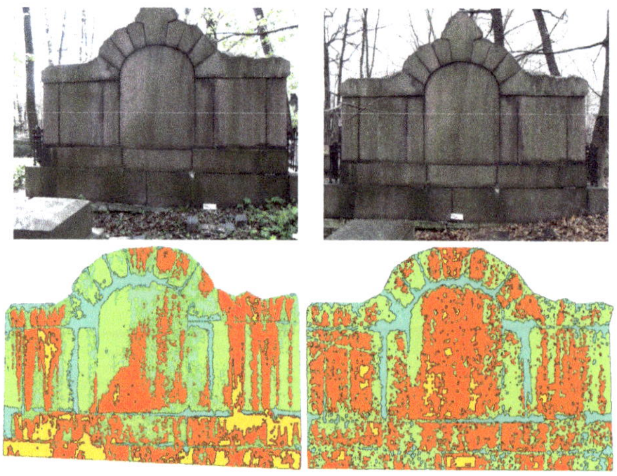

Fig. 2. Photos of memorial monument in the Saint Petersburg and maps of the current state of the monument surface in the different periods of observation.

All members of biological community occupying a monument surface have a different degree of biological activity. They have different degree of interaction with the rock monument surface. Some of them can be very dangerous and aggressive in relation to the rock substrate. They can destroy the surface level of the monument due to chemical and physical impact. They can change the mineral structure and color characteristics of the substrate.

Such biodestructors can represent the greatest danger for the monument. Other biological objects on the monument surface practically do not destroy a rock material. They have a lower aggressive degree. All members of the lithobiotic community interact with each other. Zones of such interactions can exert the specific influence on the rock surface. So we can try to mark out the different levels of biological objects aggressive on the monument surface and place them in the biofouling map using the computer technology CONSOLIDATOR-BIO.

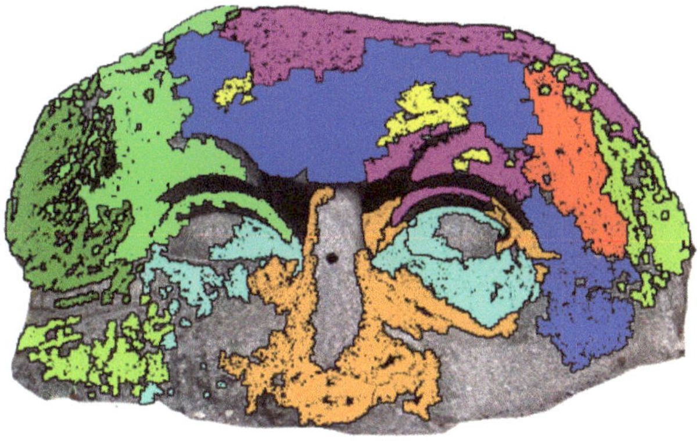

Fig. 3. The biofouling map of the granite head in the Petershoff (Saint Petersburg surburb).

The maps must be accumulated in the electronic atlas of the object. Such atlas opens the possibilities for the visualization of the monument state changes in time. The proposed computer system has the large importance for the real monitoring of the cultural heritage monuments as well as for planning of conservation works.

Dynamic mapping of migration and settlement patterns of communities - challenges and issues

P. Sumabala

Lecturer, Department of History, L.N.G. College, Ponneri
E-Mail: sumabala2k@yahoo.com

Abstract

A study of the migration of communities that took place in the history of South India, particularly of the Andhradesa and Tamilagam, between 6th to 16th centuries CE, was carried out recently. It involved a detailed investigation into the records - inscriptional, literary, and monumental - necessary to glean information on the migration of communities and on how these settlers adjusted and subsequently transformed themselves and their new environment.
Several ways of depicting these migrations are explored. A multiple-time line spanning the time zone between 6th and 12th centuries is presented. The factors associated with migrations within this time zone are described. An example of a time line of a limited geographic region that was in the flux of events throughout the time zone is presented. Two examples of maps, one dealing with migrations of Brahmanas from Andhradesa to Pallava territory and the other dealing with territorial divisions of Cholas in Andhradesa are presented.

Keywords South Indian History, Andhradesa, Tamilagam, Rock cut temples, Migrations, Time-lines, Maps, Chola, Eastern Chalukya.

Scope of the present study

A study of the migration of communities that took place in the history of South India, particularly of the Andhradesa and Tamilagam, between 6th to 16th centuries AD, was carried out recently (Sumabala, P. 2008). It involved a detailed investigation into the records - inscriptional, literary, and monumental - necessary to glean information on the migration of communities and on how these settlers adjusted and subsequently transformed themselves and their new environment.

The political contacts between Andhradesa and Tamilagam seem to have existed from 1st century onwards. The migration of common people from Andhradesa to Tamilagam and their settlement in Tamilagam appeared to have started from the period of Pallavas - especially during Nandivarman Pallavamalla (AD 731-796) (as known from the Pullur and Tandnantottam copper plates). The migration of common people from Tamilagam to Andhradesa and their settlement in Andhradesa seems to have been started from the period of Cholas as evident from the inscriptions of Virachoda's Chelluru and Pithapuram copper plates issued during 21st and 23rd regnal years of Kulottunga I, corresponding to AD 1091 and 1093. The spread of Vaishnavism following the preaching of Ramanuja further paved way for the migration of people from Tamilagam to Andhradesa in 12th century AD. The rule of Vijayanagar over the whole of South India provided an opportunity for the movement of people between these two regions. There was a widespread migration of people into Tamilagam during the Nayak rule in Madurai, Tanjore and Gingee in 16th-17th century AD. There are evidences to indicate the movement of people of all walks of life from Andhradesa to Tamilagam through the period of Pallavas to the Nayak period. However, the movement of people from Tamilagam to Andhradesa appears to be intermittent.

Historical migrations in the context of the present study span several decades or centuries and extend spatially over large geographical regions. The possibility of depicting historical migrations through time lines and maps is explored.

Time line for the 6th - 12th century AD

As migrations are extended events in time they can be represented with the help of time lines. Figure 1 is an example of a time line through the time zone between 6th - 12th centuries AD that attempts to capture migrations that took place in the southern peninsular India. Broadly, these migrations have taken place during the reigns of two dynasties, namely, the Pallavas and the Cholas. The various time divisions presented in Figure 1 are based on available records. While it is expected that these time divisions would be unequal, it is interesting to observe that there are several entries recording events spanning every two decades!

Figure 1 attempt to give a comprehensive picture by bringing together time lines corresponding to political, religious and cultural events that took place within the time zone. The backdrop against which some of these migrations took place is discussed below. Factors influencing migrations and the changes brought about by migrations are multi-layered information which when associated with the time line representation can provide an attractive descriptor of the dynamics in History.

Time	Imperial/Noble families	Brahmanas	Craftsmen/Weavers	Merchants
c. AD 6th century	Pallavas moved from Andhradesa and settled at Tondaimandalam with their capital at Kanchipuram			
c. AD 6th-7th century	Telugu Chodas in Andhra and in Andhra - Tamil border regions			
c.AD 7th century			Special craftsmen migrated from the lower Krishna valley for the construction of rock-cut monuments	
AD 764, 789		Migrations from different parts of Andhra Pradesh to Nelli, Pullur and Kudiyur in Polur taluk of Tiruvannamalai (North Arcot) district; and Tandantottam in Kumbakonam taluk of Tanjavur district		
c. AD 9th century	Migration of Kumaramurti, chief of Kaluru and Vayaluru (near Kanchipuram) to Undi in Vengi kingdom			
c. AD 1000	Matrimonial alliances between the royal families of Cholas from Tanjavur and Chalukyas from Vengi			
AD 1019 - 1022		Migrations from different parts of Andhradesa to villages in Chola kingdom		
AD 1076 – 1126				Settlements of merchants of aruvala nagaramu and aruvala veedhi in Peruru in Western Chalukyan region
AD 1093		Tamil Vaishnava brahmana settlements in several places of Andhradesa		
AD 1121	Migration of Royal officials to Draksharama			
AD 1128			Migrations of twenty weaver families in Tirukkanapuram in Coimbatore district	
AD 1151	Migration of Brahma-Kshatra Kula to Bapatla. from the village of Brihadankura in Tondaimandalam	.		
AD 11th -12th century	Territorial divisions and place names		Construction of temples in Chola style in Andhradesa	Migrations from Uraiyurkurram in Cholaimandalam to Ravada in Chola-Chalukya region

Fig. 1 .Calendar of Events - Migrations/ Settlements between Andhradesa and Tamilagam

Migration of Imperial families during 6th century AD

The time line starts with 6th century AD and portrays the movement of Pallavas from the lower Krishna valley to Tondaimandalam with their capital at Kanchipuram. Pallavas, before coming to Tamil country and establish their hegemony with their capital at Kanchipuram, were ruling parts of Andhra region. It is believed that they were the feudatories of Satavahanas and became independent after the fall of Satavahana empire. The Pallavas regarded themselves as the political successors of the Satavahanas. The Pallava king Simhavishnu of the Manchikallu inscription (*Epigraphia Indica*, Vol. XXXII, p89) first dealt with the Ikshvakus of Vijayapuri (Nagarjunakonda) at lower Krishna valley. The Pallavas remained the undisputed rulers of Andhrapatha with Dhanakada (Dhanyakataka – present day Amaravati of Guntur district) as its capital till Samudra Gupta's invasion of southern regions. They left their home and reached the Tondaimandalam where they started their political career afresh with Kanchipuram as their capital.

Another important dynasty, which ruled the regions around Cuddapah, Kurnool, Anantapur, Krishna and Godavari belt in 6th-8th centuries AD and later the Nellore region in AD 11th -12th centuries AD, was that of the Telugu Chodas. They claim their descent from Karikala Chola and maintain that they hail from Uraiyur, the ancestral home of the Cholas, in Tiruchirapalli district. The migration to the Telugu country by these Cholas seems to have been taken place around 6th century AD, probably when the Tamil regions were occupied by the Kalabhras. The Telugu Chodas were mainly feudatories. The Renati Cholas, feudatories of Pallavas bore the names and titles, which resembled those of Pallavas. The Telugu Choda ruler Srikantha's daughter Aggalanimmati was the queen of the Pandya ruler Srimara Srivallabha who gave birth to his second son Parantaka Viranarayana. Under Jata Choda Bhima the Telugu Choda kingdom recovered much of their power and glory. However, he was hostile towards Rajaraja Chola and the Vengi Chalukya Saktivarman. During Kulottunga III the Telugu Choda Nallasiddha rose in revolt against the central authority and even brought Kanchi under sway for a short period. Later, the Telugu Chodas became independent during Rajaraja III and Rajendrea III and were friendly with the Chola monarchs. The Telugu Choda Gandagopala was finally defeated by the Pandya king Jatavarma Sundara Pandya.

Migration of Craftsmen during 7th century AD

Beginning from the 7th century, the Tondaimandalam region witnessed a significant cultural change due to the migration of special craftsmen from the lower Krishna valley (C. Meenakshi, *Administration and Social Life Under the Pallavas*, University of Madras, 1977, p350). They were appointed by Mahendravarman for the creation of cave temples. These temples excavated from solid rocks were first introduced in the Tamil land by him. Earlier to him there were no stone temples in Tamilagam as evidenced by his Mandagapattu inscription (*Epigraphia Indica*, Vol. XVIII p.17). Such rock-cut temples had already come into vogue during the reigns of Vishnukundin kings in Undavalli, Mogalrajapuram etc. It was believed that young Mahendra who had spent his princely years in the region around Undavalli had observed the shaping of rocks into temples by the Andhra workmen. The craftsmen of Mahendravarman created nearly twenty cave temples in different parts of Tamil land.

Migrations during 11th and 12th centuries AD

The Pallavas were defeated by the Cholas, who came to power with their capital at Tanjavur. Rajaraja became prominent with his conquests. The Chola kingdom grew under him into an extensive empire efficiently organized and administered. The contemporary rulers at Andhradesa were the Eastern Chalukyas of Vengi (the present Peda Vegi in Eluru Taluk in West Godavari district). Their kingdom comprised of the whole coastal region extending from modern Visakhapatnam in the north to Manneru in the south.

The cordial relationship between the Andhra and Tamil countries started when the Vengi Chalukyas entered into matrimonial alliance with the Cholas. During these regimes the Andhradesa and Tamilagam came much closer by a series of dynastic alliances culminating in the succession of Kulottunga I (AD 1070-1120) to the Chalukya-Chola empire, embracing both the regions. The inscriptional and other sources pertaining to Chola-Chalukya period in Andhradesa disclose the Chola influence on political, administrative, social, cultural and religious aspects between 10th and 13th centuries AD. The usage of Tamil words in the temple rituals, in the religious practices, the names given to the administrative divisions and the personal names, the art and architecture further reveal the Chola influence on Andhradesa.

Saktivarman I and Vimaladitya, the two sons of Danarnava of Eastern Chalukya dynasty of Vengi, sought political asylum in the Chola country after the death of Danarnava in the hands of Telugu Choda Jatachoda Bhima. Rajaraja Chola, who was an ambitious monarch with a commendable foresight, realized the strategic importance of the Vengi kingdom and supported the cause of the Eastern Chalukya princes. By defeating Jata Choda Bhima he restored the Vengi throne to Saktivarman I, reducing in that process, from an autonomous state of Vengi to one of dependence on the Chola monarchy to some extent.

The Chola-Chalukya relationship was further strengthened by the marriage of Kundavai, daughter of Rajaraja I with Vimaladitya (AD 1011-18) who succeeded his brother Saktivarman. Vimaladitya's son Rajarajanarendra (AD 1019-61) married Ammangadevi, daughter of the Chola king Rajendra I (AD 1012-44). His rule ended with the annexation of the Andhra country to the Chola Empire by Rajendra II alias Kulottunga I, son of Rajarajanarendra and Ammangadevi. Since Kulottunga I, (who married Madhurantaki (the daughter of Rajendra II), had spent many years in the Chola country and he chose to remain in that country itself due to various political reasons and appointed his sons to rule over the Vengi kingdom as his viceroys.

Migration of Brahmanas during 11th century AD

The period between AD 1019 and 1022 witnessed large-scale migration of the Brahmanas from Andhra region to Chola country. Migration of a specific community on such a large scale had taken place earlier during the Pallava period from Andhradesa. This migration was recorded very graphically in the Karandai Plates of Rajendra I. This inscription records the lands given by the king to the *Brahmanas*, most of them being migrated from Andhra region. The practice of establishing Brahmadeyas, which started during the Pallava period, was adopted by the Cholas. The phenomenal rise of Brahmadeyas during this period is a clear indication of the fact that the ruling kings patronized the Brahmana settlers from the Andhra country almost continuously. For, as evidenced by the present charter, no less than 775 *Brahmanas* among the 1083 bore names of villages in the Andhra country while it is possible, that in the case of the rest they had arrived in the Tamil country much earlier and therefore had adopted the local village. In the case of Tandantottam plates of Nandivarman II, the name of the place pertaining to the donee is given whereas in the present charter there are names of two places in respect of each donee. The place names of the *Brahmanas*, who hailed from Andhra country, as mentioned in the Karandai plates, are mostly common to those of the Tandantottam grant. The forms of these names have, however, undergone some considerable change attesting to the influence of Tamil on them. Pinukkipparru, Viraparru, Vangipparru, Attamparru, Mudaparru, Karipparru, and Ponnamparru etc., of the Pallava grant have become Pinukkippuram, Virapuram, Vangipuram, Attampuram, Mudappuram, Karippuram and Ponnampuram etc. With the efflux of time between the Pallava grant and the present one it is possible that they had commenced to stand merely for the house or family since those persons or their descendants might have ceased to have anything to do with those far off places from where their ancestors migrated during the reign of Pallava kings before Pallavmalla. There are some other names pointing to their location in the Tamil country and these places had already become permanent places of residence of the persons stated to hail from them. In view of the absence of the names of places in the Telugu country in their cases, it may be surmised that these persons might have belonged to a far earlier stream of migrants from the Andhra country or might have taken to their new places sooner than the rest. (Krishnan, K.G. pp. 55-57, pp. 230-265).

Migration of merchants during 11th century AD

Several communities such as enterprising merchants migrated to Andhra country during early medieval period. A subordinate of Kalyani Chalukya Vikramaditya VI (1076 – 1126 CE), named Kanduri Bhima Choda issued an inscription, which mentions a merchant guild, namely, *Aruvala* (Tamil) *nagaramu*, in Peruru (Nalgonda Dt.). It also refers to the street of the Tamil merchants as *Aruva nagarapu vidhi* and refers to their active participation in the temple festivals. These Aruva merchants constructed a temple in Perur and called it *Aruva-nagaresvara* temple. As the term Aruva indicates to the Tamils, it may be presumed that the Tamil merchant communities migrated to Perur and successfully established themselves in the village of Perur. [It may be pointed out in this context that the ancient Tondaimandalam (present Chittoor, Kanchipuram, Tiruvallur and Vellore districts) was called Aruva Vadatalainadu or Aruva nadu during Sangam age. The Tamil-speaking people were referred to as *Aravavaru* (or *Aruvalar* - people of *Aruvanadu* or *Aruva vadatalinadu*) and the Tamil language was referred to as *Aravam* by the Telugu-speaking people].

Migration of Tamil Brahmanas during 11th century AD

This period witnessed large-scale migration of the Tamil Brahmanas from Chola country to Andhra region. (*Epigraphia Indica* Vol.V, p.70 ff). In the days of Chalukya-Chola rulers, there was a migration of several families from the Chola country. Their mother tongue was Tamil. Some of them settled in the rich and fertile delta of the Godavari as is illustrated by the Pithapuram plates of Vira Choda (AD 1093). These plates record the royal gift of lands to 554 *Dravida Desa Brahmanas* who hailed from the Tamil country. These *Brahmanas* seem to have migrated to Andhra, when Vira Choda was sent to that country as viceroy by his father. These Dravidian *Brahmanas* and other communities were encouraged to migrate from Tamil country to coastal Andhra by the Chola emperors for propagation of religion, for military assistance and for revenue administration. This inscription referred to the personal names of the *Brahmanas* who settled at Andhra region. The king, who was a *Parama Bhattaraka* made an endowment to these *Brahmanas* who bore Tamil names like *Veneya Kuttan*, *Ulahamandan Bhatta*, *Karimanikya*, *Chentaamarai Kannan*. These names indicate that Vaishnavism penetrated deeply into Andhra region and gained popularity.

The gradual shift from Saivism to Vaishnavism by some of the rulers of this period was mainly due to the visits and efforts put forth by Sri Ramanuja and his disciples. Anantacharya in his *Prapannamrita* recorded the visits of Ramanuja to places like Sri Kurmam and his attempts at converting the rulers of the region to Vaishnavism. Sri Kurmam, a famous temple in Srikakulam district, is said to have been originally a Saivite was changed into a Vaishnavite by Sri Ramanujacharya. Some of the Sri Vaishnava families like *Bhattas* and *Kandadais* from *Tamilagam* settled in several parts of *Andhra Desa*. Most of the inscriptions of the Vaishnavite temples end with the epithet "*Sri Vaishnavula Raksha*". Ramanujacharya, Ananda Tirtha, Narahari Tirtha and Padmanabha have done much for the growth of Vaishnavism in the northern Coastal Andhra.

Krishnamacharya composed a number of sacred songs and spread Vaishnavism in and around Simhachalam.

During Chalukya-Chola period the strongholds of Vaishnavism were Sarpavaram, Bhimavaram, Simhachalam, Srikurmam, Bapatla, Macherla, Mangalagiri and Srikakulam. The Vaishnavism of Andhra was influenced by the southern school of Vaishnavism. The use of Tamil words like *Amritapadi, Tirumanjanam, and Devakarmi*, which occur in the inscriptions of the period, shows how profoundly the Tamils influenced the Vaishnavite religion of Andhras. An inscription at the Rajanarayana temple at Bhimavaram records the several shares to the *Brahmanas* who were mentioned as *Tiruvaradhanaseyvaru Tirugakarchiseyu Brahmana, Tiruvadai vinnapamu cheyuvaru*, and *Tiruvamudipadi* Rajasamudra Brahma Narayana. By this evidence it is apparent that the religious book of songs *Tiruvaymoli,* which was popular in Tamil country was introduced in the temples of Andhradesa.

Migration of weaver communities in 12th century AD

The relative importance of the weavers in early medieval economy can be gauged by the attitude and policies of the state towards them. Not only did the state adopt a conciliatory stance in the face of protests by weavers, but it also offered them economic privileges and large scale concessions. Thus the state reduced taxes in the newly settled areas, to attract weavers, and for the first two or three years granted total tax remission. During the period of Vikrama Chola in AD 1128 twenty families of weavers, at the rate of four families from five villages, were newly settled in Tirukkanapuram in modern Coimbatore district and were granted special privileges, including tax remission. (*Annual Report on Indian Epigraphy*, no. 508 of 1922-23).

Migration of Brahma - Kshatriyas in 12th century AD

Two inscriptions of Rajaraja II at Bapatla refer to a family of settlers of the Brahma-Kshatra Kulas. (*South Indian Inscriptions* Vol.VI, No.127, 156). They migrated from the village of Brihadankura in Tondaimandalam. *Brahmanas* who abandoned the traditional occupation of their caste and took to Kshatriya occupation were called Brahma Kshatriyas. The Kadambas, the Pallavas and the Vishnukundins, who ruled different parts of Andhra region styled themselves as belonging to Brahma-Kshatra vamsa or family. The Pallavas claim that they belong to the family of Dronacharya, the guru of Pandava and Kaurava princes of Mahabharata. Members of this sub-caste obtained distinction as physicians, poets and philosophers in later periods.

Migration of architects during 11th-12th centuries AD

During the 11th-12th century AD Andhradesa witnessed the construction of numerous temples dedicated to Siva and Vishnu. These were constructed in the Chola style. For example, Sri Kurmam a famous temple in Srikakulam district, said to have been originally a Saivite temple, was changed into a Vaishnavite temple by Sri Ramanuja. The main *vimana* over the *garbhagriha* with the peculiar pyramidal tapering design, are characteristic of the temples built in the days of Rajaraja Chola. The Bhimesvara temple at Draksharama, recognized as one of the five *Arama kshetras* of Andhradesa, is a perfect example of a harmonious blend of the Chalukyan and Chola styles of architecture. The Bhismeswara temple at Samarlakota, one of the Pancharama temples, also contains the Chola architectural features. The Ananta Bhogeswara temple at Kolluru in Tenali Taluk is a small temple built strictly according to the conventional Chola style of architecture in 10th-11th centuries A.D. Most of the temples in Chittoor district such as those found at the Bokkisampalem, Jogimallavaram, Kalavagunta, Kattamanchi, Kanipakkam, Mogili, Kapilatirtham, Laddigam, Vaviltota, Yatavakili, Sri Kalahasti, Tondamanadu etc are Chola style temples simple in plan and modest in scale. But most of the Chola temples which were originally small in size, were amplified by the additions like the *mandapas*, small shrines and *gopuras* in the Vijayanagara period as we see at the temples of Tirumala, Tirupati, Tiruchanur, Kapilatirtham, Sri Kalahasti, Kattamanchi, Kanipakkam, Kalavagunta, Velkuru etc. The Tamil inscriptions in all these temples provide ample evidence for the Tamil influence in these regions.

Evolution of a specific geographic region in time

Time lines are useful when the geographical extent under study is relatively small and of reasonably definite boundary. For example, consider the place Kanchi or Kanchipuram. Kanchipuram presently the headquarters of the Kanchipuram district, very near Chennai is well-known religious centre and a sacred place of pilgrimage for people of all faiths, like Buddhism, Jainism, Saivism and Vaishnavism. It is one of the cities in South India, which has played a dominant role in the political and cultural history of our country from as early as the third century B.C. Kanchi has a chequered history right from the Puranic times and all through the historical period. It enjoyed a pre-eminent position as a city of political, military and cultural consequence. Hieun Tsang, the Chinese pilgrim, who visited this place around 7th century AD, described the condition of Buddhism in Kanchipuram in his travelogue. Kanchipuram came into historical limelight for the first time during the period of Sangam literature. This was the capital of Tondaimandalam and was ruled around 2nd century AD by Tondaiman Ilam Tiraiyan, believed to be a contemporary of Karikala Chola of Sangam age. Subsequently it gained importance as the capital of the Pallavas (6th-9th centuries AD). From Rajaraja Chola's reign (AD 985 - 1014), this region was named Jayangonda chola mandalam, after his title. Even from very early times it was particularly renowned as the seat of learning attracting scholars from far off places and belonging to different faiths and several sectarian creeds and castes. Broadly divided into three sectors - Vishnu Kanchi, Siva Kanchi and Jina Kanchi, it draws adherents of the respective faiths in large numbers even today. When the Pallavas occupied Tondaimandalam with their capital at Kanchipuram, this city was a flourishing cultural centre. This region assumed significance during the Pallava reign for the emergence of rock cut temples and structural temples. During the Chola reign, structural temples continued to be built but on a larger scale. Different stages of south Indian architecture from Pallavas (8th century) to Nayak period (18th century) can be seen in the temples of Kanchipuram.

There was a movement of weaver communities into south India from 11th century AD. Kanchipuram became popular as a silk centre from 18th century AD. This caused migration of several communities of weavers like Salis, Padma salis, and Devangas into this region.

People migrated to Kanchi, as it was an educational centre, pilgrim centre, religious centre, and trading centre. Different communities and followers of different religions settled here to pursue their interests.

Fig. 2 .Migration of Brahmanas from Andhradesa to Pallava territory.

Maps as spatio-temporal descriptors of History

In the context of migrations over extended geographical regions, time lines convey little. Maps, on the other hand, can depict large-scale features naturally. Two examples of maps, one dealing with migrations of Brahmanas from Andhradesa to Pallava territory and the other dealing with territorial divisions of Cholas in Andhradesa is presented here.

Figure 2 attempts to capture the migrations of Brahmanas from Guntur and Godavari regions during the reign of Nandivarman Pallavamalla (AD 731-796). The Pullur copper plate charter (AD 764) states that the king granted four villages to one hundred and eight *Brahmanas* as *brahmadeya*. Some of the recipients of this endowment were originally belonged to some places in Andhra country, such as Kavanuru (Kanuru, West Godavari Dt.), Karambichedu (Karamchedu) and Vengi etc. (*South Indian Inscriptions* Vol. II, Pt. V, pp. 519, 520).

The Tandantottam charter (AD 789) records the gift of a village by Dayamukha with the permission of Nandivarman Pallavamalla. This record provides some of the names of the donees along with their native villages, which are located in Andhra country. These are Karambichedu (Karamchedu), Tanukkeri (Tanuku), Irakkambarru (Itikampadu), Karanjai (Kranja or Kaja), Penukamparru (Penukamparru), Veliparru (Velpuru), Nutilaparru (Nutulaparru), Gunturu (Guntur) etc. All these places are situated in the present Guntur, Krishna, and Godavari districts of Andhra Pradesh. (*Epigraphia Indica,* Vol XXXVI, pp144-163). The present name of the places are given in the brackets.

These place name prefixes to their personal names denote that these brahmana scholars migrated to the Tamil country, with all probability when the Pallavas transferred their political activities from Andhra to Tamil country. The migration of *Brahmana* to the Tamil country must have commenced from the early Pallava rule in Kanchipuram and was continuously encouraged by the rulers by the system of establishing *brahmana* settlements by land endowments. The large numbers of Telugu village names in the later Pallava copper plates such as Urupputtur, Karambichedu, Nellur, Chettalur, Kumandur, Pattangi etc. are even now held by some Vaishnava families in Tamil country.

Figure 3 represents the influence of Chola names on the territorial divisions of Eastern Chalukyan territory in Andhradesa. During 11th-12th centuries AD, the Vengi region was dominated by the Cholas. The change in the political status of this Eastern Chalukyan kingdom had a great impact on the toponomy of the Andhra country. The territorial divisions and the place names of the region

Fig. 3. Territorial divisions of Cholas in Vengi Kingdom.

disclose the supremacy and influence of the Cholas in the Eastern Chalukyan territory. It is interesting to note that some of the epithets of the Chola monarch Rajaraja were given as second names to the towns included in the Vengi country. Thus Draksharama, Melpadi, Kalahasti, Tummuru and Sulluru came to be known also as Jananathapuram, Rajasrayapuram, Mummudisolapuram, Jayangondasola Chaturvedimangalam and Singalantapuram respectively. During the reign of Kulottunga I the Kalinga kings invaded the Andhra country twice. Kulottunga I successfully retaliated and to commemorate his victory he renamed Visakhapattanam as *Kulottunga chola pattanamu.* (*South Indian Inscriptions* Vol. V No.66, Vol.VI, No.200).

It is very interesting to note that the Chola kings had the practice of naming the alien territories, which they brought under their sovereignty after their own favourite epithets. It is known from inscriptions in Tamil Nadu that the kings of this family named a number of divisions in the Tamil country after the epithets of their choice.

The following are a few examples –

Jayangondasola-mandalam - named after the epithet *Jayamgondachola* (the victorious Chola king) of the Chola king Rajaraja I (985-1016 A.D.). This major division was once called Tondaimandalam (the southern parts of the present Andhra Pradesh and also parts of Tamil Nadu). (*South Indian Inscriptions*, Vol.XVIII, No.157).

Rattappadikondasola-mandalam - The famous Chola emperor Rajaraja I and as also Rajendra I (1012-1044A.D.) are known to have conquered *Irattapadi-seven-and-a-half lakhs*, a division which corresponded to the entire Chalukya empire, in order to commemorate their triumph, assumed the epithet *Irattappadikondachola*. It was this epithet that gave its name to the division, which covered the region around Punganuru in Andhradesa (the present Chittoor district). (*Epigraphia Indica* Vol.XXV, pp.241ff).

Nilagirisola-mandalam - Rajaraja I annexed the Nolambavadi to the Chola empire and renamed it as Nilagirisola-mandalam after his own epithet *Nilagirisola*. This division corresponds to modern Anantapur district in Andhra Pradesh and Bellari, Kolar, and Tumkur districts in Karnataka.(*Epigraphia Indica* Vol.XXXI,pp.269ff).

Ganggagondachoda-valanadu - This division seems to have been named after the epithet *Gangaikonda* of the Rajendra I who is known to have brought the holy water of Ganga to his capital. This division corresponded more or less to the entire East Godavari district of Andhra Pradesh. (*South Indian Inscriptions*,Vol.V, No.66)

Uttamachoda - valanadu - This division owes its name to the epithet *Uttamachoda* borne by the Chalukya-Chola king Kulottunga I (1070-1122 A.D.). It was a fairly big division spread over parts of the present Guntur district. (*South Indian Inscriptions,* Vol.VI.Nos.135,179,200).

Adhirajendra-mandalam - This division is obviously named after the Chola king Adhirajendra who ruled between 1067 and 1070 A.D. It corresponds more or less to modern Rajampet taluk in the Cuddapah district. (*Annual Reports on Indian Epigraphy* Nos. 572, 600, 581 of 1907).

Kulottungachola-mandalam - This division is named after another Chola king Kulottunga I (1070-1122). Since it is stated that *Ganggagondachoda-valanadu* and *Uttamchoda-valanadu* formed parts of this division we may conclude that this division included the entire Andhra territory that was under the rule of Kulottunga. (*South Indian Inscriptions,* Vol. V No.66).

Figure 2 and Figure 3 are examples of conventional maps modified for the purposes of depicting migrations. However, conventional maps provide only a snap shot of history such as the spatial extent of a dynasty. It would be of greater value to have a sequence of maps marking time depicting the changes in the spatial domain associated with the rise and fall of a dynasty. Migration of specific communities and the consequences, e.g. evolution of art/architecture due to migration of artisans, can be described effectively through spatio-temporal maps. Such maps can also depict the evolution of migratory patterns and migratory routes. It is believed that the creation of a digital cultural atlas, with multi-layered cultural information presented through dynamic maps, would have a significant impact moving well beyond what was possible with conventional means in forging newer perspectives.

Issues and Challenges

The creation of dynamic maps bringing the cultural facets together with the political aspects has several issues and challenges. For one, cultural facets are seldom chronicled as much as political aspects are leading to lack of temporal continuity in the former. Sources bearing on cultural facets are wide and varied since documenting culture has not been an academic pursuit. For another, the various geographical regions in south India have distinct languages and idioms requiring a multi-lingual approach.

However, the creation of community related databases could be a step towards collecting and collating historic-geographic information essential for all further work. Tracing the movement of art and craft forms through space and time within the context of the south Indian history would constitute a parallel effort. In particular, visual rendering of the art and craft forms would greatly facilitate one of the objectives of dynamic maps, namely, to disseminate cultural history effectively. Finally, documenting agricultural practices and the changes brought about in these in space and time would help address an important feature of the dynamics of migration.

It is hoped that the study of migrations begun recently would evolve into and merge with concurrent efforts made elsewhere towards the creation of a cultural atlas for south India.

References

ANNUAL REPORTS ON INDIAN EPIGRAPHY, 1919-20, 1922-23, 1939, 1959. *Archaeological Survey of India*, New Delhi.

EPIGRAPHIA INDICA, Archaeological Survey of India, New Delhi.

KRISHNAN, K.G. 1984.*The Karandai Tamil Sangam Plates of Rajendra Chola* I, Archaeological Survey of India, New Delhi.

MEENAKSHI, C. 1977. *Administration and Social Life Under the Pallavas*, University of Madras.

NILKANTA SASTRI, K.A. 1955. *The Colas*, University of Madras, Madras.

RAMACHANDRA RAO C.V. *Administration and Social Life Under the later Eastern Gangas and Suryavamsa Gajapatis.*

SOUTH INDIAN INSCRIPTIONS, Archaeological Survey of India, New Delhi.

SUMABALA,P. 2008. *Confluences in South India History*, Parker, Chennai.

Management of material heritage information : structured repository for travelers annotations over a qualified caravanserais inventory

C. Tavernari, N. Boutros and A. Del

EVCAU Research Team, National School of Architecture of Paris Val-de-Seine

Abstract

This paper wishes to present an innovative project which aims at using new technologies developed for web applications in order to renew with the tradition of travel accounts, so as to retrieve valuable information on caravanserais. The objective is to create a repository service displaying high added-value. The repository service is intended to be public and free. The added-value of the repository lies in the submission form, which will permit to give immediate valuable meaning to the data by identifying authors and annotating uploaded data with concerted ontology concepts. This kind of processing of the data will make them functional and valid to be used for scientific purposes. The expected result is twofold. Firstly, while collecting information on caravanserais that are veritably needed and will be freely available, people will become aware of this material heritage, till now frequently neglected, as they can annotate and add their own contributions. Secondly, these open internet archives will rely on highly qualified academic standards which will allow for a scientific use of the information.

Keywords : caravanserais, inventory, travel accounts, repository, semantic web

Introduction

The idea of the ICRP (Information on Caravanserais Repository Project) developed from the analysis of different situations the EVCAU team had to deal with during the research on caravanserais it carried out over the last ten years: namely the scarcity of information and, globally, of research on these edifices, the relevance of travel notes to obtain information on caravanserais and the spread of scientific and academic projects whose goal was to share data and knowledge through the use of new technologies.

State of the research on caravanserais

Caravanserais are a characteristic element of the Central Asian and Mediterranean landscapes: they dotted the roads and cities, functioning as simple sheltered stopovers, markets or warehouses. These edifices have been built over an impressive span of time, more than one thousand years, from the 8th century AD till the modern transports rendered them obsolete.

Nonetheless, these buildings, although well-known elements of the oriental landscape, and despite their broad scattering, failed, till recent years, to attract much of the scholars' attention, thus remaining poorly studied and insufficiently understood. If the gap between the scale of the caravanserais' phenomenon and the research done on it is noteworthy, the situation can be different from region to region: Turkey and Iran being the countries that have most developed research in this field, supported by significant government renovation programs due to their great number of preserved caravanserais.

Among the most important regional studies carried out from the beginning of the 20th century it is worth noting the works of Jean Sauvaget, who concentrated on the Syrian caravanserais (Sauvaget, 1939 and 1940), Kurt Erdmann, who studied the Turkish caravanserais of the Seljuk period (Erdmann, 1953), followed by Gabriele Mandel-Sugana, (Mandel-Sugana, 1988) and for the Iranian region the research led by Maxime Siroux (Siroux, 1949 and 1971) and more recently Wolfram Kleiss and Mohammad Youssef Kiani, (Kleiss and Kiani, 1995). Other studies focused on single sites (Hawari, 2001; Hillenbrand, 1992; Petersen, 1997) or, in a more comprehensive manner, on the caravanserais' architecture (Hillenbrand, 2000; Lebigre, 1996).

A renewal of interest in caravanserais has seemed to appear in these last few years as the international community has become aware that these edifices had been neglected and were rapidly disappearing. Conscious of the importance of these edifices, both as material heritage bearer of a great symbolic value and as historical witnesses aiding the reconstruction of the ancient road networks, in 1998 UNESCO organized the first seminar devoted to caravanserais with the aim of launching an analytic and systematic regional inventory of caravanserais in Central Asia. This project, in which members of the EVCAU research team participated, permitted them to collect a large amount of information and to establish a first basic list of records. It lasted six years and reached its conclusion during ICANAS 37, which was held in Moscow in 2004.

Subsequently, EVCAU continued to promote research on caravanserais and collaborated temporarily with the Japanese National Institute of Informatics within the framework of the "Digital Silk Road" project (ASPICO-DSR, Advanced Scientific Portal for International Cooperation on Digital

Silk Roads), which aimed at extending and expanding the UNESCO initiative.

Since the launch of the Japanese project other significant scientific programs focusing on Central Asia and mainly related to the Silk Road have been set up. It is worth mentioning the AISCA (Archaeological Information System of Central Asia) which plans to create a digital atlas based on a WebGIS and regroup all the archaeological data on Central Asia and the IDP (International Dunhuang Project) whose aim is to make information and images related to artifacts found on Silk Road sites freely available on the web.

The spread and success of these projects show that they meet a real demand from the public and from scholars alike; significantly, however, caravanserais are only marginally concerned. A research program devoted to caravanserais is still lacking and the authors deem one to be necessary as these edifices constituted one of the main and most visible apparatus along the roads, a visible trace of journeys and commercial exchanges and, consequently, fundamental to understand them.

Travel books: a fundamental source for the caravanserais' research

It is no wonder that travel books, meaning accounts founded on real-life experiences, came to constitute the fundamental source for the study of caravanserais: for travelers these buildings represented a promise of protection and deserved rest. Besides travel literature, only geographic treatises supply valuable information on these edifices, while other kinds of texts are often sparing with information on caravanserais, the authors hardly noticing or reporting their existence.

Used as a primary source, nonetheless, travel accounts prove to be particularly difficult to handle and they present some specific problems, the first of them being their quantity in relation to the considered geographic and historic framework. If travel accounts as a literary genre started to be successful in the Middle Ages, by the 15th century they already represented a considerable amount of sources. As an example, Stéphane Yerasimos (Yerasimos, 14) shows that between *c.* 1300 and *c.* 1600 AD more than 440 travelers left an account of their travels in the Ottoman Empire. The impressive number of travel accounts is due to their inherent characteristics, as they do not constitute a codified genre but are multiform and almost every text reporting a comment on a territory visited by the text's author came to be considered a travel book even if it did not report a precise and dated itinerary.

A further limit to a simple management of travel accounts is that information are broadly scattered in the text and precise data on caravanserais is not easy to retrieve.

The reasons evoked above transform historical research in travel books into an extremely time-consuming task in a context where the state of the situation asks for a faster work and reaction.

As the purpose of the caravanserais inventory is also to help to preserved this disregarded heritage, contemporary travel accounts should be considered as well. If it appears by now unnecessary to prove the relevance of contemporary descriptions and notes as historical sources, the problem which has to be confronted about this kind of material is its form. Paper books can be analyzed as more ancient texts, but new forms of travel diaries, such as the blogs which are spreading considerably on the internet, do not allow for the usual methods of study, because in these cases information is not organized. The need to renew the information on caravanserais, converging with the consideration that travel books are the most valuable source for this kind of edifices, and that people travel more and more widely, convinced the EVCAU team that contemporary sources should nonetheless be exploited.

Sharing knowledge: ICT and new possibilities

Digital technologies are impacting on the academic research in the field of humanities, challenging the traditional methods and working processes and, above all, influencing the final production and spread of the results. Universities and research centers are more and more willing to share their cultural and scientific information via new available web applications, thus allowing the public to have a free access to their knowledge. As evidence of this tendency many projects on digital atlases, digital and multimedia archives and historical GIS have been set up over the last few years.

Also in keeping with the emerging trends calling for a knowledge society, institutions (e.g. UNESCO) and government departments broadly promote cultural applications on the web which aim to be available to the widest number of users. Most of these projects are directed at promoting cultural heritage, particularly through the creation of digital collections.

Among the most representative projects, and which do not deal with the Silk Road, it is possible to mention:

- the CHGIS (China Historical GIS – http://www.fas.harvard.edu/~chgis/), promoted by the Harvard University;
- the DAHL (Digital Atlas of the Holy Land – http://daahl.ucsd.edu/DAAHL/), which offers a great amount and variety of data and images, historical information, freely downloadable elements;
- the "Valley of the Shadow" project (http://valley.vcdh.virginia.edu/), an example of

digital archive of different digitized primary sources, such as letters, newspapers, diaries and a varied set of official records;
- Europeana (http://www.europeana.eu/portal/);
- Michael (http://www.michael-culture.org/en/home);
- the Strabon project, which differentiates from the previous because it integrates a geographical information system and is mostly intended to promote ethical tourism around the Mediterranean sea (http://www.strabon.org/portal/).

Finally it is worth recalling that ECAI developed and supported several programs and initiatives in these fields.

All these initiatives aim at representing effective instruments for information and education but, if some of them are simply digitized libraries the user can explore, others point toward a deeper interaction with the public. Actually, in the light of the recent web and informatics tools evolution it seems necessary to re-think applications so as to offer keen user-centered digital cultural applications.

The project

In accordance with the situations described above, the EVCAU research team wished to develop a different approach to the caravanserais' research. Previous experiences and achievements in this field, in fact, showed the extreme difficulty of creating a comprehensive inventory of these edifices but, at the present state of the research, such a tool is necessary, both to help to preserve this kind of heritage and for the study of caravanserais as a network, considering their relations and interactions in order to shed new light on their distribution patterns.

The research team concentrated on the possibility of collecting information on caravanserais calling on the public experience of these buildings and basically the idea lies in appealing to the lively tradition of travel accounts. This mean will allow the citizen to become a data producer. At the same time, the application of the protocols developed for the semantic web will make it possible to enhance the standard of the submitted data so as they can be considered as being at a good scientific level. The structure of the repository will qualify and organize the information from the earliest stage of the deposit procedure, thus reducing the need for control operations. This deeper interaction with the user appears as one of the main and most innovative feature of this project and it constitutes the core of the ICRP.

If the purpose of this project is to collect information and new data on caravanserais in order to organize an up-to-date inventory, its fundamental goal is more ambitious, as it consists in increasing the public awareness towards this material heritage and to have it realize that this edifices are important witnesses of past cultural and geographical interconnections between East and West.

The practical set up of the project

From the inception of the project the EVCAU research team was confronted to a main problem: quality of the information.
The first challenge consisted in finding a balanced relation between the control of the submitted data and the difficulty of the deposit procedure. In order to improve its contribution the citizen will be provided, from the very beginning, with a precise but not too restrictive framework which will guide him through the deposit process (Figure 1). The citizen will be allowed to fill in the metadata sheet using a determined set of tools to query the academic database. The latter comprises three complementary elements:

- Caravanserais Inventory
- Caravanserais Ontology
- Toponyms Database

It will be possible to submit images, texts related to the travel experience connected to caravanserais but also historical or geographical information if known. Both contemporary records and older material will be accepted but the latter will follow a different deposit process, which will have been adapted to its peculiar characteristics.

Following the qualification process of its data, the citizen will be induced to confront its information, what he has seen and noted, to the academic knowledge which can be associated to its deposit.

The project's team is aware of the difficulty of managing such a project as, in order to secure it a scientific value, an academic control is essential and unavoidable and the control operations represent the most expensive part of the project. Nonetheless, the user's involvement arising from the registration process could represent one the essential keys in order to reduce the control needs and the time they require. A light control device could represent a pledge for the project durability.

Concluding Remarks

To ensure the ICRP success and viability the EVCAU can count on some important features. Firstly the research team can already reckon a real experience in the field of the caravanserais' research and, consequently, has by now at its disposal a rich database of these buildings comprising an important amount of different records: images, historical sources and also geographical coordinates. Subsequently, by way of the CIERAM (International Center for the Study of the Modern and Ancient Roads) the team is in contact and

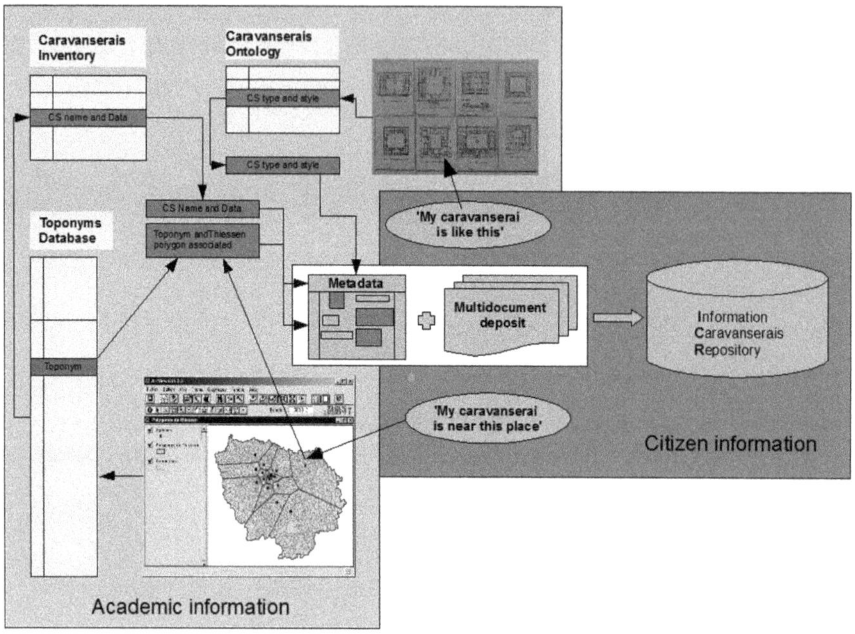

Fig. 1. The scheme of the ICRP structure.

can rely on a wide network of specialists who conducted studies on caravanserais or related arguments.

The implement of the project intends to achieve three main goals: the establishing of a comprehensive inventory of caravanserais or, at any rate, to gather information to preserve and manage this material heritage, the spreading of these data so as studies on caravanserais could develop and shed new light on this multi-factorial phenomenon. Finally, the user involvement in a valuable project on caravanserais will expectantly awake his awareness as regards these relics of the past.

References

ERDMANN, K. 1961. *Das Antolische Karavansaray des 13. Jahrhunderts*. Berlin, G. Mann.

HAWARI, M. 2001. Khan al-Lubban : a caravansarai on the Damascus-Jerusalem road. *Levant*, 33, 7-20.

HILLENBRAND, R. *ET AL.*1992. Mamluk caravansarais in Galilee. *Levant*, 24, 55-94.

HILLENBRAND, R. 2000, Caravanserai. In R. Hillenbrand, *Islamic architecture: form, function and meaning,* 331-376, Edinburgh, Edinburgh University Press.

KLEISS, W. AND KIANI M. Y. 1995. *Iranian Caravanserais*. Teheran, Iranian Cultural Heritage Organization.

LEBIGRE, P. 1996. Les caravansérails routiers du Moyen Orient XIème – XIXème siècle construction d'un regard découverte d'un langage. *Annales archéologiques arabes syriennes* 42, 217-231.

MANDEL-SUGANA, G. 1988. *I caravanserragli turchi*, Bergamo, Lucchetti.

PETERSEN, A. 1997. JALJULYA: a Village on the Cairo-Damascus Road. *Levant*, 29, 95-114.

SAUVAGET, J. 1939. Caravansérails syriens du Moyen - Âge. Caravansérails Ayyoubides. *Ars Islamica* 6, 48-55.

SAUVAGET, J. 1940. Caravansérails syriens du Moyen - Âge. Caravansérails Mamelouks. *Ars Islamica* 7, 1-18.

SIROUX, M. 1949. *Caravansérails d'Iran*, Le Caire, IFAO.

SIROUX, M. 1971. *Anciennes voies et monuments routiers de la région d'Ispahân*, Le Caire, IFAO

YERASIMOS, S. 1985. *Les voyageurs dans l'empire ottoman, XIV-XVI siècle : bibliographie, itinéraires et inventaire des lieux habités*, Unpublished PhD thesis, University of Paris VIII.

SITA (South Indian Temples Archives)
A digital data-base for iconography, architecture and epigraphy in the Pondicherry Centre of the École française d'Extrême-Orient

Valérie Gillet

École française d'Extrême-Orient, Pondicherry

Abstract

A digital photo-archive named SITA (South Indian Temples Archives) has been created in the Pondicherry Centre of the École française d'Extrême-Orient. This archive has two branches, one concentrated on the iconographical and architectural aspect of the early South Indian temples (6th-10th centuries AD) and the other on early South Indian epigraphy (2nd century BC-10th century AD). A data-base has been started for each branch, with various fields displaying the relevant information for each subject. Constituting such archives became necessary to cope with the accumulation of pictures collected during field-trips. We wanted also to document the galloping loss of valuable information regarding the temples because of ill-considered renovations and erosion. But our principal aim in building a well documented digital collection is to make the data related to different aspects of an early South Indian temple available to any scholar. In order to facilitate access to these information, we are planning to put this data-base on the internet and thereby to broaden the possibility for scholars to work from anywhere in the world. Such documents will help in developing studies on South Indian art and history.

Keywords Digital photographs – Archives – Iconography – Archaeology – Epigraphy – South Indian art – temples – Tamil Nadu.

Constituting photo-archives has been a long tradition in the École française d'Extrême-Orient (EFEO). An archive started in the year 1956 in the Centre of Pondicherry, under the supervision of the eminent member of the EFEO, P. Z. Pattabiramin. He was succeeded by another well-known and industrious EFEO scholar, Françoise L'Hernault, until her untimely demise in 1999, upon which the growth of the archives was suspended. Their "photothèque" is now conserved in the premises of the French Institute of Pondicherry. The goal of these scholars was to keep records of their field-trips and allow art historians and other scholars access to an innumerable quantity of data, mainly archaeological. But at that time only prints were available and anybody who wanted to consult this data-base needed to come to Pondicherry in person. A little more than 50 years later, a new type of data-base has been created along the lines of the valuable old one, but using the new means at our disposal, and has been named SITA (South Indian Temple Archives).

The purpose of this paper is to present a two-branched digital Archive, SITA, which is under my supervision, in the Pondicherry Centre of the EFEO. One branch of the project, the photo-archive focussing on iconography and architecture, will be described in the first part of this article and will be followed by a detailed account of the second branch of the project, which concentrates on the inscriptions.

I am indebted to Dr Charlotte Schmid who has made many suggestions regarding the organization of the data-base. I should also like to thank here N. Ramaswamy, better known as Babu, the well-travelled driver and field-assistant of the Pondicherry Centre, who has made himself invaluable for locating the sites as well as during the field-trips, and two trainees sent by the École du Louvre in Paris (Aude Berthomier in 2008 and Raphaëlle Mauger in 2009) to work for a few months on the iconographical part of the Archive. Their active participation gave a real push to the evolution of the project.

The Iconographical and Architectural branch of SITA

The first branch of the project was launched in 2007 in Pondicherry. The need for constituting this Archive had become evident because of the accumulation of thousands of digital images collected during field-trips (Figure 1). These images were not being documented and the information related to them was getting lost, rendering these pictures useless. We therefore decided to enter all the data related to each and every photo in a data-base-form supplied by the EFEO in Paris.

Fig. 1. Team at work, Paramesvaramangalam, Ceyyur taluk, Kancipuram district (picture taken by G. Ravindran, EFEO, 2007).

We are currently working on a temporary Access-type database; but by the end of 2009, it will be replaced by a more powerful one, Actimuséo, which is used by the Louvre museum, and which offers many search-facilities.

Our plan is to put it on the Internet as soon as possible in order make our data accessible from anywhere in the world, making it a valuable tool for all scholars.

The reality on the ground in Tamil Nadu has equally pushed us to start this Archive. The original aspect of most temples is disappearing at a fast pace because of well-meant but ill-considered renovations, heavy plastering, painting, insufficiently careful anastylosis. People in charge of a temple won't hesitate to transform a monument even from the 7th century in order to make it look "new". As a result, archaeologists and art historians struggle to keep records of the temples in their oldest shape, and the only way at our disposal is to take pictures before their renovation. This is why, as soon as an article is published in newspapers about a new finding or announcing a renovation, our team makes its way to the site to keep traces of the original shape of the monument. Apart from deliberate alterations that a temple can undergo, we should not forget natural forces (wind, rain, age, causing the erosion of the stone), which are also important factors in the change and sometimes loss of monuments, or parts of them.

The choice of the sites to photograph depends on our academic activities. The temples are indeed too numerous to include them all in our Archive. They are therefore selected according to the projects of the members of the team, which is why we are concentrating at the moment on the oldest sites in Tamil Nadu, i. e. the sites of the Pallava and Pandyan dynasties (Figures 2 and 3), as well as on early Cola temples. To date, our focus ends generally with 11th century temples.

Fig. 2. Entrance to the excavated Shiva temple at Kunrantarkoyil, Kulattur taluk, Pudukkottai district (7th-9th centuries) (picture taken by V. Gillet, 2009).

The documentation of the pictures

The documentation of the pictures requires the filling in of many fields described here below. After having given a reference number to each picture, the information related to the location of the temple, its GPS coordinates, as well as its name, are entered into the succeeding fields. The name of a village is always accompanied by the name of its Taluk and District, two types of geographical divisions employed in Tamil Nadu, which enable us to differentiate localities bearing the same name.

Fig. 3. Dancing Shiva, Kailasanatha temple, Kancipuram (beginning of the 8th century) (picture taken by V. Gillet, 2007).

The next field displays the subject of the picture and is followed by one for the exact location of the subject. A field is reserved for the measurements of the object. Two distinct fields present two different types of descriptions: the first one is a simple and basic description which can help the viewer looking at the main features of an icon; the second is titled "scientific description" and contains elements drawn from our own researches. For instance, an uncommon hand-gesture is simply described in the first description but is interpreted in the second one if our studies have enabled us to throw light upon its significance.

Then comes the difficult question of the dating of an object. Here we can enter a year (which is almost impossible to define in earlier times), a century or a period (defined by a span of few years or a few centuries). We may also mention the name of the king or of the dynasty, if known, under whose period and territory the temple has been built or the image carved.

The viewer will have access in a separate field to a brief bibliography stating a few studies related to the subject of the photograph. Another field entitled "Notes and remarks" will gather all the supplementary information which is valuable but which couldn't be entered in one of the specific fields. The name of the author and the date of the shot are also mentioned in their respective fields. The picture's documentation ends with a field for key-words, where different orthographies or different names of a deity will find their place, especially in order to facilitate the search of a subject.

Our aim is to cover as many fields as possible, even if it slows down the process of documentation, in order to provide a complete range of information regarding any given object. Such information on a sculpture, for instance, provides the necessary elements for anyone attempting to understand the iconographic program of a temple.

The Epigraphical branch of SITA

The Indian inscriptional record is one of the most valuable resources for the understanding and reconstruction of Indian history, much of which is still shrouded in darkness and uncertainties. Inscriptions are found sometimes on copper-plates and stelae, but we find them engraved mainly on the walls of temples. They provide information about the names and clues to the successions of kings, about property transactions, social organization, geographical divisions, etc. They often fail to furnish firm dates, but even in such cases a span of half a century can sometimes be tentatively proposed for a given inscription, based on one of the elements given by the text when used in conjunction with evidence from other inscriptions. The proportion of Tamil inscriptions in the Tamil country is the highest; nevertheless, Sanskrit is not uncommon, especially in temples excavated or built under the Pallava and Pandyan dynasties (6th-9th centuries).

If taking pictures of the sculptures and the architecture of a temple was presented as a pressing need because of their fast disappearance, photographing inscriptions seems even more urgent. The engraving of a text on walls or on stelae is usually necessarily shallow, and when such engraved texts are placed outside, they stand exposed to the vagaries of a severe climate and to the natural processes of erosion. This is without taking into account the even more dangerous "renovations" made by people in charge of the temples. Those responsible for temples today might often not hesitate, for instance, to build a brand new flight of stairs in concrete on top of a 9th century Pandyan inscription in order to facilitate the entry to the temple for pilgrims. And this is an example taken from among hundreds. Publications cannot be produced fast enough to compensate the loss of so much of valuable information. Furthermore, we can consider publications as partial and subjective if they don't provide an image of the inscription itself or an estampage of it, which is often the case, making them difficult to rely on. Estampages are indeed extremely useful, providing an excellent base for deciphering the inscriptions. Unfortunately, it is not always straightforward to get permission from making estampages while on site. The easiest and fastest way we have found is therefore to take digital pictures of the epigraphs. Focussing on the same period, i. e. mainly on epigraphs engraved during the Pallava, Pandya and early Cola dynasties, we have decided to gather these photographs in an Archive. The documentation to be entered for each inscription has been carefully defined in order that all essential information be included.

The documentation of the pictures

We come now to the fields the data-base is going to display. The primary elements, such as the reference number of the picture, the name of the author, the date of the picture, the name of the village where the inscription can be found, together with its GPS coordinates, the name of the temple and the precise location of the epigraph on the monument itself will be followed by some specifics regarding epigraphy. Language and script will be stated. The language is generally either Tamil or Sanskrit for the inscriptions included in our Archive, but the range of scripts is much wider: Tamil-Brahmi, Tamil, Grantha (Figure 4), Nagari, florid Nagari, Vatteluttu (Figure 5). The number of lines will also be given, as well as a brief summary of the contents. To decide if an inscription is worth spending the time poring over it, one needs to know which kind of epigraph it is and what type of information it records (the foundation of a temple, renovation, donation, etc.).

Fig. 4. Sanskrit musical inscription, grantha script, lower cave in Malaiyakoyil, Tirumeyyam taluk, Pudukkottai district (6th-9th centuries) (picture taken by V. Gillet, 2009).

Fig. 5. Tamil Jaina inscription, vatteluttu script, cave in the village of Arittapatti, Melur taluk, Madurai district (9th century) (picture taken by V. Gillet, 2009).

A specific field will be kept for giving the name of the king, if any, or chieftain or general, mentioned in the inscription. This often helps when assigning a date to the concerned inscription. In case the inscription has been published, a separate field will give a list of the publications where it appears.

Two fields will not be systematically filled in: transliteration and translation. The main reason is that transliterating an inscription, even without translating it, is a long process, which requires taking a lot of interpretative choices. What we intend to offer here is not a sort of digital publication where we can justify our choices but the access to the source material. The purpose of such an Archive providing a maximum amount of primary sources would be lost or at least diluted if we spent long hours transliterating the epigraphs. However, these fields should appear in our data-base. If a member of our team, which is currently composed of the epigraphist Prof G. Vijayavenugopal assisted by Dr R. Sathyanarayanan, Babu N. Ramaswamy, and myself, wishes to work on an inscription or if the transliteration or the translation has already been done, we want to have the possibility to insert it in the data-base, signed with the name of the author. Even a preliminary or subjective transliteration can of course be a valuable help for those consulting the database.

I have chosen to begin this Epigraphical branch of the Archive with the large and complex corpus of the inscriptions of the Pandyan dynasty during the first Empire, from their first mention in epigraphy until the 10th century AD, which is the core of a project Prof Vijayavenugopal and I have started in collaboration. But many inscriptions found during our field-trips will also find their place in this data-base and will be treated with equal care. We hope our efforts will contribute to a much clearer idea of the History of South India by providing the necessary tools to understand the mechanisms of medieval times.

ATS
archaeolandscapes tech & survey
Spin Off accademico

A.T.S. s.r.l. - Archaeolandscapes Tech & Survey

spin-off company

University of Siena
Department of Archaeology and History of Art

Technological Services for Archaeology and Cultural Heritage

From air to underground, from site to landscape, from reality to virtual reality, ATS brings together a wide range of specialist expertise in cultural and landscape interpretation through the application of advanced technology in land management in the fields of the cultural and environmental heritage.

ATS, combining innovative methodology with proven technological expertise, offers services in:

- recording and survey in advance of development (remote sensing and geophysics)
- 3D recording and reconstruction
- environmental monitoring and topographical survey
- archaeological and historical mapping (GIS)
- landscape recording and interpretation

The methods employed by ATS ensure a high level of recorded detail in combination with speed of acquisition, employing whatever method is most appropriate to the particular situation, based on many years of research technological experience in landscape archaeology using the most innovative methodologies available within the sector.

www.atsenterprise.com
A.T.S. s.r.l. spin-off company (University of Siena) - Strada della Coroncina 8, 53100 Siena (ITALIA)
phone +39 0564 441212/13 mobile ph. +39 3316458042 - + 393316458378 info@atsenterprise.com

ATS was founded in January 2009 as the initiative of a group of research workers at the University of Siena with the aim of making available in a commercial context the combined expertise of the partners in innovative methods of archaeological survey and prospection, management of the cultural heritage and the study of ancient landscapes.
The creation of this university spin-off company brought about a merger between three scientific laboratories active for a decade and more in Italian and international projects at university level:

- The Laboratory of Landscape Archaeology and Remote Sensing (LAP&T)
- The Geographical Laboratory
- The Laboratory of Applied Informatics and Spatial Analysis (ASIA Lab)

Examples of recent work by ATS include the following:

- Conservation archaeology in the BreBeMi Highway Project

Survey and interpretation along the route of a new motorway between Brescia, Bergamo and Milan (BreBeMi). This project gave ATS the opportunity to apply the most innovative multi-technology approach to such a project yet attempted in Italy.

The project involved the use of the most advanced archaeological expertise in the field of pre-development archaeology, reducing to a minimum the degree of uncertainty about archaeological remains within the area affected by the motorway and its related construction works so as to reduce the duration and cost of any subsequent rescue work along a track 75km long with a total 'footprint' requiring 750 acres of pre-development investigation.

- The investigation of individual archaeological sites: ancient settlements of Altino (Venice) and Sofiana (Sicily)

ATS has been commissioned to carry out large-scale geophysical prospection on archaeological sites in Italy and elsewhere to define the character and extent of the sub-surface remains. In these two examples survey was undertaken with use of magnetometer SYSTEMS GEM Overhauser (GSM-19GW, 2 sensors) and magnetometer MULTICAT Foerster (4 sensors mounted on a fibreglass trolley).

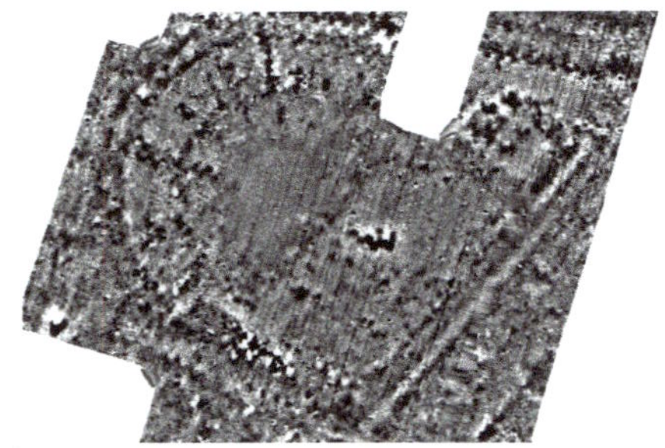

- Landscape monitoring through aerial survey in the field of environmental and cultural heritage

Monitoring and investigations of landscape areas have been undertaken through aerial survey and photography in both archaeological industrial contexts, as for instance in areas of quarrying to determine the potential impact on the surrounding countryside. It proved possible within these projects to collect air-photogrammetric data and to achieve 3D modeling, notably through the use of the company's radio-controlled drone VTOL AUMAV (Microdrones GmbH).

www.atsenterprise.com
A.T.S. s.r.l. spin-off company (University of Siena) - Strada della Coroncina 8, 53100 Siena (ITALIA)
phone +39 0564 441212/13 mobile ph. +39 3316458042 - + 393316458378 info@atsenterprise.com

www.ingramcontent.com/pod-product-compliance
Lightning Source LLC
LaVergne TN
LVHW072246300926
849160LV00018B/1606